THE OXFORD ENGINEERING SCIENCE SERIES

THE OXFORD ENGINEERING SCIENCE SERIES

1. D. R. Rhodes: *Synthesis of Planar Antenna Sources*
2. L. C. Woods: *Thermodynamics of Fluid Systems*
3. R. N. Franklin: *Plasma Phenomena in Gas Discharges*
4. G. A. Bird: *Molecular Gas Dynamics*
5. H.-G. Unger: *Optical Wave Guides and Fibres*
6. J. Heyman: *Equilibrium of Shell Structures*
7. K. H. Hunt: *Kinematic Geometry of Mechanisms*
8. D. S. Jones: *Methods in Electromagnetic Wave Propagation*
9. W. D. Marsh: *Economics of Electric Utility Power Generation*
10. P. Hagedorn: *Non-linear Oscillations*
11. R. Hill: *Mathematical Theory of Plasticity*
12. D. J. Dawe: *Matrix and Finite Element Analysis of Structures*
13. N. W. Murray: *Introduction to the Theory of Thin-Walled Structures*
14. R. I. Tanner: *Engineering Rheology*
15. M. F. Kanninen and C. H. Popelar: *Advanced Fracture Mechanics*
16. R. H. T. Bates and M. J. McDonnell: *Image Restoration and Reconstruction*
17. A. Holgate: *The Art of Structural Design*

Advanced Fracture Mechanics

MELVIN F. KANNINEN
Southwest Research Institute
CARL H. POPELAR
The Ohio State University

OXFORD UNIVERSITY PRESS · *New York*
CLARENDON PRESS · *Oxford*
1985

OXFORD UNIVERSITY PRESS

Oxford New York Toronto
Delhi Bombay Calcutta Madras Karachi
Petaling Jaya Singapore Hong Kong Tokyo
Nairobi Dar es Salaam Cape Town
Melbourne Auckland

and associated companies in
Beirut Berlin Ibadan Nicosia

Published by Oxford University Press, Inc., 200 Madison Avenue,
New York, New York 10016

Library of Congress Cataloging in Publication Data
Kanninen, Melvin F.
Advanced fracture mechanics.
(Oxford engineering science series)
Includes index.
1. Fracture mechanics. I. Popelar, C. H.
II. Title. III. Series.
TA409.K36 1985 620.1′126 84-27379
ISBN 0-19-503532-1

Printing (last digit): 9 8

Printed in the United States of America
on acid-free paper

PREFACE

Prospective authors of a technical book are faced with a dilemma. If their subject is well established—the theory of elasticity, for example—then it is likely that they have been preceded by someone who has taken far more pains to produce a book than they possibly could. On the other hand, if they have not been so pre-empted, it may well be because their subject is still evolving. Their "snapshot" of a subject in its infancy might then rapidly become out of date.

Fracture mechanics is a subject that has not yet fully matured. Yet, it has existed long enough, and its practical applications are important enough, that a great deal of information is already available. Nevertheless, we wrote this book believing that fracture mechanics is currently at a unique stage. Enough research has been performed to provide a solid foundation upon which future progress will build. At the same time, societal dictates for optimum uses of energy and materials are increasingly forcing structural integrity assessments to be made in the more realistic way afforded by a fracture mechanics approach. Accordingly, a book offering its readers a unifying treatment of the subject for a wide variety of structural materials and application areas is one that should be of value, even though much work remains to be done.

While a number of excellent books on fracture mechanics have already been written—many by our friends and colleagues—we do not feel these offer the particular perspective we have sought in this book. We have addressed the subject from the point of view of applied mechanics. At the same time we feel that some fundamental aspects have not been made as clear in the existing books as they perhaps should be for the newcomer to this field. We hope that we can also improve on this aspect for our readers.

To those not well acquainted with it, the subject of fracture mechanics may appear to be rather exotic and mysterious. But it should not. Any reader who understands the basic concepts of stress and strain, as might be acquired in an undergraduate course on the strength of materials, should find little conceptual difficulty with it. In essence, fracture mechanics circumvents the difficulty arising from the presence of a sharp crack in a stress analysis problem (where there would be an infinite stress, and fracture, under any load) by providing a parameter that characterizes the propensity of the crack to extend. This parameter, which can be generally referred to as the crack extension force, can be calculated knowing the stress-strain behavior of the material, the crack/structure geometry, and the boundary conditions. A critical value of the crack extension force is generally taken as a property of the material. This property, which can be inferred from simple tests, constitutes the only additional information needed.

Fracture mechanics, be it for elastic-brittle, ductile, time-dependent, or heterogeneous materials, is simply based on equating the calculated crack extension force for a cracked structure to the fracture property for the structural material. The result is an explicit relation for crack extension for prescribed applied load, crack size, component dimensions, and material. Applying this method in any particular circumstance may not be obvious. This book would not be necessary if it were. But the basic approach is both simple and widely applicable.

We now state more definitely what we mean by the term "fracture mechanics." In common with most researchers in the field, we define the term in the following way: Fracture mechanics is an engineering discipline that quantifies the conditions under which a load-bearing body can fail due to the enlargement of a dominant crack contained in that body. This definition is obviously quite general. Accordingly, what it does not include is perhaps equal in importance to what it does.

First, this definition does not restrict the size, shape, or location of the crack. Nor does it limit the direction or the rate at which it enlarges. Hence, relatively slow crack growth rates as in stress corrosion and fatigue are included along with dynamic processes such as rapid "brittle" crack propagation. Second, no constraint is placed on the constitutive relation obeyed by the cracked body. It follows that elastodynamic, elastic-plastic, and viscoplastic continuum material behavior, along with heterogeneous and atomistically viewed materials, are equally admissible with the conventional (and most widely used) linear elastic continuum view. Third, the causation of crack extension is not specified in our definition. Mechanical and thermal stresses that vary arbitrarily in time along with environmental agents, separately or in combination, can be considered. As a final point the definition leaves the nature of failure itself unspecified. Thus, any condition from the mere appearance of a detectable crack to catastrophic fracture can be considered within the domain of fracture mechanics.

What should follow from the definition just given is the vacuousness of statements that imply that fracture mechanics does not work in some given area. As an example, not too many years ago many people concerned with the use of fiber composites for aircraft structures undoubtedly would have subscribed to the sentiment expressed by one of them: "Fracture mechanics will work only for a composite structure that someone has attacked with a hatchet." The interpretation of such a remark is this: linear elastic fracture mechanics techniques developed for high strength metals are not directly applicable to a composite unless a through-wall crack exists that is large in comparison to the scale of the micromechanical failure events that precede fracture. Linear elastic fracture mechanics, to be sure, is by far the most highly developed and widely applied version of fracture mechanics. But, it is just that—a specialization of the general subject that must not be considered as synonymous with the subject as a whole. Thus, when "fracture mechanics doesn't work," it is very likely because the methodology has been applied at too simple a level.

A key feature of any fracture mechanics definition is the explicit requirement of a dominant crack. This is the essential difference between fracture mechanics and other kinds of structural analysis. That cracks can and do appear in every type of structure is, of course, the *raison d'etre* of fracture mechanics. But, the requirement that at least one identifiable crack exist can be troublesome. For example, fracture mechanics cannot predict failure in a simple tensile test. No engineering structure can be assessed via fracture mechanics unless at least one crack is either observed (or postulated) to exist in that structure.

Another drawback to fracture mechanics is a subtle one that even many people with long experience in the field do not always recognize. Fracture property values cannot be directly measured. Such values can only be inferred—via the interposition of some assumed analysis model—from quantities that can be experimentally determined. The reason is that there is no instrument that can be made to provide fracture property values for all materials in all testing conditions to the extent that a strain gage measures a change in a length or a thermocouple measures a change in temperature. To "measure" a material fracture property, the theoretical crack driving force is calculated for the crack length and load level at the observed point of crack extension. The fracture property is just the critical value of this crack driving force. While this is true even under linear elastic conditions, there is little difficulty in that regime. But, in nonlinear and dynamic fracture mechanics, serious consequences can result from not recognizing that the fracture "property" can be strongly affected by the analysis method used with the measurement process.

The foregoing requirements suggest a constraint on the definition of fracture mechanics. To qualify as a true fracture mechanics approach, the measured fracture properties must be broadly applicable and not restricted only to the special conditions in which the characterizing experiments are performed. Approaches in which a specific structural component is closely simulated are therefore not in this spirit. Even though such tests are performed on cracked materials, if a basic fracture parameter is not correctly involved, the results are limited to an interpolative function; that is, reliable predictions can only be made for conditions that correspond to those in which the experiments were performed.

The hallmark of a true fracture mechanics approach is that it has an extrapolative function. It should be possible to obtain reliable predictions even for conditions that differ significantly from those in which crack growth measurements were made. In accord with this constraint, fracture mechanics makes possible the use of small-scale laboratory tests (e.g., compact tension specimens) to provide material crack growth and fracture property data for integrity assessments of large-scale structures. Of course, a properly founded analysis approach provides the critical link needed to make such a transition possible.

Our basic definition of fracture mechanics may also help readers of this book to appreciate just how broad the subject of fracture mechanics is. Far from being a specialized subject, it underlies all structural analysis and materials science. No structural material is exempt from a defected condition, and, if it could not fail because of such defects, it would be pointless to analyze it in any other way. Consequently, each and every structural component is, or could be, a candidate for treatment by fracture mechanics. While all applications obviously do not now receive such scrutiny, it is clear from present trends that the years to come will see fracture mechanics assessments become more and more commonplace.

We have sought to satisfy two general groups of readers. In the first group are those who may have had little or no association with fracture mechanics, but possess a background in stress analysis and/or materials science equivalent to that acquired in an undergraduate engineering program. The second group contains those who have worked, perhaps extensively, in a particular aspect of fracture, but who have not been exposed to the variety of application areas covered. Our presentation can be likened to a paraphrase of a remark on the nature of science attributed to the French mathematician Poincaré: a technical book is built of facts the way a house is built of bricks, but an accumulation of facts is no more a book than a pile of bricks is a house. That is, we have sought to provide more than just a haphazard collection of analysis approaches and results. We want instead to show the essential unity of fracture mechanics and the basic commonality of its many specializations. Simply put, our goal is to demonstrate principles rather than recount details. Thus, we want our book to be judged on whether it enables its readers to understand fracture mechanics, not on its worth as a source of up-to-the-minute data and problem-solving techniques.

This book is partly based on lecture notes for a two-quarter course on fracture mechanics taught in the Department of Engineering Mechanics at the Ohio State University. The introductory course for advanced undergraduate and beginning graduate students is confined primarily to linear elastic or small-scale yielding fracture mechanics. It draws upon material from Chapters 1 through 3, supplemented with selected topics from Chapter 5. The more advanced topics in Chapters 2 and 4 through 7 form the subject matter for the second course. Since experience has demonstrated that the book contains more material than can conceivably be covered in a two-quarter course, the book should also be suitable for use in a two-semester course. Chapter 1 evolved from notes developed for short courses designed to introduce fracture mechanics to practicing engineers interested in structural integrity and nondestructive evaluation.

In common with most engineering-oriented subjects, fracture mechanics practitioners have had to face the problems arising from the use of different sets of units. We are convinced that the SI system will eventually become universally accepted and, accordingly, have tried to use it to the extent possible in this book. However, a great amount of data has been collected and reported in English units. We do not feel obliged to convert these data, and, in fact, because the English system is still far from obsolete, feel that we would not be providing a service in so doing. Dual systems are tedious and tend to become much more of a hindrance than a help to understanding. We have provided a conversion chart at the front of the book to assist the reader with a need to have particular results in a system other than the one in which we have reported it.

In writing this book we have been able to draw upon a vast amount of published material. This is of course not an unmixed blessing. There are simply too many worthwhile reports of research activities in fracture mechanics for us to report on but a fraction of them. For example, the two primary journals exclusively devoted to the subject—The *International Journal of Fracture* and *Engineering Fracture Mechanics*—contained some 3100 pages between them in 1983. Added to this are perhaps two dozen other technical journals that regularly contain papers on some aspect of fracture mechanics together with countless volumes of conference proceedings and other compilations. Accordingly, we make no pretense of completeness in covering the subject. We believe that the approximately 800 references we have cited will provide ready access to the remaining literature in any particular specialized area. Furthermore, we have selected references to reflect the main contributors to the subject, thereby identifying the people from whom important work in each area of interest to our readers can be expected in the future. In so doing, we have provided citations that are readily obtainable in English and would be available in most technical libraries. Our apologies to those whose major contributions we have unintentionally (and inevitably) overlooked, and to those whose claims of historical priority—particularly in non-English-language papers—we have thereby violated.

We have found it possible to embark upon the preparation of this book because of the wide diversity of the research we have been involved in. For this, both of us must primarily credit our associations with the Battelle Memorial Institute. Each of us could also compile a long list of colleagues and co-workers who have in some way contributed to extending our knowledge of fracture mechanics. That we have not named them individually does not, we hope, suggest that our debt to these associations is a small one. It is not. There are, however, four individuals whose influence on the first author have been such that he would be extremely remiss not to acknowledge them specifically. These are Mr. Eugene Eschbach, who guided his first professional work while both were employed by the General Electric Company in Richland, Washington; the late Professor Norman Goodier, his teacher, advisor and friend at Stanford University; Dr. George Hahn, his co-worker for many years at Battelle's Columbus Laboratories; and the one foremost in his affections, his wife, Jean. The second author would like to acknowledge his friend and mentor, the late

CONTENTS

CONVERSION FACTORS FOR STRESS

	MPa	ksi	kg mm^{-2}
1 MPa (N mm^{-2}) =	1	0.1450	0.1019
1 ksi =	6.895	1	0.7031
1 kg mm^{-2} =	9.807	1.4223	1

CONVERSION FACTORS FOR THE STRESS INTENSITY FACTOR

	MPa $m^{1/2}$	ksi $in^{1/2}$	N $mm^{-3/2}$	kg $mm^{-3/2}$
1 MPa $m^{1/2}$ =	1	0.910	31.62	3.224
1 ksi $in^{1/2}$ =	1.099	1	34.75	3.542
1 N $mm^{-3/2}$ =	0.03162	0.02878	1	0.1019
1 kg $mm^{-3/2}$ =	0.3102	0.2823	9.807	1

1

INTRODUCTION AND OVERVIEW

The existence of crack-like flaws cannot be precluded in any engineering structure. At the same time, increasing demands for energy and material conservation are dictating that structures be designed with smaller safety margins. Consequently, accurate quantitative estimates of the flaw tolerance of structures is increasingly becoming of direct concern for the prevention of fracture in load-bearing components of all kinds. This has not always been so. Prudent design procedures that avoided large stress concentrations—together with immediate repair or retirement from service of components that exhibited cracks—have been reasonably effective in preventing catastrophic failures. However, two important factors have now emerged to negate this traditional strategy.

First, improved nondestructive evaluation (NDE) procedures have enabled defects to be found that would have gone unnoticed earlier. Second, the presence of a crack-like defect does not necessarily mean that a structural component is at (or even near) the end of its useful service life. The cost of the repair or replacement of a flawed component can therefore be balanced against the possibility that continued service could lead to a failure. The new engineering concept known as *damage tolerance* has been developed to provide quantitative guidance for this purpose. It, in turn, is largely based upon the technology of *fracture mechanics.* While not the only ingredient of structural integrity assessments, as this book will make clear, it plays a central role.

Concern for fracture has surely existed back to antiquity. While much of this concern is unrecorded, some evidence of scholarly study that substantially predates our times does exist; see for examples Gordon's books (1.1). As described in Timoshenko's history of the strength of materials (1.2)—see also Irwin's review paper (1.3)—da Vinci performed experiments to determine the strength of iron wires in the fifteenth century. He found an inverse relationship between the wire length and the breaking load for constant diameter wires. Because this result would otherwise imply that strength is dependent upon the wire length, it can be surmised that the presence of cracks dictated the fracture stress; that is, the larger the volume of material tested, the more likely it is that a large crack exists. Considering the wire quality available at that time, this is highly plausible. Nevertheless, little of a quantitative nature could be done with this possibility. Fracture theories based on crack extension require the mathematical concepts of stress and strain that were not forthcoming until given by Cauchy and the other great French mathematician/engineers of the nineteenth century (1.4).

A. A. Griffith was the first to make a quantitative connection between strength and crack size (1.5). Hence, as one possibility, fracture mechanics could be dated from 1922. However, many would agree that fracture mechanics became largely an engineering discipline, as opposed to one of mere scientific curiosity, as a result of George Irwin's basic contributions in the years following the Second World War. Accordingly, we feel that fracture mechanics should be dated from 1948, the year of publication of the first of Irwin's classic papers (1.6).

Because the developments that directly followed from Irwin's work were almost entirely focused on linear elastic fracture mechanics, there appears to be a second distinct demarcation point in the history of fracture mechanics. This point coincides with the introduction of the basic ideas necessary for the treatment of nonlinear problems. This time can be taken as 1968. In that year J. R. Rice presented his J-integral (1.7) and J. W. Hutchinson (1.8) showed how such a concept could be used to obviate the need for a direct description of the discrete and nonlinear events involved in crack extension. We will refer to the methodology that evolved subsequently for the treatment of nonlinear and dynamic problems beyond linear elastic fracture mechanics as *advanced fracture mechanics.*

As our title suggests, this book will primarily address the subject in terms of the nonlinear and dynamic aspects that require analysis techniques beyond those now in common use. To set the stage for these presentations, this overview chapter is intended as an introduction to fracture mechanics for those not previously acquainted with the subject. It is written on a level that should be readily accessible to readers familiar with the basic concepts of stress and strain—that is, as might be acquired in an undergraduate course in the strength of materials. The first section of this chapter presents applications of current techniques. The four sections that follow step back to introduce fracture mechanics from a historical point of view. The penultimate section then presents some practical problems of current interest where conventional linear elastic fracture mechanics techniques will not entirely suffice. A short philosophical section giving our personal views on fracture mechanics and its uses concludes the chapter.

1.1 Current Fracture Mechanics and Its Applications

Fracture mechanics is an engineering discipline that primarily draws (in roughly equal proportions) upon the disciplines of applied mechanics and materials science. In its most basic form it can be applied to relate the maximum permissible applied loads acting upon a structural component to the size and location of a crack—either real or hypothetical—in the component. But, it can also be used to predict the rate at which a crack can approach a critical size in fatigue or by environmental influences, and can be used to determine the conditions in which a rapidly propagating crack can be arrested. Current damage tolerance assessment procedures are now available that make effective use of these capabilities for materials that otherwise behave in an essentially linear elastic manner.

In applications where either extensive elastic-plastic or time-dependent deformation might be experienced prior to fracture, linear elastic fracture mechanics methods are generally inadequate. But, procedures are now becoming available for such conditions. Also on the horizon are treatments of the even more complicated conditions involved in the cracking of welds and other areas where residual stresses are present, of heterogeneous materials such as fiber reinforced composite materials, of adhesives and other viscoelastic materials, and the like. However, it is unlikely that one could have an appreciation for such advanced work without understanding the manner in which fracture mechanics applications are currently being made. Accordingly, we begin by describing current fracture mechanics and its applications. Consistent with the approach that we will take throughout this book, we will do so from the point of view of applied mechanics.

1.1.1 The Consequences of Fracture

Readers of this book will certainly already have a definite interest in fracture mechanics and, no doubt, a general appreciation for the consequences of fracture in practice. Nevertheless, some graphic examples might usefully be provided. To begin, Figures 1.1, 1.2, and 1.3, taken from Burdekin (1.9), show instances where a catastrophic fracture in a structural component could be traced to the existence of a crack-like flaw. While pictures of failures such as these are not uncommon, these pictures are unusual in one respect—the flaw that triggered the fracture was specifically identified. These flaws are shown along with the fractured component in Figures 1.1, 1.2, and 1.3. In all three cases it can be seen that the initiating flaws were not overly large. Conceivably, they could well have been detected prior to failure whereupon applications of fracture mechanics would presumably have revealed that the structure was in jeopardy. Obviously, they were not.

The costs of fracture associated with industrial accidents such as those pictured are not easily determined. They would likely be dominated by the replacement cost and the loss of revenue in the interim. But, there are other instances in which these are dwarfed by another possible aspect: the loss of life and property in the neighborhood of the facility. Perhaps the most spectacular instance of this kind is the catastrophic rupture of a liquified natural gas (LNG) storage tank that took place in Cleveland in 1944. Figure 1.4 shows the attendant devastation in the neighborhood of the plant. The ruptured vessel and two of the other four originally at the site, can be seen near the center of the photograph.

According to Atallah (1.10), 79 houses, 2 factories, and some 217 automobiles were totally destroyed with another 35 houses and 13 factories being heavily damaged. The extent of the combined property damage was

Figure 1.1 A nuclear plant boiler failure precipitated by a weld crack.

estimated at 6 to 7 million (1944) dollars. The sequence of events evidently involved an initial rupture of the vessel that allowed a substantial amount of liquefied natural gas to escape. The liquid then vaporized and was somehow ignited. When the gas ignited, according to Atallah,

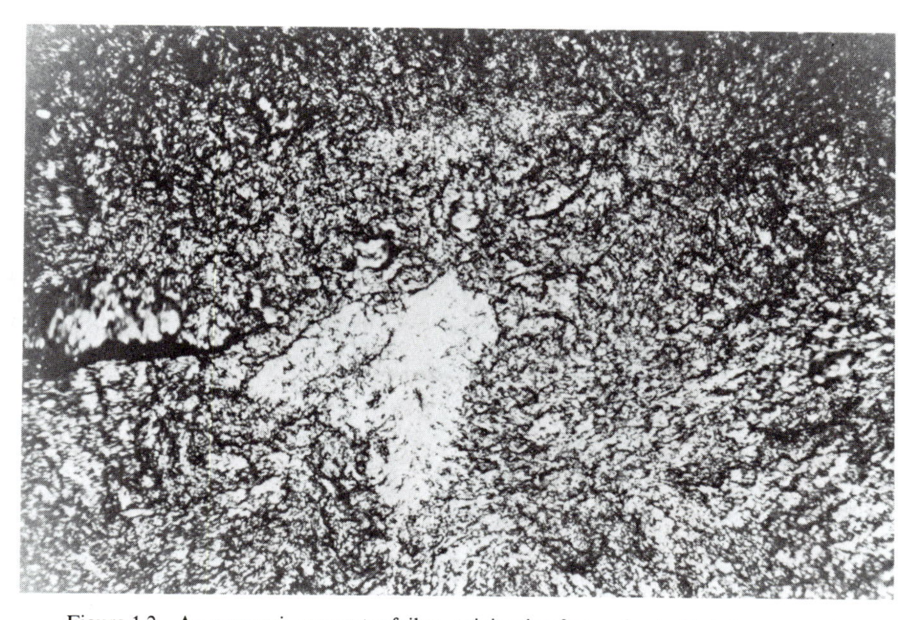

Figure 1.2 An ammonia converter failure originating from a heat-affected-zone crack.

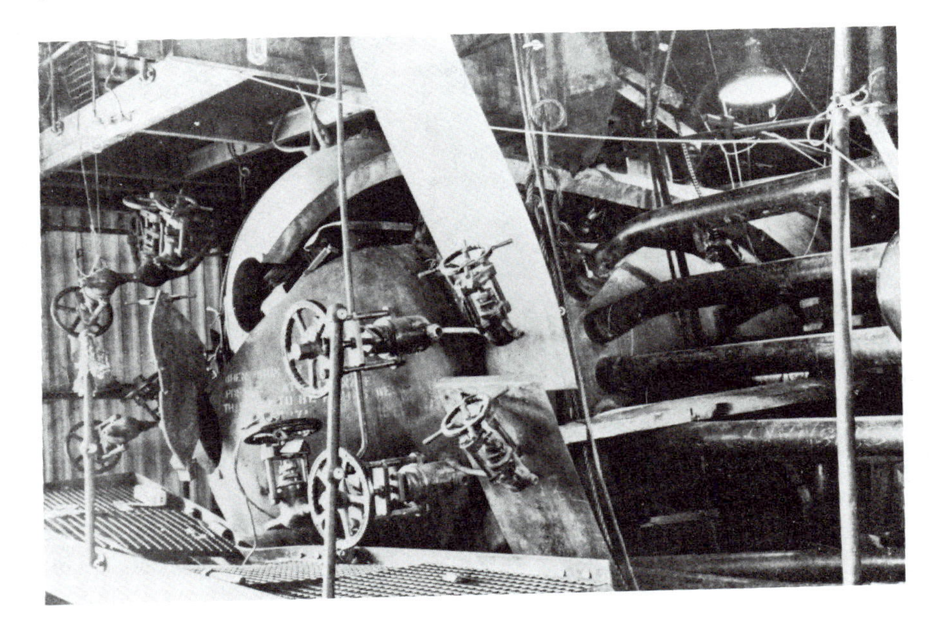

Figure 1.3 A power station boiler failure caused by a surface crack.

Figure 1.4 Devastated area in the vicinity of a ruptured liquefied natural gas storage tank.

> sewers exploded, propelling manhole covers into the air. Heavy underground blasts lifted entire street pavements, demolished houses, ruptured water and gas mains, and broke hundreds of store plate-glass and residence windows. ... the height of the ball of flame (was estimated) at 2800 feet. Roasted birds fell out of the sky.

Most serious of all was that the LNG tank fires and the explosions that accompanied them killed 130 people and seriously injured another 300. These figures are understandable in view of the devastation that can be seen in Figure 1.4.

The origin of the Cleveland LNG tank failure was never unequivocally determined. But, it is thought to have been due to a welding defect with subsequent fatigue crack growth caused by vibrations and shocks from heavy train traffic and from the many stamping mills in the vicinity (1.10). Coupled with this was the likelihood that the material used for the tank—a low-carbon 3.5 percent nickel-alloy steel—was too low in toughness at the service temperature of -250°F.

In 1982 the National Bureau of Standards (NBS) commissioned a study of the total direct and indirect costs of fracture in the economy of the United States. This study, conducted by Duga et al. (1.11), not unexpectedly reveals that the costs are high indeed. Along with the direct losses and imputed costs associated with fracture-related accidents of all kinds, they have included estimates encompassing the necessity to overdesign structures because of nonuniform material quality and to perform inspection, repair, and replacement on materials that have degraded in service. The grand total is some 120 billion dollars annually. One can only wonder about the costs worldwide, and, with the ever increasing reliance on structural integrity as our society becomes ever more complex, what these costs will become in the future.

An interesting feature of the NBS study is the estimates that have been made on possible savings. It was suggested that some 35 billion dollars annually (30 percent of the grand total) could be saved if all known best fracture control technology were applied today. Another 28 billion dollars (23 percent) could be saved as a result of applying the new knowledge that they expect to be generated in the future. Regardless of how literally one takes these estimates, it is certainly clear that fracture is a serious problem and that much more could be done to resolve it than is currently being done. In view of the relative newness of fracture mechanics as an engineering discipline, it is not too surprising that many structural designers, metallurgists, nondestructive evaluators, and others concerned with structural integrity do not employ fracture mechanics as the engineering tool as it can and should be used.

1.1.2 Fracture Mechanics and Strength of Materials

In the strength of materials approach that is certainly familiar to all structural engineers, one typically has a specific structural geometry (assumed to be defect free!) for which the load carrying capacity must be determined. To accomplish this, a calculation is first made to determine the relation between the load and the maximum stress that exists in the structure. The maximum stress so determined is then compared with the material's strength. An acceptable design is achieved when the maximum stress is less than the strength of the material, suitably reduced by a factor of safety. The similarity between this approach and that of fracture mechanics can be illustrated with the help of the example shown in Figure 1.5.

In the simple structure shown in Figure 1.5(a), a built-in cantilever beam of length L, depth H, and thickness B is required to support a weight W at its free end. As indicated in Figure 1.5(b), the maximum tensile stress acts in the outermost fibers of the beam at its built-in end and is related to the load and the beam dimensions by the relation

$$\sigma_{\max} = \frac{6WL}{BH^2} \tag{1.1-1}$$

It can be assumed that failure will not occur unless $\sigma_{\max}$ exceeds the yield strength of the material, σ_Y. Then, for fixed beam dimensions, W must be small enough that the right-hand side of Equation (1.1-1) is less than σ_Y. To assure this, a factor of safety S can be introduced to account for material variability and/or unanticipated greater service loadings. Using Equation (1.1-1), $\sigma_{\max}$ will be less than σ_Y/S if

$$W < \frac{BH^2}{6SL}\sigma_Y \tag{1.1-2}$$

whereupon the structure will be safe, at least from the viewpoint of strength of materials.

Now, consider that the beam, instead of being defect-free, contains a crack. Further suppose that, as shown in Figure 1.5(c), the crack is located where the maximum stress is anticipated. As will be thoroughly discussed below, the governing structural mechanics parameter when a crack is present, at least in the linear approach, is an entity called the stress intensity factor. This parameter, which is conventionally given the symbol K, can be determined from a mathematical analysis like that used to obtain the stresses in an uncracked component. For a relatively small crack, an analysis of the flawed beam shown in Figure 1.5(c) would give to a reasonable approximation

$$K = 1.12\,\sigma_{\max}\sqrt{\pi a} \tag{1.1-3}$$

where a is the depth of the crack and $\sigma_{\max}$ is the stress that would occur at the crack location in the absence of the crack.

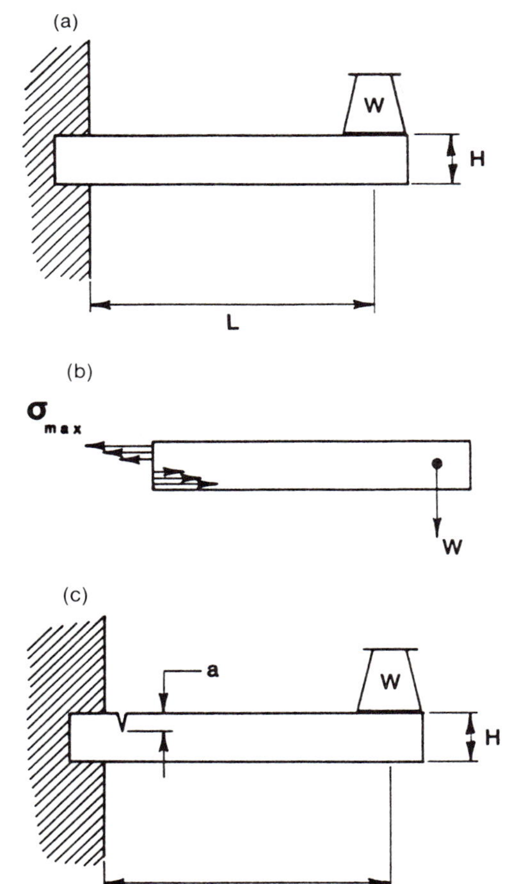

Figure 1.5 Basis for the comparison of strength of materials and fracture mechanics approaches: (1) unflawed cantilever beam, (b) tensile stresses in an unflawed beam, (c) cracked beam.

The basic relation in fracture mechanics is one that equates K to a critical value. This critical value is often taken as a property of the material called the plane strain fracture toughness, conventionally denoted as K_{Ic}. When an equality is achieved between K and K_{Ic}, the crack is presumed to grow in an uncontrollable manner. Hence, the structure can be designed to be safe from fracture by assuring that K is less than K_{Ic}. Further assurance can be obtained by having $K < K_{Ic}/S$, where, just as in the strength of materials approach, the number S is a factor of safety. Using Equation (1.1-1) to replace σ_{max} in Equation (1.1-3) then leads to

$$W < \frac{BH^2}{6SL} \frac{K_{Ic}}{1.12\sqrt{\pi a}} \tag{1.1-4}$$

which is the fracture mechanics estimate of the safe operating load.

A comparison between inequalities (1.1-2) and (1.1-4) is instructive. It can be seen that the structural geometry and the factor of safety enter both relations in exactly the same way—that is, through the multiplicative parameter $(BH^2/6SL)$. Also, both relations contain a basic, albeit different, material property. The essential difference is that the fracture mechanics approach explicitly introduces a new physical parameter: the size of a (real or postulated) crack-like flaw. In fracture mechanics the size of a crack is the dominant structural parameter. It is the specification of this parameter that distinguishes fracture mechanics from conventional failure analyses.

1.1.3 Basic Uses of Linear Elastic Fracture Mechanics

The generalization of the basis for engineering structural integrity assessments that fracture mechanics provides is portrayed in terms of the failure boundary shown in Figure 1.6. Clearly, fracture mechanics considerations do not obviate the traditional approach. Structures using reasonably tough materials (high K_{Ic}) and having only small cracks (low K) will lie in the strength of materials regime. Conversely, if the material is brittle (low K_{Ic}) and strong (high σ_Y), the presence of even a small crack is likely to trigger fracture. The fracture mechanics assessment is then the crucial one.

However, fracture is not the only way that a structure can fail. In the example of Figure 1.5, an excessive deflection at the end of the beam, even though the beam still adequately supports some load, could be judged as a failure in some applications. Nonetheless, in the presence of a crack-like flaw, fracture is likely to be of most concern.

Note here that the words "crack" and "flaw" tend to be used interchangeably. But, while all cracks can be considered to be flaws (or defects), not all flaws are cracks. The distinction is in the sharpness of the tip, a crack being a flaw with a very small radius of curvature at its tip. Volumetric defects are clearly not cracks. But, unless specific information is available to the contrary, prudence dictates that all flaws be so considered.

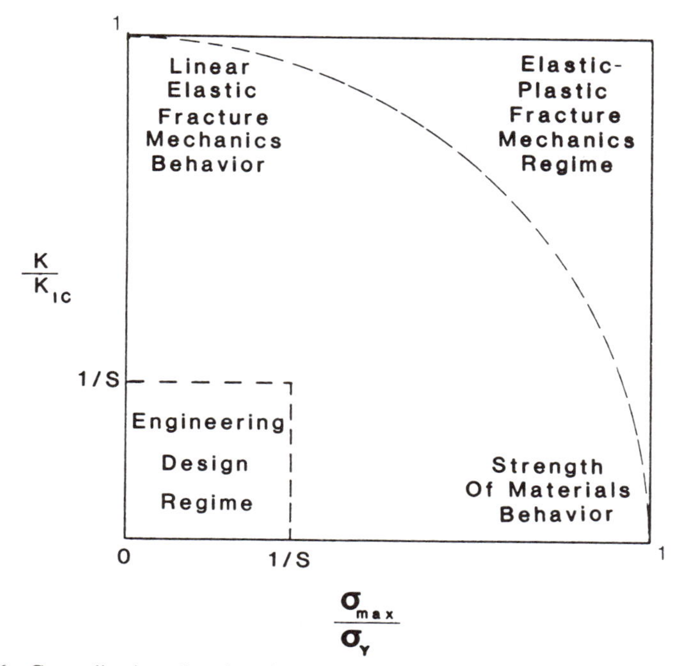

Figure 1.6 Generalization of engineering structural integrity assessments provided by fracture mechanics.

Although it is tangential to our main objective here, the special circumstances that would be called into play in the upper right-hand corner of the diagram shown in Figure 1.6 are worth noting. In this regime a cracked structure would experience large-scale plastic deformation prior to crack extension. Linear elastic treatments of the kind discussed so far are then invalid. As discussed later in this chapter, this fact necessitates the use of nonlinear fracture mechanics treatments and requires more precise definitions of the fracture parameters. However, for introductory purposes, it will suffice to simply accept the specialization of the general subject known as "linear elastic fracture mechanics" (or, simply, LEFM) as applicable for all crack/structure/loading conditions where the inherent inelastic deformation surrounding the crack tip is small.

The problem category in which LEFM is valid is also known as one satisfying "small-scale yielding" conditions. Its validity requires that the applied stresses be small enough that general plastic yielding does not occur. Figure 1.7 shows an example of the center cracked panel data collected by Fedderson (1.12), which illustrates that fracture can indeed occur well below net section yielding. These results, made on an aircraft material, typify the applications for which LEFM has so effectively been made.

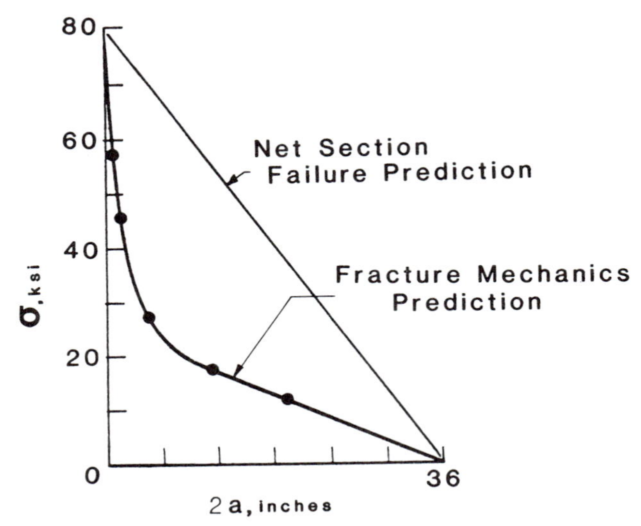

Figure 1.7 Comparison of net-section failure and fracture mechanics prediction for center-cracked tension panels of 7075-T6 aluminum alloy.

There are three general ways in which a flaw can appear in a structure. These are, (1) inherent defects that occur in the material (e.g., inclusions in a metal, debonded regions in a composite), (2) defects introduced during the fabrication of a structural component (e.g., lack of fusion in a weld, welding arc strikes), and (3) damage incurred during the service life of the component (e.g., dents and cuts, fatigue, and environmentally assisted cracking). Within the confines of linear elastic fracture mechanics, it makes no difference how a flaw is introduced. A crack-like flaw of a given size and position in a body is assumed to obey the same fracture rules regardless of its origins. The basic capability that fracture mechanics provides can then be employed in either of two general ways. First, the maximum safe operating loads that an engineering structure can sustain for the sizes and locations of *existing flaws* can be determined. Such cracks might be those actually found during an inspection, whereupon the continued safe operation of the structure is in question. Second, for *given loads*, the largest crack size that can exist without fracture can be determined. This will provide specifications to be set in advance of an inspection.

Of most importance in regard to linear elastic fracture mechanics are applications where weight is a primary concern. When the resistance to yielding is high (e.g., in a high strength steel), the fracture toughness tends to be low. Combined with the necessity to use highly stressed components, it follows that fracture mechanics analyses are essential to achieve a proper balance between performance and reliability. In fact, the use of fracture mechanics for structural integrity assessments is largely due to the importance of considering small flaws in aerospace, off shore, and other applications where high strength materials are used to minimize weight.

1.1.4 Linear Elastic Fracture Mechanics Relations

Linear elastic fracture mechanics relates the size of a crack with the loading that will fracture a given component by linking two separate activities: (1) a mathematical stress analysis of the loaded structure, and (2) experimental measurements of the material's fracture properties. Expressed in quantitative terms, fracture will occur when

$$K(a, D, \sigma) = K_c(T, \dot{\sigma}, B) \tag{1.1-5}$$

where K is a calculated parameter that, as indicated in Equation (1.1-5), depends on crack size, a, component dimensions, D, and applied stress, σ. It will not depend on the material. In contrast, K_c is a material parameter called the fracture toughness that depends on the temperature at the crack tip, T, the rate of loading, $\dot{\sigma}(\equiv d\sigma/dt)$, and B, the thickness of the cracked section. It is an experimentally measured quantity that is independent of the crack/structure geometry, of the loading imposed on the structure, and of the crack size.*

It is particularly important to understand that, in order to perform a fracture mechanics assessment, both K and K_c are needed: neither parameter is meaningful by itself. To recall the analogy between fracture mechanics and strength of materials given in Section 1.1.2, this parallels the basic distinction that exists in the latter subject. The strength of a material is the stress required to break a specimen of that material—a number usually determined in a uniaxial tension test. The strength of a structure is the force that makes the maximum stress acting in the structure equal to the strength of the material, the maximum stress being independent of the material. Thus, the relation $\sigma_{max} = \sigma_Y$ is a direct counterpart of Equation (1.1-5), including the fact that the values of the two quantities are not meaningful by themselves. These values are only relevant in a relative sense.

Table 1.1 displays a set of representative stress intensity factors for some simple load/crack/structure combinations of interest. This list is far from exhaustive (n.b., catalogs of stress intensity factors have been compiled—see Chapter 9). Figure 1.8 shows typical ranges of the fracture toughness data for aluminum, titanium, and steel alloys. These data illustrate the inverse relation between yield strength and fracture toughness that generally exists. Collections of fracture toughness data also can be found and these too are cited in Chapter 9.

The pronounced dependence of the fracture toughness upon the degree of triaxial constraint at the crack tip is of some importance. Triaxial constraint is primarily manifested by the plate thickness in a through-wall cracked plate test. This effect is illustrated by the data of Jones and Brown (1.13) on 4340 steel that are shown in Figure 1.9. It can be seen that the lowest value of K_c is that which occurs for large thicknesses—that is, the plane strain fracture toughness value, K_{Ic}. Here, plane strain conditions hold and the triaxial

* When the loading rate becomes significant, the symbol K_d is often used to denote the fracture toughness. Similarly, when Equation (1.1-5) is used to characterize the arrest of a rapidly propagating crack, the symbol K_a is used.

Table 1.1 Approximate Stress Intensity Factors for Selected Crack/Structure Geometries

Structure	Crack	Load	Stress Intensity Factor
Very large body subjected to a tensile stress	Centrally located crack of length $2a$	Remote tension σ normal to crack	$K = \sigma\sqrt{\pi a}$
Very large body subjected to a tensile stress	Edge crack of length a normal to free edge	Remote tension σ normal to crack	$K = 1.12\sigma\sqrt{\pi a}$
Strip of width $2W$ subjected to a tensile stress	Centrally located crack of length $2a$	Remote tension σ normal to crack	$K = \sigma\sqrt{\pi a}\sec^{\frac{1}{2}}\left(\frac{\pi a}{2W}\right)$
Vessel having diameter D and wall thickness h	Through-wall axial crack of length $2a$	Internal pressure p	$K = \frac{pD}{2h}\sqrt{\pi a}\left[1 + 3.22\frac{a^2}{Dh}\right]^{\frac{1}{2}}$
Vessel having diameter D and wall thickness h	Through-wall circumferential crack of length $2a$	Internal pressure p	$K = \frac{pD}{4h}\sqrt{\pi a}\left[1 + 1.12\frac{a}{\sqrt{Dh}} \cdot \left(1 - \exp\left(-1.54\frac{a}{\sqrt{Dh}}\right)\right)\right]$
Very large body with circular hole of radius R	Two symmetrical cracks at edge of hole of length a	Remote tension σ normal to the cracks	$K = \sigma\sqrt{\pi a}\left[1 + 2.365\left(\frac{R}{R+a}\right)^{2.4}\right]$

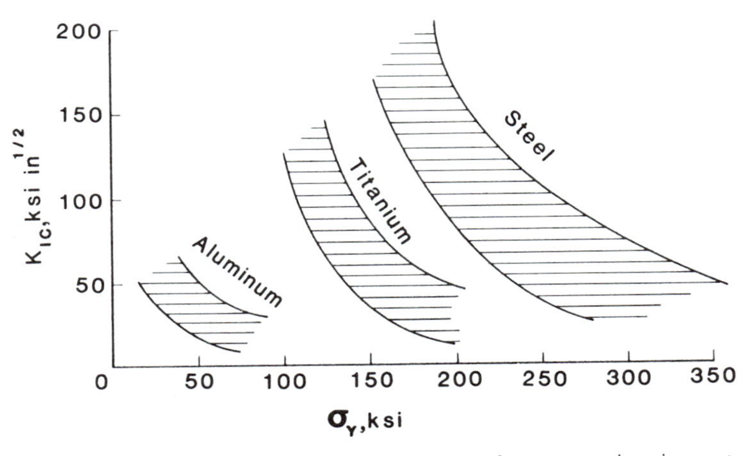

Figure 1.8 Fracture toughness and yield strength ranges for some engineering materials at ambient temperature.

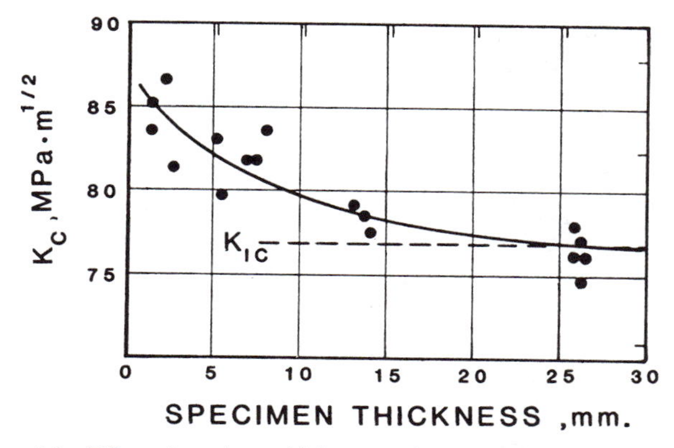

Figure 1.9 Effect of specimen thickness on fracture toughness of 4340 steel.

constraint is the greatest. The plane strain fracture toughness determinations for metallic materials are made under precisely defined procedures dictated by the American Society for Testing and Materials (ASTM) Standard E399.

A convenient relation between the plane strain fracture toughness K_{Ic} and the somewhat greater values under less constraint is the semiempirical equation developed by Irwin (1.14). This is

$$K_c = K_{Ic}\left[1 + \frac{1.4}{B^2}\left(\frac{K_{Ic}}{\sigma_Y}\right)^4\right]^{\frac{1}{2}} \tag{1.1-6}$$

where B is the plate thickness and σ_Y is the yield stress of the material. Note that this equation can be used either, (1) to estimate toughness values for conditions that approximate plane stress conditions, knowing the plane strain value, or (2) to remove the thickness dependence in small size specimen testing to obtain the plane strain value as Merkle has done (1.15).

The effects of temperature and loading rate on the fracture toughness values of engineering materials are also of importance. Typically, a pronounced difference exists between the lower toughness values at low temperatures, generally resulting from cleavage fracture behavior, and the higher toughness values at high temperatures characteristic of ductile fracture. Although the transition occurs over a range of temperatures, a material-dependent temperature is conventionally used to delineate the two regimes. This is known as the ductile-brittle transition temperature and as the nil-ductility transition (NDT) temperature. The opposite effect of loading rate above and below the ductile-brittle transition temperature might be noted. Generally, increasing the loading rate diminishes the fracture toughness in the low toughness region while increasing it in the high toughness region.

The reference temperature designations have been a source of confusion. It should be recognized that an ASTM standard, E208, exists for determining the NDT temperature. This standard, which specifies the use of the drop-weight test developed at the Naval Research Laboratory, is referenced in several ASTM specifications and the American Society of Mechanical Engineers

(ASME) Boiler and Pressure Vessel Code; see Section 1.1.8. An alternative reference parameter is RT_{NDT}, the reference temperature for nil-ductility transition. This is an approximation to the NDT that is obtained by Charpy specimen testing. The RT_{NDT} can either exceed the NDT temperature, or be equal to it, but cannot be less. The RT_{NDT} is most useful when insufficient material exists to perform either a fracture experiment or a sufficient number of drop-weight tests to properly determine the NDT.

An early fracture control measure was simply to assure that the component operated above the ductile-brittle transition temperature as determined, for example, by Charpy tests. However, such a procedure does not preclude fracture. As emphasized in the foregoing, fracture can always occur if the combination of the load and crack size gives rise to a high enough crack driving force; that is, even at "upper shelf" conditions, the material's fracture toughness is finite. Furthermore, fracture control measures based only on operation above the transition temperature do not provide quantitative connections between crack sizes and applied loads. Nevertheless, Charpy testing retains an important role in fracture mechanics.

A Charpy test is usually much less expensive to perform than a fracture toughness characterization experiment. In addition, there are instances where it is very difficult to obtain fracture toughness data directly—for example, in irradiated capsules and weldments where only a small volume of material is available for testing. Consequently, it is not surprising that a number of empirical correlations between Charpy values and fracture toughness have been established. Perhaps the best known is the upper shelf correlation developed by Rolfe and Novak (1.16). Their relation was based on results obtained on 11 steels having yield strengths ranging from 110 to 246 ksi. This relation is

$$\left(\frac{K_{Ic}}{\sigma_Y}\right)^2 = 5\left[\frac{CVN}{\sigma_Y} - .05\right] \tag{1.1-7}$$

where CVN is the upper shelf Charpy energy in ft-lbs, σ_Y is in ksi and K_{Ic} is in ksi in.$^{1/2}$ Many other correlations of this type also exist; the book of Barsom and Rolfe (see reference in Chapter 9) provides a good source for these. Needless to say, all such empirical relations have a limited range of applicability and should be used with caution.

The work of Oldfield (1.17) in developing fracture toughness reference curves should also be mentioned in this regard. He has been successful with the choice of the sigmoidal function

$$K_c = A + B\tanh\left(\frac{T - T_0}{C}\right) \tag{1.1-8}$$

where T is the temperature while A, B, C, and T_0 are arbitrary constants that fit Charpy test data. This relation gives "lower shelf" behavior for $T \ll T_0$ and "upper shelf" behavior for $T \gg T_0$ with a transition region between. Note that the physical mechanisms that give rise to these differences (i.e., the brittle cleavage fracture that generally occurs at low temperatures versus the ductile fracture at higher temperatures) are not of concern (nor need they be) in this type of formulation. We will return to consider this representation in Section 1.1.8.

1.1.5 Some Illustrative Applications of Fracture Mechanics

To demonstrate the application of fracture mechanics to practical structures, consider a cylindrical pressure vessel. The design of such a structure could be based simply upon the hoop stress, $\sigma_h = pD/2h$, where p is the internal pressure, D is the mean vessel diameter, and h is the wall thickness. Hence, following the strength of materials procedure of Section 1.1.2, a safe operating pressure would be given by equating σ_h to σ_Y to obtain

$$p = \left(\frac{2h}{SD}\right)\sigma_Y \tag{1.1-9}$$

where σ_Y is the material yield stress and, again, the factor of safety S is introduced to reflect common engineering practice.

Now, suppose that a shallow axial surface crack of depth a could exist in the vessel. Just as in the beam example given above, the fracture mechanics approach to evaluating the critical crack depth requires the appropriate stress intensity factor. Typically, one would make the conservative assumption that the crack length along the surface is much greater than its depth into the wall. An approximate relation for such a crack is that for an edge-cracked plate in tension; see Table 1.1. Equating this K value to K_{Ic}/S then gives a safe operating pressure for a flawed vessel. The result is

$$p = \left(\frac{2h}{SD}\right)\frac{K_{Ic}}{1.12\sqrt{\pi a}} \tag{1.1-10}$$

The commonality of the geometric parameters and the factor of safety in the grouping $(2h/SD)$ can again be seen in both estimates.

A use to which Equation (1.1-10) could be put is to determine the maximum vessel pressure, given the minimum flaw size that could be reliably detected. But, knowing the depth of a crack that could cause fracture in operation is of equal importance. For example, the critical flaw depth could be determined for operation at a design pressure given by Equation (1.1-9). It is easily shown by combining Equations (1.1-9) and (1.1-10) that this depth is approximately $a = \alpha(K_{Ic}/\sigma_Y)^2$, where α is a geometry-dependent constant that is equal to 0.25 in this example. Note that rough "back of the envelope" estimates are often made by taking α to be this. Thus, as this simple calculation shows, fracture mechanics estimates can be obtained using concepts not much more complex than those used routinely in engineering structural integrity assessments.

It is important to recognize that the location of a flaw is just as significant as its size. For example, consider a large tension panel of 4340 steel (a material used extensively in airframes) with a yield stress of 240 ksi and a K_{Ic} value of 50 ksi-in.$^{\frac{1}{2}}$. Suppose that the crack is a through-wall crack in the center of the panel and its length is small relative to the overall dimensions of the panel. The first entry in Table 1.1 is therefore appropriate. If the applied stress is taken as 60 ksi (i.e., $S = 4$), the critical length of such an *internal* crack is then given by

$$2a_c = \frac{2}{\pi}\left(\frac{K_{Ic}}{\sigma}\right)^2 = \frac{2}{\pi}\left(\frac{50}{60}\right)^2 = 0.44 \text{ in.}$$

For comparison, suppose that the crack instead exists at the edge of the panel, but with all other conditions being the same. Then, using the second entry in Table 1.1, the critical length of an *edge* crack is

$$a_c = \frac{1}{1.25\pi}\left(\frac{K_{Ic}}{\sigma}\right)^2 = \frac{1}{1.25\pi}\left(\frac{50}{60}\right)^2 = 0.18 \text{ in.}$$

Thus, because the critical crack size is smaller, the crack in the edge of a plate is considerably more dangerous than the one in the interior.

To both reveal the influence of the material and to demonstrate a possible remedial action when the critical flaw sizes are too small, consider a lower toughness grade of 4340 steel. Such a material might have a yield strength of 180 ksi and a K_{Ic} value of 105 ksi-in.$^{\frac{1}{2}}$ (cf. Figure 1.8). Consider again an edge crack in a large tension panel subjected to a tensile loading equal to 60 ksi. This condition leads to a critical length of

$$a_c = \frac{1}{1.25\pi}\left(\frac{K_{Ic}}{\sigma}\right)^2 = \frac{1}{1.25\pi}\left(\frac{105}{60}\right)^2 = 0.78 \text{ in.}$$

This marked difference with the high strength grade of 4340 steel clearly points up the importance of considering trade-offs in the selection of the component material.

For an example illustrating how a stress intensity factor handbook might be used, consider the cracked cantilever beam shown in Figure 1.5. A more precise expression for the stress intensity factor can be obtained from a handbook in the typical form

$$K = \frac{6WL}{bH^2}\, a^{\frac{1}{2}}\left[1.99 - 2.47\left(\frac{a}{H}\right) + 12.97\left(\frac{a}{H}\right)^2 - 23.17\left(\frac{a}{H}\right)^3 + 24.8\left(\frac{a}{H}\right)^4\right] \tag{1.1-11}$$

where L now denotes the distance between the crack and the loaded end of the beam, with other parameters as given in Figure 1.5. Note that in the earlier example the value of a/H was assumed to be negligible in comparison to unity; an assumption that is reasonable only for crack sizes that are in the order of a few percent of the beam depth. As Equation (1.1-11) indicates, for very deep cracks, this would be a nonconservative assumption. To demonstrate this, Equations (1.1-3) and (1.1-11) can be combined to give

$$K = (K)_0\left[1 - 1.24\left(\frac{a}{H}\right) + 6.52\left(\frac{a}{H}\right)^2 - 11.64\left(\frac{a}{H}\right)^3 + 12.46\left(\frac{a}{H}\right)^4\right] \tag{1.1-12}$$

where $(K)_0$ denotes here the value of the stress intensity factor as $a/H \rightarrow 0$; that is, its value from Equation (1.1-3). For example, at $a/H = 0.5$, the stress intensity factor would be some 33 percent greater than this limiting value. The stress intensity factor expressions contained in the catalogs listed in Chapter 9 are very often given in polynomial form, as in this instance.

1.1.6 Approaches for Complex Crack/Structure Geometries

The examples given in Section 1.1.5 were for rather simple crack geometries. To illustrate how a fracture mechanics approach can relate to more realistic geometries, consider the part-through-wall surface flaw shown in Figure 1.10. Here the depth of the flaw in the thickness direction is denoted by a, its length along the wall by $2c$, and the wall thickness is h. The wall is subjected to a uniform tensile stress that acts remotely in the direction normal to the crack plane. A "worse case" analysis would consider that $c \gg h > a$. The assumed crack front would then be the horizontal line in Figure 1.10 whereupon the analysis problem is one of plane strain with the only crack dimension of concern being a. To determine the critical crack depth a_c for a given applied stress σ, the approximate relation $K = 1.12\sigma\sqrt{\pi a}$ (see Table 1.1) can therefore be equated to K_{Ic} and solved for $a = a_c$ as described in Section 1.1.5.

Unstable crack growth in the thickness direction presents the structure with a through-wall crack. But, this crack will not necessarily continue to propagate. That is, because the speed at which the crack propagates after instability occurs is generally much higher than the loading rate, a safe assumption is that a through-wall crack appears at the instant of fracture instability. The length of the through-wall crack will then probably be equal to $2c$; see the vertical lines in Figure 1.10. The computational process required to determine the critical crack length c_c would then use the relation $K = \sigma\sqrt{\pi c}$. The appropriate fracture toughness value would depend upon the degree of lateral constraint at the crack tip, which, in turn, depends upon the thickness of the cracked component; see Figure 1.9. While the use of K_{Ic} for the through-wall crack would then be a conservative assumption, a more accurate procedure would be to use Equation (1.1-6) to obtain the K_c value corresponding to the wall thickness $h = B$.

The results of the part-through-wall and through-wall analyses can be coupled to give a conservative estimate of the critical dimensions of an initial flaw. Figure 1.11 compares three possible flaw shapes with these critical dimensions. Flaw Type A is benign even though its surface dimension is

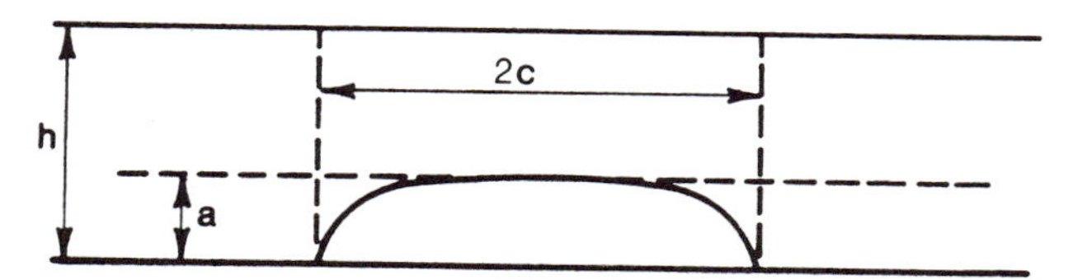

Figure 1.10 Part-through-wall surface crack.

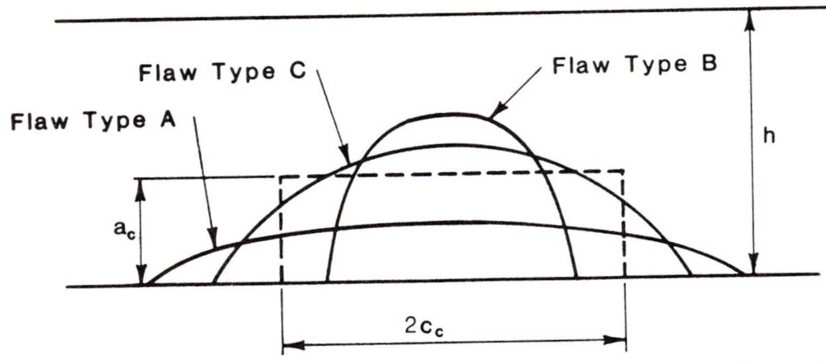

Figure 1.11 Basis for a conservative estimate of the critical dimensions of a surface crack in a loading-bearing component.

greater than the critical surface dimension because its depth dimension is less than a_c. (Recall that a_c was established on the basis of an infinitely long surface length.) Similarly, flaw Type B is benign. Despite the fact that its depth dimension exceeds a_c, because its surface dimension is less than $2c_c$, the through-wall crack resulting from the initiation of flaw Type B will not be critical. Only flaw Type C, where *both* dimensions exceed the corresponding critical values would lead to complete fracture.

In the important class of applications where Figure 1.11 represents a fluid containment boundary, flaw Type B would give rise to the so-called "leak-before-break" condition. That is, the crack would grow through the wall, allowing a contained fluid to escape but, presumably, at only a limited rate. Hence, the presence of the crack could be detected in time to take remedial action. Flaw Type C would not exhibit this desirable condition. Crack growth through and along the wall would take place almost simultaneously, allowing no time for effective operator intervention to halt the process. Both of these statements, of course, are based on the tacit assumption that the applied stress does not change during the crack growth process. In most instances this assumption would be conservative in that loss of a contained fluid is likely to be accompanied by a decrease in pressure. However, this may not always be so. In certifying that a given crack shape will exemplify leak-before-break behavior, one ought to consider the effect of crack growth on the applied stresses. This is sometimes overlooked in practice, however.

As a further thought on the subject of leak-before-break, suppose that the approximate analysis procedure just outlined suggests that fracture will occur in some instances. Because of the built-in conservatism, which could be sizable, the particular conditions may actually be fracture-safe. Remedial action is always expensive and, in some cases (e.g., an improper weld repair), may actually exacerbate the problem. Thus, analyzing more accurately before taking action is often worthwhile. This can be done by removing some of the simplifications, one of which is the assumption of linear elastic behavior. The use of advanced fracture mechanics techniques for this purpose is discussed in Section 1.5.3. Another is to consider the actual crack shape.

An early approach to the analysis of part-through wall cracks that is still often used is that given by Irwin (1.18). Starting from the exact solution for an elliptical crack embedded in an infinitely large elastic body, Irwin introduced approximate correction factors to account for the free surface and for crack-tip plasticity. His result can be written

$$K = \frac{1.12\sigma\sqrt{\pi a}}{\left[\Phi^2 - 0.212\left(\frac{\sigma}{\sigma_Y}\right)^2\right]^{\frac{1}{2}}} \tag{1.1-13}$$

where $\Phi = \Phi(a, c)$ is a factor that depends on the crack shape. It can be expressed in terms of the elliptic integral

$$\Phi = \int_0^{\pi/2} \left(\sin^2\phi + \left(\frac{a}{c}\right)^2 \cos^2\phi\right)^{\frac{1}{2}} d\phi$$

Irwin suggested that Equation (1.1-13) should be valid for crack depths up to half the thickness. This approach was subsequently extended by Kobayashi and Moss (1.19) through the use of magnification factors designed to account for back surface and other effects. Perhaps the most accurate procedure now available is the approach of Newman and Raju (1.20).

Using the results of three-dimensional finite element analyses, Newman and Raju developed an empirical stress intensity factor equation for semielliptical surface cracks. The equation applies for cracks of arbitrary shape factor in finite sized plates for both tension and bending loads. For simplicity, and because such conditions will cover the majority of all applications, only the specialized form of their equation applicable for tension loading of a wide plate will be considered here. This is

$$K = \sigma\sqrt{\pi a}\left[M_1 + M_2\left(\frac{a}{h}\right)^2 + M_3\left(\frac{a}{h}\right)^4\right]\cdot\left[1 + 1.464\left(\frac{a}{c}\right)^{1.65}\right]^{-\frac{1}{2}}$$
$$\cdot\left[\left(\frac{a}{c}\right)^2\cos^2\phi + \sin^2\phi\right]^{\frac{1}{4}}\cdot\left\{1 + \left[0.1 + 0.35\left(\frac{a}{h}\right)^2\right](1 - \sin\phi)^2\right\} \tag{1.1-14}$$

where ϕ denotes the angle between the plate surface and a generic point on the crack front. The dimensions a, c, and h are as shown in Figure 1.10, with the functions M_1, M_2, and M_3 expressable as follows:

$$M_1 = 1.13 - 0.09\left(\frac{a}{c}\right)$$

$$M_2 = 0.89\left[0.2 + \left(\frac{a}{c}\right)\right]^{-1} - 0.54$$

$$M_3 = 0.5 - \left[0.65 + \left(\frac{a}{c}\right)\right]^{-1} + 14\left[1.0 - \left(\frac{a}{c}\right)\right]^{24}$$

Newman and Raju indicate that Equation (1.1-14) is accurate to within ± 5 percent, provided $0 < a/c \leqslant 1.0$ and $a/h \leqslant 0.8$.

The Newman-Raju equation shows that, for small values of a/c the maximum value of K is at $\phi = \pi/2$, the point of deepest penetration. For a/c, about equal to 0.25, K is roughly independent of ϕ. At larger values of a/c, K exhibits a maximum on the plate surface. Because it is likely that the first-mentioned case occurs most often in practice, specific results for the point $\phi = \pi/2$ are of interest. First, for a shallow crack where $a \ll h$, Equation (1.1-14) reduces to

$$K = 1.13\sigma\sqrt{\pi a}\left[1 - .08\left(\frac{a}{c}\right)\right]\left[1 + 1.464\left(\frac{a}{c}\right)^{1.65}\right]^{-\frac{1}{2}} \qquad (1.1\text{-}15a)$$

This result can be compared with Equation (1.1-13). Second, for a very long crack where $a \ll c$, Equation (1.1-14) becomes

$$K = 1.13\sigma\sqrt{\pi a}\left[1 + 3.46\left(\frac{a}{h}\right)^2 + 11.5\left(\frac{a}{h}\right)^4\right] \qquad (1.1\text{-}15b)$$

which should be valid for $a/h \leqslant 0.8$.

1.1.7 Damage Tolerance Assessments

Fracture mechanics is not limited to determining critical crack size/load combinations for fracture instability. It can also be applied to determine the rate of progression of a crack from a defect of a benign size to a critical condition. The crack growth rate is clearly dependent upon the mechanism involved. There are two distinct types that are of most practical concern: fatigue and environmentally assisted cracking. Included in environmentally assisted cracking are corrosion, stress corrosion, and corrosion fatigue. Because even specialists in the subject cannot always agree on the precise distinctions between them, the all-inclusive name will be used herein.

The quantitative relations that have so far been developed for fatigue and environmentally assisted cracking both draw upon linear elastic fracture mechanics considerations. Specifically, for fatigue, the expression commonly used to relate the change in crack length with the number of applied load cycles is widely known as the "Paris Law." For an applied load that is cycled uniformly between K_{max} and K_{min}, this relation is given by

$$\frac{da}{dN} = C(\Delta K)^m \qquad (1.1\text{-}16)$$

where $\Delta K \equiv K_{max} - K_{min}$, while C and m are taken as material-dependent constants that also depend upon load frequency, environment, and mean load. Typical data for some steel, titanium, and aluminum alloys, taken from Bates and Clark (1.21) and Mackay et al. (1.22) are shown in Figure 1.12. These data further illustrate that the mechanical properties of the material can also affect the fatigue crack growth rates. Note the existence of the so-called threshold

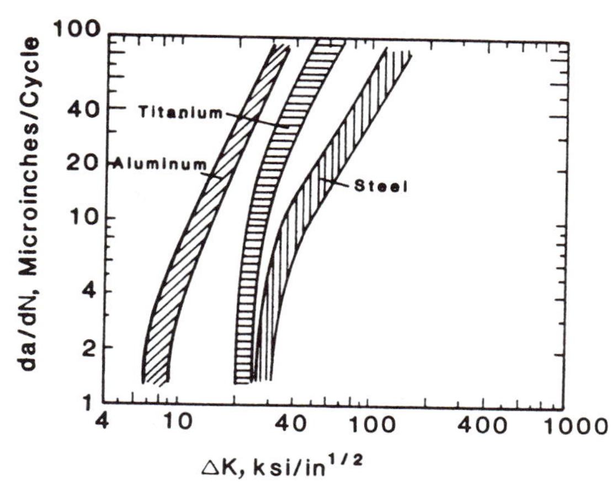

Figure 1.12 Fatigue crack growth rate data for some engineering materials at ambient temperatures.

stress intensity, $(\Delta K)_{\text{th}}$, below which no fatigue crack occurs. This parameter can depend upon $R = K_{\min}/K_{\max}$ and the mechanical properties.

Crack growth under an environmental influence can occur even under a constant or moderately varying loading. In these conditions, the relation

$$\frac{da}{dt} = DK^n \tag{1.1-17}$$

is often used. Here, a denotes the crack length, t is time, K is the LEFM stress intensity factor, and D and n are empirical constants for the material/environment system of concern. The existence of a threshold level for environmental crack growth, generally denoted as K_{Iscc}, is also important. Note that in subcritical crack growth by fatigue, at least for the particular circumstances where the load is cycled from zero to a maximum at a constant rate (i.e., dN/dt = constant), the form of Equation (1.1-16) will be identical to (1.1-17). While the values of the constants would be dissimilar, the latter form can therefore be used to illustrate the damage tolerance assessment procedure for both stress corrosion and fatigue crack growth.

Consider stress corrosion crack growth in a plate under constant applied tension. As subcritical crack growth proceeds, K will usually increase and the crack will grow more and more rapidly. Consider that the initial crack length is a_0 and, further, that a_0 is less than a_d, the minimum crack size that can be reliably detected. Referring to Figure 1.13, nondestructive examination (NDE) would not reveal such a crack prior to the time t_d. But, detection of the crack prior to the time t_c, at which the critical crack length a_c would be achieved, is essential. Consequently, there must be at least one inspection in the time interval $t_d < t < t_c$ if the crack is to be found before it reaches a critical size.

When a_0 is unknown, t_d and t_c cannot be known. Counteracting this uncertainty is one key plausibility: the time required for the crack to grow from

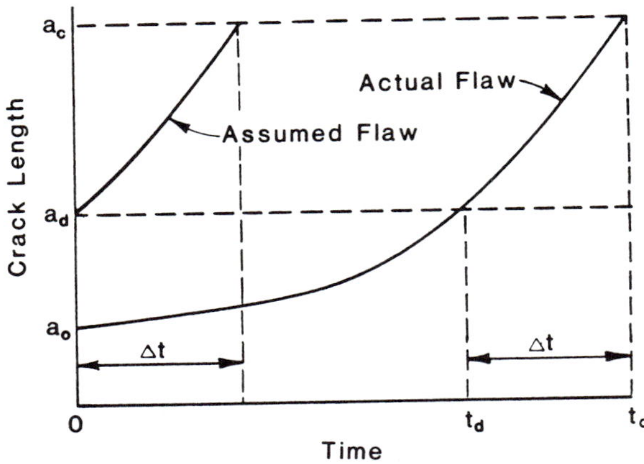

Figure 1.13 Use of fracture mechanics in the damage tolerance approach to structural integrity.

detectible size to the critical size is independent of the initial crack size. This makes it possible to design an NDE program based upon $(\Delta t)_{max}$ as the maximum permissible inspection interval. To obtain this quantity, Equation (1.1-17) can be rewritten as

$$(\Delta t)_{max} = \int_{a_d}^{a_c} \frac{da}{DK^n} \tag{1.1-18}$$

For an edge crack the relation $K = 1.12\sigma\sqrt{\pi a}$ can be used to perform the integration whereupon

$$(\Delta t)_{max} = \frac{2}{D(n-2)} \frac{a_d^{1-n/2} - a_c^{1-n/2}}{(1.12\sigma\sqrt{\pi})^n} \tag{1.1-19}$$

From this result $(\Delta t)_{max}$ can be determined from a_c and a_d (knowing of course the applied load and the material/environment constants) independently of the initial crack size. It should be recognized that similar results can be obtained for other load/crack/structure conditions as well as for fatigue.

An appropriate design would be one in which the inspection procedure need not be carried out too frequently. Typically, the interval that would be selected is $(\Delta t)_{max}/2$, which reflects the fact that NDE crack detection methods, as well as our ability to anticipate the future loads and other conditions, is not completely reliable. Clearly, despite these elements of uncertainty, a predictive methodology of this type is definitely required for assuring structural integrity by precluding fracture.

1.1.8 Code Requirements

The discussions to this point have focused on deterministic approaches that can be used to assure the integrity of cracked structures. It has been tacitly assumed that the shape and location of the crack and the type and magnitude of the load applied to the component are precisely known. Unfortunately, such certainty is seldom, if ever, reached in practice. Reflecting this uncertainty, codes providing conservative approaches have been developed to assist the designers and operators of structures where fracture is a concern. Perhaps the most widely known of these is the American Society of Mechanical Engineers (ASME) Boiler and Pressure Vessel Code, strictures that in many localities have the force of law. To illustrate the use of such a code, we will here briefly outline the procedures prescribed in Section XI of the ASME Code: "Rules for Inservice Inspection of Nuclear Power Plant Components." Of specific interest are the procedures for determining critical crack sizes that are given in Appendix A of Section XI of the ASME code along with the material fracture property data given in Appendix G.

Section XI of the ASME Code distinguishes two types of operating conditions: "normal" and "faulted." Normal conditions are supposed to occur often enough that continued operation with a minimum of interruption would be expected. Faulted conditions are very severe, but would occur so seldom that it would only be necessary to terminate operation safely. The critical crack sizes corresponding to these two sets of conditions are called a_c and a_i, respectively. The crack length a_f is defined as the crack depth at the end of the component's design life. Then, if either

$$a_f \geqslant 0.1\mathrm{a_c}$$

or

$$a_f \geqslant 0.5a_i$$

the requirements of Section XI mandate that the flaw be repaired or the component be withdrawn from service.

The analysis procedures given in Section XI to implement its rules are those of linear elastic fracture mechanics. For surface flaws the first step is to linearize the stress distribution into a membrane stress σ_m and a bending stress σ_b. Then, K can be calculated from the relation

$$K = (\sigma_m M_m + \sigma_b M_b)\left(\frac{\pi a}{Q}\right)^{\frac{1}{2}} \tag{1.1-20}$$

where M_m, M_b, and Q are geometry-dependent factors. Because concentric crack growth is assumed, only the crack depth a appears explicitly in determining the stress intensity factor.

Fatigue crack growth rates are usually calculated from Equation (1.1-16). The constants C and m are material properties chosen for the frequency, environment and R ($\equiv K_{\min}/K_{\max}$) value of concern for the particular application while $K_{\max}$ and $K_{\min}$ are obtained from Equation (1.1-20) using the maximum and minimum cyclic load levels. The crack growth rate data for nuclear pressure vessel steel for water and air environments from which these parameters can be determined are shown in Figure 1.14. The factors appearing in Equation (1.1-20) are approximate values taken from flat plate

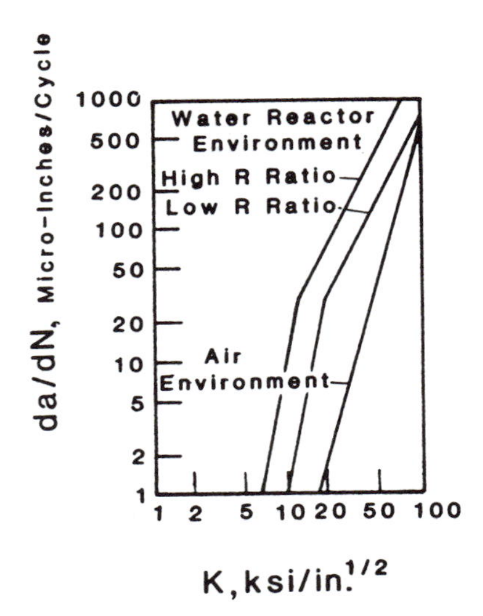

Figure 1.14 Fatigue crack growth rate data for pressure vessel steel used in the ASME code.

analyses. As suggested by McGowan (1.23), the more exact results of Newman and Raju presented in Section 1.1.6 appear in this same form and may eventually be mandated instead. A comparison is shown in Figure 1.15.

The critical crack size determinations are obtained by equating the stress intensity factor given by Equation (1.1-20) to an appropriate fracture toughness value for the temperature of interest. Recognizing that the fracture toughness of pressure vessel steel also varies with the loading rate, two reference curves are provided in Appendix XI. The K_{Ic} curve is the lower bound of all available quasi-static crack initiation test data while K_{Ia} is the lower bound of all available dynamic crack initiation and crack arrest test data. These data are shown in Figures 1.16 and 1.17 as a function of the service temperature minus the index reference temperature, RT_{NDT}. For normal operating conditions critical crack sizes are determined by using K_{Ia}. For

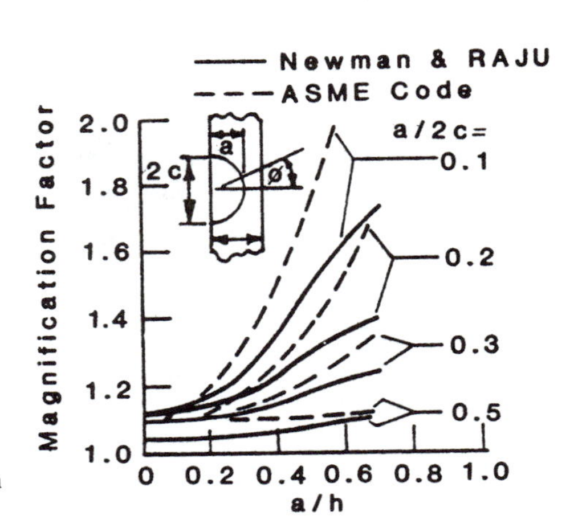

Figure 1.15 Membrane correction factors for surface flaws at $\phi = 0$.

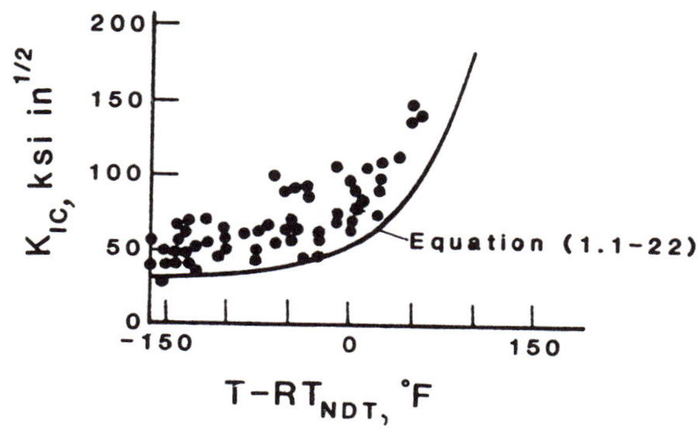

Figure 1.16 Fracture toughness data for pressure vessel steel used in the ASME code.

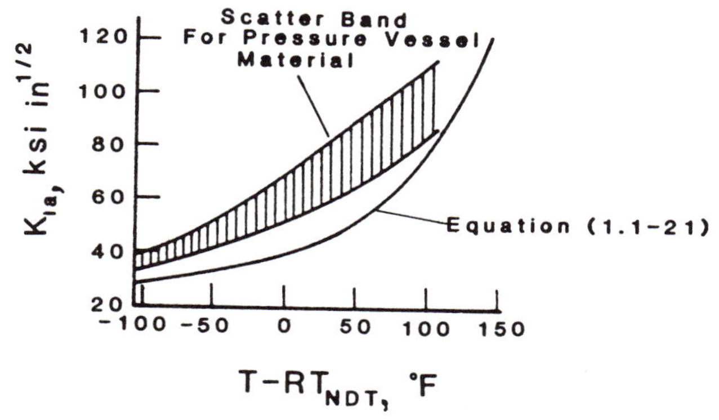

Figure 1.17 Crack arrest toughness values for pressure vessel steel used in the ASME code.

faulted conditions, crack growth initiation is supposed to be governed by K_{Ic} with the possibility of arrest using K_{Ia} also being allowed.

In initial versions of Section III of the ASME code, a lower bound curve of static, dynamic, and crack arrest data from specimen tests on A533B and A508 steel was used that is designated as the K_{IR} (R for reference) curve. This curve was supposed to apply for all ferritic materials approved for nuclear pressure vessel boundaries having a specified yield strength of 50 ksi (345 MPa) or less at room temperature.* This relation is

$$K_{IR} = 26.8 + 1.223 \exp\{.0145(T - RT_{\mathrm{NDT}} + 160)\} \qquad (1.1\text{-}21)$$

where T is the temperature at the crack tip and RT_{NDT} is the reference temperature for nil-ductility transition. For completeness, the corresponding

* The K_{IR} curve apparently originated from the work done under the auspices of the Pressure Vessel Research Committee that appeared in 1972 in Welding Research Council Bulletin No. 175. The related curve for initiation in A533B Steel is sometimes referred to as the "million dollar curve."

relation for K_{Ic} in the ASME code is

$$K_{Ic} = 33.2 + 2.806 \exp[.02(T - RT_{NDT} + 100)] \qquad (1.1\text{-}22)$$

In both Equation (1.1-21) and (1.1-22), T is in °F while K_{IR} and K_{Ic} are in ksi-in.$^{\frac{1}{2}}$. It might be noted that the dynamic fracture toughness values K_{Id} and K_{Ia}, being always less than K_{Ic}, determined the K_{IR} curve. Hence, Equation (1.1-21) provides a good approximation to K_{Ia} for nuclear pressure boundary materials.

The essential idea in Equations (1.1-21) and (1.1-22) is that heat-to-heat variations in material properties can be taken into account by determining RT_{NDT}. This has been criticized by Oldfield (1.17) on both statistical and physical grounds. In particular, the notion that the toughness increases without limit as the temperature increases is incorrect and anti-conservative.* One might look for relations such as that given by Equation (1.1-8) to eventually be embodied in the code, therefore. Dougan (1.24) in turn, has pointed out some inconsistancies in that type of relation. It can only be concluded that this type of development is still in an evolutionary state.

1.2 The Origins of Fracture Mechanics

The preceding section introduced the subject of fracture mechanics as it is now most often being applied. Because fracture mechanics is a continually growing discipline, the new concepts and techniques that will surely evolve in the years to come may be better understood if one has an appreciation of the origins of the subject. Accordingly, in this and the following two sections we step back to trace the early development of the subject.

1.2.1 The Evolution of Structural Design

As depicted in Figure 1.18, the evolution of structural design to include fracture mechanics has proceeded through a series of stages. The earliest work (e.g., the pyramids, the great cathedrals of Europe) simply relied upon previous attempts, proceeding in an essentially trial and error manner. It was not until the development of the concepts of stress and strain and their incorporation in the mathematical theory of elasticity during the nineteenth century that quantitative design procedures became possible; see references (1.1)–(1.4). The application of elasticity concepts to determine the strength of a material is shown as the second stage in the chronology of Figure 1.18. As nicely described by Gordon (1.1), the logical extension of these ideas to treat stress concentrations, however, led to a severe dilemma—the existence of singular behavior and, hence, infinite stresses—that was only resolved by the invention of fracture mechanics.

* To describe the lower bound behavior conveniently, an exponential function was fit to the test data envelope. This was understandable in that, given the lack of upper shelf data, no high toughness limit was evident. However, in practice, an upper bound of 200 ksi-in.$^{\frac{1}{2}}$ has been invoked.

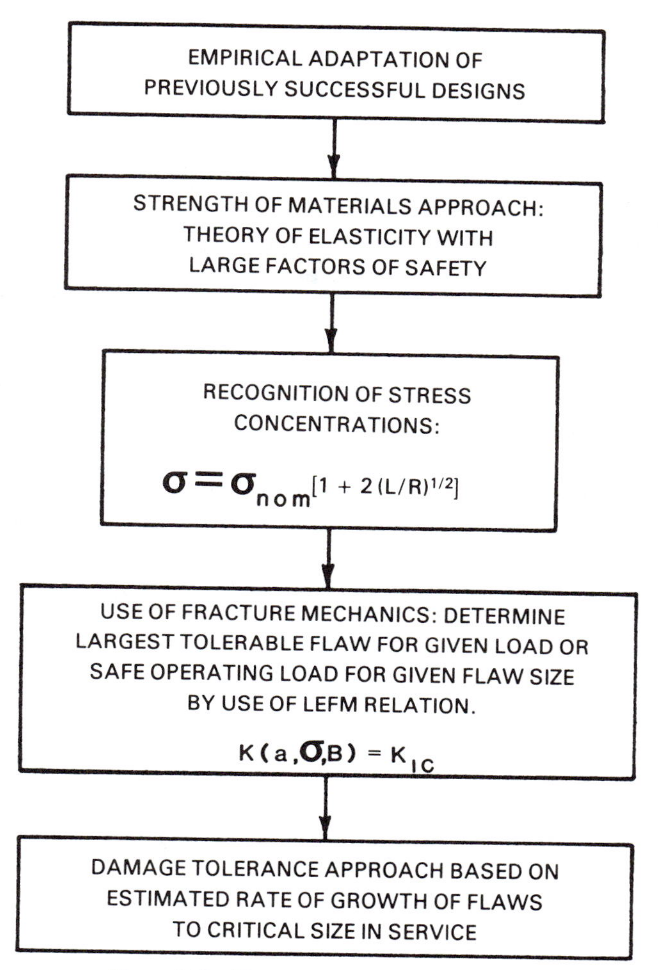

Figure 1.18 The evolution of structural design.

Shown as the third stage in Figure 1.18 is the result obtained by Inglis (1.25) in the early years of this century for the stress concentration at the end of an elliptical hole in a plate in tension.* This result relates σ, the stress acting at the most highly stressed point on the edge of the hole, to the remote or nominal stress, σ_{nom}. It can be seen that this relation depends on the ratio of L, the half length of the hole, to R, the radius of curvature at the point of interest. Note that for a circular hole, where $L = R$, $\sigma = 3\sigma_{nom}$. But, because $R = 0$ for a sharp crack, an infinite stress would then be expected. This result suggests that a body with a crack could sustain no applied load whatever!

The paradox arising from the application of Inglis' result to a crack was resolved by Griffith (1.5) whose landmark work on glass fibers first appeared

* Similar results were obtained at about the same time by Kolosov in Russia.

around 1920. However, while some additional work was subsequently contributed—for example, by Obriemoff (1.26) and Westergaard (1.27)—fracture remained for some time a scientific curiosity that did not permeate engineering design. One reason may be the apparent non-applicability of the Griffith Theory to engineering materials with fracture resistance values orders of magnitude greater than that of glass. Regardless, this state existed until sometime after the Second World War. Severe instances of fracture in Liberty ships (1.28), missile casings, and other structures (1.29) then gave the impetus to more intensive studies of fracture in the U.S. with the Comet aircraft disasters (1.30) additionally spurring work in Europe. This gave rise to the next major contribution to the subject that was given by Irwin (1.6) when he generalized Griffith's ideas for applicability to metals and other engineering materials—an idea that was suggested independently by Orowan (1.31).

A key subsequent step, to connect the stress intensity factor to Griffith's energy balance, was also performed by Irwin (1.32). Couched in the terminology of Griffith-Irwin or linear elastic fracture mechanics, this accomplishment is shown as the fourth stage of the evolution depicted in Figure 1.18. To a large extent, it represents currently applicable technology, as described in the preceding section of this chapter.

The fifth stage shown in Figure 1.18 represents the kind of activity that has been initiated only within the past few years. This is the explicit recognition that cracks do exist in every engineering structure, whether arising from initial defects in the material, from fabrication flaws, or from service conditions. Because more refined and intensive nondestructive evaluation (NDE) procedures are increasingly being applied, the integrity of the structure must be addressed by taking account of cracks. This is done by a combination of NDE and fracture mechanics calculations that (1) assume an initial crack size, (2) estimate the rate of subcritical crack growth (e.g., by fatigue, stress corrosion), and (3) determine the critical crack size for fracture instability. In this approach, it is generally assumed that a crack exists in the structure having a size that would be just missed by NDE. The growth rate calculation then allows an inspection interval to be set that would allow that crack to be found prior to its achieving the critical size. This procedure, generally known as a "damage tolerance" approach, was illustrated in Figure 1.13 and described in Section 1.1.7.

The damage tolerance concept is now mainly used in the aerospace industry. However, it seems likely that its use will be broadened in the future. Efforts are now underway to implement it for the prevention of cracking in railroad rails and offshore platforms, for example. Clearly, such a procedure can be applied to any engineering structure where, first, the growth of small flaws to a critical size is a concern and, second, NDE methods that can reliably detect subcritical cracks are available.

1.2.2 Griffith's Theory

As already mentioned, it was Griffith's fundamental contributions (1.5) that resolved the infinite crack-tip stress dilemma inherent in the use of the theory

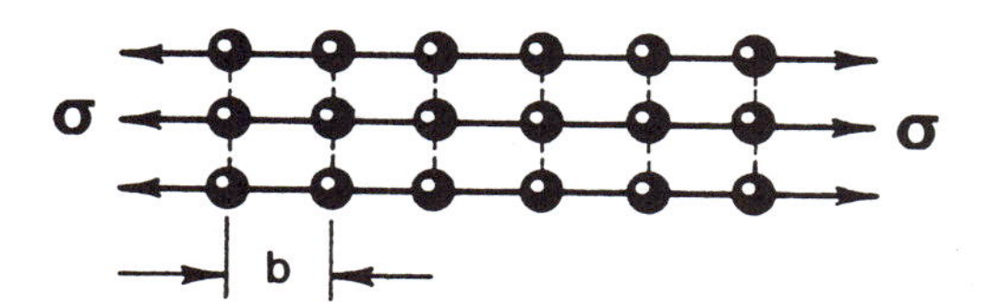

Figure 1.19 Atomic model for theoretical strength calculations.

of elasticity for cracked structures. While this ultimately led to the development of fracture mechanics as an engineering discipline, Griffith was apparently motivated by other considerations (1.1). That is, as illustrated in Figure 1.19, simple estimates can be made for the strength of a crystalline solid based on its lattice properties. But, this results in a relation for the theoretical tensile strength that is not attained in actuality. Griffith's work was primarily focused on resolving this dichotomy.

The interatomic force-separation law can be approximated by a function that exhibits three properties: (1) an initial slope that corresponds to the elastic modulus E, (2) a total work of separation (i.e., area under the curve) that corresponds to the surface energy γ, and (3) a maximum value that represents the interatomic cohesive force. Because the exact form that is selected makes little difference, it is convenient to use a sine function. As can readily be verified, the appropriate relation is then

$$\sigma(x) = \left(\frac{E\gamma}{b}\right)^{\frac{1}{2}} \sin\left[\left(\frac{Eb}{\gamma}\right)^{\frac{1}{2}}\left(\frac{x}{b}\right)\right] \qquad (1.2\text{-}1)$$

where b represents the equilibrium interatomic spacing and x denotes the displacement from the equilibrium separation distance. It follows that the theoretical strength—the maximum value exhibited by this relation—is

$$\sigma_{\text{th}} = \left(\frac{E\gamma}{b}\right)^{\frac{1}{2}} \qquad (1.2\text{-}2)$$

For many materials $\gamma \simeq Eb/40$ so that $\sigma_{\text{th}} \simeq E/6$. But, such a prediction is clearly much in excess of the observed strengths; a result that was explained by Griffith who, drawing upon the mathematical development of Inglis (1.25), traced the discrepancy to the existence of crack-like flaws. It may be of interest to note that it was the analogous discrepancy between the theoretical shear strength of a crystalline solid and the observed values that led to the identification of the dislocation as the fundamental element in metal plasticity in the mid 1930s.

Figure 1.20 shows the results of Griffith's series of experiments on glass fibers having different thicknesses. As the fiber thickness decreased, the breaking stress (load per unit area) increased. At the limit of large thickness, the strength is that of bulk glass. But, of considerably more interest, at the opposite limit of vanishingly small thickness, the theoretical strength is approached. This observation led Griffith to suppose that the apparent thickness effect was actually a crack size effect. Figure 1.21 illustrates his observation. It is worth mentioning here that this "size effect" is responsible for the usefulness of materials like glass and graphite in fiber composites; that is,

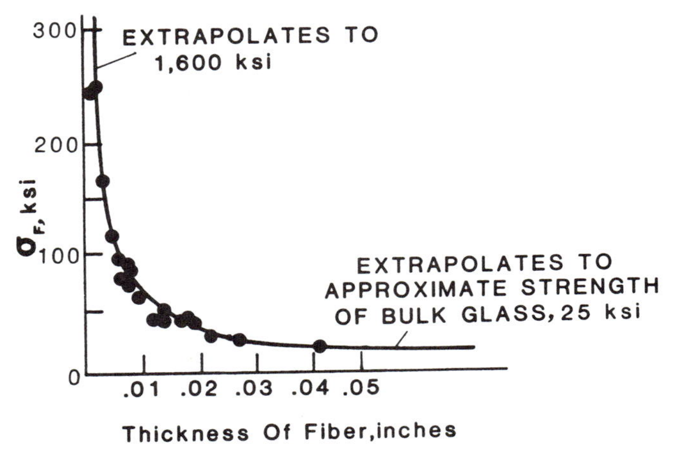

Figure 1.20 Results of Griffith's experiments on glass fibers.

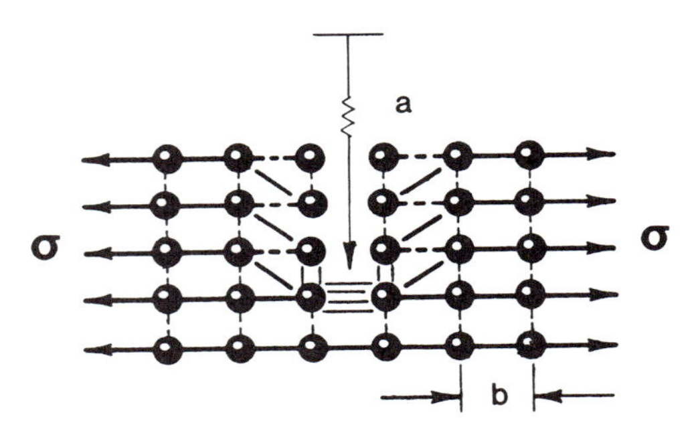

Figure 1.21 Atomic model with defect for fracture calculations.

the inherent defects can be considerably reduced by using such materials in fiber form bound together by a resin.

The basic idea in the Griffith fracture theory is that there is a driving force for crack extension (that results from the release of potential energy in the body) along with an inherent resistance to crack growth. The resistance to crack growth, in glass at least, is associated with the necessity to supply surface energy for the newly formed crack surfaces. By using the existing mathematical development of Inglis (1.26), Griffith was able to formulate an energy balance approach. This leads to a critical condition for fracture that can be written as an equality between the change in potential energy due to an increment of crack growth and the resistance to this growth. For an elastic-brittle material like glass, this is

$$\frac{dW}{dA} - \frac{dU}{dA} = \gamma \tag{1.2-3}$$

where W and U, respectively, are the external work done on the body and its internal strain energy, γ is the surface energy of the material, and $A = 4Ba$ is the crack surface area for an internal crack in a body of thickness B.

For a crack in an infinite body subjected to a remote tensile loading normal to the crack (the problem considered by Griffith), only the net change in elastic strain energy needs to be evaluated. This is today most conveniently accomplished by making use of a procedure developed by Bueckner (1.33). Bueckner recognized that the strain energy due to a finite crack is equal to one-half of the work done by stresses (of equal magnitude but opposite in sign to the applied stress) acting on the crack faces. The crack face opening for the Griffith problem is therefore given by Westergaard's solution for plane stress; this is

$$v = \frac{2\sigma}{E}(a^2 - x^2)^{\frac{1}{2}} \tag{1.2-4}$$

where σ is the applied stress and the origin of the coordinates is taken at the center of the crack. Consequently, for an internal crack, work is done at four separate surface segments. As a result, it is readily shown using Equation (1.2-4) that

$$W - U = 4B \int_0^a \tfrac{1}{2}\sigma v(x)\,dx = \frac{\pi a^2 \sigma^2 B}{E} \tag{1.2-5}$$

Equation (1.2-3) then gives

$$\sigma_f = \left(\frac{2}{\pi}\frac{E\gamma}{a}\right)^{\frac{1}{2}} \tag{1.2-6}$$

where σ_f denotes the applied stress that would lead to fracture.

While Equation (1.2-6) was derived for constant applied stress conditions, the same result is also obtained for fixed displacement conditions. Following Erdogan (1.34), if the crack is introduced after the load σ is applied with the grips then being fixed, then the total strain energy of the body will be (plane stress)

$$U = \frac{1}{2}\frac{\sigma^2}{E}V - \pi\frac{\sigma^2 a^2}{E}B \tag{1.2-7}$$

where V is the volume of the body under consideration. It can readily be seen that, because $W = 0$ under fixed grip conditions, use of Equation (1.2-3) with $dA = 4B\,da$ will again lead to Equation (1.2-6).

The influence of the local crack/structure geometry on the critical applied stress might be noted. For example, if plane strain conditions were taken, the factor of E appearing in Equation (1.2-6) would be replaced by $E/(1 - \nu^2)$. Further, as shown first by Sneddon (1.35), the axisymmetric case of a penny-shaped crack of radius a leads to an expression given by

$$\sigma_f = \left(\frac{\pi}{2}\frac{E}{(1-\nu^2)}\frac{\gamma}{a}\right)^{\frac{1}{2}} \tag{1.2-8}$$

which differs by a factor of $(2/\pi)^2$ from the plane strain version of Equation (1.2-6). Thus, the results for different conditions result in the same form of equation and differ only in the value of the numerical factor that appears in them. Recognition of this fact later provided Irwin with the key to generalizing crack problems to fracture mechanics.

Returning to Griffith's work, the theoretical strength given by Equation (1.2-2) can be combined with the fracture stress given by Equation (1.2-6) to obtain a relation between the theoretical strength and the fracture stress in the presence of a crack. This is

$$\frac{\sigma_f}{\sigma_{\text{th}}} \simeq \left(\frac{b}{a}\right)^{\frac{1}{2}} \tag{1.2-9}$$

Substituting values for bulk glass leads to a value of an inherent crack size of about .001 inches. Referring to Figure 1.20, it can be seen that when the fiber thicknesses are reduced below this value (whereupon the fibers are likely to be flaw-free), the theoretical strength begins to be approached. This strongly indicates that cracks are indeed the source of the discrepancy between theoretical and observed strength and that quantitative predictions involving them can be made.

1.2.3 Some Difficulties with the Griffith Theory

The astute reader will notice that Equation (1.2-6) was derived for a body with dimensions much greater than the crack size. Yet, it was applied to fibers having thicknesses comparable to the crack size. Such bold applications are not atypical in fracture mechanics, the rationale being that, in view of the many uncertainties that exist, order of magnitude estimates are often all that can be expected. But, there are even more fundamental difficulties.

As pointed out by Goodier (1.36), in calculating the elastic strain energy in the plate, Griffith neglected the stresses due to the surface tension that are implied by the existence of surface energy on the crack faces. If surface energy is specified, the boundary value problem to be solved is one that would include a normal traction γ/ρ, where γ is the surface tension and ρ is the radius of curvature on the "stress free" crack. This problem was solved by Rajapakse (1.37) who found that the nature of the crack tip singularly then becomes radically different. Furthermore, because the specification of an energy balance then becomes redundant (i.e., it is guaranteed by the solution of a properly posed boundary value problem), some alternative criterion for fracture must be sought.

Griffith himself found that the energy balance criterion was not useful for fracture under combined stress conditions. For combined stress states, Griffith (1.5) found it necessary to employ an alternative to the energy balance, stating that:

> the general condition for rupture will be the attainment of a specific tensile stress at the edge of one of the cracks.

To avoid the singularity, he employed a thin ellipse with a finite radius of curvature ρ_0 at the crack tip. Rajapakse emulated this by equating the normal stress at the crack tip from his model to the theoretical strength. He thereby obtained for the fracture criterion

$$2\sigma\left(\frac{a}{\rho_0}\right)^{\frac{1}{2}} - 0.273\,\frac{\gamma}{\rho_0} = \sigma_{th} \tag{1.2-10}$$

By setting $\rho_0 = b$ (the interatomic spacing) Rajapakse concluded that the contribution of the surface energy term is negligible in comparison to that of the applied stresses. While he left his development at that point, it can be taken a step further by using Equation (1.2-2) to replace σ_{th} in Equation (1.2-10). This substitution gives

$$\sigma_f = \frac{1}{2}\left(\frac{E\gamma}{a}\right)^{\frac{1}{2}}\left[1 + 0.273\left(\frac{\gamma}{Eb}\right)^{\frac{1}{2}}\right] \simeq 0.52\left(\frac{E\gamma}{a}\right)^{\frac{1}{2}} \tag{1.2-11}$$

where the approximate result was obtained using the estimate $\gamma = Eb/40$. It is readily seen that Rajapakse's result is very nearly the same as Equation (1.2-6), differing in only a minor way from the constant $(2/\pi)^{\frac{1}{2}} = 0.80$ that appears in the latter equation. The phoenix-like character of Griffith's result is evident.

It should be recognized that Equation (1.2-3) describes a condition necessary for the initiation of crack growth, but it is not a sufficient condition for fracture. For continued (unstable) crack growth, it is clear that

$$\frac{d}{dA}\left(\frac{dW}{dA} - \frac{dU}{dA}\right) > \frac{d\gamma}{dA} \tag{1.2-12}$$

But, this restriction is not a detriment to the theory. A more practical consideration in the application of the Griffith Theory is its apparent restriction to ideally brittle materials. The fracture resistances of engineering structural materials are greater than γ by several orders of magnitude. Moreover, it is possible to achieve stable crack growth even where the left-hand side of inequality (1.2-12) is positive—an impossibility if γ is a material constant independent of crack length.

1.2.4 Origins of Linear Fracture Mechanics

Despite the theoretical questions that Griffith's work engenders, it unquestionably was an important first step. Nevertheless, the subject was relatively dormant during the two decades following his work. As quoted in the prologue to a paper by Tipper (1.38), Mott stated in 1949 that:

> quite apart from its practical importance, fracture is the most interesting property of solids to the theoreticians because it is the least understood property, no progress having been made beyond the 1924 Griffith Crack Theory. It is not known how cracks exist nor what causes them. Experimented work has made clear what happens during fracture but not how it occurs.

More than any other single factor, the large number of sudden and catastrophic fractures that occurred in welded merchant ships during and following the Second World War, gave the impetus for the development of fracture mechanics. Out of approximately 5000 welded ships constructed during the war, over 1000 suffered structural damage with 150 of these being seriously damaged. Ten ships fractured into two parts; eight of these were lost.

The puzzling feature of the welded ship fracture problem was the fact that the normally ductile ship steel fractured with only limited plastic flow and, hence, little energy absorption. While this observation might have immediately suggested that use be made of Griffith's theory, other approaches were tried first. The concept of flow and fracture as competing mechanisms is one such approach. But, because of its inability to account for crack size, it was eventually found to be vacuous. Irwin and his associates at the U.S. Naval Research Laboratory were left to make use of Griffith's ideas and thereby set the foundations for fracture mechanics.

Two key contributions were required to turn Griffith's theory into an engineering discipline. These were, first, by Irwin (1.6) and Orowan (1.31) independently, to extend Griffith's theory to metals and, second, by Irwin (1.32), to connect the global concepts of Griffith to a more readily calculable crack-tip parameter. Because of the leading roles played in its development, linear elastic fracture mechanics is also known as Griffith-Irwin fracture mechanics and, somewhat less often (generally by materials scientists), as the Griffith-Irwin-Orowan theory.

The extension of Griffith's elastic-brittle fracture concepts to metals by introducing the plastic energy dissipation to supplement the surface energy involves no new mathematical developments. The fracture stress for the Griffith problem, Equation (1.2-6), was simply rewritten in the form

$$\sigma_f = \left[\frac{E(2\gamma + \gamma_p)}{\pi a}\right]^{\frac{1}{2}} \tag{1.2-13}$$

where γ_p denotes the plastic energy per unit of crack extension per crack tip. Orowan estimated that γ_p is some three orders of magnitude greater than γ in a metal, whereupon inclusion of the latter is superfluous. Equation (1.2-13) clearly then is indistinguishable from Equation (1.2-6), provided that a more general interpretation is put on γ in the latter.

In current terminology, the left-hand side of Equation (1.2-3) can be used to define a parameter called the strain energy release rate or, alternatively, the crack driving force. This is conventionally given the symbol G in honor of Griffith. Thus, by definition

$$G = \frac{dW}{dA} - \frac{dU}{dA} \tag{1.2-14}$$

Alternatively, as shown by Irwin (1.3), it is convenient to work with the compliance of the cracked component to calculate this quantity. This can be expressed as

$$G = \tfrac{1}{2} P^2 \frac{dC}{da} \tag{1.2-15}$$

where P is the applied load and C is the compliance. In either instance Griffith's energy balance statement of the critical condition for crack extension is then

$$G = G_c \tag{1.2-16}$$

where G_c represents the material's resistance to crack growth.*

Irwin (1.32) later referred to this approach as the "modified Griffith theory" for "somewhat brittle materials" and, still later (1.39), as the "old modified Griffith theory of Irwin and Orowan." The need to make this distinction will be made clear later. Regardless, the application of Equation (1.2-16) led to the solution of a number of important practical problems. As an especially pertinent example, the fracture of a generator rotor, described by Schabtach et al. (1.40), was resolved by application of Griffith-Irwin fracture mechanics in work reported by Winne and Wundt (1.41). This work provided a very significant confirmation of fracture mechanics and its utility for engineering analyses that was recognized even at that time as a landmark achievement; see the discussion to referrence (1.41).

1.2.5 The Stress Intensity Factor

Irwin's masterstroke was to provide a quantitative relation between the sometimes mathematically awkward strain energy release rate, a global parameter, and the stress intensity factor, a local crack-tip parameter. Analogous to Griffith's use of the existing mathematical development of Inglis, Irwin was able to utilize the cracked body solutions of Westergaard (1.27). Specifically, Irwin (1.32) needed two specific relations: for σ_y, the normal stress on the crack line, and v, the opening displacement of the crack faces. In current notation, these can be written for either plane stress or plane strain by introducing the material parameter κ, defined in terms of Poisson's ratio by

$$\kappa = \begin{cases} \dfrac{3 - \nu}{1 + \nu} & \text{plane stress} \\ 3 - 4\nu & \text{plane strain} \end{cases} \tag{1.2-17}$$

Then, the normal stress ahead of the crack and the displacement on the crack surface are given for the Griffith Problem by

$$\sigma_y = \frac{x\sigma}{(x^2 - a^2)^{\frac{1}{2}}}, \qquad x > a \tag{1.2-18}$$

and

$$\frac{Ev}{(1 + \nu)(\kappa + 1)} = \frac{\sigma}{2}(a^2 - x^2)^{\frac{1}{2}}, \qquad x < a \tag{1.2-19}$$

---

* While the quantity G_c was originally called the fracture toughness, despite the fact that they are not numerically equal, it is now the quantity K_c that is generally so called. This is a consequence of the predominance of K-based expressions for crack growth in engineering work.

where x is taken from an origin at the center of the crack. More convenient relations that are valid very near the crack tip can be obtained by taking $|x - a| \ll a$. Equations (1.2-18) and (1.2-19) can then be written to a close approximation as

$$\sigma_y \doteq K[2\pi(x - a)]^{-\frac{1}{2}}, \qquad x > a \tag{1.2-20}$$

and

$$v \doteq \frac{(1 + \nu)(\kappa + 1)}{E} K\left(\frac{a - x}{2\pi}\right)^{\frac{1}{2}}, \qquad x < a \tag{1.2-21}$$

where K (for Kies, one of Irwin's collaborators) is the stress intensity factor. K is a geometry-dependent quantity that here has the value $\sigma\sqrt{\pi a}$.

Imagining that the crack has extended by an amount Δa, Irwin calculated the work required to close it back up to its original length. This amount of work can be equated to the product of the energy release rate and the crack extension increment. Thus,

$$G\Delta a = 2 \int_a^{a+\Delta a} \tfrac{1}{2}\sigma_y(x)\, v(x - \Delta a)\, dx \tag{1.2-22}$$

Or, upon substituting Equations (1.2-20) and (1.2-21) into (1.2-22), Irwin obtained

$$G = \frac{1}{4}(1 + \nu)(\kappa + 1)\frac{K^2}{E} = \frac{K^2}{E'} \tag{1.2-23}$$

where $E' = E$ for plane stress and $E' = E/(1 - \nu^2)$ for plane strain. Equation (1.2-23) thus provides a replacement for finding the derivative with respect to crack length of the total strain energy in a cracked body.*

Initially, Equation (1.2-23) was perceived only as a convenient means for evaluating G. This is the reason for the ubiquitous appearance of the factor $\sqrt{\pi}$ in expressions for K. That is, the parameter that characterizes the singular behavior at a crack tip, using Equation (1.2-18), can be found to be

$$\lim_{x \to a} [2(x - a)]^{\frac{1}{2}} \sigma_y(x, 0) = \sigma\sqrt{a} \tag{1.2-24}$$

which does not include π. Thus, the awkward $\sqrt{\pi}$, incorporated artificially into the definition of K to simplify the calculation of G, is completely unnecessary. In retrospect, because by far the greatest number of fracture mechanics applications now employ K rather than G—and do so because of its connection with the local crack-tip field as shown by Equation (1.2-24)—this was rather unfortunate.

Because the analysis problem was made tractable for practical problems by the use of Equations (1.2-23), it can be said that fracture mechanics as an

* Irwin has referred to the procedure embodied in Equation (1.2-22) as the crack closure method. While this is a properly descriptive phrase, it should be recognized that the current use of the term "crack closure" more often refers to the contact of the crack faces during the unloading portion of a cyclic load in fatigue; see Chapter 8. There is no connection between these two ideas.

engineering discipline had its origins in this procedure. Through the use of complex variable methods, Sih, Paris, and Erdogan (1.42), for example, provided the first collection of stress intensity factors for practitioners of fracture mechanics. Sih, Paris, and Irwin (1.43) also provided a generalization of Equation (1.2-23) for anisotropic materials. For a homogeneous orthotropic material with the crack parallel to a plane of symmetry, this relation can be written

$$G = \left(\frac{a_{11}a_{22}}{2}\right)^{\frac{1}{2}} \left[\left(\frac{a_{11}}{a_{22}}\right)^{\frac{1}{2}} + \frac{2a_{12} + a_{66}}{2a_{22}}\right]^{\frac{1}{2}} K^2 \qquad (1.2\text{-}25)$$

where the a_{ij} are elements of the compliance matrix written in the form $\varepsilon_i = a_{ij}\sigma_j, j = 1, \ldots, 6$.

The introduction of the stress intensity factor did much to establish the basis of fracture mechanics. Nevertheless, some fundamental conceptual difficulties still existed. These centered on the necessity to reconcile the linear elastic nature of the derivation of Equations (1.2-23) and (1.2-25) with the fact that nonlinear deformation will engulf the crack tip in most engineering materials. As stated by Irwin and Paris (1.39):

> In terms of the old modified Griffith theory, the condition critical for the onset of rapid fracture was a point of stable balance, between stress field energy release rate and rate of plastic work near the crack, to be followed by a regime of unstable rapid crack propagation.

However, as they later came to recognize:

> the point of onset of rapid crack fracture was an abrupt instability point followed by a stable regime in which work rate and loss of stress field energy were balanced through a considerable range of crack speeds. Indeed, the instability point could be preceded by a slow regime of crack extension in which the crack-extension process was also stable. From these facts, it was not clear that the assumption of an equality between rate of strain energy release and plastic work rate would be helpful for understanding (the) sudden onset of fast crack extension.

This view, strongly influenced by the studies of the plasticity at the crack tip developed by McClintock and Irwin (1.44), had the effect of shifting the emphasis in fracture mechanics away from the energy balance to crack-tip characterization. This viewpoint is paramount in modern fracture mechanics.

Further reinforcement for the crack-tip characterization viewpoint emerged from the work of Barenblatt (1.45). He found it necessary to invoke two postulates: that (1) intense cohesive forces act over a small interval at the ends of the crack, and (2) the local distribution of these cohesive forces is always the same for a given material under given conditions. Goodier (1.36) suggested that a further feature of Barenblatt's approach be considered as a third postulate: the stress singularity arising from the cohesive forces is such that it cancels that due to the applied stresses. As a direct consequence of abolishing the stress singularity, the two crack faces close smoothly in a cusp shape (rather than opening to an ellipse with a finite radius of curvature as in the Griffith problem). A material constant emerges that is called the modulus of cohesion.

Barenblatt's theory supposes that, as the applied loads increase from zero, the cohesive stresses at the crack tip will also increase. But, there is a limit to the material's ability to restrain the crack faces from opening. Then, an "immobile equilibrium crack" becomes a "mobile equilibrium crack" and dynamic crack propagation ensues. The transition point was then expressed quantitatively using the complex variable representation of the theory of elasticity formulated by Muskhelishvili (1.46) and Barenblatt's postulates. This leads to an expression for the modulus of cohesion K_0 given by

$$K_0 = \int_0^d \frac{g(\xi)\,d\xi}{\xi^{\frac{1}{2}}} \tag{1.2-26}$$

where $g(\xi)$ denotes the intensity of the cohesive force and d is the (small) interval at the crack tips where it acts. While he was unable to evaluate the integral in Equation (1.2-26), Barenblatt was able to show that

$$K_0 = (\pi E \gamma)^{\frac{1}{2}} \tag{1.2-27}$$

which eliminates the necessity to perform calculations on the molecular scale. Finally, use of the singularity canceling procedure leads to the fracture criterion

$$K = \frac{K_0}{\pi} \tag{1.2-28}$$

where K is the stress intensity factor. Using Equation (1.2-27) and taking $K = \sigma(\pi a)^{\frac{1}{2}}$ for the Griffith problem recovers Equation (1.2-6) to within a constant.

1.2.6 Atomic Simulation of Fracture

In this section we will digress to look at the various analysis models that have been proposed to address fracture in the most fundamental way possible—by the rupture of the interatomic bonds that keep a solid intact. It is not difficult to recognize that such modeling efforts can only proceed with sweeping assumptions and a degree of computational complexity that essentially obviates any practical applications. Yet, such work is nonetheless of interest for the insight that it can bring to a basic understanding of the origins of fracture toughness.

Implicit in fracture analyses from their inception is the idea that atomic bonds must be ruptured to allow a crack to propagate; for example, see Figure 1.21. The first quantitative treatment that explicitly considered interatomic bond rupture may have been that of Elliott (1.47). Using the linear elastic (continuum!) solution for a cracked body under uniform tension, Elliott evaluated the normal stress and displacement values along a line parallel to the crack plane (see Figure 1.22), but situated at a small distance $b/2$ into the body. He then plotted the stress at each position as a function of the

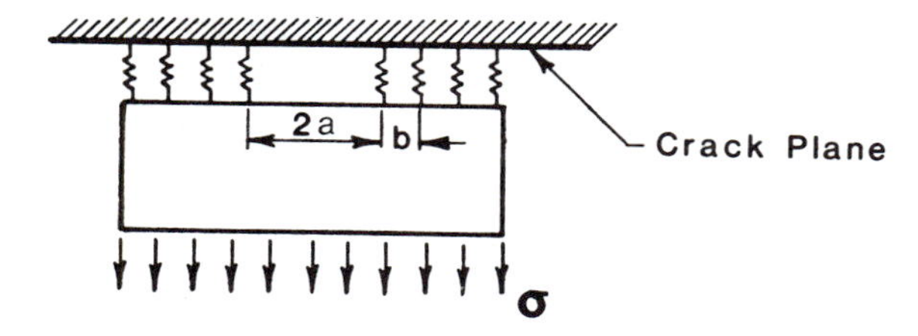

Figure 1.22 Pseudo atomic model for fracture mechanics calculations.

displacement at that point. The shape of this function turns out to be consistent with what one would intuitively expect for interatomic force-separation law—that is, one that obeys Hooke's law for small separations, exhibits a maximum, and diminishes monotonically to zero at very large separations [cf. Equation (1.2-1)].

On the basis of this interesting finding, Elliott formulated a model in terms of two semi-infinite blocks that attract each other with interatomic forces. This is shown in Figure 1.22. In this model the distance b between the blocks is taken as the equilibrium interatomic separation distance, and the area under the force-separation curve is set equal to the surface energy γ. The final step is to equate the maximum stress to the critical rupture stress for the material. This gives a relation that can be written in a form like that of Griffith. For plane strain conditions and $\nu = 0.25$, this is

$$\sigma_f = \frac{8}{7}\left(\frac{E\gamma}{2a}\right)^{\frac{1}{2}} \tag{1.2-29}$$

which can be compared with Equation (1.2-6). Remarkably, it can be seen that, apart from a difference in the numerical constants of only about 1 percent, these results are identical. Nevertheless, because the approach is based completely on linear elastic continuum theory (as Elliott certainly recognized), such a result might be regarded as fortuitous. Cribb and Tomkins (1.48) subsequently performed a more direct analysis of the interatomic cohesive forces at the tip of a crack in a perfectly brittle solid, nevertheless finding a result in essential agreement with Elliott's.

Elliott was concerned also with another aspect of the problem, one that troubles all atomistic and energy balance approaches. Because there is only one equilibrium point, the crack should close up at all applied stress values less than the critical value! Consequently, some physical agency not present in the model (e.g., a missing layer of atoms, gas pressure, a nonadhering inclusion) must be postulated to assure the existence of the crack prior to fracture instability. Elliott argued that such a deus ex machina would be compatible with his approach. Subsequent investigators simply took the initial displacements of the atoms on the crack plane to be beyond the separation distance corresponding to the maximum cohesive force.

Improvements on continuum-based analyses of discrete atomic-scale events were forthcoming only with the advent of large-scale numerical computation. The first of these may have been that of Goodier and Kanninen; see reference (1.36). While actually motivated by Barenblatt's postulates, they in effect extended Elliott's model. Specifically, they considered two linear elastic semi-infinite solids connected by an array of discrete nonlinear springs spaced a distance b apart (b was taken as the interatomic separation distance). The crack-tip region in this model is shown in Figure 1.23.

Goodier and Kanninen selected four different analytical forms to represent the interatomic force separation law. In each of these the initial slope was made to correspond to the elastic modulus E with the area under the curve, representing the work of separation, equated to the surface energy γ [cf. Equation (1.2-1)]. For any of these "laws," the solution for the resulting mixed nonlinear boundary condition problem was obtained numerically by monotonically loading the body (with a finite length $2a$ over which the atoms were already supposed to be out-of-range of their counterparts) until the maximum cohesive strength was achieved. Computations performed for a range of crack lengths led to a relation for the fracture stress that can be written as

$$\sigma_f = \alpha \left(\frac{E\gamma}{a} \right)^{\frac{1}{2}} \tag{1.2-30}$$

where α was a number on the order of unity that depend modestly on the interatomic force law that was used in the calculation. The similarity with Equations (1.2-6) and (1.2-29) is evident.

An obvious shortcoming in the model of Goodier and Kanninen is the limitation to the atomic pairs bridging the crack plane. As later shown by Rice (1.7), the application of the J-integral (see Section 1.4.4) to such a problem results in $G = 2\gamma$ regardless of the force law that is used—the difference found for the various choices simply reflecting the discreteness of the model, not the nonlinearity. Recognizing this, Gehlen and Kanninen (1.49) extended the Goodier-Kanninen treatments by directly considering the crystal structure at the crack tip. Specifically, equilibrium atomic configurations at the tip of a

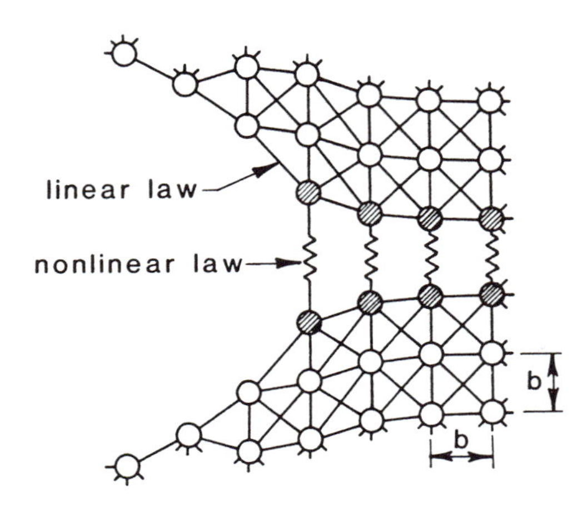

Figure 1.23 Crack tip in a pseudo-atomic fracture model.

crack in alpha-iron, a body-centered-cubic structure for which interatomic potentials are reasonably well known, were determined for different load levels. But, owing to the limited sizes of the model that they were able to employ, the crack growth condition had to be deduced in an artificial way. This work nevertheless also produced a relation of the type given by Equation (1.2-30), again with a constant about equal to unity. A similar approach pursued by Chang (1.50) led to the same type of result.

Because the number of "free" atoms that Kanninen and Gehlen could admit into their computation was small (typically about 30), the atomic positions were highly constrained by the linear elastic continuum displacement field in the vicinity of the crack tip. Consequently, the coincidence between their result and that of Griffith is not surprising. Nonetheless, this effort was valuable in that the fundamental process responsible for cleavage crack extension was for the first time confronted in a realistic way and, simplicity aside, no artificial postulates were required to support it. Of more importance, such a model allows the process of dislocation nucleation—the origin of crack-tip plasticity—to occur naturally. Consequently, it should be possible to delineate the mechanical properties of a material that dictate whether brittle or ductile fracture will occur. The later work of Gehlen et al. (1.51), where larger models with less rigid constraints in the boundary of the computational model were used, did indeed permit bond rupture to be possible; see Figure 1.24. Their work also showed the origins of dislocation nucleation. Subsequently, Markworth et al. (1.52) showed that the influence of a foreign atom (e.g., hydrogen) on the rupture of atomic bonds was also admissable in this treatment. They found that the iron atoms in the proximity of the hydrogen atom were attracted to it. This elongated the Fe-Fe bonds which then ruptured more readily.

Work of current note in atomic fracture simulation was initiated by Weiner and Pear (1.53), who first addressed rapidly propagating cracks. A somewhat

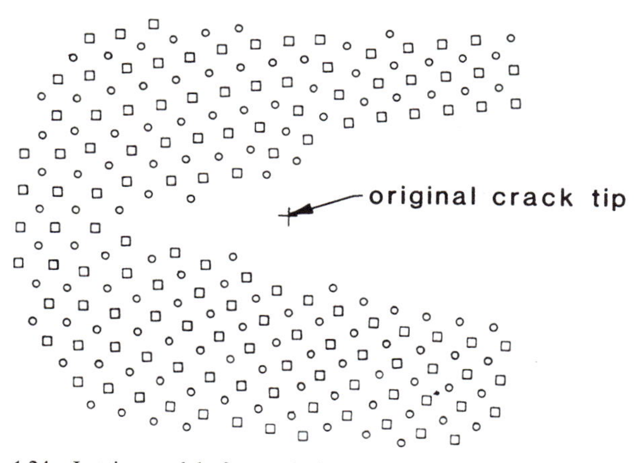

Figure 1.24 Lattice model of a crack tip in bcc iron showing crack extension.

tangential line of development that has much in common with the finite element method is one reflected by Ashhurst and Hoover (1.54). The current state-of-the-art in large-scale atomistic computer simulation of crack tips is probably represented by the work of Markworth et al. (1.55), Mullins (1.56), and Paskin et al. (1.57).

The calculations of Markworth et al. (1.55) were based upon body-centered-cubic iron. Reasonably reliable interatomic force-separation laws (conventionally expressed in terms of two-body potential functions) are available for this system and for its interactions with hydrogen and helium atoms. A typical computation is performed by inserting a hydrogen (or helium) atom into the lattice ahead of the crack tip. In contrast to their earlier result (n.b., different potential functions were used), the computation that was carried out shows that the presence of the hydrogen atom causes severe local distortion of the iron crystal, whereupon a relatively small applied stress can bring about a unit of crack advance by bond rupture. In this sense, the iron crystal was indeed "embrittled" by the hydrogen. While such a result is intuitively reasonable and probably to be expected, quantitative results can be obtained only through computations such as these.

While results such as those of Markworth et al. are encouraging, the prospects for further progress in atomistic simulation of fracture are daunting. The most serious barrier would seem to be the paucity of reliable multi-body interatomic force-separation laws (for both like as well as unlike atoms) that can account for temperature effects. Work on the atomic scale will always be handicapped by computer limitations. Compounding this constraint is the ultimate necessity to treat corrosion fatigue that involves low-level but repeated loadings. Nevertheless, formidable as the computational problems are, we feel that work in this area is now much more constrained by inadequacies in the physics of the problem (i.e., in deriving reliable potential functions) than in the mechanics aspects.

In concluding this section, it might be recognized that simpler models also exist. As a prime example, Kelly, Tyson, and Cottrell (1.58) have devised a rough indication of whether a crystal will fail in a ductile or in a brittle mode. Their criterion was that, if the ratio of the largest tensile stress to the largest shear stress at the crack tip exceeds the ratio of the ideal cleavage strength to the ideal shear strength, then a fully brittle crack will be sustained. Otherwise, dislocations are presumed to be generated whereupon crack extension will be accompanied by significant plastic flow and the fracture mode will be ductile. The simple approach of Kelly, Tyson, and Cottrell was later extended by Rice and Thomson (1.59).

1.3 The Establishment of Fracture Mechanics

The essential feature of fracture mechanics is the delineation of the contributions of the load and the structural geometry from that of the material. This procedure permits a material's fracture property to be determined in a convenient laboratory-scale test and, in conjunction with a mathematical analysis, to be used to predict the behavior of full-scale engineering structures. As just described, the introduction of the stress intensity factor was instrumental in a practical approach for the initiation of crack propagation under conditions where predominantly elastic material response can be expected. In this section the extensions of fracture mechanics to subcritical crack growth, nonmetals, and to rapid crack propagation that employ this parameter are described.

1.3.1 Subcritical Crack Growth

The failures in welded ships discussed by Williams and Ellinger (1.28), the molasses tank and other vessel failures reviewed by Shank (1.29), the Comet aircraft disasters described by Bishop (1.30), the collapse of the Point Pleasant bridge over the Ohio River reported by Bennett and Mindlin (1.60), and the generator rotor fractures described by Schabtach et al. (1.40, 1.41), reveal clearly the necessity of quantifying the processes by which a crack achieves a critical size. We here turn to the extension of the Griffith-Irwin concepts to treat these aspects of fracture mechanics. As a prime example, in work that appeared in 1961, Paris, Gomez, and Anderson (1.61) first showed how the stress intensity factor could be effectively applied to describe fatigue crack growth. As discussed later by Paris and Erdogan (1.62), a number of competing fatigue "laws" existed at that time that could adequately describe data obtained from specific specimen geometries and loading types. But, by applying a discriminating test in which fatigue crack growth rates obtained under different conditions could be correlated, Paris and Erdogan clearly showed the superiority of relations based upon K. Figure 1.25 shows this key result. Here, data for center cracked tensile panels under remote tension (K increasing with crack length) and under a wedge loading on the crack faces (K decreasing with crack length) very nicely consolidate. This result fairly well established the type of relation that is now commonly known as the "Paris law"—that is, Equation (1.1-16).

Fatigue crack growth presents a number of complications beyond the simple behavior exemplified by a Paris-law equation. The load history to be considered must not substantially deviate from uniformity. At one extreme, it is known that crack growth does not occur if ΔK is less than some threshold value. At the opposite extreme, fatigue cracking will be superseded by rapid crack growth. In addition, there is usually an effect of the mean stress in the load cycle. As described in Chapter 8, a plethora of empirical fatigue relations exist to account for one or more of these complications. For example, a variant of Equation (1.1-16) that accounts for the lower limit to ΔK for fatigue cracking that can be seen in Figure 1.25 is

$$\frac{da}{dN} = C(\Delta K - (\Delta K)_{\text{th}})^m \tag{1.3-1}$$

where $(\Delta K)_{\text{th}}$ is known as a threshold value. An extension of Equation (1.1-16)

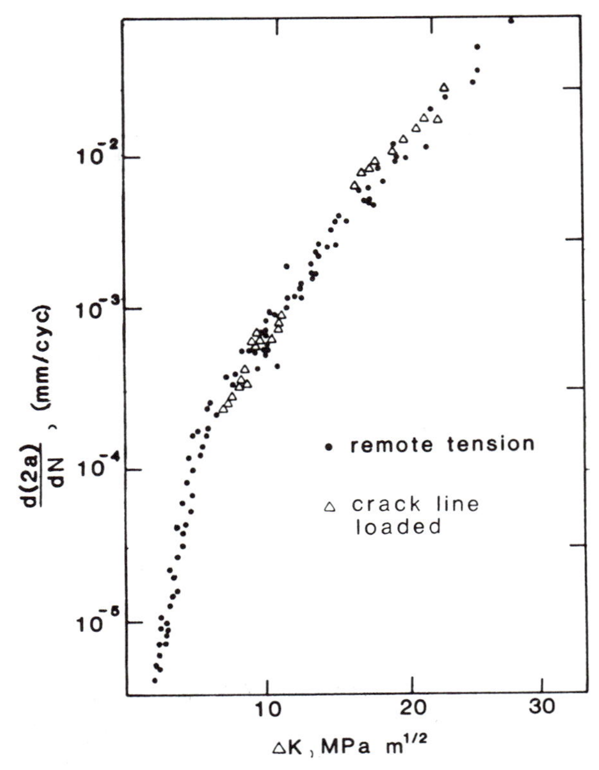

Figure 1.25 Fatigue crack propagation results in 7075-T6 aluminum showing correspondence between remote load and crack-line load results.

for the higher ΔK crack growth regime was suggested by Foreman et al. (1.63). Their result can be written as

$$\frac{da}{dN} = \frac{C(\Delta K)^n}{(1 - R)(K_{Ic} - K_{\max})} \tag{1.3-2}$$

where $R = K_{\min}/K_{\max}$. Like Equation (1.1-16), this equation has a limited theoretical basis. But, owing to its ability to accurately represent observed results, it is very widely used in structural integrity assessments.

A semiempirical formulation of great importance in fatigue crack propagation is the crack closure concept advanced by Elber (1.64) in 1971. He recognized that, because a fatigue crack leaves a wake of plastic deformation behind it, the crack faces can close on each other under a tensile applied stress. Hence, even if the minimum load level on the unloading portion of the load cycle has not been reached, no further damage occurs at the crack tip during that load cycle. Equivalently, during the loading portion of the cycle, no effect will occur until the crack has opened. On this basis, Elber proposed the idea of an effective stress, $(\Delta K)_{\text{eff}} \equiv K_{\max} - K_{0p}$, where K_{0p} corresponds to the opening stress level. The quantity $(\Delta K)_{\text{eff}}$ can then be used to replace ΔK in Equations (1.1-16), (1.3-1), or (1.3-2). While this substitution is logically defensible, unfortunately, K_{0p} is a difficult quantity to obtain. Nevertheless, as discussed in Chapter 8, Elber's idea is clearly a cornerstone in modern attempts to develop nonlinear fatigue crack propagation models.

In an analogous manner to fatigue crack growth characterization, environmentally assisted crack growth (i.e., corrosion cracking, corrosion fatigue) has also come to employ the stress intensity factor. In contrast, a widely accepted counterpart of the Paris-law type of relation does not exist. Where agreement generally exists is on the threshold level required for environmental crack growth, K_{Iscc}. That is, it is supposed that there is a material and environment-dependent constant such that if $K < K_{Iscc}$, an existing crack cannot propagate. The origin of this notion is probably due to Brown and Beacham (1.65).

Much less of a consensus exists on relations that are appropriate to determine crack growth rates when K exceeds K_{Iscc}. Perhaps the most widely accepted relation is the power law form suggested by Evans and Johnson (1.66), Equation (1.1-17). This formulation was based upon earlier work by Charles (1.67) who formulated an energy release rate relation for environmentally assisted crack growth. As formulated in terms of K by Evans and Johnson, environmentally assisted crack growth consists of three characteristic regions. This is shown in Figure 1.26. Just above K_{Iscc} there is region I—a region in which the crack velocity varies rapidly with K. A nearly constant velocity region, termed Region II, exists at higher K levels. At K levels approaching K_c, there is Region III; another region in which the crack velocity varies rapidly. The values of A and n, of course, will differ depending upon which region is being considered.

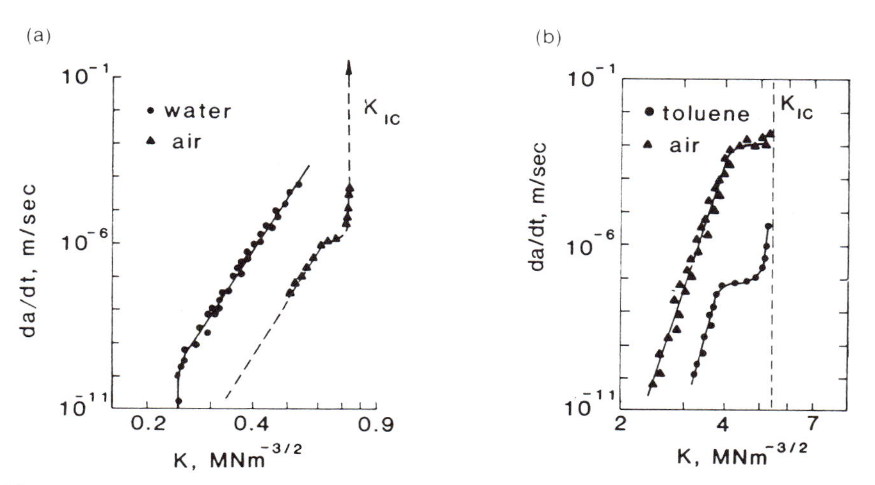

Figure 1.26 Environmentally assisted crack growth data for two materials: (a) glass and (b) alumina.

Alternatives to the power law form also exist. For example, a relation that has been used to treat crack growth in ceramics is

$$\frac{da}{dt} = V_0 \exp\left(\frac{CK - E_0}{RT}\right) \tag{1.3-3}$$

where V_0 and C are empirical material-dependent constants while E_0, R, and T denote the activation energy, the universal gas constant and the absolute temperature, respectively. It can nevertheless be seen that the stress intensity factor retains a key role in this type of relation.

1.3.2 Fiber Reinforced Composites

By transcending the initially micromechanical approaches to take a complete reliance on continuum mechanics, early work on the fracture of fiber-reinforced composites paralleled the development of relations for their deformation behavior. Prominent among the primary contributors to formulating micromechanical models for the various energy absorbing processes involved in the fracture of fiber composites (e.g., fiber breakage, matrix cracking, fiber pull-out, fiber-matrix debonding, inter-ply delamination) were Cooper and Kelly (1.68). However, while useful for a basic understanding of composite materials, such isolated models do not suffice for structural integrity assessments. At the same time, the use of the rule of mixtures and lamination theory that has been so successful for deformation behavior in fiber composite laminates will not suffice for fatigue and fracture. The trend has therefore been towards semiempirical adaptations of the LEFM formulations with disposable parameters being introduced to account for the additional complexities that arise.

An early application of continuum fracture mechanics to composites was made by Wu (1.69), who developed a failure surface approach for mixed mode fracture. Wu established a relation that can be generalized to the form

$$\left(\frac{K_{\mathrm{I}}}{K_{1c}}\right)^m + \left(\frac{K_{\mathrm{II}}}{K_{2c}}\right)^n = 1 \tag{1.3-4}$$

where K_{I} and K_{II}, respectively, are the Mode I and Mode II stress intensity factors,* K_{1c} and K_{2c} are their critical values, while m and n are material-dependent constants. Using experimental data on wood, Wu found the particular values $m = 1$ and $n = 2$ to be appropriate. Different values have been found to be necessary when applying Equation (1.3-4) to other types of composites.

* Work in fracture mechanics considers three distinct fracture modes: Mode I, the opening mode, Mode II, the sliding mode, and Mode III, the out-of-plane mode. These encompass all possible ways in which the crack tip can deform. Mode I predominates for isotropic materials because crack growth tends to occur in planes that are normal to the maximum tensile stress where $K_{\mathrm{II}} = 0$. Hence, only this condition has been considered to this point. See Chapter 3 for a further description.

Because Equation (1.3-4) cannot completely account for the diversity of crack growth processes in a fiber composite laminate, Zweben (1.70), Konish et al. (1.71) and others developed alternate forms. In essence, by allowing the fracture toughness to be direction-dependent, their models allowed for crack growth to occur in a non-self similar manner. A similar approach was formulated by Harrison (1.72). He delineated "cracking" from "splitting" by assigning separate toughness values for the two processes. Recognizing that the crack driving force is also direction-dependent, the crack growth criterion is then given by the particular equality (i.e., for cracking or for splitting) that is satisfied at the lower load level.

Together with the several crack length adjustment models that have been offered, the models of Zweben, Konish et al., and Harrison can be categorized as two-parameter relations. Although having broader utility than only for fiber composites, the strain energy density theory of Sih can be contrasted with these. As advanced for composites by Sih and Chen (1.73), the approach centers on the strain energy density quantity given by

$$S = \alpha_{11} K_{\mathrm{I}}^2 + 2\alpha_{12} K_{\mathrm{I}} + \alpha_{22} K_{\mathrm{II}}^2 \qquad (1.3\text{-}5)$$

where α_{11}, α_{12}, and α_{22} are known functions of the elastic constants and of the direction of crack extension. The strain energy density theory is based upon two hypotheses: that crack extension occurs, (1) in a direction determined by the stationary value of S, and (2) at a load level such that S equals a critical value, S_c. It can be seen that, while the strain energy density model depends only on a single experimentally determined parameter, it can also determine the direction of crack growth. For further details, see Chapter 6.

While the models that have been developed are appropriate for uni-directional composites, they are not as well suited for multi-directional laminates. Also, because of the usual requirement in fracture mechanics for an identifiable dominant crack, the small defects that are usually associated with the initiation of fracture in fiber composite laminates pose difficulties. Current work is therefore becoming focused on hybrid continuum/micromechanical models and probabilistic treatments. The fracture mechanics models that have been developed for fiber composite materials are discussed in Chapter 6.

1.3.3 Elastomeric Materials

Because of their importance as engineering materials, most fracture mechanics developments have at least tacitly focused on metals. Consequently, a classic paper in the field has not received the attention that it deserves. This is the extension of the Griffith theory to the rupture of rubber, contributed by Rivlin and Thomas (1.74) in 1953. This work is of interest because it initiated an entire branch of the subject—the fracture mechanics of elastomers—and it extended the basis of the general subject as a whole.

The essence of the Rivlin-Thomas approach was to demonstrate the existence of a characteristic energy for the tearing of thin sheets of rubber that is independent of the test piece configuration. To do this, they introduced a quantity called the tearing energy, defined for fixed grip loading conditions as

$$T = -\frac{\partial U}{\partial A} \tag{1.3-6}$$

where U is the stored elastic energy and A is the slit area (e.g., for an edge-cracked constant width specimen, $A = aB$, where a is the slit length and B is the specimen thickness). While this is similar to the Griffith approach, Rivlin and Thomas recognized that the elastic energy given up to support crack extension need not only be transferred to surface energy. Like Irwin and Orowan, they expected that energy changes would be proportional to the amount of growth and, hence, would be related to the crack-tip deformation state, independent of the shape of the test specimen and the manner in which it is loaded. As they put it:

> It is therefore to be anticipated that the energy which must be expended at the expense of the elastically stored energy, in causing a given small increase in the cut length at constant overall deformation of the test-piece will be substantially independent of the shape of the test-piece and the manner in which the deforming forces are applied to it. This energy will therefore be a characteristic energy for tearing of thin sheets of the material, although it may depend on the shape of the tip of the cut.

As Rivlin and Thomas later recognized, the critical tearing energy can also be affected by the temperature and the loading rate, as of course are the fracture properties of most materials.*

Rivlin and Thomas performed experiments using three different test specimen geometries to determine critical (or characteristic) tearing energy values numerically via Equation (1.3-6). For several different vulcanizates, they found essentially constant values of T_c for catastrophic tearing. These experiments apparently involved a significant amount of stable crack growth prior to instability and their values for the initiation of growth were substantially lower. Regardless, of most interest here, they demonstrated that a simple energy balance analysis could be developed to evaluate T for a given specimen that accounts for the large deformations and the incompressible nature of rubber. Then, using T_c values from measurements using an entirely different test specimen geometry, reasonably accurate predictions were made of the onset of crack growth.

Interestingly, these results suggest that, in rubber, T_c is independent of the fracture mode. But, of most importance to the foundations of fracture mechanics in general, was the implicit connection between a global energy

* There is some ambiguity in the parameter designation used by Rivlin and Thomas who denoted both the crack driving force and its critical value by the symbol T. In accord with more recent fracture mechanics approaches, we will preserve the distinction through the use of the equality $T = T_c$ for elastomeric crack growth. Note that, while the symbol T is also currently being used to denote the tearing modulus in elastic-plastic fracture mechanics (see Section 1.4.8), these concepts are totally dissimilar.

balance and the autonomous state of the crack tip at the onset of fracture that Rivlin and Thomas demonstrated for an elastomeric material. This work spawned an entire branch of fracture mechanics—for example, the work of Gent, Lindley, and Thomas (1.75) and Lindley and Stevenson (1.76) on the fatigue of rubber. For recent work, see Rivlin and Thomas (1.77) and the publications of the Malaysian Rubber Producers' Association of the U.K.

In common usage are expressions for the tearing energy valid for the large deformations experienced in rubber having the typical form

$$T = 2kwa \tag{1.3-7}$$

which Rivlin and Thomas (1.74) developed for an edge-cracked tension specimen. In Equation (1.3-7), w is the strain energy density and k is a strain-dependent geometrical factor. For linear elastic conditions where $w = \sigma^2/2E$, it can be seen that Equation (1.3-7) is consistent with the LEFM relations provided $k = (\pi/4)(1 + \nu)(\kappa + 1)$; [cf. Equation (1.2-23)]. But, for large deformations and small shape factors (i.e., the ratio of the loaded area to the stress free surface of a rubber component), this relation will differ.

1.3.4 Numerical Methods in Fracture Mechanics

Returning to linear elastic fracture mechanics, one of our primary contentions is that Irwin's introduction of the stress intensity factor, by superseding a global energy balance approach, constituted the origin of fracture mechanics. This view is supported by the proliferation of mathematical solutions that have been developed for the stress intensity factors in various crack/structure geometries and by the relative ease for which numerical solutions can be made for situations where these do not exist. Adding vital support for the practical utilization of this idea was the detailed development of the stress state in the vicinity of a crack tip provided by Williams (1.78) and, later, by Karp and Karal (1.79). This work has been particularly important in the variety of numerical techniques that have been used.

One of the first uses of the finite element method to determine stress intensity factors was that of Chan et al. (1.80). They used conventional elements at the crack tip and were therefore forced to use an awkward procedure in which K was inferred by extrapolating crack opening displacement values back to the crack tip. Shortly thereafter a number of investigators devised special elements that explicitly contained the elastic singularity. The first of these may have been that of Byskov (1.81). However, even this procedure became unnecessary owing to the discovery by Henshell and Shaw (1.82) of the so called quarter-point element. This is a conventional finite element in which the mid-side nodes have been shifted whereupon, by serendipity, the element appears to reflect the $r^{-\frac{1}{2}}$ singular behavior needed for elastic crack problems. Because of its simplicity, the quarter-point element became the usual choice of analysts for LEFM problems. A further contribution was made by Barsoum (1.83) who developed analysis procedures using isoparametric finite elements.

Several alternative numerical analysis methods have been used in fracture mechanics applications. These include the boundary collocation method used by Kobayashi et al. (1.84), the boundary integral method of Cruse (1.85), the finite difference method advocated by Wilkins and coworkers (1.86), and the alternating method. A review of numerical methods in fracture mechanics with particular emphasis on their application to the three-dimensional corner crack problem has been given recently by Akhurst and Chell (1.87). As a general guideline, the tendency among numerical analysts currently is to use a finite element code with conventional elements and to evaluate the crack driving force through a path-independent contour integral such as J (see Section 1.4.4). This technique allows the crack-tip region to be modeled with much less mesh refinement than is necessary to model the singular behavior directly and eliminates the need for a library of special elements.

1.3.5 Rapid Crack Propagation

The first quantitative assessment of dynamic fracture may be the paper by Mott (1.88) in 1948. By the use of a dimensional analysis procedure to determine the kinetic energy of a crack moving quasi-statically in an infinite domain, Mott deduced an expression suggesting the existence of a material-dependent limiting crack propagation speed. His limiting speed result was proportional to C_0, through an undetermined constant, where $C_0 = \sqrt{E/\rho}$ is the elastic bar wave speed for the material. Note that this result is independent of the fracture resistance of the material. However, there is a basic error in Mott's derivation that invalidates his crack length history result (see below).

The landmark paper of Yoffe (1.89) provided the first dynamic solution—that is, with inertia forces included in the equations of motion—albeit by considering the rather unrealistic case of a constant length crack translating at a constant speed in an infinite domain. Her paper, which appeared in 1951, focused on determining the stress field ahead of a propagating crack in order to develop a criterion for crack branching. This work led to a limiting crack speed estimate of $0.6C_2$, where $C_2 = \sqrt{\mu/\rho}$ is the shear wave speed and $\mu = E/2(1 + \nu)$ is the shear modulus.

The next subsequent noteworthy effort was the enlargement of Mott's quasi-static approach that was made by Roberts and Wells in 1954 (1.90). In particular, Roberts and Wells were able to estimate the undetermined constant in Motts' work—a result that suggests that $0.38C_0$ is the limiting crack speed. At about the same time, Hall (1.91) began to consider the origins of fracture toughnesses via crack-tip plasticity. Although the idea is implicit in the earlier papers of Irwin—see reference (1.6), for example—Hall appears to be the first to include the effect of plastic deformation in rapid crack propagation. His results suggested that the fracture requirement may depend upon crack speed, a now generally accepted idea.

A definite escalation in interest in the analysis of rapid crack propagation came in 1960 with the appearance of four major papers. Included are the quasi-static analyses of Dulaney and Brace (1.92) and of Berry (1.93), and the dynamic analyses of Craggs (1.94) and Broberg (1.95). Dulaney-Brace and Berry independently derived a crack growth history relation that corrected the error in Mott's result. By incorporating the Roberts and Wells limiting speed estimate, this result can be written as

$$V = 0.38\, C_0\left(1 - \frac{a_0}{a}\right) \tag{1.3-8}$$

where $a = a(t)$ is the instantaneous crack length and $a_0 = a(0)$ is the initial crack length. This result is widely quoted. However, it must be recognized that its derivation employed several assumptions that are seldom met in actual applications. Specifically, Equation (1.3-8) is based on quasi-static crack motion in an infinite domain subjected to a constant applied stress and assumes that the material's fracture energy is independent of the crack speed. Nevertheless, the limiting speed estimate of 0.38 C_0 that arises in this result is entirely reasonable. Broberg's 1960 paper improved Yoffe's solution by considering a uniformly expanding crack—a result that suggested that the limiting speed was C_R, the Rayleigh wave speed.*

A key subsequent step was contributed by Atkinson and Eshelby (1.96), who considered the flow of energy to the tip of a dynamically propagating crack and thereby introduced the contour integral idea into dynamic fracture mechanics. Subsequently, in a series of papers appearing in 1972 through 1974, Freund (1.97) provided results for unrestricted crack growth in an infinite domain. Coupled with the 1973 work of Nilsson (1.98), this work has completed the picture for elastodynamic crack propagation problems when wave reflections from component boundaries are not important. In so doing it has also set the basis for the quasi-static approach to crack arrest by providing what is now called the Freund-Nilsson relation. This key result, which is the dynamic generalization of Equation (1.2-23), can be expressed as

$$G = A(V)\frac{K^2}{E'} \tag{1.3-9}$$

where $A(V)$ is a monotonically increasing, geometry-independent, function of the crack speed that has a value of unity at $V = 0$ and increases monotonically to become infinite at the Rayleigh speed; see Chapter 4. While this result appears to have been first obtained by Craggs, the universality of Equation (1.3-9) stems more directly from the work of Freund and Nilsson and is therefore more properly attributed to them.

On the experimental side, the earliest reported work may be that of Schardin that was conducted in the 1930's; see reference (1.99). An early instance of an actual crack speed measurement was the value of 40,400 in./sec (in steel) reported by Hudson and Greenfield in 1947 (1.100). Later, the

* For a Poisson's ratio of 0.25, $C_R/C_2 = 0.91$ (plane stress). Thus, the limiting speed estimates of Roberts and Wells and of Yoffe, are nearly the same. For comparison with Broberg, their limiting speed value is $0.66C_R$.

photoelastic investigations of Wells and Post (1.101) in 1958 and the early crack growth measurements of Carlsson (1.102) reported in 1963 are perhaps most noteworthy. There of course have been many large-scale testing programs such as those conducted at the University of Illinois, in Japan, and the large-scale pipe fracture experiments conducted at Battelle's Columbus Laboratories.

Irwin (1.103) identified six particular research areas—crack speed, crack division, crack arrest, the onset of rapid fracturing, minimum toughness, and the direction of cracking—and introduced the concept of "progressive fracturing" to link them. Thus, one has the slow stable growth of small flaws to near critical size, the rapid spreading of the worst flaw (i.e., the one that first achieves the critical condition), and, possibly, the arrest of such a crack. Each of these events must be governed by a criterion that connects the load/geometry of the cracked component to an appropriate material property for the particular event. At that precipitous time, these characterizations, which had been largely limited to linear elastic material behavior and static conditions, could begin to include nonlinear and dynamic effects. This development ushered in the modern approach to dynamic fracture mechanics that is discussed in Chapter 4.

1.3.6 Dynamic Crack Arrest

The subject of crack arrest has been a virtual battleground throughout much of its history. Two mutually exclusive points of view for the arrest of a rapidly propagating crack have been vigorously advanced. One stems from the suggestion, apparently made first by Irwin and Wells (1.104), that crack arrest can be considered as the reverse in time of crack initiation. This proposition means that crack arrest should be characterizable in terms of a distinct material constant (commonly, the plane strain linear elastic fracture mechanics parameter K_{Ia}) and the crack driving force that corresponds to the crack length at the arrest point. Since there is no necessity to consider the rapid crack propagation process preceding arrest—indeed, there is not even the possibility of so doing—this approach must be expressed in static terms.

The opposing point of view is one based upon an equality that governs the moving crack. Crack arrest occurs as the special case when this equality can no longer be satisfied. Such an approach has previously been known as the dynamic view of crack arrest (because of the dynamic effects in the equations of motion of the cracked body) to contrast it with the static view of Irwin and Wells. But, refering to it as a kinetic approach, as we will do in this book, is more accurate because the focal point is on the crack growth process itself regardless of the importance of dynamic (inertia) forces.

An understanding has now been reached between the two opposing viewpoints. Dynamic effects generally exist in crack arrest. Nevertheless, statically determined arrest values can be determined experimentally that often suffice for practical purposes. Specifically, the quasi-static approach requires a small crack jump length in the experimentation being conducted to determine the arrest property. The reason is that a dynamic analysis for crack propagation in an infinite medium provides the legitimization for the quasi-static view. In actual structures, the return of kinetic energy to the crack tip prior to crack arrest can make the quasi-static approach invalid. As this condition is often not inhibiting, the bulk of current state-of-the-art applications of dynamic fracture mechanics to rapid crack propagation and arrest are based upon the use of quasi-static crack arrest concepts.

For the initiation and termination of rapid crack propagation under linear elastic fracture mechanics (LEFM) conditions, one has Equation (1.1-5) for crack initiation, and

$$K = K_{Ia}(T) \tag{1.3-10}$$

for crack arrest where T denotes the temperature at the crack tip. Note that, in both instances, K is calculated just as if the crack were stationary (quasi-static growth) while K_{Ic} and K_{Ia} are taken as temperature-dependent material properties. Again, this point of view does not include (nor could it) any consideration of the unstable crack propagation process that links the initiation and arrest points.

The kinetic point of view gives direct consideration to crack propagation with crack arrest occurring only when continued propagation becomes impossible. Within the confines of elastodynamic behavior, unstable crack propagation occurs in such an approach under the condition that

$$K = K_{ID}(V, T) \tag{1.3-11}$$

where K_{ID}, the dynamic fracture toughness, is a temperature-dependent function of the crack speed, V. Note that this function generally does not include an initation value [i.e., $K_{ID}(0, T) \neq K_{Ic}(T)$] so that Equation (1.1-5) is common to both approaches. It also does not include an arrest value as such. In the kinetic approach arrest occurs at the position and time for which K becomes less than the minimum value of K_{ID} and remains less for all greater times.*

In Equation (1.3-11), K generally is the dynamically calculated value of the stress intensity factor. It therefore must be determined as part of the solution of an initial value, moving boundary value problem. In early work an energy balance was adopted in which Equation (1.2-14) was generalized to

$$G = \frac{dW}{dA} - \frac{dU}{dA} - \frac{dT}{dA} \tag{1.3-12}$$

where T denotes the kinetic energy in the body. Equation (1.3-9) was then used to obtain the dynamic stress intensity factor in order to employ Equation (1.3-11) as the crack growth criterion. Within this framework the kinetic

* As Barenblatt (1.45) has noted, crack propagation has a reasonably close analogy with a block pushed along a rough surface. The difference between static and sliding friction is similar to the difference between the initiation and propagation fracture toughness. Note that there is no "arrest" friction coefficient per se. The block motion terminates (arrests) when it is not pushed firmly enough to overcome the sliding friction.

counterpart of Equation (1.3-10) is the inequality

$$K < \min_{0 \leq V < C_R} \{K_{ID}\} \tag{1.3-13}$$

The minimum value of K_{ID} has at various times been designed as K_{IM} and as K_{IA} with an emerging preference for the latter. This is obviously a key parameter. In fact, the legitimacy of the quasi-static approach rests upon the notion that K_{Ia} is a good approximation to K_{IA}.

Ripling and his associates at the Materials Research Laboratory have provided extensive data on crack arrest using a quasi-static interpretation of this process (1.105). Interestingly, while their original purpose was to obtain K_{Ic} data for rapidly applied loading (i.e., K_{Id} values), this work initiated quantitative treatments of crack arrest; see, for example, Figure 1.17. The work of Hahn and co-workers at Battelle's Columbus Laboratories beginning shortly thereafter was foremost in developing the kinetic approach to crack arrest (1.106). While a considerable number of data and analyses were developed on crack arrest in the Battelle and MRL programs, decisive evidence for one or the other point of view was lacking. This was eventually forthcoming from the experiments on photoelastic materials performed by Kalthoff et al. (1.107) and by Kobayashi et al. (1.108). These were instrumental in resolving the issue in favor of the kinetic point of view. A key result is that of Kalthoff et al. shown in Figure 1.27.

The experimental results shown in Figure 1.27, obtained using the method of caustics combined with high-speed photography, enabled a direct measure-

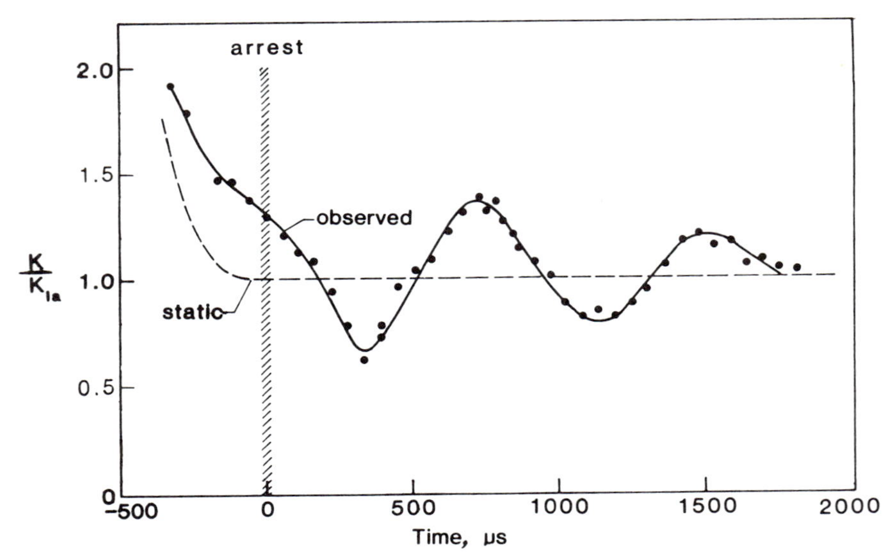

Figure 1.27 Comparison of static stress intensity factors and actual values during rapid crack propagation and following arrest in a DCB test specimen as determined by the method of caustics.

ment of the crack driving force for a fast moving crack to be obtained for the first time. It can be seen that there is a distinct difference between the actual arrest value and the static value that exists at a later time. Of some importance, these results provided direct confirmation of the computational results obtained earlier by Kanninen (1.109) that indicated the incompatibility of Equation (1.3-10) and inequality (1.3-13) in certain conditions. In addition, by performing experiments in different crack speed regimes, Kalthoff and his co-workers were also able to establish K_{ID} functions that were in general accord with those developed indirectly by Hahn et al. A relation constructed from tests like that shown in Figure 1.27 is shown in Figure 1.28.

Ironically, results of the type shown in Figure 1.28, while undermining the conceptual basis for the quasi-static approach to crack arrest, also provide a pragmatic basis for the use of Equation (1.3-10). Specifically, results of this type can generally be well-correlated by an empirical relation of the form

$$K_{ID} = \frac{K_{IA}}{1 - \left(\frac{V}{V_l}\right)^m} \tag{1.3-14}$$

where K_{IA}, V_l and m are temperature-dependent material constants. As described more completely in Chapter 4, if a K_{Ia} value is obtained under the proper conditions, it will be identical to K_{IA}; that is, crack jump lengths short

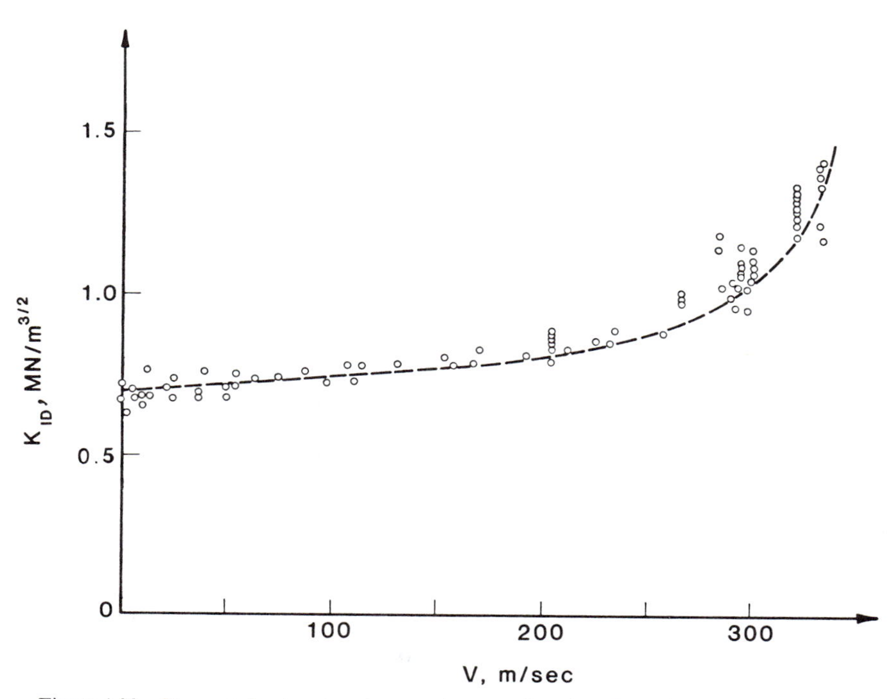

Figure 1.28 Dynamic fracture toughness values as a function of crack speed in Araldite-B.

enough that reflected stress waves do not return to interfere with the moving crack tip together with moderate crack speeds. Then, a kinetic prediction of crack arrest will be indistinguishable from the quasi-static prediction. As a consequence, it now appears that the controversy over the use of quasi-static or kinetic treatments of rapid crack propagation, which dominated the field over the decade of the 1970's, is now fairly well resolved.

In essence, it is now generally understood that, while kinetic treatments are necessary for certain geometries (e.g., where the crack propagation path parallels a nearby free surface), the simpler quasi-static point of view suffices for most engineering structures in many types of applications. Replacing this controversy at the forefront of research interest is one involving the appropriate nonlinear characterization for propagating cracks. Key work in this area is being focused on the question of the correct crack-tip characterizing parameter for inelastic conditions. These efforts are employing the crack opening angle parameter and other new concepts to make crack propagation/arrest predictions in dynamic-plastic conditions and in dynamic viscoplastic conditions. As described in Chapter 4, research in the decade of the 1980's will clearly see progress that will generalize dynamic fracture mechanics to address such nonlinear problems.

1.3.7 *Probabilistic Fracture Mechanics*

Probabilistic fracture mechanics treatments fall into one of two general categories. In the first category are applications where a dominant flaw does not exist and, hence, fracture initiates from one of a distribution of flaws. The prime application areas are to highly brittle materials like ceramics and to fiber composite materials. The key variable is then the time to failure. The second category in which probabilistic treatments are used is for applications involving a chain of events, each of which has some degree of uncertainty involved in it. The simplest example is where the existence of a flaw of a given length, the variation in the material toughness properties, and the magnitude of the applied stresses are all uncertain and must be treated in a statistical manner. In this instance the key variable is the hypothetical failure probability—a number that has significance only in a relative sense.

A cynical view might be that all probabilistic approaches are expressions of ignorance of one kind or another. As more intimate knowledge is acquired for a better quantitative understanding, the need for statistical treatments diminshes. Indeed, all too often, statistics are applied to experimental results when basic results exist that would alleviate the observed "scatter." Nevertheless, it cannot be denied that a lack of precise information exists in many practical applications where fracture is a concern. These situations therefore are candidates for probabilistic treatments.

The origins of probabilistic fracture mechanics as a quantitative approach to structural integrity can be directly connected to the work of Weibull (1.110). Yet, in a qualitative sense, probabilistic phenomena have played a prominent role since the beginnings of serious thinking about fracture. As reviewed by Irwin and Wells (1.104), da Vinci, the prototypical Renaissance man, conducted a series of experiments on iron wires in which he found that short lengths of wire were noticeably stronger than longer lengths. From the mechanics point of view, advanced somewhat later by Galileo, the breaking strength should of course only depend upon the cross-sectional area, and be independent of the length. The explanation lies in the flaw probability.

Considering the quality of the material that was probably available in the sixteenth century, the wires tested by da Vinci would have contained a significant number of imperfections whereupon the fracture strength would be dictated by the most critical flaw in a distribution of flaws. A longer length of wire would simply be more likely to contain a large flaw in a highly stressed location than would a shorter length. Hence, its breaking point would be lower. Irwin and Wells estimated on this basis that the fracture strength should be roughly in inverse propagation to the length of the wires—an effect that could be readily recognized even with the load measuring equipment available to da Vinci. This size effect was noted in many different studies, including one concerning the wires used in British aircraft during World War I. Such results, in fact, apparently led Griffith to his cornerstone fracture theory.

Weibull supposed that a structural component behaves as if composed of many subunits, each possessing an intrinsic fracture strength. By assuming a specific distribution of strengths, he was able to predict a variation in strength with component volume that corresponded at least qualitatively with observed fracture behavior. In so doing, the Griffith theory was taken as a base whereupon the approach is sometimes referred to as the Weibull-Griffith theory. Weibull's two-parameter distribution relation can be expressed as

$$R(t) = \exp\left[-\left(\frac{t}{\beta}\right)^{\alpha}\right] \tag{1.3-15}$$

where α and β are stress and temperature-dependent parameterss known as the shape factor and the scale factor, respectively. More general forms of the Weibull distribution also exist that can be used. An early use of probabilistic fracture mechanics made by Besuner and Tetleman (1.111) is illustrated in Figure 1.29. Example applications of a probabilistic fracture mechanics approach are given by Rau and Besuner (1.112), Gamble and Strosnider (1.113), and by Harris and Lim (1.114).

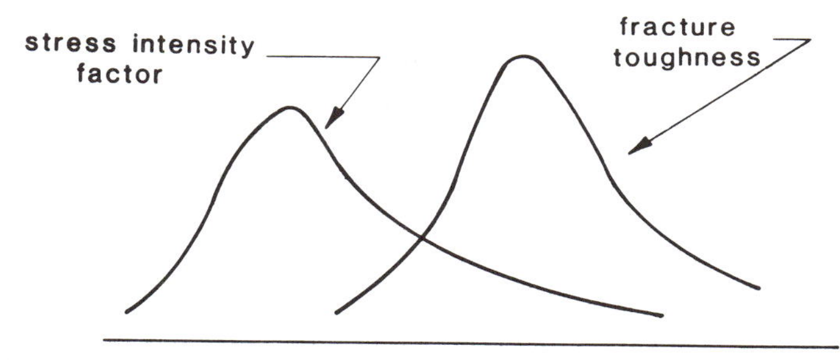

Figure 1.29 Overlap of fracture toughness and stress intensity factors in a probabilistic fracture mechanics approach.

1.4 Nonlinear Considerations

The original motivation for the development of fracture mechanics was (and largely still is) to be able to account for materials that fracture with limited plastic deformation—that is, at applied stress levels less than those producing net section yielding. The discipline was initially focused exclusively on essentially linear elastic-brittle behavior. But, with the successes that were achieved with LEFM, materials for which such an approximation would be invalid also became of interest. Primary concern was for conditions in which the cracked structural component itself would likely obey the dictates of LEFM, but the small-scale laboratory experiments needed to provide the necessary fracture properties would not; that is, for materials sufficiently ductile and tough that the extent of plastic yielding accompanying the crack growth would be comparable to the specimen dimensions. Interest in direct consideration of the plastic deformation attending crack growth has also been pursued for the purpose of achieving a better understanding of the metallurgical origins of fracture toughness in metals. Most recently, attention has been focused on structural components (e.g., stainless steel nuclear plant piping) for which LEFM is inappropriate. The origins of the nonlinear fracture mechanics techniques that have been developed for these circumstances are addressed next.

1.4.1 Simple Crack-Tip Plasticity Models

The first quantitative accounting for the effect of the plastic zone at the crack tip seems to be that suggested by Irwin, Kies, and Smith (1.115). On the basis that a plastically deformed region cannot support the same level of stress that it could if yielding did not intervene, they argued that a cracked body is somewhat weaker than a completely elastic analysis would suggest. To account for this within the framework of linear elasticity, they supposed that the effect would be the same as if the crack length were slightly enlarged. Thus, a plasticity-modified stress intensity factor for a crack in an infinite medium can be written as

$$K = \sigma\sqrt{\pi\,(a + r_y)} \tag{1.4-1}$$

where r_y is supposed to be a measure of the plastic zone size—for example, the radius of a circular zone.

Estimates of r_y can be obtained in a variety of ways. Perhaps the simplest is to take $2r_y$ as the point on the crack line where $\sigma_y = \sigma_Y$, the yield stress. Using Equation (1.2-20), this simple argument gives

$$r_y = \begin{cases} \dfrac{K^2}{2\pi\sigma_Y^2} & \text{plane stress} \\[2ex] \dfrac{K^2}{6\pi\sigma_Y^2} & \text{plane strain} \end{cases} \tag{1.4-2}$$

The distinction arises because σ_Y is taken as the uniaxial tensile yield stress for plane stress while for plane strain it is appropriate to use $\sqrt{3}\sigma_Y$ as the parameter governing the plastic zone size. Thus, r_y is a factor of 3 smaller in plane strain than in plane stress. Combining Equations (1.4-1) and (1.4-2) gives (plane stress)

$$K = \sigma\sqrt{\pi a}\left(1 - \frac{1}{2}\frac{\sigma^2}{\sigma_y^2}\right)^{-\frac{1}{2}} \tag{1.4-3}$$

From Equation (1.4-3) it can be seen that this approximate plastic zone correction will be negligible when $\sigma \ll \sigma_Y$, but will increase K by about 40 percent at applied stresses that are of yield stress magnitude. In plane strain, the correction will generally be less than 10 percent, however.

Through the use of Equations (1.2-23), Equation (1.4-3) can be written in terms of the energy release rate. For plane stress conditions and assuming that $\sigma \ll \sigma_Y$, this relation is

$$G = \frac{\pi\sigma^2 a}{E}\left[1 + \frac{1}{2}\left(\frac{\sigma}{\sigma_y}\right)^2\right] \tag{1.4-4}$$

which is a form once in common use for a plasticity-corrected crack driving force. It will be useful here only for comparisons with the results obtained using the COD approach described below.

1.4.2 Origins of the COD Approach

At about the same time as Irwin and his associates were developing the plasticity-enhanced stress intensity factor to broaden the applicability of the linear elastic approach, Wells (1.116) advanced an alternative concept in the hope that it would apply even beyond general yielding conditions. This concept employs the crack opening displacement (COD) as the parameter governing crack extension. Wells evaluated this parameter using Irwin's plastic zone estimate and the equations for a center crack in an infinite elastic body. Specifically, substituting Equation (1.4-2) into Equation (1.2-21) to obtain $\delta = 2v(r_y)$ gives

$$\delta = \alpha\frac{K^2}{E\sigma_Y} \tag{1.4-5}$$

Here, α is a numerical factor that in Wells' work was equal to $4/\pi$.

Wells recognized that the factor $4/\pi$ is inconsistent with an energy balance approach (which would require a factor of unity) and subsequently adopted $\alpha = 1$. Other investigators later found other values of α to be appropriate. Regardless, Equation (1.4-5) shows that the COD approach is entirely consistent with LEFM where the latter applies. Note from Equations (1.4-2) and (1.4-5) that

$$r_y = \frac{\delta}{2\pi e_Y} \tag{1.4-6}$$

where $e_Y = \sigma_Y/E$ is the uniaxial yield strain. Then, it can be shown that

$$\frac{\delta}{2\pi e_Y a} = \left[2\left(\frac{\sigma_Y}{\sigma}\right)^2 - 1\right]^{-1} \tag{1.4-7}$$

which Wells felt would be acceptable up to $\sigma/\sigma_Y = 0.8$. While there may be little that is remarkable about this result, Wells' next step certainly was. This was to convert Equation (1.4-7) to general yielding conditions. Wells argued that

> It is appropriate to assume, although it is not thereby proven, that the crack opening displacement δ will be directly proportional to overall tensile strain e after general yield has been reached.

By assuming that $r_y/a = e/e_Y$, Wells' intuitive argument led to

$$\frac{\delta}{2\pi e_Y a} = \frac{e}{e_Y}, \qquad e > e_Y \tag{1.4-8}$$

which he suggested would be an approximate fracture criterion for the post-yield regime. This set the basis for the widely used COD method, as will be seen in the following section.

While elastic-plastic analyses to determine the plastic region at a crack tip were available, an explicit relation was needed for δ in order to advance the COD concept. This was provided in a key paper published in 1960 by Dugdale (1.117) in which he developed a closed-form solution applicable for plane stress conditions. Using methods of the complex variable theory of elasticity developed by Muskhelishvili (1.46), Dugdale supposed that for a thin sheet loaded in tension, the yielding will be confined to a narrow band lying along the crack line. Mathematically, this idea is identical to placing internal stresses on the portions of the (mathematical) crack faces near its tips; the physical crack being the remaining stress-free length.

The magnitude of the internal stresses in Dugdale's model are taken to be equal to the yield stress of the material. In order to determine the length over which they act, Dugdale postulated that the stress singularity must be abolished.* For a crack of length $2a$ in an infinite medium under uniform tension σ, this led Dugdale to the relation

$$\frac{a}{c} = \cos\left(\frac{\pi}{2}\frac{\sigma}{\sigma_Y}\right) \tag{1.4-9}$$

Here, $c = a + d$, where d denotes the length of the plastic zone at each crack tip; see Figure 1.30(a). This can also be written as

$$d = 2a\sin^2\left(\frac{\pi}{4}\frac{\sigma}{\sigma_Y}\right) \simeq \frac{\pi}{8}\left(\frac{K}{\sigma_Y}\right)^2 \tag{1.4-10}$$

---

* Goodier and Field (1.118) have pointed out the resemblance to the finiteness condition invoked in Joukowski's hypothesis in airfoil theory.

(a)

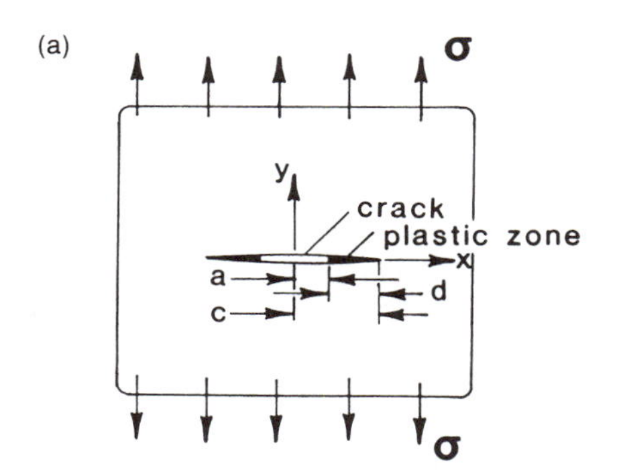

(b)

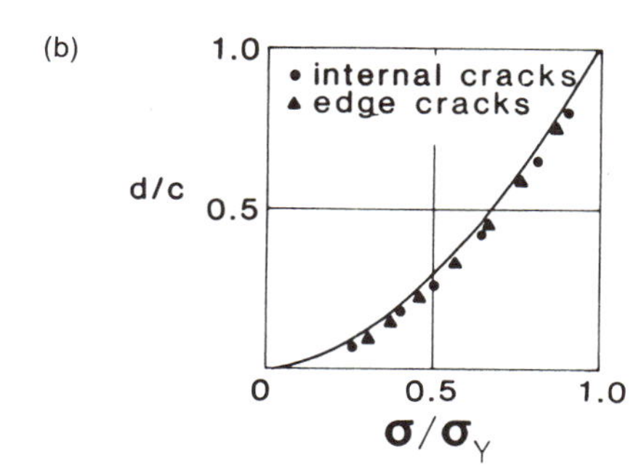

Figure 1.30 (a) The Dugdale model; (b) Dugdale's results for the plastic zone size and a comparison with analysis.

where the approximation is valid for small-scale yielding conditions. This can be compared with the expression for r_y given by Equation (1.4-2). Note that the obvious mathematical similiarities between this approach and that of Barenblatt (1.45) has led a number of writers to refer to the "Barenblatt-Dugdale" crack theory. We believe that this is misleading on the basis that the physical bases of the two approaches (i.e., one representing molecular cohesion, the other macroscopic plasticity) differ so markedly.

Dugdale obtained experimental results that could be compared with the plastic-zone-size predictions of Equation (1.4-10) by strain etching steel sheets having both internal and edge slits. The comparison is shown in Figure 1.30(b). Despite this good agreement, his approach was not immediately accepted. This could have been due to doubt that yielding could actually occur in this fashion (n.b., Dugdale did not provide pictorial evidence in his classic paper). It remained for Hahn and Rosenfield (1.119) to show that zones in plane stress conditions do, in fact, conform to the Dugdale model.

1.4.3 Extension of the COD Approach

The crack face displacements in the Dugdale model were first worked out in 1963 by Goodier and Field (1.118). The form of their result is rather cumbersome. But, as shown, for example, by Kanninen et al. (1.120), the normal displacement on the crack faces can be written in the reasonably convenient form

$$v(x) = \frac{2}{\pi}\frac{a\sigma_Y}{E}\left\{\log\left|\frac{\sqrt{c^2 - a^2} + \sqrt{c^2 - x^2}}{\sqrt{c^2 - a^2} - \sqrt{c^2 - x^2}}\right| + \frac{x}{a}\log\left|\frac{x\sqrt{c^2 - a^2} - a\sqrt{c^2 - x^2}}{x\sqrt{c^2 - a^2} + a\sqrt{c^2 - x^2}}\right|\right\}, \quad 0 \leq x \leq c \qquad (1.4\text{-}11)$$

It may be of passing interest to recognize that $v(x)$ has a vertical slope at $x = a$, as can be found by differentiating Equation (1.4-11). However, conceptual sketches of the Dugdale model seldom show this [cf. Figure 1.30(a)].

Of more interest is the displacement at the crack tip given by the Dugdale model. This can be obtained via a limiting procedure applied to Equation (1.4-11). The result is

$$v(a) = \frac{4}{\pi}\frac{a\sigma_Y}{E}\log\frac{c}{a} \qquad (1.4\text{-}12)$$

Noting that the COD in Wells' approach is $\delta = 2v(a)$, and using Equation (1.4-9) to eliminate c, then gives the widely used form first obtained by Goodier and Field. This is

$$\delta = \frac{8}{\pi}\frac{a\sigma_Y}{E}\log\left[\sec\left(\frac{\pi}{2}\frac{\sigma}{\sigma_Y}\right)\right] \qquad (1.4\text{-}13)$$

This key result was later obtained independently by Burdekin and Stone (1.121) who used it directly to advance Wells' COD concept. They first showed, by expanding the right-hand side of Equation (1.4-13) in a Taylor's series, that

$$\delta = \frac{K^2}{E\sigma_Y}\left[1 + \frac{\pi^2}{24}\frac{\sigma^2}{\sigma_Y^2} + \cdots\right] \qquad (1.4\text{-}14)$$

whereupon Equation (1.4-14) demonstrates the compatibility of the Dugdale model with LEFM for $\sigma \ll \sigma_Y$. This can also be seen by comparing Equation (1.4-14) with (1.4-4), recognizing that $G = \sigma_Y\delta$.

Burdekin and Stone also demonstrated the plausibility of the notion that fracture could be governed by critical δ values by experimental results on mild steel in tension and bending. This work provided the basis for the semiempirical "COD Design Curve" approach that is now used extensively in the United Kingdom for fracture under contained yielding conditions. As evolved by Dawes (1.122) and his co-workers at the UK Welding Institute, the COD design curve approach is based upon empirical correlations that currently take the form

$$\Phi \equiv \frac{\delta_c}{2\pi e_Y \bar{a}_{\max}} = \begin{cases} \left(\dfrac{e}{e_Y}\right)^2, & \dfrac{e}{e_Y} \leq 0.5 \\ \dfrac{e}{e_Y} - 0.25, & \dfrac{e}{e_Y} \geq 0.5 \end{cases} \qquad (1.4\text{-}15)$$

where e is the local applied strain that would exist in the vicinity of the crack if the crack were absent, $e_Y = \sigma_Y/E$ is the yield strain, δ_c is the critical COD, and $\bar{a}_{\max}$ is a conservative (an underestimate) of the critical flaw size. By comparing Equation (1.4-15) with Equations (1.4-7) and (1.4-8), it can be seen that the origin of the COD approach is directly traceable to Wells (1.116) although, clearly, semiempirical improvements have been made.

Applications of the COD design curve approach are usually focused on the "yielding fracture mechanics" regime—that is, one between linear elastic fracture mechanics behavior and fully plastic behavior where failure occurs by plastic instability. As described by Harrison et al. (1.123), these applications are intended to be conservative. This can readily be seen by comparing Equation (1.4-15) with (1.4-7). This comparison shows that for small values of $\sigma/\sigma_Y = e/e_Y$, there is a built-in factor of safety of two on the crack size—a factor that is assumed to hold for the entire loading range. Corrections to account for residual stresses and geometric stress concentrations, as might be found in welded structures, have been developed, primarily through the efforts of The Welding Institute of the U.K.

1.4.4 The J-Integral

An alternative to the COD design curve for applications of fracture mechanics where elastic-plastic deformation must be taken into account is one based upon the path-independent contour integral introduced by Rice (1.7), coupled with the crack characterization studies of Hutchinson (1.8) and Rice and Rosengren (1.124). While similar concepts were advanced independently by Sanders (1.125), Eshelby (1.126), and Cherepanov (1.127), Rice's approach has clearly carried the day. In what is very likely one of the two most quoted papers in all of the fracture mechanics literature (the other is Griffith's), Rice laid the ground work for the bulk of the applications in elastic-plastic fracture mechanics and for crack-tip characterization in a variety of other applications. The basic relation is

$$J = \int_\Gamma W\,dy - \mathbf{T}\frac{\partial \mathbf{u}}{\partial x}\,ds \tag{1.4-16}$$

where Γ is a curve that surrounds the crack-tip, $\mathbf{T}$ is the traction vector, $\mathbf{u}$ is the displacement vector, W is the strain energy density and the y direction is taken normal to the crack line. Of most importance is that, for deformation plasticity (i.e., nonlinear elastic behavior), J is "path-independent" and will have the same value for all choices of Γ.

The use of J in elastic-plastic fracture mechanics will be taken up in Chapter 5 where it will be shown that the interpretation of J as the rate of change of the potential energy for nonlinear constitutive behavior plays a key role for the analysis of fracture in elastic-plastic conditions. Here, we will focus on the role played by J in unifying linear elastic fracture mechanics. Specifically, by taking Γ to be a contour that just circumscribes the cohesive zone in Barenblatt's model (1.45), Rice readily found that

$$J = \int_0^{\delta_t} \sigma(\delta)\,d\delta \tag{1.4-17}$$

where σ denotes the cohesive stress and δ_t is the separation distance at the crack tip. At the onset of fracture, δ_t must be equal to δ_c, the out-of-range interatomic separation distance. Then, the right-hand side of Equation (1.4-17) would be twice the surface energy. Thus, for fracture, $J_c = 2\gamma$. This relation strongly suggests that, for linear elastic conditions, J and G are equivalent.

This equivalence can also be shown directly through an energy release rate interpretation of J, which results in

$$J = -\frac{\partial \Pi}{\partial a} \equiv G \tag{1.4-18}$$

where Π denotes the potential energy of the cracked body. From this finding Rice was able to conclude that:

> the Griffith theory is identical to a theory of fracture based on atomic cohesive forces, regardless of the force-attraction law, so long as the usual condition is fulfilled that the cohesive zone be negligible in size compared to characteristic dimensions

Finally, Rice also applied the J-integral to the Dugdale model. The result is just the same as Equation (1.4-17) provided $\sigma(\delta)$ is taken equal to σ_Y. The result is simply

$$J = \sigma_Y \delta_t \tag{1.4-19}$$

where δ_t is the crack-tip opening displacement. Equations (1.4-18), and (1.4-19) taken together with (1.2-23) show the equivalence of all of the popular fracture mechanics parameters under linear elastic conditions.

That J is based on deformation plasticity should not be viewed as an extreme deficiency for, as Budiansky (1.128) has shown, provided proportional loading exists, deformation plasticity and incremental plasticity are equivalent. However, this will not be true for a growing crack. Crack advance in an elastic-plastic material involves elastic unloading and nonproportional loading around the crack tip. Neither of these processes is adequately accommodated by deformation theory. This fact has led Hutchinson (1.129) to state that

> Tempting though it may be, to think of the criterion for initiation of crack growth based on J as an extension of Griffith's energy balance criterion, it is nevertheless incorrect to do so. This is not to say that an energy balance does not exist, just that it cannot be based on the deformation theory J.

Nonetheless, as the work reported in Chapter 5 will show, the energy-based definition of J given by Equation (1.4-18) has been very useful in mathematical analyses, both to determine critical J values from experimental load-deflection records and for component fracture predictions.

1.4.5 The Collinear Strip Yield Model

The term "strip yield model" refers to the introduction of crack-tip plasticity as a line or strip element that emanates from the crack tip. The Dugdale or collinear strip yield model discussed above is the simplest and best known of these. It has a physical justification for thin sections where the through-the-thickness plastic relaxation characteristic of plane stress conditions occurs. However, for thicker sections where plane strain deformation might occur, the Dugdale model would clearly not represent the plastic deformation that would ensue. A mathematical model has been developed for these conditions by assuming that the plastic yielding occurs on lines inclined to the crack plane. Like the Dugdale model, this model can be formulated by linear elastic analysis methods whereupon the principle of linear superposition is valid. Because this possibility admits a great deal of mathematical convenience, many problems can be solved, albeit at the expense of some physical reality. These models are discussed in this section.*

Lying as it does completely within the realm of linear elasticity, the collinear strip yield model is highly amenable to mathematical analyses (n.b., the use of linear superposition is valid). It can also be physically realistic. A research program carried out at Battelle's Columbus Laboratories in 1967 illustrates these facts. The experiments, performed by C. R. Barnes, used rectangular, center-cracked coupons of steel foil that displayed elongated plastic zones. One such experiment is shown in Figure 1.31, where the plastic regions can be seen at a series of monotonically increasing loads.

Because the extent of the yielding can clearly become comparable to the specimen dimensions, the infinite plane solution would be inadequate to predict the results shown in Figure 1.31. In addition, there will be an effect of misaligning the load. This issue was investigated by Kanninen (1.130) who solved the Dugdale problem for a linearly varying remote tensile loading $\sigma_\infty = \alpha x$, where x is the direction parallel to the crack. For small values of α, this result can be used to find a correction to the crack opening displacement at the more highly strained crack tip. This is

$$\Delta\delta = \frac{4\alpha a^2}{E} \tan\left(\frac{\pi}{2}\frac{\sigma}{\sigma_Y}\right) \tag{1.4-20}$$

which would be added to Equation (1.4-13) to obtain the true crack opening displacement. Unfortunately, one would not necessarily know the appropriate value of α, particularly if it characterizes an unwanted misalignment. But, for foil at least, what could be observed is the discrepancy in the plastic zone lengths; see Figure 1.31. An approximate expression obtained by Kanninen from his exact result is

$$\frac{\Delta\delta}{\delta} = \frac{\Delta d}{d} \cos\left(\frac{\pi}{2}\frac{\sigma}{\sigma_Y}\right) \tag{1.4-21}$$

* Within applied mechanics the terms "plane stress" and "plane strain" have very precise meanings. However, these terms are applied in somewhat looser ways in fracture mechanics. Specifically, while plane stress rigorously means that the principal stress acting in the direction normal to the plane of interest is negligibly small, the plane stress condition in fracture mechanics is commonly taken to characterize thin components with in-plane loading and the surface layer of thicker components. But, for a state of plane stress to occur, the stress gradients in the direction normal to the plane must also be negligibly small—a condition that applies only approximately to a thin plate but certainly not to the surface of a thick body.

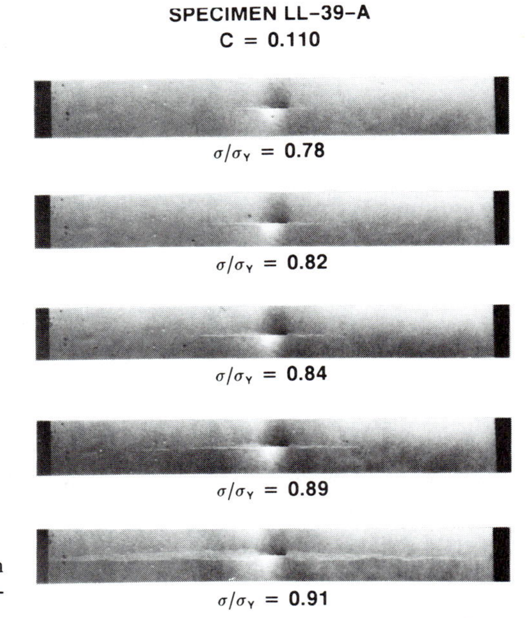

Figure 1.31 Plastic zones observed in center-cracked steel foil sheets subjected to tension loading.

where d is the plastic zone length given by Equation (1.4-10) and $2\Delta d$ is equal to the difference between the plastic zone lengths at the two crack tips. Interestingly, Equation (1.4-21) reveals that the plastic zone lengths are more sensitive to load misalignment than are the crack opening displacements. Hence, they are a conservative indicator of the propensity of a skewed loading to produce a premature fracture.

A strip yield zone model generally does not provide a good description of the crack-tip plastic zone per se. It has a further disadvantage in that the extent of strain hardening can only be related to a local strain in an arbitrary manner—that is, because a line zone has a zero gage length. Kanninen et al. (1.120) generalized the collinear strip yield model to permit a variable flow stress along the plastic zone. The flow stress at each point was then associated with the displacement at that point through a local necking model that associated the displacement with a strain. But, the complexity so introduced, together with the attendant arbitrariness, makes such an approach lose its prime virtue of simplicity. Hence, such approaches are useful indeed for relating the COD to component dimensions and applied loads (see above), but cannot be expected to provide a faithful representation of the details of the crack-tip deformation.

One of the drawbacks to the COD method for fracture problems is the difficulty of calculating such values for arbitrary crack/structure geometries and loading conditions. Elastic-plastic finite element methods are cumbersome and, in addition, because the crack-tip in a finite element model will have a zero normal displacement, they require an arbitrary definition of COD. For this reason, the Dugdale model has been of central importance for the implementation of this approach. The computational procedures that have been applied to obtain solutions for Dugdale cracks in various geometries have recently been reviewed by Petroski (1.131) who also provides a solution (by a weight function method) for edge cracks. To illustrate, for an edge crack in a semi-infinite plate subject to remote uniform tension, he has found that

$$0.06952\left(\frac{d}{a+d}\right)^{\frac{5}{2}} + 0.7099\left(\frac{d}{a+d}\right)^{\frac{3}{2}} + 3.1724\left(\frac{d}{a+d}\right)^{\frac{1}{2}} = 1.258\pi\,\frac{\sigma}{\sigma_Y} \tag{1.4-22}$$

While this method does not necessarily give good estimates of the crack opening displacement, Chell (1.132) has shown that Equation (1.4-12) is very nearly geometry independent. Hence, to a good approximation, d values determined from Equation (1.4-22) can be used to obtain $\delta(=2v)$ values for edge cracks.

1.4.6 Other Strip Yield Models

Dugdale formulated the collinear strip yield model using the complex variable formulation of the theory of elasticity. Shortly thereafter, Bilby, Cottrell, and Swinden (1.133) presented an alternative approach based on the use of linear dislocation arrays. This approach is sometimes referred to as the BCS model in honor of the authors of that paper. Mathematically, their approach replaces the solution of the usual differential equations of the theory of elasticity by an integral equation solution. Like Dugdale's approach, a singularity canceling equation is introduced.

A dislocation pileup model gives a unique value of the crack-tip opening displacement, but appears to be very limited in the boundary value problems that can be handled. Consequently, it has been largely superseded by the continuum approach. The more enduring idea that emerged from this work is the subsequent development by Bilby and Swinden (1.134) of a model in which plastic relaxation occurs by dislocation pileups on slip planes that are inclined symmetrically from the crack plane. This approach is known both as the Bilby-Swinden model and as the inclined strip yield model.

The Bilby-Swinden model was originally proposed in terms of a continuous distribution of edge dislocations along both the crack line (i.e., as in the BCS model) and along the assumed slip lines. However, this problem proved to be intractable. Only some preliminary numerical results (of uncertain validity) were obtained by a finite difference scheme. Atkinson and Kay (1.135) neatly circumvented the mathematical difficulties by introducing a superdislocation to represent the net effect of the plastic zone dislocation array. This approach was later extended by Atkinson and Kanninen (1.136), who had in mind the use of this model to address problems in which fatigue crack growth leaves behind a wake of plasticity.

The basic equations of the Atkinson-Kanninen inclined strip yield model are for the equilibrium of the superdislocation representing the crack-tip plasticity and, like all strip yield models, for the cancellation of the crack-tip singularity. In the equilibrium equation, the superdislocation is "pushed out" from the crack-tip by the applied stress and resisted by a "friction stress" related to the yield stress. The unknowns are the strength of the superdislocation and its position relative to the crack-tip. Such a model clearly does not give an accurate representation of the plastic zone size. But, this is tolerable because it does give a good estimate of the crack-tip opening displacement. Other uses of the inclined strip yield zone model have also been made. These include the contributions of Riedel (1.137), Vitek (1.138), and Evans (1.139).

A third type of strip yield model has been proposed recently by Weertman et al. (1.140). Their "double slip plane crack model" consists of a Griffith crack with slip planes parallel to the crack on both sides. The model is supposed to represent either Mode II (in-plane shear) or Mode III (anti-plane shear). They report that, when the crack tip advances, the stress intensity factor becomes smaller whereupon the residual plasticity left behind a growing crack can be represented. Their work, like that of Kanninen and Atkinson (1.141), is aimed at fatigue crack growth.

Strip yield zone models have also been useful in aspects other than elastic-plastic conditions. Knauss (1.142) and Schapery (1.143) have used this idea in developing a viscoelastic fracture mechanics model. Propagating crack models employing a dynamic generalization of the Dugdale model were contributed by Kanninen (1.144), who adopted Yoffe's approach, and Atkinson (1.145), who emulated Broberg. Embley and Sih (1.146) later provided a complete solution for this class of problems. These approaches are described more fully in Chapter 4.

1.4.7. Origins of Elastic-Plastic Fracture Mechanics

While the initial work in fracture mechanics was based upon an energy balance criterion, later work identified alternate fracture parameters: principally, the stress intensity factor, the crack opening displacement, and the J-integral parameter. In linear elastic fracture mechanics (LEFM), these are all interrelated. Specifically, for plane strain conditions in the "opening" mode, it has been shown that

$$G = J = \frac{1 - \nu^2}{E} K^2 = \delta\sigma_Y \tag{1.4-23}$$

In view of Equation (1.4-23), which of the four basic parameters involved in LEFM is the "most basic" may appear to be a purely academic question. However, it is considerably more important when it becomes necesssary to select the basis of nonlinear fracture mechanics for elastic-plastic conditions. Many possibilities exist, all of which have their origins in one of the LEFM parameters. But, a set of equalities like (1.4-23) does not exist beyond LEFM. Consequently, a considerable amount of research has been focused on developing inelastic fracture parameters.

The basic work in elastic-plastic fracture mechanics was that contributed from a completely theoretical point of view, primarily by Rice (1.7), and by Hutchinson (1.8). Hence, in what can be seen as a common pattern in the development of fracture mechanics (cf. Inglis and Griffith, Westergaard and Irwin, Muskhelishvili and Dugdale), the mathematical framework already existed that could be exploited by experimentalists to make a significant advance in the subject. In the case of elastic-plastic fracture mechanics, it was the perceptiveness of Begley and Landes (1.147, 1.148) that accomplished this. Their key papers appeared in 1972. While aided by an element of good fortune—see (1.149)—they were able to establish the J-integral parameter as the premier criterion in elastic-plastic fracture mechanics, as follows.

Begley and Landes, in seeking a failure criterion that could predict fracture for both small- and large-scale plasticity, recognized that J provides three distinct attractive features: (1) for linear elastic behavior it is identical to G, (2) for elastic-plastic behavior it characterizes the crack-tip region and, hence, would be expected to be equally valid under nonlinear conditions, and (3) it can be evaluated experimentally in a convenient manner. The third of these follows from the path-independent property of J and its energy release rate interpretation.

Specifically, Rice has shown that, for deformation plasticity, J can be interpreted as the potential energy difference between two identically loaded bodies having neighboring crack sizes; see Equation (1.4-18). Along with all subsequent uses of the approach, Begley and Landes recognized that the energy interpretation of J, because deformation plasticity becomes invalid when unloading occurs, is not strictly valid for an extending crack. J cannot therefore be identified with the energy available for crack extension in elastic-plastic materials. But, as they argued, because it is a measure of the characteristic crack-tip elastic-plastic field, Equation (1.4-18) nevertheless provides a physically relevant quantity.

Subsequent work of note was contributed by Bucci et al. (1.150), who first showed how simple engineering estimates could be made using the energy-based definition of the J-integral. A classic use of this approach was made by Rice, Paris, and Merkle (1.151) to determine critical J values from experimental load-deflection records for a bend specimen. By assuming the specimen to be so deeply cracked that the only relevant specimen dimension is the ligament length b, they were led to the relation

$$J = \frac{2}{b}\int_0^{\phi} M \, d\phi \tag{1.4-24}$$

where M is the applied moment per unit thickness and ϕ is the rotation through which M works, minus the corresponding rotation in the absence of the crack. This can be put into the more convenient form

$$J = \eta \frac{A}{bB} \tag{1.4-25}$$

where A is the area under a load-displacement curve, B is the specimen thickness, b is the remaining ligament size, and η is a constant that has a value of two for a deeply cracked bend specimen. This approach was subsequently extended by Merkle and Corten (1.152). The use of the η-factor as a geometry-dependent parameter to extend the applicability of Equation (1.4-25) was introduced by Turner (1.153). This factor currently forms the basis for generalizing the energy-based approach to work-hardening and other complicating aspects; see Chapter 5.

Applications of elastic-plastic fracture mechanics using J clearly also require the ability to compute it for an engineering structure. In large structures this computation is often done through the use of plasticity-corrected strain energy release rate expressions. This method is acceptable for load levels that do not produce full-scale yielding. For more extreme conditions, resort must usually be made to finite elements or other numerical methods. But, some simple estimation methods do exist. For example, for power-law stress-strain behavior of the form $\varepsilon/\varepsilon_0 = \sigma/\sigma_0 + \alpha(\sigma/\sigma_0)^n$, a solution can be obtained relative to the limit load solution for perfectly plastic behavior. If P here denotes a load parameter, this result can be written

$$J = \alpha\sigma_0\varepsilon_0 a h_n \left(\frac{P}{P_0}\right)^{n+1} \tag{1.4-26}$$

where P_0 refers to the limit load solution. In Equation (1.4-26), h_n is a dimensionless function of n and the component geometry that has been compiled for a number of cases. Further details are discussed in Chapter 5.

1.4.8 Tearing Instability Theory

Very prominent in current nonlinear fracture mechanics applications, particularly for nuclear plants, is the tearing modulus concept. This concept is based upon the fact that fracture instability can occur after some amount of stable crack growth in tough and ductile materials with an attendant higher applied load level at fracture. To take account of such a process, the resistance curve concept is useful. The idea of a resistance curve seems to have first been suggested by Irwin around 1960; see Srawley and Brown (1.154). It was extensively developed and used for thin section materials in LEFM terms by Heyer and McCabe (1.155).

In the LEFM version of the resistance curve approach, the material fracture toughness is expressed by the function $K_R = K_R(\Delta a)$, where Δa denotes the extent of stable crack growth. This function is taken as a material property in the same sense as the initiation toughness K_c is a thickness-dependent material property. In fact, $K_c = K_R(0)$. The condition for fracture of a cracked component is obviously not synonymous with the conditions for achieving crack initiation. The fracture point is instead determined through a stability analysis. This focuses on the point in the stable crack growth process at which the rate of change of the crack driving force exceeds the rate of change of the material's resistance to continued crack growth. Thus, fracture instability occurs when

$$\frac{dK}{da} \geq \frac{dK_R}{da} \tag{1.4-27}$$

from which it can be seen that the point of fracture is dictated by the compliance of the loading system. It will be different for load-control than for displacement-control, for example.

The use of J as the crack driving force parameter in a resistance curve approach was a natural idea that was adopted soon after the establishment of J as a elastic-plastic fracture parameter by Begley and Landes. But, it was clearly the contributions of Paris and his co-workers (1.156) that led to the widespread acceptance of this concept. In essence, the resistance curve concept was simply reformulated as $J_R = J_R(\Delta a)$, where again Δa denotes the extent of stable crack growth. Fracture instability then occurs when dJ/da exceeds dJ_R/da. Paris formalized this concept by defining the parameters

$$T = \frac{E}{\sigma_0^2}\frac{dJ}{da} \tag{1.4-28}$$

and

$$T_R = \frac{E}{\sigma_0^2}\frac{dJ_R}{da} \tag{1.4-29}$$

where σ_0 is the flow stress of the material. The dimensionless parameter T is known as the tearing modulus with its critical value $T_R = T_R(\Delta a)$ taken to be a property of the material. Paris' concept (1.157) is illustrated in Figure 1.32.

Shown in Figure 1.32(a) is a typical J-resistance curve. It is important to recognize that all such relations have a finite range of applicability. The limit is denoted by the value $(\Delta a)_{\text{lim}}$, which can be estimated from the ω parameter

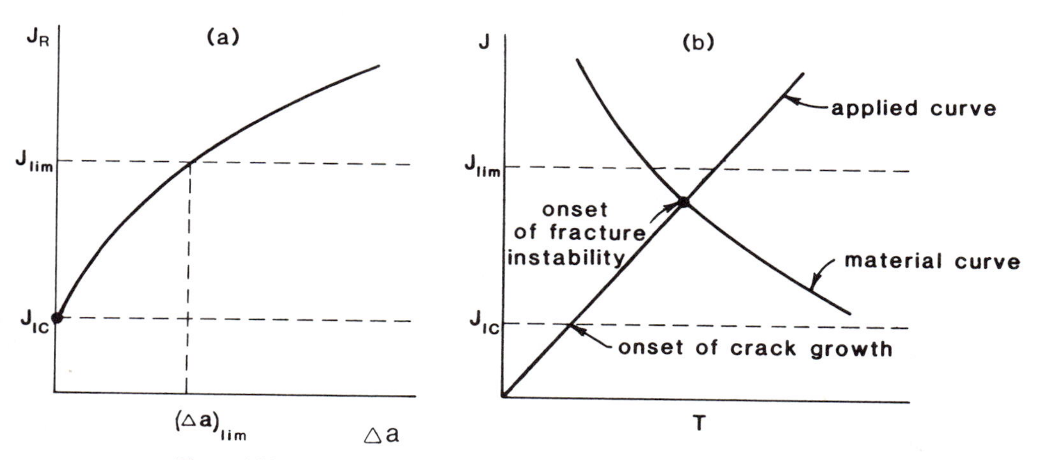

Figure 1.32 Basis of the tearing modulus prediction of fracture instability.

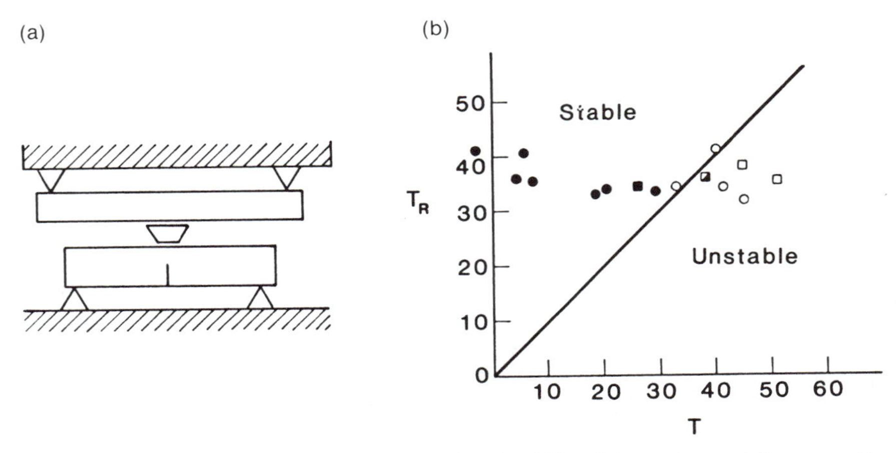

Figure 1.33 Experimental results of Paris showing the ability of the tearing modulus concept to delimit stable and unstable crack growth behavior for a ductile material. (a) Spring-loaded beam specimen. (b) Comparison of experiment and theory.

introduced in the work of Hutchinson and Paris (1.158). This is

$$\omega = \frac{b}{J}\frac{dJ}{da} \tag{1.4-30}$$

where b denotes the smallest relevant dimension from the crack tip to the boundary of the cracked component—for example, the remaining ligament size. The work of Hutchinson and Paris shows that $\omega \gg 1$ for the theory to be valid whereupon there will be some value of $(\Delta a)_{\text{lim}}$ that designates the largest amount of crack growth (and J value) for which the theory is valid.* Assuming that the fracture instability point would occur before $(\Delta a)_{\text{lim}}$ is reached, its determination can be readily found via the J/T diagram shown as Figure 1.32(b). Clearly, to use this approach, one needs the J-resistance curve (and some means for accurately determining its slope) together with estimates of J and T for the crack/structure/load conditions of interest.

The incentive for the use of the tearing modulus approach is that it offers a convenient way to estimate the often substantial increase in load carrying capacity over that associated with initiation that accrues when small amounts of crack growth are permitted in ductile materials. The fracture prediction then centers on crack growth stability. The first demonstration that such predictions could be accurately made was that performed by Paris and co-workers (1.156). They employed a spring-like loading system and a material for which $T_R \simeq 36$ just following initiation. As shown in Figure 1.33, test results obtained by varying the compliance of the loading system to change T systematically did indeed produce stable and unstable behavior in accord with the theory.

---

* The appropriate value of ω is geometry-dependent, and cannot be stated precisely for any configuration. Currently, a value of perhaps $\omega > 5$ to 7 is thought to be realistic.

1.4.9 Criteria for Crack Growth in Nonlinear Conditions

Particularly for nuclear plant components, the tearing modulus approach will generally give a more accurate assessment of the margin of safety than will one based only upon crack initiation. Nevertheless, this approach to nonlinear fracture mechanics (along with every alternative) has drawbacks as well as virtues. One of these is obviously the restriction to small amounts of stable crack growth reflected by the $(\Delta a)_{\lim}$ parameter. A possible way of overcoming this limitation is through the use of the crack-tip opening angle (CTOA) parameter that after an initial transient, appears to remain constant for extended amounts of stable crack growth. However, while an intriguing possiblity, effective use of the CTOA parameter in nonlinear fracture mechanics remains in the future. A more important consideration currently is the lack of verification that has been achieved for actual structures. The first of relatively few such attempts was the analysis of circumferentially cracked nuclear plant piping developed by Zahoor and Kanninen (1.159), which was assessed by experiments conducted by Wilkowski et al. (1.160).

A point made early in this chapter was that fracture mechanics considerations do not obviate the need for considering failure due to plastic yielding. Especially for the tough, ductile materials used in nuclear plant piping, ordinary plastic collapse analyses will possibly suffice. Kanninen et al. (1.161), for example, demonstrated this possibility by showing that a net section collapse load analysis procedure predicts the failure point of circumferentially cracked stainless steel piping.

The key to developing an analysis procedure for plastic fracture is to identify an apropriate crack-tip fracture criterion. Work performed by Kanninen et al. (1.162) and Shih et al. (1.163) in a cooperative effort encompassed three main stages. First, steel and aluminum test specimens were tested to obtain detailed data on crack growth initiation and stable growth. Second, "generation-phase" analyses were performed in which the experimentally observed applied stress/stable crack growth behavior was reproduced in a finite element model. In each such computation, the critical values of each of a number of candidate crack initiation and stable growth criteria were determined for the material tested. In the third stage, "application-phase" finite element analyses were performed for another crack/structure geometry using one of the candidate criteria. These computations determined the applied stress/crack growth behavior that could be compared with experimental results for the given specimen geometry.

The fracture criteria examined included the J-integral, the local and average crack opening angles, the conventional LEFM R curve, and various generalized energy release rates. Because each of the candidate criteria was attractive in one way or another, the task of selecting the best criterion for application to nuclear steels was difficult. Clearly, geometry-independence is a crucial test of the acceptability with practicality being another. The advantages of the J-integral are its virtual independence of finite element type and element size, the computational ease involved in evaluating it, and, because of its history-independence, its catalogability. However, it is valid only for a limited amount of stable crack growth and is unable to cope with large amounts of stable crack growth attended by large-scale plasticity.

Use of the crack opening angle parameter as a stable crack growth criterion is appealing because of its readily grasped physical significance and the opportunity that it offers for direct measurement. The fact that after an initial transient, stable crack growth appears to proceed with a virtually constant crack shape provides a particularly simple criterion. This was first reported in the experimental work of Green and Knott (1.164), Berry and Brook (1.165), and, in the integrated experimentation/analysis approach of deKoning (1.166). However, there are two different definitions of the crack opening angle: a crack-tip value that reflects the actual slope of the crack faces (CTOA), and an average value based on the original crack position (COA). While the critical value of the COA can be measured, how its value has any direct connection with the fracture process is difficult to see. Conversely, while the critical value of the CTOA can likely be associated with the fracture process, it presents a formidable measurement task. In addition, there are clearly some difficulties in making either value apply to mixed character shear/flat crack growth.

A proper stable crack growth criterion must differentiate between the energy dissipated in direct fracture-related processes near the crack-tip and energy dissipated in geometry-dependent plastic deformation remote from the crack-tip. With this in mind, a number of investigators have opted for a generalization of the LEFM energy release rate as the basic plastic fracture methodology. But inherent in this approach is a basic difficulty: There is a theoretical basis for expecting a computational step size dependence in the energy release rate parameter that is based on the work of separating the crack faces. This consequence can be handled by appealing to micromechanical considerations, as Kfouri and Miller (1.167) have argued. Regardless, it appears that the necessity to arbitrarily circumvent the inherent step size difficulty with any energy release rate parameter makes its use somewhat unattractive.

There is also the basic point, generally credited to Rice (1.168), in connection with the use of the energy release rate in a nonlinear analysis. This is, for a material with a stress-strain curve that saturates at infinity, the solution of an elastic-plastic boundary value problem (which automatically includes an energy balance) will have no surplus of energy that can be assigned to the energy release rate. An energy balance calculation will then produce a value of G that is identically zero. This point was also recognized by Goodier and Field (1.118). A tabulation of the various attempts to produce generalized energy release rate formulations for nonlinear material behavior can be found in reference (1.161).

It would be inappropriate to conclude this account of elastic-plastic fracture mechanics without mention of the pioneering work of McClintock (1.169), who provided an approach to ductile fracture via the growth of holes. Also of importance is the two-parameter (fracture and yielding) criterion first advanced by Milne (1.170). The key advance provided by Shih (1.171) that connected J and the CTOD for power law hardening materials should also be cited. By defining an effective crack-tip opening displacement by the separation at the points where 45° lines emanating from the crack-tip intersect the crack faces, Shih found that

$$\delta = d_n \frac{J}{\sigma_Y} \tag{1.4-31}$$

where d_n is a constant of order unity that depends primarily upon the strain hardening exponent n.

The pioneering work of M. L. Williams on viscoelastic fracture mechanics (1.172) was followed by a number of investigators, his students being prominent among them. These include Knauss (1.142) and Schapery (1.143), who developed their ideas by extending the strip yield zone concept originated by Dugdale for elastic-plastic materials. Following a suggestion made by Goldman and Hutchinson (1.173), Landes and Begley (1.174) introduced the C^* parameter by extending the J-integral to apply to time-dependent material behavior. Other recent work of note for crack growth in viscoelastic materials is that of Christensen and Wu (1.175) and Bassani and McClintock (1.176). It is safe to say that the characterization of high-temperature creep crack growth, for which these formulations are needed, is not yet resolved. This subject is taken up in Chapter 7.

The proliferation of fracture mechanics parameters may be somewhat daunting to one who, perhaps with the help of the earlier sections of this chapter, has become comfortable with the basically simple and straightforward uses of LEFM. The situation might be likened to atomic physics. After becoming accustomed to concepts based on neutrons, protons, and electrons, we have had quarks and other more exotic entities thrust upon us. Yet, as may eventually also be the case in physics, we would like to conclude this section by suggesting that a basic order will eventually be restored to fracture mechanics. Our personal feeling is that focusing on the local crack opening displacement may offer the way to accomplish this. The investigations of Rice and his co-workers—see reference (1.177), for example—have tacitly assumed that such a parameter does govern the crack growth process. Other possibilities also exist—for example, the approach being developed by Andrews (1.178). Regardless, the most important issue that now confronts the subject of fracture mechanics is the identification and development of appropriate crack extension criterion for nonlinear and dynamic fracture mechanics.

1.5 The Necessity for Nonlinear and Dynamic Treatments

The preceding sections have discussed the development of fracture mechanics from a historical point of view, beginning with Griffith, but concentrating on the time period from the end of the Second World War to the early 1970s. The development of that era was strongly influenced by the energy balance concept with the stress intensity factor originally receiving its legitimacy by being connected to the energy release rate. However, there were both conceptual difficulties with the basis of the energy release rate concept and with the generally cumbersome mathematical procedures needed for its application. The focus in fracture mechanics therefore shifted to crack-tip characterization with K replacing G as the working parameter. But, as inelastic applications became more important, the inadequacies of K have become clear. The more modern point of view is one that blends a crack-tip characterizing parameter with an energy release rate formulation for its implementation.

Before considering the specific nonlinear and dynamic research areas that constitute advanced fracture mechanics, it may be useful to touch on a few significant application areas where fracture mechanics techniques beyond those of LEFM would appear to be required. The examples we have selected are applications to nuclear reactor power plant pressure vessels and piping. While probably no more susceptible to subcritical cracking and fracture than other types of engineering structures, because of the catastrophic consequences of a failure, nuclear plant systems have been subjected to an unprecedented degree of scrutiny. Such scrutiny has explored many situations in which applications of linear elastic fracture mechanics (as conservatively permitted by code procedures) would indicate that failure should occur when in fact experience has demonstrated otherwise. Such observations have led to a great amount of research focused on the development of nonlinear (e.g., elastic-plastic) and dynamic fracture mechanics methods to obtain more realistic assessments of the risk of fracture in nuclear plant components. With the possible exception of the aerospace industry, the advanced fracture mechanics treatments that have evolved since roughly 1975 onwards have been primarily motivated by the concerns of the nuclear power industry. We review this application area as an introduction to the more detailed fracture mechanics that are contained in the remaining chapters of this book.

1.5.1 The Thermal Shock Problem

The two most interesting fracture mechanics applications for nuclear plant systems lie in the reactor pressure vessel and in its attendant piping systems. In the first area is the so-called "thermal shock" problem of nuclear reactor power plant pressure vessels. This application is currently of profound concern to the nuclear power industry (1.179), to the U.S. Nuclear Regulatory Commission (1.180), and has received attention in the public press (1.181). The scenario referred to as thermal shock involves the possibility of the fracture of a nuclear reactor pressure vessel during a loss of coolant accident (LOCA) under circumstances that are most likely to occur in a pressurized water reactor (PWR) plant. Three conditions appear to be necessary for such an incident to occur: (1) a large upward shift in the nil ductility transition (NDT) temperature due to a combination of nuclear irradiation during service and the presence of high copper and nickel content in the vessel welds, (2) the existence of an initial flaw on the inner surface of the vessel, and (3) a severe over-cooling transient caused by the splashing of cold water on the inner surface of the vessel—for example, by the activation of the emergency core cooling system (ECCS) during a LOCA.

It should be recognized that, while LOCA's have occurred during reactor operation (e.g., Three Mile Island II), because all three of the above conditions were not simultaneously satisfied, no catastrophic fractures of nuclear plant pressure vessels have been experienced. Current data indicate that the problem is confined to those few plants in which the welding procedures have since been disallowed. Furthermore, even in those particular plants, the extent of neutron irradiation will not soon reach a danger point. Nonetheless, the application is clearly significant and must be taken seriously. This in turn calls for the utmost in fracture mechanics analysis.

The most significant research currently addressing the thermal shock problem is being conducted at the Oak Ridge National Laboratory (ORNL) by Cheverton et al. (1.182) under the auspices of the U.S. Nuclear Regulatory Commission. On the experimental side, a series of medium-size vessel tests is in progress (planning is scheduled through 1986) to study the key events that might follow a LOCA in a nuclear plant. It is important to recognize that an exact simulation is not being sought. The impossibility of achieving a toughness reduction by neutron irradiation would alone preclude this. Rather, the idea is to obtain measurements on crack propagation/arrest behavior under conditions similar to an actual event and to use these observations to validate an analysis procedure. If the analysis is then applied to the actual conditions of interest, its predictions can be accepted with confidence.

The analysis effort developed in the ORNL research program relies upon linear elastic, quasi-static fracture mechanics considerations. That is, Equation (1.1-5) is used for crack growth initiation and (1.3-10) for arrest. A time element enters because the temperature distribution in the vessel wall during the transient period following a thermal shock is a function of time.* This causes time-varying thermal stresses and, in turn, time-varying stress intensity factors. Figure 1.34 shows the types of temperature-dependent material fracture property data that are needed for an analysis. The calculation of a run/arrest event at a specific time during a transient is illustrated in Figure 1.35. Finally, Figure 1.36 shows the crack propagation/arrest history that was predicted by Cheverton et al. for one of their experiments (TSE-5A). The key results from three of their experiments are provided in Table 1.2.

Figure 1.36 shows several interesting features of the ORNL predictions for the thermal shock experiment TSE-5A. Of most importance is that, in accord with the experiment itself, crack arrest occurs far before the complete

* Because the equations of heat transmission are parabolic, even if the vessel surface is subjected to a step change in temperature, there will be no shock in the usual solid mechanics sense. Shock wave propagation can occur only in systems governed by hyperbolic equations. This misnomer is nevertheless common usage and for this reason will be used in this discussion.

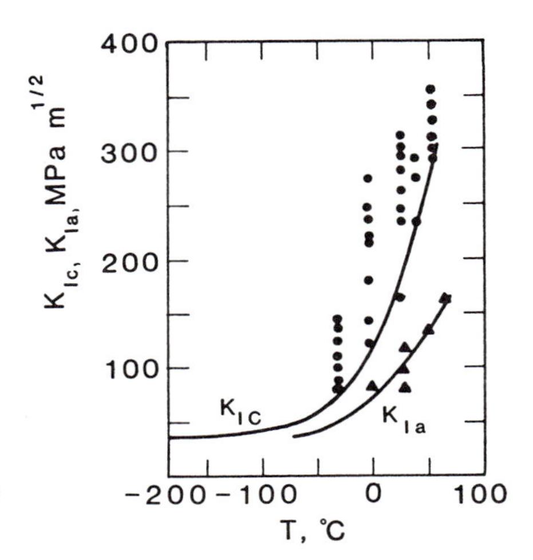

Figure 1.34 Fracture toughness data used in thermal shock analyses.

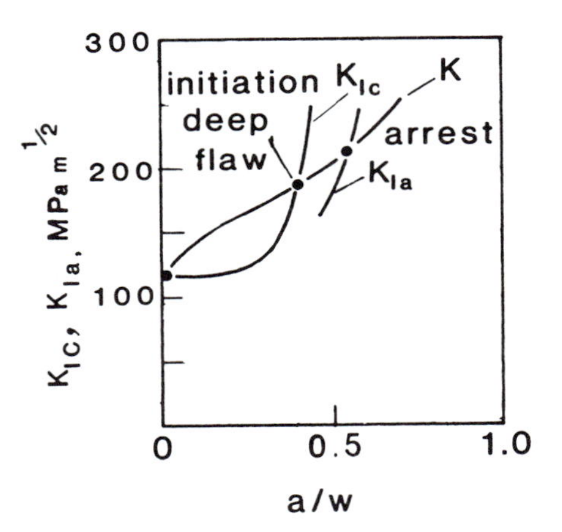

Figure 1.35 Possible crack initiation and arrest points at one time in a thermal shock analysis.

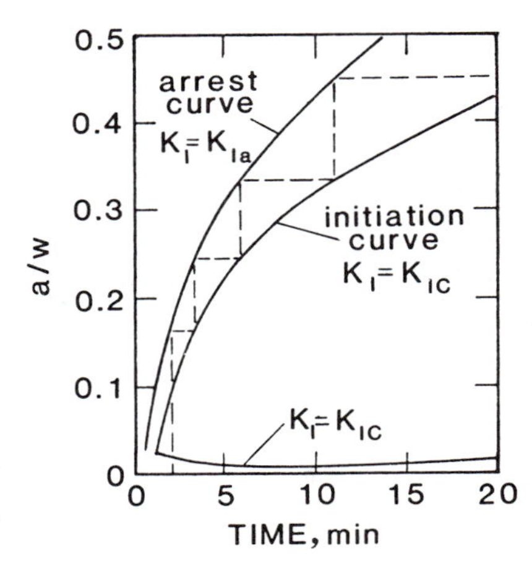

Figure 1.36 Crack propagation/arrest prediction for a thermal shock event.

Table 1.2 Summary of Events for ORNL Thermal Shock Experiments with Long Axial Cracks in A508 Steel Cylinders

Time (sec)	Event	Temperature (°C)	a/w	Δa (mm)	K_I ($\mathrm{MPam^{\frac{1}{2}}}$)
Experiment No. TSE-5[a]					
105	Initiation	−9	0.10		79
	arrest	36	0.20	15	86
177	Initiation	−3	0.20		111
	arrest	82	0.63	65	104
205	Initiation	79	0.63		115
	arrest	89	0.80	26	92
Experiment No. TSE-5A[b]					
79	Initiation	−11	0.076		70
	arrest	22	0.138	9	76
91	Initiation	12	0.138		85
	arrest	38	0.199	9	86
123	Initiation	13	0.199		108
	arrest	51	0.316	18	107
185	Initiation	21	0.316		135
	arrest	67	0.535	33	130
Experiment No. TSE-6[c]					
69	Initiation	−12	0.10		46
	arrest	34	0.27	13	63
137	Initiation	−28	0.27		87
	arrest	64	0.93	50	104

[a] Wall thickness = 152 mm, outer wall diameter = 991 mm, initial wall temperatures = 93°C, thermal shock temperature (inner wall) = −197°C; experiment conducted August, 1979.
[b] Vessel dimensions and test temperatures same as in TSE-5; experiment conducted September, 1980.
[c] Wall thickness = 76 mm, outside diameter = 991 mm, initial wall temperature = 96°C, inner wall temperature = −196°C; experiment conducted December, 1981.

penetration of the vessel wall. It can be seen that four separate crack jumps are predicted, and this too is consistent with the results of this particular experiment; see Table 1.2. One key feature associated with the final arrest is the "warm prestress" (WPS) effect. WPS is invoked in the analysis to preclude reinitiation whenever $dK/dt < 0$ on the basis that the prior crack-tip blunting reduces the actual crack driving force. Key experimental work supporting this concept has been contributed by Loss et al. (1.183); see also the recent review by Pickles and Cowan (1.184).

The WPS effect is of course a nonlinear effect that cannot be directly treated within the context of a linear analysis. The WPS effect is analogous to that of crack growth retardation in fatigue, which also arises because of prior plastic deformation that alters the crack driving force; see Chapter 8. It too is a nonlinear problem not effectively treatable by linear methods.

In a large-break LOCA a step change would occur in the coolant temperature that is in contact with the inner surface of the pressure vessel. The change would typically be from 288°C to 21°C. Simultaneously, the internal pressure would be expected to fall from the operating pressure to one atmosphere. (If the pressure remains high, the event is known as a pressurized thermal shock (PTS) event.) In the actual ORNL experiments a more severe thermal shock is administered to partly compensate for the absence of radiation damage; see Table 1.2. Additional control over the fracture toughness is achieved through the use of differing tempering temperatures for the test vessel materials.

A long axial inner surface crack was considered in all three of the experiments shown in Table 1.2. Cladding on the inner wall was omitted. The times at which events occurred during the experiments were indicated by step changes in the crack opening displacement (COD) measurements made on the vessel surface. The crack sizes at these times were inferred from ultrasonic measurements that connected the COD data to crack depth via a post-test finite element computation. The measured temperature distributions and the inferred crack depths at the times of crack initiation and arrest events were then used to calculate critical stress intensity values for those events. These are the K_{Ic} and K_{Ia} values given in Table 1.2.

One objective of the ORNL experimentation was to compare their test results for K_{Ic} and K_{Ia} with the corresponding laboratory data—for example, with Figure 1.34. These comparisons were generally favorable. Cheverton et al. were able to offer several additional key conclusions. First, complete penetration of the vessel will not occur under thermal stresses alone. Second, quasi-static LEFM procedures appear to be valid for the analysis of thermal shock conditions. Third, crack arrest will occur under the condition that $K = K_{Ia}$ even though dK/da is increasing (n.b., K_{Ia} data are generally obtained under conditions where $dK/da < 0$). Finally, crack initiation is precluded if dK/dt is decreasing even if $K \gg K_{Ic}$. This is the warm prestress phenomenon.

In addition to the warm prestress effect, there are aspects of the thermal shock problem that, as effective as the ORNL analysis has been for thermal shock, suggest the need for nonlinear treatments. One is the possibility of stable tearing after the arrest of a crack that penetrates deeply into the vessel. This problem must be handled by elastic-plastic fracture mechanics. Similarly, as the inner surface of an actual vessel has a weld-deposited cladding, the small flaws that are known to exist under the cladding could be affected by the residual stress state that exists there. As revealed by Figure 1.35, such flaws can be critical and, moreover, because they would likely give rise to longer crack jumps, they would be more dangerous than deeper flaws. Finally, elastic-viscoplastic fracture mechanics procedures appear to be needed to address dynamic crack arrest at the high toughness upper shelf material conditions that would be experienced for a deeply penetrating crack.

1.5.2 Degraded Nuclear Plant Piping

One of the most likely causes of a loss of coolant accident in a nuclear power plant is a rupture in the piping system caused by stress corrosion. Many incidents of stress corrosion cracking have been reported, particularly in smaller diameter stainless steel pipes. Consequently, there is also a great need for a quantitative understanding of the behavior of cracked pipes under normal operating and postulated accident conditions—for example, a seismic event. In most instances, concern is for surface cracks that initiate at the inner surface of the pipe in the heat affected zone around a girth weld. Assisted by the weld-induced residual stress, these cracks tend to grow circumferentially and radially, sometimes attaining a size that is a significant fraction of the pipe wall area. Figure 1.37 shows an example of the cracks that were discovered in

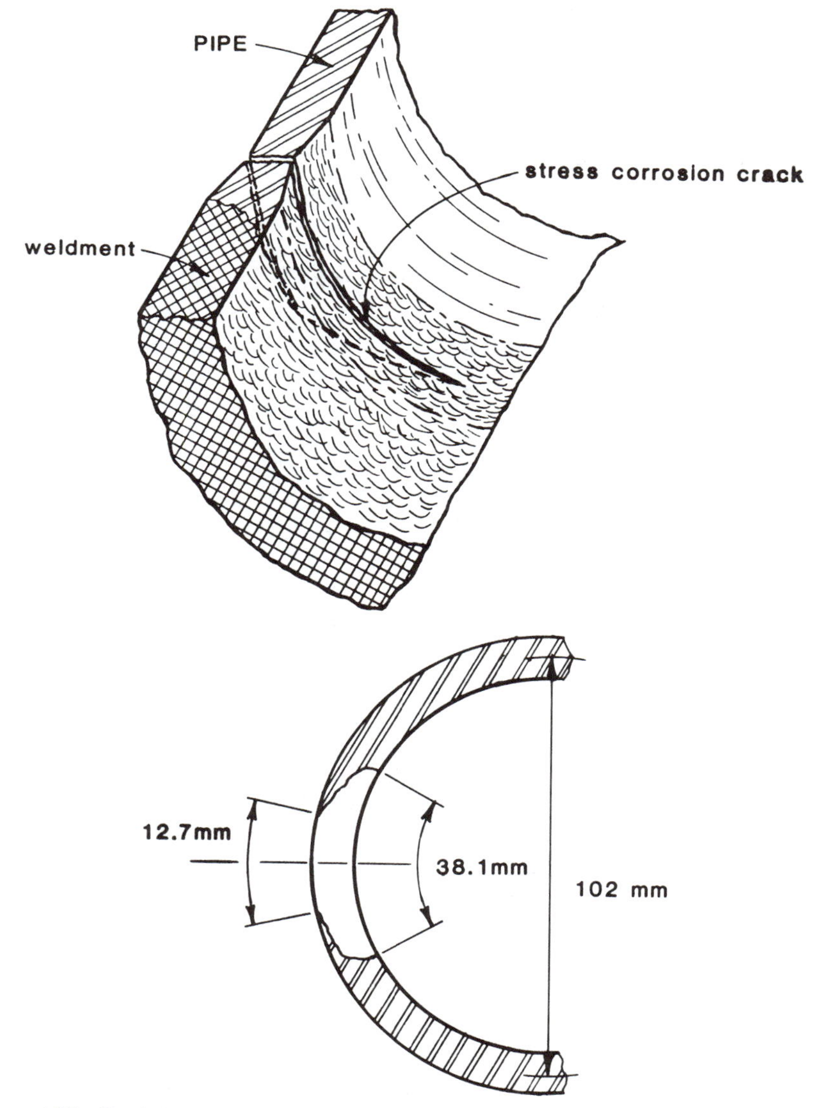

Figure 1.37 Crack detected in a 4-in. diameter recirculation by-pass of a boiling water reactor plant.

the wall of 4-in. diameter stainless steel pipes in a boiling water reactor (BWR) plant in 1974. This example illustrates that substantial crack sizes can be achieved before detection.

A prime consideration in analyzing cracked nuclear plant pipes is to determine, if failure actually occurs, whether it will lead to a "leak-before-break" condition. As described in Section 1.1.6, the leak-before-break concept generally refers to a pressure containment system failure event in which a part-through-wall crack extends to become a through-wall crack, thus allowing the contained fluid to escape. If no further crack growth occurs, then the loss of fluid can presumably be detected in one way or another and the system can be shut down safely. The alternative—where the through-wall crack propagates along the wall—is very likely to lead to a catastrophic event. Obviously, it must be avoided. If failure occurs, it is very desirable to be assured that it will be confined to the leak-before-break mode.

In nuclear power plant applications it is necessary to show that the leak-before-break concept is the applicable failure mode in critical piping systems where cracking has occurred or even could occur. Specifically, it must be established that a pipe crack will be revealed by leak detection techniques before it reaches a condition where fracture could occur under normal operating or postulated accident conditions. Anticipating the failure mode in connection with subsequent events triggered by a pipe failure is also important. The design basis accident used in nuclear plant regulations around the world is the so-called full guillotine offset break (i.e., an instantaneous circumferential fracture). This extreme condition has resulted in the incorporation of massive pipe whip restraints into nuclear piping systems. These restraints are not only very expensive to design and install, but they can reduce the reliability of inservice inspection while increasing the radiation hazard in the inspection process. The necessity for such devices in the design stage and as modifications in operating plants would be substantially relieved if leak-before-break conditions could be demonstrated. Consequently, there is currently a great deal of research interest in developing more precise fracture mechanics analysis methods.

Nuclear plant piping materials are very ductile and tough and this is the essential difficulty in the application of fracture mechanics. In the materials selected for such service, crack growth is generally preceded by substantial crack-tip blunting while significant amounts of stable crack growth can occur prior to fracture instability. Linear elastic fracture mechanics techniques usually provide very conservative predictions in such circumstances, conservatism that often prompts unnecessary remedial action. Compounding the complexity of the analysis problem is the fact that pipe cracks tend to be located within weld-induced residual stress and deformation fields. Recent results have indicated that the fracture toughness values of nuclear piping welds can be substantially less than that of the base material. At the time of this writing, generally accepted analysis procedures for these conditions do not exist.

1.5.3 The Leak-Before-Break Condition

Leak-before-break can be simply viewed as one possible outcome of a sequence of crack extension events. For a ductile material, a general sequence of events might be one in which a weld defect or other intrinsic flaw enlarges in service through the following series of events: (1) subcritical crack growth by fatigue and/or corrosion to a critical crack depth, (2) stable crack growth under operating or accident loads, (3) fracture instability and subsequent rapid crack growth through the wall, (4) arrest of the through-thickness crack (leak), (5) reinitiation of the through-thickness crack and subsequent stable growth along the wall, and (6) fracture instability and rapid crack growth (break). Leak-before-break occurs when events 4 and 5 are well separated in time or when events 5 and 6 are precluded.

The above list describes several factors affecting the occurrence of leak-before-break. These include: (1) the orientation of the initial flaw, (2) the size and shape of the crack at the onset of stable crack growth, (3) the type, intensity, and duration of the applied loads, (4) the distribution of any residual stresses that might be present, and (5) the mechanical and fracture properties of the pressure boundary material. To illustrate these effects, consider that a part-through-wall axial crack exists in a pressurized pipe under conditions such that a linear elastic fracture mechanics approach is applicable. Further, suppose that the length of the crack along the surface is long in comparison to the crack depth (n.b., this will lead to a conservative prediction for all crack aspect ratios). Then, the initiation of unstable crack propagation will occur in the radial direction when, approximately (see Table 1.1)

$$1.12 \frac{p_0 R}{h} \left[\pi a \sec\left(\frac{\pi}{2}\frac{a}{h}\right)\right]^{\frac{1}{2}} = K_{Ic} \tag{1.5-1}$$

where p_0 is the internal pressure, K_{Ic} is the plane strain fracture toughness, while a, c, R, and h are the geometric parameters shown in Figure 1.38.

The nature of the unstable crack propagation event that follows the satisfaction of Equation (1.5-1) is likely to result in a through-wall crack of length equal to the surface length of the original part-through crack. It can be assumed that this state will exist, momentarily at least, before the crack continues to grow along the wall. The critical condition for reinitiation of crack growth to occur is given by (see Table 1.1)

$$\frac{p_1 R}{h}\left[\pi c \left(1 + 1.61 \frac{c^2}{Rh}\right)\right]^{\frac{1}{2}} = K_c \tag{1.5-2}$$

where p_1 is the internal pressure at the onset of longitudinal crack instability and K_c is the fracture toughness that corresponds to the wall thickness.

A relation that provides the boundary between leak and break behavior can be obtained by combining Equations (1.5-1) and (1.5-2):

$$\left[1 + 1.61 \frac{h}{R}\left(\frac{c}{h}\right)^2\right]\frac{c}{h} = 1.25 \left(\frac{K_c}{K_{Ic}}\right)^2 \left(\frac{p_0}{p_1}\right)^2 \frac{a}{h} \sec\left(\frac{\pi}{2}\frac{a}{h}\right) \tag{1.5-3}$$

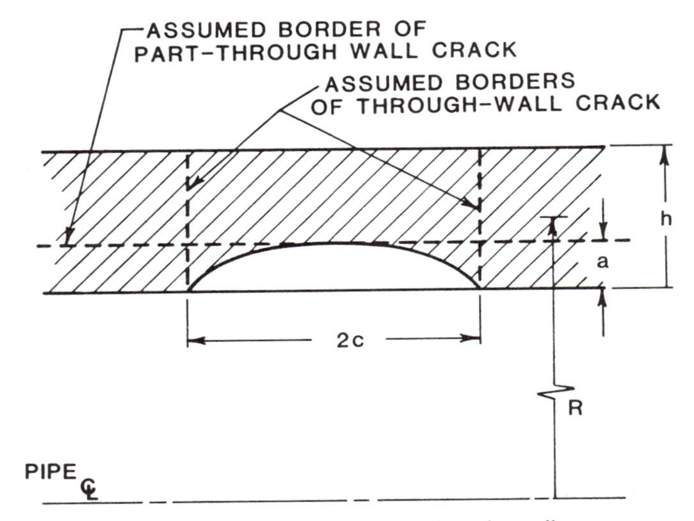

Figure 1.38 Axial surface crack in a pipe wall.

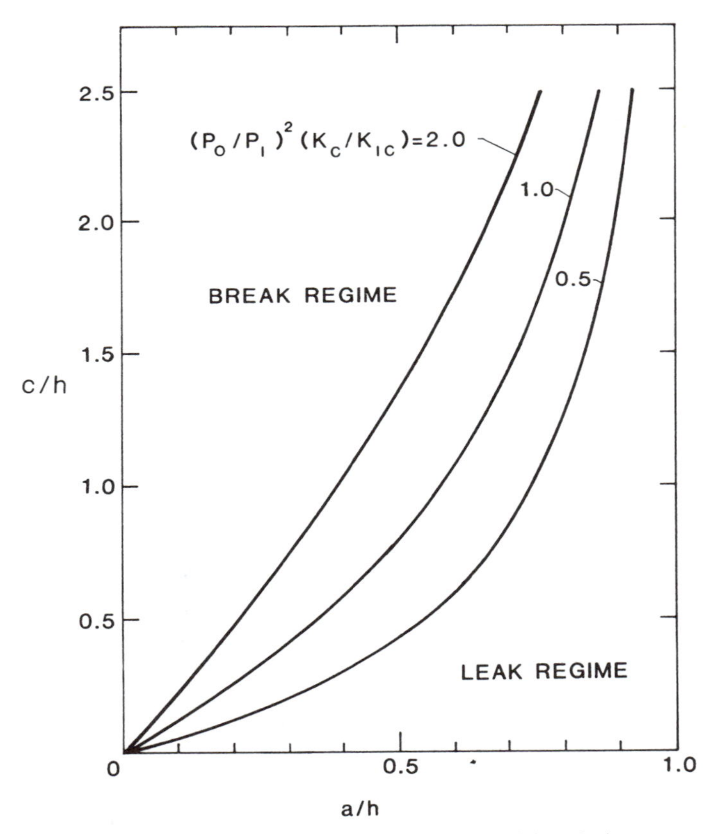

Figure 1.39 LEFM leak-before-break assessment diagram for axial cracks in a pressurized pipe.

which provides a leak-before-break delimitation for axial cracks in pressurized pipes. Equation (1.5-3) can be solved numerically (e.g., by Newton's method) for c/h as a function of a/h, provided the remaining parameters are specified. Equation (1.1-6) demonstrates that K_c will generally differ from K_{Ic}. Of equal importance to this discussion, p_1 also can differ from p_0. In the case of fluid leakage, the inequality $p_1 < p_0$ is expected. However, if the fracture is caused by a waterhammer or other dynamic loading, $p_1 > p_0$ is entirely conceivable. Thus, it is of interest to explore the effect of variations in these parameters. Figure 1.39 shows the set of results obtained for $R/h = 10$.

Figure 1.39 reveals the important (albeit intuitive) conclusion that the differences in the fracture properties and the change in applied stress during the fracture event can have a significant effect on the leak-before-break delimitation. For example, accounting for the greater toughness that generally confronts a through-wall crack and admitting a reduction of the pressure due to fluid leakage would shift the boundary so as to enlarge the "leak" zone. But, in contrast, should an escalation in the pressure overcome the increased toughness, it would be the "break" region that enlarges. Clearly, considerations of this kind should enter into any leak-before-break assessment procedure. Further information on leak-before-break and its applications to nuclear plant piping can be found in reference (1.185).

1.6 Status and Prospects of Fracture Mechanics

A definition of fracture mechanics was offered in the preface to this book. At this transition point between the introductory aspects of the subject and the more advanced treatments of the chapters to come, it may be worth restating.

> Fracture mechanics is an engineering discipline that quantifies the conditions under which a load-bearing solid body can fail due to the enlargement of a dominant crack contained in that body.

The consequences of this definition—primarily its generality—have been broadly amplified in this chapter. Here, it may be worthwhile to emphasize an aspect that has been implicit in the foregoing. This aspect is one that can be called "transferability". This term refers to the use of a measurement taken from a simple test specimen for a prediction of failure in a structural component. That is, many different quantities can be measured in a fracture mechanics experiment. The question is, which of these will have the same value when that material is used in a different geometric configuration and loading system. Only those that do can be said to possess transferability.

For a metal, if small-scale yielding conditions occur in both a characterization test and in the application, then the parameter K_{Ic} will exhibit transferability. If both the characterization test and the application occur under net section yielding conditions, it may be that a critical flow stress instead is transferable. But, there are many important instances where a similar desirable correspondence does not occur. In fact, it can be said that advanced fracture mechanics is necessitated by the many practical instances where the conveniently performed crack growth tests do not adequately mirror the conditions expected for the application. The development of a physically sound basis for selecting a transferable parameter is then the hallmark of an advanced fracture mechanics treatment.

It should be recognized that, while the concept of transferability is implicit in all fracture mechanics applications, it is often used with little physical justification. Consequently, transferability is not exclusively vested in one particular parameter. Provided the characterizing test and the intended application are sufficiently similar, virtually any measured value taken from the former will suffice to predict the latter. Indeed, much of the progress in applications of fracture mechanics to ductile materials has evolved from empirical correlations of full-scale experimental data—for example, gas transmission pipes. Less obvious are fatigue crack growth characterizations in terms of stress intensity factors where the necessity to closely approximate the application conditions is known as the "similitude" requirement.

Reliance on large-scale component testing for basic crack growth data is a rather inefficient and restrictive use of fracture mechanics. Clearly, the firmer the physical basis for a given fracture mechanics application, the less constraining will be the requirements on the testing needed to obtain values of the crack growth parameter and the more reliable will be its predictions. The capability provided by a sound fracture mechanics technique to allow laboratory scale test data to be used to predict the behavior of engineering components is the most effective possible usage of the technology.

The necessity to focus on a dominant crack in fracture mechanics is a result of inherent limitations in computational techniques and in inspection and measurement equipment. However, this focus does not mean to exclude conditions other than those specific instances with only one potentially dangerous crack. There is a growing area of interest that addresses problems when a dominant crack does not exist. This field is coming to be called "damage mechanics"; for example, see Chaboche (1.186). Damage mechanics primarily applies to brittle bodies like ceramics that contain a fairly dense population of defects with the largest defects being more or less equal in size. As such, it may be viewed as a precursor to the fracture mechanics regime; the demarcation (obviously, a somewhat arbitrary one) being when one member of the population either grows more rapidly than the others or some other reason exists for it to be singled out. In either instance, a deterministic treatment, via the fracture mechanics techniques introduced in this chapter, is then appropriate.

The linear elastic fracture mechanics techniques outlined in Section 1.1 cover the great majority of all applications of fracture mechanics at present. While nonlinear techniques beyond those of LEFM are increasingly being developed and used, it is unlikely that this trend will obviate LEFM. Simplicity of application and a conservative prediction are generally associated with a LEFM approach. If nothing else, one will often find that LEFM techniques are useful for a first cut at a given problem before resorting to the generally more complex procedures inherent in a nonlinear fracture mechanics treatment. It should be clear from this chapter that the LEFM methodology is far from being the entirety of fracture mechanics, however.

One of our basic contentions is the universality of fracture mechanics. Rather than being a narrow specialized discipline, fracture mechanics is as broad as materials science and structural mechanics together. That is, for any circumstances in which a structural material may be said to fail, the presence of a crack-like defect can only intensify the risk of failure. We note here in passing that fracture need not always be synonymous with failure. Achieving a fracture well could be the objective—in comminution processes, for example. The principles that govern the crack growth process supplied by fracture mechanics are neverthless the same.

Just as it should not be assumed that fracture mechanics is confined to metals, it should not be assumed that it is useful only for engineering structures. Applications to bone have been made by a number of investigators—for example, by Bonfield et al. (1.187). Applications have also been made to paper; see, for example, Seth and Page (1.188). Finally, Suh and co-workers have employed fracture mechanics in developing their innovative approach to wear (1.189).

The process of writing this book has confirmed a notion about the progression of fracture mechanics that can be expressed in words attributed to the French writer Alphonse Karr, "plus ca change, plus c'est la meme chose": the more things change, the more they remain the same. Fracture mechanics evolved because of the paradox recognized by C. E. Inglis, G. V. Kolsov, and others at the beginning of this century when the theory of elasticity became refined enough to treat a sharp crack. Because of its singular nature, the natural idea of relating failure to the existence of a finite critical stress or strain value is inapplicable when a sharp crack is admitted.

A. A. Griffith's work resolved the dilemma via an energy balance approach and, with the key contributions provided later by G. R. Irwin, fracture mechanics was initiated. This development provided the crack-tip characterization view. However, their work relied upon several important assumptions, among them that the cracked body is essentially linear elastic. When the extent of inelastic deformation attending a crack becomes large enough, or significant amounts of crack growth occur, the Griffith-Irwin linear elastic fracture mechanics approach must be superseded by nonlinear approaches. Currently, for many such applications, the J-integral is being used as the crack extension force parameter with its evaluation being drawn from energy balance considerations. Because of the inability of J to characterize extended crack growth, measures such as a critical strain in the nonlinear deformation region ahead of an advancing crack-tip are also being invoked. Thus, as Karr might have anticipated, structural integrity assessments have, to some extent, come full circle.

This observation should not be taken in any sense as a condemnation of fracture mechanics. The likely correct interpretation is that both present computational capabilities and the pressing practical needs for nonlinear analyses have simply out paced our ability to model the details of the fracture process. This is precisely where progress in fracture mechanics will come in the next several years. The objective of this book, by reflecting the commonality that underlies the wide ranging applications of the methodology, is to further this progress.

1.7 References

(1.1) Gordon, J. E., *The New Science of Strong Materials, or Why You Don't Fall Through the Floor*, Penguin, New York (1976); and *Structures or Why Things Don't Fall Down*, Penguin, New York (1978).

(1.2) Timoshenko, S. P., *History of the Strength of Materials*, McGraw-Hill, New York (1953).

(1.3) Irwin, G. R., "Structural Aspects of Brittle Fracture," *Applied Materials Research*, **3**, pp. 65–81 (1964).

(1.4) Bell, E. T., *Men of Mathematics*, Simon and Schuster, New York (1937).

(1.5) Griffith, A. A., "The Phenomena of Rupture and Flow in Solids," *Philosophical Transactions of the Royal Society of London*, **A221**, pp. 163–197, 1921; and "The Theory of Rupture," *Proceedings of the First International Conference of Applied Mechanics*, Delft (1924).

(1.6) Irwin, G. R., "Fracture Dynamics," *Fracturing of Metals*, American Society for Metals, Cleveland, pp. 147–166 (1948).

(1.7) Rice, J. R., "A Path Independent Integral and the Approximate Analysis of Strain Concentrations by Notches and Cracks," *Journal of Applied Mechanics*, **35**, pp. 379–386 (1968).

(1.8) Hutchinson, J. W., "Singular Behavior at the End of a Tensile Crack in a Hardening Material," *Journal of the Mechanics and Physics of Solids*, **16**, p. 13–31 (1968).

(1.9) Burdekin, F. M., "The Role of Fracture Mechanics in the Safety Analysis of Pressure Vessels," *International Journal of the Mechanical Sciences*, **24**, pp. 197–208 (1982).

(1.10) Atallah, S., "U.S. History's Worst LNG Disaster," *Firehouse*, January (1979), p. 29 et seq.

(1.11) Duga, J. J. and others, *The Economic Effects of Fracture in the United States*, Battelle's Columbus Laboratories Report to the National Bureau of Standards, March (1983).

(1.12) Feddersen, C. E., "Evaluation and Prediction of the Residual Strength of Center Cracked Tension Panels," *Damage Tolerance in Aircraft Structures*, ASTM STP 486, American Society for Testing and Materials, Philadelphia, pp. 50–78 (1971).

(1.13) Jones, M. H. and Brown, W. F., Jr., "The Influence of Crack Length and Thickness in Plane Strain Fracture Toughness Tests," *Review of Developments in Plane Strain Fracture Toughness Testing*, ASTM STP 463, American Society for Testing and Materials, Philadelphia, pp. 63–101 (1970).

(1.14) Irwin, G. R., "Fracture Mode Transition for a Crack Traversing a Plate," *Journal of Basic Engineering*, **82,** pp. 417–425 (1960).

(1.15) Merkle, J. G., "New Method for Analyzing Small Scale Fracture Specimen Data in the Transition Zone," *Tenth Water Reactor Safety Meeting*, **4**—*Materials Engineering Research*, U.S. Nuclear Regulatory Commission, Washington, D.C. (1982).

(1.16) Rolfe, S. T. and Novak, S. R., "Slow Bend K_{Ic} Testing of Medium-Strength High-Toughness Steels," *Review of Developments in Plane Strain Fracture Toughness Testing*, ASTM STP 463, American Society for Testing and Materials, Philadelphia, pp. 124–159 (1970).

(1.17) Oldfield, W., "Development of Fracture Toughness Reference Curves," *Journal of Engineering Materials and Technology*, **102,** pp. 107–117 (1980).

(1.18) Irwin, G. R., "Crack Extension Force for a Part-Through Crack in a Plate," *Journal of Applied Mechanics*, **29**, pp. 651–654 (1962).

(1.19) Kobayashi, A. S. and Moss, W. L., "Stress Intensity Magnification Factors for Surface-Flawed Tension Plate and Notched Round Tension Bar," *Fracture 1969*, Chapman and Hall, London, pp. 31–45 (1969).

(1.20) Newman, J. C., Jr. and I. S. Raju, "An Empirical Stress Intensity Factor Equation for Surface Cracks," *Engineering Fracture Mechanics*, **15**, pp. 185–192 (1981).

(1.21) Bates, R. C. and Clark, W. G., Jr., "Fractography and Fracture Mechanics," *Transactions of the ASM*, **62**, pp. 380–389 (1969).
(1.22) Mackay, T. L., Alperin, B. J., and Bhatt, D. D., "Near-Threshold Fatigue Crack Propagation of Several High Strength Steels," *Engineering Fracture Mechanics*, **18**, pp. 403–416 (1983).
(1.23) McGowan, J. J., "An Overview of Current Methods Used for Assessing Surface Flaws in Nuclear Reactor Vessels," *Nuclear Engineering and Design*, **73**, pp. 275–281 (1982).
(1.24) Dougan, J. R., "Relationships Between Charpy V-Notch Impact Energy and Fracture Toughness," Oak Ridge National Laboratory Report ORNL/TM-7921, NUREG/CR-2362, U.S. Nuclear Regulatory Commission, Washington, D.C. (1982).
(1.25) Inglis, C. E., "Stresses in a Plate Due to the Presence of Cracks and Sharp Corners," *Transactions of the Institute of Naval Architects*, **55**, pp. 219–241 (1913).
(1.26) Obriemoff, J. W., "The Splitting Strength of Mica," *Proceedings of the Royal Society of London*, **127A**, pp. 290–297 (1930).
(1.27) Westergaard, H. M., "Bearing Pressures and Cracks," *Transactions of the American Society of Mechanical Engineers*, **61**, pp. A49–A53 (1939).
(1.28) Williams, M. L. and Ellinger, G. A., "Investigation of Structural Failures of Welded Ships," *Welding Journal*, **32**, pp. 498s–528s (1953).
(1.29) Shank, M. E., "Brittle Failure of Steel Structures—A Brief History," *Metal Progress*, **66**, pp. 83–88 (1954).
(1.30) Bishop, T., "Fatigue and the Comet Disasters," *Metal Progress*, **67**, pp. 79–85 (1955).
(1.31) Orowan, E., "Fracture and Strength of Solids," *Reports on Progress in Physics*, **XII**, p. 185 (1948).
(1.32) Irwin, G. R., "Analysis of Stresses and Strains Near the End of a Crack Traversing a Plate," *Journal of Applied Mechanics*, **24**, pp. 361–364 (1957).
(1.33) Bueckner, H. F., "The Propagation of Cracks and the Energy of Elastic Deformation," *Transactions of the American Society of Mechanical Engineers*, **80**, pp. 1225–1230 (1958).
(1.34) Erdogan, F., "Stress Intensity Factors," *Journal of Applied Mechanics*, **50**, pp. 992–1002 (1983).
(1.35) Sneddon, I. N., "The Distribution of Stress in the Neighborhood of a Crack in an Elastic Solid," *Proceedings of the Royal Society of London*, **A187**, pp. 229–260 (1946).
(1.36) Goodier, J. N., "Mathematical Theory of Equilibrium Cracks," *Fracture*, H. Liebowitz (ed.), Vol. II, Academic, New York, pp. 1–66 (1968).
(1.37) Rajapakse, Y.D.S., "Surface Energy and Surface Tension at Holes and Cracks," *International Journal of Fracture*, **11**, pp. 57–69 (1975).
(1.38) Tipper, C. F., "The Fracture of Metals," *Metallurgia*, **39**, pp. 133–137 (1949).
(1.39) Irwin, G. R. and Paris, P. C., "Fundamental Aspects of Crack Growth and Fracture," *Fracture*, H. Liebowitz (ed.), Vol. III, Academic, New York, pp. 1–46 (1971).
(1.40) Schabtach, C., Fogleman, E. L., Rankin, A. W., and Winne, D. H., "Report of the Investigation of Two Generator Rotor Fractures," *Transactions of the American Society of Mechanical Engineers*, **78**, pp. 1567–1584 (1956).
(1.41) Winne, D. H. and Wundt, B. M., "Application of the Griffith-Irwin Theory of Crack Propagation to the Bursting Behavior of Disks, Including Analytical and Experimental Studies," *Transactions of the American Society of Mechanical Engineers*, **80**, pp. 1643–1655 (1958).
(1.42) Sih, G. C., Paris, P. C., and Erdogan, F., "Crack Tip, Stress-Intensity Factors for Plane Extension and Plate Bending Problems," *Journal of Applied Mechanics*, **29**, pp. 306–312 (1962).
(1.43) Sih, G. C., Paris, P. C., and Irwin, G. R., "On Cracks in Rectilinearly Anisotropic Bodies," *International Journal of Fracture Mechanics*, **1**, pp. 189–203 (1965).
(1.44) McClintock, F. A. and Irwin, G. R., "Plasticity Aspects of Fracture Mechanics," *Fracture Toughness Testing and Its Applications*, ASTM STP 381, American Society for Testing and Materials, Philadelphia, pp. 84–113 (1965).
(1.45) Barenblatt, G. I., "The Mathematical Theory of Equilibrium of Crack in Brittle Fracture," *Advances in Applied Mechanics*, **7**, pp. 55–129 (1962).
(1.46) Muskhelisvili, N. I., *Some Basic Problems in the Mathematical Theory of Elasticity*, Nordhoff, The Netherlands (1954).
(1.47) Elliott, H. A., "An Analysis of the Conditions for Rupture Due to Griffith Cracks," *Proceedings of the Physical Society*, **59**, pp. 208–223 (1947).
(1.48) Cribb, J. L. and Tomkins, B., "On the Nature of the Stress at the Tip of a Perfectly Brittle Crack," *Journal of the Mechanics and Physics of Solids*, **15**, pp. 135–140 (1967).

(1.49) Gehlen, P. C. and Kanninen, M. F., "An Atomic Model for Cleavage Crack Propagation in Alpha Iron," *Inelastic Behavior of Solids*, M. F. Kanninen et al. (ed.), McGraw-Hill, New York, pp. 587–603 (1970).
(1.50) Chang, R., "An Atomistic Study of Fracture," *International Journal of Fracture Mechanics*, **6**, pp. 111–125 (1970).
(1.51) Gehlen, P. C., Hahn, G. T., and Kanninen, M. F., "Crack Extension by Bond Rupture in a Model of BCC Iron." *Scripta Metallurgica*, **6**, pp. 1087–1090 (1972).
(1.52) Markworth, A. J., Kanninen, M. F., and Gehlen, P. C., "An Atomic Model of an Environmentally Affected Crack in BCC Iron," *Stress Corrosion Cracking and Hydrogen Embrittlement of Iron Base Alloys*, R. W. Staehle et al. (eds.), National Association of Corrosion Engineers, Houston, Texas, pp. 447–454 (1977).
(1.53) Weiner, J. H., and Pear, M., "Crack and Dislocation Propagation in an Idealized Crystal Model," *Journal of Applied Physics*, **46**, pp. 2398–2405 (1975).
(1.54) Ashurst, W. T., and Hoover, W. G., "Microscopic Fracture Studies in the Two-Dimensional Triangular Lattice," *Physical Review B*, **14**, pp. 1465–1473 (1976).
(1.55) Markworth, J. A., Kahn, L. R., Gehlen, P. C., and Hahn, G. T., "Atomistic Computer Simulation of Effects of Hydrogen and Helium on Crack Propagation in BCC Iron," *Res. Mechanica*, **2**, pp. 141–162 (1981).
(1.56) Mullins, M., "Molecular Dynamics Simulation of Propagating Cracks," *Scripta Metallurgica*, **16**, pp. 663–666 (1982).
(1.57) Paskin, A., Som, D. K., and Dienes, Q. J., "The Dynamic Properties of Moving Cracks," *Acta Metallurgica*, **31**, pp. 1841–1848 (1983).
(1.58) Kelly, A., Tyson, W. R., and Cottrell, A. H., "Ductile and Brittle Crystals," *Philosophical Magazine*, **15**, pp. 567–586 (1967).
(1.59) Rice, J. R. and Thomson, R., "Ductile versus Brittle Behavior of Crystals," *Philosophical Magazine*, **29**, pp. 73–97 (1974).
(1.60) Bennett, J. A., and Mindlin, H., "Metallurgical Aspects of the Failure of the Point Pleasant Bridge," *Journal of Testing and Evaluation*, **1**, pp. 152–161 (1973).
(1.61) Paris, P. C., Gomez, M. P., and Anderson, W. P., "A Rational Analytic Theory of Fatigue," *The Trend in Engineering*, **13**, pp. 9–14 (1961).
(1.62) Paris, P. and Erdogan, F., "A Critical Analysis of Crack Propagation Laws," *Journal of Basic Engineering*, **85**, pp. 528–534 (1963).
(1.63) Foreman, R. G., Kearney, V. E., and Engle, R. M., "Numerical Analysis of Crack Propagation in Cyclic-Loaded Structures," *Journal of Basic Engineering*, **89**, pp. 459–464 (1967).
(1.64) Elber, W., "The Significance of Fatigue Crack Closure," *Damage Tolerance in Aircraft Structures*, ASTM STP 486, American Society for Testing and Materials, Philadelphia, pp. 230–242 (1971).
(1.65) Brown, B. F. and Beachem, C. D., "A Study of the Stress Factor in Corrosion Cracking by Use of the Pre-Cracked Cantilever Beam Specimen," *Corrosion Science*, **5**, pp. 745–750 (1965).
(1.66) Evans, A. G. and Johnson, H., "The Fracture Stress and its Dependence on Slow Crack Growth," *Journal of Materials Science*, **10**, pp. 214–222 (1975).
(1.67) Charles, R. J., "Dynamic Fatigue of Glass," *Journal of Applied Physics*, **29**, pp. 1657–1662 (1958).
(1.68) Cooper, G. A. and Kelly, A., "Tensile Properties of Fibre-Reinforced Materials: Fracture Mechanics," *Journal of the Mechanics of Physics and Solids*," **15**, pp. 279–297 (1967).
(1.69) Wu, E. M., "Application of Fracture Mechanics to Anisotropic Plates," *Journal of Applied Mechanics*, **34**, pp. 967–974 (1967).
(1.70) Zweben, C., "On the Strength of Notched Composites," *Journal of the Mechanics and Physics of Solids*, **19**, pp. 103–116 (1971).
(1.71) Konish, J. J., Swedlow, J. L., and Cruse, T. A., "Fracture Phenomena in Advanced Fibre Composite Materials," AIAA *Journal*, **11**, pp. 40–43 (1973).
(1.72) Harrison, N. L., "Strain Energy Release Rate for Turning Cracks," *Fibre Science and Technology*, **5**, pp. 197–212 (1972).
(1.73) Sih, G. C. and Chen, E. P., "Fracture Analysis of Unidirectional Composites," *Journal of Composite Materials*, **7**, pp. 230–244 (1973).
(1.74) Rivlin, R. S. and Thomas, A. G., "Rupture of Rubber. I. Characteristic Energy for Tearing," *Journal of Polymer Science*, **X**, pp. 291–318, 1953; "II. The Strain Concentration at an Incision," **XVIII**, pp. 177–188 (1955).
(1.75) Gent, A. N., Lindley, P. B., and Thomas, A. G., "Cut Growth and Fatigue of Rubbers:

I. the Relationship Between Cut Growth and Fatigue," *Journal of Applied Polymer Science*, **8**, pp. 455–466 (1964).
(1.76) Lindley, P. B., and Stevenson, A., "Fatigue Resistance of Natural Rubber in Compression," *Rubber Chemistry and Technology*, **55**, pp. 337–351 (1982).
(1.77) Rivlin, R. S. and Thomas, A. G., "The Incipient Characteristics Tearing Energy for an Elastomer Crosslinked under Strain," *Journal of Polymer Science, Polymer Physics Edition*, **21**, pp. 1807–1814 (1983).
(1.78) Williams, M. L., "On the Stress Distribution at the Base of a Stationary Crack," *Journal of Applied Mechanics*, **24**, pp. 109–114 (1957).
(1.79) Karp, S. N. and Karal, F. C., Jr., "The Elastic-Field Behavior in the Neighborhood of a Crack of Arbitrary Angle," *Communications on Pure and Applied Mathematics*, **XV**, pp. 413–421 (1962).
(1.80) Chan, S. K., Tuba, I. S. and Wilson, W. K., "On the Finite Element Method in Linear Fracture Mechanics," *Engineering Fracture Mechanics*, **2**, pp. 1–17 (1970).
(1.81) Byskov, E., "The Calculation of Stress Intensity Factors Using the Finite Element Method With Cracked Elements," *International Journal of Fracture Mechanics*, **6**, pp. 159–167 (1970).
(1.82) Henshell, R. D. and Shaw, K. G., "Crack-Tip Finite Elements are Unnecessary," *International Journal for Numerical Methods in Engineering*, **9**, pp. 495–507 (1975).
(1.83) Barsoum, R. S., "On the Use of Isoparametric Finite Elements in Linear Elastic Fracture Mechanics," *International Journal of Numerical Methods in Engineering*, **10**, pp. 25–37 (1976).
(1.84) Kobayashi, A. S., Cherepy, R. D., and Kinsel, W. C., "A Numerical Procedure for Estimating the Stress Intensity Factor for a Crack in a Finite Plate," *Journal of Basic Engineering*, **86**, pp. 681–684 (1964).
(1.85) Cruse, T. A., "Lateral Constraint in a Cracked Three-Dimensional Elastic Body," *International Journal of Fracture Mechanics*, **6**, pp. 326–328 (1970).
(1.86) Wilkins, M. L. and Streit, R. D., "Computer Simulation of Ductile Fracture," *Nonlinear and Dynamic Fracture Mechanics*, N. Perrone and S. N. Atluri (eds.), American Society of Mechanical Engineers, AMD, **35**, pp. 67–77 (1979).
(1.87) Akhurst, K. N. and Chell, G. G., "Methods of Calculating Stress Intensity Factors for Nozzle Corner Cracks," *International Journal of Pressure Vessels and Piping*, **14**, pp. 227–257 (1983).
(1.88) Mott, N. F., "Fracture of Metals: Theoretical Considerations," *Engineering*, **165**, pp. 16–18 (1948).
(1.89) Yoffe, E. H., "The Moving Griffith Crack," *Philosophical Magazine*, **42**, pp. 739–750 (1951).
(1.90) Roberts, D. K. and Wells, A. A., "The Velocity of Brittle Fracture," *Engineering*, **178**, pp. 820–821 (1954).
(1.91) Hall, E. O., "The Brittle Fracture of Metals," *Journal of the Mechanics of Physics and Solids*, **1**, pp. 227–233 (1953).
(1.92) Dulaney, E. N. and Brace, W. F., "Velocity Behavior of a Growing Crack," *Journal of Applied Physics*, **31**, pp. 2233–2236 (1960).
(1.93) Berry, J. P., "Some Kinetic Considerations of the Griffith Criterion for Fracture," *Journal of the Mechanics of Physics and Solids*, **8**, pp. 194–216 (1960).
(1.94) Craggs, J. W., "On the Propagation of a Crack in an Elastic-Brittle Material," *Journal of the Mechanics of Physics and Solids*, **8**, pp. 66–75 (1960).
(1.95) Broberg, K. B., "The Propagation of a Brittle Crack," *Arkiv for Fysik*, **18**, pp. 159–192 (1960).
(1.96) Atkinson, C. and Eshelby, J. D., "The Flow of Energy Into the Tip of a Moving Crack," *International Journal of Fracture Mechanics*, **4**, pp. 3–8 (1968).
(1.97) Freund, L. B., "Crack Propagation in an Elastic Solid Subjected to General Loading," *Journal of the Mechanics of Physics and Solids*, **20**, pp. 129–140, pp. 141–152 (1972); **21**, pp. 47–61 (1973); **22**, pp. 137–146 (1974).
(1.98) Nilsson, F., "A Note on the Stress Singularity at a Non-Uniformly Moving Crack Tip," *Journal of Elasticity*, **4**, pp. 73–75 (1974).
(1.99) Schardin, H., "Velocity Effects in Fracture," *Fracture*, M.I.T. Press, Cambridge, Mass., pp. 297–330, 1959.
(1.100) Hudson, G. and Greenfield, M., "Speed of propagation of Brittle Cracks in Steel," *Journal of Applied Physics*, **18**, pp. 405–408 (1947).
(1.101) Wells, A. A. and Post, D., "The Dynamic Stress Distribution Surrounding a Running

Crack—A Photoelastic Analysis," *Proceedings of the Society for Experimental Stress Analysis*, **16**, pp. 69–92, 1958.
(1.102) Carlsson, A. J., "On the Mechanism of Brittle Fracture Propagation," *Transactions of the Royal Institute of Technology (Sweden)*, **205**, pp. 2–38, 1963.
(1.103) Irwin, G. R., "Basic Concepts for Dynamic Fracture Testing," *Journal of Basic Engineering*, **91**, pp. 519–524 (1969).
(1.104) Irwin, G. R. and Wells, A. A., "A Continuum-Mechanics View of Crack Propagation," *Metallurgical Reviews*, **10**, pp. 223–270 (1965).
(1.105) Crosley, P. B. and Ripling, E. J., "Dynamic Fracture Toughness of A533 Steel," *Journal of Basic Engineering*, **91**, pp. 525–534 (1969).
(1.106) Hahn, G. T., Hoagland, R. G., Kanninen, M. F., and Rosenfield, A. R., "A Preliminary Study of Fast Fracture and Arrest in the DCB Test Speciman," *Dynamic Crack Propagation*, G. C. Sih (ed.), Noordhoff, Leyden, The Netherlands, pp. 649–662 (1973).
(1.107) Kalthoff, J. F., Beinert, J., and Winkler, S., "Measurements of Dynamic Stress Intensity Factors for Fast Running and Arresting Cracks in Double-Cantilever-Beam Specimens," *Fast Fracture and Crack Arrest*, G. T. Hahn and M. F. Kanninen, (ed.), ASTM STP 627, American Society for Testing and Materials, Philadelphia, pp. 161–176 (1977).
(1.108) Kobayashi, A. S., Seo, K., Jou, J. Y., and Urabe, Y., "A Dynamic Analysis of Modified Compact Tension Specimens Using Homolite-100 and Polycarbonate Plates," *Experimental Mechanics*, **20**, pp. 73–79 (1980).
(1.109) Kanninen, M. F., "An Analysis of Dynamic Crack Propagation and Arrest for a Material Having a Crack Speed Dependent Fracture Toughness," *Prospects of Fracture Mechanics*, G. C Sih et al. (ed.), Noordhoff, Leyden, The Netherlands, pp. 251–266 (1974).
(1.110) Weibull, W., "A Statistical Distribution Function of Wide Applicability," *Journal of Applied Mechanics*, **18**, pp. 293–297 (1951).
(1.111) Besuner, P. M. and Tetelman, A. S., "Probabilistic Fracture Mechanics," *Nuclear Engineering and Design*, **43**, pp. 99–114 (1977).
(1.112) Rau, C. A., Jr. and Besuner, P. M., "Risk Analysis by Probabilistic Fracture Mechanics," *Product Engineering*, **50**, No. 10, pp. 41–47 (1979).
(1.113) Gamble, R. M. and Strosnider, J., Jr., *An Assessment of the Failure Rate for the Beltline Region of PWR Pressure Vessels During Normal Operation and Certain Transient Conditions*, U.S. Nuclear Regulatory Commission Report NUREG-0778 (1981).
(1.114) Harris, D. O. and Lim, E. Y., "Applications of a Probabilistic Fracture Mechanics Model to the Influence of In-Service Inspection on Structural Reliability," *Probabilistic Fracture Mechanics and Fatigue Methods: Applications for Structural Design and Maintenance*, J. M. Bloom and J. C. Ekvall (ed.), ASTM STP 798, American Society for Testing and Materials, Philadelphia, pp. 19–41 (1983).
(1.115) Irwin, G. R., Kies, J. A., and Smith, H. L., "Fracture Strengths Relative to Onset and Arrest of Crack Propagation," *Proceedings of the American Society for Testing Materials*, **58**, pp. 640–657 (1958).
(1.116) Wells, A. A., "Application of Fracture Mechanics at and Beyond General Yielding," *British Welding Journal*, **10**, pp. 563–570 (1963).
(1.117) Dugdale, D. S., "Yielding of Steel Sheets Containing Slits," *Journal of the Mechanics and Physics of Solids*, **8**, pp. 100–108 (1960).
(1.118) Goodier, J. N. and Field, F. A., *Fracture of Solids*, D. C. Drucker and J. J. Gilman Wiley, New York, pp. 103–118 (1963).
(1.119) Hahn, G. T. and Rosenfield, A. R., "Local Yielding and Extension of a Crack Under Plane Stress," *Acta Metallurgica*, **13**, pp. 293–306 (1965).
(1.120) Kanninen, M. F., Mukherjee, A. K., Rosenfield, A. R., and Hahn, G. T., "The Speed of Ductile Crack Propagation and the Dynamics of Flow in Metals," *Mechanical Behavior of Materials Under Dynamic Loads*, U.S. Lindholm (ed.), Springer-Verlag, New York, pp. 96–133 (1969).
(1.121) Burdekin, F. M. and Stone, D. E. W., "The Crack Opening Displacement Approach to Fracture Mechanics in Yielding Materials," *Journal of Strain Analysis*, **1**, pp. 145–153 (1966).
(1.122) Dawes, M. G., "Fracture Control in High Yield Strength Weldments," *Welding Journal Research Supplement*, **53**, pp. 369S–379S (1974).
(1.123) Harrison, J. D., Dawes, M. G., Archer, G. L., and Kamath, M. S., "The COD Approach and Its Application to Welded Structures," *Elastic-Plastic Fracture*, J. D. Landes et al. (ed.), ASTM STP 668, American Society for Testing and Materials, Philadelphia, pp. 606–631 (1979).

(1.124) Rice, J. R. and Rosengren, G. F., "Plane Strain Deformation Near a Crack Tip in a Power-Law Hardening Material," *Journal of the Mechanics of Physics and Solids*, **16**, pp. 1–12 (1968).
(1.125) Sanders, J. L., "On the Griffith-Irwin Fracture Theory," *Journal of Applied Mechanics*, **27**, pp. 352–353 (1960).
(1.126) Eshelby, J. D., "Energy Relations and the Energy-Momentum Tensor in Continuum Mechanics," *Inelastic Behavior of Solids*, M. F. Kanninen et al. (ed.), McGraw-Hill, New York, pp. 77–115 (1969).
(1.127) Cherepanov, G. P., "On Crack Propagation in Solids," *International Journal of Solids and Structures*, **5**, pp. 863–871 (1969).
(1.128) Budiansky, B., "A Reassessment of Deformation Theories of Plasticity", *Journal of Applied Mechanics*, **26**, pp. 259–264 (1959).
(1.129) Hutchinson, J. W., "Fundamentals of the Phenomenological Theory of Nonlinear Fracture Mechanics," *Journal of Applied Mechanics*, **50**, pp. 1042–1051 (1983).
(1.130) Kanninen, M. F., "A Solution for a Dugdale Crack Subjected to a Linearly Varying Tensile Loading," *International Journal of Engineering Science*, **8**, pp. 85–95 (1970).
(1.131) Petroski, H. J., "Dugdale Plastic Zone Sizes for Edge Cracks," *International Journal of Fracture*, **15**, pp. 217–230 (1979).
(1.132) Chell, G. G., "The Stress Intensity Factors and Crack Profiles for Centre and Edge Cracks in Plates Subjected to Arbitrary Stresses," *International Journal of Fracture*, **12**, pp. 33–46 (1976).
(1.133) Bilby, B. A., Cottrell, A. H., and Swinden, K. H., "The Spread of Plastic Yield from a Notch," *Proceedings of the Royal Society*, **A272**, pp. 304–314 (1963).
(1.134) Bilby, B. A. and Swinden, K. H., "Representation of Plasticity at Notches by Linear Dislocation Arrays," *Proceedings of the Royal Society*, **A285**, pp. 22–33 (1965).
(1.135) Atkinson, C. and Kay, T. R., "A Simple Model of Relaxation at a Crack Tip," *Acta Metallurgica*, **19**, pp. 679–683 (1971).
(1.136) Atkinson, C. and Kanninen, M. F., "A Simple Representation of Crack Tip Plasticity: The Inclined Strip Yield Superdislocation Model," *International Journal of Fracture*, **13**, pp. 151–163 (1977).
(1.137) Riedel, H., "Plastic Yielding on Inclined Slip-Planes at a Crack Tip," *Journal of the Mechanics of Physics and Solids*, **24**, pp. 277–289 (1976).
(1.138) Vitek, V., "Yielding on Inclined Planes at the Tip of a Crack Loaded in Uniform Tension," *Journal of the Mechanics and Physics of Solids*, **24**, pp. 263–275 (1976).
(1.139) Evans, J. T., "Reverse Shear on Inclined Planes at the Tip of a Sharp Crack," *Journal of the Mechanics of Physics and Solids*, **27**, pp. 73–88 (1979).
(1.140) Weertman, J., Lin, I. H., and Thomson, R., "Double Slip Plane Crack Model," *Acta Metallurgica*, **31**, pp. 473–482 (1983).
(1.141) Kanninen, M. F. and Atkinson, C., "Application of an Inclined-Strip-Yield Crack Tip Plasticity Model to Predict Constant Amplitude Fatigue Crack Growth," *International Journal of Fracture*, **16**, pp. 53–69 (1980).
(1.142) Knauss, W. G., "Delayed Failure—The Griffith Problem for Linearly Viscoelastic Materials," *International Journal of Fracture Mechanics*, **6**, pp. 7–20 (1970).
(1.143) Schapery, R. A., "A Theory of Crack Initiation and Growth in Viscoelastic Media," *International Journal of Fracture*, **11**, pp. 141–159, pp. 369–388 pp. 549–562 (1975).
(1.144) Kanninen, M. F., "An Estimate of the Limiting Speed of a Propagating Ductile Crack," *Journal of the Mechanics of Physics and Solids*, **16**, pp. 215–228 (1968).
(1.145) Atkinson, C., "A Simple Model of a Relaxed Expanding Crack," *Arkiv for Fysik*, **26**, pp. 469–476 (1968).
(1.146) Embley, G. T. and Sih, G. C., "Plastic Flow Around an Expanding Crack," *Engineering Fracture Mechanics*, **4**, pp. 431–442 (1972).
(1.147) Begley, J. A. and Landes, J. D., "The *J*-integral as a Fracture Criterion," *Fracture Toughness*, ASTM STP 514, American Society for Testing and Materials, Philadelphia, pp. 1–20 (1972).
(1.148) Landes, J. D. and Begley, J. A., "The Effect of Specimen Geometry on J_{Ic}," *Fracture Toughness*, ASTM STP 514, American Society for Testing and Materials, Philadelphia, pp. 24–29 (1972).
(1.149) Begley, J. A. and Landes, J. D., "Serendipity and the *J*-Integral," *International Journal of Fracture*, **12**, pp. 764–766 (1976).
(1.150) Bucci, R. J., Paris, P. C., Landes, J. D., and Rice, J. R., "*J*-Integral Estimation Procedures," *Fracture Toughness*, ASTM STP 514, American Society for Testing and Materials, Philadelphia, pp. 40–69 (1972).

(1.151) Rice, J. R., Paris, P. C., and Merkle, J. G., "Some Further Results of J-Integral Analysis and Estimates," *Progress in Flaw Growth and Fracture Toughness Testing*, ASTM STP 536, American Society for Testing and Materials, pp. 231–245 (1973).

(1.152) Merkle, J. G. and Corten, H. T., "A J-Integral Analysis for the Compact Specimen, Considering Axial Force as Well as Bending Effects," *Journal of Pressure Vessel Technology*, **96**, pp. 286–292 (1974).

(1.153) Turner, C. E., "The Ubiquitous η Factor," *Fracture Mechanics: Twelfth Conference*, ASTM STP 700, American Society for Testing and Materials, Philadelphia, pp. 314–337 (1980).

(1.154) Srawley, J. E. and Brown, W. F., Jr., "Fracture Toughness Testing Methods," *Fracture Toughness Testing and Its Applications*, ASTM STP 381, American Society for Testing and Materials, Philadelphia, pp. 133–196 (1965).

(1.155) Heyer, R. H. and McCabe, D. E., "Crack Growth Resistance in Plane Stress Fracture Testing," *Engineering Fracture Mechanics*, **4**, pp. 413–430 (1972).

(1.156) Paris, P. C., Tada, H., Zahoor, A., and Ernst, H. A., "Instability of the Tearing Mode of Elastic-Plastic Crack Growth," *Elastic-Plastic Fracture*, J. D. Landes et al. (ed.), ASTM STP 668, American Society of Testing and Materials, Philadelphia, pp. 5–36 and pp. 251–265 (1979).

(1.157) Paris, P. C. and Johnson, R. E., "A Method of Application of Elastic-Plastic Fracture Mechanics to Nuclear Vessel Analysis," *Elastic-Plastic Fracture: Second Symposium*, ASTM STP 803, C. F. Shih and J. P. Gudas (ed.), American Society for Testing and Materials, Philadelphia, Vol. II, pp. 5–40 (1983).

(1.158) Hutchinson, J. W. and Paris, P. C., "Stability Analysis of J-Controlled Crack Growth," *Elastic-Plastic Fracture*, ASTM STP 668, American Society for Testing and Materials, Philadelphia, pp. 37–64 (1979).

(1.159) Zahoor, A. and Kanninen, M. F., "A Plastic Fracture Mechanics Prediction of Fracture Instability in a Circumferentially Cracked Pipe in Bending—Part I: J-Integral Analysis," *Journal of Pressure Vessel Technology*, **103**, pp. 352–258 (1981).

(1.160) Wilkowski, G. M., Zahoor, A., and Kanninen, M. F., "A Plastic Fracture Mechanics Prediction of Fracture Instability in a Circumferentially Cracked Pipe in Bending—Part II: Experimental Verification on a Type 304 Stainless Steel Pipe," *Journal of Pressure Vessel Technology*, **103**, pp. 359–365 (1981).

(1.161) Kanninen, M. F., Broek, D., Hahn, G. T., Marschall, C. W., Rybicki, E. F., and Wilkowski, G. M., "Towards an Elastic-Plastic Fracture Mechanics Predictive Capability for Reactor Piping," *Nuclear Engineering and Design*, **48**, pp. 117–134 (1978).

(1.162) Kanninen, M. F., Rybicki, E. F., Stonesifer, R. B., Broek, D., Rosenfield, A. R., Marschall, C. W., and Hahn, G. T., "Elastic-Plastic Fracture Mechanics for Two-Dimensional Stable Crack Growth and Instability Problems," *Elastic-Plastic Fracture*, ASTM STP 668, J. D. Landes, et al. (ed.), American Society for Testing and Materials, Philadelphia, pp. 121–150 (1979).

(1.163) Shih, C. F., Delorenzi, H. G., and Andrews, W. R., "Studies on Crack Initiation and Stable Crack Growth," *Elastic-Plastic Fracture*, ASTM STP 668, J. D. Landes et al. (ed.), American Society for Testing and Materials, Philadelphia, pp. 65–120(1979).

(1.164) Green, G. and Knott, J. F., "On Effects of Thickness on Ductile Crack Growth in Mild Steel," *Journal of the Mechanics and Physics of Solids*, **23**, pp. 167–183 (1975).

(1.165) Berry, G. and Brook, R., "On the Measurement of Critical Crack-Opening-Displacement When Slow Crack Growth Precedes Rapid Fracture," *International Journal of Fracture*, **11**, pp. 933–938 (1975).

(1.166) deKoning, A. U., "A Contribution to the Analysis of Quasistatic Crack Growth in Sheet Materials," *Fracture 1977; Proceedings of the Fourth International Conference on Fracture*, Taplin (ed.), **3**, pp. 25–31 (1977).

(1.167) Kfouri, A. P. and Miller, K. J., "Crack Separation Energy Rates for Inclined Cracks in an Elastic-Plastic Material," *Three-Dimensional Constitution Relations and Ductile Fracture*, S. Nemat-Nasser (ed.), North Holland, pp. 79–105 (1981).

(1.168) Rice, J. R., *Proceedings of the First International Conference on Fracture*, Vol. I, Japanese Society for Strength and Fracture of Materials, Tokyo, pp. 283–308 (1966).

(1.169) McClintock, F. A., "A Criterion for Ductile Fracture by the Growth of Holes," *Journal of Applied Mechanics*, **35**, pp. 363–371 (1968).

(1.170) Milne, I., "Failure Analysis in the Presence of Ductile Crack Growth," *Materials Science and Engineering*, **39**, pp. 65–79 (1979).

(1.171) Shih, C. F., "Relationships Between the J-Intergral and the Crack Opening Displacement

for Stationary and Extending Cracks," *Journal of the Mechanics and Physics of Solids*, **29**, pp. 305–326 (1982).
(1.172) Williams, M. L., "The Fracture of Viscoelastic Materials," *Fracture of Solids*, D. C. Drucker and J. J. Gilman (eds.), Gordon and Breach, New York, pp. 157–188 (1963).
(1.173) Goldman, N. L. and Hutchinson, J. W., "Fully Plastic Crack Problems: The Center-Cracked Strip Under Plane Strain," *International Journal of Solids and Structures*, **11**, pp. 575–591 (1975).
(1.174) Landes, J. D. and Begley, J. A., "A Fracture Mechanics Approach to Creep Crack Growth," *Mechanics of Crack Growth*, ASTM STP 590, American Society for Testing and Materials, Philadelphia, pp. 128–148 (1976).
(1.175) Christensen, R. M. and Wu, E. M., "A Theory of Crack Growth in Viscoelastic Materials," *Engineering Fracture Mechanics*, **14**, pp. 215–225 (1981).
(1.176) Bassani, J. L. and McClintock, F. A., "Creep Relaxation of Stress Around a Crack Tip," *International Journal of Solids and Structures*, **17**, pp. 479–492 (1981).
(1.177) Drugan, W. J., Rice, J. R., and Sham, T. L., "Asymptotic Analysis of Growing Plane Strain Tensile Cracks in Elastic-Ideally Plastic Solids," *Journal of the Mechanics of Physics and Solids*, **30**, pp. 447–473 (1982).
(1.178) Andrews, E. H. et al., "Generalized Fracture Mechanics," *Journal of Materials Science*: *Part 1*, **9**, pp. 887–894 (1974); *Part 2*, **11**, pp. 1354–1361 (1976); *Part 3*, **12**, pp. 1307–1319 (1977).
(1.179) Marston, T. U., Smith, E., and Stahlkopf, K. E., "Crack Arrest in Water-Cooled Reactor Pressure Vessels During Loss-of-Coolant Accident Conditions," *Crack Arrest Methodology and Applications*, G. T. Hahn and M. F. Kanninen (ed.), ASTM STP 711, American Society for Testing and Materials, Philadelphia, pp. 422–431 (1980).
(1.180) Serpan, C. Z., Jr., "USNRC Materials Research for Evaluation of Pressurized Thermal Shock in RPV of PWR's," *Nuclear Engineering and Design*, **72**, pp. 53–64 (1982).
(1.181) Wald, M. L., "Steel Turned Brittle by Radiation Called a Peril at 13 Nuclear Plants," *New York Times*, September 27, 1981; see also D. L. Basdekos, "The Risk of a Meltdown," *New York Times*, March 29, 1982; and Bill Paul, "High Strength Steel is Implicated as Villain in Scores of Accidents," *The Wall Streel Journal*, January 16, 1984.
(1.182) Cheverton, P. D., Canonico, D. A., Iskander, S. K., Bolt, S. E., Holtz, P. P., Nanstad, R. K., and Stelzman, W J., "Fracture Mechanics Data Deduced from Thermal Shock and Related Experiments with LWR Pressure Vessel Material," *Journal of Pressure Vessel Technology*, **105**, pp. 102–110 (1983).
(1.183) Loss, F. J., Gray, R. A., Jr., and Hawthorne, J. R., "Investigation of Warm Prestress for the Case of Small ΔT During a Reactor Loss-of-Coolant Accident," *Journal of Pressure Vessel Technology*, **101**, pp. 298–304 (1979).
(1.184) Pickles, B. W. and Cowan, A., "A Review of Warm Prestressing Studies," *International Journal of Pressure Vessels and Piping*, **14**, pp. 95–131 (1983).
(1.185) Strosnider, J. R., Jr. et al. (eds.), *Proceedings of the CSNI Specialists Meeting on Leak-Before-Break in Nuclear Reactor Piping*, NUREG/CP 51, CSNI Report 82, U.S. Nuclear Regulatory Commission, Washington, D.C. (1984).
(1.186) Chaboche, J. L., "Continuous Damage Mechanics—A Tool to Describe Phenomena Before Crack Initiation," *Nuclear Engineering and Design*, **64**, pp. 233–247 (1981).
(1.187) Bonfield, W., Grynpass, M. D., and Young, R. J., "Crack Velocity and the Fracture of Bone," *Journal of Biomechanics*, **11**, pp. 473–479 (1978).
(1.188) Seth, R. S. and Page, D. H., "Fracture Resistance of Paper," *Journal of Materials Science*, **9**, pp. 1745–1753 (1974).
(1.189) Suh, N. P., "An Overview of the Delamination Theory of Wear," *Wear*, **44**, pp. 1–16 (1977).

2

ELEMENTS OF SOLID MECHANICS

In this chapter the fundamental concepts and the field equations of solid mechanics that provide the basis for work in advanced fracture mechanics are reviewed. No pretense of completeness is made. Rather, attention is focused upon those particular elements of elasticity, viscoelasticity, plasticity, and viscoplasticity that we believe are essential for the understanding of the subject matter in the following chapters. The material is presented from the viewpoint of continuum mechanics and, for the most part, within the confines of small (infinitesimal) strains. Cartesian tensors and the indicial notation are employed because of the convenience, compactness, and simplicity they offer.

The reader who is well versed in these theories may choose to pass over this chapter, yet much of the notation used in subsequent chapters is set forth here. For this reason it may be worthy of at least a casual reading. On the other hand, the reader who finds the present treatment too terse or who desires even more detail is referred to references (2.1)–(2.10) and to the references contained therein.

2.1 Analysis of Stress

In general a body can experience two types of external loadings: body forces acting directly on volume elements of the body and surface forces acting over elements of surface area. For the body B illustrated in Figure 2.1, these forces are assumed to be prescribed functions of the fixed Cartesian coordinates

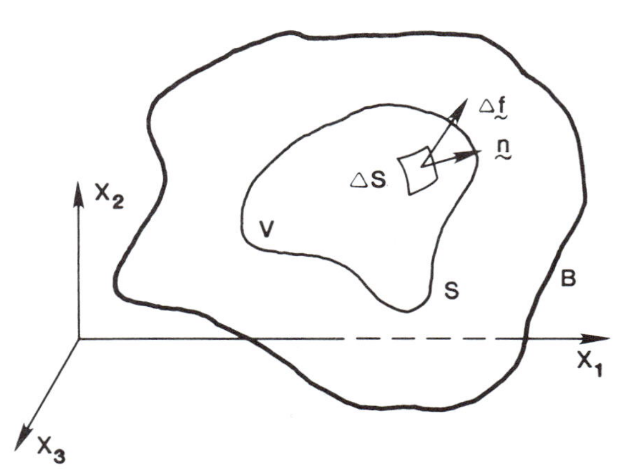

Figure 2.1 Internal forces in a solid body.

x_1, x_2, x_3. Now, consider an elemental surface area ΔS of a closed surface S in the body and let $\mathbf{n}$ be the local outward normal to ΔS. The material exterior to S produces a force $\Delta \mathbf{f}$ over the area ΔS. The stress vector or traction $\mathbf{T}^{(n)}$ at the point (x_1, x_2, x_3) is defined as

$$\mathbf{T}^{(n)} \equiv \lim_{\Delta S \to 0} \frac{\Delta \mathbf{f}}{\Delta S} = \frac{d\mathbf{f}}{dS} \qquad (2.1\text{-}1)$$

where the limit is assumed to exist. The superscript n has been used to emphasize the dependence of the stress vector on the orientation of the elemental surface area.

If the special case for which the normal to the surface element is parallel to a coordinate axis x_i $(i = 1, 2, 3)$ is considered, then

$$\mathbf{T}^{(i)} = T_1^{(i)}\mathbf{e}_1 + T_2^{(i)}\mathbf{e}_2 + T_3^{(i)}\mathbf{e}_3 \equiv T_j^{(i)}\mathbf{e}_j \qquad (2.1\text{-}2)$$

where $T_j^{(i)}$ are the scalar components of $\mathbf{T}^{(i)}$ and the unit base vectors $\mathbf{e}_i$ are directed along the positive coordinate axis x_i. The last equality introduces the summation convention; that is, a repeated minuscule index (subscript) in a term denotes a summation with respect to that index over its range. Unless otherwise noted, the range of a roman index will be from 1 to 3.

It is convenient and customary to write

$$\sigma_{ij} \equiv T_j^{(i)} \qquad (2.1\text{-}3)$$

where the nine scalar quantities σ_{ij} are the components of the stress tensor. The first subscript of σ_{ij} corresponds to the direction of the outward normal to the element of area. The second subscript indicates the direction of the component of the stress vector. The stress component is considered positive if *both* its sense and the outer normal to the surface are in the same coordinate direction, whether the direction be positive or negative. Otherwise, a negative value is assigned to the stress component. The positive sign convention is illustrated in Figure 2.2. The components σ_{ij} with $i = j$ (i.e., $\sigma_{11}, \sigma_{22}, \sigma_{33}$) are

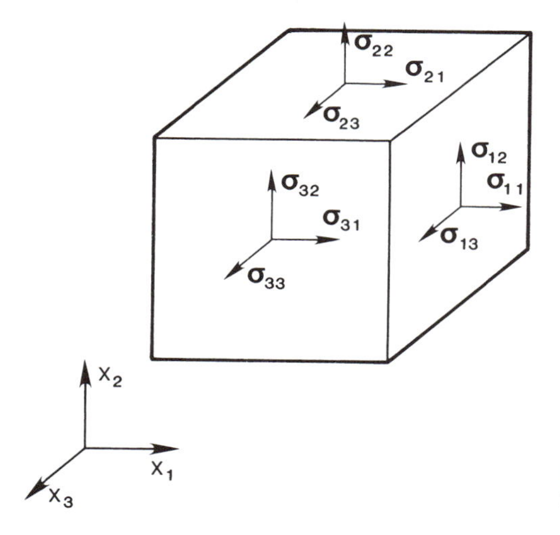

Figure 2.2 Positive sign convention for stress components.

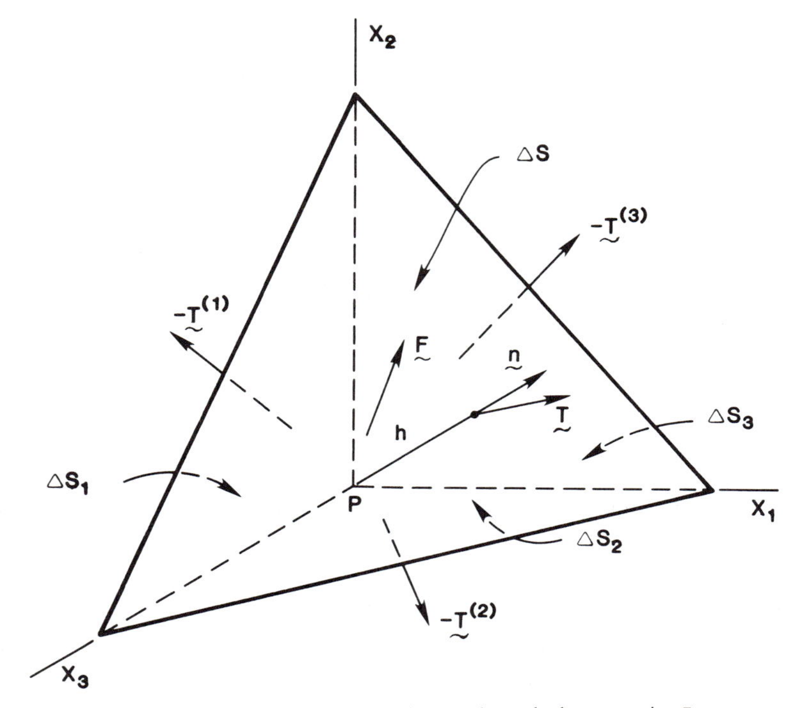

Figure 2.3 Forces acting on an elemental tetrahedron at point P.

called normal stresses and the remaining components $\sigma_{ij}, i \neq j$, are referred to as shear stresses. These stress components define the state of stress at a point.

2.1.1 Equilibrium Equations

Once the components of the stress tensor are known at a point P, then the stress vector $\mathbf{T}$ acting on any other plane through P having a unit normal $\mathbf{n}$ can be determined. With $\mathbf{F} = F_i\mathbf{e}_i$ denoting a body force per unit of volume, the equilibrium of the small tetrahedron of Figure 2.3 requires that*

$$\mathbf{T}\,\Delta S - \mathbf{T}^{(1)}\,\Delta S_1 - \mathbf{T}^{(2)}\,\Delta S_2 - \mathbf{T}^{(3)}\,\Delta S_3 + \mathbf{F}h\frac{\Delta S}{3} = 0 \tag{2.1-4}$$

where $\Delta S_i = \Delta S n_i$ and n_i are the direction cosines of $\mathbf{n}$. After introducing the combination of Equations (2.1-2) and (2.1-3) into Equation (2.1-4) and taking the limit as $h \rightarrow 0$, we obtain

$$T_i = \sigma_{ji} n_j \tag{2.1-5}$$

for the components of the traction $\mathbf{T} = T_i\mathbf{e}_i$. Hence, knowing the components of the stress tensor, one can determine from Equation (2.1-5) the components

* Other terms that tend to zero in the limit as $h \rightarrow 0$ have been neglected in writing Equation (2.1-4).

of the traction on an element of surface whose orientation is defined by its unit normal.

If the body B is in equilibrium, then any volume V of material bounded by a closed surface S within B (see Figure 2.1) must also be in equilibrium. The forces acting on this material consist of the body force $\mathbf{F}$ distributed throughout V and the stress vector or traction $\mathbf{T}$ distributed over the boundary S. Force equilibrium of this volume of material requires that

$$\int_V \mathbf{F}\,dV + \int_S \mathbf{T}\,dS = 0 \tag{2.1-6}$$

or, equivalently,

$$\int_V F_i\,dV + \int_S T_i\,dS = 0 \tag{2.1-7}$$

Noting Equation (2.1-5), we may rewrite the latter as

$$\int_V F_i\,dV + \int_S \sigma_{ji} n_j\,dS = 0 \tag{2.1-8}$$

Assuming that the stress components σ_{ji} are continuous and possess continuous first derivatives, one may use the divergence theorem to transform the surface integral in Equation (2.1-8) to a volume integral.* Hereby, Equation (2.1-8) becomes

$$\int_V (\sigma_{ji,j} + F_i)\,dV = 0 \tag{2.1-9}$$

Since Equation (2.1-9) must hold for an arbitrary volume V, then

$$\sigma_{ji,j} + F_i = 0 \tag{2.1-10}$$

throughout the body.

Moment equilibrium demands that

$$\int_V \mathbf{r} \times \mathbf{F}\,dV + \int_S \mathbf{r} \times \mathbf{T}\,dS = 0 \tag{2.1-11}$$

where $\mathbf{r} = x_i \mathbf{e}_i$ is the position vector from the origin to the point (x_1, x_2, x_3). The scalar form of Equation (2.1-11) is

$$\int_V \varepsilon_{ijk} x_j F_k\,dV + \int_S \varepsilon_{ijk} x_j \sigma_{lk} n_l\,dS = 0 \tag{2.1-12}$$

in which Equation (2.1-5) has been used and where the alternating tensor ε_{ijk} is defined by

$$\varepsilon_{ijk} = \begin{cases} 0 \text{ if any two of } i, j, k \text{ are equal} \\ 1 \text{ if } i, j, k \text{ is a cyclic permutation of } 1, 2, 3 \\ -1 \text{ if } i, j, k \text{ is a cyclic permutation of } 1, 3, 2 \end{cases} \tag{2.1-13}$$

* According to the divergence theorem $\int_S \phi_i n_i\,dS = \int_V \phi_{i,i}\,dV$, where $\phi_{i,i} \equiv \partial\phi_i/\partial x_i$.

The divergence theorem permits writing Equation (2.1-12) as

$$\int_V \varepsilon_{ijk}[x_j(\sigma_{lk,l} + F_k) + \sigma_{lk}x_{j,l}]\, dV = 0 \tag{2.1-14}$$

Due to Equation (2.1-10) the quantity within the parentheses of this integrand vanishes. Furthermore, $x_{j,l} = \delta_{jl}$, where the Kronecker delta δ_{jl} is defined as

$$\delta_{jl} = \begin{cases} 1 & \text{if} \quad j = l \\ 0 & \text{if} \quad j \neq l \end{cases} \tag{2.1-15}$$

Consequently, $\sigma_{lk}x_{j,l} = \sigma_{lk}\delta_{jl} = \sigma_{jk}$. Since Equation (2.1-14) must hold for any material volume V, then the integrand must vanish and, with the foregoing, this condition reduces to

$$\varepsilon_{ijk}\sigma_{jk} = 0$$

Because $\varepsilon_{ijk} = -\varepsilon_{ikj}$, the latter equation implies that the stress tensor is symmetric; that is,

$$\sigma_{ij} = \sigma_{ji} \tag{2.1-16}$$

Thus, Equations (2.1-10) and (2.1-16) form the equations of equilibrium.

For the sake of generality, the equation of motion can be readily obtained from Equation (2.1-10) by interpreting the body force as due to inertia. Then

$$\sigma_{ji,j} = \rho\ddot{u}_i \tag{2.1-17}$$

where u_i denote the components of the displacement in the direction of x_i, ρ is the mass density of the material, and a superposed dot indicates a partial differentiation with respect to time, t.

2.1.2 *Principal Stresses*

At every point in a body there exists a plane, called a principal plane, such that the stress vector lies along the normal **n** to this plane. That is,

$$T_i = \sigma n_i = \sigma\delta_{ij}n_j \tag{2.1-18}$$

where σ is the normal stress acting on this plane. The implication is that there is no shear stress acting on a principal plane. The direction of **n** is referred to as the principal direction. The introduction of Equation (2.1-18) into Equation (2.1-5) yields

$$(\sigma_{ji} - \sigma\delta_{ij})n_j = 0 \tag{2.1-19}$$

which is a set of three homogeneous equations for the direction cosines n_i defining the principal direction. Since $n_in_i = 1$, then to avoid the trivial solution

$$|\sigma_{ji} - \sigma\delta_{ij}| = 0 \tag{2.1-20}$$

which in the unabridged form is

$$\begin{vmatrix} \sigma_{11} - \sigma & \sigma_{12} & \sigma_{13} \\ \sigma_{21} & \sigma_{22} - \sigma & \sigma_{23} \\ \sigma_{31} & \sigma_{32} & \sigma_{33} - \sigma \end{vmatrix} = 0 \qquad (2.1\text{-}21)$$

This is a cubic equation in σ that can be written as

$$\sigma^3 - I_1\sigma^2 + I_2\sigma - I_3 = 0 \qquad (2.1\text{-}22)$$

where I_1, I_2, and I_3 are scalar quantities that are independent of the coordinate system in which the stress components are expressed. They are called stress invariants and are expressed as

$$\begin{aligned} I_1 &= \sigma_{ii} \\ I_2 &= \tfrac{1}{2}(\sigma_{ii}\sigma_{jj} - \sigma_{ij}\sigma_{ij}) \\ I_3 &= \tfrac{1}{6}\varepsilon_{ijk}\varepsilon_{pqr}\sigma_{ip}\sigma_{jq}\sigma_{kr} \end{aligned} \qquad (2.1\text{-}23)$$

or, in unabridged form,

$$\begin{aligned} I_1 &= \sigma_{11} + \sigma_{22} + \sigma_{33} \\ I_2 &= (\sigma_{11}\sigma_{22} + \sigma_{22}\sigma_{33} + \sigma_{33}\sigma_{11}) - \sigma_{12}^2 - \sigma_{23}^2 - \sigma_{31}^2 \\ I_3 &= \begin{vmatrix} \sigma_{11} & \sigma_{12} & \sigma_{13} \\ \sigma_{21} & \sigma_{22} & \sigma_{23} \\ \sigma_{31} & \sigma_{32} & \sigma_{33} \end{vmatrix} \end{aligned} \qquad (2.1\text{-}24)$$

Due to the symmetry of the stress tensor, there are three real roots $(\sigma_1, \sigma_2, \sigma_3)$, referred to as principal stresses, of Equation (2.1-21). Associated with each principal stress is a principal direction satisfying Equation (2.1-19) and $n_i n_i = 1$. The three principal directions and the associated principal planes are mutually orthogonal. It can be shown that the principal stresses correspond to the maximum, intermediate, and minimum normal stresses at a point. Moreover, the maximum shear stress at this point is equal to one-half of the difference between the maximum and minimum principal stresses and acts on a plane making an angle of 45 degrees with the direction of these stresses. A knowledge of the principal stresses is important because they form the basis of failure theories of materials.

2.2 Analysis of Strain

Strain is induced in a continuous body when a physical action causes its configuration to change or deform. The change is normally assumed to be continuous with a one-to-one correspondence existing between the deformed and undeformed states. An analysis of the kinematics of the deformation under these conditions is presented in this section. It is noteworthy that the one-to-one correspondence will be lost for points lying on a prospective fracture plane when a crack passes through these points. Consideration will be given to this anomaly in the subsequent chapters.

2.2.1 Strain Tensor

Let the position of two neighboring points P and Q, separated by a distance dS in an undeformed body, be defined by the coordinates x_i and $x_i + dx_i$, respectively (see Figure 2.4). When the body is loaded, P and Q are displaced to points p and q, defined by coordinates ξ_i and $\xi_i + d\xi_i$, respectively. Let the separation between these points in the deformed body be ds.

It can be assumed for our purposes that a one-to-one correspondence exists between x_i and ξ_i and, in addition, that the displacement components u_i at P are functions of x_i; that is,

$$\xi_i = \xi_i(x_1, x_2, x_3) \tag{2.2-1}$$

and

$$u_i = u_i(x_1, x_2, x_3) \tag{2.2-2}$$

It follows that

$$dS^2 = dx_i \, dx_i \tag{2.2-3}$$

and

$$ds^2 = d\xi_i \, d\xi_i \tag{2.2-4}$$

where

$$d\xi_i = (\delta_{ij} + u_{i,j}) \, dx_j$$

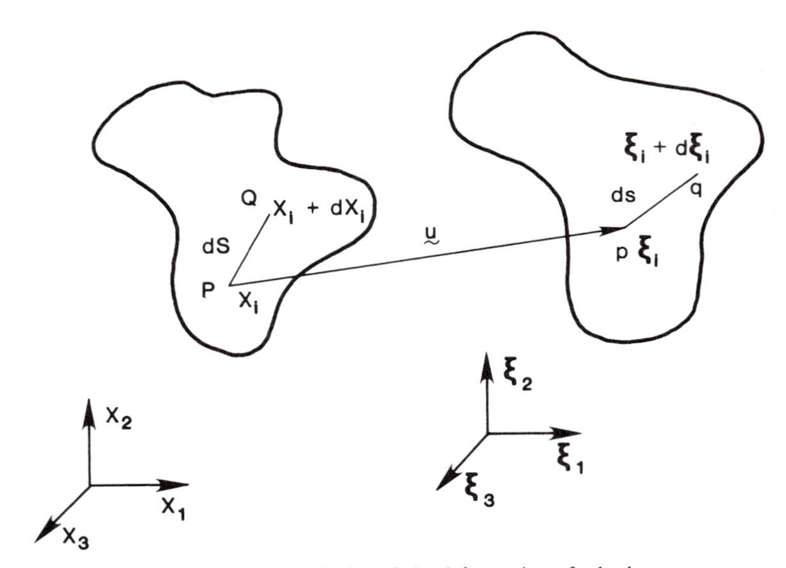

Figure 2.4 Description of the deformation of a body.

Hence, Equation (2.2-4) becomes

$$ds^2 = (\delta_{ij} + u_{i,j} + u_{j,i} + u_{k,i}u_{k,j})\, dx_i\, dx_j \tag{2.2-5}$$

Clearly, there will be no relative change in position of points P and Q if $dS^2 = ds^2$. Conversely, straining occurs whenever $dS^2 \neq ds^2$. This straining can be expressed by the components γ_{ij} of the Green strain tensor defined by

$$ds^2 - dS^2 = 2\gamma_{ij}\, dx_i\, dx_j \tag{2.2-6}$$

It follows with Equations (2.2-3) and (2.2-5) that the finite-deformation strain components are given by

$$\gamma_{ij} = \tfrac{1}{2}(u_{i,j} + u_{j,i}) + \tfrac{1}{2}u_{k,i}u_{k,j} \tag{2.2-7}$$

and, furthermore, that the strain tensor is symmetric. For sufficiently small displacement gradients $u_{k,i}$ (i.e., $|u_{k,i}| \ll 1$), their products in Equation (2.2-7) are negligible and γ_{ij} reduce to

$$\varepsilon_{ij} = \tfrac{1}{2}(u_{i,j} + u_{j,i}) \tag{2.2-8}$$

This simplification yields the strain tensor for infinitesimal strains.

2.2.2 *Compatibility Equations*

Equation (2.2-8) represents six independent equations relating six strain components to only three displacement components. To ensure single-valued displacements u_i, the strain components ε_{ij} cannot be assigned arbitrarily but must satisfy certain integrability or *compatibility conditions*. These conditions can be obtained by eliminating u_i between the equations of Equation (2.2-8) through differentiating the latter with respect to x_k and x_l and interchanging the order of differentiation. Thus, the compatibility equations are

$$\varepsilon_{ij,kl} + \varepsilon_{kl,ij} - \varepsilon_{ik,jl} - \varepsilon_{jl,ik} = 0 \tag{2.2-9}$$

Of the 81 equations included in Equation (2.2-9), only six are independent. The remainder are either identities or repetitions due to the symmetry of ε_{ij}. The six independent compatibility equations are

$$\begin{aligned}
2\varepsilon_{12,12} - \varepsilon_{11,22} - \varepsilon_{22,11} &= 0 \\
2\varepsilon_{23,23} - \varepsilon_{22,33} - \varepsilon_{33,22} &= 0 \\
2\varepsilon_{31,31} - \varepsilon_{33,11} - \varepsilon_{11,33} &= 0 \\
\varepsilon_{11,23} + \varepsilon_{23,11} - \varepsilon_{31,12} - \varepsilon_{12,13} &= 0 \\
\varepsilon_{22,31} + \varepsilon_{31,22} - \varepsilon_{12,23} - \varepsilon_{23,21} &= 0 \\
\varepsilon_{33,12} + \varepsilon_{12,33} - \varepsilon_{23,31} - \varepsilon_{31,32} &= 0
\end{aligned} \tag{2.2-10a–f}$$

Finally, the displacement gradient can be written as

$$\begin{aligned}
u_{i,j} &= \tfrac{1}{2}(u_{i,j} + u_{j,i}) + \tfrac{1}{2}(u_{i,j} - u_{j,i}) \\
&= \varepsilon_{ij} + \omega_{ij}
\end{aligned} \tag{2.2-11}$$

where ω_{ij} is the skew-symmetric rotation tensor. The compatibility equations ensure that the displacements in a simply connected region will be single-valued functions of the coordinates when evaluated by integrating the displacement gradients along any path in this region. If the region is multiply connected—for example, a plane body with an interior crack—then the displacements may be multiple-valued.

2.3 Elasticity

The equations of equilibrium and the kinematic relations are independent of the type of material. However, the connection between the stress and strain components depends upon the specific type of material behavior. In this section constitutive relations for elastic behavior are considered. Constitutive relations for viscoelastic, elastoplastic, and elastic-viscoplastic responses appear in the following sections.

2.3.1 Strain Energy Density

With only a slight modification we adopt Green's (2.11) definition of an elastic material as one for which there exists a positive-definite, single-valued, potential function $W = W(\varepsilon_{kl})$ of the strains ε_{kl} defined by

$$W = \int_0^{\varepsilon_{kl}} \sigma_{ij}\, d\varepsilon_{ij} \tag{2.3-1}$$

This function is referred to as the strain energy density. It is further required that W be a convex function of the strains in the sense that for two strain fields ε_{ij} and ε''_{ij}

$$W(\varepsilon''_{ij}) - W(\varepsilon_{ij}) - (\varepsilon''_{kl} - \varepsilon_{kl}) \left.\frac{\partial W}{\partial \varepsilon_{kl}}\right|_{\varepsilon_{ij}} \geq 0$$

Convexity of W is a sufficient condition for its positive definiteness and assures that the material is stable.

For W to be independent of the loading path and a function of the final strains only,

$$dW = \sigma_{ij}\, d\varepsilon_{ij} \tag{2.3-3}$$

in which

$$\sigma_{ij} = \frac{\partial W}{\partial \varepsilon_{ij}} \tag{2.3-4}$$

must be a perfect differential. This condition is satisfied if

$$\frac{\partial^2 W}{\partial \varepsilon_{ij}\, \partial \varepsilon_{kl}} = \frac{\partial^2 W}{\partial \varepsilon_{kl}\, \partial \varepsilon_{ij}} \tag{2.3-5}$$

or, equivalently,

$$\frac{\partial \sigma_{kl}}{\partial \varepsilon_{ij}} = \frac{\partial \sigma_{ij}}{\partial \varepsilon_{kl}} \tag{2.3-6}$$

Equation (2.3-4) is the constitutive relation connecting the stress and strain components while Equation (2.3-5) ensures symmetry in this relation. It is assumed that the stress-strain relation has a unique inverse; that is, the Jacobian $|\partial^2 W/\partial \varepsilon_{ij}\,\partial \varepsilon_{kl}| \neq 0$.

2.3.2 *Linear Elastic Materials*

For example, if W is the quadratic function

$$W = \tfrac{1}{2} C_{ijkl}\varepsilon_{ij}\varepsilon_{kl} \tag{2.3-7}$$

then Equation (2.3-4) yields the generalized Hooke's law

$$\sigma_{ij} = C_{ijkl}\varepsilon_{kl} \tag{2.3-8}$$

for a linear elastic material. The tensor of elastic constants or moduli of the material is denoted by C_{ijkl}. Symmetry of the stress and strain tensors requires that $C_{ijkl} = C_{jikl}$ and $C_{ijkl} = C_{ijlk}$, respectively. The condition [Equation (2.3-5)] for the existence of a strain energy function also requires that $C_{ijkl} = C_{klij}$. These latter conditions reduce the number of elastic constants from 81 to 21. The existence of material planes of elastic symmetry further decreases the number of independent constants until there are only two for an isotropic material. Consequently, for an isotropic material

$$\sigma_{ij} = 2\mu\left(\varepsilon_{ij} + \frac{v}{1 - 2v}\,\delta_{ij}\varepsilon_{kk}\right) \tag{2.3-9}$$

where μ is the shear modulus, v is Poisson's ratio, and the elastic modulus is $E = 2(1 + v)\mu$.

The combination of Equations (2.3-7) and (2.3-8) yields

$$W = \tfrac{1}{2}\sigma_{ij}\varepsilon_{ij} \tag{2.3-10}$$

for a linear elastic material. The introduction of Equation (2.3-9) into Equation (2.3-10) gives

$$W = \mu\left(\varepsilon_{ij}\varepsilon_{ij} + \frac{v}{1 - 2v}\,\varepsilon_{ii}\varepsilon_{jj}\right) \tag{2.3-11}$$

for an isotropic material.

2.3.3 *Complementary Strain Energy Density*

The existence of a unique inverse of the constitutive relation, Equation (2.3-4), subject to

$$\frac{\partial \varepsilon_{ij}}{\partial \sigma_{kl}} = \frac{\partial \varepsilon_{kl}}{\partial \sigma_{ij}} \tag{2.3-12}$$

assures the existence of the complementary strain energy density $W^* = W^*(\sigma_{ij})$ defined by

$$W^* = \sigma_{ij}\varepsilon_{ij} - W \tag{2.3-13}$$

From chain-rule differentiation of Equation (2.3-13) and the introduction of Equation (2.3-4) we find that

$$\varepsilon_{ij} = \frac{\partial W^*}{\partial \sigma_{ij}} \tag{2.3-14}$$

It is a straightforward task to show that convexity of W^* follows from convexity of W.

For a linear elastic material the combination of Equation (2.3-10) and (2.3-13) yields

$$W = W^* = \tfrac{1}{2}\sigma_{ij}\varepsilon_{ij} \tag{2.3-15}$$

One can also write for this case that

$$W^* = \tfrac{1}{2}C^*_{ijkl}\sigma_{ij}\sigma_{kl} \tag{2.3-16}$$

where the elastic tensor C^*_{ijkl} is the inverse of the tensor C_{ijkl} and $C^*_{ijkl} = C^*_{jikl} = C^*_{ijlk} = C^*_{klij}$. It follows from Equations (2.3-14) and (2.3-16) that

$$\varepsilon_{ij} = C^*_{ijkl}\sigma_{kl} \tag{2.3-17}$$

For an isotropic material Equation (2.3-17) reduces to

$$\varepsilon_{ij} = \frac{1+\nu}{E}\sigma_{ij} - \frac{\nu}{E}\delta_{ij}\sigma_{kk} \tag{2.3-18}$$

and W^* becomes

$$W^* = \frac{1+\nu}{2E}\sigma_{kl}\sigma_{kl} - \frac{\nu}{2E}\sigma_{kk}\sigma_{ll} \tag{2.3-19}$$

If a power law relationship between stress and strain exists such that the strain is a homogeneous function of degree n of the stress, then Equation (2.3-14) implies that W^* must be a homogeneous function of the stress components of degree $n + 1$. It follows from Euler's theorem on homogeneous functions [e.g., see (2.12)] and Equation (2.3-14) that

$$W^* = \frac{1}{n+1}\frac{\partial W^*}{\partial \sigma_{ij}}\sigma_{ij} = \frac{1}{n+1}\sigma_{ij}\varepsilon_{ij} \tag{2.3-20}$$

Furthermore, the combination of Equations (2.3-13) and (2.3-20) permits writing

$$W = \frac{n}{n+1}\sigma_{ij}\varepsilon_{ij} \tag{2.3-21}$$

When stress is proportional to strain ($n = 1$), Equations (2.3-20) and (2.3-21) become identical to Equation (2.3-15).

2.3.4 Elastic Boundary Value Problems

For an elastic material the governing field equations are the equilibrium equations [Equations (2.1-10) and (2.1-16)], the kinematic relations

[Equation (2.2-8)] and the compatibility equations [Equation (2.2-10)]. These must be supplemented by a constitutive relation whose most general form is expressed by either Equation (2.3-4) or Equation (2.3-14). For the case of a linear material, the latter assume the forms of Equations (2.3-8) and (2.3-17), which reduce, respectively, to Equations (2.3-9) and (2.3-18) for an isotropic material.

The preceding equations must hold within the volume V of the body and on its boundary S. On the portion S_T of the boundary where the tractions are prescribed we require that T_i be equal to the specified values $\bar{T}_i$. Over the remainder of the boundary $S_u = S - S_T$, the displacements u_i must assume their prescribed values $\bar{u}_i$.

2.3.5 Rubber Elasticity

The discussion to this point has been largely confined to small strain behavior—a limitation that is not particularly severe for most engineering materials. However, for rubber and other elastomers, a small strain assumption would be highly inappropriate. In this section the generally accepted theoretical models embodying the large strain behavior characteristic of rubber elasticity will be outlined. The discussion will focus on the constitutive relations developed by Rivlin and Mooney, following the treatments of the subject given by Treloar (2.13) and Gent (2.14).

In a manner that is analogous to the development of Section 2.1.2 above, the homogeneous deformation of an elastic body can be expressed in terms of three principal extension ratio $\lambda_1, \lambda_2, \lambda_3$ (i.e., ratio of the deformed dimension to the undeformed dimension) and, in turn, to three invariants given by

$$\begin{aligned} I_1 &= \lambda_1^2 + \lambda_2^2 + \lambda_3^2 \\ I_2 &= \lambda_1^2\lambda_2^2 + \lambda_2^2\lambda_3^2 + \lambda_3^2\lambda_1^2 \\ I_3 &= \lambda_1^2\lambda_2^2\lambda_3^2 \end{aligned} \tag{2.3-22}$$

Rubber is very nearly an incompressible material whereupon it is usually appropriate to set $I_3 = 1$. This allows Equation (2.3-22) to be simplified to

$$\begin{aligned} I_1 &= \lambda_1^2 + \lambda_2^2 + \lambda_3^2 \\ I_2 &= \lambda_1^{-2} + \lambda_2^{-2} + \lambda_3^{-2} \end{aligned} \tag{2.3-23}$$

Rivlin has developed a general treatment of rubber-like solids by introducing the basic assumption that the material is elastically isotropic in the undeformed state. More specifically, he argued that the strain energy density function W must be a function of I_1 and I_2 that vanishes in the undeformed state. The most general form satisfying these conditions is

$$W = \sum C_{mn}(I_1 - 3)^m(I_2 - 3)^n \tag{2.3-24}$$

where the upper limit of the summation is arbitrary. Taking the limit to be unity gives the first-order expression

$$W = C_1(I_1 - 3) + C_2(I_2 - 3) \tag{2.3-25}$$

which is sometimes referred to as the Mooney-Rivlin equation.

Because of the assumption of incompressibility only the differences of principal stresses can be determined. Expressed in terms of the strain energy density function, the expressions are

$$\sigma_1 - \sigma_2 = 2(\lambda_1^2 - \lambda_2^2)\left[\frac{\partial W}{\partial I_1} + \lambda_3^2 \frac{\partial W}{\partial I_2}\right] \tag{2.3-26}$$

with similar expressions for $\sigma_2 - \sigma_3$ and $\sigma_3 - \sigma_1$. Thus, for a Mooney-Rivlin material

$$\sigma_1 - \sigma_2 = 2(\lambda_1^2 - \lambda_2^2)(C_1 + \lambda_3^2 C_2) \tag{2.3-27}$$

2.4 Energy Principles

Because of the conservative nature of many loadings and the reversible nature of elastic materials, energy principles represent important concepts in the theory of elasticity. As described in Chapter 1, these principles have played (and continue to play) a fundamental role in the development of many of the concepts in fracture mechanics. They frequently form the bases of finite element methods used to obtain numerical solutions to boundary value problems. Due to their importance we review some of the most pertinent principles in this section. The reader can find additional treatments of energy principles and methods in references (2.2) and (2.3).

2.4.1 Principle of Virtual Work

Consider a body that occupies the volume V and is bounded by the surface $S = S_T + S_u$ to be in static equilibrium under the action of prescribed body forces F_i and surface tractions $\bar{T}_i$ on S_T. Over the remaining portion of the boundary S_u the displacements $\bar{u}_i$ are specified. A statically admissible stress field σ'_{ij} is defined as one that satisfies the equations of equilibrium, Equations (2.1-10) and (2.1-16), and whose stress vector $T'_i = \sigma'_{ij} n_j$ takes on the prescribed values on S_T. A kinematically admissible displacement field u''_i is one that is three times continuously differentiable and assumes the specified values on S_u. There need not be any connection between a statically admissible stress field and a kinematically admissible displacement field.

For a statically admissible stress field σ'_{ij} and a kinematically admissible displacement field u''_i,

$$\int_S T'_i u''_i \, dS + \int_V F_i u''_i \, dV = \int_V \sigma'_{ij} \varepsilon''_{ij} \, dV \tag{2.4-1}$$

expresses the principle of virtual work. The strain field ε''_{ij} is derivable from u''_i through Equation (2.2-8). The proof of Equation (2.4-1) follows directly upon substituting $T_i' = \sigma_{ij}' n_j$ into the left-hand side, employing the divergence theorem and noting that σ_{ij}' satisfies the equilibrium equations. If this virtual work statement holds for *all* kinematically admissible displacement fields, then the stress field σ_{ij}' must necessarily be statically admissible.

If σ_{ij}' and u_i'' are the actual stress σ_{ij} and displacement u_i fields in a body obeying a power law relationship between stress and strain, then Equations (2.4-1), (2.3-20), and (2.3-21) yield

$$\int_S T_i u_i \, dS + \int_V F_i u_i \, dV = \frac{n+1}{n} \int_V W \, dV \equiv \frac{n+1}{n} U$$

$$= (n+1) \int_V W^* \, dV \equiv (n+1) U^* \qquad (2.4\text{-}2)$$

The strain energy and complementary strain energy of the body are denoted by U and U^*, respectively. For a linear elastic material ($n = 1$), Equation (2.4-2) becomes an expression of Clapeyron's theorem, which states that the work done by the body forces and tractions acting through the displacements from the unstressed state to the final equilibrium configuration is equal to twice the strain energy or the complementary strain energy of the body.

2.4.2 Potential Energy

For a kinematically admissible displacement field u_i'', let the potential energy Π of a body under the action of conservative body forces F_i and prescribed surface tractions $\bar{T}_i$ on S_T be defined by

$$\Pi(u_i'') = \int_V W(\varepsilon_{ij}'') \, dV - \int_{S_T} \bar{T}_i u_i'' \, dS - \int_V F_i u_i'' \, dV \qquad (2.4\text{-}3)$$

where the compatible strain field ε_{ij}'' is related to u_i'' by Equation (2.2-8). If σ_{ij}, ε_{ij}, and u_i are the actual stress, strain, and displacement fields, respectively, then

$$\Pi(u_i'') - \Pi(u_i) = \int_V \left[W(\varepsilon_{ij}'') - W(\varepsilon_{ij}) - (\varepsilon_{ij}'' - \varepsilon_{ij}) \frac{\partial W}{\partial \varepsilon_{ij}} \right] dV$$

$$+ \int_V \sigma_{ij}(\varepsilon_{ij}'' - \varepsilon_{ij}) \, dV - \int_S T_i(u_i'' - u_i) \, dS - \int_V F_i(u_i'' - u_i) \, dV$$

$$= \int_V \left[W(\varepsilon_{ij}'') - W(\varepsilon_{ij}) - (\varepsilon_{ij}'' - \varepsilon_{ij}) \frac{\partial W}{\partial \varepsilon_{ij}} \right] dV \qquad (2.4\text{-}4)$$

In arriving at Equation (2.4-4) the surface integral has been extended over the entire surface S since $u_i'' - u_i$ vanishes on S_u. Furthermore, the principle of virtual work has been invoked since $u_i'' - u_i$ is also a kinematically admissible displacement field. Due to the assumed convexity of W it follows that

$$\Pi(u_i'') \geqslant \Pi(u_i) \qquad (2.4\text{-}5)$$

Equation (2.4-5) expresses the principle of minimum potential energy. Among all the kinematically admissible displacement fields the actual displacement field, which is also statically admissible, minimizes the potential energy.

2.4.3 Complementary Potential Energy

For a statically admissible stress field σ'_{ij} the complementary potential energy Π^* is defined by

$$\Pi^*(\sigma'_{ij}) = \int_V W^*(\sigma'_{ij})\, dV - \int_{S_u} T'_i u_i\, dS \tag{2.4-6}$$

where $T'_i = \sigma'_{ij} n_j$ is the reaction on S_u. If σ_{ij}, ε_{ij}, and u_i are the actual stress, strain, and displacement fields, then

$$\begin{aligned}\Pi^*(\sigma'_{ij}) - \Pi^*(\sigma_{ij}) = \int_V \left[W^*(\sigma'_{ij}) - W^*(\sigma_{ij}) - (\sigma'_{ij} - \sigma_{ij}) \frac{\partial W^*}{\partial \sigma_{ij}} \right] dV \\ + \int_V \varepsilon_{ij}(\sigma'_{ij} - \sigma_{ij})\, dV - \int_S (T'_i - T_i) u_i\, dS \end{aligned} \tag{2.4-7}$$

where the surface integral has been extended to include S_T since $T'_i - T_i = 0$ there. The contribution of the last two integrals is zero due to the principle of virtual work. Recalling the convexity of W^* we conclude from Equation (2.4-7) that

$$\Pi^*(\sigma'_{ij}) \geqslant \Pi^*(\sigma_{ij}) \tag{2.4-8}$$

which is the basis of the principle of minimum complementary potential energy. Of all the statically admissible stress fields, the actual one which is also kinematically admissible (compatible) minimizes the complementary potential energy.

For the actual stress, strain and displacements fields, upon adding Equations (2.4-3) and (2.4-6) and employing Equation (2.3-13), we can write

$$\begin{aligned}\Pi^*(\sigma_{ij}) + \Pi(u_i) &= \int_V (W^* + W)\, dV - \int_S T_i u_i\, dS - \int_V F_i u_i\, dV \\ &= \int_V \sigma_{ij}\varepsilon_{ij}\, dV - \int_S T_i u_i\, dS - \int_V F_i u_i\, dV \end{aligned} \tag{2.4-9}$$

But, the right-hand side of Equation (2.4-9) vanishes due to the principle of virtual work so that

$$\Pi^*(\sigma_{ij}) = -\Pi(u_i) \tag{2.4-10}$$

Furthermore, the minimal properties of Π and Π^* permit writing

$$-\Pi(u''_i) \leqslant -\Pi(u_i) = \Pi^*(\sigma_{ij}) \leqslant \Pi^*(\sigma'_{ij}) \tag{2.4-11}$$

which provides upper and lower bounds to the potential energy. If $u_i = 0$ on S_u (i.e., the reactions do no work), then the introduction of Equation (2.4-6) into Equation (2.4-10) yields

$$\Pi^*(\sigma_{ij}) = -\Pi(u_j) = U^*(\sigma_{ij}) \tag{2.4-12}$$

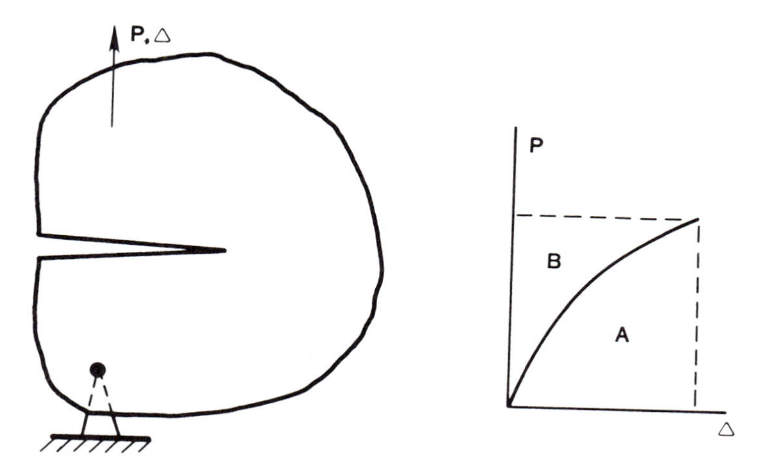

Figure 2.5 Typical structure and load-displacement diagram.

In this case Equation (2.4-11) also provides upper and lower bounds to the complementary strain energy. When the material obeys a power law, Equation (2.4-11) also provides bounds to the strain energy.

It is noteworthy that the preceding developments and principles are applicable to structural analysis by viewing the stresses, strains, and displacements in the foregoing as generalized forces, strains, and displacements, respectively. As an example consider the structure and its load versus load-point displacement depicted in Figure 2.5. Here P represents a generalized force and Δ is the conjugate generalized displacement such that the product $P\Delta$ represents the work done. The area A under the curve is equal to the strain energy U and the area B between the curve and the load axis is the complementary strain energy U^*. The potential energy is

$$\Pi = U - P\Delta = -U^* \tag{2.4-13}$$

in agreement with Equation (2.4-12).

2.5 Viscoelasticity

As the name reflects, a viscoelastic material is one that possesses both elastic and viscous properties. Under a step loading in time (see Figure 2.6) such a material may exhibit an initial elastic response followed by creep where the material continues to strain at a rate that depends upon the intensity of the stress. Upon removal of the load there may be a partial instantaneous elastic recovery followed by a period of decreasing strain. When the strain recovery is complete, it is referred to as delayed elasticity. On the other hand, if the material is subjected to a sustained constant strain, the material will relax in the sense that the stress decreases with time from its initial value.

Viscoelastic materials display a pronounced sensitivity to the rate of straining or stressing and possess time-dependent material properties. This

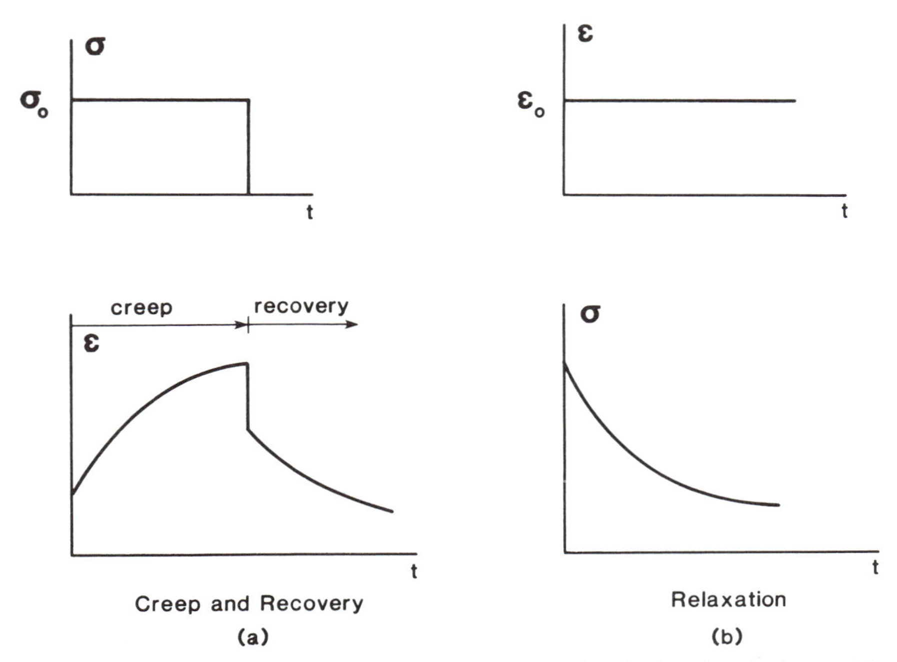

Figure 2.6 Examples of (a) creep and recovery and (b) stress relaxation in a viscoelastic material.

type of behavior is readily observable at room temperature in many polymers and adhesives that are increasingly being found in load-bearing members. Metals in high temperature applications such as gas-turbines, nuclear power plants, and space vehicles can also exhibit similar characteristics.

2.5.1 Linear Viscoelastic Materials

Suppose a time-dependent material is subjected to a stress history $\sigma(t) = c_1\sigma_1(t) + c_2\sigma_2(t - \tau)$, where c_1 and c_2 are constants. If the strain $\varepsilon[\sigma(t)]$ satisfies the basic property of a linear functional that

$$\varepsilon[c_1\sigma_1(t) + c_2\sigma_2(t - \tau)] = c_1\varepsilon[\sigma_1(t)] + c_2\varepsilon[\sigma_2(t - \tau)] \qquad (2.5\text{-}1)$$

then the material is referred to as being linear viscoelastic. For such materials a simple uniaxial creep test can be used to establish the relationship between stress and strain. In this test the material is subjected to a uniaxial stress, $\sigma = \sigma_0 H(t)$, where $H(t)$ is the Heaviside step function defined as $H(t) = 1$ for $t > 0$ and $H(t) = 0$ for $t < 0$. The axial strain is observed to obey the law

$$\varepsilon(t) = \sigma_0 C(t) \qquad (2.5\text{-}2)$$

Equation (2.5-2) defines the creep compliance $C(t)$, which is a monotonically increasing function of t for $t \geqslant 0$. Without loss in generality it is understood that $C(t) \equiv 0$ for $t < 0$. Alternatively, in a relaxation test the material is subjected to a step strain, $\varepsilon = \varepsilon_0 H(t)$, and the stress is measured. In the case of a linear material

$$\sigma(t) = \varepsilon_0 G(t) \tag{2.5-3}$$

where $G(t)$ is the relaxation modulus that is a monotonically decreasing function of t for $t \geqslant 0$ and $G(t) \equiv 0$ if $t < 0$.

Knowing the creep compliance we can now express the strain for an arbitrary stress history depicted in Figure 2.7. The strain at time t can be visualized as resulting from a sequence of infinitesimal step stresses occurring at times τ for $-\infty < \tau < t$. For example, the incremental strain $d\varepsilon$ due to the infinitesimal stress $d\sigma(\tau)H(t-\tau)$ illustrated in Figure 2.7 is

$$d\varepsilon = C(t-\tau)\, d\sigma(\tau)$$

Integrating the foregoing expression from $\tau = -\infty$ to t, we obtain

$$\varepsilon(t) = \int_{-\infty}^{t} C(t-\tau) \frac{d\sigma(\tau)}{d\tau}\, d\tau \tag{2.5-4}$$

Equation (2.5-4) signifies that the current strain is a function of the previous stress history. Consequently, the convolution integral in this equation is referred to as a hereditary integral.

Equivalent forms of Equation (2.5-4) also exist. For example, since $C(t-\tau) \equiv 0$ for $\tau > t$, then the upper limit of integration in this equation can be replaced by ∞. In this manner the strain can be expressed in terms of the Stieljes integral

$$\varepsilon(t) = \int_{\tau=-\infty}^{\tau=\infty} C(t-\tau)\, d\sigma(\tau) \tag{2.5-5}$$

If the loading commences at time $t = 0$ so that $\sigma_{ij} = \varepsilon_{ij} = 0$ for $t < 0$, then Equation (2.5-4) can be written as

$$\varepsilon(t) = \sigma(0^+)C(t) + \int_{0}^{t} C(t-\tau) \frac{d\sigma(\tau)}{d\tau}\, d\tau \tag{2.5-6}$$

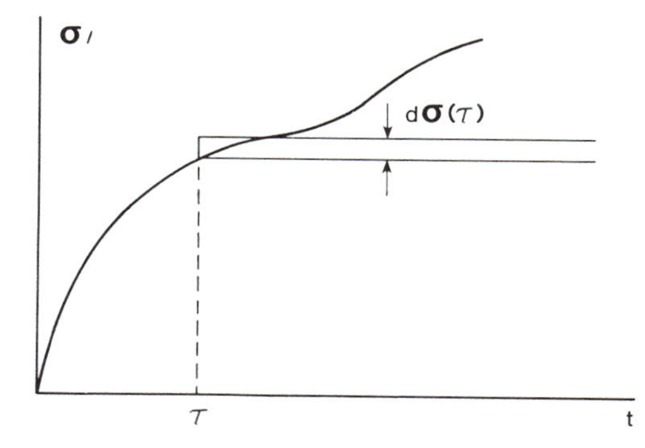

Figure 2.7 Stress-time history.

The first term reflects the strain due to the initial stress and the integral represents the strain due to the ensuing stress. An integration by parts of Equation (2.5-6) leads to

$$\begin{aligned} \varepsilon(t) &= \sigma(t)C(0) + \int_0^t \sigma(\tau)\frac{dC(t-\tau)}{d(t-\tau)}\,d\tau \\ &= \sigma(t)C(0) + \int_0^t \sigma(t-\tau)\frac{dC(\tau)}{d\tau}\,d\tau \end{aligned} \tag{2.5-7}$$

If the material is subjected to a prescribed strain history rather than a stress history, a similar analysis yields

$$\sigma(t) = \int_{-\infty}^{t} G(t-\tau)\frac{d\varepsilon(\tau)}{d\tau}\,d\tau \tag{2.5-8}$$

$$= \int_{\tau=-\infty}^{\tau=\infty} G(t-\tau)\,d\varepsilon(\tau) \tag{2.5-9}$$

$$= \varepsilon(0^+)G(t) + \int_0^t G(t-\tau)\frac{d\varepsilon(\tau)}{d\tau}\,d\tau \tag{2.5-10}$$

$$= \varepsilon(t)G(0) + \int_0^t \varepsilon(t-\tau)\frac{dG(\tau)}{d\tau}\,d\tau \tag{2.5-11}$$

When either the creep compliance or the relaxation modulus is known, the other can be determined. This connection is readily demonstrated with the Laplace transform method. Let the Laplace transform $\hat{f}(s)$ of the function $f(t)$ be defined by

$$\hat{f}(s) = \int_0^\infty e^{-st} f(t)\,dt \tag{2.5-12}$$

It follows from an integration by parts that the Laplace transform of df/dt is $s\hat{f}(s) - f(0^+)$. The Laplace transform of the convolution integral

$$I(t) = \int_0^t f(t-\tau)g(\tau)\,d\tau$$

is $\hat{I}(s) = \hat{f}(s)\hat{g}(s)$; see reference (2.15).

Operating on Equation (2.5-6) with the Laplace transform, we obtain

$$\hat{\varepsilon}(s) = s\hat{C}(s)\hat{\sigma}(s) \tag{2.5-13}$$

Following the same procedure for Equation (2.5-10) one finds that

$$\hat{\sigma}(s) = s\hat{G}(s)\hat{\varepsilon}(s) \tag{2.5-14}$$

We note in passing that Equation (2.5-14) in the transformed space has the same form as the elastic law if $s\hat{G}(s)$ is associated with E. Equations (2.5-13) and (2.5-14) can be combined to yield

$$\hat{C}(s)\hat{G}(s) = \frac{1}{s^2} \tag{2.5-15}$$

The left-hand side of Equation (2.5-15) is the Laplace transform of the convolution integral whereas the right-hand side is the Laplace transform of t. Hence,

$$\int_0^t G(t-\tau)C(\tau)\,d\tau = \int_0^t G(\tau)C(t-\tau)\,d\tau = t \tag{2.5-16}$$

is the connection between the creep compliance and the relaxation modulus. In addition, the limiting properties (2.15) of the Laplace transforms as $s \to 0$ and $s \to \infty$ provide that

$$C(0)G(0) = C(\infty)G(\infty) = 1 \tag{2.5-17}$$

When generalizing the constitutive relations, Equations (2.5-4) and (2.5-8), it is convenient to introduce the deviatoric stress components s_{ij} and strain components e_{ij} defined by

$$s_{ij} = \sigma_{ij} - \tfrac{1}{3}\sigma_{kk}\,\delta_{ij}, \qquad s_{ii} = 0 \tag{2.5-18}$$

$$e_{ij} = \varepsilon_{ij} - \tfrac{1}{3}\varepsilon_{kk}\,\delta_{ij}, \qquad e_{ii} = 0 \tag{2.5-19}$$

For a linear, isotropic viscoelastic material

$$e_{ij} = \int_{-\infty}^{t} C_1(t-\tau)\frac{ds_{ij}(\tau)}{d\tau}\,d\tau \tag{2.5-20}$$

$$\varepsilon_{ii} = \int_{-\infty}^{t} C_2(t-\tau)\frac{d\sigma_{ii}(\tau)}{d\tau}\,d\tau \tag{2.5-21}$$

or, alternatively,

$$s_{ij} = \int_{-\infty}^{t} G_1(t-\tau)\frac{de_{ij}(\tau)}{d\tau}\,d\tau \tag{2.5-22}$$

$$\sigma_{ii} = \int_{-\infty}^{t} G_2(t-\tau)\frac{d\varepsilon_{ii}(\tau)}{d\tau}\,d\tau \tag{2.5-23}$$

where $C_1(t)$ and $G_1(t)$ are the creep compliance and relaxation modulus in shear, respectively, and $C_2(t)$ and $G_2(t)$ are their dilatational counterparts. The integrals in these constitutive relations can also be expressed in any of the equivalent forms discussed earlier for the uniaxial case. Carrying the analogy an additional step, one obtains

$$\int_0^t G_\alpha(t-\tau)C_\alpha(\tau)\,d\tau = \int_0^t G_\alpha(\tau)C_\alpha(t-\tau)\,d\tau = t \tag{2.5-24}$$

where no sum over $\alpha = 1, 2$ is intended.

2.5.2 *Thermorheologically Simple Materials*

The creep compliance and the relaxation modulus of many viscoelastic polymers depend upon the temperature T. There is a class of materials known as thermorheologically simple materials for which a change of temperature from a reference temperature T_0 is equivalent to changing the time scale. Temperatures above (below) the reference temperature effectively increase (decrease) the real time. For these materials it was first demonstrated empirically and later shown theoretically that

$$C(T,t) = C(T_0, t/a_T(T)) \tag{2.5-25}$$

where the shift factor $a_T(T)$ is a measured function of temperature. Consequently, for temperatures above T_0, where $a_T(T)$ is less than unity, the process appears to be accelerated relative to the reference temperature. The opposite effect occurs for temperatures below T_0, where $a_T(T)$ is greater than unity. Williams, Landel, and Ferry (2.16) have found that for many polymers $a_T(T)$ is given by the WLF equation

$$\log a_T(T) = -\frac{c_1(T - T_0)}{c_2 + T - T_0} \tag{2.5-26}$$

where c_1 and c_2 are material properties. A relation analogous to Equation (2.5-25) can also be written for the relaxation modulus.

Not only does this time-temperature superposition principle permit the inclusion of the influence of temperature upon the constitutive law, but it also provides an efficient technique for establishing master relaxation modulus and creep compliance curves. In developing a master curve, the relaxation modulus or the creep compliance is measured over one or two decades of time for a range of temperatures. The logarithm of the relaxation modulus or creep compliance is plotted against the logarithm of time for each temperature. All these curves except the one at the reference temperature are shifted parallel along the abscissa until they form a single continuous master curve extending over perhaps ten or more decades of time. Examples of the use of this principle to determine master curves for polymers together with further discussion can be found in the texts of Tobolsky (2.17) and Ferry (2.18).

2.5.3 *Correspondence Principle*

The governing equations for a linear, isotropic viscoelastic material are Equations (2.1-10), (2.1-16), (2.2-8), (2.5-20), (2.5-21), and the boundary conditions

$$T_i = \bar{T}_i \quad \text{on} \quad S_T \tag{2.5-27}$$

$$u_i = \bar{u}_i(t) \quad \text{on} \quad S_u \tag{2.5-28}$$

where the overbar is used to denote prescribed quantities. Alternatively, Equations (2.5-20) and (2.5-21) can be replaced with Equations (2.5-22) and (2.5-23). Operating on these governing equations with the Laplace transform, we obtain

$$\begin{aligned}
&\hat{\sigma}_{ij,j} + \hat{F}_i = 0, \qquad \hat{\sigma}_{ij} = \hat{\sigma}_{ji} \\
&\hat{\varepsilon}_{ij} = \tfrac{1}{2}(\hat{u}_{i,j} + \hat{u}_{j,i}) \\
&\hat{\varepsilon}_{ij} = s\hat{C}_1\hat{\sigma}_{ij} + \frac{s}{3}(\hat{C}_2 - \hat{C}_1)\delta_{ij}\hat{\sigma}_{kk} \\
&\hat{\sigma}_{ij} = s\hat{G}_1\hat{\varepsilon}_{ij} + \frac{s}{3}(\hat{G}_2 - \hat{G}_1)\delta_{ij}\hat{\varepsilon}_{kk} \\
&\hat{T}_i = \hat{\hat{T}}_i \quad \text{on} \quad S_T \\
&\hat{u}_i = \hat{\hat{u}}_i \quad \text{on} \quad S_u
\end{aligned} \tag{2.5-29a–f}$$

These transformed equations have the same form as the corresponding equations of linear, isotropic elasticity if the transformed variable is associated with the equivalent elastic variable and if $s\hat{G}_1, s\hat{G}_2, s\hat{C}_1$, and $s\hat{C}_2$ are associated with the corresponding elastic constants. It follows that the solution of the viscoelastic problem in the transformed plane corresponds to the solution of the equivalent elasticity problem. Therefore, if the solution to the latter problem is known, the solution to the viscoelastic problem reduces to effecting the inverse Laplace transform. The association of the solution of the transformed viscoelastic equations with the solution of the elasticity equations forms the basis of the correspondence principle of linear viscoelasticity. While only isotropic materials have been considered here, the correspondence principle also holds for anisotropic materials.

At first hand, the correspondence principle may appear to be quite general, but a little reflection reveals two important limitations associated with mixed boundary value problems. First, in transforming the boundary conditions it is assumed that at a point on the boundary either the traction or the conjugate displacement is prescribed for all $t > 0$. This is not the case, for example, in contact or indentation problems where the boundary conditions on the prospective contact surface change with time from prescribed traction to prescribed displacement. The second limitation, which has direct bearing on the subject of this text, assumes that the boundaries S_T and S_u do not vary with time. This is not, of course, the case during crack propagation where new surfaces are being created.

Subject to additional restrictions, Graham (2.19) has extended the correspondence principle to include a class of mixed boundary value problems in isotropic viscoelasticity involving time-dependent boundaries. Included in this class of problems are the aforementioned contact and fracture problems. The specific restrictions required for the problem of a propagating crack are: (1) the crack front is not permitted to retreat; (2) the crack plane stresses for the equivalent elastic problem must be independent of the elastic constants; and (3) the elastic crack plane displacement v must have the separable form $v = f(E, \nu)V(x_i, t)$, where $V(x_i, t)$ is independent of the elastic constants whereas f is a function of only these constants.

The first restriction does not permit any crack closure or crack healing. With respect to the second restriction we note that if only stresses are prescribed on the boundary of the body, then the stress field will be independent of the modulus of elasticity but not necessarily independent of Poisson's ratio. However, for simply connected bodies and for multiply connected bodies in plane stress or plane strain for which the resultant force on each boundary vanishes, the dependence of the stresses upon Poisson's ratio also disappears. The final restriction is perhaps the most exacting. But, if the viscoelastic materials has a constant Poisson's ratio, a condition approximately satisfied by many polymers, then Graham (2.20) has further shown that the correspondence principle is still applicable even though Poisson's ratio cannot be included in the separable factor. It is also applicable to anisotropic viscoelastic materials with a single relaxation function.

The emphasis in this section has been on linear viscoelastic materials. The treatment of nonlinear viscoelastic materials and extensive references can be found in the treatises by Christensen (2.4) and by Findley, Lai, and Onaran (2.5).

2.6 Elastoplasticity

When a material is loaded beyond its elastic limit, plastic deformation, an irreversible process, ensues. Consequently, the final state of deformation depends not only upon the final loading, but also upon the loading path. In contrast to viscoelastic behavior, plastic deformation, at least for quasi-static loading, tends to be rate-independent. At a minimum the constitutive relation for plastic deformation must reflect these characteristics. The theory that satisfies this criterion and that has found widespread application is the incremental or flow theory. Deformation or total strain theory, which has a more restricted range of application but often possesses the redeeming feature of reduced mathematical complexity, is also worthy of consideration. In the following we develop the constitutive relations based on these two theories and then compare them. Finally, the slip-line theory for plane strain, perfectly plastic behavior is summarized.

2.6.1 Yield Criteria

It is convenient to write the strain ε_{ij} in an elastic-plastic material as

$$\varepsilon_{ij} = \varepsilon_{ij}^e + \varepsilon_{ij}^p \qquad (2.6\text{-}1)$$

where ε_{ij}^e and ε_{ij}^p are the elastic and plastic contributions, respectively. The elastic component of the strain is assumed to be related to the stress σ_{ij} by one of the Hookean laws discussed in Section 2.3. If the stress level is such that no yielding occurs, then the plastic strain component is identically zero. When yielding does occur, the plastic deformation is assumed to be incompressible or, equivalently,

$$\varepsilon_{ii}^p = 0 \qquad (2.6\text{-}2)$$

Consequently, there is no need to make a distinction between the plastic strain ε_{ij}^p and its deviatoric counterpart e_{ij}^p.

A yield criterion is required to assess whether or not yielding or plastic deformation is imminent or has occurred. We postulate the existence of a yield function

$$f = f(\sigma_{ij}) \tag{2.6-3}$$

which at the very minimum is a function of the current state of stress σ_{ij}. The yield surface $f = 0$ represents a hypersurface in the nine-dimensional Euclidean stress space σ_{ij}. In general, the yield surface can expand (isotropic hardening), translate (kinematic hardening) or both during plastic deformation. However, all current stress states must lie either on or inside ($f \leqslant 0$) this surface, but never outside ($f > 0$) it. All stress states or variations falling within the surface are elastic whereas during plastic deformation the current stress state must be on the yield surface.

We note that $\partial f/\partial \sigma_{ij}$ are the components of the gradient to the yield surface and consider a current state of stress that lies on the yield surface. A small change in the loading giving rise to stress increments $d\sigma_{ij}$ produces elastic unloading if the incremental stress vector having components $d\sigma_{ij}$ is directed towards the interior of the surface (i.e., elastic unloading from a plastic state occurs when $f = 0$ and $(\partial f/\partial \sigma_{ij})\, d\sigma_{ij} < 0$). On the other hand if $f = 0$ and $(\partial f/\partial \sigma_{ij})\, d\sigma_{ij} > 0$, then the incremental stress vector is directed outward from the yield surface and plastic loading occurs. For the intermediate case when the incremental stress vector lies in the tangent plane to the yield surface [i.e., $f = 0$ and $(\partial f/\partial \sigma_{ij})\, d\sigma_{ij} = 0$], the loading is neutral.

For isotropic material behavior the yield function must be an isotropic function of the stress. This requires that

$$f = f(I_1, I_2, I_3) \tag{2.6-4}$$

where I_i are the stress invariants of Equation (2.1-23). Since these invariants can be expressed in terms of the principal stresses, then alternatively

$$f = f(\sigma_1, \sigma_2, \sigma_3) \tag{2.6-5}$$

Experimental evidence indicates that yielding of most metals is not influenced by a moderate hydrostatic pressure

$$p = \sigma_{ii}/3 = I_1/3 \tag{2.6-6}$$

For such cases the yield function is independent of the first stress invariant and depends only upon the deviatoric stresses s_{ij}. Accordingly, for isotropic behavior

$$f = f(J_2, J_3) \tag{2.6-7}$$

where

$$\begin{aligned} J_2 &= \tfrac{1}{2} s_{ij} s_{ij} \\ J_3 &= \tfrac{1}{3} s_{ij} s_{ik} s_{jk} \end{aligned} \tag{2.6-8}$$

are the second and third invariants of the deviatoric stress tensor. As a consequence the yield surface $f(\sigma_1, \sigma_2, \sigma_3) = 0$ in the principal stress space will be normal to the octahedral Π-plane, $\sigma_1 + \sigma_2 + \sigma_3 = 0$. If the material manifests the same yield stress in tension and compression (i.e., no Bauschinger effect), then $f(s_{ij}) = f(-s_{ij})$ and f must be an even function of J_3.

Two of the most widely used yield criteria are the von Mises and the Tresca yield conditions. The von Mises yield function is

$$f = J_2 - k^2 \tag{2.6-9}$$

where k depends upon the strain history for a work hardening material and assumes a constant value for a perfectly plastic material. In terms of the principal stresses the von Mises yield surface is defined by

$$f = \tfrac{1}{6}[(\sigma_1 - \sigma_2)^2 + (\sigma_2 - \sigma_3)^2 + (\sigma_3 - \sigma_1)^2] - k^2 = 0 \tag{2.6-10}$$

For a material in pure shear, $\sigma_1 = -\sigma_3 = \tau$ and $\sigma_2 = 0$, and it follows from Equation (2.6-10) that $k = \tau$ is the yield stress in shear. For yielding in uniaxial tension, $\sigma_1 = \sigma_y$ and $\sigma_2 = \sigma_3 = 0$; whereupon $k = \sigma_y/\sqrt{3}$. In the principal stress space Equation (2.6-10) defines a circular cylindrical surface whose axis is the ray $\sigma_1 = \sigma_2 = \sigma_3$.

The Tresca criterion postulates that yielding occurs when the maximum shear stress in the material reaches the yield stress k in shear. In terms of the principal stresses this criterion can be expressed as

$$\max[|\sigma_1 - \sigma_2|, |\sigma_2 - \sigma_3|, |\sigma_3 - \sigma_1|] = 2k \tag{2.6-11}$$

or in the symmetric form

$$f = [(\sigma_1 - \sigma_2)^2 - 4k^2][(\sigma_2 - \sigma_3)^2 - 4k^2][(\sigma_3 - \sigma_1)^2 - 4k^2] = 0 \tag{2.6-12}$$

The Tresca yield surface is a regular hexogonal cylinder with the axis $\sigma_1 = \sigma_2 = \sigma_3$ in the principal stress space. When the principal stresses are not known a priori, then the form,

$$4J_2^3 - 27J_3^2 - 36k^2J_2^2 + 96k^4J_2 - 64k^6 = 0 \tag{2.6-13}$$

due to Reuss can be used.

2.6.2 *Incremental Plasticity*

The phenomenon of stable work hardening in a material under uniaxial tension is associated with a positive increment of stress $d\sigma$ for a plastic strain increment $d\varepsilon^p$ or, equivalently, $d\sigma\, d\varepsilon^p > 0$, which is equally valid for compressive loadings. Drucker (2.21) generalized this simple concept of work hardening to include all states of stress and loading paths by considering the work done by an external agency as it quasi-statically applies additional stresses to an already stressed medium and then removes them. After such a cycle the original configuration may or may not be recovered. It is to be understood that the external agency is independent of the one producing the existing states of stress and strain. A stable work hardening material is postulated to be one for which the work done by an external agency during the application of additional stresses is positive and for which the net work performed by the external agency during a complete cycle of loading and unloading is nonnegative.

Assume the initial stress field σ_{ij}^0 to lie either on or inside the yield surface. Allow the external agency to add stresses to this state by following a path entirely within the yield surface until a state of stress σ_{ij} lying on this yield surface is reached (see Figure 2.8). Up to this point only elastic deformations have occurred. Now, further suppose the external agency increments these latter stresses by $d\sigma_{ij}$ and produces an increment of plastic strain $d\varepsilon_{ij}^p$. Next, permit the external agency to unload along an elastic path to the state of stress σ_{ij}^0. Since the elastic work is recoverable, the net work done by the external agency according to Drucker's postulate is

$$(\sigma_{ij} - \sigma_{ij}^0)\, d\varepsilon_{ij}^p + d\sigma_{ij}\, d\varepsilon_{ij}^p \geqslant 0 \tag{2.6-14}$$

Since σ_{ij}^0 can include σ_{ij}, then

$$d\sigma_{ij}\, d\varepsilon_{ij}^p \geqslant 0 \tag{2.6-15}$$

On the other hand the difference $\sigma_{ij} - \sigma_{ij}^0$ can be made arbitrarily large compared to $d\sigma_{ij}$ so that Equation (2.6-14) also yields

$$(\sigma_{ij} - \sigma_{ij}^0)\, d\varepsilon_{ij}^p \geqslant 0 \tag{2.6-16}$$

To understand the significance of Equation (2.6-16), let P denote an arbitrary point on the yield surface and consider a hyperplane at this point normal to the incremental strain vector having components $d\varepsilon_{ij}^p$ (see Figure 2.8). The inner product of Equation (2.6-16) implies that the stress vector, $\sigma_{ij} - \sigma_{ij}^0$, cannot make an obtuse angle with the strain vector.

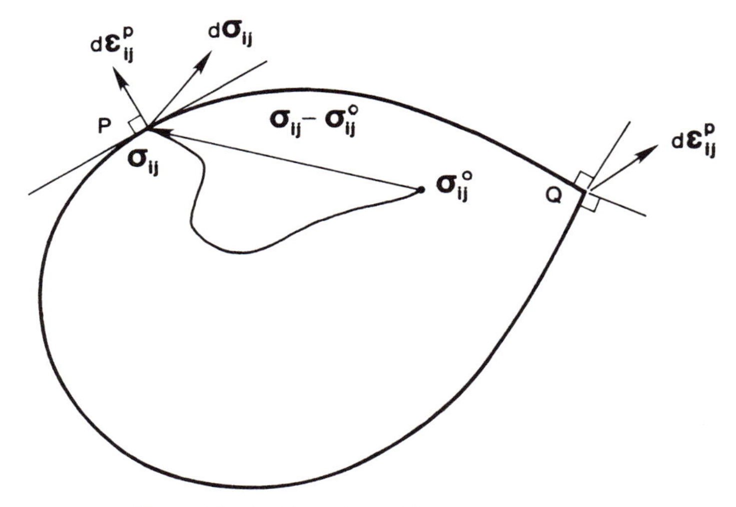

Figure 2.8 A yield surface with a corner at point Q.

Consequently, this stress vector must lie to one side of the hyperplane. Since P can be any point on the yield surface, then the surface must be convex.

During neutral or plastic loading the stress vector $d\sigma_{ij}$ is directed outward from the yield surface and is limited by the tangent plane to the yield surface. If P is a smooth or regular point on the yield surface, then Equation (2.6-15) implies that the incremental plastic strain vector $d\varepsilon_{ij}^p$ must be normal to the yield surface and has a well defined direction. However, at a corner Q where the normal to the yield surface is undefined the vector $d\varepsilon_{ij}^p$ can assume any direction bounded by the limiting normals to the yield surface at the corner.

The normality of the plastic strain increment to the yield surface implies that at a smooth point

$$d\varepsilon_{ij}^p = d\lambda \frac{\partial f}{\partial \sigma_{ij}} \tag{2.6-17}$$

where $d\lambda > 0$ during plastic loading. A comparison of the incremental form of Equation (2.6-17) and Equation (2.3-14) suggests that f can also be viewed as a plastic potential.

Consider isotropic hardening wherein the yield surface does not change shape or translate in the stress space, but only expands uniformly with plastic deformation. Suppose there exists a universal stress-strain curve, $\bar{\sigma} = h(\int d\bar{\varepsilon}^p)$, expressing an effective stress $\bar{\sigma}$ as a function h of an effective plastic strain increment $d\bar{\varepsilon}^p$ integrated over the strain history. With the effective stress defined by

$$\bar{\sigma} = (3J_2)^{\frac{1}{2}} = (\tfrac{3}{2} s_{ij} s_{ij})^{\frac{1}{2}} \tag{2.6-18}$$

the size of the von Mises yield surface is measured by $\bar{\sigma}$. The coefficient in Equation (2.6-18) is chosen such that $\bar{\sigma} = |\sigma_{11}|$ for a uniaxial state of stress. The corresponding effective plastic strain increment is defined by

$$d\bar{\varepsilon}^p = (\tfrac{2}{3}\, d\varepsilon_{ij}^p\, d\varepsilon_{ij}^p)^{\frac{1}{2}} \tag{2.6-19}$$

where the numerical factor is chosen such that for a uniaxial state of stress

$$d\bar{\varepsilon}^p = d\varepsilon_{11}^p = -2d\varepsilon_{22}^p = -2d\varepsilon_{33}^p$$

The introduction of von Mises yield criterion, Equation (2.6-9), into Equation (2.6-17) yields

$$d\varepsilon_{ij}^p = d\lambda s_{ij} \tag{2.6-20}$$

whence it follows that

$$d\varepsilon_{ij}^p\, d\varepsilon_{ij}^p = (d\lambda)^2 s_{ij} s_{ij}$$

or, equivalently,

$$\tfrac{3}{2}(d\bar{\varepsilon}^p)^2 = \tfrac{2}{3}(d\lambda)^2 \bar{\sigma}^2$$

Consequently,

$$d\lambda = \frac{2}{3}\frac{d\bar{\varepsilon}^p}{\bar{\sigma}} = \frac{2}{3}\frac{d\bar{\sigma}}{\bar{\sigma} h'}, \qquad h' \neq 0 \tag{2.6-21}$$

where h' is the slope of the universal stress-strain curve at the current effective stress $\bar{\sigma}$. After substituting Equation (2.6-21) into Equation (2.6-20) we have

$$d\varepsilon_{ij}^{p} = \frac{3}{2}\frac{s_{ij}}{\bar{\sigma}}\, d\bar{\varepsilon}^{p} \tag{2.6-22}$$

or

$$d\varepsilon_{ij}^{p} = \frac{3}{2}\frac{s_{ij}}{\bar{\sigma}h'}\, d\bar{\sigma} \qquad h' \neq 0 \tag{2.6-23}$$

When there is no work hardening ($h' = 0$), Equation (2.6-23) is no longer valid and Equation (2.6-22) must be used. However, the latter equation only permits the determination of the strain increments within an arbitrary multiplicative constant. The plastic deformation is governed by the amount of plastic work

$$dW_p = \bar{\sigma}\, d\bar{\varepsilon}^{p} \tag{2.6-24}$$

done on a unit volume of material. Therefore, Equation (2.6-22) can be rewritten as

$$d\varepsilon_{ij}^{p} = \frac{3}{2}\frac{dW_p}{\bar{\sigma}^2}\, s_{ij} = \frac{1}{2}\frac{dW_p}{J_2}\, s_{ij} \tag{2.5-25}$$

which is equally valid for hardening and nonhardening materials.

Finally, the incremental constitutive relations based upon von Mises yield criterion or J_2 and incompressible plastic flow are

$$d\varepsilon_{ij} = \frac{1+\nu}{E}\, d\sigma_{ij} - \frac{\nu}{E}\,\delta_{ij}\, d\sigma_{kk} + \frac{3}{2}\,\alpha\,\frac{s_{ij}}{\bar{\sigma}h'}\, d\bar{\sigma} \tag{2.6-26}$$

where $\alpha = 1$ if $f = 0$ and $d\bar{\sigma} > 0$ and $\alpha = 0$ if either $f < 0$ or $d\bar{\sigma} < 0$. These relations are frequently referred to as the Prandtl-Reuss equations.

2.6.3 Deformation Plasticity

In addition to the incremental theory of plasticity, there is the deformation or total strain theory due to Hencky. Whereas the incremental theory is quite general, the deformation theory, which is actually a nonlinear elasticity theory, has very definite limitations. Nevertheless, it offers certain mathematical simplifications that makes its use attractive.

In the deformation theory it is assumed that

$$\varepsilon_{ij}^{p} = \phi s_{ij} \tag{2.6-27}$$

where ϕ is a scalar function of the invariants of stress and plastic strains. It follows from Equation (2.6-27) that

$$\varepsilon_{ij}^{p}\varepsilon_{ij}^{p} = \phi^2 s_{ij}s_{ij} \tag{2.6-28}$$

$$\phi = \frac{3}{2}\frac{\bar{\varepsilon}^{p}}{\bar{\sigma}} \tag{2.6-29}$$

where the effective plastic strain $\bar{\varepsilon}^p$ is defined by

$$\bar{\varepsilon}^p = (\tfrac{2}{3}\, \varepsilon_{ij}^p \varepsilon_{ij}^p)^{\frac{1}{2}} \tag{2.6-30}$$

Combining Equations (2.6-27) and (2.6-30) one obtains

$$\varepsilon_{ij}^p = \frac{3}{2}\frac{\bar{\varepsilon}^p}{\bar{\sigma}} s_{ij} = \frac{3}{2}\left(\frac{E - E_s}{EE_s}\right) s_{ij} \tag{2.6-31}$$

where E_s is the secant modulus in uniaxial tension. Once again, the existence of a universal stress-strain curve $\bar{\sigma} = h(\bar{\varepsilon}^p)$ is assumed. When these plastic strains are added to the elastic strains, then

$$\varepsilon_{ij} = \frac{1 + \nu}{E} \sigma_{ij} - \frac{\nu}{E} \delta_{ij}\sigma_{kk} + \frac{3}{2}\frac{\bar{\varepsilon}^p}{\bar{\sigma}} s_{ij} \tag{2.6-32}$$

for an isotropic material.

The principal limitation of the deformation theory of plasticity is reflected by Equation (2.6-27). That is, the plastic strains depend only on the current state of stress and are independent of the path leading to this state. This is contrary to the usually observed behavior of plastic deformation. However, for proportional or radial loading (loading along a radial line in the stress space), one may write [see Hill (2.7)]

$$s_{ij} = cS_{ij} \tag{2.6-33}$$

where S_{ij} is constant nonzero reference state of stress and c is a monotonically increasing function, say, of λ. The introduction of Equation (2.6-33) into Equation (2.6-20) gives

$$d\varepsilon_{ij}^p = cS_{ij}\, d\lambda$$

which upon integration yields

$$\varepsilon_{ij}^p = S_{ij} \int c d\lambda = \left[\frac{1}{c} \int c d\lambda\right] s_{ij}$$

If ϕ is identified with $(\int c\, d\lambda)/c$, then the latter equation assumes the form of Hencky's Equation (2.6-27). Consequently, for proportional loading, the incremental or flow theory and the deformation or total strain theory of plasticity coincide.

Proportional loading is a sufficient condition for the incremental and deformations theories to agree. This is not to say that for some other nonproportional loadings that the two theories cannot agree approximately. Examples for which this can occur are discussed in reference (2.6). In many cases the deformation theory has been used because of the relative mathematical simplicity it embraces.

2.6.4 Rigid Plastic Materials

In regions of rather extensive plastic deformation the elastic strains may be much smaller than the plastic strains and can be neglected. With the neglect of the elastic strains the material is considered incompressible. If work hardening is insignificant, then the material can be modeled as being rigid-perfectly plastic. The slip-line theory then offers an effective method of solution for plane strain problems satisfying these conditions. A brief discussion of the theory is presented here. The reader interested in more detail is referred to Hill (2.7), Kachanov (2.6), or Mendelson (2.8).

Under plane strain loading $u_1 = u_1(x_1, x_2)$, $u_2 = u_2(x_1, x_2)$, and $u_3 \equiv 0$, which implies that $\varepsilon_{i3} = 0$. Under this condition Equation (2.6-20) provides that $s_{i3} = 0$ and whence $\sigma_{13} = \sigma_{23} = 0$. Moreover,

$$\sigma_{33} = \tfrac{1}{2}(\sigma_{11} + \sigma_{22}) = p \tag{2.6-34}$$

is a principal stress. The other two principal stresses are

$$\left.\begin{matrix}\sigma_1\\ \sigma_2\end{matrix}\right\} = \frac{\sigma_{11} + \sigma_{22}}{2} \pm \left[\left(\frac{\sigma_{11} - \sigma_{22}}{2}\right)^2 + \sigma_{12}^2\right]^{\frac{1}{2}} \tag{2.6-35}$$

which upon the introduction of the von Mises yield criterion

$$\left(\frac{\sigma_{11} - \sigma_{22}}{2}\right)^2 + \sigma_{12}^2 = k^2 \tag{2.6-36}$$

can be written as

$$\left.\begin{matrix}\sigma_1\\ \sigma_2\end{matrix}\right\} = p \pm k \tag{2.6-37}$$

where k can be identified with the maximum shear stress. Hence, the state of stress at a point can be specified in terms of the hydrostatic pressure p and the yield stress k in shear. There remains to specify the principal directions or, equivalently, the direction of maximum shear stress.

It is convenient to introduce a set of orthogonal slip lines α and β that are tangent at every point to the directions of maximum shear stress. Since the principal directions of stress and strain increments coincide, the rate of extension along a slip line is zero for an incompressible material. Let ϕ (see Figure 2.9) be the angle measured positive counterclockwise from the x_1-axis to the local tangent of an α-line. It follows for the stress state shown that

$$\begin{aligned}\sigma_{11} &= p - k\sin 2\phi\\ \sigma_{22} &= p + k\sin 2\phi\\ \sigma_{12} &= k\cos 2\phi\end{aligned} \tag{2.6-38}$$

The substitution of Equation (2.6-38) into the equilibrium equations [Equation (2.1-10)] yields

$$\begin{aligned}\frac{\partial p}{\partial x_1} - 2k\left(\cos 2\phi\,\frac{\partial \phi}{\partial x_1} + \sin 2\phi\,\frac{\partial \phi}{\partial x_2}\right) &= 0\\ \frac{\partial p}{\partial x_2} - 2k\left(\sin 2\phi\,\frac{\partial \phi}{\partial x_1} - \cos 2\phi\,\frac{\partial \phi}{\partial x_2}\right) &= 0\end{aligned} \tag{2.6-39}$$

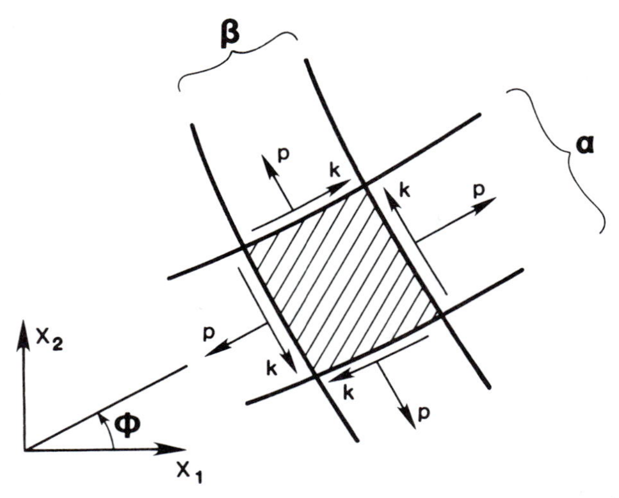

Figure 2.9 Stresses on a small curvilinear element bounded by pairs of α- and β-slip lines.

in the absence of body forces. If a local Cartesian coordinate system (s_1, s_2, s_3) is introduced with s_1 and s_2 tangent to the α- and β-lines, respectively, then Equation (2.6-39) is still applicable if x_i is replaced by s_i and $\phi = 0$. Hence, Equation (2.6-39) assumes the simple form

$$\frac{\partial}{\partial s_1}(p - 2k\phi) = 0$$
$$\frac{\partial}{\partial s_2}(p + 2k\phi) = 0 \tag{2.6-40}$$

Integrating Equation (2.6-40) we obtain the Hencky equations

$$p - 2k\phi = \text{constant on an } \alpha\text{-line}$$
$$p + 2k\phi = \text{constant on a } \beta\text{-line} \tag{2.6-41}$$

If the stresses are prescribed on the entire boundary $S = S_T$, then Equation (2.6-41) is sufficient to determine p and ϕ, and hence the state of stress, throughout the body. On the other hand, if velocities are prescribed on a part of the boundary S_u, then Equation (2.6-41) is insufficient to obtain a solution and it must be supplemented with kinematic relations.

Let u and v be velocity components along the α- and β-lines, respectively. Within second-order terms the increment of velocity (see Figure 2.10) along an α-line is $(u + du - v\,d\phi) - u$. A similar relation can also be written for the β-line. Since the rates of extension along the slip lines vanish, then

$$du - v\,d\phi = 0 \text{ along an } \alpha\text{-line}$$
$$dv + u\,d\phi = 0 \text{ along a } \beta\text{-line} \tag{2.6-42}$$

The simultaneous solution of Equations (2.6-41) and (2.6-42) is generally a

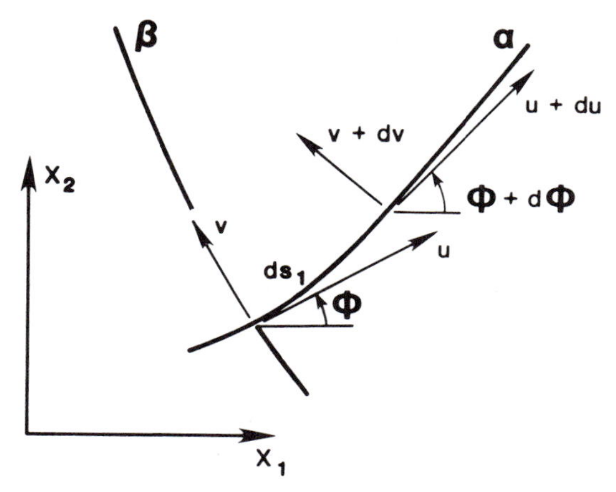

Figure 2.10 Velocity components along α- and β-slip lines.

formidable task and is usually done by trial and error. A slip line field satisfying the boundary conditions on S_T is constructed. The boundary conditions on S_u are checked and the slip line field is modified accordingly. The procedure is repeated until all the boundary conditions are satisfied.

We list below some of the properties of slip lines and refer the reader to Kachanov (2.6) for their proofs.

1. Along a slip line the hydrostatic pressure p is proportional to the angle ϕ between the line and the x_1-axis.
2. The change in the angle ϕ and hydrostatic pressure p is constant in passing from one slip line to another of the same family along any intersecting slip line.
3. If the value of p is known at one point in a given slip line field, it can be determined throughout the field.
4. If a segment of a slip line is straight, then the state of stress is constant along this segment.
5. If both the α- and β-lines are straight in a region, then a uniform state of stress exists in the region.
6. If a segment of an $\alpha(\beta)$-line cut by two $\beta(\alpha)$-lines is straight, then any other $\alpha(\beta)$-line between these two $\beta(\alpha)$-lines is straight.
7. Straight segments of slip lines between two slip lines of the other family have the same lengths.
8. The radii of curvature of $\alpha(\beta)$-lines where they intersect a $\beta(\alpha)$-line decreases in proportion to the distance traveled in the positive direction along a $\beta(\alpha)$-line.
9. The center of curvature of the $\beta(\alpha)$-lines at points of intersection with $\alpha(\beta)$-lines form an involute of the $\beta(\alpha)$-line.
10. If the radius of curvature of an $\alpha(\beta)$-line is discontinuous as it crosses a $\beta(\alpha)$-line, then all $\alpha(\beta)$ lines crossing the $\beta(\alpha)$-line are discontinuous as are the stress gradients.

These properties will be exploited in Chapter 5 to construct and interpret slip fields near the crack tip in an elastic-perfectly plastic material.

2.7 Elastic-Viscoplasticity

As the name implies, elastic-viscoplasticity characterizes the response of materials that are not only elastic but also exhibit strain rate dependent permanent deformation. The primary usefulness of such a theory is for the representation of the constitutive behavior of materials that are strain rate sensitive in the inelastic range. Figure 2.11 shows, for titanium, the behavior of interest.

In contrast to the well-established character of elastic, viscoelastic, and elastoplastic behavior, viscoplasticity is still in a formative stage. A number of competing constitutive models currently exist. From a structural mechanics point of view at least, there are three requirements of any such model: (1) a physical basis, (2) material property values that can be obtained in a straightforward manner, and (3) ease of application. Which model will prove to be able to satisfy these requirements best is uncertain now. Nevertheless, the model developed by Bodner and his associates (2.22, 2.23) clearly has a number of advantages for structural mechanics applications in general and for fracture

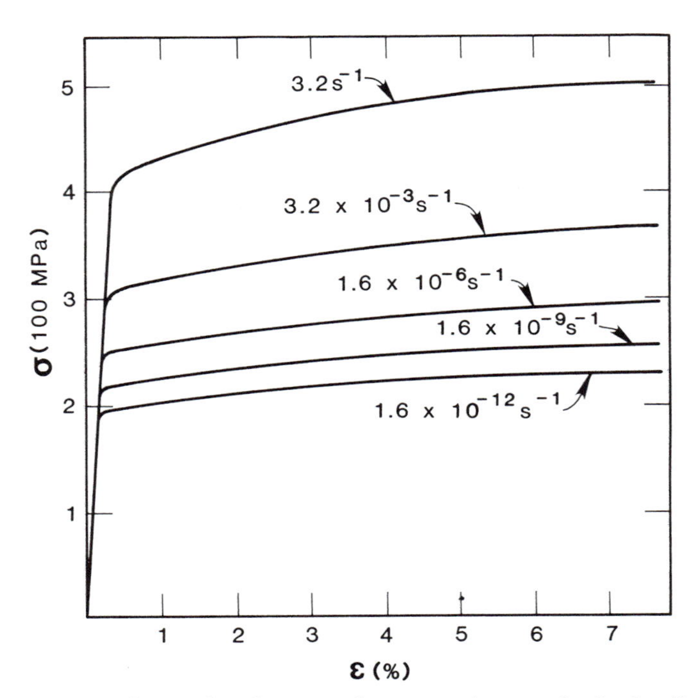

Figure 2.11 Influence of strain-rate on the stress-strain curves for titanium (2.21).

mechanics in particular. Accordingly, and also because a full discussion of this subject is beyond the scope of this book, we will confine our discussion to Bodner's approach.

The constitutive relations for an elastic-viscoplastic work hardening material proposed by Bodner and Partom (2.22) offer two important advantages for fracture mechanics problems. Neither a separate specific yield criterion is required nor is it necessary to consider loading and unloading separately. A disadvantage is that both elastic and inelastic deformations are present at all stages of loading and unloading. However, the plastic contribution is very small when the material behavior should be essentially elastic. The model does allow anisotropic (directional) hardening—the Bauschinger effect—to be included in the material characterization.

Consistent with the usual assumption of elastoplastic material behavior, the Bodner-Partom treatment assumes that the elastic and plastic strain rate components are additive at all stages of deformation [cf. Equation (2.6-1)]. Thus, the total rate of strain can be written as

$$\dot{\varepsilon}_{ij} = \dot{\varepsilon}_{ij}^{e} + \dot{\varepsilon}_{ij}^{p} \tag{2.7-1}$$

The elastic strain rates are related to the stress-rates by Hooke's law. Thus, from Equation (2.3-18)

$$\dot{\varepsilon}_{ij}^{e} = \frac{1}{2\mu}\left(\dot{\sigma}_{ij} - \frac{\nu}{1+\nu}\,\delta_{ij}\dot{\sigma}_{kk}\right) \tag{2.7-2}$$

It is assumed that the plastic deformation is incompressible, whereupon Equation (2.6-2) is applicable, and that the Prandtl-Ruess flow rule, Equation (2.6-20), holds. Introducing the second invariant of the plastic deformation rate deviator

$$D_2^p = \tfrac{1}{2}\dot{e}_{ij}^{p}\dot{e}_{ij}^{p} \tag{2.7-3}$$

then, using a dislocation dynamics argument, it is assumed that D_2^p is a function of J_2. That is, since D_2^p is a measure of the effective inelastic shear deformation rate and J_2 is the effective shear stress, the relation

$$D_2^p = f(J_2) \tag{2.7-4}$$

can be considered to be a multidimensional generalization of the uniaxial result.

The particular form selected by Bodner and Partom is one that they believe has both a physical basis and allows for flexibility in modeling actual material response. This is

$$D_2^p = D_0^2 \exp[-(A^2/J_2)^n] \tag{2.7-5}$$

where

$$A^2 = \tfrac{1}{3}Z^2\left(\frac{n+1}{n}\right)^{1/n} \tag{2.7-6}$$

and D_0^2, n, and Z are material constants; each of which has a physical interpretation.* Specifically, D_0^2 is the limiting value of D_2^p for very high stresses, n is related to the steepness of the D_2^p versus J_2 curve and to the rate-sensitivity, while Z is a history-dependent internal state variable that represents the overall resistance to plastic flow. For isotropic hardening, Z is a scalar that corresponds in a general way to the level of the flow stress. An empirically based form must be postulated with $\dot{Z}$ being a function of some measure of hardening—for example, the rate of plastic work.

Bodner and Aboudi (2.23) offer a rationale for arriving at the relation selected more arbitrarily by Bodner and Partom (2.22). Their argument is based on the plastic work W_p being the controlling factor in hardening. Then

$$\dot{Z} = \left(\frac{dZ}{dW_p}\right)\left(\frac{dW_p}{dt}\right) \tag{2.7-7}$$

where W_p can be obtained from Equation (2.6-24). The form that has been used in the Bodner-Partom model is

$$\frac{dZ}{dW_p} = \frac{m}{Z_0}(Z_1 - Z) \tag{2.7-8}$$

On the basis of Equation (2.7-8), Equation (2.7-7) can be integrated to give

$$Z = Z_1 - (Z_1 - Z_0)\exp(-mW_p/Z_0) \tag{2.7-9}$$

where Z_0, Z_1, and m are temperature-dependent material constants. Specifically, m indicates the rate of work hardening, Z_0 represents the initial hardness while Z_1 is the upper limit of Z. Note that the hardness must have an upper limit to preclude D_2^p from approaching zero for large W_p. This would imply fully elastic behavior at large strains—behavior that would obviously not be realistic.

The relations given above are applicable to multi-dimensional deformation conditions, subject to the limitations of small strains and isotropic behavior. But, to demonstrate more clearly the material response that they predict, they can be specialized to uniaxial conditions. Then, taking σ_{11} as the only nonzero stress component, use of Equations (2.6-18) and (2.7-3) give $J_2 = \sigma_{11}^2/3$ and $D_2^p = 3(\dot{\varepsilon}_{11}^p)^2/4$. Substituting these values into Equation (2.7-5) then leads to

$$\dot{\varepsilon}_{11}^p = \frac{2D_0}{\sqrt{3}}\exp\left[-\frac{1}{2}\left(\frac{\sqrt{3}A}{\sigma_{11}}\right)^{2n}\right] \tag{2.7-10}$$

It is convenient to define

$$x = \frac{\sqrt{3}}{2}\frac{\dot{\varepsilon}_{11}^p}{D_0} \tag{2.7-11}$$

* In recent papers, Bodner and his associates have redefined A as simply $Z/\sqrt{3}$ and m as n/Z_0. These changes do not affect the concepts that are involved in this approach.

and

$$y = \frac{\sigma_{11}}{\sqrt{3}A} \tag{2.7-12}$$

so that Equation (2.7-10) can be expressed in the simpler form

$$y = \left[2 \ln \left(\frac{1}{x} \right) \right]^{-1/2n} \tag{2.7-13}$$

This result is shown in Figure 2.12 for selected values of n. Note that Bodner suggests that a temperature dependence can be incorporated in this formulation via

$$n = B + C/T \tag{2.7-14}$$

where B and C are material constants. Hence, Figure 2.12 also illustrates the behavior predicted by the Bodner-Partom model as a function of both strain rate and temperature.

Values of the parameters for the Bodner-Partom model have been determined for a number of materials including titanium, copper, and the

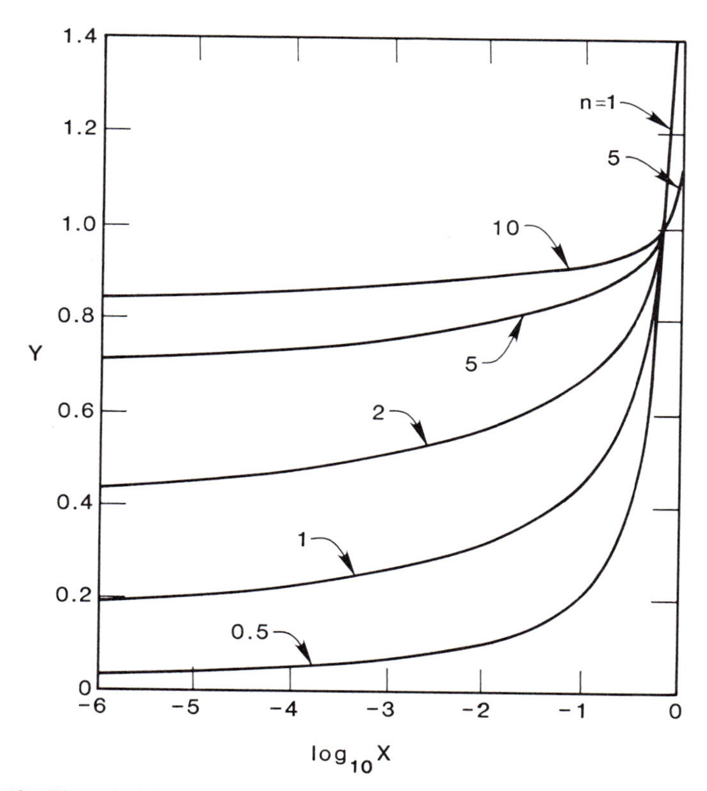

Figure 2.12 The uniaxial stress and plastic strain rate relation for the Bodner-Partom model.

engine material Rene' 95. Parameters for a particular material can be obtained by fitting test results at two constant rates of extension. For example, values for commercially pure titanium at room temperature are $E = 1.18 \times 10^5$ MPa, $\mu = 0.44 \times 10^5$ MPa, $\rho = 4.87$ gm/cm^3, $Z_0 = 1.15 \times 10^3$ MPa, $Z_1 = 1.40 \times 10^3$ MPa, $D_0 = 10^4$ sec^{-1}, $n = 1$, and $m = 100$. These values were used to calculate the family of curves shown in Figure 2.11 (2.23).

Note that the model can be based upon material parameter values determined at moderate strain rates, but it will still reflect the effects anticipated at very high strain rates. This can be seen from Figure 2.12. This is important to the subject of dynamic fracture mechanics where the strain rates in the near vicinity of a crack tip are too high to be accessible to commonly used test procedures. As the results of Abuodi and Achenbach (2.24) for rapid crack propagation and of Merzer (2.25) for adiabatic shear demonstrate, predictions based upon the Bodner-Partom model can offer realistic results. Further discussions of this topic are given in Chapter 4.

2.8 References

(2.1) Sokolnikoff, I. S., *Mathematical Theory of Elasticity*, 2nd ed., McGraw-Hill, New York (1956).

(2.2) Washizu, K., *Variational Methods in Elasticity and Plasticity*, 2nd ed., Pergamon, Oxford (1975).

(2.3) Dym, C. L. and Shames, I. H., *Solid Mechanics: A Variational Approach*, McGraw-Hill, New York (1973).

(2.4) Christensen, R. M., *Theory of Viscoelasticity*, Academic, New York (1971).

(2.5) Findley, W. N., Lai, J. S., and Onaran, K., *Creep and Relaxation of Nonlinear Viscoelastic Materials*, North-Holland, Amsterdam (1976).

(2.6) Kachanov, L. M., *Foundations of the Theory of Plasticity*, North-Holland, Amsterdam (1971).

(2.7) Hill, R., *The Mathematical Theory of Plasticity*, Oxford University Press, Oxford (1956).

(2.8) Mendelson, A., *Plasticity: Theory and Application*, Macmillan, New York (1968).

(2.9) Fung, Y. C., *Foundations of Solid Mechanics*, Prentice-Hall, Englewood Cliffs, New Jersey (1965).

(2.10) Malvern, L. E., *Introduction to the Mechanics of a Continuous Medium*, Prentice-Hall, Englewood Cliffs, New Jersey (1969).

(2.11) Green, G., "On the Propagation of Light in Crystallized Media," *Transactions of the Cambridge Philosophical Society*, **7**, pp. 121–140 (1839).

(2.12) Courant, R. and Fritz, J., *Introduction to Calculus and Analysis*, Vol. 2, Wiley, New York (1974).

(2.13) Treloar, L. R. G., "The Elasticity and Related Properties of Rubbers," *Reports on Progress in Physics*, **36**, pp. 755–826 (1973).

(2.14) Gent, A. N., "Rubber Elasticity: Basic Concepts and Behavior," *Science and Technology of Rubber*, F. R. Eirich (ed.), Academic, New York, pp. 1–21 (1978).

(2.15) Churchill, R. V., *Operational Mathematics*, 3rd ed., McGraw-Hill, New York (1972).

(2.16) Williams, M. L., Landel, R. F., and Ferry, J. D., "The Temperature Dependence of Relaxation Mechanism in Amorphous Polymers and Other Glass-Forming Liquids," *Journal of the American Chemical Society*, **77**, pp. 3701–3707 (1955).

(2.17) Tobolsky, A. V., *Properties and Structure of Polymers*, Wiley, New York (1960).

(2.18) Ferry, J. D., *Viscoelastic Properties of Polymers*, 3rd ed., Wiley, New York (1980).

(2.19) Graham, G. A. C., "The Correspondence Principle of Linear Viscoelastic Theory for Mixed Boundary Value Problems Involving Time-Dependent Boundary Regions," *Quarterly of Applied Mathematics*, **26**, pp. 167–174 (1968).

(2.20) Graham, G. A. C., "The Solution of Mixed Boundary Value Problems that Involve Time-

Dependent Boundary Regions for Viscoelastic Materials with One Relaxation Function," *Acta Mechanics*, **8**, pp. 188–204 (1969).

(2.21) Drucker, D. C., "A More Fundamental Approach to Plastic Stress-Strain Relations," Proceedings of the First U.S. National Congress of Applied Mechanics, Edward Brothers, Ann Arbor, Mich., pp. 487–491 (1952).

(2.22) Bodner, S. R. and Partom, Y., "Constitutive Equations for Elastic-Viscoplastic Strain Hardening Materials," *Journal of Applied Mechanics*, **42**, pp. 385–389 (1975).

(2.23) Bodner, S. R. and Aboudi, J., "Stress Wave Propagation in Rods of Elastic-Viscoplastic Material," *International Journal of Solids and Structures*", **19**, pp. 305–314 (1983).

(2.24) Aboudi, J. and Achenbach, J. D., "Numerical Analysis of Fast Mode-I Fracture of a Strip of Viscoplastic Work-Hardening Material," *International Journal of Fracture*, **21**, pp. 133–147 (1983).

(2.25) Merzer, A. M., "Modelling of Adiabatic Shear Band Development from Small Imperfections," *Journal of the Mechanics and Physics of Solids*, **30**, pp. 323–338 (1982).

3

LINEAR ELASTIC FRACTURE MECHANICS

In this chapter fracture mechanics within the confines of the linear theory of elasticity—that is, linear elastic fracture mechanics (LEFM)—is treated. It is not the intent of this chapter to present a complete exposé on LEFM. Not only would it be impossible to do so in a single chapter, but it is unnecessary due to the numerous texts (3.1–3.7) devoted almost entirely to this subject. Rather, the objective is to set forth the theory, principles, and concepts of LEFM in sufficient detail so that this and the preceding chapter can serve as a springboard for addressing the advanced topics in the subsequent chapters. In so doing, it is hoped that the limitations inherent in LEFM that must be relaxed in order to treat the advanced subjects will become more readily apparent. An equally important goal is to identify those principles and concepts of LEFM that can be extended or generalized to nonlinear fracture mechanics.

This chapter commences with an analysis of the linear elastic crack tip stress and displacement fields whose strengths are measured by the stress intensity factor K. The concept of small-scale yielding and the existence of a "K-dominant" region that form the modern view of LEFM and that lead naturally to a fracture criterion are presented. Because of its importance, methods for determining the stress intensity factor for a cracked body and its loading are discussed. Next, the energetics of cracked bodies are examined. This gives rise to the energy release rate parameter G and the energy balance criterion for fracture. The path-independent J-integral for nonlinear elastic materials and other invariant integrals are treated. The equivalence between various LEFM fracture characterizing parameters—the stress intensity factor, the energy release rate, and the crack-tip opening displacement—is established. The plastic zone attending the crack tip and its relationship to the fracture toughness is examined. This chapter concludes with a treatment of stable crack growth in a material whose fracture resistance increases with crack extension.

3.1 Linear Elastic Crack-Tip Fields

Aside from ideally brittle materials, any loading of a cracked body is accompanied by inelastic deformation in the neighborhood of the crack tip due to stress concentrations there. Consequently, the ultimate utility of an elastic analysis of a real cracked body must necessarily depend upon the extent of the region of inelastic deformation being small compared to the size of the crack and any other characteristic length. Inelastic deformation satisfying this

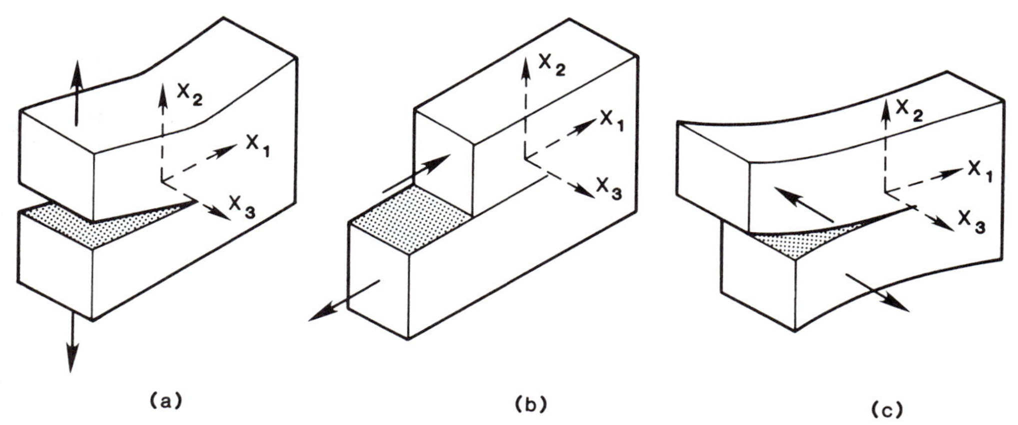

Figure 3.1 Three basic loading modes for a cracked body: (a) Mode I, opening mode; (b) Mode II, sliding mode; (c) Mode III, tearing mode.

condition is referred to as small-scale yielding and is embodied in the concept of linear elastic fracture mechanics (LEFM).

Other essential concepts of LEFM are most readily assimilated and demonstrated for plane elasticity problems. Let the crack plane lie in the x_1x_3-plane and take the crack front to be parallel to the x_3-axis. For plane problems the stress and displacement fields are functions of x_1 and x_2 only. The deformations due to the three primary modes of loading are illustrated in Figure 3.1. Mode I is the opening or tensile mode where the crack faces separate symmetrically with respect to the x_1x_2- and x_1x_3-planes. In Mode II, the sliding or in-plane shearing mode, the crack faces slide relative to each other symmetrically about the x_1x_2-plane, but antisymmetrically with respect to the x_1x_3-plane. In the tearing or antiplane mode, Mode III, the crack faces also slide relative to each other but antisymmetrically with respect to the x_1x_2- and x_1x_3-planes. In the vernacular of dislocation theory these three modes correspond, respectively, to wedge, edge, and screw dislocations.

An investigation of crack-tip stress and displacement fields is important because these fields are typically the ones that govern the fracture process occurring at the crack tip. In the following sections, the crack-tip fields are developed for the three primary modes of loading in a homogeneous, isotropic, linear elastic material.

3.1.1 The Antiplane Problem

Because of its relative simplicity, the antiplane, Mode III problem wherein $u_1 = u_2 \equiv 0$ and $u_3 = u_3(x_1, x_2)$ is considered first. Equation (2.2-8) yields the following nonzero strain components

$$\varepsilon_{3\alpha} = \tfrac{1}{2}u_{3,\alpha} \tag{3.1-1}$$

where Greek subscripts are understood to have the range 1, 2. Therefore, according to Equation (2.3-9) the nontrivial stress components are

$$\sigma_{3\alpha} = 2\mu\varepsilon_{3\alpha} \tag{3.1-2}$$

Finally, the only relevant equation of equilibrium in the absence of body forces is

$$\sigma_{3\alpha,\alpha} = 0 \tag{3.1-3}$$

Equations (3.1-1)–(3.1-3) can be combined to yield Laplace's equation

$$u_{3,\alpha\alpha} = \nabla^2 u_3 = 0 \tag{3.1-4}$$

where $\nabla^2 = \partial^2/\partial x_1^2 + \partial^2/\partial x_2^2$ is the two-dimensional Laplacian operator.

The complex variable method provides a powerful technique for establishing the solution of Equation (3.1-4) and other plane elasticity problems. Let the complex variable z be defined by $z = x_1 + ix_2$ or, equivalently, in polar coordinates $z = re^{i\theta}$, where $i = \sqrt{-1}$. The overbar is used to denote the complex conjugate; for example, $\bar{z} = x_1 - ix_2 = re^{-i\theta}$. It follows that

$$\begin{aligned} x_1 &= \mathrm{Re}(z) = (z + \bar{z})/2 \\ x_2 &= \mathrm{Im}(z) = (z - \bar{z})/2i \end{aligned} \tag{3.1-5}$$

where Re and Im denote the real and imaginary parts, respectively. By chain rule differentiation one can write

$$2\frac{\partial}{\partial z} = \frac{\partial}{\partial x_1} - i\frac{\partial}{\partial x_2}, \qquad 2\frac{\partial}{\partial \bar{z}} = \frac{\partial}{\partial x_1} + i\frac{\partial}{\partial x_2} \tag{3.1-6}$$

and, therefore,

$$4\frac{\partial^2}{\partial z\,\partial \bar{z}} = \frac{\partial^2}{\partial x_1^2} + \frac{\partial^2}{\partial x_2^2} = \nabla^2 \tag{3.1-7}$$

Let $f(z)$ be a holomorphic function* of the complex variable z, which can be written as

$$f(z) = u(x_1, x_2) + iv(x_1, x_2) \tag{3.1-8}$$

where u and v are real functions of x_1 and x_2. It is permissible to write

$$\begin{aligned} \frac{\partial f}{\partial x_1} &= \frac{\partial f}{\partial z}\frac{\partial z}{\partial x_1} = f'(z) \\ \frac{\partial f}{\partial x_2} &= \frac{\partial f}{\partial z}\frac{\partial z}{\partial x_2} = if'(z) \end{aligned} \tag{3.1-9}$$

where the prime is used to denote a differentiation with respect to the argument of the function. It follows that

$$f'(z) = \frac{\partial f}{\partial x_1} = -i\frac{\partial f}{\partial x_2}$$

* A complex function is said to be holomorphic or analytic and regular in a region if it is single valued and if its complex derivative df/dz exists in the region.

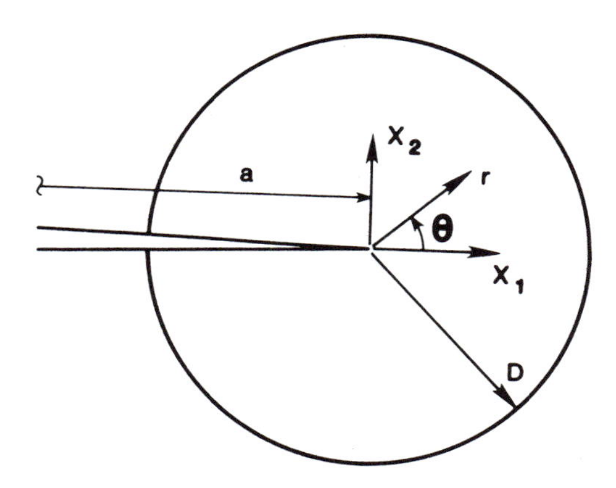

Figure 3.2 Crack-tip region and coordinate system.

whence upon the substitution of Equation (3.1-8) yields

$$\frac{\partial u}{\partial x_1} + i\frac{\partial v}{\partial x_1} = \frac{\partial v}{\partial x_2} - i\frac{\partial u}{\partial x_2}$$

Equating real and imaginary parts we obtain the Cauchy-Riemann equations

$$\frac{\partial u}{\partial x_1} = \frac{\partial v}{\partial x_2}, \qquad \frac{\partial v}{\partial x_1} = -\frac{\partial u}{\partial x_2}$$

which may be combined to yield

$$\nabla^2 u = \nabla^2 v = 0$$

Thus, the real and imaginary parts of any holomorphic function are solutions to Laplace's equation.

Therefore, the solution of Equation (3.1-4) can be written as

$$u_3 = \frac{1}{\mu}[f(z) + \bar{f}(\bar{z})] \tag{3.1-10}$$

where $\bar{f}(\bar{z}) = u(x_1, x_2) - iv(x_1, x_2)$ is the complex conjugate of $f(z)$.* Introducing Equation (3.1-10) into Equation (3.1-1) and employing Equation (3.1-9) we find that

$$\begin{aligned} \varepsilon_{31} &= \frac{1}{2\mu}[f'(z) + \bar{f}'(\bar{z})] \\ \varepsilon_{32} &= \frac{i}{2\mu}[f'(z) - \bar{f}'(\bar{z})] \end{aligned} \tag{3.1-11}$$

Combining Equations (3.1-2) and (3.1-11) one can write

$$\sigma_{31} - i\sigma_{32} = 2f'(z) \tag{3.1-12}$$

Let the origin of the x_1, x_2, x_3 coordinate system be located at the tip of a crack lying along the negative x_1-axis as shown in Figure 3.2. Attention is focused upon a small region D containing the crack tip and no other

* Frequently, the notation $\overline{f(z)}$ is also used to denote the complex conjugate of $f(z)$.

singularities. The dominant character of the stress and displacement fields in D is sought. Consider the holomorphic function

$$f(z) = Cz^{\lambda+1}, \qquad C = A + iB \tag{3.1-13}$$

where A, B, and λ are real undetermined constants. For finite displacements at the crack tip ($|z| = r = 0$), $\lambda > -1$. The substitution of Equation (3.1-13) into Equation (3.1-12) yields

$$\sigma_{31} - i\sigma_{32} = 2(\lambda + 1)Cz^{\lambda} = 2(\lambda + 1)r^{\lambda}(A + iB)(\cos \lambda\theta + i \sin \lambda\theta)$$

whence,

$$\begin{aligned} \sigma_{31} &= 2(\lambda + 1)r^{\lambda}(A \cos \lambda\theta - B \sin \lambda\theta) \\ \sigma_{32} &= -2(\lambda + 1)r^{\lambda}(A \sin \lambda\theta + B \cos \lambda\theta) \end{aligned} \tag{3.1-14}$$

The boundary condition that the crack surfaces be traction free requires that $\sigma_{32} = 0$ on $\theta = \pm\pi$. Consequently,

$$\begin{aligned} A \sin \lambda\pi + B \cos \lambda\pi &= 0 \\ A \sin \lambda\pi - B \cos \lambda\pi &= 0 \end{aligned}$$

To avoid the trivial solution the determinant of the coefficients of the foregoing equations must vanish. This leads to

$$\sin 2\lambda\pi = 0$$

which for $\lambda > -1$ has the roots

$$\lambda = -\tfrac{1}{2}, n/2, \qquad n = 0, 1, 2, \ldots$$

Of the infinite set of functions of the form of Equation (3.1-13) that yield traction-free crack surfaces within D, the function with $\lambda = -\frac{1}{2}$ for which $A = 0$ provides the most significant contribution to the crack-tip fields. For this case Equations (3.1-14) and (3.1-10) become, respectively,

$$\begin{Bmatrix} \sigma_{31} \\ \sigma_{32} \end{Bmatrix} = \frac{K_{\text{III}}}{(2\pi \text{r})^{\frac{1}{2}}} \begin{Bmatrix} -\sin(\theta/2) \\ \cos(\theta/2) \end{Bmatrix} \tag{3.1-15}$$

and

$$u_3 = \frac{2K_{\text{III}}}{\mu} \left(\frac{r}{2\pi}\right)^{\frac{1}{2}} \sin(\theta/2) \tag{3.1-16}$$

where B has been chosen such that

$$K_{\text{III}} \equiv \lim_{r \to 0} \{(2\pi r)^{\frac{1}{2}} \sigma_{32}|_{\theta=0}\} \tag{3.1-17}$$

The quantity K_{III} is referred to as the Mode III stress intensity factor, which is established by the far field boundary conditions and is a function of the applied loading and the geometry of the cracked body. Whereas the stresses associated with the other values of λ are finite at the crack tip, the stress components of Equation (3.1-15) have an inverse square root singularity at the crack tip. It is clear that the latter components will dominate as the crack tip is approached. In this sense Equations (3.1-15) and (3.1-16) represent the asymptotic forms of the elastic stress and displacement fields.

3.1.2 The Plane Problem

Before discussing methods for determining the stress intensity factor and the role it plays in LEFM, the asymptotic fields for the plane strain problem, wherein $u_1 = u_1(x_1, x_2)$, $u_2 = u_2(x_1, x_2)$, and $u_3 = 0$, will be developed. According to Equation (2.2-8) the strain components ε_{3i} will vanish. It follows from Equation (2.3-18) that $\sigma_{3\alpha} = 0$ and

$$\varepsilon_{\alpha\beta} = \frac{1+\nu}{E}\left[\sigma_{\alpha\beta} - \nu\delta_{\alpha\beta}\sigma_{\gamma\gamma}\right]$$
$$\sigma_{33} = \nu\sigma_{\alpha\alpha} \tag{3.1-18}$$

where $\sigma_{\alpha\beta} = \sigma_{\alpha\beta}(x_1, x_2)$. In the absence of body forces the equilibrium equations, Equation (2.1-10), reduce to

$$\sigma_{\alpha\beta,\alpha} = 0 \tag{3.1-19}$$

and the nontrivial compatibility equation [Equation (2.2-10a)] becomes

$$\varepsilon_{\alpha\beta,\alpha\beta} - \varepsilon_{\alpha\alpha,\beta\beta} = 0 \tag{3.1-20}$$

The equilibrium equations will be identically satisfied if the stress components are expressed in terms of the Airy stress function, $\Psi = \Psi(x_1, x_2)$, such that

$$\sigma_{\alpha\beta} = -\Psi_{,\alpha\beta} + \Psi_{,\gamma\gamma}\,\delta_{\alpha\beta} \tag{3.1-21}$$

After the introduction of Equation (3.1-21) into Equation (3.1-18), the compatibility equation requires that the Airy function satisfy the biharmonic equation

$$\Psi_{,\alpha\alpha\beta\beta} = \nabla^2(\nabla^2\Psi) = 0 \tag{3.1-22}$$

Noting that $\nabla^2\Psi$ satisfies Laplace's equation, one can write analogous to the antiplane problem that

$$\nabla^2\Psi = 4\frac{\partial^2\Psi}{\partial z\,\partial\bar{z}} = f(z) + \bar{f}(\bar{z}) \tag{3.1-23}$$

where $f(z)$ is a holomorphic function. Equation (3.1-23) can be integrated to yield the real function

$$\Psi = \tfrac{1}{2}\left[\bar{z}\Omega(z) + z\bar{\Omega}(\bar{z}) + \omega(z) + \bar{\omega}(\bar{z})\right] \tag{3.1-24}$$

where $\Omega(z)$ and $\omega(z)$ are holomorphic functions.

The substitution of Equation (3.1-24) into Equation (3.1-21) permits writing

$$4\frac{\partial^2\Psi}{\partial z\,\partial\bar{z}} = \sigma_{11} + \sigma_{22} = 2[\Omega'(z) + \bar{\Omega}'(\bar{z})] \tag{3.1-25}$$

$$\begin{aligned} 4\frac{\partial^2\Psi}{\partial\bar{z}^2} &= \sigma_{22} - \sigma_{11} - 2i\sigma_{12} \\ &= 2[z\bar{\Omega}''(\bar{z}) + \bar{\omega}''(\bar{z})] \end{aligned} \tag{3.1-26}$$

and

$$\sigma_{22} - i\sigma_{12} = \Omega'(z) + \bar{\Omega}'(\bar{z}) + z\bar{\Omega}''(\bar{z}) + \bar{\omega}''(\bar{z}) \tag{3.1-27}$$

Let

$$D = u_1 + iu_2 \tag{3.1-28}$$

define the complex displacement. Consequently,

$$2\frac{\partial D}{\partial \bar{z}} = \varepsilon_{11} - \varepsilon_{22} + 2i\varepsilon_{12} \tag{3.1-29}$$

and

$$\frac{\partial D}{\partial z} + \frac{\partial \bar{D}}{\partial \bar{z}} = \varepsilon_{11} + \varepsilon_{22} \tag{3.1-30}$$

The introduction of the stress-strain relation, Equation (3.1-18), into the preceding equations and the employment of Equations (3.1-25)–(3.1-27) provide

$$2\mu\frac{\partial D}{\partial \bar{z}} = -[z\bar{\Omega}''(\bar{z}) + \bar{\omega}''(\bar{z})] \tag{3.1-31}$$

and

$$\frac{2\mu}{1-2\nu}\left(\frac{\partial D}{\partial z} + \frac{\partial \bar{D}}{\partial \bar{z}}\right) = 2[\Omega'(z) + \bar{\Omega}'(\bar{z})] \tag{3.1-32}$$

Integrating Equations (3.1-31) and (3.1-32) we obtain within a rigid body displacement

$$2\mu D = \kappa\Omega(z) - z\bar{\Omega}'(\bar{z}) - \bar{\omega}'(\bar{z}) \tag{3.1-33}$$

for the complex displacement where

$$\kappa = 3 - 4\nu \tag{3.1-34}$$

This complex variable formulation is equally valid for generalized plane stress [see Green and Zerna (3.8)] if

$$\kappa = \frac{3-\nu}{1+\nu} \tag{3.1-35}$$

is used.

To examine the character of the Mode I stress and displacement fields, we again position the origin of the coordinate system at the crack tip. Due to symmetry with respect to the crack plane a solution of the form

$$\Omega = Az^{\lambda+1}, \qquad \omega' = Bz^{\lambda+1} \tag{3.1-36}$$

where A, B, and λ are real constants is chosen. For nonsingular displacements at the crack tip, $\lambda > -1$. The introduction of Equation (3.1-36) into

Equation (3.1-27) yields

$$\sigma_{22} - i\sigma_{12} = (\lambda + 1)r^{\lambda}\{A[2\cos\lambda\theta + \lambda\cos(\lambda - 2)\theta] + B\cos\lambda\theta \\ -i[A\lambda\sin(\lambda - 2)\theta + B\sin\lambda\theta]\} \quad (3.1\text{-}37)$$

which must vanish for $\theta = \pm\pi$. Consequently,

$$A(2 + \lambda)\cos\lambda\pi + B\cos\lambda\pi = 0$$
$$A\lambda\sin\lambda\pi + B\sin\lambda\pi = 0$$

for which a nontrivial solution exists if

$$\sin 2\lambda\pi = 0$$

or, equivalently,

$$\lambda = -\tfrac{1}{2}, \tfrac{n}{2}, \qquad n = 0, 1, 2, \ldots$$

Again the dominant contribution to the crack-tip stress and displacement fields occurs for $\lambda = -\frac{1}{2}$ for which $A = 2B$. As in the antiplane problem, an inverse square root singularity in the stress field exists at the crack tip. Substituting Equation (3.1-36) with $A = 2B$ and $\lambda = -\frac{1}{2}$ into Equations (3.1-25), (3.1-33), and (3.1-37), we find that

$$\begin{Bmatrix} \sigma_{11} \\ \sigma_{12} \\ \sigma_{22} \end{Bmatrix} = \frac{K_{\mathrm{I}}}{(2\pi r)^{\frac{1}{2}}}\cos(\theta/2)\begin{Bmatrix} 1 - \sin(\theta/2)\sin(3\theta/2) \\ \sin(\theta/2)\cos(3\theta/2) \\ 1 + \sin(\theta/2)\sin(3\theta/2) \end{Bmatrix} \quad (3.1\text{-}38)$$

and

$$\begin{Bmatrix} u_1 \\ u_2 \end{Bmatrix} = \frac{K_{\mathrm{I}}}{2\mu}\left(\frac{r}{2\pi}\right)^{\frac{1}{2}}\begin{Bmatrix} \cos(\theta/2)[\kappa - 1 + 2\sin^2(\theta/2)] \\ \sin(\theta/2)[\kappa + 1 - 2\cos^2(\theta/2)] \end{Bmatrix} \quad (3.1\text{-}39)$$

The Mode I stress intensity factor K_{I} is defined by

$$K_{\mathrm{I}} = \lim_{r \to 0}\{(2\pi r)^{\frac{1}{2}}\sigma_{22}|_{\theta=0}\} \quad (3.1\text{-}40)$$

When the foregoing is repeated with A and B being pure imaginary, the Mode II fields

$$\begin{Bmatrix} \sigma_{11} \\ \sigma_{12} \\ \sigma_{22} \end{Bmatrix} = \frac{K_{\mathrm{II}}}{(2\pi r)^{\frac{1}{2}}}\begin{Bmatrix} -\sin(\theta/2)[2 + \cos(\theta/2)\cos(3\theta/2)] \\ \cos(\theta/2)[1 - \sin(\theta/2)\sin(3\theta/2)] \\ \sin(\theta/2)\cos(\theta/2)\cos(3\theta/2) \end{Bmatrix} \quad (3.1\text{-}41)$$

and

$$\begin{Bmatrix} u_1 \\ u_2 \end{Bmatrix} = \frac{K_{\mathrm{II}}}{2\mu}\left(\frac{r}{2\pi}\right)^{\frac{1}{2}}\begin{Bmatrix} \sin(\theta/2)[\kappa + 1 + 2\cos^2(\theta/2)] \\ -\cos(\theta/2)[\kappa - 1 - 2\sin^2(\theta/2)] \end{Bmatrix} \quad (3.1\text{-}42)$$

are obtained where

$$K_{\mathrm{II}} = \lim_{r \to 0}\{(2\pi r)^{\frac{1}{2}}\sigma_{12}|_{\theta=0}\} \quad (3.1\text{-}43)$$

is the Mode II stress intensity factor. For plane strain $\sigma_{33} = \nu(\sigma_{11} + \sigma_{22})$ whereas $\sigma_{33} = 0$ for plane stress.

3.1.3 Fracture Criterion

It bears repeating that the foregoing stress and displacement fields for the three modes of loading represent the asymptotic fields as $r \to 0$ and may be viewed as the leading terms in the expansions of these fields about the crack tip. The applied loading σ, the crack length a, and perhaps other dimensions of the cracked body will affect the strength of these fields only through the stress intensity factor; that is, $K = K(\sigma, a)$. When using these expressions, attention must be confined to a sufficiently small neighborhood of the crack tip where only the leading terms are dominant. In Figure 3.3, a measure of the characteristic size of this "K-dominant" neighborhood is represented by D.

Because of the singular nature of the elastic stress field, there exists an inelastic (plastic) region surrounding the crack tip where the processes of void nucleation, growth, and coalescence that constitute ductile fracture occur. Let R be a representative dimension of this inelastic region. An estimate for R can be obtained, say, for Mode I by equating σ_{22} to the yield stress σ_y at $r = R$ and $\theta = 0$, so that

$$R = \frac{1}{2\pi}\left(\frac{K_I}{\sigma_y}\right)^2 \tag{3.1-44}$$

Within this region the linear elastic solution is invalid. It is not possible, therefore, to characterize directly the fracture process with a linear elastic formulation. This is not essential provided the inelastic region is confined to the K-dominant region. The situation where R is small compared to D and any other geometrical dimension is referred to as small-scale yielding.

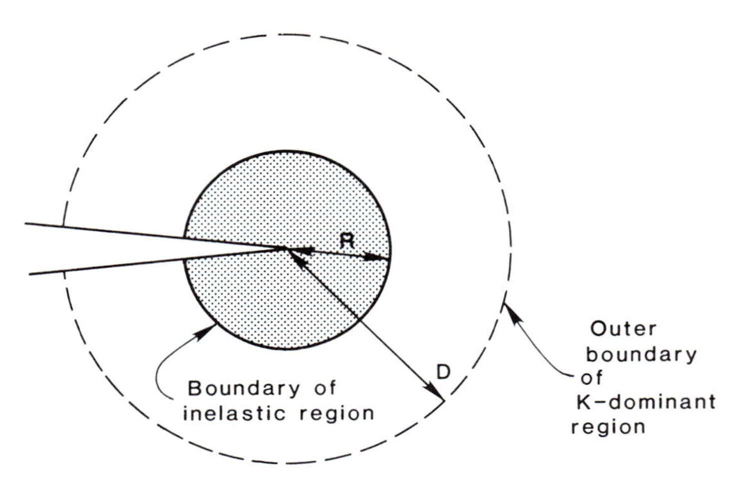

Figure 3.3 Basis of linear elastic fracture mechanics.

The elastic analysis indicates that the distributions of stresses and strains within the K-dominant region are the same regardless of the configuration and loading. Thus, given two bodies with different size cracks and different loadings of the same mode, but otherwise identical, then the near tip stress and deformation fields will be the same if the stress intensity factors are equal. Consequently, the stress intensity factor characterizes the load or deformation experienced by the crack tip and is a measure of the propensity for crack extension or of the crack driving force. That is, if crack growth is observed to initiate in the first body at a certain critical stress intensity factor, then crack extension in the second body can be expected when its stress intensity factor attains the same critical value. Therefore, within the confines of small-scale yielding the LEFM fracture criterion for incipient crack growth can be expressed as

$$K(\sigma, a) = K_c \tag{3.1-45}$$

where K_c is the critical value of the stress intensity factor K and is a measure of the material's resistance to fracture.

In general the assessment of the structural integrity of a cracked component requires a comparison of the crack driving force—for example, as measured by the stress intensity factor K, and the material's fracture toughness, say, K_c. An assessment involves either determining the critical loading to initiate growth of a prescribed crack or establishing the critical size of a crack for a specified loading. The determination of K and K_c for use in an evaluation within LEFM is considered in the following sections.

3.2 The Stress Intensity Factor

The task of determining the stress intensity factor is by no means a simple one. Because of the difficulties in satisfying the boundary conditions for finite bodies, only a limited number of closed-form solutions exist. Nevertheless, when the size of the crack is small compared to other dimensions of the body, the crack can be viewed as being in an infinite body. In this case there are standard techniques for establishing the stress intensity factor.

3.2.1 Closed Form Solutions

Based upon the analysis of the preceding section, the asymptotic form of $\Omega(\zeta)$ for the plane strain and plane stress problems can be written as

$$\Omega(\zeta) = \frac{K_{\mathrm{I}} - iK_{\mathrm{II}}}{(2\pi)^{\frac{1}{2}}} \zeta^{\frac{1}{2}} \tag{3.2-1}$$

whence

$$\Omega'(\zeta) = \frac{K_{\mathrm{I}} - iK_{\mathrm{II}}}{2(2\pi)^{\frac{1}{2}}} \zeta^{-\frac{1}{2}} \tag{3.2-2}$$

For convenience z has been replaced by ζ where the latter is understood to be measured from the crack tip. It follows from Equation (3.2-2) that

$$K_{\mathrm{I}} - iK_{\mathrm{II}} = \lim_{\zeta \to 0} \{2(2\pi\zeta)^{\frac{1}{2}}\Omega'(\zeta)\} \tag{3.2-3}$$

Hence, the task of determining the stress intensity factor reduces to establishing $\Omega'(\zeta)$ and proceeding to the limit in Equation (3.2-3).

Consider an infinite body with n collinear planar cracks lying along the segments L_r, $a_r \leqslant x_1 \leqslant b_r$, $r = 1, 2, \ldots, n$, of the real axis. Suppose that for the prescribed loading the stress field is known for the uncracked body where the stress components on the surface $x_2 = 0$ are

$$\sigma_{2i} = p_i(x_1)$$

It is now necessary to superimpose the solution to the residual problem where the remote boundary is traction free and where the prescribed stresses on the crack faces are now

$$\sigma_{2i} = -p_i(x_1) \quad \text{on} \quad L \tag{3.2-4}$$

where L is the union of L_r. Only the stress field of the latter problem is singular at the crack tip and, therefore, is instrumental in determining the stress intensity factor.

Outside of the cracks on $x_2 = 0$ there is continuity of the displacements and the stress components σ_{2i}. For both plane strain and plane stress, continuity of the displacements [see Equation (3.1-33)] requires that

$$\kappa\Omega^+(x_1) - x_1\overline{\Omega'^+(x_1)} - \overline{\omega'^+(x_1)} = \kappa\Omega^-(x_1) - x_1\overline{\Omega'^-(x_1)} - \overline{\omega'^-(x_1)}$$

where the + and − denote evaluations for $x_2 \to 0^+$ and $x_2 \to 0^-$, respectively. Since $\lim_{x_2 \to 0^+} \Omega(z) = \lim_{x_2 \to 0^-} \Omega(\bar{z})$, the above equation can be rewritten as

$$\lim_{x_2 \to 0^+} \{\kappa\Omega(z) + z\overline{\Omega'(\bar{z})} + \overline{\omega'(\bar{z})}$$

$$= \lim_{x_2 \to 0^-} \{\kappa\Omega(z) + z\overline{\Omega'(\bar{z})} + \overline{\omega'(\bar{z})}\}$$

Consequently, the function

$$\phi(z) = \kappa\Omega(z) + z\overline{\Omega'(\bar{z})} + \overline{\omega'(\bar{z})} \tag{3.2-5}$$

is continuous outside of the cracks on $x_2 = 0$ and, moreover, is holomorphic in the whole plane cut along L.

Similarly, continuity of the stress vector $\sigma_{22} - i\sigma_{12}$ [see Equation (3.1-27)] demands that

$$\Omega'^+(x_1) + \overline{\Omega'^+(x_1)} + x_1\overline{\Omega''^+(x_1)} + \overline{\omega''^+(x_1)}$$

$$= \Omega'^-(x_1) + \overline{\Omega'^-(x_1)} + x_1\overline{\Omega''^-(x_1)} + \overline{\omega''^-(x_1)}$$

or, alternatively,

$$\lim_{x_2 \to 0^+} \{\Omega'(z) - \overline{\Omega'(\bar{z})} - z\overline{\Omega''(\bar{z})} - \overline{\omega''(\bar{z})}\}$$

$$= \lim_{x_2 \to 0^-} \{\Omega'(z) - \overline{\Omega'(\bar{z})} - z\overline{\Omega''(\bar{z})} - \overline{\omega''(\bar{z})}\}$$

Therefore,

$$\psi(z) = \Omega(z) - z\overline{\Omega'(\bar{z})} - \overline{\omega'(\bar{z})} \tag{3.2-6}$$

is continuous outside of the cracks on $x_2 = 0$ and is holomorphic in the whole plane cut along L. Equations (3.2-5) and (3.2-6) further imply that $\Omega(z)$ is also holomorphic in this same cut plane. Equation (3.2-6) can be used to eliminate $\omega'(z)$ in favor of $\psi(z)$ so that the complex stress and displacement can be written as

$$\sigma_{22} - i\sigma_{21} = \Omega'(z) + \Omega'(\bar{z}) + (z - \bar{z})\overline{\Omega''(z)} - \psi'(\bar{z}) \tag{3.2-7}$$

and

$$2\mu D = \kappa\Omega(z) - \Omega(\bar{z}) + (\bar{z} - z)\overline{\Omega'(z)} + \psi(\bar{z}) \tag{3.2-8}$$

Equation (3.2-7) can be used to express the boundary conditions on the crack surfaces as

$$-p_2(x_1) + ip_1(x_1) = \Omega'^{+}(x_1) + \Omega'^{-}(x_1) - \psi'^{-}(x_1)$$

and

$$-p_2(x_1) + ip_1(x_1) = \Omega'^{+}(x_1) + \Omega'^{-}(x_1) - \psi'^{+}(x_1)$$

on L. Adding and subtracting the foregoing equations we have

$$2\Omega'^{+}(x_1) - \psi'^{+}(x_1) + 2\Omega'^{-}(x_1) - \psi'^{-}(x_1) = -2p_2(x_1) + 2ip_1(x_1) \tag{3.2-9}$$

and

$$\psi'^{+}(x_1) - \psi'^{-}(x_1) = 0 \tag{3.2-10}$$

for x_1 on L.

The problem of finding a sectionally holomorphic function in the whole plane subject to the boundary condition of the form of Equation (3.2-9) is known as a Hilbert problem. An accounting of the solution of the Hilbert problem can be found in references (3.8)–(3.11). For the present problem the solution can be written as

$$\Omega'(z) - \tfrac{1}{2}\psi'(z) = \frac{X(z)}{2\pi i}\int_L \frac{p_2(t) - ip_1(t)}{X^{+}(t)(t - z)}\,dt + X(z)P(z) \tag{3.2-11}$$

where $X(z)$ is the Plemelj function

$$X(z) = \prod_{k=1}^{n} (z - a_k)^{-\frac{1}{2}}(z - b_k)^{-\frac{1}{2}} \tag{3.2-12}$$

where the branch is selected such that $z^n X(z) \to 1$ as $|z| \to \infty$. The undetermined function $P(z)$ is holomorphic in the whole plane. Because the stresses must vanish as $|z| \to \infty$ and because $X(z)$ is of the order of z^{-n} for large $|z|$, then $P(z)$ is a polynominal of degree $n - 1$ with n undetermined coefficients. These coefficients are chosen such that the displacement field is single valued; that is,

$$\int_C \frac{\partial D}{\partial z}\,dz = 0 \tag{3.2-13}$$

where C is any closed contour within the body and, in particular, a contour enclosing a crack. When Equation (3.2-13) is written for contours enclosing each crack, then n linear, complex algebraic equations for determining the n coefficients are obtained. For the homogeneous Hilbert problem defined by Equation (3.2-10) with vanishing stresses as $|z| \to \infty$,

$$\psi'(z) = 0 \tag{3.2-14}$$

If contrary to assumed here, there is a nonzero resultant of the initial tractions on the crack surfaces, then a nontrivial $\psi'(z)$ exists. The development for the latter can be found in England (3.9) with examples given by Paris and Sih (3.12).

If a similar analysis is repeated for the antiplane strain loading where $\sigma_{32} = p_3(x_1)$ on L for the uncracked body, then

$$f'(z) = \frac{X(z)}{2\pi} \int_L \frac{p_3(t)}{X^+(t)(t-z)}\, dt + X(z)P(z) \tag{3.2-15}$$

where $P(z)$ is again a polynomial of degree $n - 1$. The coefficients of this polynomial are determined such that

$$\int_C \frac{\partial u_3}{\partial z}\, dz = 0 \tag{3.2-16}$$

for closed contours enclosing each of the n cracks.

Consider a single crack extending from $x_1 = -a$ to $x_1 = a$. The Plemelj function for this case is

$$X(z) = (z + a)^{-\frac{1}{2}}(z - a)^{-\frac{1}{2}} = (z^2 - a^2)^{-\frac{1}{2}}$$

Letting $z - a = r_1 e^{i\theta_1}$ and $z + a = r_2 e^{i\theta_2}$, we conclude that

$$X^+(x_1) = -X^-(x_1) = -i(a^2 - x_1^2)^{-\frac{1}{2}}, \qquad |x_1| \leqslant a$$

For single-valued displacements, $P(z) = 0$ and Equation (3.2-11) becomes

$$\Omega'(z) = -\frac{(z^2 - a^2)^{-\frac{1}{2}}}{2\pi} \int_{-a}^{a} \frac{(a^2 - t^2)^{\frac{1}{2}}[p_2(t) - ip_1(t)]}{t - z}\, dt \tag{3.2-17}$$

In general the usual approach to determine $\Omega'(z)$ is to express the foregoing integral in terms of a contour integral and to evaluate the latter by Cauchy's integral theorem. However, for the present purposes this is not necessary. Instead, write $z = \zeta + a$ and substitute Equation (3.2-17) into Equation (3.2-3) to obtain

$$K_{\mathrm{I}} - iK_{\mathrm{II}} = \frac{1}{(\pi a)^{\frac{1}{2}}} \int_{-a}^{a} \left(\frac{a + t}{a - t}\right)^{\frac{1}{2}} [p_2(t) - ip_1(t)]\, dt \tag{3.2-18}$$

for the stress intensity factors at the crack tip $x_1 = a$. A similar analysis for the antiplane strain problem yields

$$K_{\mathrm{III}} = \frac{1}{(\pi a)^{\frac{1}{2}}} \int_{-a}^{a} \left(\frac{a + t}{a - t}\right)^{\frac{1}{2}} p_3(t)\, dt \tag{3.2-19}$$

If $p_i(t)$ is an even function of t, then it is convenient to introduce the change of variable $t = a \sin \phi$ so that Equations (3.2-18) and (3.2-19) become

$$K_{\mathrm{I}} - iK_{\mathrm{II}} = 2\left(\frac{a}{\pi}\right)^{\frac{1}{2}} \int_0^{\pi/2} [p_2(a \sin \phi) - ip_1(a \sin \phi)] \, d\phi$$
$$K_{\mathrm{III}} = 2\left(\frac{a}{\pi}\right)^{\frac{1}{2}} \int_0^{\pi/2} p_3(a \sin \phi) \, d\phi \tag{3.2-20}$$

For an infinite body subjected to uniform remote tractions σ_{ij}^{∞} (the Griffith problem), then $p_i = \sigma_{2i}^{\infty}$ and Equation (3.2-20) leads to

$$K_{\mathrm{I}} = \sigma_{22}^{\infty}\sqrt{(\pi a)}, \qquad K_{\mathrm{II}} = \sigma_{21}^{\infty}\sqrt{(\pi a)}, \qquad K_{III} = \sigma_{23}^{\infty}\sqrt{(\pi a)} \tag{3.2-21}$$

If equal and opposite forces per unit thickness are applied to the crack surfaces at $x_1 = b$ such that $p_2(x_1) = P\delta(x_1 - b)$ and $p_1(x_1) = Q\delta(x_1 - b)$, where $\delta(x_1 - b)$ is the Dirac delta function, then Equation (3.2-18) yields

$$K_{\mathrm{I}} - iK_{\mathrm{II}} = \frac{P - iQ}{(\pi a)^{\frac{1}{2}}} \int_{-a}^{a} \left(\frac{a + t}{a - t}\right)^{\frac{1}{2}} \delta(t - b) \, dt$$
$$= \frac{P - iQ}{(\pi a)^{\frac{1}{2}}} \left(\frac{a + b}{a - b}\right)^{\frac{1}{2}}$$

Koiter (3.13), using the method of this section, investigated the problem depicted in Figure 3.4 of a plane body that contains an infinite periodic array of collinear cracks and that supports uniform remote tractions $\sigma_{\alpha\beta}^{\infty}$. While

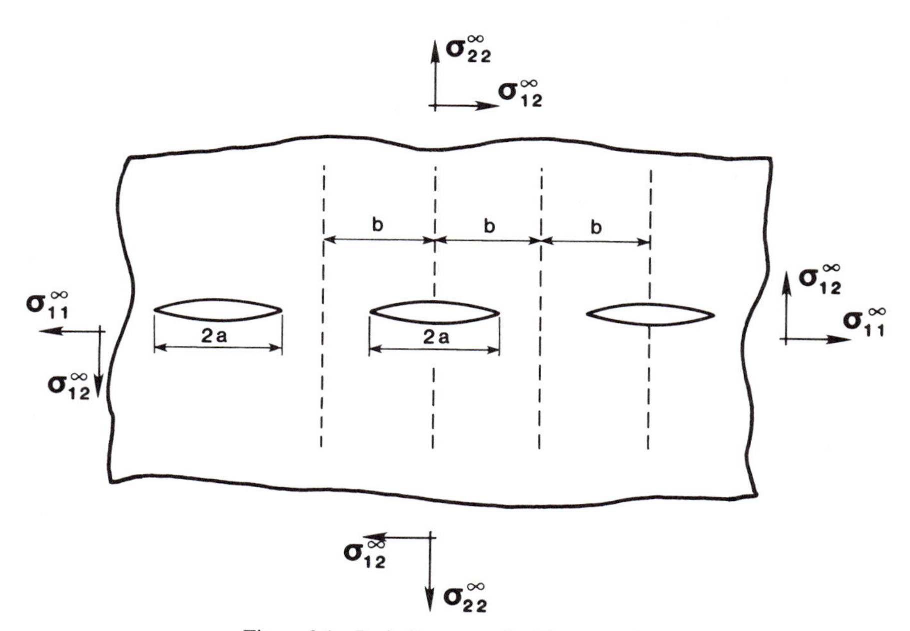

Figure 3.4 Periodic array of collinear cracks.

Koiter computed the change in the strain energy due to the presence of the cracks, he stopped short of determining the stress intensity factors. Nevertheless, by exploiting the relationship between the rate of decrease in the strain energy and the stress intensity factors of Section 3.3 [cf. Equation (3.3-17)], it is a simple manner to show that

$$\begin{aligned} K_{\mathrm{I}} &= \sigma_{22}^{\infty}(\pi a)^{\frac{1}{2}}\left[\frac{2b}{\pi a}\tan\left(\frac{\pi a}{2b}\right)\right]^{\frac{1}{2}} \\ K_{\mathrm{II}} &= \sigma_{21}^{\infty}(\pi a)^{\frac{1}{2}}\left[\frac{2b}{\pi a}\tan\left(\frac{\pi a}{2b}\right)\right]^{\frac{1}{2}} \end{aligned} \tag{3.2-22}$$

The latter were also obtained by Irwin (3.14) using the Westergaard stress function. In the limit as $b \rightarrow \infty$, Equation (3.2-21) for Modes I and II is recovered from Equation (3.2-22). Equation (3.2-22) indicates that there is insignificant interference between the cracks when the periodic spacing is greater than about twice the crack length.

From Mode I loading the shear stress σ_{12} vanishes on the planes of symmetry represented by the dashed lines in Figure 3.4. If the body is cut along a pair of these planes, a finite width strip supporting a uniform remote stress σ_{22}^{∞} and containing variously a central crack, a single edge crack or double edge cracks is obtained. In addition a nonuniform stress σ_{11} acts on the exposed planes. However, if the crack tip is sufficiently far removed from these edges, then the influence of the latter stress on the stress intensity factor will be relatively small. For example, Paris and Sih (3.12) demonstrate that the difference between Equation 3.2-22 and Isida's (3.15) numerical analysis of the remotely stressed center-cracked strip is less than 7 percent for $a/b < \frac{1}{2}$. The difference for single-edge- and double-edge-cracked strips is less than 3 percent for $a/b < \frac{1}{2}$. On the other hand, for $a/b \ll 1$

$$K_{\mathrm{I}} = 1.12\sigma_{22}^{\infty}\sqrt{(\pi a)} \tag{3.2-23}$$

and Equation (3.2-22) underestimates the stress intensity factor by 12 percent.

Until now only cracks with straight fronts have been considered. Sneddon (3.16) analyzed the axisymmetric problem of a circular (penny-shaped) crack of radius a in an infinite body subjected to a uniform remote uniaxial stress. Green and Zerna (3.8) generalized the loading to permit the crack to be opened by a normal stress $\sigma_{33} = -p(\rho)$ depending only upon the radial distance ρ from the center of the crack. Again, there is an inverse square root stress singularity at the crack tip and the forms of the near crack-tip stress and displacement fields are the same as those for plane strain. The stress intensity factor is given by

$$K_{\mathrm{I}} = \frac{2}{(\pi a)^{\frac{1}{2}}}\int_0^a \frac{\rho p(\rho)}{(a^2 - \rho^2)^{\frac{1}{2}}}\,d\rho \tag{3.2-24}$$

which yields for a uniform remote stress σ_{33}^{∞}

$$K_{\mathrm{I}} = \frac{2}{\pi}\sigma_{33}^{\infty}\sqrt{(\pi a)} \tag{3.2-25}$$

Starting with the crack opening displacement (COD) of Green and Sneddon (3.17) for an elliptical crack embedded in an infinite solid loaded by a uniform remote stress σ_{33}^{∞} normal to the crack plane, and noting that the near crack-tip stress and displacement fields relative to a local coordinate system at the crack front are the same as the plane strain fields, Irwin (3.18) found that

$$K_{\mathrm{I}} = \frac{\sigma_{33}^{\infty}(\pi a)^{\frac{1}{2}}}{E(k)}\left[\sin^2\phi + \left(\frac{a}{c}\cos\phi\right)^2\right]^{\frac{1}{4}} \tag{3.2-26}$$

In Equation (3.2-26), $2a$ and $2c$ are, respectively, the minor and major dimensions of the elliptical crack, $E(k)$ is the complete elliptic integral of the second kind and $k^2 = (c^2 - a^2)/c^2$. The coordinates of a point on the boundary of the crack are expressed in parametric form by $x_1 = c\cos\phi$ and $x_2 = a\sin\phi$. In this case the stress intensity factor varies along the crack front from its minimum value

$$(K_{\mathrm{I}})_{\min} = \frac{\sigma_{33}^{\infty}(\pi a^2/c)^{\frac{1}{2}}}{E(k)} \tag{3.2-27}$$

at the extremities of the major diameter to its maximum value

$$(K_{\mathrm{I}})_{\max} = \frac{\sigma_{33}^{\infty}(\pi a)^{\frac{1}{2}}}{E(k)} \tag{3.2-28}$$

at the ends of the minor diameter. Consequently, if crack extension is governed by the stress intensity factor attaining its critical value, then the elliptical crack will have a tendency to grow into a circular one. Kassir and Sih (3.19) have determined the stress intensity factors when the direction of the remote uniaxial stress makes an arbitrary angle with the plane of the elliptical crack. In this case the crack front experiences all three modes of loading. For mixed mode loading such as this there is a lack of agreement on what the appropriate fracture criterion ought to be.

Until now primary consideration has been given to cracks in elastic bodies of infinite extent because it is frequently possible in these instances to obtain closed form expressions for the stress intensity factors. Paris and Sih (3.12) present a relatively extensive listing of these factors that can be superposed for more complicated loadings. It has been demonstrated by way of examples that if an uncracked boundary or another crack is of the order of a crack length or more from the crack tip, then these solutions provide reasonable approximations for the stress intensity factors. When the interference between the crack tip and an uncracked boundary is significant, the task of satisfying the boundary conditions on the latter can be sufficiently difficult to preclude closed-form expressions for the stress intensity factors. Under such circumstances one must resort to numerical methods of solution.

3.2.2 Numerical Methods

In the following some of the more popular methods that have been used to determine the stress intensity factors catalogued in references (3.20)–(3.23) are summarized. The boundary condition that the crack surfaces be stress free [cf. Equation (3.2-9) with $p_1(x_1) = p_2(x_1) = 0$] and Equation (3.2-10) will be satisfied by

$$\Omega'(z) = (z^2 - a^2)^{-\frac{1}{2}}F(z) + G(z) \qquad (3.2\text{-}29)$$
$$\psi'(z) = 2G(z)$$

for an internal crack and by

$$\Omega'(z) = z^{-\frac{1}{2}}F(z) + G(z) \qquad (3.2\text{-}30)$$
$$\psi'(z) = 2G(z)$$

for an edge crack. The holomorphic functions $F(z)$ and $G(z)$ must be determined such that the boundary conditions on the uncracked portion of the boundary are satisfied.

In principle, these functions can always be expanded in Laurent series with the undetermined coefficients chosen to satisfy the boundary conditions point by point. Because this is impractical, truncated series

$$F(z) = \sum_{n=0}^{N} a_n z^n, \qquad G(z) = \sum_{m=0}^{M} b_m z^m \qquad (3.2\text{-}31)$$

are used. The condition for finite displacements at the crack tip requires that the exponents in these series be limited to nonnegative integers. The boundary collocation method requires that the boundary conditions be satisfied at an appropriate number of selected points. The minimum number of points must be sufficient to generate enough independent algebraic equations to determine the coefficients. Frequently, an improved solution is attained if a number in excess of this minimum is chosen. The resulting overdeterminate system of equations is satisfied in a least square sense. It is only necessary to determine a_0 in order to establish the stress intensity factor from Equation (3.2-3).

The stress intensity factors for the single edge notch, three point bend, and compact tension fracture specimens were computed by Gross et al. (3.24–3.26) using the boundary collocation method. Kobayashi et al. (3.27) used this method to determine the stress intensity factor of a central crack in a finite width strip.

The Schwarz-Neumann alternating method [see references (3.28) and (3.29)] has been used to study the interference of a crack with a free surface. In principle the method is simple, but its application to the numerical solution of the crack problem can be tedious. Basically, the method yields a solution in the overlapping region of interest formed by two intersecting regions B_1 and B_2 by alternately solving a sequence of separate, but related, boundary value problems in each region. In the limit the solutions in the common region converge to the desired solution.

For example, consider a crack near the free surface of a half space B_1 and assume the surfaces of the crack are subjected to a specified normal pressure. Next suppose that the solution for this same crack and pressure loading in an infinite medium B_2 is known. This solution will give rise to tractions at the site of the free surface of B_1. A residual problem, whereby these tractions are negated, is solved for an uncracked half-space. This in turn will produce normal tractions at the location of the crack surfaces which must be canceled by solving once more the problem of a crack in an infinite medium subjected to this new pressure loading. The process of alternately establishing solutions in the two regions is continued until the residual tractions on the free surface and the crack surfaces are deemed insignificant.

This approach was used by Smith et al. (3.30) to determine the stress intensity factor for a semicircular edge crack in a semi-infinite solid. Smith and Alavi (3.31) used this method to establish the stress intensity factor for a penny-shaped crack near the free surface of a half-space subjected to uniaxial tension. If the crack tip is removed from the free surface by a distance equal to at least the radius of the crack, then the stress intensity factor is nearly equal to that for a crack in an infinite medium. Shah and Kobayashi (3.32–3.34) used the alternating method to study the interaction of an elliptical crack and the free surface of a half-space.

The boundary integral equation method follows from Betti's reciprocal theorem, which states that, for a linear elastic body subjected to two different loadings, the work done by the first loading acting through the displacements produced by the second loading equals the work done by the second loading acting through the displacements due to first loading. That is,

$$\int_S T_i^{(1)} u_i^{(2)}\, dS + \int_V F_i^{(1)} u_i^{(2)}\, dV = \int_S T_i^{(2)} u_i^{(1)}\, dS + \int_V F_i^{(2)} u_i^{(1)}\, dV \qquad (3.2\text{-}32)$$

where the superscripts identify the quantities of the given loading. The proof of the theorem follows from the principle of virtual work [cf. Equation (2.4-1)] and the fact that $\sigma_{ij}^{(1)}\varepsilon_{ij}^{(2)} = \sigma_{ij}^{(2)}\varepsilon_{ij}^{(1)}$ for a linear elastic material.

Let the first loading be the one of interest and for the second loading consider a system of orthogonal unit loads in the x_j direction acting at point p. Define $T_{ij}(p, Q)$ and $U_{ij}(p, Q)$ to be the tractions and displacements in the x_i direction at point Q on the boundary due to the unit loads at p. Then, in the absence of body forces, Equation (3.2-32) yields

$$u_i(p) = -\int_S T_{ij}(p, Q) u_j(Q)\, dS + \int_S U_{ij}(p, Q) T_j(Q)\, dS \qquad (3.2\text{-}33)$$

for an interior point p and where for convenience the superscript 1 has been dropped. In arriving at Equation (3.2-33) a cut is introduced from the boundary to a spherical surface enclosing the point p and then the radius of the sphere is allowed to approach zero. In the limit as p tends to a point P on the boundary [see Rizzo (3.35)] the boundary integral equation becomes

$$\tfrac{1}{2} u_i(P) = -\int_S T_{ij}(P, Q) u_j(Q)\, dS + \int_S U_{ij}(P, Q) T_j(Q)\, dS \qquad (3.2\text{-}34)$$

On the portion S_u of the boundary where the displacements $u_j(Q)$ are prescribed, the tractions $T_j(Q)$ are unknown and vice versa on the portion S_T of the boundary. In order to solve for these unknowns, Equation (3.2-34) is reduced to a system of algebraic equations. This is done by representing the boundary by a set of boundary segments over which interpolation functions are written in terms of the unknown nodal values; for example, see Cruse (3.36). Equation (3.2-34) is written for each nodal point. The resulting algebraic equations are solved for the nodal quantities. Once the boundary data has been determined, Equation (3.2-33) can be used to determine the displacement at any interior point p. Having established the displacement field in the proximity of the crack tip, one can compute the crack opening displacement and/or the stresses and numerically extrapolate estimates for the stress intensity factor; for example, for Mode I, from

$$K_1 = \frac{2\mu}{\kappa + 1} \lim_{r \to 0} \left\{ \left(\frac{2\pi}{r} \right)^{\frac{1}{2}} u_2 \big|_{\theta = \pi} \right\} \tag{3.2-35}$$

or

$$K_1 = \lim_{r \to 0} \left\{ (2\pi r)^{\frac{1}{2}} \sigma_{22} \big|_{\theta = 0} \right\} \tag{3.2-36}$$

Equation (3.2-35) usually gives more consistent and precise estimates.

Cruse (3.37) has compared solutions using the boundary integral equation method with other solutions for a variety of cracked configurations. When both crack surfaces of a planar crack are modeled, the coefficient matrix becomes singular. This problem can be circumvented if the surfaces are modeled as distinct with a small nonzero distance of separation; that is, the crack is represented by a notch. Blanford et al. (3.38) used multidomain discretization to eliminate the problem of a singular matrix without modeling the crack as a notch. A principal advantage of the boundary integral equation method is that only discretization on the boundary but not in the interior is required. A computational disadvantage is that the method leads to a nonsymmetrical, nearly fully populated system of equations.

Rice (3.39) in 1968 noted that the finite element method of numerical analysis of fracture was in its infancy and possessed great potential for handling the crack-tip singularity. Since that time there have been literally hundreds of publications devoted to finite element analysis of fracture. Perhaps the unique feature that makes this method so attractive is its potential for handling fracture problems well beyond the limits of LEFM. The method will be discussed only from an elementary point of view, since to do otherwise would lead to a significant departure from the central purpose of this chapter. Fortunately, a number of excellent tests on the finite element method—for example, see references (3.40)–(3.43)—and several reviews, references (3.44)–(3.48), of its application to fracture mechanics analysis are at the disposal of the reader desiring a more sophisticated and detailed treatment of the subject.

In the finite element method the domain of interest is divided into a finite number of subdomains called elements. The stiffness finite element method consists of assuming for an element a local displacement field

$$\{u_i\} = [N(x_i)]\{u_i^e\} \tag{3.2-37}$$

where $[N(x_i)]$ denotes the assumed displacement shape of interpolation function and $\{u_i^e\}$ is the vector of nodal displacements of the element. When Equation (3.2-37) is introduced into Equation (2.2-8), the elemental strain field can be written as

$$\{\varepsilon\} = [N'(x_i)]\{u_i^e\} \tag{3.2-38}$$

where $[N'(x_i)]$ is the matrix obtained after performing the required differentiation. For the assumed displacement field satisfying relevant interelement continuity the strain energy U_e of the element computed from Equations (2.3-7) and (2.4-2) can be written as

$$U_e = \tfrac{1}{2}\{u_i^e\}^T[k^e]\{u_i^e\} \tag{3.2-39}$$

The superscript T denotes the transpose and the elemental stiffness matrix $[k^e]$ is

$$[k^e] = \int_{V_e} [N'(x_i)]^T[C][N'(x_i)]\, dV \tag{3.2-40}$$

in which $[C]$ is the matrix of elastic coefficients C_{ijkl}. The elemental stiffness matrices are summed to form a global stiffness matrix $[K]$. The equations of equilibrium that are derivable from the principle of minimum potential energy may be written as

$$[K]\{u\} = \{F\} \tag{3.2-41}$$

where $\{u\}$ is the vector of global nodal displacements and $\{F\}$ is the vector of generalized nodal forces. Standard numerical techniques may be used to invert Equation (3.2-41) to obtain the nodal displacements.

Two important considerations in the development of a finite element analysis for fracture mechanics are the proper modeling of the crack-tip singularity and the interpreting of the results in terms of a stress intensity factor or a crack driving force. These considerations have been addressed in the review by Gallagher (3.44) not only for the stiffness method but also for the hybrid method (assumed displacement and stress method).

Initially conventional (nonsingular) elements were used to model the crack-tip singularity. This requires a large number of small elements in the proximity of the crack tip. Consequently, the number of equations to be solved can be very large and, therefore, obtaining a solution can be expensive. This is partially offset by the fact that the global stiffness matrix tends to be strongly banded, a condition which can be exploited by efficient equation-solving algorithms.

In order to account explicitly for the singularity, singular elements have been developed. Singular elements based upon using the classical singular solutions for a crack in an infinite plate as the interpolation functions have been established. These are usually circular elements that interface with conventional elements. Special techniques are usually required to handle the displacement discontinuity between the two different types of elements. Such techniques are unnecessary if triangular elements in conjunction with polynomial interpolation functions that yield a $1/\sqrt{r}$ strain singularity are used. Since a few singular elements could replace a large number of conventional elements, increased computational efficiency might be expected. Certainly the number of equations to be solved is reduced, but generally at the expense of an increased bandwidth for the stiffness matrix, and the apparent savings can evaporate.

It is well known that the transformation from physical to isoparametric coordinates will be singular if the nodes along the sides of the element assume certain positions. When the midside nodes nearest to the crack tip node of the eight-noded quadrilateral isoparametric element and the 20-noded three-dimensional isoparametric brick are moved to the quarter point, an inverse square root singularity develops only along the edges of these elements. Barsoum (3.49) showed that the singularity can be made to occur along all rays emanating from the vertex of a triangular element formed by collapsing a quadrilateral element. The modeling of the singularity by judicious placement of side nodes in an isoparametric element is particularly appealing because these elements are usually present in general purpose finite element codes.

Having modeled the crack-tip singularity and having determined the nodal displacements, it is necessary to compute the stress intensity factor. This can be done by using the extrapolation procedure discussed in connection with the boundary integral equation method [cf. Equations (3.2-35) and (3.2-36)]. Because the precision of the displacements obtained from the stiffness method is greater than for the stresses, the extrapolation is usually based upon the displacements. As will be shown in the next section the stress intensity factor can be related to the rate of decrease of the strain energy with crack extension, the crack closure integral, and the path independent J-integral. The implementation of the latter into the finite element method is addressed by Gallagher (3.44) and the references therein. The J-integral method seems to be the most efficient alternative for interpreting the finite element computations.

The treatment of other analytical techniques—for example, conformal mapping, integral transform methods, and singular integral equations—as well as the preceding ones can be found in Sih (3.50).

3.3 Energetics of Cracked Bodies

Griffith (3.51) approached the fracture of an ideally brittle material from a thermodynamic viewpoint. He postulated that during an increment of crack extension da there can be no change in the total energy E composed of the sum of the potential energy of deformation Π and the surface energy S; that is,

$$dE = d\Pi + dS = 0 \tag{3.3-1}$$

For a crack in a two-dimensional deformation field it is convenient to define these energies as being per unit thickness of the body. If γ denotes the surface energy density per unit area, then, $dS = 2\gamma\, da$ for the two increments of fracture surfaces formed during the extension. At incipient or during crack growth Equation 3.3-1 yields

$$G = 2\gamma \tag{3.3-2}$$

where

$$G = -\frac{d\Pi}{da} \tag{3.3-3}$$

is known as the energy release rate. Because G is derivable from a potential function in much the same way as a conservative force can be, it is often referred to as a crack driving force. The right-hand side of Equation (3.3-2) represents the fracture resistance that must be overcome by the driving force in order to produce a unit of crack extension. This resistance is a characteristic of the material whereas G depends upon the loading and geometry of the cracked body.

3.3.1 *The Energy Release Rate*

Consider the cracked linear elastic body depicted in Figure 3.5. Let P represent a generalized force per unit thickness and let Δ be the corresponding generalized load-point displacement through which P does work. The potential energy [cf. Equation (2.4-13)] for a prescribed P (dead loading) is

$$\Pi = U - P\Delta = -U^* \tag{3.3-4}$$

where U and U^* are, respectively, the strain and complementary strain

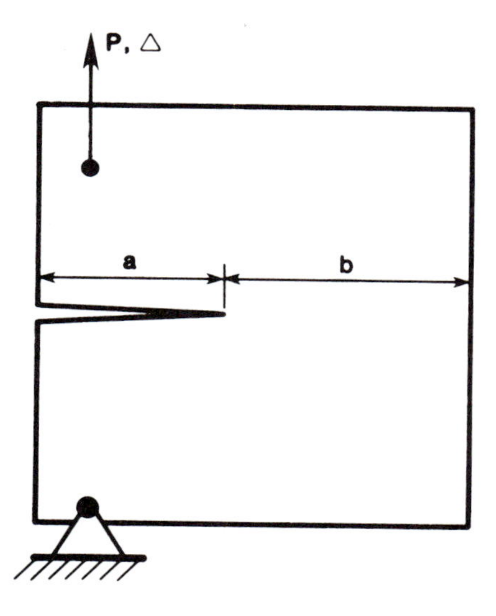

Figure 3.5 A flawed body.

energies. By Clapeyron's theorem $U = P\Delta/2$ and Equation (3.3-4) becomes

$$\Pi = -P\Delta/2 = -U^* \tag{3.3-5}$$

Substitution of Equation (3.3-5) into Equation (3.3-3) yields

$$G = \left(\frac{\partial U^*}{\partial a}\right)_P = \tfrac{1}{2}P\left(\frac{\partial \Delta}{\partial a}\right)_P \tag{3.3-6}$$

Note that in this development only a single crack tip has been considered. When there exist two equally loaded crack tips—for example, the Griffith problem—it will be necessary to replace the differentiation with respect to a by $2a$ in order to obtain the energy release rate per crack tip. The lack of proper consideration of this point has been the source of confusion and error.

For a linear elastic body

$$\Delta = CP \tag{3.3-7}$$

where C denotes the compliance of the body of unit thickness and is a function of the geometry of the body and its elastic constants. Combining Equation (3.3-6) and (3.3-7) one obtains

$$G = \frac{P^2}{2}\frac{dC}{da} = \frac{1}{2}\frac{\Delta^2}{C^2}\frac{dC}{da} \tag{3.3-8}$$

If the compliance of the body is known, then Equation (3.3-8) yields the rate of energy that would be release during a virtual extension of the crack tip. The energy release rate can also be established experimentally from measuring the compliance of the cracked body. The crack is then extended a small increment Δa, say, by cutting, and the change in compliance ΔC is measured. The ratio $\Delta C/\Delta a$ is used to approximate the derivative in Equation (3.3-8). Of course, care must be exercised in these measurements because one is dealing with differences of quantities of approximately the same magnitude.

If the load-point displacement Δ is prescribed (fixed-grip loading), then the potential energy is

$$\Pi = U = P\Delta/2$$

and it follows from Equation (3.3-3) that

$$G = -\left(\frac{\partial U}{\partial a}\right)_\Delta = -\tfrac{1}{2}\Delta\left(\frac{\partial P}{\partial a}\right)_\Delta \tag{3.3-9}$$

Substituting Equation (3.3-7) into Equation (3.3-9) we arrive again at Equation (3.3-8) and conclude that the energy release rate is independent of the type of loading. Due to the form of Equation (3.3-9), G is also known as the strain energy release rate. The introduction of a spring, representing the stiffness of a compliant loading device, between the specimen and the load will not change Equation (3.3-8) since the energy stored in the spring is independent of a.

Equations (3.3-6) and (3.3-9) can be given a graphical interpretation. Consider the load versus load-point displacement curve for a cracked body

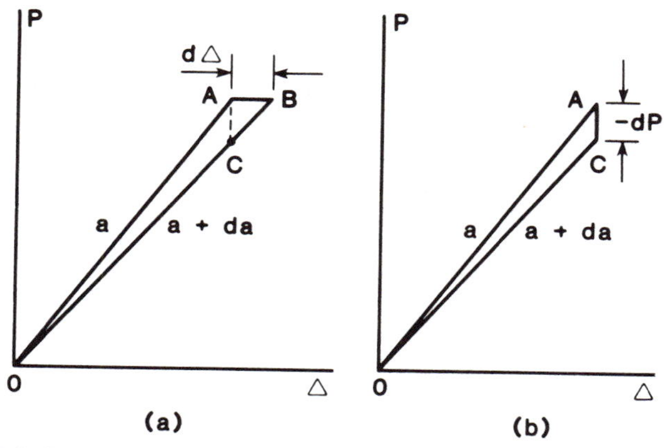

Figure 3.6 Load-deflection curves for (a) dead loading and (b) fixed-grip loading.

(see Figure 3.6). Now extend the crack (e.g., by cutting) an amount da under constant load. The area of the triangle OAB in Figure 3.6a is the increase in the complementary energy at fixed load. According to Equation (3.3-6), this area is Gda. If this procedure is repeated but with the displacement held fixed, then the area OAC in Figure 3.6b is the decrease in the strain energy and by Equation (3.3-9) equals Gda. Since the difference between the areas OAB and OAC is the infinitesimal area ABC of higher order, the area between the two curves for loadings intermediate to fixed load and fixed displacement is also Gda. This interpretation offers an alternative for determining experimentally the energy release rate of a cracked body by measuring the area between P-Δ curves for slightly different crack lengths.

To establish the connection between the stress intensity factor and the energy release rate, consider duplicate loading of two linear elastic bodies that are otherwise identical except the first has a crack length a whereas the second has a crack length $a + \Delta a$. Both bodies can be envisioned as having a crack of length $a + \Delta a$ except that the crack of the first one is closed by an amount Δa by stresses σ_{2i} acting on the crack surfaces over $a < x_1 < a + \Delta a$.

By means of Clapeyron's theorem the potential energy per unit thickness of the first body in the absence of body forces is

$$\Pi(a) = -\tfrac{1}{2} \int_{S_T} T_i u_i(a)\, dS$$

where S_T is the portion of the boundary of the body on which the tractions are prescribed. In the above, $u_i(a)$ is used to denote the displacement field associated with a crack of length a. In a similar manner for the second body

$$\Pi(a + \Delta a) = -\tfrac{1}{2} \int_{S'_T} T_i u_i(a + \Delta a)\, dS$$

where S'_T is the union of S_T and the additional traction-free crack surfaces associated with the increment of crack length Δa. Since $T_i = 0$ on the latter surfaces, then

$$\Delta\Pi = \Pi(a + \Delta a) - \Pi(a) = -\tfrac{1}{2}\int_S T_i[u_i(a + \Delta a) - u_i(a)]\,dS \quad (3.3\text{-}10)$$

The above integral has been extended over the entire boundary S of the first body since $u_i(a + \Delta a) = u_i(a)$ on S_u.

The displacement field $u_i(a)$ of the first body is a kinematically admissible field for the second body. If the stress components σ_{2i} on $a < x_1 < a + \Delta a$ and $x_2 = 0$ are included as part of the tractions, then the field $u_i(a + \Delta a)$ of the second body is kinematically admissible for the first body. The principle of virtual work permits writing

$$\int_S T_i u_i(a + \Delta a)\,dS + \int_a^{a+\Delta a} [u_i^-(a + \Delta a) - u_i^+(a + \Delta a)]\sigma_{2i}|_{x_2=0}\,dx_1$$
$$= \int_A \sigma_{ij}(a)\varepsilon_{ij}(a + \Delta a)\,dA \quad (3.3\text{-}11)$$

where $u_i^+(a + \Delta a)$ and $u_i^-(a + \Delta a)$ are the displacement components of the upper and lower crack faces, respectively. Betti's reciprocal theorem and the principle of virtual work provide that

$$\int_A \sigma_{ij}(a)\varepsilon_{ij}(a + \Delta a)\,dA = \int_A \sigma_{ij}(a + \Delta a)\varepsilon_{ij}(a)\,dA$$
$$= \int_S T_i u_i(a)\,dS \quad (3.3\text{-}12)$$

The combination of Equations (3.3-11) and (3.3-12) leads to

$$\int_S T_i[u_i(a + \Delta a) - u_i(a)]\,dS$$
$$= \int_a^{a+\Delta a} [u_i^+(a + \Delta a) - u_i^-(a + \Delta a)]\sigma_{2i}(a)|_{x_2=0}\,dx_1 \quad (3.3\text{-}13)$$

Substitute Equation (3.3-13) into Equation (3.3-10) and make the change of variable $X_1 = x_1 - a$ to obtain

$$-\Delta\Pi = \tfrac{1}{2}\int_0^{\Delta a} [u_i^+(a + \Delta a) - u_i^-(a + \Delta a)]\sigma_{2i}(a)|_{x_2=0}\,dX_1 \quad (3.3\text{-}14)$$

The right side of this equation is the work that must be done during quasi-static application of the crack-plane stresses σ_{2i} on $a < x_1 < a + \Delta a$ and $x_2 = 0\pm$ to close a crack of length $a + \Delta a$ by an amount Δa. The integral of Equation (3.3-14) is referred to as the crack closure integral (3.52). Because of the reversible nature of elastic bodies, this is also the energy that would be released during a quasi-static virtual crack extension Δa. Thus,

$$G = \lim_{\Delta a \to 0}\left\{\frac{1}{2\Delta a}\int_0^{\Delta a} [u_i^+(a + \Delta a) - u_i^-(a + \Delta a)]\sigma_{2i}(a)|_{x_2=0}\,dX_1\right\} \quad (3.3\text{-}15)$$

For Mode I, $u_2^+ = -u_2^-$ and $\sigma_{21} = \sigma_{23} = 0$ on $x_2 = 0$. The asymptotic forms,

$$\sigma_{22}(a)\big|_{x_2=0} = \frac{K_\mathrm{I}(a)}{(2\pi X_1)^{\frac{1}{2}}}$$

and

$$u_2^+(a + \Delta a) = \frac{\kappa + 1}{2\mu} K_\mathrm{I}(a + \Delta a)\left(\frac{\Delta a - X_1}{2\pi}\right)^{\frac{1}{2}}$$

from Equations (3.1-38) and (3.1-39) are sufficient for evaluating the right side of Equation (3.3-15). The stress intensity factors for the prescribed loading and crack lengths a and $a + \Delta a$ are denoted, respectively, by $K_I(a)$ and $K_I(a + \Delta a)$. Hence,

$$\begin{aligned} G &= \lim_{\Delta a \to 0} \left\{ \frac{(\kappa + 1)K_I(a + \Delta a)K_I(a)}{4\pi\mu\Delta a} \int_0^{\Delta a} \left(\frac{\Delta a - X_1}{X_1}\right)^{\frac{1}{2}} dX_1 \right\} \\ &= \frac{(\kappa + 1)K_\mathrm{I}^2}{8\mu} = \frac{K_\mathrm{I}^2}{E'} \end{aligned} \tag{3.3-16}$$

where $E' = E$ for plane stress and $E' = E/(1 - \nu^2)$ for plane strain. This relationship between G and K_I was established by Irwin (3.52). When all three modes of deformation are present, Equation (3.3-15) yields

$$G = \frac{1}{E'}(K_\mathrm{I}^2 + K_\mathrm{II}^2) + \frac{1}{2\mu} K_\mathrm{III}^2 \tag{3.3-17}$$

When K_I, for example, attains its critical value, then G must also reach its critical value and Equation (3.3-16) implies that

$$G_c = K_c^2/E' \tag{3.3-18}$$

Consequently, for linear elastic bodies, the stress intensity factor and the energy balance approaches to fracture are equivalent.

If Equation (3.3-18) were used to compute G_c from K_c, say, for a metal, the value of G_c so determined would be several orders of magnitude greater than the surface energy 2γ. Or, conversely, if G were equated to the surface energy, then unrealistically small failure loads would be predicted for metals. Until about 1950 it was thought that the Griffith energy balance theory of fracture was only applicable to brittle materials such as glass. About that time Irwin (3.53) and Orowan (3.54) independently recognized that the most significant part of the released energy went not into surface energy, but was dissipated in the plastic flow around the crack tip and in the creation of a new plastic zone as the crack tip extends. The energy balance approach to fracture is known as the Griffith-Orowan-Irwin theory and the fracture criterion is frequently written as

$$G = G_c = R = 2\Gamma \tag{3.3-19}$$

where R and 2Γ are variously referred to as the plastic energy dissipation rate, work of fracture, or simply fracture toughness. Literal interpretations of the latter have been the source of much confusion. It is perhaps best to consider them as simply the critical value of the energy release rate parameter at incipient crack extension; cf., K and K_c.

3.3.2 *The J-Integral*

Equation (3.3-15) for the energy release rate is only valid for linear elastic material behavior. A generalization to nonlinear elastic materials leads to the path independent J-integral, which plays an important role not only in elastic but also elastic-plastic fracture mechanics. Although Eshelby (3.55) was the first to derive this integral, Rice (3.39) was apparently the first to recognize its potential use in fracture mechanics.

Consider the two-dimensional deformation of the nonlinear elastic body shown in Figure 3.7. In the absence of body forces the potential energy of the body is

$$\Pi(a) = \int_A W\,dA - \int_{\Gamma_T} T_i u_i\,ds \tag{3.3-20}$$

where Γ_T denotes the contour of the body on which the tractions are prescribed and A is the total area of the two-dimensional body. The tractions are assumed to be independent of a and the crack surfaces are taken to be traction free. Differentiating Equation (3.3-20), one obtains

$$\frac{d\Pi}{da} = \int_A \frac{dW}{da}\,dA - \int_{\Gamma_0} T_i \frac{du_i}{da}\,ds \tag{3.3-21}$$

The contour of the line integral can be extended along the boundary Γ_0 of the body in the counterclockwise direction from the lower crack face to the upper one since $du_i/da = 0$ on the boundary Γ_u where the displacements are prescribed independently of a. In performing the differentiation it is convenient to introduce a coordinate system $X_i = x_i - a\delta_{i1}$ attached to the crack tip.

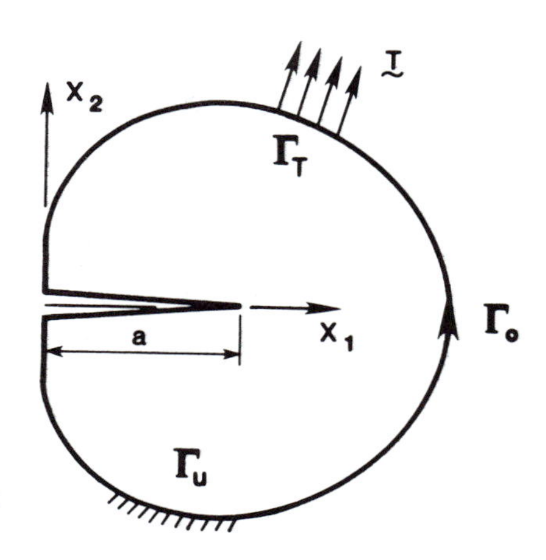

Figure 3.7 A plane, cracked nonlinear elastic body.

It follows that

$$\frac{d}{da} = \frac{\partial}{\partial a} + \frac{\partial X_1}{\partial a}\frac{\partial}{\partial X_1} = \frac{\partial}{\partial a} - \frac{\partial}{\partial X_1} = \frac{\partial}{\partial a} - \frac{\partial}{\partial x_1}$$

since $\partial X_1/\partial a = -1$ and $\partial/\partial X_1 = \partial/\partial x_1$. Therefore, Equation (3.3-21) becomes

$$\frac{d\Pi}{da} = \int_A \left(\frac{\partial W}{\partial a} - \frac{\partial W}{\partial x_1}\right) dA - \int_{\Gamma_0} T_i \left(\frac{\partial u_i}{\partial a} - \frac{\partial u_i}{\partial x_1}\right) ds \tag{3.3-22}$$

But

$$\frac{\partial W}{\partial a} = \frac{\partial W}{\partial \varepsilon_{ij}}\frac{\partial \varepsilon_{ij}}{\partial a} = \sigma_{ij}\left(\frac{\partial u_i}{\partial a}\right)_{,j}$$

where the constitutive relation, Equation (2.3-4), and symmetry of the stress tensor have been used. Since $\partial u_i/\partial a$ is a kinematically admissible displacement, then the principle of virtual work permits writing

$$\int_A \frac{\partial W}{\partial a} dA = \int_A \sigma_{ij}\left(\frac{\partial u_i}{\partial a}\right)_{,j} dA = \int_{\Gamma_0} T_i \frac{\partial u_i}{\partial a} ds$$

Therefore, Equation (3.3-22) reduces to

$$-\frac{d\Pi}{da} = \int_A \frac{\partial W}{\partial x_1} dA - \int_{\Gamma_0} T_i \frac{\partial u_i}{\partial x_1} ds$$

which upon application of the divergence theorem becomes

$$\begin{aligned} -\frac{d\Pi}{da} &= \int_{\Gamma_0} \left(W n_1 - T_i \frac{\partial u_i}{\partial x_1}\right) ds \\ &= \int_{\Gamma_0} \left(W\, dx_2 - T_i \frac{\partial u_i}{\partial x_1} ds\right) \end{aligned} \tag{3.3-23}$$

in which $n_1\, ds = dx_2$ has been used to write the last integral.

Rather than considering the contour of the body for the path of integration, Rice (3.39) considered an arbitrary contour Γ starting from the lower crack face extending counterclockwise around the crack tip to a point on the upper face (see Figure 3.8) and defined

$$J = \int_{\Gamma} \left(W n_1 - T_i \frac{\partial u_i}{\partial x_1}\right) ds \tag{3.3-24}$$

It can be shown that the J-integral is path independent. Let J_1 denote J obtained for any other contour Γ_1. One can write

$$J_1 - J = \int_{\Gamma_1 + \Gamma + S_1 + S_2} \left(W n_1 - T_i \frac{\partial u_i}{\partial x_1}\right) ds$$

where the contour has been closed by including S_1 and S_2 on which $T_i = n_1 = 0$. This contour is traversed such that the enclosed area A is on the left.

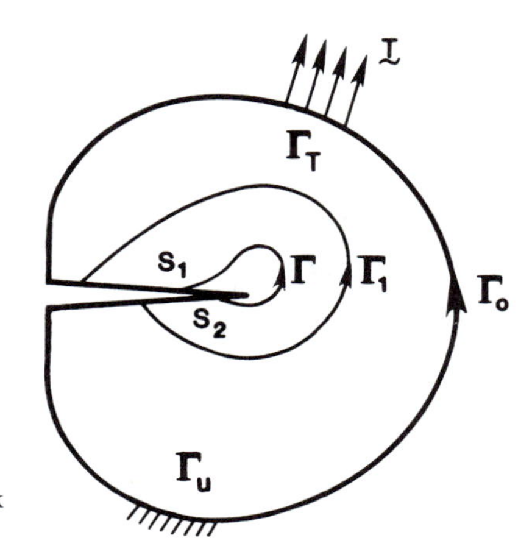

Figure 3.8 Contours enclosing a crack tip.

Once more the divergence theorem is invoked to yield

$$J_1 - J = \int_A \left[\frac{\partial W}{\partial x_1} - \frac{\partial}{\partial x_j}\left(\sigma_{ij}\frac{\partial u_i}{\partial x_1}\right)\right] dA$$

$$= \int_A \left[\frac{\partial W}{\partial \varepsilon_{ij}}\frac{\partial \varepsilon_{ij}}{\partial x_1} - \sigma_{ij}\frac{\partial}{\partial x_1}(u_{i,j})\right] dA$$

$$= \int_A \left[\sigma_{ij}\frac{\partial \varepsilon_{ij}}{\partial x_1} - \sigma_{ij}\frac{\partial}{\partial x_1}(u_{i,j})\right] dA = 0$$

Hence, $J = J_1$ and J is independent of the path. Since the J-integral is path independent, then according to Equations (3.3-23) and (3.3-24) the J-integral is equal to the rate of decrease of the potential energy; that is,

$$J = -\frac{d\Pi}{da} \tag{3.3-25}$$

Consequently, for a linear elastic material J and G are synonymous and

$$J = G = \frac{1}{E'}(K_{\mathrm{I}}^2 + K_{\mathrm{II}}^2) + \frac{1}{2\mu}K_{\mathrm{III}}^2 \tag{3.3-26}$$

The path independence of J can be exploited at times to determine the stress intensity factor without the necessity of involved computations. For example, consider the semi-finite crack in the infinite strip of Figure 3.9. Assume the lateral edges ($x_2 = |h|$) are clamped and then symmetrically displaced u_{20}. For the contour Γ shown, $n_1 = \partial u_i/\partial x_1 = 0$ on $x_2 = |h|$ and $\sigma_{ij} = 0$ at $x_1 = -\infty$. Therefore, the only contribution to J for the contour Γ must occur at $x_1 = \infty$, where $\partial u_i/\partial x_1 = 0$ and $\varepsilon_{22} = u_{20}/h$. Hence, J reduces to

$$J = \int_{-h}^{h} W\big|_{x_1 = \infty}\, dx_2$$

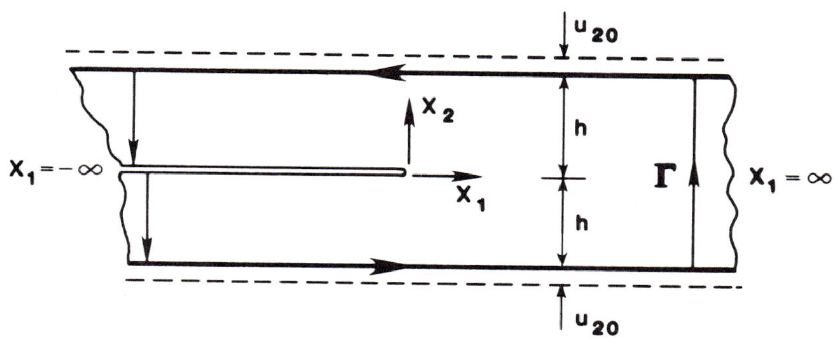

Figure 3.9 A semi-infinite crack in an infinite strip.

where by Equation (2.3-11)

$$W\big|_{x_1=\infty} = \frac{(1-\nu)E\varepsilon_{22}^2}{2(1+\nu)(1-2\nu)}$$

for plane strain. Therefore,

$$J = \frac{(1-\nu)Eu_{20}^2}{(1+\nu)(1-2\nu)h}$$

or, equivalently, from Equation (3.3-26)

$$K_{\text{I}} = \frac{Eu_{20}}{[(1+\nu)^2(1-2\nu)h]^{\frac{1}{2}}} \tag{3.3-27}$$

In an effort to eliminate the unrealistic prediction of singular stresses that accompany an elastic analysis, Dugdale (3.56) and Barenblatt (3.57) independently introduced yielded or cohesive strip zones extending from the crack tip, as depicted in Figure 3.10, to model the inelastic response of real materials in this region. In this model the opening of prospective fracture surfaces ahead of the crack tip is assumed to be opposed by a cohesive stress that Dugdale took to be the yield stress of the material. The extent d of the cohesive zone is determined by the condition that the stresses be nonsingular.

Let σ be the cohesive stress that in general can depend upon the separation $\delta = u_2^+ - u_2^-$ of the upper and lower prospective crack surfaces. For a contour shrunk to the boundary of the right yield or cohesive zone $n_1 = T_1 = 0$ and J becomes

$$J = -\int_0^d \sigma\left[\frac{\partial u_2^+}{\partial x_1} - \frac{\partial u_2^-}{\partial x_1}\right] dx_1 = -\int_0^d \sigma \frac{\partial \delta}{\partial x_1}\, dx_1$$

$$= \int_0^{\delta_t} \sigma(\delta)\, d\delta \tag{3.3-28}$$

where δ_t is the crack-tip opening displacement. If the cohesive stress is taken to be the yield stress—that is, $\sigma = \sigma_y$—then Equation (3.3-28) yields

$$J = \sigma_y\, \delta_t \tag{3.3-29}$$

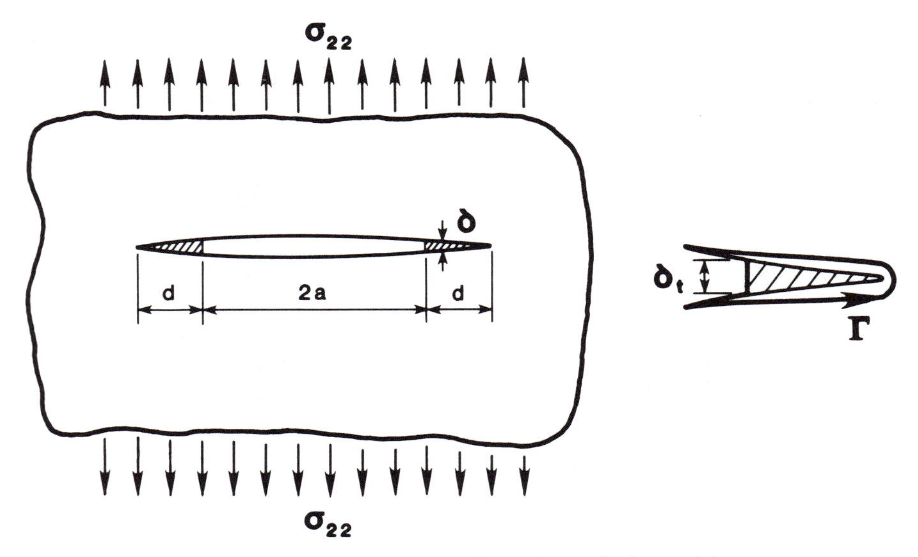

Figure 3.10 Dugdale yielded strip or cohesive model.

If the extent of the cohesive zone is small compared to any other characteristic dimension of the body, then sufficiently remote from these zones the deformation field will differ only imperceptibly from the elastic solution that ignores these zones. Since J is path independent, then $J = G$ for a remote contour. Consequently,

$$J = G = K_{\mathrm{I}}^2/E' = \sigma_y \, \delta_t \tag{3.3-30}$$

Hence, the J-integral, the energy release rate, the stress intensity factor, and the crack-tip opening displacement are all equivalent fracture parameters within the confines of small-scale yielding. Consequently, when any one of the four attains its critical value, then the others must also reach simultaneously their critical values. Therefore, the fracture criterion can be expressed in terms of any of the four parameters. Which parameter is the more basic is primarily an academic question. From an application viewpoint the choice revolves about which one is more convenient to compute and to extract from measurements. In this regard, the J-integral or, equivalently, the energy release rate and the stress intensity factor are about equal and offer distinct advantages over the crack-tip opening displacement. For example, the J-integral can be readily incorporated into finite element codes and thereby eliminate the need for extrapolated estimates of the stress intensity factor.

3.3.3 *Other Invariant Integrals*

The J-integral is not the only invariant integral or conservation law of elastostatics. For example, Knowles and Sternberg (3.58) have established a total of three such conservation laws. For a two-dimensional deformation field these integrals are

$$J_\alpha = \int_C (Wn_\alpha - T_i u_{i,\alpha})\, ds \tag{3.3-31}$$

$$L = \int_C \varepsilon_{3\alpha\beta}(Wx_\beta n_\alpha + T_\alpha u_\beta - T_k u_{k,\alpha} x_\beta)\, ds \tag{3.3-32}$$

$$M = \int_C (Wx_\alpha n_\alpha - T_k u_{k,\alpha} x_\alpha)\, ds \tag{3.3-33}$$

where C is a closed contour in the x_1x_2-plane. The J-integral is the first component of the vector J_α. Whereas the integrals J_α and L will vanish for any closed contour in an elastic (linear or nonlinear) region containing no singularities, the integral M will vanish only for contours in linear elastic regions. If the contour encloses a singularity, the integrals will, in general, differ from zero. However, the value of the integral for any other contour enclosing this singularity and no other will remain invariant. The principal advantage of these integrals is that they describe an invariant characteristic of the singularity. If the contour encloses the crack tip, it is important that the contour extending along the crack faces to the crack tip be included in order to ensure path independence of J_2.

For the general three-dimensional deformation field, these integrals generalize to

$$J_k = \int_S (Wn_k - T_i u_{i,k})\, dS \tag{3.3-34}$$

$$L_k = \int_S \varepsilon_{kij}(Wx_j n_i + T_i u_j - T_l u_{l,i} x_j)\, dS \tag{3.3-35}$$

$$M = \int_S (Wx_i n_i - T_j u_{j,i} x_i - \tfrac{1}{2} T_i u_i)\, dS \tag{3.3-36}$$

where S is a closed surface with components n_i for the outward unit normal. The integral J_k was originally established by Eshelby (3.55).

Budiansky and Rice (3.59) have interpreted J_k, L_k, and M as being the energy release rate when a cavity is translated along the x_k-axis relative to the material body, is rotated about the x_k-axis, and is expanded uniformly, respectively. With the exception of the work of King and Herrmann (3.60), there has been apparently little effort to apply L_k and M to physical fracture problems.

Fletcher (3.61) has extended these invariant integrals to linear elastodynamics. Cherepanov (3.3,3.62) has developed invariant integrals that can be used to characterize singularities in other physical fields such as electromagnetics and hydrodynamics.

The preceding two-dimensional invariant integrals involved only contour integrals. Kishimoto et al. (3.63) introduced the path-independent integral $\hat{J}_\alpha$

defined by

$$\hat{J}_\alpha = \int_{\Gamma+\Gamma_s} \left(W n_\alpha - T_i \frac{\partial u_i}{\partial x_\alpha} \right) ds + \int_A \left(\sigma_{ij} \frac{\partial \varepsilon_{ij}^*}{\partial x_\alpha} - F_i \frac{\partial u_i}{\partial x_\alpha} \right) dA \quad (3.3\text{-}37)$$

for static loading with body force F_i. The area integral extends over the region bounded by the contour Γ enclosing the crack tip and the portion Γ_s of the crack faces between Γ and the crack tip. The eigen strain components ε_{ij}^* are defined in terms of the total strain components ε_{ij} and the stress-induced elastic strain components ε_{ij}^e by

$$\varepsilon_{ij}^* = \varepsilon_{ij} - \varepsilon_{ij}^e \quad (3.3\text{-}38)$$

Examples of eigen strains are thermal strains, moisture-absorption strains, and plastic strains. The elastic strain energy density W is the same function of the elastic strain components introduced previously; that is, $W = W(\varepsilon_{ij}^e)$ such that $\sigma_{ij} = \partial W / \partial \varepsilon_{ij}^e$.

It is clear that, when eigen strains and body forces are present, the $\hat{J}_\alpha$-integral is not expressible by only a contour integral. In the absence of these quantities, $\hat{J}_\alpha$ reduces to J_α and, moreover, $\hat{J}_1$ conforms to the J-integral. The $\hat{J}_\alpha$-integral enjoys the same physical interpretation as the J_α-integral.

The $\hat{J}_\alpha$-integral has proven to be useful in studies of elastic-plastic fracture and fracture in the presence of thermal gradients. For example, the eigen strains due to a temperature change θ from the natural state are

$$\varepsilon_{ij}^* = \alpha_{ij}\,\theta \quad (3.3\text{-}39)$$

where α_{ij} are the coefficients of thermal expansion. When body forces are neglected, the introduction of Equation (3.3-39) into Equation (3.3-37) yields

$$\hat{J}_\alpha = \int_{\Gamma+\Gamma_s} \left(W n_\alpha - T_i \frac{\partial u_i}{\partial x_\alpha} \right) ds + \int_A \alpha_{ij} \sigma_{ij} \frac{\partial \theta}{\partial x_\alpha}\, dA \quad (3.3\text{-}40)$$

As expected $\hat{J}_\alpha$ reduces to J_α for an isothermal field. Furthermore, $\hat{J}_1$ is identical to the J_θ-integral introduced by Ainsworth et al. (3.64). The path-independent integral introduced by Wilson and Yu (3.65) for a linear elastic, isotropic material can be shown to be a special case of $\hat{J}_1$ and J_θ.

When thermal strains are present, one should be cognizant of the differences in the strain energy density functions used by various investigators. For example, Kishimoto et al. (3.63) and Ainsworth et al. (3.64) define

$$W = \int_0^{\varepsilon_{kl}^e} \sigma_{ij}\, d\varepsilon_{ij}^e \quad (3.3\text{-}41)$$

which for a linear elastic, isotropic material becomes

$$W = \mu \left(\varepsilon_{ij}^e \varepsilon_{ij}^e + \frac{\nu}{1 - 2\nu} \varepsilon_{ii}^e \varepsilon_{jj}^e \right) \quad (3.3\text{-}42)$$

By contrast Wilson and Yu (3.65) use

$$W = \mu \left(\varepsilon_{ij} \varepsilon_{ij} + \frac{\nu}{1 - 2\nu} \varepsilon_{ii} \varepsilon_{jj} \right) - \frac{E \alpha \theta \varepsilon_{ii}}{2(1 - 2\nu)} \quad (3.3\text{-}43)$$

and Gurtin (3.66)

$$W = \mu\left(\varepsilon_{ij}\varepsilon_{ij} + \frac{\nu}{1-2\nu}\,\varepsilon_{ii}\varepsilon_{jj}\right) \tag{3.3-44}$$

These variances in the definition of the strain energy density yield path-independent integrals that appear outwardly to be different but in reality are identical.

As an alternative to J, Blackburn (3.67) proposed the path-independent integral

$$J^* = \lim_{\rho\to 0}\int \tfrac{1}{2}\sigma_{ij}\varepsilon_{ij}\,dx_2 - T_i\frac{\partial u_i}{\partial x_1}\,ds \tag{3.3-45}$$

$$= \int_\Gamma \tfrac{1}{2}\sigma_{ij}\varepsilon_{ij}\,dx_2 - T_i\frac{\partial u_i}{\partial x_1}\,ds + \lim_{\rho\to 0}\int_A\left[\tfrac{1}{2}\sigma_{ij}\frac{\partial^2 u_i}{\partial x_1\,\partial x_j} - \tfrac{1}{2}\frac{\partial\sigma_{ij}}{\partial x_1}\frac{\partial u_i}{\partial x_j} - \frac{\partial}{\partial x_3}\left(\sigma_{i3}\frac{\partial u_i}{\partial x_1}\right)\right]dA \tag{3.3-46}$$

where ρ is the radius of a small circle centered at the crack tip and A is the area bounded by this circle, the contour Γ, and the crack surfaces. Whenever it is possible to construct a contour Γ lying wholly within a linear elastic region, then the contour integral in Equation (3.3-46) is equal to J. Hence, the difference between J and J^* is given by the area integral in Equation (3.3-46). If the material enclosed by Γ is entirely linear elastic, this integral vanishes and J^* is identical to J. While it has been purported that J^* will be closer to the true potential energy release rate, this is not the case for nonlinear elastic (power law hardening) materials for which $J^* < J$. Therefore, an energy release rate interpretation can not generally be associated with J^*.

Blackburn et al. (3.68) and Batte et al. (3.69) claim J^* can be used where creep, elastic unloading from a plastically deformed state, or thermal or inhomogenity effects render J inadmissible. Because the theoretical basis for J^* is not clear, it is not entirely certain that these claims are valid.

The number of path-independent integrals appears to be unlimited. It is always possible to generate another one by adding or subtracting two path-independent integrals. For example, if the path-independent integral

$$\oint d(u_i\sigma_{i2})$$

is subtracted from J, then one obtains

$$I = -\int_\Gamma\left[(\sigma_{ij}\varepsilon_{ij} - W)\,dx_2 - n_j u_i\frac{\partial\sigma_{ij}}{\partial x_1}\,ds\right] \tag{3.3-47}$$

The quantity within the parentheses is recognized as the complementary energy density W^*. The path-independent I-integral was originally established by Bui (3.70) using complementary energy principles. In this sense the I-integral is the dual of the J-integral. Because displacement-based finite element methods yield inherently less precise stress fields, this expression containing the derivative of the stress tensor lacks the computational efficacy of the J-integral. However, the advantages of J virtually vanish when a hybrid finite element method that yields the stress and displacement fields to the same precision is used.

The path-independent integrals considered until now have been limited to small (infinitesimal) deformation. Atluri (3.71) has generalized the conservation law, Equation (3.3-31), of Knowles and Sternberg (3.58) to finite deformations. Included in this generalization are the effects of body forces, material accelerations, and arbitrary traction and displacement conditions on the crack faces. While not explicitly included, thermal effects can be readily accommodated by replacing the body force F_i by $-\alpha E\theta_{,i}/(1\text{-}2\nu)$. Hence, Atluri's generalization includes the path independent integrals of Rice (3.39), Kishimoto et al. (3.63), Ainsworth et al. (3.64), and Wilson and Yu (3.65) as special cases.

3.4 The Plastic Zone and Fracture Toughness

Linear elastic fracture mechanics is based upon the condition that the size of the plastic zone attending the crack tip is small compared to the K-dominant region, the crack length, and any other characteristic geometric length. Within these restrictions of small-scale yielding the equivalence expressed by Equation (3.3-30) is justifiable. Based upon the near crack-tip elastic analysis, estimates of the size and shape of the plastic zone will be developed. The relationship between the size of the plastic zone and the fracture toughness will be examined.

Equation (3.1-38) yields

$$\sigma_{22} = \frac{K_{\mathrm{I}}}{(2\pi r)^{\frac{1}{2}}} \tag{3.4-1}$$

for the ideally elastic Mode I stress distribution on the crack plane ($\theta = 0$). This distribution and a hypothetical elastic-perfectly plastic distribution are shown schematically in Figure 3.11. The length

$$r_y = \frac{1}{2\pi}\left(\frac{K_{\mathrm{I}}}{\sigma_y}\right)^2 \tag{3.4-2}$$

identifies the point on the crack plane where the elastic stress σ_{22} of Equation (3.4-1) equals the uniaxial yield stress σ_y. Local yielding near the crack tip in a real material leads to a redistribution of the stress as depicted in Figure 3.11. To a first approximation this yielding causes the elastic load over the region $0 < r < r_y$ on the crack plane to be uniformly distributed over the length, r_p, the extent of yielding on this plane. That is,

$$\int_0^{r_y} \sigma_{22}\, dr = \sigma_y r_p$$

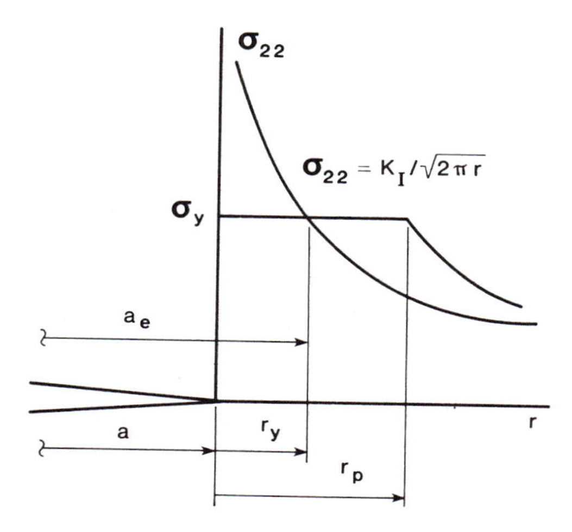

Figure 3.11 Elastic and inelastic crack plane stress distributions.

which for the distribution of Equation (3.4-1) provides

$$r_p = 2r_y = \frac{1}{\pi}\left(\frac{K_I}{\sigma_y}\right)^2 \tag{3.4-3}$$

for a state of plane stress. Irwin (3.72) estimated that the constraint introduced by the conditions of plane strain elevates the stress required to produce yielding by a factor of $\sqrt{3}$. The plane strain equivalent of Equation (3.4-3) is

$$r_p = 2r_y = \frac{1}{3\pi}\left(\frac{K_I}{\sigma_y}\right)^2 \tag{3.4-4}$$

The Dugdale model (see Figure 3.10) can also be used to provide yet another estimate of the extent of the plastic zone. In this model the opening of the crack by a uniform remote stress σ is restrained in part by a uniform stress $\sigma_{22} = \sigma_y$ in the strip cohesive or plastic zone of length d. Viewing the crack to be of length $2a + 2d$ and employing superposition one is left with the residual problem on $x_2 = 0$ of $p_2(x_1) = \sigma$ on $|x_1| < a$ and $p_2(x_1) = \sigma - \sigma_y$ on $a < |x_1| < a + d$. Making use of Equation (3.2-20) and the condition that the stress intensity factor for the Dugdale model must vanish for nonsingular stresses one obtains

$$\sigma\alpha + (\sigma - \sigma_y)\left(\frac{\pi}{2} - \alpha\right) = 0 \tag{3.4-5}$$

where

$$\alpha = \sin^{-1}[a/(a + d)] \tag{3.4-6}$$

It follows from Equation (3.4-5) that

$$d = a\left[\sec\left(\frac{\pi}{2}\frac{\sigma}{\sigma_y}\right) - 1\right] \tag{3.4-7}$$

Recalling that $K_{\mathrm{I}} = \sigma(\pi a)^{\frac{1}{2}}$ in the absence of yielding and taking σ to be small compared to σ_y, we find from Equation (3.4-7) the approximation

$$d = \frac{\pi}{8}\left(\frac{K_{\mathrm{I}}}{\sigma_y}\right)^2 \tag{3.4-8}$$

for the extent of yielding. The latter compares favorably with the plane stress estimate of Equation (3.4-3).

An effect of yielding is to increase the displacements or, equivalently, to reduce the stiffness of the body relative to the ideally elastic one. Based upon the antiplane, elastic-perfectly plastic solution of Hult and McClintock (3.73), Irwin (3.74) argued that the same effect can be approximated in the ideally elastic body by increasing the effective length of the crack. As a first approximation Irwin, who viewed the crack tip as being centered in the plastic zone, incremented the crack length by the plastic zone radius and introduced the effective crack length

$$a_e = a + r_y \tag{3.4-9}$$

The latter expresses what has become known as the Irwin's plastic zone correction for the crack length. This effective length is used in computing the stress intensity factor. Because the effective stress intensity factor is a function of a_e, which in turn depends upon the former, an iterative solution is usually required to establish the effective stress intensity factor.

Equations (3.4-3), (3.4-4), and (3.4-8) only provide estimates for the size of the plastic zone. Before the shape of the zone can be established, a yield criterion must be specified. The von Mises yield criterion of Equation (2.6-10) can be expressed as

$$(\sigma_1 - \sigma_2)^2 + (\sigma_2 - \sigma_3)^2 + (\sigma_3 - \sigma_1)^2 = 2\sigma_y^2 \tag{3.4-10}$$

For the plane problems the principal stresses are

$$\left.\begin{matrix}\sigma_1\\ \sigma_2\end{matrix}\right\} = \frac{\sigma_{11} + \sigma_{22}}{2} \pm \left[\left(\frac{\sigma_{11} - \sigma_{22}}{2}\right)^2 + \sigma_{12}^2\right]^{\frac{1}{2}}$$
$$\sigma_3 = \begin{cases} 0, & \text{plane stress} \\ \nu(\sigma_1 + \sigma_2), & \text{plane strain}\end{cases} \tag{3.4-11}$$

Strictly speaking the stress field from an elastic-plastic analysis should be used in establishing the shape of the plastic zone. We shall have to wait until Chapter 5 for such an analysis. In lieu of that, a first approximation to the shape of the zone can be obtained by using the elastic fields. The introduction of Equation (3.1-38) for the Mode I stress field into Equation (3.4-11) leads to

$$\left.\begin{matrix}\sigma_1\\ \sigma_2\end{matrix}\right\} = \frac{K_{\mathrm{I}}}{(2\pi r)^{\frac{1}{2}}}(1 \pm \sin\theta/2)\cos\theta/2$$
$$\sigma_3 = \begin{cases} 0, & \text{plane stress} \\ \dfrac{2\nu K_{\mathrm{I}}}{(2\pi r)^{\frac{1}{2}}}\cos\theta/2, & \text{plane strain}\end{cases} \tag{3.4-12}$$

The substitution of Equation (3.4-12) into Equation (3.4-10) provides, respectively,

$$r_y(\theta) = \frac{K_I^2}{4\pi\sigma_y^2}\left[(1-2v)^2(1+\cos\theta) + \tfrac{3}{2}\sin^2\theta\right] \quad (3.4\text{-}13)$$

and

$$r_y(\theta) = \frac{K_I^2}{4\pi\sigma_y^2}\left[1+\cos\theta + \tfrac{3}{2}\sin^2\theta\right] \quad (3.4\text{-}14)$$

for the plane strain and plane stress boundaries of the plastic zone. From a comparison of these boundaries for $v = 0.3$ in Figure 3.12 it is clear that the plane strain zone is significantly smaller. The coordinates x, y of points on these boundaries are normalized with respect to $r_p = (K_I/\sigma_y)^2/\pi$ for plane stress. Plastic zones based upon the Tresca yield condition can be found in

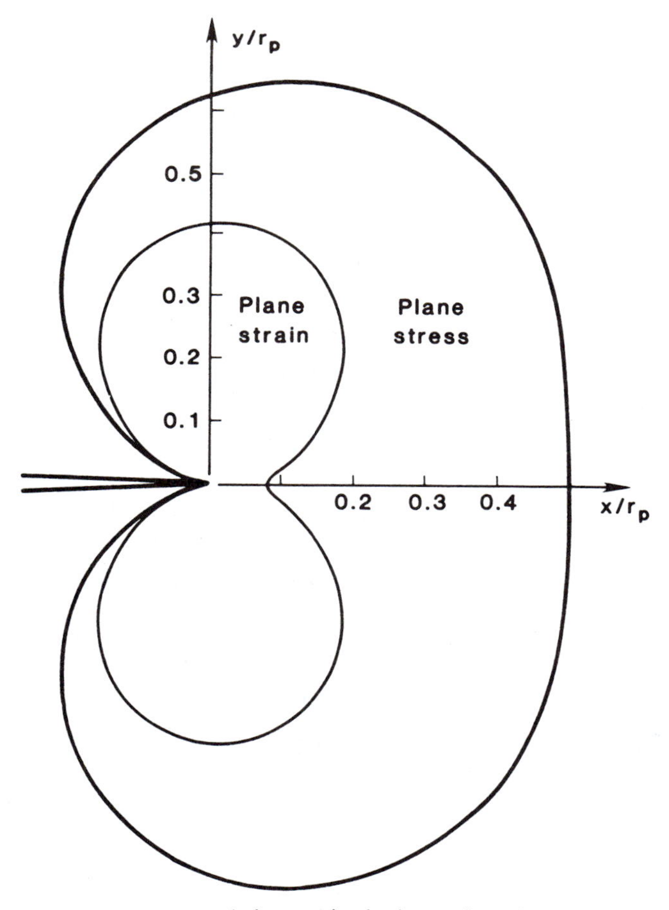

Figure 3.12 Plane stress and plane strain plastic zone boundaries for $v = 0.3$.

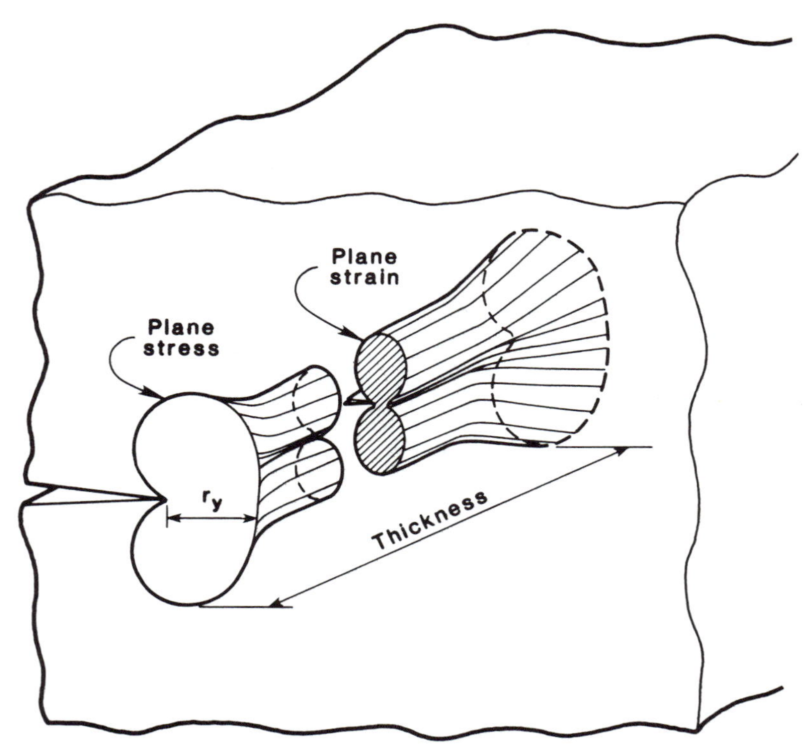

Figure 3.13 Schematic of Mode I plastic zone varying from plane stress at the lateral surfaces to plane strain at the mid-section.

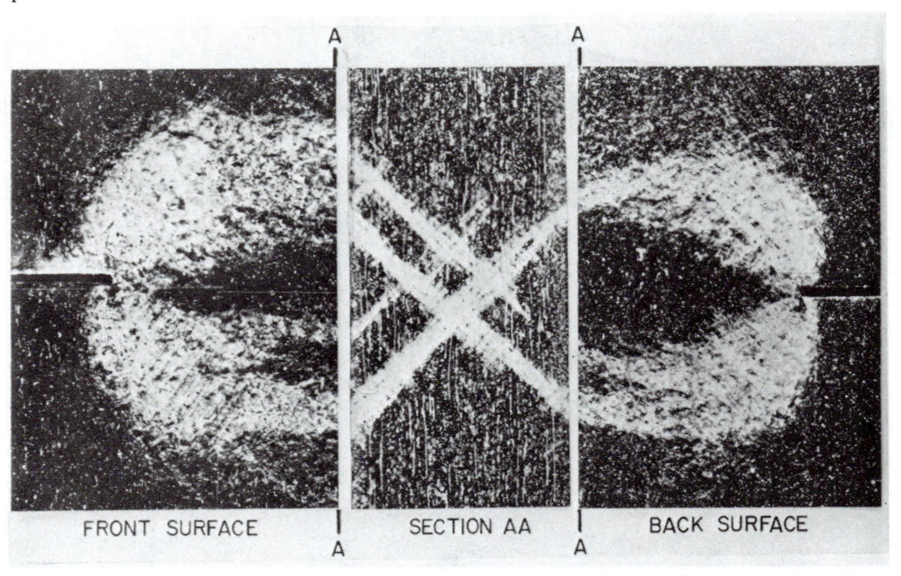

Figure 3.14 Appearance of plane stress plastic deformation at the front surface, a normal section and the back surface of a fracture specimen (3.75).

Broek (3.1), and they do not differ appreciably in appearance from these. While a more refined analysis will yield somewhat different configurations for the elastic-plastic interface, no significant change in the relative sizes of the envelopes is to be expected.

The size of the plastic zone relative to the thickness of the body influences whether the crack tip state of stress is essentially plane stress, plane strain, or a combination of the two. Conversely, the type of stress field dictates the size of the plastic zone. Clearly, the classification of the stress field is not a simple task; particularly, in light of the fact that the traction-free lateral surfaces of the body are in a state of plane stress. Ordinarily this would not be such an important issue if it were not for the fact that the fracture characteristics of a material are influenced by the stress state attending the crack tip. It is possible to establish guidelines for determining whether this state will be predominantly plane stress or plane strain.

When the size of the plastic zone as computed from Equation (3.4-3) is of the order of the thickness B of the body or greater, then insufficient material at the interior exists to prevent through-the-thickness straining. Such deformation is usually evidenced by dimpling or necking in the crack-tip region. Under these conditions the state of stress is predominantly plane stress. At the other extreme where the size of the plastic zone is small—say, $r_p < B/25$—then there is sufficient material present to prevent through-the-thickness straining. Except for the states of plane stress confined to thin boundary layers near the traction-free lateral surfaces, the state of stress is primarily one of plane strain. For intermediate sizes of the plastic zones, $1 > r_p/B > \frac{1}{25}$, the state of plane stress extends beyond the simple boundary layers near the free surfaces, but not completely through the thickness. There is a gradual transition from plane stress at the free surface to plane strain at the interior, as depicted schematically in Figure 3.13.

One can expect the plastic deformation in ductile materials to occur as slipping along planes of maximum shear stress. The principal stress σ_3 is always normal to the free lateral surfaces. For plane stress $\sigma_3 = 0$ and the planes of maximum shear stress make 45-degree angles with the free surfaces. By contrast $\sigma_2 < \sigma_3 < \sigma_1$ for plane strain and the planes of maximum shear stress are normal to the free surfaces.

By loading notched and fatigue-cracked, silicon iron fracture specimens and then etching the polished surfaces, Hahn and Rosenfield (3.75) were able to reveal the plastic zone and the accompanied slip planes. Figure 3.14 shows the appearance of the plastic zone on the front surface, a section normal to the crack plane and the back surface. In these photographs the lighter portions represent regions of rather extensive slipping. The presence of the 45-degree slip bands are consistent with the direction of maximum shear stress in a plate under plane stress. This slip produces a considerable strain in the thickness direction, which appears as localized necking. Figure 3.15 shows evidence of the plastic slip, appearing as dark regions in these photographs, at sections parallel to the specimen's lateral surfaces and demonstrates the anticipated convergence of these slip bands for plane stress as the midsection is

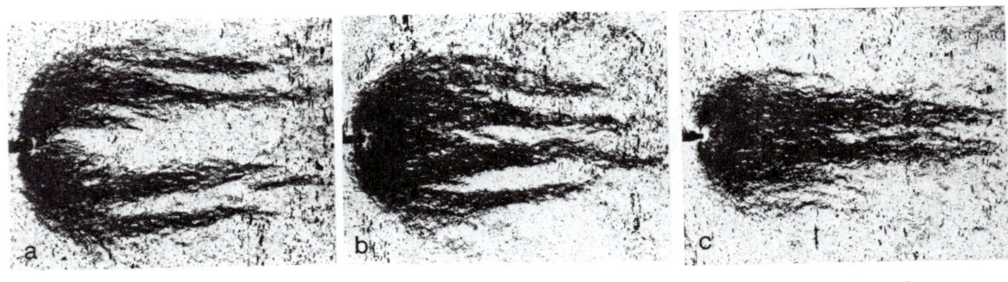

Figure 3.15 Plastic zone revealed by etching (a) the surface of the specimen, (b) a section halfway between the surface and mid-section, and (c) the mid-section (3.75).

approached. These results are for a plastic zone whose size is approximately twice the 5-mm thickness of the specimen and are compatible with the character of plane stress deformation.

On the other hand, for a relatively lower stress level for which the plastic zone is smaller than the thickness of the specimen, the character of the plastic deformation revealed by etching sections parallel to the plate surfaces remained nearly invariant. This is consistent with plane strain deformation in which the slip bands are normal to the surface of the plate.

From these observations Hahn and Rosenfield constructed the crack-tip deformation patterns shown in Figure 3.16. Under plane strain the plastic flow tends to occur around a hinge. In plane stress wedges are typically formed and along their surfaces slipping occurs to produce the rather large through-thickness straining.

It should be clear by now that the thickness of the specimen relative to the size of the plastic zone influences the state of stress and the deformation within the zone. Consequently, the fracture characteristics of the specimen can also be expected to depend upon its thickness. For example fracture tests frequently indicate that the critical stress intensity factor K_c at initiation of crack growth

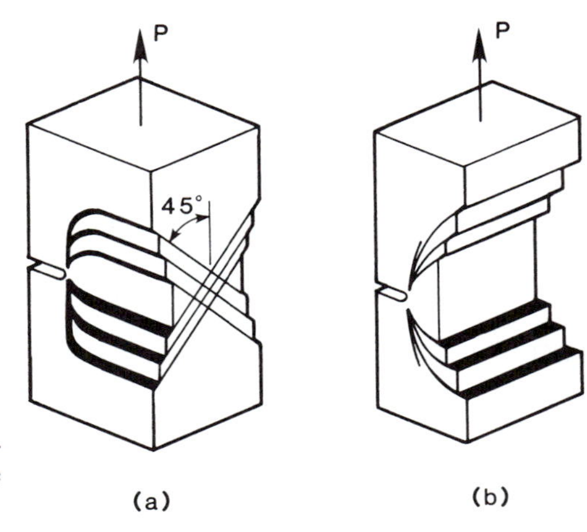

Figure 3.16 Crack tip plastic deformation in (a) plane stress and (b) plane strain.

for Mode I varies with the thickness B of the specimen as depicted in Figure 3.17. Also shown are the distinguishing features of the fracture surfaces that are typical of the various regions of the curve.

For relatively large thicknesses the fracture surface is predominantly flat with relatively small shear lips (slant fracture surfaces) occurring near the free surfaces. The former is typical of plane strain fracture in the interior, whereas the latter is a characteristic of plane stress fracture at the free surfaces. For large thicknesses sufficient constraint can exist to produce a triaxial state of stress. This triaxiality tends to reduce the apparent ductility of the material and fracture proceeds at a lower critical stress intensity factor. Because further increases of the thickness beyond a certain minimum value B_c does not change appreciably the triaxiality, the fracture resistance or toughness remains invariant. This plane strain fracture toughness is denoted by K_{Ic}, is independent of the thickness and is considered to be a material property. It is clear from Figure 3.17 that it also represents the lower limit of the critical value K_c.

At the other extreme of thickness B_m there is minimal constraint provided by the thickness and a biaxial state of stress exists. The loss of triaxiality contributes to an apparent increase in the ductility and, hence, in the fracture resistance. In this case the fracture surfaces are slanted at 45 degrees to the specimen's surfaces and are composed entirely of shear lips. This maximum fracture resistance can be several times larger than the value of K_{Ic}. For still thinner sections the curve may level off or even decrease. Which prevails likely

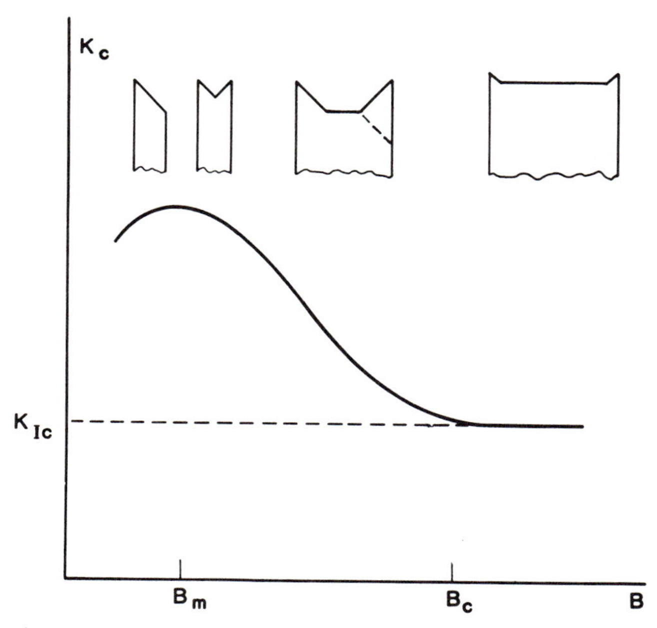

Figure 3.17 Critical Mode I stress intensity factor as a function of specimen thickness.

depends upon the extent of straining the material can withstand prior to failure.

At intermediate thicknesses between B_m and B_c some constraint is present but not sufficient to develop the triaxiality fully. The central portion of the specimen will be nearly in plane strain, whereas the remainder is closer to plane stress. The fracture surface usually consists of a central flat portion and appreciable shear lips at the edges. This combination results in a fracture resistance that is intermediate to the minimum plane strain toughness K_{Ic} and the maximum plane stress toughness.

Because predictions of quasi-static loads required to initiate crack growth based upon the fracture toughness K_{Ic} will be conservative, a great deal of effort has been devoted towards measuring K_{Ic}. The American Society for Testing and Materials (ASTM) has established Standard E-399 (3.76) for the measurement of K_{Ic}. The three point bend and compact tension specimens are frequently used in the determination of K_{Ic}. To produce a sharp crack tip the machined starter notch is fatigue cracked. The specimen is loaded quasistatically and the load required to initiate crack growth is measured. This load and the specimen's geometry are used to calculate the stress intensity factor, which will be equal to K_{Ic} if all the specifications of the Standard are met. The specifications must be rigorously followed in order to establish a valid K_{Ic} value—that is, the value for which one is assured that a larger or thicker specimen will not yield an even smaller quantity. The reader contemplating performing such tests would be well advised to study the ASTM E-399 Standard in detail.

In order to ensure that the size of the plastic zone is small compared to the thickness and other geometric dimensions, the Standard requires for a valid K_{Ic} that

$$
\begin{aligned}
a &> 2.5\left(\frac{K_{Ic}}{\sigma_y}\right)^2 \\
B &> 2.5\left(\frac{K_{Ic}}{\sigma_y}\right)^2 \qquad (3.4\text{-}15) \\
W &> 5.0\left(\frac{K_{Ic}}{\sigma_y}\right)^2
\end{aligned}
$$

where B and W denote the thickness and depth or width of the specimen, respectively. The thickness of the specimen is required to be approximately 50 times the radius of the plane strain plastic zone. For a material with relatively low or moderate toughness and high strength, only a modest thickness is required. A minimum thickness of only 5.5 mm is required for a low-temperature heat treated AISI 4340 steel with $K_{Ic} = 65$ MPa $\text{m}^{\frac{1}{2}}$ and $\sigma_y =$ 1400 Mpa. For the latter a larger thickness may be required to prevent buckling during testing. By contrast an inordinately large thickness may be needed for a high toughness-low strength material. For example, the minimum thickness required for A533B, a nuclear reactor grade steel, having a toughness of $K_{Ic} = 180$ MPa m$^{\frac{1}{2}}$ and a yield stress of $\sigma_y = 350$ MPa is approximately 0.6 m! Due to the large specimens needed in K_{Ic} testing of high toughness-low strength materials, alternative test methods with less stringent size requirements must be used. Depending upon the circumstances LEFM may not be appropriate for these materials.

Compendia of sources of fracture toughness data for metallic alloys have been prepared by Hudson and Seward (3.77, 3.78). Because the fracture toughness depends upon many variables, the data should not be used indiscriminately. The user should be sure that all factors influencing the toughness are properly considered.

Prior to the development of fracture mechanics the Charpy V-notch (CVN) impact test was used to compare the fracture resistance of materials. In the standard Charpy test (3.79) a notched three-point bend specimen is impacted with a pendulum. The fracture energy is equated to the energy lost by the pendulum during the impact. The test is relatively inexpensive, simple, easy to conduct and widely used. Consequently, a great deal of data has been generated. These data have formed the bases of empirical correlations and engineering judgments that have been used to translate Charpy energies into specifications for material toughnesses in designs.

When CVN tests are conducted on low or intermediate strength steels at different temperatures, the dependence of the Charpy energy upon temperature has the character depicted in Figure 3.18. At the lower temperatures corresponding to Charpy energies on the lower shelf or plateau of this curve the fracture surface is reminiscent of the quasi-static plane strain fracture surface in that it is flat (cleavage) with little or no shear lips. The nil-ductility temperature (NDT) is used to define the upper limit of temperature for which plane strain fracture under impact exists. At the other extreme of temperatures the fracture surface associated with the upper shelf energies exhibits general yielding and ductile failure accompanied usually, but not always, by large

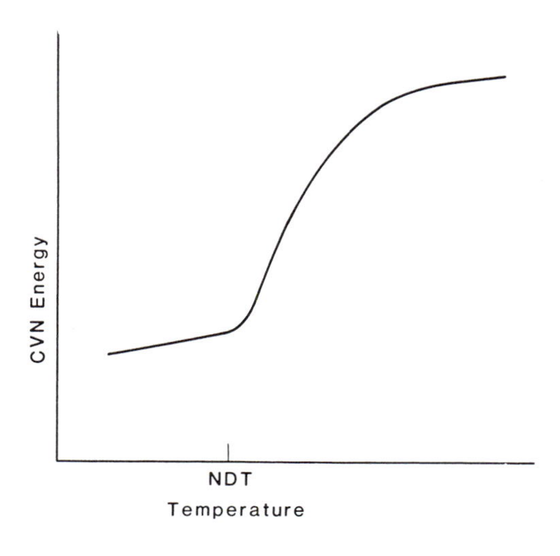

Figure 3.18 Typical Charpy V-notch energy versus temperature for low and intermediate strength steels.

shear lips. In the brittle-ductile transition between these two extremes, elastic-plastic fracture occurs and the fracture surface features a mixture of cleavage and shear lips. This sudden shift or transition in absorbed energy with temperature frequently does not occur in high strength steels and many other materials.

It has become common practice in the bridge and ship building industries to specify the minimum toughness of a material by prescribing a minimum Charpy energy at a given service temperature. As an added measure of safety the NDT is usually required to be below the lowest anticipated service temperature so that the load-bearing material is operating on the upper shelf. Since the transition for quasi-static loading frequently occurs at a lower temperature, the practice of specifying impact fracture energies instead of static values tends to be conservative.

Correlations between K_{Ic} and K_{Id}, the plane strain fracture toughness under dynamic or impact loading, and the CVN energy have been made. For rather extensive discussions of these correlations the reader is referred to Chapter 6 of Rolfe and Barsom (3.5). Because the transition temperature can depend upon the thickness of the specimen, dynamic tear (DT) and drop weight tear test (DWTT) have been designed to accommodate the full thickness of the plate of material. Correlations between K_{Ic} and the energy absorbed in these compact tests can be found in Hertzberg (3.6) and references therein.

3.5 Plane Stress Fracture and the *R*-Curve

As already noted, the fracture toughness K_{Ic} is independent of the thickness of the body and the extent of crack growth when the triaxial constraint is such that a state of plane strain is attained. On the other hand, when insufficient thickness exists to support this constraint—that is, when the size of the plastic zone is no longer small compared to the thickness—the fracture resistance depends upon the thickness. Unlike the plane strain case, the plane stress fracture resistance is frequently observed to increase with increasing crack growth. The fracture resistance may increase to several times its value at crack initiation. Hence, there exists a potential reserve toughness that may be exploited.

Aside from temperature and thickness the fracture resistance usually is a function only of the amount of crack extension Δa and independent of the crack length. The fracture resistance as a function of the crack growth is referred to as the resistance curve or R-curve. The R-curve can be expressed in terms of K, G, the crack opening displacement, or any other equivalent parameter within the context of LEFM or small-scale yielding. An analysis using any of these fracture parameters can be performed. In the following an energy approach is selected because it can be readily extended to the analysis of stable elastic-plastic crack growth. A parallel treatment based upon a K-resistance curve is given by Hutchinson (3.80).

During crack growth equilibrium between the crack driving force and the fracture resistance can be expressed by

$$G(a) = R(\Delta a) \tag{3.5-1}$$

where $a = a_0 + \Delta a$ is the current crack length and a_0 is the initial crack length. A typical R-curve is depicted in Figure 3.19(a). The current configuration is stable if a slight (infinitesimal) increase of the crack length at constant load does not give rise to a driving force that exceeds the material's resistance. This condition implies that in addition to satisfying Equation (3.5-1),

$$\left(\frac{dG}{da}\right)_P < \frac{dR}{d\Delta a} \tag{3.5-2}$$

for stable crack growth. In other words, the crack growth is stable if the rate of increase of the driving force with crack length does not exceed the rate of increase of the material's resistance to crack growth. The limit of stable growth is expressed by

$$\left(\frac{dG}{da}\right)_P = \frac{dR}{d\Delta a} \tag{3.5-3}$$

Equations (3.5-1) and (3.5-3) represent two equations for determining the driving force and the extent of crack growth at instability. In a graphical solution of these equations, G and $R(\Delta a)$ are superimposed on the same plot as shown in Figure 3.19(b). A family of G-curves can be plotted for prescribed loads $P_3 > P_2 > P_1$ as depicted there. The point of intersection of the driving force and resistance curves defines the amount of crack growth for that load.

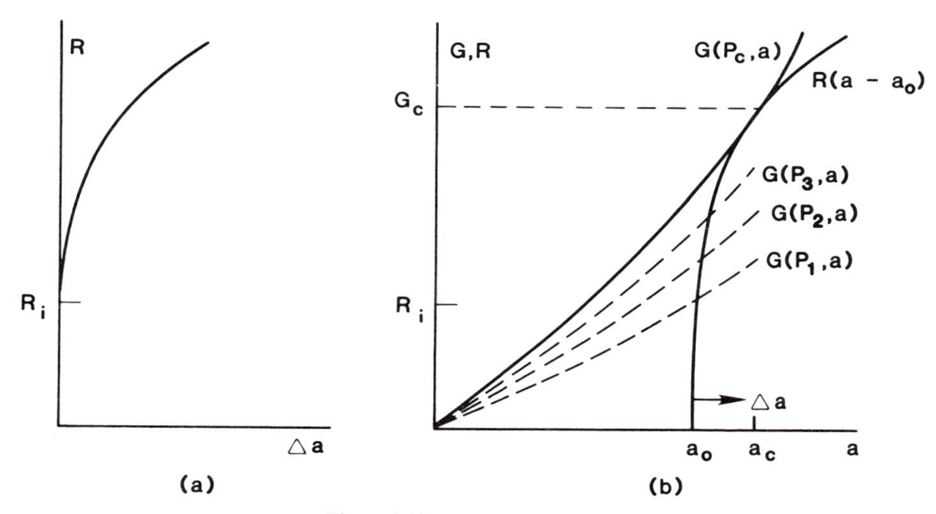

Figure 3.19 R-curve analysis.

Equations (3.5-1) and (3.5-3) imply that the limit of stable crack growth occurs when the crack driving force curve for the critical load P_c is tangent to the R-curve. It is clear from this diagram that an increase of the load above P_c will give rise to a driving force that exceeds the fracture resistance and unstable crack growth will ensue.

For a more general treatment, consider the compliant loading of the cracked component in Figure 3.20. Let P be the induced load per unit thickness of the component due to the prescribed total displacement Δ_T and let Δ be the load-point displacement. Denote by $\bar{C}_M$ the total compliance of the loading device so that the prescribed displacement can be written as

$$\Delta_T = C_M P + \Delta \tag{3.5-4}$$

where $C_M = \bar{C}_M B$. Treating G and Δ as functions of P and a, as well as other invariant parameters, one can write

$$d\Delta_T = C_M dP + \left(\frac{\partial \Delta}{\partial P}\right)_a dP + \left(\frac{\partial \Delta}{\partial a}\right)_P da = 0 \tag{3.5-5}$$

and

$$dG = \left(\frac{\partial G}{\partial P}\right)_a dP + \left(\frac{\partial G}{\partial a}\right)_P da \tag{3.5-6}$$

The combination of Equations (3.5-5) and (3.5-6) yields

$$\left(\frac{dG}{da}\right)_{\Delta_T} = \left(\frac{\partial G}{\partial a}\right)_P - \left(\frac{\partial G}{\partial P}\right)_a \left(\frac{\partial \Delta}{\partial a}\right)_P \left[C_M + \left(\frac{\partial \Delta}{\partial P}\right)_a\right]^{-1} \tag{3.5-7}$$

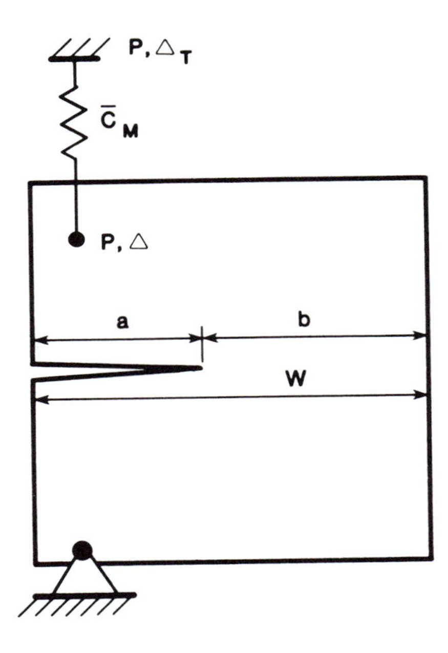

Figure 3.20 A cracked component under compliant loading.

It can be shown through manipulation of Equations (3.5-1), (3.5-4), and (3.5-6) that Equation (3.5-3) is equivalent to a stationary value of Δ_T; that is, $d\Delta_T/dG = 0$. The latter condition will usually correspond to a maximum of Δ_T (limit point) beyond which no solution to Equation (3.5-1) exists.

Since $\Delta = CP$, where $C = C(a)$ is the compliance of the component having a unit thickness, then

$$\left(\frac{\partial \Delta}{\partial P}\right)_a = C, \qquad \left(\frac{\partial \Delta}{\partial a}\right)_P = PC' \tag{3.5-8}$$

where the prime is used to denote a differentiation with respect to the argument of the function. When there exist m equally loaded crack tips, it follows that

$$\left(\frac{\partial G}{\partial P}\right)_a = \frac{PC'}{m}, \qquad \left(\frac{\partial G}{\partial a}\right)_P = \frac{P^2}{2}\frac{C''}{m} \tag{3.5-9}$$

Equations (3.5-7)–(3.5-9) lead to

$$\left(\frac{dG}{da}\right)_{\Delta_T} = \frac{P^2C''}{2m} - \frac{(PC')^2}{(C_M + C)m} \tag{3.5-10}$$

for the rate of increase in the driving force for fixed Δ_T.

While the crack driving force is independent of the compliance of the loading device, its rate of change depends upon this compliance. For dead loading ($C_M \to \infty$) Equation (3.5-10) reduces to

$$\left(\frac{dG}{da}\right)_{\Delta_T} = \frac{P^2C''}{2m} = \left(\frac{dG}{da}\right)_P \tag{3.5-11}$$

whereas for fixed grip loading ($C_M \to 0$), Equation (3.5-10) yields

$$\left(\frac{dG}{da}\right)_{\Delta_T} = \frac{P^2C''}{2m} - \frac{(PC')^2}{mC} = \left(\frac{dG}{da}\right)_\Delta \tag{3.5-12}$$

The condition for stable crack growth is

$$\left(\frac{dG}{da}\right)_{\Delta_T} < \frac{dR}{d\Delta a} \tag{3.5-13}$$

At incipient unstable crack growth,

$$\left(\frac{dG}{da}\right)_{\Delta_T} = \frac{dR}{d\Delta a} \tag{3.5-14}$$

A comparison of Equations (3.5-10) and (3.5-11) shows that

$$\left(\frac{dG}{da}\right)_P > \left(\frac{dG}{da}\right)_{\Delta_T} \tag{3.5-15}$$

Consequently, for the same driving force the prescribed load rather than displacement presents a more severe loading condition for stable crack growth.

For a uniform remote loading, $\sigma_{22}^{\infty} = P/W$, of a square plate having a width W large compared to a central crack of length $2a$, the energy release rate per crack tip is given by

$$G = \frac{\pi P^2 a}{EW^2} \tag{3.5-16}$$

When the latter equation is compared to Equation (5.3-9) with $m = 2$, then

$$C' = \frac{4\pi a}{EW^2} \tag{3.5-17}$$

Upon integrating Equation (3.5-17) and selecting the constant of integration to be the compliance of the uncracked plate, one finds that

$$C = \frac{2\pi a^2}{EW^2} + \frac{1}{E} \approx \frac{1}{E} \tag{3.5-18}$$

The substitution of Equations (3.5-16)–(3.5-18) into Equation (3.5-10) yields

$$\left(\frac{dG}{da}\right)_{\Delta_T} = \frac{G}{a}\left[1 - \frac{8\pi a^2}{W^2(1 + EC_M)}\right] \tag{3.5-19}$$

Due to the assumed smallness of $2a$ relative to W, the second term within the brackets of Equation (3.5-19) can be neglected. Consequently, within this approximation

$$\left(\frac{dG}{da}\right)_{\Delta_T} = \frac{G}{a} = \frac{\pi P^2}{EW^2} \tag{3.5-20}$$

and there is no significant difference for this example between prescribed load and prescribed displacement. The condition for stability, Equation (3.5-13), becomes

$$G < aR' \tag{3.5-21}$$

At initiation of crack growth it follows from Equations (3.5-1) and (3.5-21) that stability will be ensured if

$$a_0 > R_i/R_i'$$

where $R_i = R(0)$ and $R_i' = R'(0)$. Clearly, the growth will be unstable for all crack lengths if $R_i' = 0$ as is the case for plane strain fracture. If $R_i' > 0$, then the initial crack growth in the center cracked panel will be stable for sufficiently long cracks and unstable otherwise. For a given thickness the length $l \equiv R_i/R_i'$ is a material parameter. When the initial portion of the R-curve is approximated by its tangent at $\Delta a = 0$, the material length l represents the amount of crack growth associated with a doubling of R above R_i. This length may vary from a few millimeters for the more ductile material behavior to a few centimeters or more for brittle-like behavior.

For other types of structural components alternate expressions for $(dG/da)_{\Delta_T}$ that may be more convenient than Equation (3.5-10) exist [see Paris et al. (3.81)]. For example, the Mode I stress intensity factor can be written as

$$K_{\mathrm{I}} = \frac{P\sqrt{a}}{W} Y\left(\frac{a}{W}\right) \tag{3.5-22}$$

The dimensionless function $Y = Y(a/W)$ may also depend upon other fixed geometric parameters. Expressions for $Y(a/W)$ can be found for a number of specimens and structural components in Tada et al. (3.21). The introduction of Equation (3.5-22) into Equation (3.3-30) yields

$$G = \frac{P^2 a}{EW^2} Y^2 \tag{3.5-23}$$

Consequently, by Equation (3.5-9)

$$\left(\frac{\partial G}{\partial a}\right)_P = \frac{P^2 C''}{2m} = \frac{G}{a}\left[1 + 2\frac{a}{W}\frac{Y'}{Y}\right] \tag{3.5-24}$$

and

$$C' = \frac{2mG}{P^2} = \frac{2maY^2}{EW^2} \tag{3.5-25}$$

The substitution of Equations (3.5-24) and (3.5-25) into Equation (3.5-10) leads to

$$\left(\frac{dG}{da}\right)_{\Delta_T} = \frac{G}{a}\left[1 + 2\frac{a}{W}\frac{Y'}{Y} - 4\frac{a^2}{W^2}\frac{Y^2}{E}\frac{m}{C_M + C}\right] \tag{3.5-26}$$

The latter equation reduces to Equation (3.5-19) for the center cracked plate where $m = 2$, $Y = \sqrt{\pi}$, and $C = 1/E$. Furthermore, Equation (3.5-1) holds during crack growth so that

$$\left(\frac{dG}{da}\right)_{\Delta_T} = \frac{R}{a}\left[1 + 2\frac{a}{W}\frac{Y'}{Y} - 4\frac{a^2}{W^2}\frac{Y^2}{E}\frac{m}{C_M + C}\right] \tag{3.5-27}$$

is a function of the current crack length. When this expression is introduced into Equation (3.5-14), a transcendental equation for the critical crack length is obtained. The critical load can be determined from Equation (3.5-23) for this crack length and the associated value of $R = G$.

In determining C it is sometimes convenient to write

$$\Delta = \Delta_c + \Delta_{nc} \tag{3.5-28}$$

where Δ_{nc} is the load-point displacement of the uncracked component and Δ_c is the contribution to the load-point displacement due to the presence of the crack. It follows that

$$C = C_c + C_{nc} \tag{3.5-29}$$

where C_{nc} is the compliance of the uncracked body that can be computed using standard methods and $C_c \equiv \Delta_c/P$ is the compliance due to the crack. Tada et al. (3.21) give expressions for Δ_c for a number of cracked bodies so that C_c can be readily computed for these cases.

From known expressions, either computed or available from handbooks, for Y and C_c and for a given R-curve of the material, it is a straightforward task to evaluate the righthand side of Equation (3.5-27) and to ascertain from Equation (3.5-14) the limits of stable crack growth. In principle it makes no difference if either Equation (3.5-10) or Equation (3.5-26) is used in evaluating $(dG/da)_{\Delta_T}$; particularly, when exact expressions are available for Y and C. In practice the latter are often determined numerically and then curve fitted to obtain an analytical relation with an error of usually less than 1 percent. Each successive derivative of the fitted functions will increase the error. Since Equations (3.5-26) and (3.5-27) contain lower-order derivatives, these equations are preferred when dealing with numerically determined values for Y and C.

Further discussion of the R-curve analysis and testing can be found in Broek (3.1) and Rolfe and Barsom (3.5). While there is no standard R-curve test, ASTM (3.82) has a tentative recommended practice.

3.6 References

(3.1) Broek, D., *Elementary Engineering Fracture Mechanics*, 3 ed., Martinus Nijhoff, The Hague (1982).
(3.2) Parker, A. P., *The Mechanics of Fracture and Fatigue*, E. & F. N. Span, London (1981).
(3.3) Cherepanov, G. P., *Mechanics of Brittle Fracture*, McGraw-Hill, New York (1979).
(3.4) Knott, J. F., *Fundamentals of Fracture Mechanics*, Butterworths, London (1979).
(3.5) Rolfe, S. I. and Barsom, J. M., *Fracture and Fatigue Control in Structures*, Prentice-Hall, Englewood Cliffs, New Jersey (1977).
(3.6) Hertzberg, R. W., *Deformation and Fracture Mechanics of Engineering Materials*, Wiley, New York (1976).
(3.7) Lawn, B. R. and Wilshaw, T. R., *Fracture of Brittle Solids*, Cambridge University Press, Cambridge (1975).
(3.8) Green, A. E. and Zerna, W., *Theoretical Elasticity*, 2nd ed., Oxford University Press, London (1968).
(3.9) England, A. H., *Complex Variable Methods in Elasticity*, Wiley-Interscience, New York (1971).
(3.10) Muskhelishvili, N. I., *Singular Integral Equations*, J. R. M. Radok, trans., Noordhoff, Groningen, The Netherlands (1953).
(3.11) Muskhelishvili, N. I., *Some Basic Problems of the Mathematical Theory of Elasticity*, J. R. M. Radok, trans., Noordhoff, Groningen, The Netherlands (1963).
(3.12) Paris, P. C. and Sih, G. C., "Stress Analysis of Cracks," *Fracture Toughness Testing and Its Applications*, ASTM STP 381, ASTM, Philadelphia, pp. 30–83 (1965).
(3.13) Koiter, W. T., "An Infinite Row of Collinear Cracks in an Infinite Elastic Sheet," *Ingenieur-Archiv*, **28**, pp. 168–172 (1959).
(3.14) Irwin, G., "Fracture," *Handbook der Physik*, **6**, S. Flugge (ed.), Springer-Verlag, Berlin, pp. 551–590 (1958).
(3.15) Isida, M., "On the Tension of a Strip With a Central Elliptical Hole," *Transactions of the Japan Society of Mechanical Engineers*, **21**, pp. 507–518 (1955).
(3.16) Sneddon, I. N., "The Distribution of Stress in the Neighbourhood of a Crack in an Elastic Solid," *Proceedings of the Royal Society A*, **187**, pp. 229–260 (1946).
(3.17) Green, A. E. and Sneddon, I. N., "The Distribution of Stresses in the Neighbourhood of a Flat Elliptical Crack in an Elastic Solid," *Proceedings of the Cambridge Philosophical Society*, **46**, pp. 159–163 (1950).
(3.18) Irwin, G. R., "Crack-Extension Force for a Part-Through Crack in a Plate," *Journal of Applied Mechanics*, **29**, pp. 651–654 (1962).

(3.19) Kassir, M. K. and Sih, G. C., "Three-Dimensional Stress Distribution Around an Elliptical Crack Under Arbitrary Loadings," *Journal of Applied Mechanics*, **33**, pp. 601–611 (1966).
(3.20) Sih, G. C., *Handbook of Stress Intensity Factors*, Lehigh University, Bethleham, Pa. (1973).
(3.21) Tada, H., Paris, P. C., and Irwin, G. R., *Stress Analysis of Cracks Handbook*, Del Research Corporation, Hellertown, Pa. (1973).
(3.22) Rooke, D. R. and Cartwright, D. J., *Compendium of Stress Intensity Factors*, Hillingdon, Uxbridge, England (1976).
(3.23) Parmerter, R. R., "Stress Intensity Factors For Three Dimensional Problems," AFRPL-TR-76-30, Air Force Rocket Propulsion Laboratory, Edwards AFB, Ca., April (1976).
(3.24) Gross, B., Srawley, J. E., and Brown, W. F., "Stress Intensity Factor for a Single-Edge-Notch Tension Specimen by Boundary Collocation," NASA TN D-2395 (1964).
(3.25) Gross, B. and Srawley, J. E., "Stress Intensity Factors for Single-Edge-Notch Specimens in Bending or Combined Bending and Tension," NASA TN D-2603 (1965).
(3.26) Gross, B. and Srawley, J. E., "Stress Intensity Factors for Three Point Bend Specimens by Boundary Collocation," NASA TN D-3092 (1965).
(3.27) Kobayashi, A. S., Cherepy, R. B., and Kinsel, W. C., "A Numerical Procedure for Estimating the Stress Intensity Factor of a Crack in a Finite Plate," *Journal of Basic Engineering*, **86**, pp. 681–684 (1964).
(3.28) Sokolnikoff, I., *Mathematical Theory of Elasticity*, McGraw-Hill, Co., New York (1956).
(3.29) Kantorovich, L. V. and Krylov, V. I., *Approximate Methods of Higher Analysis*, Interscience, New York (1964).
(3.30) Smith, F. W., Emery, A. F., and Kobayashi, A. S., "Stress Intensity Factors for Semi-Circular Cracks, Part II—Semi-Infinite Solid," *Journal of Applied Mechanics*, **34**, pp. 953–959 (1967).
(3.31) Smith, F. W. and Alavi, M. J., "Stress Intensity Factors for a Penny Shaped Crack in a Half-Space," *Engineering Fracture Mechanics*, **3**, pp. 241–254 (1971).
(3.32) Shah, R. C. and Kobayashi, A. S., "Stress Intensity Factor for an Elliptical Crack Approaching the Surface of a Plate in Bending," *Stress Analysis and Growth of Cracks*, ASTM STP 513, ASTM, Philadelphia, pp. 3–21 (1972).
(3.33) Shah, R. C. and Kobayashi, A. S., "Stress Intensity Factors for an Elliptical Crack Approaching the Free Surface of a Semi-Infinite Solid," *International Journal of Fracture*, **9**, pp. 133–146 (1973).
(3.34) Shah, R. C. and Kobayashi, A. S., "Effect of Poisson's Ratio on Stress Intensity Magnification Factor," *International Journal of Fracture*, **9**, pp. 360–362 (1973).
(3.35) Rizzo, F. J., "An Integral Equation Approach to Boundary Value Problems of Classical Elastostatics," *Quarterly of Applied Mathematics*, **25**, pp. 83–95 (1967).
(3.36) Cruse, T. A., "An Improved Boundary-Integral Equation for Three Dimensional Elastic Stress Analysis," *Computers and Structures*, **4**, pp. 741–754 (1974).
(3.37) Cruse, T. A., "Boundary-Integral Equation Fracture Mechanics Analysis," *Boundary-Integral Equation Method: Computational Application in Applied Mechanics*, AMD-Vol. 11, ASME, New York, pp. 31–46 (1975).
(3.38) Blandford, G. E., Ingraffea, A. R., and Liggett, J. A., "Two-Dimensional Stress Intensity Factor Computations Using the Boundary Element Method," *International Journal for Numerical Methods in Engineering*, **17**, pp. 387–404 (1981).
(3.39) Rice, J. R., "Mathematical Analysis in the Mechanics of Fracture," *Fracture-An Advanced Treatise*, Vol. II, H. Liebowitz (ed.), Academic, New York, pp. 191–308 (1968).
(3.40) Bathe, K. J. and Wilson, E. L., *Numerical Methods in Finite Element Analysis*, Prentice-Hall, Englewood Cliffs, New Jersey (1976).
(3.41) Becker, E. G., Graham, F. C., and Oden, J. T., *Finite Elements, An Introduction*, Vol. I., Prentice-Hall, Englewood Cliffs, New Jersey (1981).
(3.42) Cook, R. D., *Concepts and Applications of Finite Element Analysis*, 2nd ed., Wiley, New York (1981).
(3.43) Zienkiewicz, O. C., *The Finite Element Method*, 3rd ed., McGraw-Hill, New York (1977).
(3.44) Gallagher, R. H., "A Review of Finite Element Techniques in Fracture Mechanics," *Numerical Methods in Fracture Mechanics*, A. R. Luxmoore and D. R. J. Owen (eds.), Pineridge, Swansea, U. K., pp. 1–25 (1978).
(3.45) Apostal, M., Jordan, S., and Marcal, P. V., "Finite Element Techniques for Postulated Flaws in Shell Structures," EPRI Report 22, Electric Power Research Institute, Palo Alto, Ca., August (1975).
(3.46) Pian, T. H. H., "Crack Elements," *Proceedings of World Congress on Finite Element Methods in Structural Mechanics*, Vol. 1, Bournemouth, U. K., pp. F. 1–F.39 (1975).

(3.47) Rice, J. R. and Tracey, D., "Computational Fracture Mechanics," *Numerical and Computer Methods in Structural Mechanics*, S. J. Fenves et al. (eds.), Academic, New York, pp. 555–624 (1973).

(3.48) Gallagher, R. H., "Survey and Evaluation of the Finite Element Method in Linear Fracture Mechanics Analysis," Proceedings of First International Conference on Structural Mechanics in Reactor Technology, Vol. 6, Part L, Berlin, North-Holland, pp. 637–648 September (1971).

(3.49) Barsoum, R. A., "Triangular Quarter Point Elements as Elastic and Perfectly-Plastic Crack Tip Elements," *International Journal for Numerical Methods in Engineering*, **11**, pp. 85–98 (1977).

(3.50) Sih, G. C. (ed.), *Methods of Analysis and Solutions of Crack Problems*, Noordhoff International, Leyden, The Netherlands (1973).

(3.51) Griffith, A. A., "The Phenomena of Rupture and Flow in Solids," *Philosophical Transactions of the Royal Society*, **A221**, pp. 163–198 (1920).

(3.52) Irwin, G. R., "Analysis of Stresses and Strains Near the End of a Crack Traversing a Plate," *Journal of Applied Mechanics*, **24**, pp. 361–364 (1957).

(3.53) Irwin, G. R., "Fracture Dynamics," *Fracturing of Metals*, American Society of Metals, Cleveland, Ohio, pp. 147–166 (1948).

(3.54) Orowan, E., "Fundamentals of Brittle Behavior of Metals," *Fatigue and Fracture of Metals*, W. M. Murray (ed.), Wiley, New York, pp. 139–167 (1952).

(3.55) Eshelby, J. D., "The Continuum Theory of Lattice Defects," *Solid State Physics*, Vol. 3, F. Seitz and D. Turnbull (eds.), Academic, New York, pp. 79–141 (1956).

(3.56) Dugdale, D. S., "Yielding of Steel Sheets Containing Slits," *Journal of the Mechanics and Physics of Solids*, **8**, pp. 100–108 (1960).

(3.57) Barenblatt, G. I., "The Mathematical Theory of Equilibrium Cracks in Brittle Fracture," *Advances in Applied Mechanics*, **7**, pp. 55–129 (1962).

(3.58) Knowles, J. K. and Sternberg, E., "On a Class of Conservation Laws in Linearized and Finite Elastostatics," *Archive for Rational Mechanics and Analysis*, **44**, pp. 187–211 (1972).

(3.59) Budiansky, B. and Rice, J. R., "Conservation Laws and Energy-Release Rates," *Journal of Applied Mechanics*, **40**, pp. 201–203 (1973).

(3.60) King, R. B. and Herrmann, G., "Nondestructive Evaluation of the J and M Integrals," *Journal of Applied Mechanics*, **48**, pp. 83–87 (1981).

(3.61) Fletcher, D. C., "Conservation Laws in Linear Elastodynamics," *Archive for Rational Mechanics and Analysis*, **60**, pp. 329–353 (1975).

(3.62) Cherepanov, G. P., "Invariant Γ-Integrals," *Engineering Fracture Mechanics*, **14**, pp. 39–58 (1981).

(3.63) Kishimoto, K., Aoki, S., and Sakata, M., "On the Path Independent Integral-$\hat{J}$," *Engineering Fracture Mechanics*, **13**, pp. 841–850 (1980).

(3.64) Ainsworth, R. A., Neale, B. K., and Price, R. H., "Fracture Behavior in the Presence of Thermal Strains," *Proceedings of the Institute of Mechanical Engineers Conference on Tolerance of Flaws in Pressurized Components*, London, pp. 171–178 (1978).

(3.65) Wilson, W. K. and Yu, L.-W., "The Use of the J-Integral in Thermal Stress Crack Problems," *International Journal of Fracture*, **15**, pp. 377–387 (1979).

(3.66) Gurtin, M. E., "On a Path-Independent Integral for Thermoelasticity," *International Journal of Fracture*, **15**, pp. R169–R170 (1979).

(3.67) Blackburn, W. S., "Path Independent Integrals to Predict Onset of Crack Instability in an Elastic-Plastic Material," *International Journal of Fracture Mechanics*, **8**, pp. 343–346 (1972).

(3.68) Blackburn, W. S., Jackson, A. D., and Hellen, T. K., "An Integral Associated with the State of a Crack Tip in a Non-Elastic Material," *International Journal of Fracture*, **13**, pp. 183–200 (1977).

(3.69) Batte, A.D., Blackburn, W. S., Elsender, A., Hellen, T. K., and Jackson, A. D., "A Comparison of the J^*-Integral with Other Methods of Post Yield Fracture Mechanics," *International Journal of Fracture*, **21**, pp. 49–66 (1983).

(3.70) Bui, H. D., "Dual Path Independent Integrals in the Boundary-Value Problems of Cracks," *Engineering Fracture Mechanics*, **6**, pp. 287–296 (1974).

(3.71) Atluri, S. N., "Path-Independent Integrals in Finite Elasticity and Inelasticity, With Body Forces, Inertia, and Arbitrary Crack-Face Conditions," *Engineering Fracture Mechanics*, **16**, pp. 341–364 (1982).

(3.72) Irwin, G. R., "Linear Fracture Mechanics, Fracture Transition and Fracture Control," *Engineering Fracture Mechanics*, **1**, pp. 241–257 (1968).

(3.73) Hult, J. A. H. and McClintock, F. A., "Elastic-Plastic Stress and Strain Distributions Around Sharp Notches Under Repeated Shear," *Proceedings of the Ninth International Congress for Applied Mechanics*, **8**, University of Brussels, pp. 51–58 (1957).

(3.74) Irwin, G. R., "Plastic Zone Near a Crack and Fracture Toughness," *Proceedings of the Seventh Sagamore Ordnance Materials Conference*, Syracuse University, pp. IV-63-IV-78 (1960).

(3.75) Hahn, G. T. and Rosenfield, A. R., "Local Yielding and Extension of a Crack Under Plane Stress," *Acta Metallurgica*, **13**, pp. 293–306 (1965).

(3.76) "Standard Test Method for Plane-Strain Fracture Toughness of Metallic Materials," *ASTM Annual Book of Standards*, Part 10, American Society for Testing and Materials, Philadelphia, E399-81, pp. 588–618 (1981).

(3.77) Hudson, C. M. and Seward, S. K., "A Compendium of Sources of Fracture Toughness and Fatigue-Crack Growth Data for Metallic Alloys," *International Journal of Fracture*, **14**, pp. R151–R184 (1978).

(3.78) Hudson, C. M. and Seward, S. K., "A Compendium of Sources of Fracture Toughness and Fatigue-Crack Growth Data for Metallic Alloys—Part II," *International Journal of Fracture*, **20**, pp. R59–R117 (1982).

(3.79) "Standard Methods for Notched Bar Impact Testing of Metallic Materials," *ASTM Annual Book of Standards*, Part 10, American Society for Testing and Materials, Philadelphia, E23-72, pp. 273–289 (1980).

(3.80) Hutchinson, J. W., *Nonlinear Fracture Mechanics*, Department of Solid Mechanics, The Technical University of Denmark (1979).

(3.81) Paris, P. C., Tada, H., Zahoor, A., and Ernst, H., "The Theory of Instability of the Tearing Mode of Elastic-Plastic Crack Growth," *Elastic-Plastic Fracture*, ASTM STP 668, American Society for Testing Materials, Philadelphia, pp. 5–36 (1979).

(3.82) "Recommended Practice for R-Curve Determination," *ASTM Annual Book of Standards*, Part 10, American Society for Testing and Materials, Philadelphia, E561-80, (1980).

4

DYNAMIC FRACTURE MECHANICS

It might be thought that the term "dynamic fracture mechanics" only applies to those fracture problems in which inertia forces must be included in the equations of motion of the body. While such problems are certainly included, we believe that the subject is actually much broader. In our view it encompasses all fracture mechanics problems where either the load or the crack size changes rapidly, regardless if inertia forces thereby become significant. It follows that any time-dependent boundary value problem addressing rapid crack growth initiation, propagation, and/or arrest lies within dynamic fracture mechanics. Note that, while the demarcation between "rapid" and "slow" crack propagation processes is difficult to fix in a formal way, the distinction is usually not difficult in practice.

The word "dynamic" has been used to connotate crack growth processes accompanied by rapidly occurring changes in the crack/structure geometry *and* where these changes are not necessarily well described by a sequence of static equilibrium states. (That latter usage, of course, identifies those boundary value problems where the inclusion of inertia forces is necessary.) This ambiguity has caused a great deal of unnecessary confusion. To avoid this, we will use the term "kinetic" to designate an analysis model that explicitly includes the propagating crack and "static" for those analyses that consider only the end points of such an event. Both the kinetic and the static models are of course part of dynamic fracture mechanics.

General background on wave propagation in solids can be found in the books of Kolsky (4.1), Bland (4.2), and Achenbach (4.3). The books of Broek and of Lawn and Wilshaw (see Chapter 9 for references) have addressed dynamic fracture mechanics specifically, but are largely limited to quasi-static concepts. Several comprehensive review articles are contained in Liebowitz's Fracture Treatise. Of these, Erdogan's (4.4) is particularly noteworthy; see also Kolsky and Rader (4.5) and Bluhm (4.6). But, of course, these do not reflect the progress that has been achieved since the late 1960s. Since that time three volumes of conference proceedings entirely devoted to dynamic fracture mechanics have appeared (4.7–4.9). In addition, there have been a number of papers reviewing specific aspects of the subject. The most notable of these are the articles of Achenbach (4.10), Freund (4.11), Rose (4.12), Francois (4.13), Kanninen (4.14), Kamath (4.15), and Nilsson and Brickstad (4.16).

A phenomenological classification of problems in dynamic fracture mechanics might include five specific focal points: the initiation of rapid crack propagation, the ensuing crack path, the crack speed, crack branching, and crack arrest. In our view, applications of the technology are best served by considering the latter four phenomena within a general theory of crack propagation. Indeed, in practice, two kinds of dynamic fracture mechanics problems have received most attention. These are:

1. bodies with stationary cracks that are subjected to a rapidly varying applied load, and
2. bodies under fixed or slowing varying loading that contain a rapidly moving crack.

By far the most attention and effort has been placed upon the second classification. Accordingly, it will be emphasized in this chapter.

4.1 Dynamic Crack Propagation and Arrest Concepts

At present, much as in all other branches of fracture mechanics, the bulk of the applications in dynamic fracture mechanics assume linear elastic conditions. The development of this aspect of the subject is now fairly complete. Current research efforts are focused on incorporating elastic-plastic and other forms of nonlinear material behavior into dynamic treatments. The full range of these activities will be addressed in this chapter. To begin, this section first introduces the basic concepts of linear elastic dynamic fracture mechanics. The development of the subject is then traced from a historical point of view. The section concludes with a discussion of inelastic considerations.

4.1.1 Basic Definitions and Terminology

Two points of view on dynamic fracture mechanics are extant: continuum-based and micromechanical-based. Except for one dominant crack-like defect, the former view generally assumes the material to be continuous. The latter, in contrast, considers the failure process to develop from the initiation, growth, and coalescence of a great many random material imperfections. This chapter emphasizes the continuum view, consistent with by far the most research and application work in the field. The micromechanical view, of interest primarily for high-energy processes such as penetration/perforation processes in armaments, is touched upon briefly for completeness and for the potential that this application area offers for future dynamic fracture mechanics applications.

In an applied mechanics analysis, fracture mechanisms enter only through their influence on the quantitative fracture resistance level. Cleavage fracture will generally be associated with a low fracture energy requirement, ductile fibrous fracture with a high fracture energy requirement. While the reasons for this varying behavior are intuitively understood, quantitative relations have yet to be evolved from metallurgical considerations in a completely satisfactory way. However important these connections may be in developing better structural materials, they are really unnecessary for the analysis of fracture in the terms presented here. What is essential is an appropriate measure of the material's fracture resistance to crack growth at the temperature, constraint, loading rate, and crack speed of interest. Dynamic fracture mechanics treatments are therefore focused on critical values of a crack-tip characterizing parameter.

It is important to recognize that cleavage fracture in a metal, which generally occurs below the ductile-brittle transition temperature for the material, involves limited plastic deformation. Cleavage fracture can therefore ordinarily be addressed by linear elastic fracture mechanics (LEFM) techniques. In contrast, ductile fracture is usually accompanied by significant plastic deformation, making the use of LEFM suspect for such conditions. Despite the fact that rapid ductile fracture is not an impossibility, one often hears rapid crack propagation referred to as "brittle fracture." While brittle materials most often do produce fast running cracks, it should not ever be forgotten that large-scale ductile fracture is also possible. Consequently, we prefer to use "brittle fracture" only in a metallurgical sense, and not as a synonym for rapid crack propagation.

Two points of notation will arise throughout this chapter. First, unlike some authors, we will not provide an explicit designation on the stress intensity factor to indicate that it is dynamically computed. We reserve the use of subscripts and superscripts for quantities that represent material properties. The context should make clear whether the crack driving force is computed dynamically or not. The second point of notation is that we will carry the subscript I in denoting a fracture property. This usage for designating a material fracture property is technically correct only when plane strain conditions are satisfied. However, the distinction is not too important for the purposes of this chapter.

As stated above, there are two general classes of dynamics fracture mechanics problems: crack initiation under rapidly applied loading and rapid crack propagation following initiation. Within the framework of linear elastic material behavior, the first classification is relatively free from controversy. There is general agreement that an appropriate quantitative form for the initiation of unstable crack growth is given by the relation

$$K = K_{Id}(\dot{\sigma}, T) \tag{4.1-1}$$

where K_{Id} is supposed to be a material property that depends upon the loading rate $\dot{\sigma}$ and the temperature T. The disagreement that does exist involves the necessity to determine K by dynamic computations for practical problems. There is general agreement in principle on the correctness of so doing even though in most instances it is not.

The second class of dynamic fracture mechanics problems, particularly as it involves the arrest of rapid crack propagation, has been a virtual battleground throughout most of the history of the subject. However, it can now be said that an understanding has been reached between two opposing viewpoints: the static and the kinetic points of view. Because dynamic effects exist at crack arrest, the static approach is a simplification. Nevertheless, reasonably constant statically determined arrest values can be determined experimentally that will suffice for many practical purposes. Such an approach has been successfully used in one important practical application—the thermal shock problem in nuclear power plant pressure vessels described in Section 1.5.1. Nevertheless, caution should always be exercised in a static approach because the neglect of dynamic effects tends to be anticonservative.

The static approach requires a small crack jump length in experimentation to determine the arrest property. The reason is that a kinetic analysis for crack propagation in an infinite plane provides the legitimization for the static view. In actual structures, the return of kinetic energy to the crack tip prior to crack arrest can make the static approach invalid. But, as this condition is often not inhibiting, the bulk of current state-of-the-art applications of dynamic fracture mechanics to rapid crack propagation and arrest are based upon the use of static crack arrest concepts. Specifically, for the initiation and termination of unstable crack propagation under linear elastic fracture mechanics (LEFM) conditions, one has

$$K = K_{Ic}(T) \tag{4.1-2}$$

and

$$K = K_{Ia}(T) \tag{4.1-3}$$

In both instances K is calculated just as if the crack were stationary (quasi-static growth) while K_{Ic} and K_{Ia} are taken as temperature-dependent material properties. It can be noted that this point of view does not include (nor could it) any consideration of the crack propagation process that links the initiation and arrest points.

The static view of crack propagation requires crack arrest to occur smoothly with an intimate connection between the slowing down process and the static deformation state long after arrest; that is, crack arrest must be the reverse, in time, of crack growth initiation. However, there is ample evidence to suggest that elastodynamic crack arrest instead occurs abruptly at a value that is unrelated to the corresponding static condition. The kinetic point of view, which gives direct consideration to crack propagation with crack arrest occurring only when continued propagation becomes impossible, is in much better agreement with the observations. Within the confines of elastodynamic behavior, rapid crack propagation occurs in such an approach under the condition that

$$K = K_{ID}(V, T) \tag{4.1-4}$$

where K_{ID}, the dynamic propagating fracture toughness, is a temperature-dependent function of the crack speed, V. Note that this function generally does not include an initiation value; that is, $K_{ID}(O, T) \neq K_{Ic}(T)$. Thus, a kinetic approach will also generally employ Equation (4.1-2). Of more significance, the kinetic approach does not include an arrest value as such. In the kinetic approach arrest occurs at the position and time t_a for which K becomes less than the minimum value of K_{ID}, and remains less, for all $t > t_a$.

The minimum value of K_{ID}, which has been designed as K_{IM} and is here called K_{IA}, is obviously a key parameter. In fact, the legitimacy of the static approach rests upon the notion that K_{Ia} is a good approximation to K_{IA}. There are indeed many practical problems where the difference between the predictions of a static and a kinetic approach is not great. These applications are typified by, (1) the use of K_{Ia} values measured from short crack jumps, whereupon K_{Ia} provides a close approximation to K_{IA}, and (2) component boundaries that do not reflect stress waves back to the running crack tip. These conditions are often satisfied in a structural component, but are less often valid in a small-scale test specimen.

4.1.2 Quasi-Static Analyses of Propagating Crack Speeds

The first quantitative prediction for the speed of a rapidly propagating crack appears to be that given by Mott (4.17) who in 1948 extended Griffith's theory to include a kinetic energy contribution. Mott considered an infinite body subjected to a remote tensile stress σ containing a propagating crack whose instantaneous length is $2a$. Assuming that the crack speed is small in comparison to the velocity of sound in the body, Mott was able to express the kinetic energy of the body, per unit thickness, through a term having the form

$$\tfrac{1}{2}\rho V^2 \iint \left(\frac{du}{da}\right)^2 dx\,dy$$

where ρ is the mass density, V is the crack speed, and $u(x, y)$ represents the quasi-static displacements in the body. On dimensional grounds, Mott asserted that the above integral must be proportional to the quantity $(a\sigma/E)^2$. The kinetic energy can then be written as $\tfrac{1}{2}k\rho V^2(a\sigma/E)^2$, where k is a numerical constant to be determined. The total energy due to the presence of the crack was then obtained by adding the kinetic energy to the diminution of the elastic energy and the surface energy increase due to the crack. Provided that no external work is supplied, conservation of energy then requires that

$$\tfrac{1}{2}k\rho a^2 V^2 \left(\frac{\sigma}{E}\right)^2 - \frac{\pi\sigma^2 a^2}{E} + 4\gamma a = \text{constant} \tag{4.1-5}$$

where γ denotes the surface energy.

Mott reasoned that, since the total energy must be constant, its derivative with respect to crack length will be zero. Making the questionable assumption that $dV/da = 0$, Mott deduced an expression for the crack speed having the form

$$V = \left(\frac{2\pi}{k}\right)^{\frac{1}{2}} \left(\frac{E}{\rho}\right)^{\frac{1}{2}} \left(1 - \frac{a_0}{a}\right)^{\frac{1}{2}} \tag{4.1-6}$$

where $a_0 = (2/\pi)(E\gamma/\sigma^2)$, the critical crack length in the Griffith Theory, was taken as the crack length at the start of the event [cf. Equation (1.2-6)]. It can readily be seen that Equation (4.1-6) predicts a limiting crack speed

$V_l = (2\pi/k)^{\frac{1}{2}}(E/\rho)^{\frac{1}{2}}$. From this result Mott concluded that

> the velocity of propagation, under uniform stress, of a crack in a material that is not ductile, will tend towards a value of the order of the velocity of sound in the material, and which is independent of the stress applied or of the atomic cohesive forces across the cleavage plane.

However, the exact value of this limiting speed, which depends upon the numerical constant k, was not specified. It might be kept in mind that a tacit assumption in Mott's work was that $V \ll C_0 = (E/\rho)^{\frac{1}{2}}$. Nevertheless, the result was carried to a point where this assumption is clearly violated.

Roberts and Wells (4.18), in a paper that appeared in 1954, advanced Mott's analysis by determining a specific value for k. This was done by numerically evaluating the kinetic energy in the body during quasi-static crack growth. They used Westergaard's solution for the displacements in an infinite domain under remote biaxial stresses with a crack of length $2a$ on the x-axis. For plane strain, these relations are

$$\begin{aligned} 2\mu u &= (1 - 2\nu)\,\mathrm{Re}\,\bar{Z} - y\,\mathrm{Im}\,Z \\ 2\mu v &= 2(1 - \nu)\,\mathrm{Im}\,\bar{Z} - y\,\mathrm{Re}\,Z \end{aligned} \tag{4.1-7}$$

For a crack in an infinite domain subjected to an applied stress σ

$$Z = \frac{\sigma z}{(z^2 - a^2)^{\frac{1}{2}}} \tag{4.1-8}$$

where μ is the shear modulus, $z = x + iy$ and

$$Z = \frac{d}{dz}\bar{Z}$$

Because the integration for the kinetic energy does not converge, Roberts and Wells related the value of the integral to an arbitrarily chosen finite limit radius of integration, r. For each such limit, values of the quantity $(2\pi/k)^{\frac{1}{2}}$ were determined numerically. To select the correct outer limit, the argued that

> If the displacements are immediately communicated to the outermost parts of the plate, then the crack may only move with a very small terminal velocity. However, it also appears that the communication of these displacements is limited by the velocity of elastic waves themselves.

From this Roberts and Wells concluded that, if the crack has grown to length a from a very small value a_0, then the farthest wavefront will only have traveled a distance equal to $r = C_0 t = a(k/2\pi)^{\frac{1}{2}}$, a result that apparently follows from integrating Equation (4.1-6) with $a_0 = 0$. This provides the result that $r/a = 2.62$ and gives a lower limit for $(2\pi/k)^{\frac{1}{2}}$ equal to 0.38 (n.b., this result was obtained for $\nu = 0.25$). Thus, Mott's equation can be put into the more definite form

$$V = 0.38 C_0 \left(1 - \frac{a_0}{a}\right)^{\frac{1}{2}} \tag{4.1-9}$$

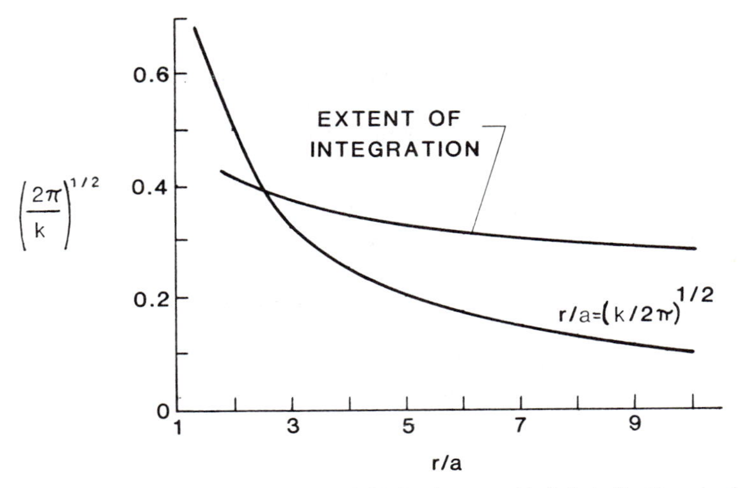

Figure 4.1 The calculation of Roberts and Wells that provided their limiting crack speed estimate of $0.38C_0$.

where, as above, $C_0 = (E/\rho)^{\frac{1}{2}}$ is the velocity of elastic waves in a thin rod (one-dimensional wave propagation). This gives the terminal or limiting crack speed $V_l = 0.38C_0$, a widely quoted result. Figure 4.1 shows their analysis procedure.

The assumptions violated in Mott's approach are, of course, also violated in arriving at Equation (4.1-9). In addition, it should be recognized that the assumption made to determine r, apparently to avoid a difficult integration, is a critical one. A more exact estimate would necessarily involve the initial crack length a_0; that is, from Equation (4.1-6), using the approximation $r = C_0 t$, it is easily shown that

$$r = \left(\frac{k}{2\pi}\right)^{\frac{1}{2}} \int_{a_0}^{a} \frac{da}{\left(1 - \dfrac{a_0}{a}\right)^{\frac{1}{2}}}$$

But, because the integral will clearly be a function of a/a_0, the entire approach is dubious. However, because the value of $(2\pi/k)^{\frac{1}{2}}$ obtained by integrating the kinetic energy is a slowly varying function of r (see Figure 4.1), it could be argued that there probably is some appropriate value of r and that it is not too important just what this value is. In addition, the result is consistent with experimental results that show that terminal velocities exist and do not appear to depend upon the manner of producing the crack nor upon the applied stress. Indeed, values of measured crack speeds for both glasses and metals typically lie in a range from about $0.2C_0$ to $0.4C_0$, in rough agreement with this estimate. Table 4.1 provides some measured values for various materials. Table 4.2 gives values of the elastic wave speed that can be used for comparison with these observations.

Table 4.1 Observed Limiting Crack Speeds (4.6)

Material	C_2 (m/sec)	V_l (m/sec)	V_l/C_2
Glass (soda–lime–silica)	3140	1540	0.49
Glass (silica)	3700	2190	0.59
Cellulose acetate	680	300	0.44
Steel	3160	1500	0.47
Columbia resin	940	550	0.59

Table 4.2 Approximate Value of Elastic Wave Speeds for Common Engineering Materials (4.1)

	Steel	Copper	Aluminum	Glass	Rubber
E (GPa)	210	120	70	70	20
ρ (kg/m^3)	7800	8900	2700	2500	900
ν	0.29	0.34	0.34	0.25	0.5
C_R (m/sec)	2980	2120	2920	3080	26
C_2 (m/sec)	3220	2250	3100	3350	27
C_0 (m/sec)	5190	3670	5090	5300	46
C_1 (m/sec)	5940	4560	6320	5800	1040

Stroh (4.19), recognizing the possibility that the neglect of "relativistic" effects in Mott's treatment could be significant when the crack speed approaches the velocity of sound, suggested in 1957 that the limiting speed must instead be the Rayleigh velocity. His argument follows from the presumption that the limiting velocity is independent of the surface energy of the material. If so, the situation is just the same as if $\gamma = 0$, whereupon the work done in creating new crack surfaces must also be zero. This is tantamount to a disturbance moving on a stress-free surface, disturbances that moves at the Rayleigh velocity. For a Poisson's ratio of 0.25 (see Table 4.3), this would give a limiting speed of $C_R = 0.58C_0$—a speed somewhat in excess

Table 4.3 Elastodynamic Wave Speeds as a Function of Poisson's Ratio

		Plane Stress		Plane Strain	
ν	C_0/C_2	C_R/C_2	C_1/C_2	C_R/C_2	C_1/C_2
0	1.414	0.874032	1.414	0.874032	1.414
0.10	1.483	0.891416	1.491	0.893106	1.500
0.20	1.549	0.905184	1.581	0.910996	1.633
0.25	1.581	0.910996	1.633	0.919402	1.732
0.30	1.612	0.916214	1.690	0.927413	1.871
0.35	1.643	0.920916	1.754	0.935013	2.082
0.40	1.673	0.925165	1.826	0.942195	2.449
0.45	1.703	0.929019	1.907	0.948960	3.317
0.50	1.732	0.932526	2.000	0.955313	—

of the observed values. Nonetheless, Stroh's intuitive conjecture is interesting in that the more precise dynamic treatments that came later definitely established the Rayleigh velocity as a theoretical limiting speed.

The next important advance in the theoretical crack propagation formulations appeared in two papers published almost simultaneously in 1960. These are the papers of Berry (4.20) and of Dulaney and Brace (4.21), both of which re-examined (and corrected) Mott's problem. While the approaches differed, they reached essentially the same result. The more straightforward approach of Dulaney and Brace takes Equation (4.1-5) as its starting point and, like Mott, introduces a definition of the initial crack length from the Griffith Theory to eliminate the surface energy. By imposing an initial condition such that $V = 0$ when $a = a_0$ and replacing γ by $(\pi/2)(\sigma^2 a_0/E)$, Equation (4.1-5) can be written as

$$-\tfrac{1}{2}k\rho a^2 V^2 \left(\frac{\sigma}{E}\right)^2 - \frac{\pi\sigma^2 a^2}{E} + \frac{2\pi\sigma^2 a a_0}{E} = \frac{\pi\sigma^2 a_0^2}{E} \tag{4.1-10}$$

Hence, by simply solving this equation for V, Dulaney and Brace readily found that

$$V = \left(\frac{2\pi}{k}\right)^{\frac{1}{2}} \left(\frac{E}{\rho}\right)^{\frac{1}{2}} \left(1 - \frac{a_0}{a}\right) \tag{4.1-11}$$

By using Roberts and Wells' result to estimate k, this becomes

$$V = 0.38 C_0 \left(1 - \frac{a_0}{a}\right) \tag{4.1-12}$$

Because of the use of the erroneous assumption $dV/da = 0$ used by Mott, Equation (4.1-12) is clearly to be preferred over Equation (4.1-9).*

Neither Dulaney and Brace nor Berry (like Mott) could provide a way to establish the constant k and had to rely upon the result of Roberts and Wells for this purpose. It might be noted that Roberts and Wells' questionable assumption (that $a_0 = 0$) could be removed by integrating Equation (4.1-12). The kinetic energy integration limit arising in Roberts and Wells' treatment would then take the form

$$\frac{r}{a} = \left(\frac{k}{2\pi}\right)^{\frac{1}{2}} \left(\frac{a_0}{a}\right) f\left(\frac{a}{a_0}\right) \tag{4.1-13}$$

From Figure 4-1 it can be clearly seen that Roberts and Wells' result corresponds to the condition that $(a_0/a)f(a/a_0) = 1$; a condition that would indeed be approached, but only for very long crack lengths.

It would be possible to improve upon the corrected form of Mott's quasi-static crack growth relation through a more accurate implementation of Roberts and Wells' results via Equation (4.1-13). However, this has been rendered somewhat superfluous by the dynamic approach of Freund (4.23–

* This was recognized by Wells and Post (4.22) in a footnote to their 1958 paper.

4.26) for crack propagation in an infinite domain that appeared in 1972. As shown in Section 4.1.6, by using a slight numerical approximation, his result can be applied to Mott's problem (i.e., crack propagation in an infinite medium under uniform tensile loading) to obtain

$$V = C_R\left(1 - \frac{a_0}{a}\right) \tag{4.1-14}$$

where C_R is the Rayleigh speed. The ratio of C_R to C_0 varies from 0.54 to 0.62, depending upon Poisson's ratio. Thus, the crack speed history predicted by Equation (4.1-14) would be qualitatively the same as that arising from Equation (4.1-12), but would be roughly 50 percent greater at each stage of growth up to and including the terminal state.

While observed crack speeds tend to agree somewhat better with Equation (4.1-12) than with (4.1-14), this is somewhat fortuitous. The latter form is essentially free of the several questionable assumptions embodied in the former. A more important consideration is that, to this point in the discussion, it has tacitly been assumed that the fracture energy of the propagating crack is not only the same as for the onset of rapid fracture, it is independent of the crack propagation speed. As will be shown later in this chapter, these assumptions are generally not true. In addition, it is possible that crack branching will intercede to provide a limiting crack speed.

The crack propagation relations that have been obtained are very definite in predicting that crack speeds cannot exceed a theoretical upper limit that is connected to an elastic wave speed. As suggested by Winkler, Shockey, and Curran (4.27), under the normal kinds of loading conditions, the upper limit must arise from a restriction on the flow of energy to the crack tip. That is, since crack propagation at speeds less than C_R must be an energy absorbing process, the rate at which it can proceed depends upon the rate at which sufficient energy can be supplied to the crack-tip process region. Usually, the load is applied at a distance from the crack tip whereupon the energy must be transmitted through the material at a rate that is restricted by the elastic wave speeds. But, as Winkler et al. showed, when the load is applied directly to the crack tip by a laser impulse, the resulting crack speeds can greatly exceed those normally observed. Moreover, they succeeded in producing crack speeds from one to two orders of magnitude above C_R by this artifice.

While the results of Winkler et al. are important for establishing the theoretical framework for crack propagation, of more practical significance is the role of plastic deformation at the crack tip in governing the crack speed. Hall (4.28) seems to have been the first to have considered the effect of the plastic energy dissipation at the tip of a propagating crack. In 1953 he reasoned that

> It is to be expected that the magnitude (of the plastic work) will depend upon the velocity of the crack front ... the faster the motion of the stress field around the tip of the crack, the less time there is for plastic deformation to spread as plastic waves into the bulk of the material on either side of the crack before the stress concentration has passed,

Hall concluded from this that an additional relation is needed to connect the crack speed and the plastic work, and, in so stating, he anticipated the modern approach to the subject.

Hall developed a simple model to estimate the effect of crack-tip plasticity based upon plastic wave propagation theory. Unfortunately, as the above statement implies, his model suggested a sharp decrease in the amount of plastic work with crack speed—a result that was not supported by later results, which, as embodied in the K_{ID} parameter, generally show a rising character with crack speed. Note that these results did not take the increase in yield stress with strain rate into account, an effect that would lower the plastic work still more. Stroh (4.29) subsequently developed a simple crack growth model in which the fracture energy was associated with crack-tip plasticity and, hence, dependent upon temperature and strain rate. From this reasoning, he suggested that a transition from brittle to ductile behavior with increasing temperature could be explained. However, only qualitative results were obtained.

It might be noted at this point in our review of early treatments of rapid crack propagation that these efforts were based upon several tacit assumptions that have since been abandoned. One is that there is always a distinct period of acceleration prior to crack propagation at a constant speed. Other assumptions were that quasi-static analysis procedures based upon an energy balance procedure that use a crack speed-independent fracture energy, would suffice. As the following will indicate, these were gradually superceded in evolving the present-day dynamic fracture mechanics analysis procedures.

4.1.3 Dynamic Crack Propagation Analyses

The first analysis treatment to include the effect of inertia forces on crack propagation was also the first to attempt a quantitative treatment of crack branching. This was the "moving Griffith crack" solution of Elizabeth Yoffe (4.30) given in 1951. She considered a crack propagating in an infinite elastic region under a uniform applied stress σ acting normal to the crack line. The special feature of her analysis was that the crack retains its original length; in essence, it is a disturbance that propagates at a constant speed without change of form. While physically unrealistic, it did at least provide an indication of the influence of crack speed on the stress state at the tip of a rapidly propagating crack.

Yoffe was motivated by a suggestion of Orowan (see reference 4.30) who anticipated that, while the stresses about a stationary crack are such that crack extension will occur in the line of the crack, there might be a tendency for the crack to curve or branch at high propagation speeds. For this idea to be quantitatively assessed, the stress state ahead of a crack propagating rapidly enough for inertia forces to be important was needed. Consideration of constant length, constant speed, crack propagation in an infinite medium allows a solution to be obtained without resort to numerical methods.

Yoffe used a Fourier method that is somewhat lengthy (a more direct approach is given in Section 4.3.4) to obtain a closed form speed-dependent solution that reduces to that of Inglis in the limit of zero speed. Of most interest in view of Orowan's suggestion was the stress component $\sigma_{\theta\theta}(r, 0)$ in the near vicinity of the crack tip. For comparison, in the static case this is

$$\sigma_{\theta\theta} = \sigma\left(\frac{a}{2r}\right)^{\frac{1}{2}} \cos^3\frac{\theta}{2} + \cdots \tag{4.1-15}$$

where the terms omitted are of higher order in r. The speed-dependent solution can be expressed in a similar way whereupon the specific value of a/r is irrelevant for the purpose of assessing the propensity of the crack to deviate from the original crack plane. Yoffe plotted the θ-dependent portion of her solution as shown in Figure 4.2. From this plot she concluded that there is a critical velocity of about $0.6C_2$ for crack curving; that is, above this speed the angle for which the maximum value of $\sigma_{\theta\theta}$ occurs changes from $\theta = 0$ to some value $\theta > 0$. Here, C_2 denotes the shear wave speed and is given by $\sqrt{\mu/\rho}$.

The compilation of limiting crack speed data given in Table 4.1 can be compared with Yoffe's prediction. It can be seen that these results are generally in the range from $0.4C_2$ to $0.6C_2$. Thus, the observed results generally lie just under Yoffe's prediction. For comparison with the quasi-static approaches described in the preceeding subsection, for an average Poisson's ratio of 0.25, $C_2 = 0.63C_0$. Thus, the range of observed values is from 0.25 to $0.38V_l/C_0$.

The next important dynamic crack propagation solutions were those contributed by Broberg (4.31) and by Craggs (4.32), both in 1960. Guided by Mott's original approach, Broberg argued that, if the surface energy is negligibly small, then the crack will nucleate from an infinitesimally small microcrack and will achieve the limiting velocity immediately [c.f. Equation (4.1-6)]. He therefore solved the dynamic problem of a crack expanding from zero length at a uniform rate. Baker (4.33) subsequently

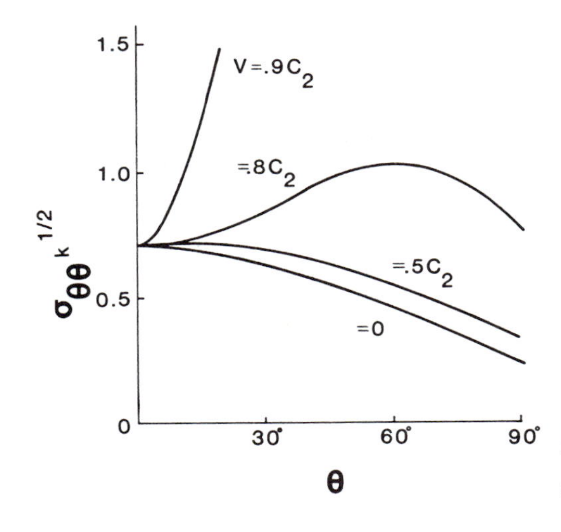

Figure 4.2 The calculated result of Yoffe that provided a crack branching speed estimate of $0.6C_2$.

generalized Broberg's solution to include a finite initial crack. Like Yoffe's solution, these are also artificial solutions that have no direct structural application. However, they also have provided useful insights and continue to be useful today for the assessment of dynamic numerical analyses; that is, by providing check cases in which the stress and displacement fields attending a moving crack are known.

An example of this usage is that of Nishioka and Atluri (4.34) who compared their moving singularity formulation with Broberg's result

$$K = \sigma\sqrt{\pi a}\, k(V) \tag{4.1-16}$$

where $k(V)$ is a monotonically decreasing function of the crack speed and the elastic wave speeds that becomes equal to zero at the Rayleigh velocity. In common with the results of other investigators, to a reasonable approximation, this function can be written as

$$k(V) \simeq 1 - \frac{V}{C_R} \tag{4.1-17}$$

where C_R denotes the velocity of Rayleigh surface waves. Specifically, Nishioka and Atluri expanded a crack in a finite element model at a selected constant speed and computed $K/\sigma\sqrt{\pi Vt}$ as a function of time. By checking their result with Equation (4.1-17) they were able to verify the suitability of their model formulation.

Craggs (4.32) considered a semi-infinite crack propagating at a constant speed under the action of surface tractions on the crack faces that move with the crack tip. Hence, like Yoffe's solution, a steady-state problem was considered. Craggs results can be interpreted to show that the instantaneous crack extension force G can be related to the dynamic stress intensity factor K through a crack-speed-dependent relation given by

$$G = \frac{1+\nu}{E}\left(\frac{V}{C_2}\right)^2\left[1-\left(\frac{V}{C_1}\right)^2\right]^{\frac{1}{2}}\frac{K^2}{D(V)} \tag{4.1-18}$$

where C_1 and C_2 are elastic wave speeds and D is a universal function of the crack speed given by

$$D(V) = 4\left[1-\left(\frac{V}{C_1}\right)^2\right]^{\frac{1}{2}}\left[1-\left(\frac{V}{C_2}\right)^2\right]^{\frac{1}{2}} - \left[2-\left(\frac{V}{C_2}\right)^2\right]^2 \tag{4.1-19}$$

This function vanishes at zero crack speed and at the Rayleigh velocity. It is positive at intermediate values. It can be shown through a limiting procedure that

$$D \rightarrow \frac{4}{\kappa+1}\left(\frac{V}{C_2}\right)^2 \quad \text{as} \quad V \rightarrow 0$$

noting that

$$\frac{C_1^2}{C_2^2} = \frac{\kappa+1}{\kappa-1} \tag{4.1-20}$$

It can be seen that Equation (4.1-18) reduces, as it must, to Equation (1.2-23) for zero crack speed.

The development of Equation (4.1-18) might be Craggs most lasting contribution. Nevertheless, in addition, he also developed results for crack branching that were more systematic than those of Yoffe, but nonetheless showed general agreement with the $V = 0.6C_2$ criterion. Crack branching was also a central concern of Carlsson (4.35). In his 1963 paper Carlsson supposed that a crack propagates through the formation of microcracks that grow ahead of the crack tip and link up with it. He was able to formulate a model using complex variable methods of the theory of elasticity. Carlsson showed that his model was qualitatively in accord with experimental observations of crack branching, but did not obtain a clear-cut quantitative prediction. Carlsson also suggested that the Dugdale model could be incorporated into a crack propagation formulation to reflect the effect of crack-tip plasticity. But, the actual accomplishment of this idea did not occur until somewhat later.

4.1.4 Crack Branching

Recall that Yoffe's solution was motivated by crack branching. Hence, before discussing the topic of crack arrest in the next subsection, we consider progress on crack branching. First, even though crack branching offers attractive research opportunities to both mathematicians and experimentalists, its practical applications are not of nearly as much significance. Comminution processes and terminal ballistics aside, the importance of crack branching in structural mechanics probably lies in its role in limiting rapid crack propagation. That is, after a sufficiently high crack speed has been attained, further increases in the crack driving force apparently only cause repeated branching with no increase in the average crack speed. This possibility was suggested by several investigators including Cotterell (4.36), who concluded that, while fracture propagation will normally proceed along a local symmetry line where the principal stress normal to the line is a maximum, two such lines will develop at high speeds causing the fracture to branch. Clark and Irwin (4.37) have stated that successful crack branching implies that a limiting speed has been nearly achieved. The origins of the latter's ideas are as follows.

Clark and Irwin, in discussing a qualitative explanation for the existence of a limiting crack speed offered by Saibel (4.38), used the fact that observed limiting crack speeds are roughly half of the Rayleigh velocity. From this observation they reasoned that

> This indicates that damping processes rather than inertia of the material dominate in setting the value of V_l. Thus, we can regard the running crack as subsonic, over-damped, low inertia disturbance moved ahead by a driving force G. Until the value of G begins to overdrive the crack toward crack division, the crack speed increases and decreases in phase with the force.

In common with the observations reported by Schardin (4.39), Clark and Irwin found that the crack speed just before crack division was nearly the same as that of the most advanced crack in a multiple division crack pattern. Hence, in contrast to Yoffe's argument, the crack does not branch solely because it has reached some critical velocity. Clark and Irwin were able to arrive at an alternative explanation through estimates (quasi-static) of the crack driving forces prior to and following branching. Upon finding them to be nearly equal, they concluded that

> It seems best therefore to regard attainment of a critical K (or G), rather than a crack-speed-induced modification of the stress pattern, as the primary factor controlling crack division.

This same conclusion was reach independently by Congleton and Petch (4.40) who offered a rationalization based on the nucleation of cracks ahead of the moving crack tip. In contrast to the observations reported by Clark and Irwin, however, their findings indicated that branching can occur at crack speeds well below the limiting value. These conclusions were subsequently reinforced by Anthony et al. (4.41).

An energy balance approach to the prediction of crack branching was proposed by Johnson and Holloway (4.42) in 1966. Starting from Mott's equation, Equation (4.1-5), and assuming that the only effect of crack division would be to double the surface energy contribution, they arrived at a condition for branching that can be expressed as

$$\sigma^2 \pi a = 4\gamma E\left[1 - \frac{k}{2\pi}\left(\frac{V}{C_0}\right)^2\right]^{-1} \tag{4.1-21}$$

This approach was criticized by Rabinovitch (4.43) on the basis that, if, as Johnson and Holloway assumed, the crack speed at branching is the limiting speed V_l, then the bracketed term in Equation (4.1-21) is identically equal to zero; that is, from Equation (4.1-5) it can be seen that $V_l \equiv (2\pi/k)^{\frac{1}{2}} C_0$. A similar conclusion was reached by Jacobson (4.44).

Anderson (4.45) has calculated the stress intensity factors for crack branches of infinitesimally small length. On the basis that crack division occurs at the angle for which K_{I} is maximum, he predicted that the angle between two branches should be about 60 degrees. However, as Kalthoff (4.46) pointed out, the experimental measurements indicate that actual angles are instead only about 30 degrees. Kalthoff suggested an alternative branching criterion based on consideration of K_{I} and K_{II}. In a numerical study, he noted that for two cracks oriented at an angle of 28 degrees, K_{II} is just equal to zero and crack growth at that angle would be stable whereas greater or lesser angles would be unstable. However, while this finding is undoubtedly connected with the development of crack division, it does not provide any quantitative insight into the initiation of branching.

To conclude this discussion we note that the mathematical study of crack bifurcation under dynamic conditions is currently a fairly active research area. As yet, no universal crack branching criterion has emerged. It can be conjectured that, like other areas of fracture mechanics, the most fruitful avenue is one that integrates experimentation with mathematical studies. This may be the reason for the progress recently achieved by Ramulu and

Kobayashi (4.47) who have arrived at a combined criterion that appears to be in good agreement with experimental observations. This involves, (1) a necessary condition for the generation of secondary cracks given by $K_I = K_{Ib}$, where K_{Ib} is a material constant, and (2) a sufficiency condition given $r_0 < r_c$, where r_0 is a characteristic distance governed by the local stress state and r_c is a material property.

4.1.5 Early Views on Crack Arrest

Broadly speaking, the behavior of the materials used in engineering structures can be classified as either ductile or brittle. In the presence of a crack, the failure point of a ductile material can be estimated reasonably well in terms of the net cross-sectional area that remains on the crack plane to support the flow stress. In contrast, a brittle material will fail well before this point, particularly when the crack is small in comparison to the remaining ligament size. A given material can exhibit both extremes of behavior. That is, a metal that is ordinarily tolerant of flaws (i.e., because it fails only after substantial plastic deformation) can also exhibit brittle behavior at low temperatures and/or high loading rates. That the risk of inducing brittle failure is increased by the presence of notches, cold, and impact was well known to construction engineers of the nineteenth century. But, as evidenced by the catalog of failures cited in Chapter 1, this phenomenon has perhaps not always been recognized by their successors in the present century.

This lack of appreciation for the dual nature of engineering materials led to compartmentalized approaches to failure analysis. Some investigators concentrated attention on fast brittle crack propagation and others on crack initiation from flaws. According to Wells (4.48),

> The proponents of each study defended their hypothesis... and a controversial battle was joined. Like other battles, it was wasteful of resources and led to casualties among scientific reputations".

It is clear from the predominance of efforts focused on fracture control by precluding crack growth initiation which viewpoint eventually carried the day. Nevertheless, the alternate viewpoint—fracture control by arrest of rapid crack propagation—has not been obviated. In fact, owing to ever more widespread recognition that crack initiation cannot ever be absolutely precluded, interest in crack arrest for structural components is growing significantly at present.

Generally speaking, if the initiation of crack propagation cannot be absolutely precluded and the consequences of fracture are sufficiently large, then a crack arrest strategy is mandated as a second line of defense against a catastrophic rupture. A prominent example of the necessity for crack arrest considerations is the thermal shock problem for nuclear pressure vessels, as described in Chapter 1. In this problem reduced toughness levels due to neutron irradiation coupled with the severe loading conditions experienced in a loss of coolant accident could well lead to initiation of crack growth from a

Figure 4.3 A pragmatic crack arrest device in a ship structure.

small flaw. But, serious consequences can be avoided if crack arrest will intercede before the moving crack can penetrate the wall. As another example, rapid crack propagation over several miles has been experienced in gas transmission pipelines. Ship structures also offer instances where crack arrest considerations are appropriate. Figure 4.3 shows a primitive application of a crack arrest approach in a ship at sea. While this application was obviously successful (no such picture would likely have survived otherwise), more quantitative approaches are certainly useful.

There are two approaches to characterizing crack arrest. These are the static and the kinetic arrest theories. The former has received its rationalization from Irwin and Wells (4.49) who stated in 1965 that

> It would now appear that the majority of running brittle cracks in mild steels are manifestations of plane-strain fracture, such that arrests are simple reversals on the time scale of possible plane-strain initiations.

As a consequence of this observation, there must exist a material property, commonly designated K_{Ia}, that governs crack arrest in the same sense as K_{Ic} governs initiation. But, the formal concept of a crack arrest toughness, as measured in a static interpretation of a post-arrest condition, had its origins in the work of Crosley and Ripling (4.50). They later suggested a formal definition of K_{Ia} as being the value of K at about one millisecond after a run-arrest segment. They assumed that a nearly static stress state would be established by that time. This, of course, avoids the difficult problem of characterizing the crack tip at the instant of crack arrest and offers the possibility of treating crack arrest by methods that are no more difficult than are used for initiation.

Interestingly, while their objective was to study the effect of strain rate on the fracture toughness of a pressure vessel steel, Crosley and Ripling found that the K_{Ia} parameter exhibited much less variability than did K_{Ic} measurements on the same materials under the same testing conditions. Moreover, these data were found to be in good agreement with the lower bound of all static and dynamic crack initiation toughness measurements. The K_{IR} reference fracture toughness curve of the American Society of Mechanical Engineers (ASME) Boiler and Pressure Vessel Code in fact uses K_{Ia} data in this spirit (see Section 1.1.8). It was only subsequently that K_{Ia} values were used together with statically computed stress intensity factors to estimate whether or not a rapidly propagating crack would arrest.

As summarized by Kanazawa (4.51), a great deal of very important research on crack arrest has been performed in Japan. But, because the formal development of the dynamic crack arrest theory owes much to the efforts of G. T. Hahn and co-workers at Battelle (4.52), we will focus on this work here. This work was based on an alternative to the static arrest approach—an approach we now refer to as kinetic. A kinetic approach focuses on the crack propagation process and considers arrest only as the termination of such a process. The quantitative development of this approach was intimately connected with the use of the double cantilever beam (DCB) test specimen, shown in Figure 4.4. Figure 4.5 shows schematically results obtained by Hahn and co-workers that revealed clearly the importance of a kinetic analysis, at least for the DCB test specimen.

Because of the blunted initial crack tip commonly used in the type of experiment shown in the upper portion of Figure 4.5, the stress intensity factor at the onset of crack growth, K_Q, can be made arbitrarily greater than K_{Ic}. Accordingly, the crack speed and the crack jump length can be systematically increased by increasing the ratio K_Q/K_{Ic}. The more blunt the initial crack, the

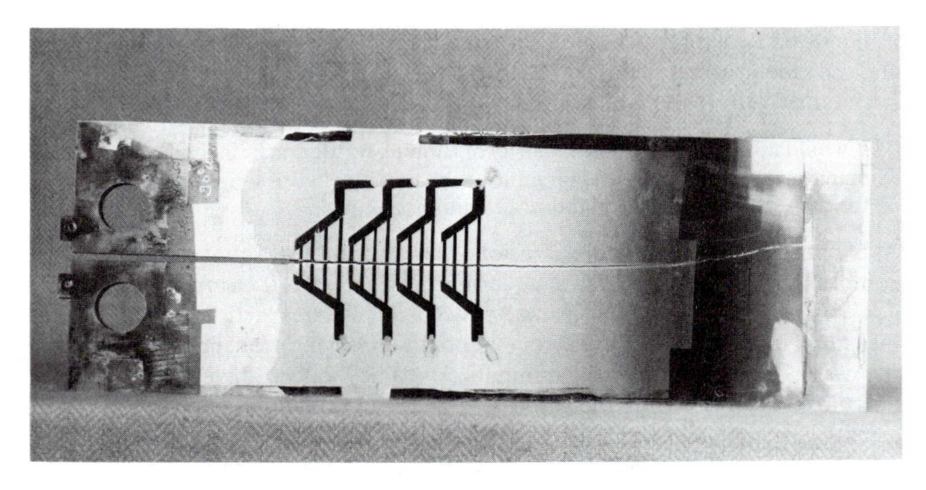

Figure 4.4 A double cantilever beam (DCB) test specimen instrumented with timing wires to determine the crack propagation history.

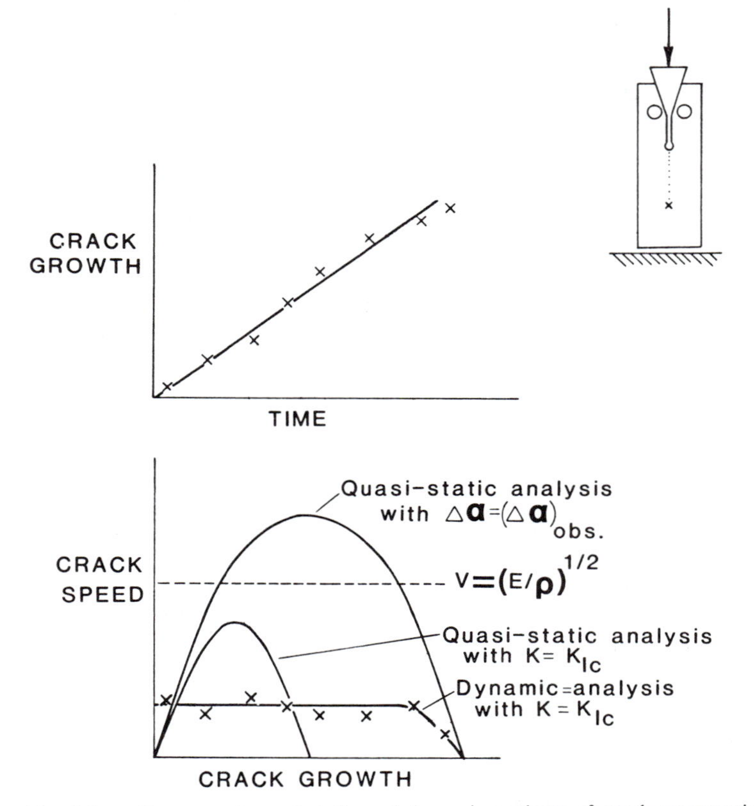

Figure 4.5 Schematic comparison of static and dynamic analyses of crack propagation and arrest in the DCB test specimen.

more strain energy can be stored in the arms of the DCB specimen whereupon the crack will be driven further and faster. Because the crack run/arrest event proceeds under essentially fixed grip conditions, the crack propagates into a diminishing stress field whereupon the arrest of a fast moving crack within a DCB specimen is possible. Also, because of the proximity of the free surfaces that parallel the crack plane, the DCB specimen is the most "dynamic" of any of the commonly used test specimens. Consequently, even though it may not be representative of service conditions, the DCB specimen is ideally suited to an elucidation of crack arrest principles.

The beam-like character of the DCB specimen was exploited by a number of investigators—for example, Benbow and Roessler (4.53) and Gilman (4.54)—who developed expressions for the energy release rate using simple built-in beam relations. Kanninen (4.55) extended this idea by modeling the uncracked portion of the specimen as a beam on an elastic foundation. From this an expression for the specimen compliance was obtained for the entire range of possible crack lengths. This result was shown to be in good agreement with measured values by Hahn et al. In terms of an applied load P, the stress intensity factor deduced for this model is

$$K = 2\sqrt{3}\,\frac{Pa}{Bh^{\frac{3}{2}}}\left[\frac{\sinh^2 \lambda c + \sin^2 \lambda c}{\sinh^2 \lambda c + \sin^2 \lambda c} + \frac{1}{\lambda a}\,\frac{\sinh \lambda c \cosh \lambda c - \sin \lambda c \cos \lambda c}{\sinh^2 \lambda c - \sin^2 \lambda c}\right] \tag{4.1-22}$$

where $2h$ and B, respectively, are the specimen height and thickness, c is the uncracked speciment length, and $\lambda = (6)^{\frac{1}{4}}/h$. Equation (4.1-22) was found to be in excellent agreement with the numerical results of Srawley and Gross (4.56). More recently, Fichter (4.57) has used a Fourier transform and the Wiener-Hopf technique to determine the stress intensity factor for plane stress conditions. Because he neglected the free end, a comparison can be made only with the special case where $c \gg h$. For completeness, the three results can be written for this case as

$$\frac{KBh^{\frac{1}{2}}}{2\sqrt{3}P} = \begin{cases} \dfrac{a}{h} + 0.69, & \text{Gross and Srawley} \\[2ex] \dfrac{a}{h} + 0.6728 + .0377\,\dfrac{h^2}{a^2}, & \text{Fichter} \\[2ex] \dfrac{a}{h} + 0.64, & \text{Kanninen} \end{cases} \tag{4.1-23}$$

For practical purposes, these results are essentially the same. Equation (4.1-22), however, provides the only closed-form expression for the complete specimen. In particular, that result indicates that free end effects cannot be neglected unless $c > 2h$.

For dynamic conditions, the energy balance condition with the kinetic energy included was taken as the crack growth criterion. This can be written, per unit thickness, as

$$G = \frac{dW}{da} - \frac{dU}{da} - \frac{dT}{da} \tag{4.1-24}$$

where, as above, W and U denote the external work done on the component and its elastic energy, per unit thickness, respectively, while T denotes the kinetic energy in the component per unit thickness. Crack growth is then supposed to proceed according to a relation $G(t) = G_c(V)$ with crack arrest occurring at a time when the criterion is no longer satisfied. Now, as indicated schematically in Figure 4.5, crack propagation in a DCB specimen from an initially blunted crack tip under slowly inserted wedge loading proceeds at an ostensibly constant velocity. This fact, albeit unexpected, made possible a decisive comparison of the various analysis approaches.

The simplest of the approaches shown in Figure 4.5 is one in which the crack propagates under quasi-static conditions with a fracture toughness that is always equal to the initiation toughness, K_{Ic}. This approach uses an energy balance, assigning the differences between the strain energy release rate and the fracture energy requirement to kinetic energy. From this a crack speed can be inferred. As shown in the lower part of Figure 4.5, for quasi-static conditions with G_c being speed-independent, a crack speed that differs markedly from the observed values is predicted. Also, the crack jump length is considerably underestimated.

An obvious improvement in the approach can be based upon the use of different initiation and running fracture energies. For heuristic purposes, one could select a value of G_c to match the observed crack arrest point. But, Figure 4-5 also reveals the inadequacy of this approach: the predicted crack speeds exceed the elastic wave speeds for the material. Clearly, therefore, the resolution of this difficulty does not lie in the choice of a fracture toughness property alone. It would instead appear that extended amounts of rapid crack propagation cannot be characterized with a static computational approach. A dynamic solution using the beam on elastic foundation model is shown in Figure 4.5 where, to be a quite good approximation, the experimental results were reproduced both qualitatively (i.e., a linear crack length-time record virtually from the onset of crack growth to just prior to arrest) and quantitatively.

This success, coupled with the inadequacies of static analyses, led to questioning of the then widely accepted static post-arrest characterization of crack arrest. For example, Kanninen (4.58) performed a series of computations for different initiation conditions in the DCB specimen which showed that the static condition following arrest was a very definite function of the crack jump length in the test. This means that the post-arrest condition characterized by K_{Ia} cannot be related to the material properties controlling the propagation event. Clearly, these two approaches are theoretically incompatible and, on the basis of the foregoing, it appears to be the kinetic approach that is correct. However, it remained for the decisive experimental work of Kalthoff et al. (4.59) and A. S. Kobayashi et al. (4.60) to resolve the issue. These results, shown as Figures 4.6 and 4.7, were obtained on photoelastic materials using optical techniques.

The results shown in Figure 4.6 were obtained on DCB specimens using the method of caustics (4.59). In contrast, the results shown in Figure 4.7 were obtained on a compact tension (CT) specimen via a photoelastic method (4.60). In both cases an estimate of the stress intensity factor of the propagating crack was made from optical images using high-speed photography (see Section 4.4.1). These results clearly show that the static interpretation of a run-arrest is vacuous. In particular, it can clearly be seen that the actual stress intensity factor at the point of arrest does not correspond to the static value. Figure 4.6, which presents the results of five individual experiments, further reveals that there is a progressive increase in the discrepancy between the static and observed values with increasing crack jump length. The constancy of the actual stress intensity factor value at crack arrest is evident in this set of results. The "ring-down" in the stress intensity

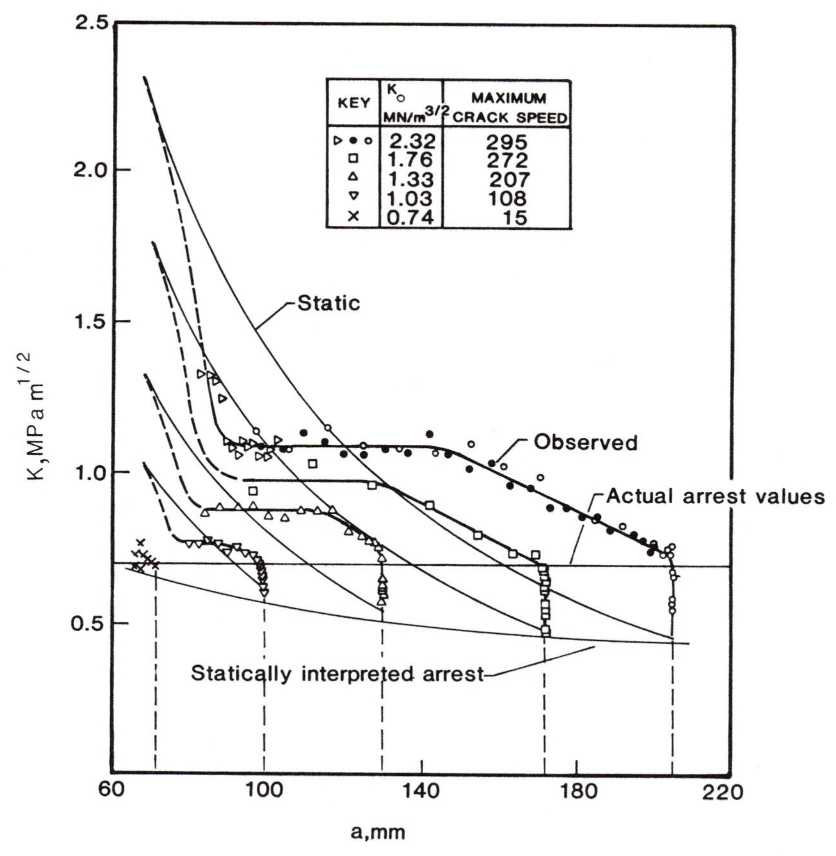

Figure 4.6 Comparison of static stress intensity factors with values observed by the method of Caustics for Araldite B DCB specimens.

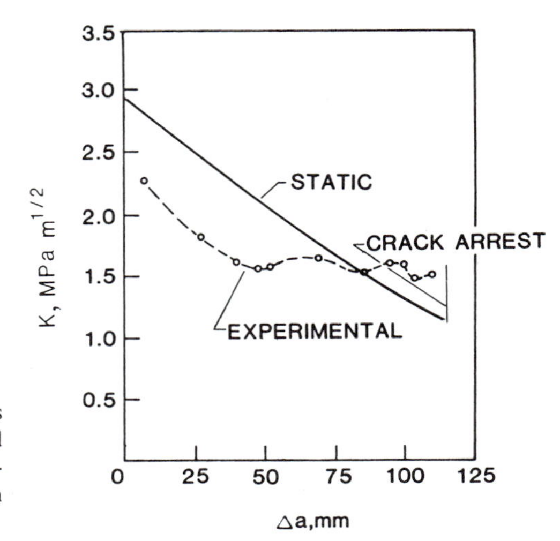

Figure 4.7 Comparison of static stress intensity factors with values observed by the photoelastic method in a polycarbonate modified compact tension specimen.

factor long after arrest, which indicates that the static value is eventually restored, was also shown in the work of Kalthoff et al. (4.59). This was shown in Chapter 1 as Figure 1.27.

Although it is tangential to the question of crack arrest, the analogous situation of static versus dynamic interpretations of rapidly loaded components might also be noted. Kalthoff et al. (4.61) have employed their method of caustics approach to study the impact loading of a three-point bend specimen. Their results reveal that there is again a complete lack of correspondence between the static interpretation and the observed stress intensity factors. Finally, in concluding this section, it should be recognized that a considerable amount of experimental work has been contributed that has served to illuminate the basic issues. This body of work would include the early contributions of Schardin and Smekal in Germany, Kanazawa in Japan, Carlsson in Sweden, Robertson in Britain, and Hall in the United States. Pertinent recent work would include that of Kalthoff et al. (4.62), Hoagland et al. (4.63), Kobayashi and Dally (4.64), Kobayashi and Mall (4.65), Crosley and Ripling (4.66), Dahlberg et al. (4.67), and Kanazawa and Machida (4.68). We will return to this subject in Section 4.1.7.

4.1.6 The Basis of Elastodynamic Fracture Mechanics

Paralleling the more pragmatic efforts to identify appropriate measures for crack arrest in practical engineering problems was mathematical work. The focus of these efforts was the determination of the dynamic counterpart of Irwin's relation between the energy release rate and the stress intensity factor—that is, Equation (1.2-23). Craggs (4.32) seems to have been the first to deduce such a relation, albeit for the artificial conditions of a constant speed, semi-infinite crack. For an expanding crack, a result for a steadily moving crack was later obtained by Sih (4.69). Freund (4.23), using the energy flow rate formulation developed by Atkinson and Eshelby (4.70), appears to be the first to develop such a relation for non-steady state motion. A rigorous derivation of this key is given in Section 4.2.2.

Broberg, Baker, and others developed relations for the dynamic stress intensity factor for an infinite medium. Their relations can be written in a common form as

$$K(t) = k(V)K(0) \tag{4.1-25}$$

where $K(t)$ denotes a dynamically computed value of the stress intensity factor, with $K(0)$ being its static counterpart, while $k(V)$ is a universal (i.e., geometry-independent) function of the crack speed. The $k(V)$ function is slightly different in the different solutions. Values of this function for two different solutions are shown in Figure 4.8.

The function $k(V)$ is defined by a complicated expression that is not readily manipulated. Accordingly, the approximate simplified form developed by Rose (4.71) is useful for practical computations. This is

$$k(V) = \left(1 - \frac{V}{C_R}\right)(1 - hV)^{-\frac{1}{2}} \tag{4.1-26}$$

where h is a function of the elastic wave speeds (having units of reciprocal crack speed) that Rose develops in several different ways, each to a different degree of approximation. The only closed-form expression is

$$h = \frac{2}{C_1}\left(\frac{C_2}{C_R}\right)^2\left[1 - \left(\frac{C_2}{C_1}\right)^2\right] \tag{4.1-27}$$

where C_1, C_2, and C_R are the elastic wave speeds defined in Section 4.2. Detailed calculations provided by Rose show that use of Equation (4.1-27) will give only a slight underestimate of k (e.g., within 5 percent) up to $V/C_2 = 0.5$, which covers the great majority of all practical applications.

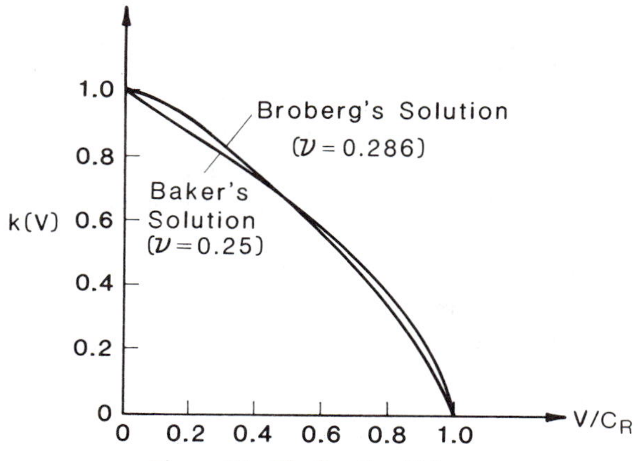

Figure 4.8 The function $k(v)$.

The result derived by Freund (4.23) for the dynamic energy release rate is

$$G(t) = g(V)G(0) \tag{4.1-28}$$

where $G(t)$ is the dynamic energy release rate, $G(0)$ is its static counterpart, and $g(V)$ is a universal function of crack speed. The latter function is given to a close approximation by

$$g(V) = 1 - \frac{V}{C_R} \tag{4.1-29}$$

This relation can be used to deduce the speed of a rapidly propagating crack in a large body. Let us suppose, like Mott (4.17), that a critical energy release rate governs crack propagation; that is that crack propagation proceeds in accord with the equality $G = G_c$. Now, consider that crack propagation has initiated from a crack of length $2a_0$ in an infinite domain subjected to a remote constant applied stress σ. Then, $G(0) = \sigma^2 \pi a$. For the special circumstances where G_c is the same for initiation and propagation, it is evident that $\sigma^2 \pi a_0 = G_c$. Substituting these quantities into Equation (4.1-28) and using Equation (4.1-29) produces Equation (4.1-14), which was introduced above to compare with the quasi-static results of Mott et seq.

As described in Section 4.1.5, two seemingly opposing views of crack arrest exist: the static approach centered on the K_{Ia} parameter and the kinetic approach based on $K_{ID} = K_{ID}(V)$ function that views arrest as the termination of propagation. Substantial amounts of evidence were accumulated in support of each point of view. Widespread acceptance of one or the other position awaited the more direct experimental evidence that was eventually forthcoming in the shadow pattern (or method of caustics) and dynamic photoelasticity work described in Section 4.1.5. Both techniques, when coupled with flash photography, enable extraction of the stress intensity factor of a fast running crack to be made. If, as assumed in the kinetic point of view, crack propagation occurs only when $K = K_{ID}(V)$, then experimental results such as these can be used to determine directly the material property K_{ID} as a function of crack speed V. An example was shown in Chapter 1 as Figure 1.28. However, because these results were obtained on a polymeric material, the subsequent adaptation of the method of caustics in reflection, which enabled it to be applied to study metals, was important. This was first accomplished Shockey et al. (4.73) and later by Rosakis and Freund (4.74). A comparison of results for a polymer and for steel showing that the same effects are present is shown in Figure 4.9.

The observations revealed in Figure 4.9 were anticipated by the results of kinetic fracture calculations—for example, see Kanninen (4.58). That the major influence causing the so-called dynamic effect is the reflection of stress waves from the specimen boundaries can be seen in the results of these calculations as illustrated in Figure 4.10. This figure compares a dynamic solution that takes into account the finite dimensions of the DCB specimen with the dynamic solution for an infinite medium given by combining

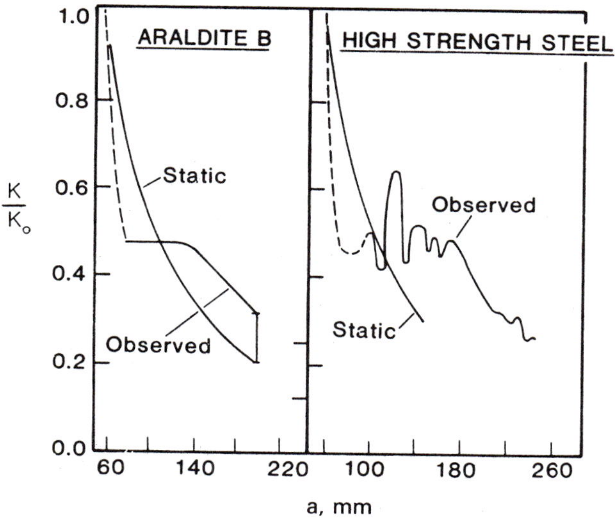

Figure 4.9 Stress intensity factors obtained by the method of Caustics for Araldite B (Transmission) and a high strength steel (reflection).

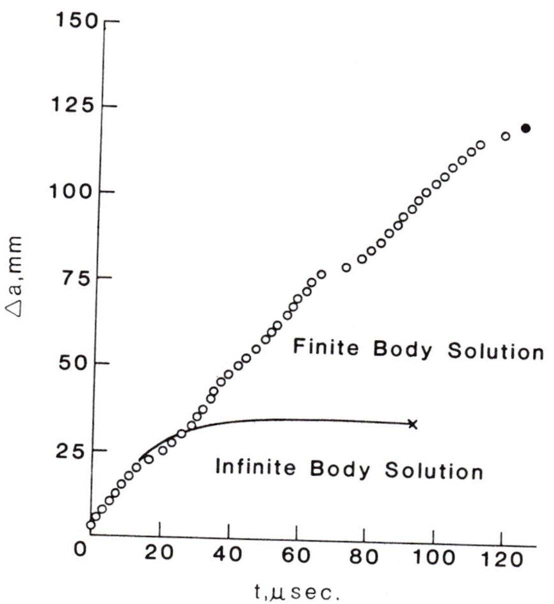

Figure 4.10 Comparison of calculated crack propagation/arrest results for a DCB specimen with an infinite medium solution.

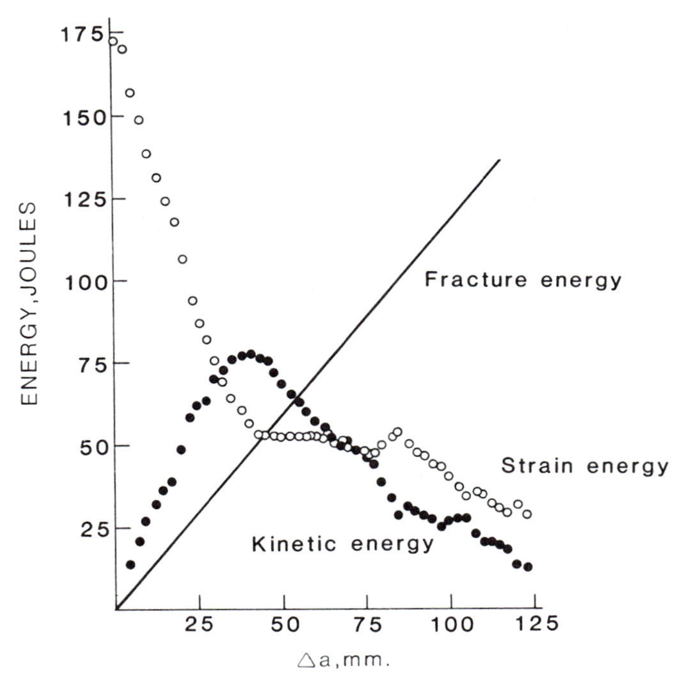

Figure 4.11 Distribution of energy during crack propagation in the DCB specimen experiment of Figure 4.10.

Equation (4.1-22) with Equation (4.1-25). This result demonstrates an important conclusion: a kinetic approach via an initial value–boundary value problem will coincide with the quasi-static crack arrest prediction when reflected stress waves are not important—see short time agreement in Figure 4-10—but will otherwise predict that the crack may penetrate further into the body.

The result presented in Figure 4.9 also reveals that kinetic energy plays a crucial role in crack propagation; see Equation (4.1-24). That is, the time required for an elastic stress wave to travel from the crack tip to the specimen boundary in a DCB test specimen and return is approximately $2h/C_0$. For the specimen dimensions considered, this time is 26 μsec. And, as Figure 4-10 shows, this is about where the infinite medium solution departs from the finite body solution. Figure 4.11, which shows the partitioning of the initial strain energy contained in the specimen during the run-arrest event, further bears this out.

Figure 4.11 shows that the kinetic energy rises to a maximum at about the statically predicted arrest point (i.e., $a - a_0 = 35$ mm). The subsequent decrease indicates the reflection of waves from the specimen boundaries. The results obtained by Kobayashi et al. (4.60), as shown in Figure 4.12, reveal that quantitatively similar behavior occurs in a compact tension specimen.

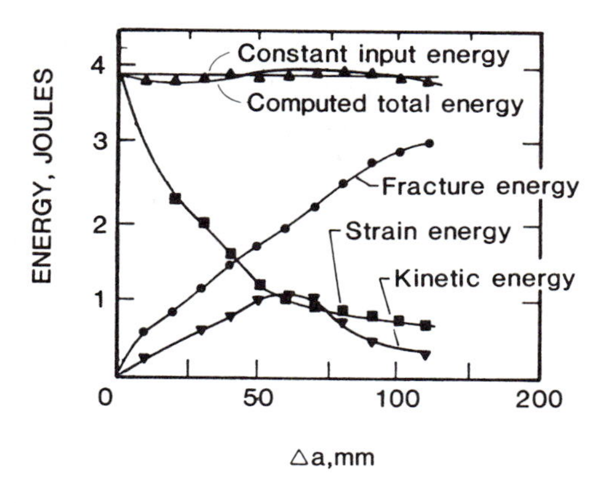

Figure 4.12 Distribution of energy during crack propagation in the CT specimen experiment of Figure 4–7.

Thus, while the DCB specimen is highly dynamic, it displays an effect that is common to all run/arrest events in the vicinity of free surfaces.

Figure 4.13 shows the result of a kinetic calculation by Popelar and Kanninen (4.75) for the oscillation of the stress intensity factor following crack arrest at a hole in the DCB specimen. Their approach utilized a dynamic viscoelastic analysis approach for the DCB specimen. Of most significance, the Popelar-Kanninen analysis indicated that the amount of viscous energy dissipation that occurs prior to crack arrest is negligible. This conclusion is at odds with the inferences made from observations by Shukla et al. (4.76) that seem to indicate that a substantial amount of energy is lost away from the crack tip. This question is currently unresolved.

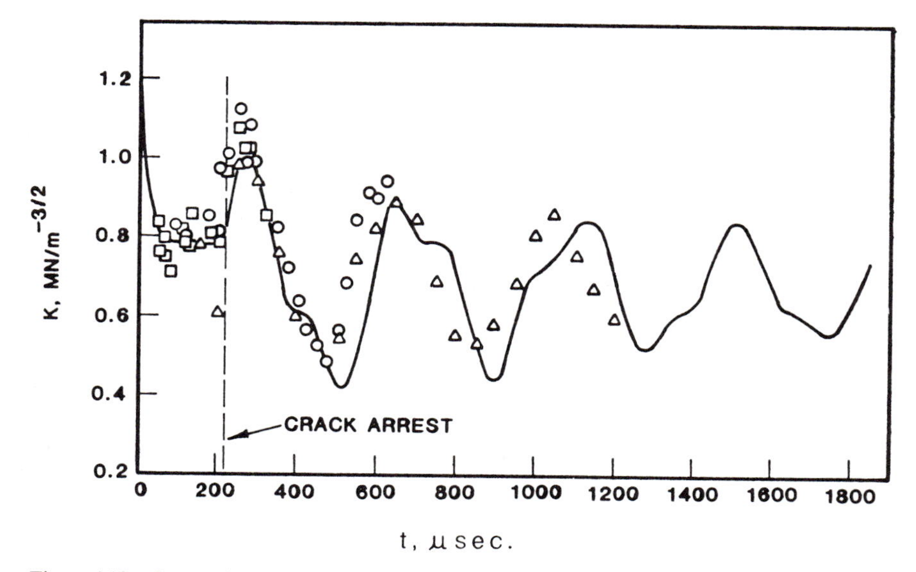

Figure 4.13 Comparison of stress intensity factors calculated by a dynamic viscoelastic analysis and values observed by the method of caustics.

4.1.7 Crack Arrest Methodology

While the kinetic analyses made in the work conducted by Hahn and his associates were based upon a dynamic energy balance and a critical energy release rate parameter, the Freund-Nilsson relation, Equation (4.1-18), shows that this would be equivalent to a crack growth criterion based on the stress intensity factor. Thus, the expression for dynamic crack propagation, Equation (4.1-4), can be generally taken in the form $K(t) = K_{ID}(V)$, where K is computed from the equations of motion for the cracked body as part of an initial value–boundary value problem. Formally, the crack arrest condition can be expressed as the inequality

$$K_I < \min_{0<V<c_R} \{K_{ID}(V)\} \tag{4.1-30}$$

Despite the apparent dichotomy posed by the two opposing crack arrest criteria offered by inequality (4.1-30) and Equation (4.1-3), an accomodation exists. This can be drawn from the infinite medium results presented in the preceding section, as follows.

Equation (4.1-18) can then be used to relate $K_{ID}(V)$ to a critical energy release rate for a rapidly propagating crack. For plane strain conditions

$$K_{ID} = \left[\frac{E}{1-\nu^2}\frac{G_c}{A(V)}\right]^{\frac{1}{2}} \tag{4.1-31}$$

where $A(V)$ is defined in an obvious way from Equation (4.1-18). Next, Equation (4.1-25) can be used to introduce the static stress intensity factor. Using a linear approximation to the $k(V)$ function—see Figure 4.8—this gives the result

$$K(0) = \frac{K_{ID}(V)}{\left(1 - \frac{V}{C_R}\right)^{\frac{1}{2}}} \tag{4.1-32}$$

where, again, $K(0)$ denotes the value of the stress intensity factor calculated *statically* from the instantaneous position and load level.

Suppose that $K_{ID}(V_M)$ is the minimum value of the running fracture toughness (n.b., this respresents the most general situation and does not exclude $V_M = 0$). Then, a material property K_{IA} can be defined such that

$$K_{IA} \equiv \frac{K_{ID}(V_M)}{\left(1 - \frac{V_M}{C_R}\right)^{\frac{1}{2}}} \tag{4.1-33}$$

Then, substitution of Equation (4.1-33) into Equation (4.1-32) would produce exactly the same form as Equation (4.1-3). Hence, if $K_{Ia} = K_{IA}$, Equation (4.1-3) would then be consistent with inequality (4.1-30). But, while this correspondence might be taken as a verification of the static crack arrest theory, there is one key point that must be recognized: Equation (4.1-32) was developed for an infinite medium. Hence, it is valid for a finite domain only

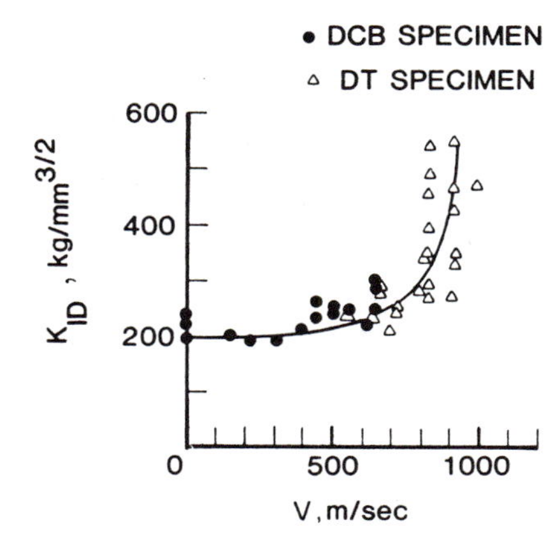

Figure 4.14 Dynamic fracture toughness values for a ship steel deduced from DCB specimen testing and wide plate experiments—test temperature −40°C.

until stress wave reflections from the nearest boundary begin to impinge upon the crack tip.* As revealed in Figure 4.6, if crack arrest occurs before this, the static and kinetic approaches will correspond. But, for more extended crack propagation, they do not.

It is commonly accepted that the use of the static crack arrest approach is conservative. Indeed, if one were to infer a value of K_{Ia} from a test like that shown in Figure 4.6, the true fracture toughness property would clearly be underestimated. However, the use of such a value in an application to a structure via a static approach will not be conservative if the dynamically computed stress intensity factor exceeds its static counterpart. As a heuristic illustration of this idea, suppose that a K_{Ia} value had been obtained from the short crack jump length experiment shown in Figure 4.6. This could give a value that underestimates the true crack arrest toughness by perhaps 10 percent. Now, if this value were used together with the static K_I curve to predict the point of crack arrest in the longest crack jump experiment shown in Figure 4.6, it can be seen that an *underestimate* of roughly 25 percent of the crack jump would be made! To be sure, such a dramatic effect might not occur in an actual structure. Nevertheless, it should be recognized that such a possibility does exist and, unless it is certain that reflected stress waves cannot return to the crack tip, at least some initial dynamic calculations may be prudent in critical applications.

To help make the foregoing somewhat more definite, consider the $K_{ID} = K_{ID}(V)$ data of the type first suggested by Eftis and Krafft (4.77) for metals and Paxton and Lucas (4.78) for polymers. Of particular interest are the data generated by Kanazawa and Machida (4.68) shown in Figures 4.14 and 4.15 and those of Rosakis et al. (4.74) in Figure 4.16. These are typical of most of the data that have been reported which suggest that, while K_{ID} is roughly speed independent at low crack speeds, it increases rapidly with increasing

* The dynamic K value computed for the infinite medium is sometimes called the "reflectionless stress intensity factor."

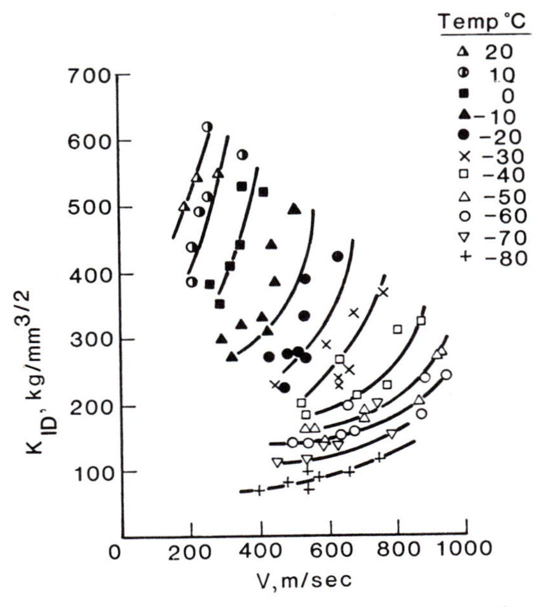

Figure 4.15 Dynamic fracture toughness values for a ship steel as a function of crack speed and temperature.

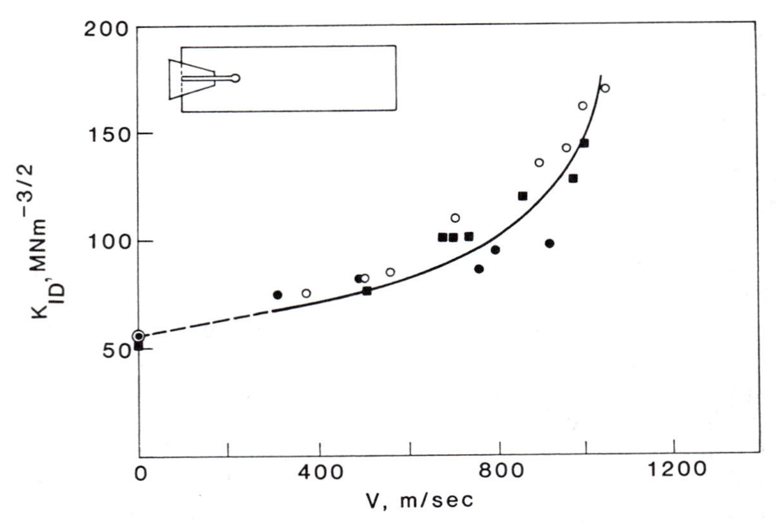

Figure 4.16 Dynamic fracture toughness values for 4340 steel.

crack speed as some material-dependent limiting speed is approached. A generic form for the functional dependence of K_{ID} might then be

$$K_{ID} = \frac{K_{IA}}{1 - \left(\frac{V}{V_l}\right)^m} \tag{4.1-34}$$

where K_{IA}, V_l and m are material constants that must be determined empirically. Using this relation, if continued crack propagation requires that $K = K_{ID}$, then it is clear that crack arrest will occur at a time when $K < K_{IA}$ for all subsequent times.

Now, as already stated, a kinetic analysis admits two complications beyond the static analysis. These are inertia forces and reflected stress waves. If K_{Ia} is the *statically* determined value of K that corresponds to the arrest condition, then, if there are no reflected stress waves, K_{Ia} will be *exactly* equal to K_{IA}. It follows that K_{Ia} is a perfectly legitimate fracture parameter *if* it is measured in a short-jump crack arrest test—that is, before reflected stress waves can impinge upon the crack tip. For larger crack jumps, K_{Ia} is still useful because it then provides a lower bound conservative estimate to the true crack arrest measure, K_{IA}.

To illustrate the use of dynamic calculations to assess the use of a static approach to crack arrest in a critical application area, let us return to the thermal shock problem for nuclear power plant pressure vessels described in Section 1.5.1. As stated there, the ORNL work reached the conclusion that dynamic effects at arrest are negligible in a thermal shock event. To assess this conclusion in a more direct way, Jung and Kanninen (4.79) performed a series of run/arrest computations, both statically and dynamically, for thermal shock experiment number TSE-5A; see Table 1.2 for the experimental details.

The immediate difficulty in any such calculation is the lack of proven $K_{ID} = K_{ID}(V, T)$ relation for the A508 steel vessel material. Since the available data typically have the general character shown in Figures 4.14 to 4.16, it might be assumed that this relation is given by Equation (4.1-34). For heuristic purposes, Jung and Kanninen reasoned that the crack speeds in a thermal shock event might be such that $V \ll V_l$ whereupon the speed-dependent term in Equation (4.1-35) can be neglected. They further assumed that the existing K_{Ia} data were collected under conditions such that they provide a sufficiently close approximation to K_{IA}. The dynamic calculations were performed to contrast the static calculations on this basis. The results are shown in Table 4.4.

Table 4.4 Comparison of Quasi-Static and Dynamic Crack Jump Lengths with ORNL Experimental Results for Thermal Shock Experiment TSE-5A

		Average Data		Lower Bound Data	
Event	Experimental (mm)	Dynamic (mm)	Quasi-static (mm)	Dynamic (mm)	Quasi-static (mm)
1st	5.78	6.60	2.14	13.3	5.19
2nd	13.7	7.25	3.81	17.8	9.15
3rd	10.7	11.0	6.10	14.3	15.2
4th	39.7	27.2	23.2	43.8	36.6
Total	69.9	52.1	35.4	89.2	66.1

The computational results of Jung and Kanninen provided in Table 4.4 were made using initial conditions that correspond to the experimentally determined times of crack initiation events in TSE 5-A, but are otherwise not connected to the results of that experiment. It can be seen that two separate sets of computations were made. In the first, the K_{Ia} and K_{ID} property values were taken as the average of the laboratory data. In the second computation, lower bound K_{Ia} and K_{ID} values, consistent with the values obtained in the thermal shock experiment, were used. These data are shown in Figure 4.17.

Two general conclusions are illustrated by the comparisions shown in Table 4.4. First, if the calculations are made on a common basis, when inertial forces and reflected stress wave effects are not too important, then the static and kinetic approaches will give essentially the same result (cf. Figure 4.10). Therefore, it is not possible that the static approach is ever correct when the kinetic approach is not. The second conclusion is that, again if the computations are made on the same basis, then the kinetic approach will predict a larger crack jump at arrest. Specifically, the kinetic results given in Table 4.4 obtained using the small specimen data account for roughly half of the discrepancy between the static prediction and the measured values. In the computation using the K_{Ia} values derived from the vessel experiment, the static result of course agrees well with the measured values, while the kinetic prediction is somewhat higher.

The foregoing was intended to demonstrate that the static and kinetic approaches to the analysis of crack arrest are compatible, the former being simply a special case of the latter that is applicable when reflected stress waves do not interfere with the moving crack tip. Nonetheless, the literature still

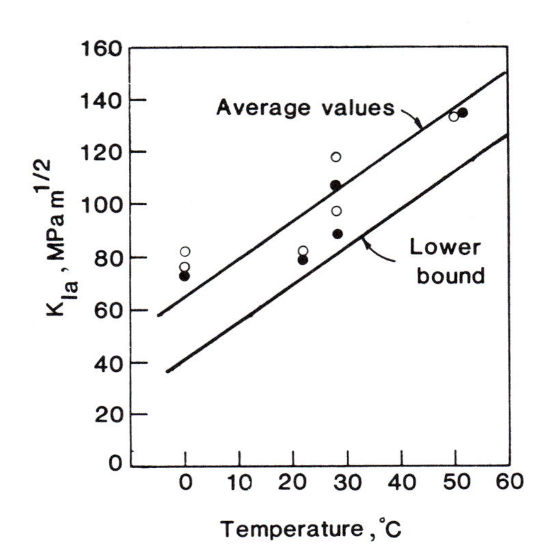

Figure 4.17 Crack arrest data used in the analyses for the thermal shock experiment TSE-5A.

provides evidence that this point is not understood. There appear to be two factors that cause this. First, there has been some confusion arising from the use of approximate relations for extracting K_{ID} values from experiments. One that has been used is

$$K_{ID} \simeq (K_Q K_{Ia})^{\frac{1}{2}} \tag{4.1-35}$$

where K_Q is the apparent K value at the initiation of rapid crack propagation. Some investigators apparently believe such a relation to be a definition of K_{ID}. Obviously, it is not. Among other shortcomings, use of this relation does not recognize the inherent speed dependence of K_{ID} and, more importantly, it can never coincide with K_{Ia} for short crack jumps, as it must. The results of the cooperative crack arrest test program conducted by Crosley et al. (4.80) suffer somewhat in this regard.

A second factor is the evidence that is cited by Crosley and Ripling (4.66) for cracks that slow down smoothly to arrest. Figure 4.18 shows a typical load-time record for a crack run/arrest experiment in a tapered DCB test specimen together with the corresponding crack length history. According to Crosley and Ripling, because the crack appears to arrest smoothly, a dynamic approach predicts a more abrupt arrest and therefore must be invalid. The two strain gage histories shown in Figure 4.18 were offered as further support of this postulate (n.b., these gages are at fixed locations and do not, of course, move with the crack tip). It seems likely that such data cannot be characterized elastically (i.e., small-scale yielding is not valid) whereupon a nonlinear (e.g.,

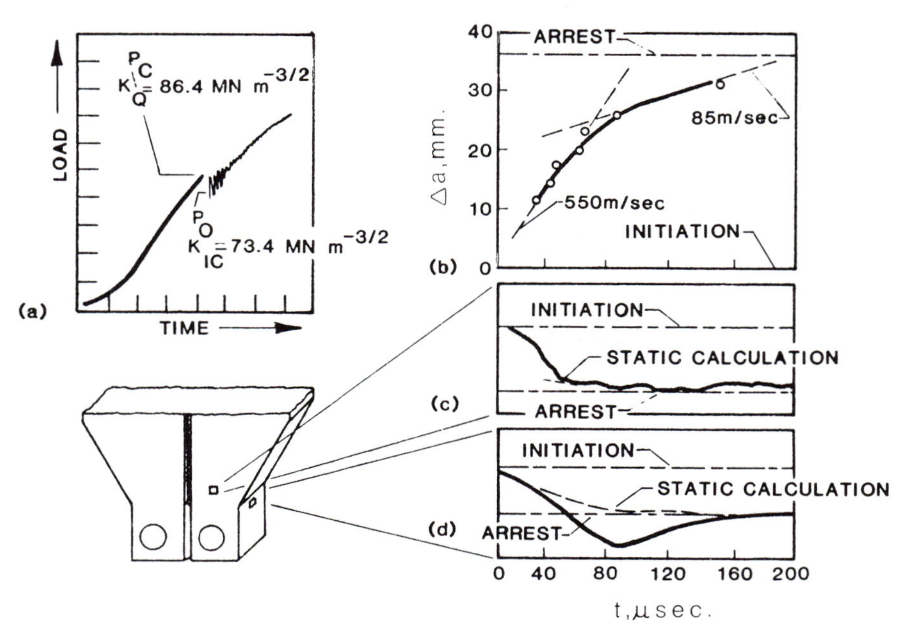

Figure 4.18 Load and strain gage readings during a run/arrest experiment in a pressure vessel steel CT specimen.

viscoplastic) dynamic analysis would be necessary. Thus, while most attention has been placed upon the conduct and interpretation of crack arrest tests by elastic considerations, it must be recognized that nonlinear interpretations may be necessary in some regimes.

4.1.8 Nonlinear Aspects of Dynamic Fracture Mechanics

As should be apparent from the discussion to this point, applications of dynamic fracture mechanics for crack arrest are largely confined to elastodynamic considerations. Nonetheless, questions that are now outstanding appear to be resolvable only by resort to nonlinear dynamic treatments. One of these bears on the uniqueness of the K_{ID} property. For example, in Figure 4-14, it can be seen that values obtained using the DCB specimen (lower toughnesses) match reasonably well the data collected in wide plate experiments (higher toughnesses). Although there is clearly scatter in these data, it can be concluded that the data are reasonably geometry-independent, as they must be for K_{ID} to qualify as a material property. This is a key point because other investigations have revealed a small, but systematic, dependence on specimen geometry. A particularly revealing result is that of Kalthoff (4.81) shown in Figure 4.19.

One explanation for a geometry dependence is that K_{ID} may not be a function only of the instantaneous crack speed but may also depend upon

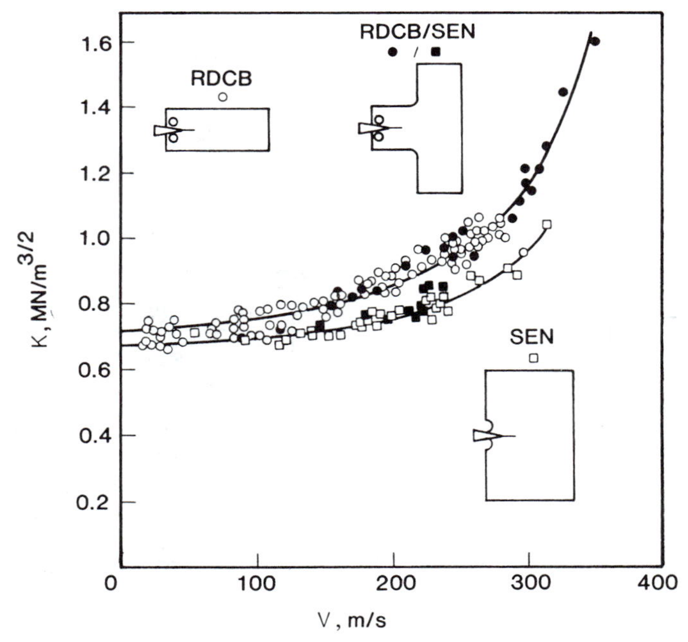

Figure 4.19 Geometry-dependence of the dynamic fracture toughness revealed by results in different specimens.

higher derivatives of crack length. An intuitive argument against this was given by Eshelby (4.82) who stated that "the crack tip has no inertia," meaning that the acceleration of the crack tip does not affect the calculated value of the crack driving force. It is generally concluded from this that the K_{ID} property also cannot depend on the acceleration. It must therefore be concluded that local nonlinear effects must be responsible for whatever nonuniqueness in K_{ID} that exists.

Except for the fact that a generally accepted crack advance criterion for such conditions does not exist, dynamic elastic-plastic and dynamic viscoplastic crack propagation computations pose no difficulty in principle. Important work on the characterization of the crack tip for dynamic elastic-plastic conditions has been given by Achenbach et al. (4.83), Freund and Douglas (4.84), Dantam and Hahn (4.85), and Ahmad et al. (4.86), among others. An early appreciation of the problem was in fact evidenced by Tetelman (4.87).

The essence of the problem is that a wake of relaxed plasticity is left behind the moving crack tip and this violates the "K-dominance" requirement described in Chapter 3. Hence, an approach based upon incremental plasticity is required. But, owing to the high rates of deformation associated with a rapidly propagating crack, viscoplastic treatments that account for strain rate effects may well be required. A start on this type of problem has been made by Aboudi and Achenbach (4.88, 4.89), Hoff et al. (4.90), Lo (4.91), and Brickstad (4.92). An illustrative example is given in Figure 4.20 that reveals the practical effect of treating viscoplastic material behavior.

Figure 4.20 shows the result obtained by Brickstad (4.92) using the viscoplastic model of Perzyna in analyzing experimental test data on crack

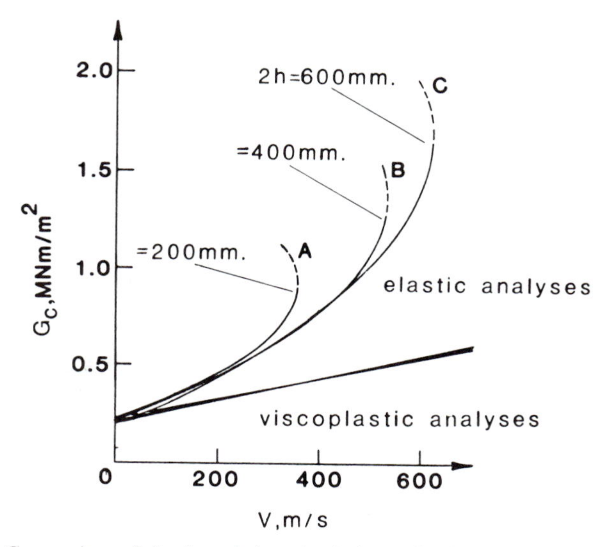

Figure 4.20 Comparison of elastic and viscoplastic determinations of the critical energy release rate for crack propagation in different sized edge-cracked tension panel experiments.

propagation in single edge notched (SEN) plates. The elastic interpretation of these results, given earlier by Dahlberg et al. (4.67), showed distinct geometry-dependence. But, by including viscoplastic effects, these have essentially been eliminated. Indeed, as Figure 4.20 shows, much of the speed-dependence is also removed. It might be noted that a number of viscoplastic formulations are available. However, the model advanced by Bodner and Partom (4.93) appears to have the most potential for use in dynamic fracture mechanics. One reason is the type of result shown in Figure 4.21. Further details on this model are given in Chapter 2.

The condition that the inelastic deformation surrounding the crack tip is "dominated" by the elastic K field (see Chapter 3) sets a requirement for contained or small-scale yielding. But, because a moving crack inevitably leaves a wake of relaxed plasticity behind, except for very short crack jumps, LEFM clearly cannot in principle offer a valid characterization of a propagating crack. We note that the same conceptual difficulty arises in fatigue (albeit at generally much lower stress levels) where it is dealt with by appealing to the idea of "similitude." For similitude the parameters governing the crack growth rate must be measured for the same type of load history as in the application of interest. When similitude is violated, the fatigue relations do

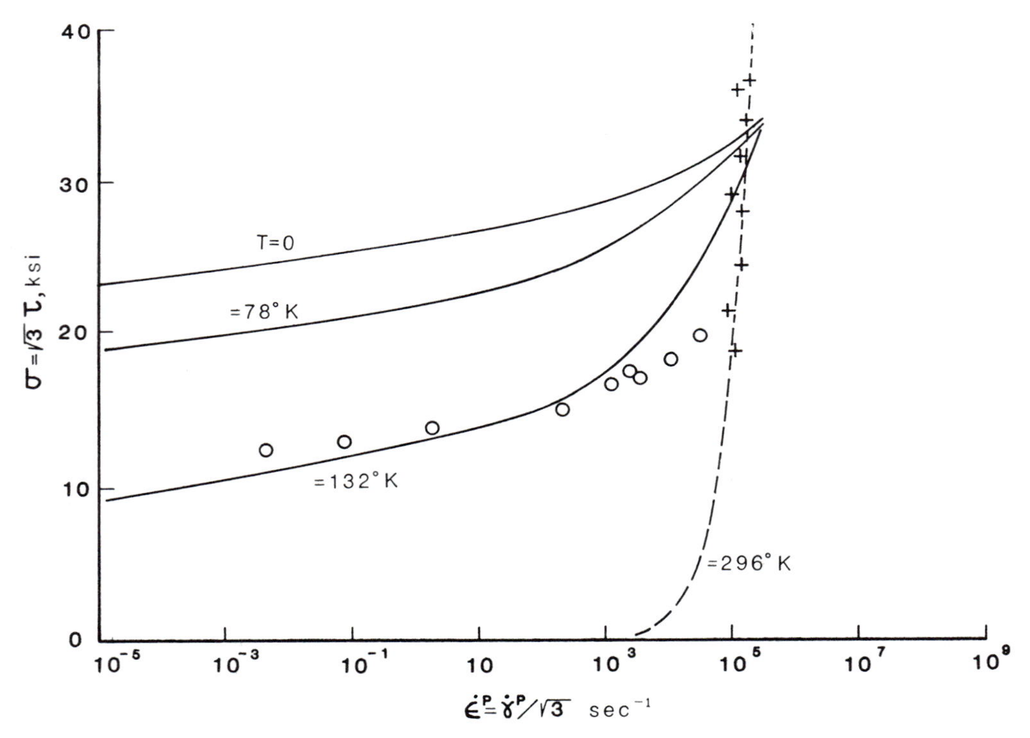

Figure 4.21 Comparison of the Bodner-Partom viscoplastic model with high rate loading experiments in aluminum.

not work—for example, in crack growth retardation following an overload. One might therefore expect the same kind of effects in dynamic fracture. Hence, just as current advanced research efforts in fatigue are now being made via direct consideration of plastic yielding and relaxation, dynamic fracture mechanics researchers are also turning to elastic-plastic analyses.

A difficulty that immediately arises in elastic-plastic analyses is in identifying the proper crack growth criterion. A critical strain at a critical distance ahead of the crack tip has been proposed, for example. Freund and Douglas (4.84) have used this criterion effectively in deducing the running fracture resistance values shown in Figure 4.22. But, such a parameter is somewhat unappealing—it cannot be measured and is not in any event palatable unless it can somehow be connected to a micromechanical picture. While results have been obtained in anti-plane strain that show a connection between ductile hole growth and a continuum strain measure ahead of a crack tip, a general result has yet to be obtained. An alternative to a critical strain criterion is the crack-tip opening angle (CTOA). This parameter is attractive from both a computational point of view and from the extensive experience garnered in elastic-plastic fracture mechanics that shows the constancy of the CTOA in stable growth; see Chapter 5. Nevertheless, the use of this parameter similarly requires a proper theoretical basis that does not now exist.

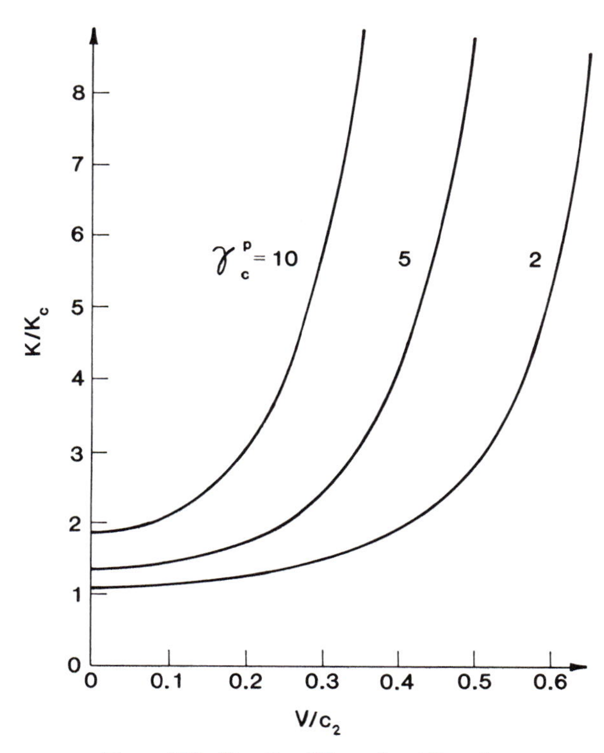

Figure 4.22 Results of Freund and Douglas.

Computations in which a crack extension criterion is specified and the crack length history determined in a given initial value–boundary value problem Kanninen (4.14) has called "application-phase" computations. As already mentioned, their counterparts, "generation-phase" computations, are those in which an actual crack propagation event is simulated in order to generate values of one or more selected crack growth criteria. Clearly, while the current lack of a well-established elastic-plastic dynamic criterion prohibits application-phase computations, generation-phase computations on appropriate experiments are still possible.

4.2 Mathematical Basis of Dynamic Fracture Mechanics

Engineering structures requiring protection against the possibility of large-scale catastrophic crack propagation (i.e., nuclear reactor pressure vessels and piping, ship hulls, gas transmission pipelines) are generally constructed of ductile, tough materials. Consequently, the procedures of linear elastic fracture mechanics (LEFM) can give only approximately correct predictions for such materials. More rigorous fully elastic-plastic and viscoplastic treatments are required to give precise results. However, because the direct incorporation of inelastic effects in dynamic fracture problems is only in its initial phases, the process of rapid unstable crack propagation and arrest in structures is therefore predominantly treated in terms of LEFM concepts and parameters. This section presents the mathematical basis of dynamic fracture mechanics with emphasis on the elastodynamic point of view.

4.2.1 Elastodynamic Crack-Tip Fields

The asymptotic crack-tip fields for self-similar crack propagation in a linear elastic isotropic material are important for a sound theoretical basis of dynamic fracture mechanics. In developing these, the crack speed is assumed to be sufficiently great to warrant retention of inertia effects. But, it will not exceed the characteristic Rayleigh wave speed of the material.

Consider a plane crack in the x_1-x_2-plane. The x_2-axis is normal to the crack plane. Crack extension occurs in the x_1 direction. Then, combining the equations of motion, Equation (2.1-17), the strain-displacement relation of Equation (2.2-8), and the constitutive relation, Equation (2.3-9), the Navier equations

$$\mu u_{i,jj} + (\lambda + \mu) u_{j,ij} = \rho \ddot{u}_i \tag{4.2-1}$$

are obtained. Each superposed dot denotes a partial derivative with respect to time. Also,

$$\lambda = 2\mu\nu/(1 - 2\nu) \tag{4.2-2}$$

where μ denotes the shear modulus and ν is Poisson's ratio.

For plane strain (i.e., $u_3 \equiv 0$). it is convenient to express the in-plane displacement components in terms of two potential functions. These are the dilatational and shear wave potentials ϕ and ψ, given respectively by

$$u_1 = \frac{\partial\phi}{\partial x_1} + \frac{\partial\psi}{\partial x_2}, \qquad u_2 = \frac{\partial\phi}{\partial x_2} - \frac{\partial\psi}{\partial x_1} \tag{4.2-3}$$

Equation (4.2-1) then reduces to

$$C_1^2\nabla^2\phi = \ddot{\phi}, \qquad C_2^2\nabla^2\psi = \ddot{\psi} \tag{4.2-4}$$

Notice that, by writing

$$C_1^2 = \frac{\kappa+1}{\kappa-1}\frac{\mu}{\rho} \qquad \text{and} \qquad C_2^2 = \frac{\mu}{\rho} \tag{4.2-5}$$

for the dilatational and shear wave speeds, Equations (4.2-4) are also valid for plane stress. Here

$$\kappa = \begin{cases} 3 - 4\nu & \text{for plane strain} \\ (3-\nu)/(1+\nu) & \text{for plane stress} \end{cases} \tag{4.2-6}$$

as in Chapter 3.

Introduce an $x - y$ coordinate system attached to the crack tip such that $x = x_1 - a(t)$ and $y = x_2$. Then, $x_1 = a(t)$ and $x_2 = 0$ define the location of the propagating tip. This change of coordinates permits writing

$$\ddot{\phi} = V^2\frac{\partial^2\phi}{\partial x^2} - 2V\frac{\partial^2\phi}{\partial x\partial t} + \frac{\partial^2\phi}{\partial t^2} - \dot{V}\frac{\partial\phi}{\partial x} \tag{4.2-7}$$

where the instantaneous crack speed is $V = \dot{a}$. A similar relation can be obtained for $\ddot{\psi}$.

Because the crack-tip fields are expected to be singular at the crack tip, the leading term on the right-hand side of Equation (4.2-7) will dominate the others there. Hence, the asymptotic fields are governed by

$$\begin{aligned} \beta_1^2\frac{\partial^2\phi}{\partial x^2} + \frac{\partial^2\phi}{\partial y^2} &= 0 \\ \beta_2^2\frac{\partial^2\psi}{\partial x^2} + \frac{\partial^2\psi}{\partial y^2} &= 0 \end{aligned} \tag{4.2-8}$$

where

$$\beta_1^2 = 1 - V^2/C_1^2 \quad \text{and} \quad \beta_2^2 = 1 - V^2/C_2^2 \tag{4.2-9}$$

Note that $\beta_1^2 > \beta_2^2 > 0$. As a final step, let

$$y_1 = \beta_1 y \quad \text{and} \quad y_2 = \beta_2 y \tag{4.2-10}$$

The use of these variables then reduces Equations (4.2-8) to equations having the form of the Laplace equation; that is,

$$\frac{\partial^2\phi}{\partial x^2} + \frac{\partial^2\phi}{\partial y_1^2} = 0, \qquad \frac{\partial^2\psi}{\partial x^2} + \frac{\partial^2\psi}{\partial y_2^2} = 0 \tag{4.2-11}$$

Therefore, ϕ can be expressed as either the real or imaginary part of an analytic function of the complex variable

$$z_1 = x + iy_1 = x + i\beta_1 y = r_1 e^{i\theta_1} \tag{4.2-12}$$

Similarly, ψ can be expressed in terms of

$$z_2 = x + iy_2 = x + i\beta_2 y = r_2 e^{i\theta_2} \tag{4.2-13}$$

It follows that

$$\phi = A_1 \operatorname{Re}[z_1^s], \qquad \psi = A_2 \operatorname{Im}[z_2^s] \tag{4.2-14}$$

with A_1, A_2, and s being real constants. In this formulation the stresses and deformations are symmetric with respect to the crack plane—that is, Mode I. Moreover, $s > 1$ for nonsingular displacements at the crack tip.

From here on the analysis follows that given for a stationary crack in Chapter 3. The condition that $\sigma_{22} = \sigma_{21} = 0$ on $\theta_1 = \theta_2 = \pi$ yields two linear homogeneous algebraic equations for A_1 and A_2. The singular solution is identified with $s = \frac{3}{2}$ and $A_2 = -2A_1\beta_2/(1 + \beta_1^2)$. With the Mode I dynamic stress intensity factor defined by

$$K(t) = \lim_{r \to 0} [(2\pi r)^{\frac{1}{2}} \sigma_{22}(r, 0, t)] \tag{4.2-15}$$

the dynamic crack-tip singular field can be written as

$$\begin{aligned}
\sigma_{11} &= \frac{K(t)B}{(2\pi r)^{\frac{1}{2}}} \left[(1 + 2\beta_1^2 - \beta_2^2)\left(\frac{r}{r_1}\right)^{\frac{1}{2}} \cos(\theta_1/2) \right. \\
&\qquad \left. - \frac{4\beta_1\beta_2}{1 + \beta_2^2}\left(\frac{r}{r_2}\right)^{\frac{1}{2}} \cos(\theta_2/2) \right] \\
\sigma_{12} &= \frac{2K(t)B\beta_1}{(2\pi r)^{\frac{1}{2}}} \left[\left(\frac{r}{r_1}\right)^{\frac{1}{2}} \sin(\theta_1/2) - \left(\frac{r}{r_2}\right)^{\frac{1}{2}} \sin(\theta_2/2) \right] \\
\sigma_{22} &= \frac{K(t)}{(2\pi r)^{\frac{1}{2}}} B \left[-(1 + \beta_2^2)\left(\frac{r}{r_1}\right)^{\frac{1}{2}} \cos(\theta_1/2) \right. \\
&\qquad \left. + \frac{4\beta_1\beta_2}{1 + \beta_2^2}\left(\frac{r}{r_2}\right)^{\frac{1}{2}} \cos(\theta_2/2) \right]
\end{aligned} \tag{4.2-16}$$

where

$$B = (1 + \beta_2^2)/D(V) \tag{4.2-17}$$

and $D(V) = 4\beta_1\beta_2 - (1 + \beta_2^2)^2$ is equivalent to that given in Equation (4.1-19).

Note that, in the limit as the crack speed V tends to zero, the stress field of Equation (3.1-38) for a stationary crack is recovered. Nilsson (4.94) and Freund and Clifton (4.95) developed Equation (4.2-16) for an arbitrarily advancing crack tip whose instantaneous speed is V. Consequently, Equation (4.2-16) includes as a special case the constant speed crack case.

The particle-velocity field near the crack tip is

$$\dot{u}_1 = -\frac{K(t)BV}{(2\pi r)^{\frac{1}{2}}\mu}\left[\left(\frac{r}{r_1}\right)^{\frac{1}{2}}\cos(\theta_1/2) - \frac{2\beta_1\beta_2}{1+\beta_2^2}\left(\frac{r}{r_2}\right)^{\frac{1}{2}}\cos(\theta_2/2)\right]$$

$$\dot{u}_2 = -\beta_1\frac{K(t)BV}{(2\pi r)^{\frac{1}{2}}\mu}\left[\left(\frac{r}{r_1}\right)^{\frac{1}{2}}\sin(\theta_1/2) - \frac{2}{1+\beta_2^2}\left(\frac{r}{r_2}\right)^{\frac{1}{2}}\sin(\theta_2/2)\right] \tag{4.2-18}$$

This velocity field also has an inverse square root singularity at the crack tip. The crack-tip stress and velocity fields can be derived in a similar way for Mode II loading.

The dynamic stress intensity factor, which in general will be a function of the loading, crack length, and geometry of the flawed body, determines the strength of the crack-tip singular field. When the size of the plastic zone attending the crack tip is small compared to the characteristic dimension of the region over which the singular field dominates, the concept of K-dominance for a stationary crack tip can be extended to a propagating crack (cf. Chapter 3). That is, when K-dominance exists, the fracture process is governed by $K(t)$. Hence, the linear elastodynamic crack propagation criterion is an equality between K and its critical value, a parameter known as the running fracture toughness. On the basis that the plane strain value of this parameter can depend only on temperature and crack speed, the crack propagation equation can be written as

$$K(t) = K_{ID}(V, T) \tag{4.2-19}$$

Equation (4.2-19), which can be viewed as the equation of motion for the crack tip, can be integrated to yield the crack growth history.

The K_{ID} relations that have so far been developed, shown in the preceding section of this chapter, contain a sharp increase at higher crack speeds. A partial explanation for this effect may be due to the effect of stress triaxiality. That is, it follows from Equation (4.2-16) on $\theta = 0$ that the ratio of the principal stresses can be written as

$$\frac{\sigma_{22}}{\sigma_{11}} = \frac{D(V)}{(1+\beta_2^2)(1+2\beta_1^2-\beta_2^2) - 4\beta_1\beta_2} \tag{4.2-20}$$

Equation (4.2-20) shows that σ_{22} decreases continuously relative to σ_{11} from equality at zero crack speed to zero at the Rayleigh speed. Because the toughness generally increases as the triaxiality decreases, the consequent reduction in the stress triaxiality as V increases could be at least partly responsible for the increasing fracture toughness at high crack speeds in rate-sensitive materials.

Achenbach et al. (4.83) have developed asymptotic analyses for dynamic crack propagation in elastic-plastic materials. Of most importance, assuming that the nature of the singularity in the stress field has the form r^{-s}, they were able to determine explicit values for s. Their solution employed a bilinear constitutive relation with E_t denoting the slope of the stress-strain curve in the plastic region. They then found that s is a strong function of E_t/E, where E is

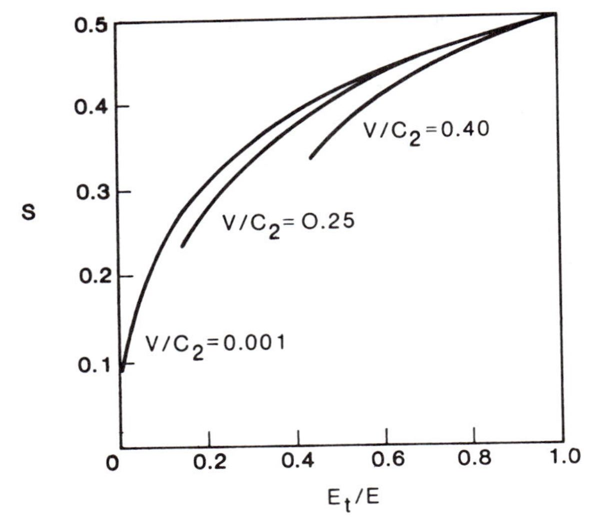

Figure 4.23 Order of the crack-tip singularity for crack propagation in a bilinear elastic-plastic material under plane stress conditions.

the elastic modulus. But, the results revealed only a modest dependence upon V/C_2. Their results for plane stress conditions are shown in Figure 4.23. It can be seen that, while the crack speed does effect the singularity, the stress-strain behavior is much more influential.

4.2.2 *The Energy Release Rate*

To establish the connection between the stress intensity factor and the energy release rate, consider an elastic body containing a propagating planar crack. This is depicted in Figure 4.24. To determine the energy released to the crack tip, consider a vanishingly small loop Γ^* surrounding the crack tip. The loop

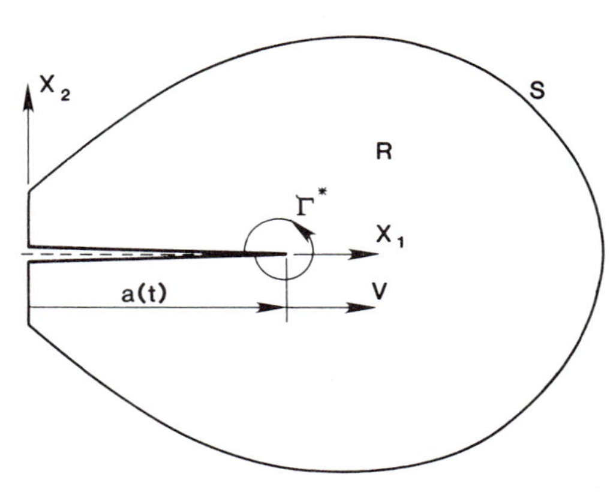

Figure 4.24 Dynamic crack propagation in a plane body.

Γ^* can have an arbitrary shape, but it must remain fixed relative to a coordinate system attached to the crack tip. The region R of interest is bounded by the outer boundary S of the body, the traction-free crack faces, and the loop Γ^*.

The rate of work P done by the tractions on S is equal to the rate of increase of the strain energy $\dot{U}$ and kinetic energy $\dot{T}$ in R and the flux F of energy into the crack-tip region. That is,

$$P = \dot{U} + \dot{T} + F \tag{4.2-21}$$

where

$$P = \int_S T_i \dot{u}_i \, ds \tag{4.2-22}$$

$$U = \lim_{\Gamma^* \to 0} \int_R W(\varepsilon_{ij}) \, dA \tag{4.2-23}$$

$$T = \lim_{\Gamma^* \to 0} \int_R \tfrac{1}{2} \rho \dot{u}_i \dot{u}_i \, dA \tag{4.2-24}$$

Since the loop Γ^* moves with the crack tip, the region R is time-dependent. Therefore, the time rate of change of the strain energy equals the integral over R of the time rate of change of the strain energy density $W(\varepsilon_{ij})$ less the flow of this density through the loop Γ^*. Hence,

$$\dot{U} = \lim_{\Gamma^* \to 0} \int_R \dot{W} \, dA - \lim_{\Gamma^* \to 0} \int_{\Gamma^*} W V_n \, ds \tag{4.2-25}$$

where $V_n = V n_1 = V \cos\theta$ is the component of the crack-tip velocity normal to Γ^* and θ is the angle that the outward unit normal to Γ^* makes with the x_1 direction. The positive direction for traversing Γ^* is in the counterclockwise direction.

With the aid of Equation (2.3-4) and the symmetry of the stress tensor, the energy rate expression can be written as

$$\dot{W} = \frac{\partial W}{\partial \varepsilon_{ij}} \dot{\varepsilon}_{ij} = \sigma_{ij} \dot{\varepsilon}_{ij} = \sigma_{ij} \dot{u}_{i,j} \tag{4.2-26}$$

Hence,

$$\dot{U} = \lim_{\Gamma^* \to 0} \int_R \sigma_{ij} \dot{u}_{i,j} \, dA - \lim_{\Gamma^* \to 0} \int_{\Gamma^*} W V_n \, ds \tag{4.2-27}$$

Similarly,

$$\dot{T} = \lim_{\Gamma^* \to 0} \int_R \rho \ddot{u}_i \dot{u}_i \, dA - \lim_{\Gamma^* \to 0} \int_{\Gamma^*} \tfrac{1}{2} \rho \dot{u}_i \dot{u}_i V_n \, ds \tag{4.2-28}$$

The introduction of Equations (4.2-22), (4.2-27), and (4.2-28) into (4.2-21) then gives an expression for the energy flux,

$$F = \lim_{\Gamma^* \to 0} \int_{\Gamma^*} (W + \tfrac{1}{2}\rho \dot{u}_i \dot{u}_i) V_n \, ds + \int_S T_i \dot{u}_i \, ds - \lim_{\Gamma^* \to 0} \int_R (\sigma_{ij} \dot{u}_{i,j} + \rho \ddot{u}_i \dot{u}_i) \, dA \tag{4.2-29}$$

After introducing $\sigma_{ij}\dot{u}_{i,j} = (\sigma_{ij}\dot{u}_i)_{,j} - \sigma_{ij,j}\dot{u}_i$, invoking the divergence theorem, and making use of the equations of motion—see Equation (2.1-17)—then Equation (4.2-29) can be written as

$$F = \lim_{\Gamma^* \to 0} \int_{\Gamma^*} [(W + \tfrac{1}{2}\rho \dot{u}_i \dot{u}_i) V_n + T_i \dot{u}_i] \, ds \tag{4.2-30}$$

Equation (4.2-30) states that the flux of energy to the crack tip is the sum of the flux of total energy (strain energy plus kinetic energy) through Γ^* as the crack tip moves through the material, plus the rate of work done by the material outside of Γ^* on that within Γ^*.

The dynamic energy release rate G is related to the energy flux by $F = VG$. Therefore, using Equation (4.2-30), G can be expressed as

$$G = \lim_{\Gamma^* \to 0} \frac{1}{V} \int_{\Gamma^*} [(W + \tfrac{1}{2}\rho \dot{u}_i \dot{u}_i) V_n + T_i \dot{u}_i] \, ds \tag{4.2-31}$$

Usually the symbol G is considered to pertain only to linear elastic material behavior. However, it should be noted that Equation (4.2-31) is actually valid for nonlinear elastic behavior.

Consider another contour Γ extending from the lower crack face counterclockwise around the tip to the upper crack face. The application of the divergence theorem to the region R^* bounded by Γ, Γ^*, and the crack faces can be shown to give

$$VG = \int_{\Gamma} [(W + \tfrac{1}{2}\rho \dot{u}_i \dot{u}_i) V_n + T_i \dot{u}_i] \, ds - \lim_{\Gamma^* \to 0} \int_{R^*} \left[\rho \left(\ddot{u}_i + V \frac{\partial \dot{u}_i}{\partial x_1} \right) \dot{u}_i + \sigma_{ij} \left(\dot{u}_{i,j} + V \frac{\partial u_{i,j}}{\partial x_1} \right) \right] dA \tag{4.2-32}$$

In contrast to the static problem, the energy release rate can not be expressed by a path-independent contour integral. That this is to be expected is readily demonstrated following an argument given originally by Eshelby. If a wave front intercepts one contour, but not the second one, then two different values for G would be expected if it were represented by a contour integral only. It is the area integral in Equation (4.2-32) that preserves the invariant character of G. Equation (4.2-32) and its variations have proven to be useful in numerical investigations—for example, the finite element method, where, due to numerical difficulties, it is not feasible to evaluate the limiting contour integral of Equation (4.2-31). An alternative to Equation (4.2-32) is developed in the next section.

Under conditions of steady-state crack propagation where $\dot{u}_i = -V\,\partial u_i/\partial x_1$, Equation (4.2-32) reduces to

$$G = \int_\Gamma \left[\left(W + \tfrac{1}{2}\rho V^2 \frac{\partial u_i}{\partial x_1}\frac{\partial u_i}{\partial x_1} \right) n_1 - T_i \frac{\partial u_i}{\partial x_1} \right] ds \qquad (4.2\text{-}33)$$

Hence, as first shown by Atkinson and Eshelby (4.70), G is path-independent for steady-state crack growth. It is also clear that Equation (4.2-33) reduces to the J-integral when $V = 0$ [cf. Equation (3.3-24)].

The introduction of the linear elastic stress and velocity fields of Equations (4.2-16) and (4.2-18) into Equation (4.2-31) gives

$$G = \frac{V^2\beta_1}{C_2^2 D(V)} \frac{K^2(t)}{2\mu} \qquad (4.2\text{-}34)$$

where $D(V)$ is given by Equation (4.1-19). This relationship establishes the generality of Equation (4.1-18). It follows that, whenever $K(t)$ is at its critical value, then G is necessarily at its critical value. Hence, as in the static case, the dynamic fracture criterion for a linear elastic material can be expressed in terms of either a critical stress intensity factor or a critical energy release rate.

It is important to recognize that, while Equation (4.2-34) is geometry-independent, the individual values of K and G are distinctly geometry-dependent. Consider Yoffe's constant crack length solution discussed in Section 4.1.3. Because K is independent of crack speed in the Yoffe problem, it follows from Equation (4.2-34) that G in this case increases monotonically with crack speed and becomes unbounded at the Rayleigh wave speed. In Broberg's expanding crack solution, K is a monotonically decreasing function of crack speed that becomes zero at the Rayleigh wave speed. This produces a finite value of G throughout. Figure 4.25, taken from Cotterell (4.36), contrasts the results of Yoffe and Broberg.

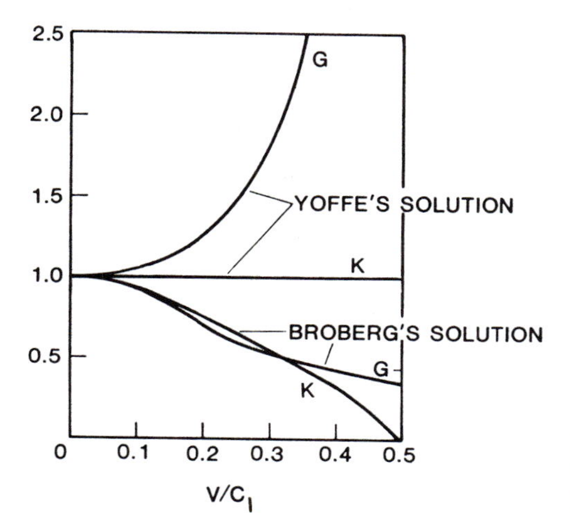

Figure 4.25 Comparison of constant-length and uniformly expanding crack propagation models.

As an illustration of how the path independence of G for steady-state crack growth can be exploited, consider the steady-state crack growth of a semi-infinite crack in an infinite linear elastic strip. The static analog of this problem was treated in Chapter 3; for example, see Figure 3.9. The lateral edges, $|x_2| = h$, are displaced uniformly and symmetrically by an amount u_{20} without tangential motion. If the same contour Γ as in the static problem is used, then $\partial u_i/\partial x_1 = 0$ on it. Hence, the kinetic energy density in Equation (4.2-33) vanishes on Γ. The dynamic and static energy release rates are then identical; for example, for plane strain see the development leading to Equation (3.3-27). This result is

$$G = \frac{(1 - \nu)Eu_{20}^2}{(1 + \nu)(1 - 2\nu)h} \tag{4.2-35}$$

Therefore, the energy release rate is independent of the crack speed for $V < C_R$. This expression for G is simply the strain energy per unit cross-sectional area far ahead of the crack tip. Bergkvist (4.96) used this interpretation to establish G. When the lateral edges are shear-free, Popelar and Atkinson (4.97) have shown that such an interpretation is no longer valid.

The introduction of Equation (4.2-35) into Equation (4.2-34) leads to

$$K = \left(\frac{(1 - \nu)D(V)}{\beta_1(1 - \beta_2^2)}\right)^{\frac{1}{2}} \frac{Eu_{20}}{[(1 + \nu)^2(1 - 2\nu)h]^{\frac{1}{2}}} \tag{4.2-36}$$

Again, as dimensional analysis requires, the dynamic stress intensity factor is the product of the equivalent static stress intensity factor and a function of crack speed. Nilsson (4.98) used the Wiener-Hopf technique to obtain this result among others. Since $D(V) \to 0$ as $V \to C_R$, then $K \to 0$ as $V \to C_R$.

4.2.3 Elastodynamic Contour Integrals

Using Equation (4.2-31) to evaluate the dynamic energy release rate presents no particular difficulties when closed form solutions for the stress and velocity fields are available. Unfortunately, only a few such solutions for dynamic crack propagation and crack arrest exist. And, these solutions are often for highly idealized problems. In practice, the stress and velocity fields are typically determined numerically. In many instances the greatest imprecision in a numerical technique (e.g., the finite element method) is associated with modeling the singular behavior in the crack-tip region. Specifically, it is impossible to proceed numerically to the limit that is required in Equation (4.2-31). Therefore, we develop an equivalent representation for the energy release rate that is less sensitive to numerical inaccuracies in the crack-tip region.

For a convective coordinate system attached to the crack tip the particle velocity can be written as

$$\dot{u}_i = -V\frac{\partial u_i}{\partial x_1} + \frac{\partial u_i}{\partial t} \tag{4.2-37}$$

The leading convective term in Equation (4.2-37) will be dominant in the limiting crack-tip region bounded by the crack faces and the contour Γ^* in Figure 4.24. Hence, it is permissible to rewrite Equation (4.2-31) as

$$G = \lim_{\Gamma^* \to 0} \int_{\Gamma^*} \left[(W + \tfrac{1}{2}\rho \dot{u}_i \dot{u}_i) n_1 - T_i \frac{\partial u_i}{\partial x_1} \right] ds \tag{4.2-38}$$

As in the preceding section let Γ denote a contour extending from the lower crack face counterclockwise around the crack tip to the upper crack face. Invoking the divergence theorem for the region R^* bounded by the union of Γ, Γ^*, and the traction-free crack faces permits writing

$$\begin{aligned} G = &\int_{\Gamma} \left(W n_1 - T_i \frac{\partial u_i}{\partial x_1} \right) ds + \lim_{\Gamma^* \to 0} \int_{\Gamma^*} \tfrac{1}{2} \rho \dot{u}_i \dot{u}_i n_1 \, ds \\ &+ \lim_{\Gamma^* \to 0} \int_{R^*} \rho \ddot{u}_1 \frac{\partial u_i}{\partial x_1} \, dA \end{aligned} \tag{4.2-39}$$

In obtaining Equation (4.2-39) we have also introduced the equations of motion, Equation (2.1-17), and the constitutive relation, Equation (2.3-4). This expression for the energy release rate is independent of the contour Γ selected. Moreover, it agrees with the energy release rate obtained by Aoki et al. (4.99) who also developed the dynamic analogues of the L- and M-integrals—that is, Equations (3.3-32) and (3.3-33).

It is also possible through use of Equation (4.2-37) and the divergence theorem to write

$$\begin{aligned} \lim_{\Gamma^* \to 0} \int_{\Gamma^*} \tfrac{1}{2} \rho \dot{u}_i \dot{u}_i n_1 \, ds = &\int_{\Gamma} \tfrac{1}{2} \rho V^2 \frac{\partial u_i}{\partial x_1} \frac{\partial u_i}{\partial x_1} n_1 \, ds \\ &- \lim_{\Gamma^* \to 0} \int_{R^*} \rho V^2 \frac{\partial^2 u_i}{\partial x_1^2} \frac{\partial u_i}{\partial x_1} \, dA \end{aligned} \tag{4.2-40}$$

Therefore, Equation (4.2-39) can be written as

$$\begin{aligned} G = &\int_{\Gamma} \left[\left(W + \tfrac{1}{2} \rho V^2 \frac{\partial u_i}{\partial x_1} \frac{\partial u_i}{\partial x_1} \right) n_1 - T_i \frac{\partial u_i}{\partial x_1} \right] ds \\ &+ \lim_{\Gamma^* \to 0} \int_{R^*} \rho \left(\ddot{u}_i - V^2 \frac{\partial^2 u_i}{\partial x_1^2} \right) \frac{\partial u_i}{\partial x_1} \, dA \end{aligned} \tag{4.2-41}$$

This form for the energy release rate is less sensitive to numerical inaccuracies in the crack-tip region than Equation (4.2-31) or, equivalently, Equation (4.2-38). Equation (4.2-41) obviously reduces to Equation (4.2-33) for steady-state crack propagation.

The path independent $\hat{J}$-integral developed by Kishimoto et al. (4.100) is based upon writing the energy balance equation as

$$P = \lim_{\Gamma^* \to 0} \int_{R} (\dot{W} + \rho \ddot{u}_i \dot{u}_i) \, dA + V \hat{J} \tag{4.2-42}$$

When this equation is compared with Equation (4.2-21) et seq., it follows that

$$\hat{J} = G - \lim_{\Gamma^* \to 0} \int_{\Gamma^*} (W + \tfrac{1}{2}\rho \dot{u}_i \dot{u}_i) n_1 \, ds \tag{4.2-43}$$

It is therefore clear that $\hat{J}$ differs from the energy release rate by the flux of the strain and kinetic energies through the contour Γ^*. It might be noted that Kishimoto and co-workers have provided a more comprehensive relation that contains thermal stresses and incremental plasticity.

4.3 Analyses of Some Simple Configurations

Although it is seldom made on a rigorous basis, the commonly used assumption that dynamic effects can be neglected in the arrest of rapid crack propagation in engineering structures is probably a reasonable one in most instances. Freund's analysis for crack propagation in an infinite medium provides the legitimacy for this assumption. Unfortunately, the requirement that stress waves not be reflected from boundaries to the propagating crack tip is less likely to be met in laboratory test specimens. Even with modest crack jump lengths, the finite size of the specimen must be taken into account if the experiment is to be properly interpreted. While this can usually be done with large-scale finite element models, simpler analysis approaches are always useful. Two of the specimen geometries commonly used for dynamic crack arrest experimentation lend themselves to such treatments. These are the "beam-like" configurations known as the double cantilever beam (DCB) specimen and the double torsion (DT) specimen. Their geometries can be effectively exploited to produce one-dimensional (spatial) analysis models. Analogous to these is crack propagation in an infinite stretched strip and in a pressurized pipeline. Also discussed are steady-state crack propagation and the use of strip-yield zone models in dynamic crack propagation.

4.3.1 The Double Cantilever Beam Specimen

A laboratory test specimen used effectively by many investigators is the double cantilever beam (DCB) specimen; see Figure 4.26. A model for dynamic crack propagation in the DCB specimen has been developed by Kanninen and co-workers in a series of papers culminating with the work of Gehlen et al. (4.101). In the latter paper Reissner's (4.102) variational principle combined with assumed forms for the stress and displacement fields was used to develop the governing equations. This approach has an advantage in that, once the forms of these fields are selected, an entirely consistent formulation follows without additional ad hoc assumptions. Let

$$L = \int_{t_1}^{t_2} \int_{\Omega} (\tfrac{1}{2}\rho \dot{u}_i \dot{u}_i - \sigma_{ij}\varepsilon_{ij} + W^*) \, d\Omega \, dt \tag{4.3-1}$$

where the integration is over the volume Ω of the specimen and W^* is the

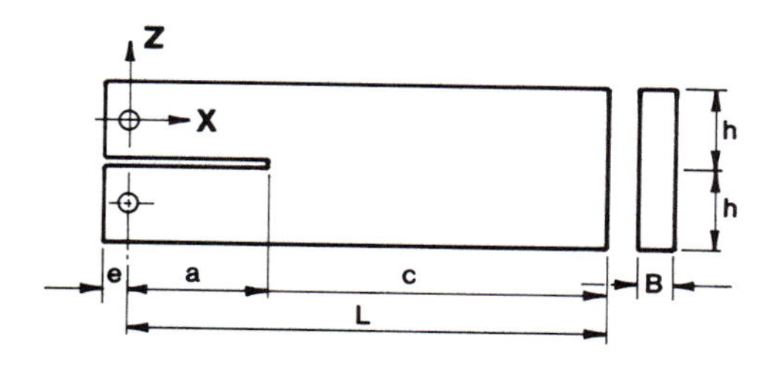

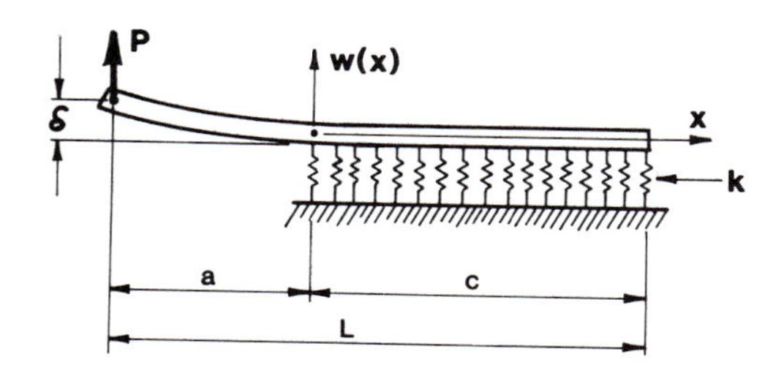

Figure 4.26 Double cantilever beam specimen and beam on an elastic foundation model.

complementary strain energy density; see Equation (2.3-19). Reissner's principle is then embodied in the variational equation

$$\delta L = 0 \tag{4.3-2}$$

for arbitrary variations of the stress and displacement fields subject to $\delta u_i(t_1) = \delta u_i(t_2) = 0$.

Due to the symmetry with respect to the crack plane, only the upper half $|z| \leqslant h/2$ of the specimen depicted in Figure 4.26 needs to be considered. Exploiting the beam-like character of this specimen, the nonzero stress components can be assumed to have the forms

$$\begin{aligned}\sigma_{11} &= \frac{Mz}{I} \\ \sigma_{13} &= \frac{S}{2I}\left(\frac{h^2}{4} - z^2\right) \\ \sigma_{33} &= \frac{p}{B} H(-z)H(x-a)\end{aligned} \tag{4.3-3}$$

where M, S, and p (the bending moment, transverse shear and crack-plane tension) are undetermined functions of x and t, H is the Heaviside step function, and I is the centroidal moment of inertia. The displacement components are then written as

$$\begin{aligned}u_1 &= -z\psi, \qquad u_2 = 0 \\ u_3 &= w - \frac{\nu M z^2}{2EI} + 2\frac{w}{h} zH(-z)H(x-a)\end{aligned} \tag{4.3-4}$$

where $w = w(x, t)$ is the vertical displacement of the centerline ($z = 0$) and $\psi = \psi(x, t)$ is its rotation. The nonzero strain components become

$$\varepsilon_{11} = -z\frac{\partial \psi}{\partial x}, \qquad \varepsilon_{13} = \frac{1}{2}\left(\frac{\partial w}{\partial x} - \psi\right)$$
$$\varepsilon_{33} = -\frac{\nu M z}{EI} + \frac{2w}{h}H(-z)H(x - a) \tag{4.3-5}$$

where the z dependence of ε_{13} has been neglected.

With the introduction of Equations (4.3-3)–(4.3-5) into Equation (4.3-1), Equation (4.3-2) leads to the equations of motion

$$\frac{\partial M}{\partial x} - S = -\rho I \ddot{\psi}$$
$$\frac{\partial S}{\partial x} - pH(x - a) = \rho A \ddot{w} \tag{4.3-6}$$

and the constitutive relations

$$M = -EI\frac{\partial \psi}{\partial x}$$
$$S = \frac{5}{6}\mu A\left(\frac{\partial w}{\partial x} - \psi\right) \tag{4.3-7}$$
$$p = \frac{2EB}{h}wH(x - a)$$

where $A = Bh$ is the cross-sectional area. Note that, while a rectangular DCB specimen is considered here (i.e., h = constant), the analysis is readily adapted to an arbitrarily contoured specimen [i.e., $h = h(x)$].

As shown in the preceding sections of this chapter, the energy release rate approach and the stress intensity factor approach are equivalent. Because the crack-tip singularity is not modeled explicitly, the former is more useful here. Accordingly, the fracture criterion can more conveniently be written as

$$G = R(V) \tag{4.3-8}$$

where the fracture resistance $R(V)$ in terms of the fracture toughness is

$$R(V) = A(V)K_{ID}^2(V)/E' \tag{4.3-9}$$

The energy release rate is given by Equation (4.1-24) or, equivalently, by

$$G = -\frac{1}{V}\frac{d}{dt}(\Pi + T) \tag{4.3-10}$$

where Π is the potential energy per unit thickness; see Equation (2.4-13). When Equations (4.3-3) and (4.3-6) are introduced into the strain energy per unit thickness,

$$U = \frac{1}{B}\int_{\Omega}(\sigma_{ij}\varepsilon_{ij} - W^*)\,d\Omega \tag{4.3-11}$$

then it becomes

$$U = \frac{1}{B} \int_{-e}^{L} \left[EI\left(\frac{\partial \psi}{\partial x}\right)^2 + \frac{5}{6}\mu A\left(\frac{\partial w}{\partial x} - \psi\right)^2 + \frac{2EB}{h} w^2 H(x-a) \right] dx \quad (4.3\text{-}12)$$

The kinetic energy per unit thickness is

$$T = \frac{\rho}{B} \int_{-e}^{L} (I\dot{\psi}^2 + A\dot{w}^2)\, dx \quad (4.3\text{-}13)$$

Under wedge opening (fixed-grip) loading, $\Pi = U$. The introduction of Equations (4.3-12) and (4.3-13) into Equation (4.3-10) then leads to

$$G = 2E \left[\frac{w^2}{h} \right]_{x=a} \quad (4.3\text{-}14)$$

It can be shown that Equation (4.3-14) is equally valid for compliant loading.

Equations (4.3-6) and (4.3-7) can be combined to obtain the form of a Timoshenko beam on an elastic foundation. This alternative formulation is

$$\frac{\partial^2 w}{\partial x^2} - \frac{\partial \psi}{\partial x} - \frac{6}{h^2} H(x-a) w = \frac{3}{C_0^2} \frac{\partial^2 w}{\partial t^2} \quad (4.3\text{-}15)$$

and

$$\frac{\partial^2 \psi}{\partial x^2} + \frac{4}{h^2} \frac{\partial w}{\partial x} - \psi = \frac{1}{C_0^2} \frac{\partial^2 \psi}{\partial t^2} \quad (4.3\text{-}16)$$

where C_0 is the elastic bar wave speed. Under static loading it is always possible to obtain closed-form solutions to these equations for a uniform specimen and to use Equation (4.3-14) to establish a static energy release rate. The stress intensity factor—for example, see Equation (4.1-22)—follows from Equation (3.3-16) relating G and K. As described in connection with Equation (4.1-23), the predictions of this model are in excellent agreement with the more exact approaches.

In order to achieve the supercritical condition necessary for extended rapid crack propagation in the DCB specimen, the initial crack tip is intentionally blunted. The degree of blunting can be characterized by K_Q, the static stress intensity factor existing at initiation of crack extension. Commonly, the view is taken that the blunting inherently alters the intrinsic fracture resistance of the material in the neighborhood of the tip such that initiation of crack growth is associated with K_Q. After initiation, however, further growth is governed by the dynamic fracture toughness $K_{ID}(V)$.

For a rapidly propagating crack, Equations (4.3-15) and (4.3-16) must be numerically integrated—for example, using a finite difference method. Equation (4.3-14) is then used to compute the energy release rate at each time step. This value of G is compared with R using a virtual crack speed based upon the time since the last increment of growth. When $G = R$, the crack is permitted to advance the next increment. If G remains less than R for an arbitrarily long period, the crack is considered to have arrested. In this manner it is possible to determine the crack history.

A comparison of the predictions of this model with the crack propagation and arrest phenomena observed by Kalthoff et al. (4.59) for Araldite B was made by Gehlen et al. (4.101). The dynamic fracture toughness used in the analysis is that presented in Figure 1.28 for Araldite B. A further complicating characteristic of Araldite B, common among other polymers, is that it exhibits rate effects. For example, its dynamic elastic modulus is approximately 10 percent greater than its static value. While this difference is relatively small, it can have an appreciable influence upon the crack history. A viscoelastic analysis along the lines performed by Popelar and Kanninen (4.75) would be required to accommodate such rate effects. Unfortunately, this requires further material characterization. In lieu of such, the dynamic properties were used in the dynamic analysis. Gehlen et al. found that using the static properties generally resulted in less favorable comparisons.

Figure 4.27 shows the predicted crack history and the energy composition as a function of crack growth for a wedge-loaded DCB specimen of the photoelastic material Araldite B with $K_Q = 1.34$ MPa m$^{\frac{1}{2}}$. The measured crack history is also shown for comparison. Both the measured and predicted crack histories reveal that crack extension over a substantial portion of the event occurs at a constant speed. The analysis underpredicts slightly the crack speed and overestimates the extent of crack growth in this test. It can also be seen that the strain energy decreases at a greater rate during the initial stages than during the final portion while the kinetic energy increases initially and then

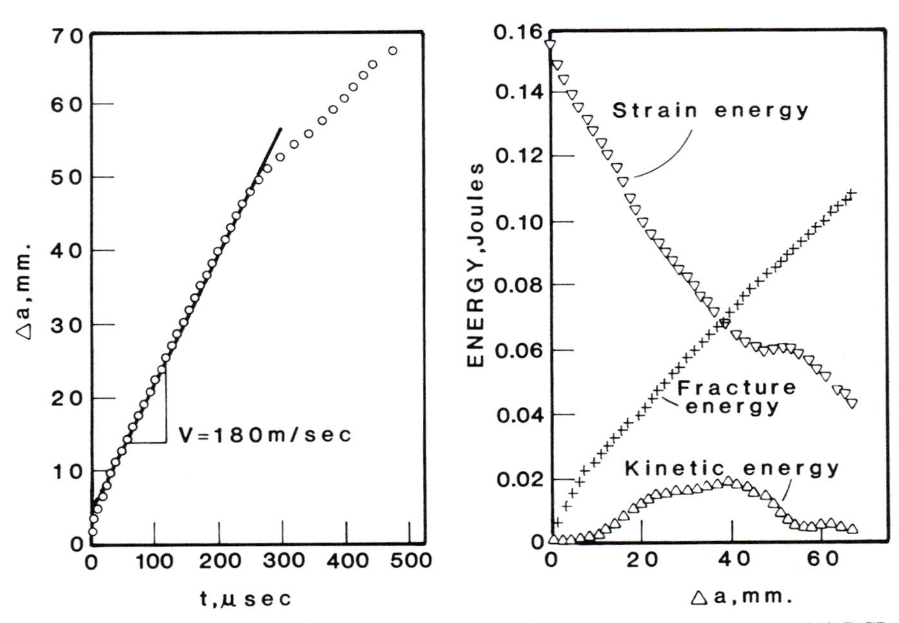

Figure 4.27 Crack propagation and energy composition history for a wedge-loaded DCB specimen of Araldite-B.

decreases later. The maximum kinetic energy in this case is approximately 10 percent of the initial stored energy. When $K_Q = 2.35$ MPa m$^{\frac{1}{2}}$, the maximum kinetic energy increases to approximately 25 percent of the initial stored energy.

Figure 4.28 compares the measured and predicted dynamic stress intensity factors with crack growth. Except near the end of the event the agreement is quite good. It would be impossible to achieve this kind of correlation without the inclusion of dynamic effects; for example, consider the simple static predictions displayed in Figure 4.6.

In contrast to the DCB model presented here, Burns and Chow (4.102) have devised a model based on the representation of the specimen as a simple built-in cantilever beam. In order to obtain a dynamic solution for this model (in which crack propagation occurs by increasing the length of the beam), it is necessary to restrict the solution to the case where the load points are continually displaced and the crack length is initially zero. However, these conditions approximate reasonably well the impact loading experimental procedure used by Burns.

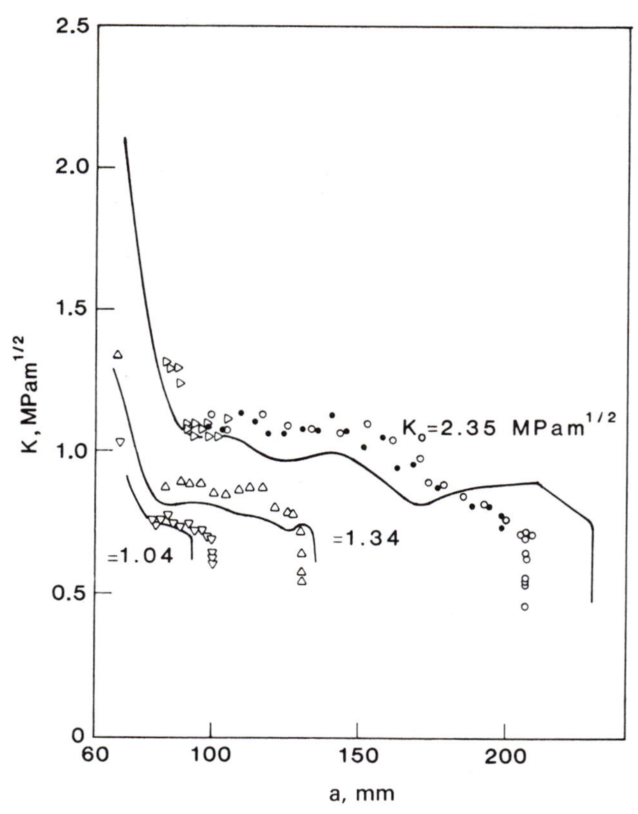

Figure 4.28 Comparison of measured and predicted dynamic stress intensity factors for wedge-loaded DCB specimens.

As Malluck and King (4.103) have shown, the results obtained by Burns and Chow are virtually identical to those obtained when the early version of model devised by Kanninen et al. (4.104) is suitably specialized. Further comparisons with the closed-form result obtained by Freund (4.105) using an approximate shear beam model for the DCB specimen also give good qualitative agreement with Equations (4.3-15) and (4.3-16). A beam model formulation has also been used by Steverding and Lehnigk (4.106) and by Bilek and Burns (4.107).

4.3.2 The Double Torsion Specimen

Perhaps the simplest illustration of the role of stress waves in dynamic crack propagation and crack arrest is given by the one-dimensional analysis of the double torsion (DT) specimen. As shown in Figure 4.29 this specimen consists of a rectangular plate with an initial blunted precrack of length a_0. A four-point loading system subjects the arms of the specimen to equal and opposite torques until a sharp crack emanates from the blunted tip. During the subsequent dynamic crack propagation event no further rotation of the end $z = 0$ is usually permitted; that is, fixed grip loading is considered.

Following the approach developed by Popelar (4.108), several simplifications are introduced to achieve a one-dimensional (spatial) model. They are, (1) the crack front is straight and normal to the plane of the plate, (2) elastic torsion theory describes the deformation, and (3) the stiffness of the plate ahead of the crack tip is such that the deformation in this region can be neglected. For rectangular sections, the motion in the region $0 < z < a(t)$ is governed by the torsional wave equation

$$\mu K \phi'' = \rho I \ddot{\phi} \tag{4.3-17}$$

where $\phi(z, t)$ is the rotation of the cross section about the centroidal axis z, μK is the torsional rigidity, ρI is the mass moment of inertia, and a prime is used to denote a partial derivative with respect to z.

The introduction of the rate of twist, $\alpha(z, t) \equiv \phi'$, and the angular velocity, $\omega(z, t) \equiv \dot{\phi}$, permits replacing Equation (4.3-17) by a system of first order

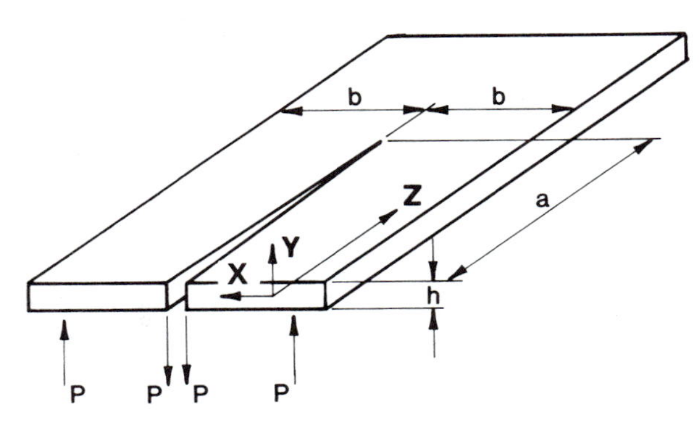

Figure 4.29 Mathematical model for double torsion specimen.

equations. They are

$$C^2\alpha' - \dot{\omega} = 0$$
$$\dot{\alpha} - \omega' = 0 \tag{4.3-18}$$

where $C = (\mu K/\rho I)^{\frac{1}{2}} = (K/I)^{\frac{1}{2}}C_2$ is the torsional wave speed. By differentiating the boundary conditions $\phi(0,t) = \phi_0$ and $\phi[a(t),t] = 0$, it is found that $\omega = 0$ at $z = 0$ and $\omega + V\alpha = 0$ at $z = a(t)$. When $t = 0$, $\alpha = -\phi_0/a_0 \equiv \alpha_0$ and $\omega = 0$.

For fixed grip loading, the energy release rate for the DT specimen can be written as

$$G = -(\dot{U} + \dot{T})/Vh \tag{4.3-19}$$

where

$$U = \mu K \int_0^a \alpha^2\, dz$$
$$T = \rho I \int_0^a \omega^2\, dz \tag{4.3-20}$$

Following an integration by parts and the introduction of Equations (4.3-18) and the boundary conditions, it can be shown that Equation (4.3-19) gives

$$G = \frac{\mu K}{h}\left(1 - \frac{V^2}{C^2}\right)\alpha^2[a(t),t], \qquad V < C \tag{4.3-21}$$

The energy release rate at incipient crack extension is

$$G_Q = \mu K\, \alpha_0^2/h \tag{4.3-22}$$

During crack extension $G = R$. While the fracture resistance R can in general depend upon the crack speed, a speed-independent resistance will be used in the following.

The method of characteristics provides an efficient solution technique. The solution domain is defined by $0 < z < a(t)$ and $t > 0$. The characteristics of Equation (4.3-18) are defined by $z \pm Ct = \text{constant}$ and are depicted in Figure 4.30. On these characteristics $\omega \pm C\alpha = \text{constant}$, respectively. At $t = 0$, when a sharp crack emanates with a speed V_A from the blunted tip (point A in Figure 4.30), there is a precipitous decrease in α at $z = a_0$. This produces an unloading wave that propagates along the characteristic AB towards B corresponding to $z = 0$. There the wave is reflected and propagates along the characteristic BC where it overtakes the crack tip at C.

In the domain AOB, $\omega = 0$ and $\alpha = \alpha_0$. Within ABC the method of characteristics gives for a typical point P

$$\alpha_P = \frac{\alpha_0}{(1 + V_N/C)}, \qquad \omega_P = \frac{-\alpha_0 V_N}{(1 + V_N/C)} \tag{4.3-23}$$

where the subscript is used to denote the point in the solution domain at which the variable is to be evaluated. When the crack tip is at point N intermediate to

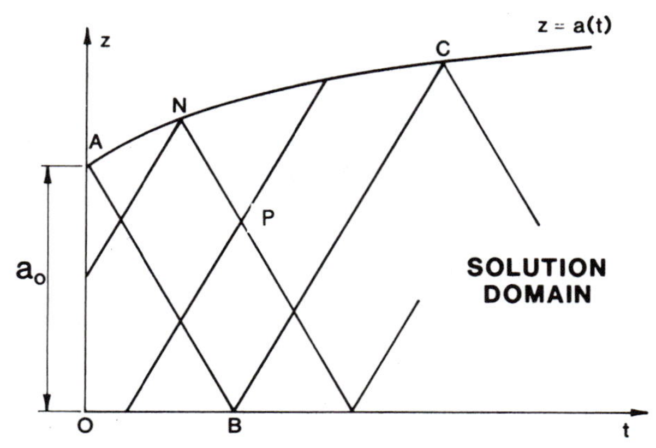

Figure 4.30 Method of characteristics solution.

A and C, then the combination of Equations (4.3-21)–(4.3-23) leads to

$$\frac{V_N}{C} = \frac{(1 - R/G_Q)}{(1 + R/G_Q)} \tag{4.3-24}$$

Therefore, from the time of initiation of crack growth until the reflected unloading wave overtakes the crack tip, the crack speed V is constant. If the fracture resistance is a function of crack speed, the crack will also propagate during this interval with a constant speed. Equation (4.3-24) then becomes a transcendental equation for the crack speed.

By the time the reflected unloading wave overtakes the crack tip, it has restored the angular velocity of the specimen to zero and reduced the rate of twist in $0 < z < a_c$ to

$$\alpha = \alpha_0 \frac{(1 - V/C)}{(1 + V/C)} \tag{4.3-25}$$

This rate of twist is insufficient to support further crack growth. Crack arrest then occurs. In this time interval the crack tip will have advanced from A ro C at a speed V. Hence, the crack length at arrest is

$$\frac{a_c}{a_0} = \frac{1 + V/C}{1 - V/C} = \frac{G_Q}{R} \tag{4.3-26}$$

Just before crack arrest $G = R$, while after arrest,

$$G_a = R \frac{(1 - V/C)}{(1 + V/C)} \tag{4.3-27}$$

or, equivalently,

$$G_a = \frac{Ra_0}{a_c} = \frac{R^2}{G_Q} \tag{4.3-28}$$

Equation (4.3-28) demonstrates that G_a depends upon the initial condition G_Q

and is not a material property. Only if the crack jump $\Delta a = a_c - a_0$ is small compared to a_0, will G_a be approximately equal to R.

The method of characteristics may be used to study how the total energy is partitioned into strain, kinetic, and fracture F energies during the dynamic crack propagation and arrest event. The partitioning of these energies, normalized with respect to the initial strain energy, for $G_Q/R = 4$, is shown in Figure 4.31. As the sharp crack emerges from the blunted notch, G decreases abruptly from G_Q to R. Because of this sudden reduction, the specimen is no longer in static equilibrium and an unloading wave propagates with the speed C towards the loaded end. As it does so it reduces the strain energy and increases the kinetic energy as more of the specimen gathers momentum. When the unloading wave reaches the loaded end, the crack has increased its length by 60 percent. The reflected wave begins to bring the specimen to rest as it speeds towards overtaking the crack tip. After the crack has quadrupled in length, the reflected wave overtakes the crack tip, the specimen is at rest, the driving force decreases to less than the fracture resistance, and crack arrest occurs.

Because of dispersive effects that the present model does not exhibit, the specimen will not normally be quiescent when crack arrest occurs. Hence, the driving force will generally ring down from R to G_a and will not take place instantaneously as this model predicts. The dashed curves in Figure 4.31 are the results of a prediction based upon a static interpretation of this dynamic

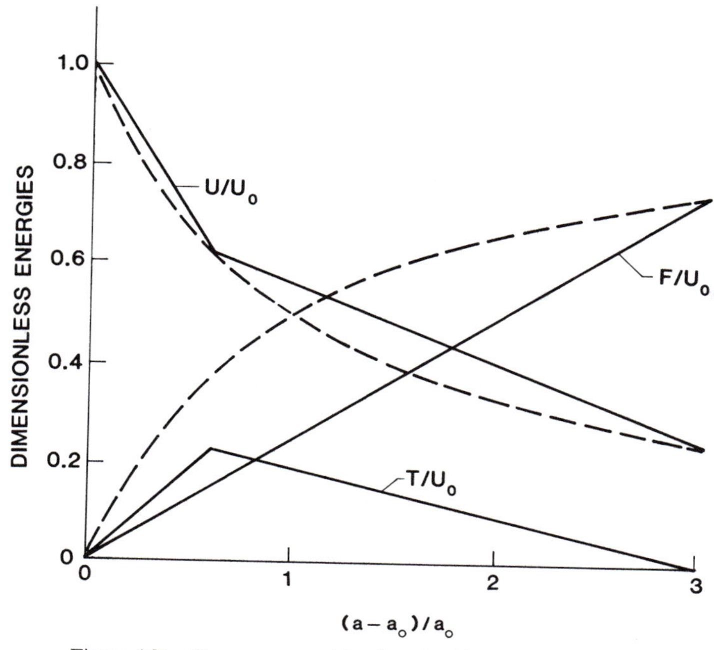

Figure 4.31 Energy composition in a double torsion specimen.

event. The static analysis overestimates the crack driving force for $a_0 < a < 2a_0$ and underestimates it for greater crack lengths associated with the latter stages of the event. These results are analogous to those obtained by Freund (4.105) for a shear model of the double cantilever beam specimen.

4.3.3 Axial Crack Propagation in a Pressurized Pipeline

Crack propagation in pressurized pipelines, as determined by full-scale tests, generally occurs at an essentially constant speed. And, when arrest takes place, it does so in a fairly abrupt manner. Typically, the ductile (or shear) crack speeds observed in full-scale tests range from 100 to 300 m/sec, brittle crack propagation speeds from 600 to 1000 m/sec. Note that, for typical gas transmission pipe dimensions (e.g., $R/h = 40$), these are considerably smaller than the Rayleigh wave speed, C_R.

Kanninen (4.109) provided a limiting speed prediction for rapid crack propagation in a gas transmission pipeline using a beam-like model. This approach was based on an elastic-plastic extension of Yoffe's branching criterion and a model based upon the analogy between the deformation of a circular cylindrical shell under axisymmetric loading and the deflection of a beam on an elastic foundation. Under the assumptions that, if the crack speed exceeds the gas decompression speed, the pressure behind the propagating crack tip is zero and the presence of the crack introduces a step change in the shell stiffness, the limiting crack speed was found to be

$$V_l = \tfrac{3}{4} C_0 \left(\frac{h}{R} \right)^{\frac{1}{2}} \tag{4.3-29}$$

where h and R are the pipe wall thickness and radius, respectively. As shown in Figure 4.32, comparisons with measured speeds of cleavage crack propagation in full-scale tests conducted by Maxey et al. (4.110) agree well with this result.

Most experimental work has been focused on obtaining empirical guidelines for the toughness necessary to insure crack arrest in the ductile regime. Empirical results based on a minimum value of the Charpy upper-shelf energy are available from the results of Maxey et al. (4.110) and others. While a decisive comparison of the different relations might seem possible experimentally, this is not the case. The difficulty lies in the fact that no experiment has yet been able to determine a specific value of an arrest parameter for given operating conditions. It can only determine whether the crack has propagated or not. Consequently, while qualitative comparisons are possible, direct quantitative verification is not. A theoretical analysis would appear to offer the only way to resove this dilemma. Accordingly, the dynamic fracture mechanics analysis procedure of Kanninen et al. (4.111) for pipelines results in a one-dimensional representation for crack propagation. Under steady-state conditions, explicit consideration was given both to the principal features of the gas-filled pipe problem and the important elements of dynamic crack propagation. Their model is shown in Figure 4.33.

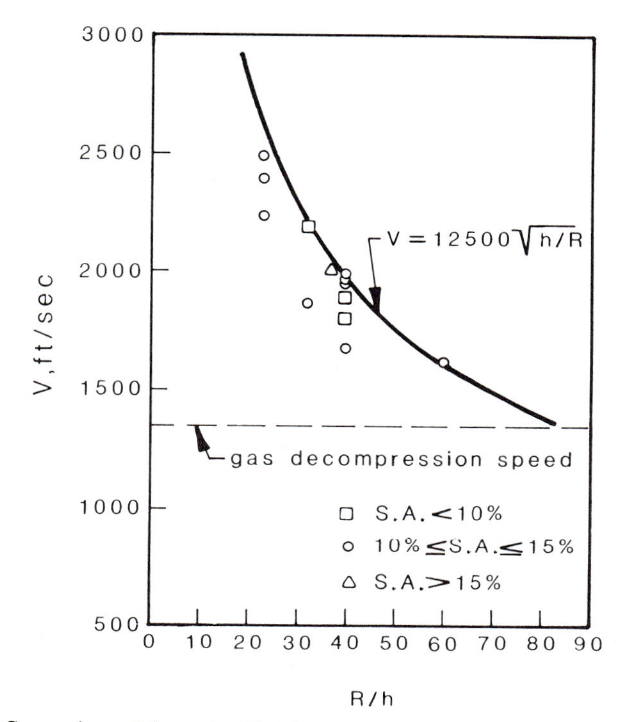

Figure 4.32 Comparison of theoretical limiting crack speed in pressurized steel pipelines with measured speeds.

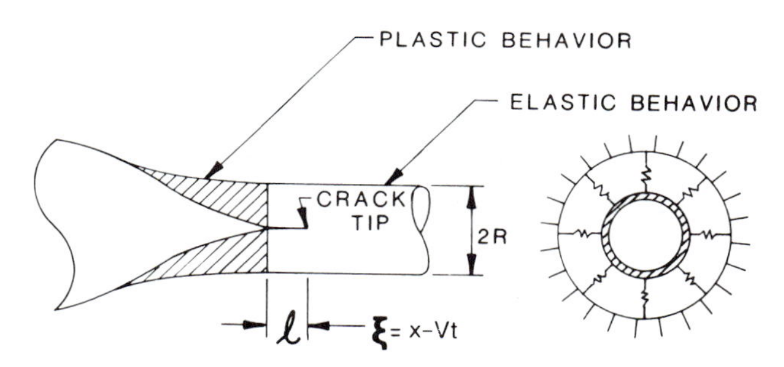

Figure 4.33 Fracture propagation analysis model for a buried pressurized pipeline.

Starting from the equations for a circular cylindrical shell, four key assumptions were introduced to make the analysis tractable: (1) radial deformations predominate, (2) circumferential variations in pressure can be neglected, (3) the crack-opening displacement is equal to the circumferentially integrated radial displacement $\bar{w}$ at any cross section in the cracked region, and (4) a plastic yield hinge is developed behind the crack tip. Further simplification can be introduced by specializing to steady-state conditions.

This was accomplished with the coordinate transformation $\xi = x - Vt$, where x is the axial coordinate, t is time, and V is the crack speed. This led to the ordinary differential equation

$$\frac{d^4\bar{w}}{d\xi^4} + 12(1-\nu^2)\frac{V^2}{h^2C_0^2}\frac{d^2\bar{w}}{d\xi^2} + \frac{24(1-\nu^2)\mu_s\bar{w}}{Eh^3R} + \frac{12(1-\nu^2)}{R^2h^2}H(\xi)\bar{w}$$
$$= 24\pi\frac{1-\nu^2}{Eh^3}p \qquad (4.3\text{-}30)$$

where μ_s is the shear modulus of the soil. Because the governing equation involves the pressure exerted on the pipe walls, a fluid mechanics treatment is also required. By assuming a predominately axial flow problem and accounting for the change of pipe cross-sectional area plus gas leakage behind the crack tip, an equation governing the pressure distribution $p = p(\xi, V)$ was obtained—a result that has been verified by the more detailed analyses of Emery et al. (4.112).

The next step involved the development of an expression for the crack driving force G as a function of $\bar{w}$ via Equation (4.3-30). Omitting the details, a relation $G = G(V)$ was obtained as a function of the pipe diameter and wall thickness; the elastic modulus, Poisson's ratio, and yield stress of the pipe material; the specific heat ratio, speed of sound, and pressure of the undisturbed gas; and the shear modulus of the soil surrounding the pipe. A typical result is shown in Figure 4.34. Of particular importance is the existence of a maximum crack driving force value for any given set of pipe geometry and operating conditions. This was used to estimate the minimum crack arrest toughness value. Also of interest in this solution is that the limiting crack speed agrees with Equation (4.3-29).

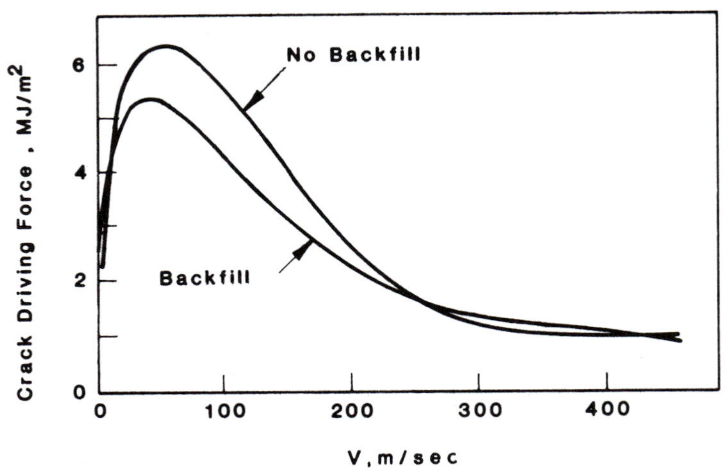

Figure 4.34 Calculated crack driving force for crack propagation in a pressurized pipeline ($R = .457$ m, $h = 8.4$ min, $p = 8$ M Pa).

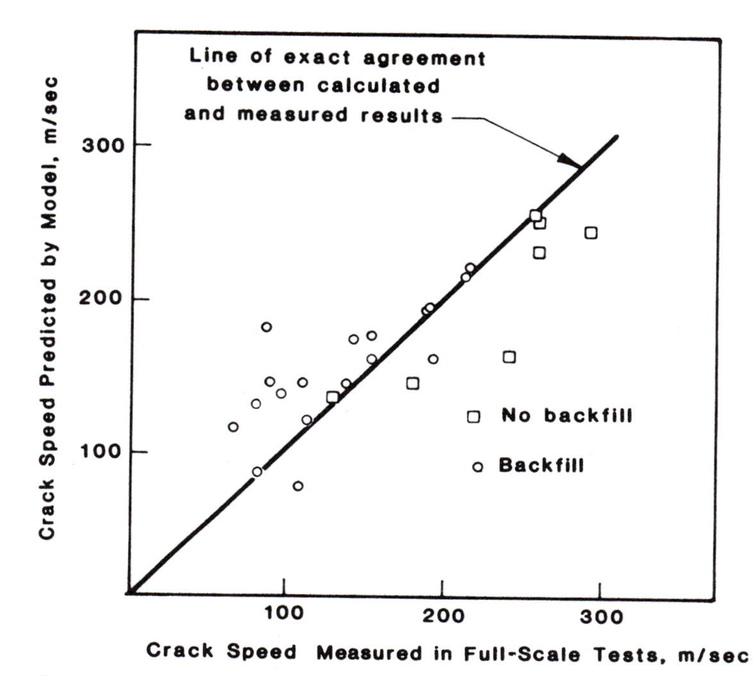

Figure 4.35 Comparison of predicted and measured crack speeds in a pressurized pipeline.

To assess the accuracy of the model, crack speeds were calculated for comparison with the experimental results. This was done using a value of the dynamic fracture energy requirement obtained from a drop-weight tear test (the best available alternative) in conjunction with curves typified by those of Figure 4.34. In this way the steady-state equilibrium point was determined. As shown in Figure 4.35, the crack speeds so determined were found to compare well with observed speeds in full-scale pipeline tests. Note that, while other pipe fracture models have been developed—for example, Freund et al. (4.113)—the model just described is so far the only one that offers a prediction of crack speeds that can be compared with experimental results.

4.3.4 Steady-State Crack Propagation

Significant analytical progress can be made by specializing the theory to admit only disturbances that propagate at a constant speed V in the positive x_1 direction. This can be accomplished by introducing the transformation $\xi = x_1 - Vt$ and using a stress function $\Phi = \Phi(\xi, y, t)$. The result is generalized biharmonic equation given by

$$\left\{\beta_1^2 \frac{\partial^2}{\partial \xi^2} + \frac{\partial^2}{\partial y^2}\right\}\left\{\beta_2^2 \frac{\partial^2}{\partial \xi^2} + \frac{\partial^2}{\partial y^2}\right\}\Phi = 0 \tag{4.3-31}$$

where β_1 and β_2 are defined in Equation (4.2-9).

If $V < C_2 < C_1$, Equation (4.3-31) is elliptic and has the real solution

$$\Phi = F_1(z_1) + F_2(z_2) + \overline{F_1(z_1)} + \overline{F_2(z_2)} \tag{4.3-32}$$

where now

$$z_1 = \xi + i\beta_1 y, \qquad z_2 = \xi + i\beta_2 y \tag{4.3-33}$$

and, as usual, a bar over a quantity denotes its complex conjugate. Relations for the in-plane stress and displacement components, known as the Sneddon-Radok equations (4.114), are as follows:

$$\sigma_{22} = (1 + \beta_2^2)\,\mathrm{Re}[F_1''(z_1) + F_2''(z_2)] \tag{4.3-34}$$

$$\sigma_{11} + \sigma_{22} = -2(\beta_1^2 - \beta_2^2)\,\mathrm{Re}\,F_1''(z_1) \tag{4.3-35}$$

$$\sigma_{12} = 2\,\mathrm{Im}\left[\beta_1 F_1''(z_1) + \frac{(1 + \beta_2^2)^2}{4\beta_2} F_2''(z_2)\right] \tag{4.3-36}$$

$$Gu_1 = -\mathrm{Re}[F_1'(z_1) + \tfrac{1}{2}(1 + \beta_2^2)F_2'(z_2)] \tag{4.3-37}$$

$$Gu_2 = \mathrm{Im}\left[\beta_1 F_1'(z_1) + \frac{1 + \beta_2^2}{2\beta_2} F_2'(z_2)\right] \tag{4.3-38}$$

It must of course be recognized that these results are applicable only to steady-state conditions; that is, an observer moving with crack tip can sense no change whatever in the deformation field from one time (or position) to another. While this is a rather unrealistic condition, because the mathematics is so much more amenable than otherwise, there is a definite benefit to proceeding in this manner.

To demonstrate the utility of the Sneddon-Radok equations, the problem originally solved numerically by Yoffe using the Fourier method can be obtained much more directly as follows. The appropriate boundary conditions for a constant length crack of length $2a$ propagating at a constant speed in an infinite medium under remote biaxial tension are, on $y = 0$, $\sigma_{22} = \sigma_{12} = 0$, $|\xi| < a$ and $u_2 = \sigma_{12} = 0$, $|\xi| < a$. At infinity, $\sigma_{22} = \sigma$, $\sigma_{11} = \lambda\sigma$, and $\sigma_{12} = 0$. The dynamic solution that satisfies these boundary conditions can be written as

$$\begin{aligned} F_1''(z_1) &= \sigma\left\{A_1^*\left[1 - \frac{z_1}{(z_1^2 - a^2)^{\frac{1}{2}}}\right] + B_1^*\right\} \\ F_2''(z_2) &= \sigma\left\{A_2^*\left[1 - \frac{z_2}{(z_2^2 - a^2)^{\frac{1}{2}}}\right] + B_2^*\right\} \end{aligned} \tag{4.3-39}$$

where

$$A_1^* = \frac{1 + \beta_2^2}{D}, \qquad A_2^* = -\frac{4\beta_1\beta_2}{D(1 + \beta_2^2)}$$

$$B_1^* = -\frac{1 + \lambda}{2(\beta_1^2 - \beta_2^2)}, \qquad B_2^* = \frac{1}{1 + \beta_2^2} + \frac{1 + \lambda}{2(\beta_1^2 - \beta_2^2)}$$

and, as above, $D = 4\beta_1\beta_2 - (1 + \beta_2^2)^2$.

Of most interest in any crack problem are the normal stresses ahead of the crack and the normal displacement on the crack faces. First, by substituting Equations (4.3-39) into Equation (4.3-34), the normal stress acting ahead of the propagating crack tip is

$$\sigma_{22} = \frac{\sigma\xi}{(\xi^2 - a^2)^{\frac{1}{2}}} \tag{4.3-40}$$

Similarly, substituting Equations (4.3-39) into Equation (4.3-38) gives the crack opening displacement as

$$\delta = \frac{2\beta_1(1 - \beta_2^2)}{4\beta_1\beta_2 - (1 + \beta_2^2)^2} \frac{\sigma}{\mu} (a^2 - \xi^2)^{\frac{1}{2}} \tag{4.3-41}$$

From these it can be seen that σ_{22} is independent of the crack speed in Yoffe's solution while, in contrast, δ depends upon the crack speed. In particular, it can be shown (by a limit process) that, while Equation (4.3-41) reduces to that of the static problem when $V = 0$, for $V > 0$, δ is always greater than its static counterpart, becoming infinite at the Rayleigh velocity C_R.

4.3.5 Use of Strip Yield Models

The Sneddon-Radok equations are readily usable to incorporate a collinear strip yield zone. As shown by Kanninen (4.115), the crack opening displacements for a constant-length moving crack then have the same form as in static conditions—see Equation (1.4-11)—but are escalated by a multiplicative factor $L = L(V)$ defined as

$$L(V) = \frac{4}{\kappa + 1} \frac{1}{D} \left(\frac{V}{C_2}\right)^2 \left[1 - \left(\frac{V}{C_1}\right)^2\right] \tag{4.3-42}$$

In particular, for plane stress, the crack-tip opening displacement at the crack tip is

$$\delta = \frac{8}{\pi} \frac{\alpha\sigma_Y}{E} L(V) \log \frac{c}{a} \tag{4.3-43}$$

Because the singularity canceling equation in the Yoffe/Dugdale model is the same as in the static case, Equation (1.4-9) can be inserted in the above to give the dynamic COD in terms of the applied stress. Then,

$$\delta = \frac{8}{\pi} \frac{a\sigma_Y}{E} L(V) \log\left[\sec\left(\frac{\pi}{2} \frac{\sigma}{\sigma_Y}\right)\right] \tag{4.3-44}$$

Because $L(0) = 1$, this result reduces to that of Equation (1.4-13) for stationary cracks, as it must, for $V = 0$.

On the basis that the crack opening displacement is constant during rapid crack propagation, Kanninen used Equation (4.3-44) to predict crack speeds. These predictions were made to compare with measured speeds observed in steel foil as a function of crack length at five different load levels. The comparison is shown in Figure 4.36. The results are reasonable, particularly in view of the fact that the analysis did not attempt to account for the finite dimensions of the test specimens. This approach was among the earliest quantitative approaches that employed a critical crack opening displacement criterion—a procedure that has recently become of particular interest for use in elastic-plastic dynamic crack propagation analyses.

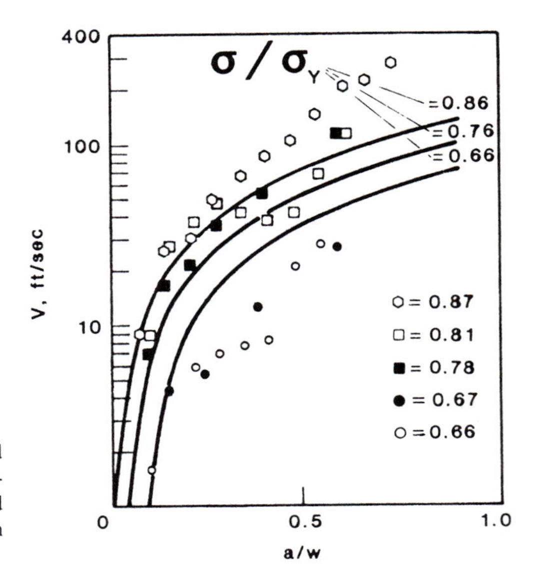

Figure 4.36 Comparison of observed crack propagation results with prediction of a propagating strip yield zone model for steel foil sheets in tension.

Also of interest in this solution was the development of a plasticity-based limiting crack speed prediction, which was based on the point at which the two in-plane principal stresses become equal. Estimates of the crack speed dependence of the flow stress were also incorporated in this relation. While these results are perhaps not too realistic, they are of interest in that they anticipated a current trend in dynamic fracture mechanics—the use of dynamic elastic-plastic analyses to estimate the resistance of the material to crack growth.

Atkinson (4.116) has provided a strip yield zone generalization of Broberg's model. His Broberg/Dugdale propagating crack solution was confined to the elucidation of the finiteness condition (i.e., the singularity canceling equation). Embley and Sih (4.117) subsequently broadened this treatment to include the crack face displacements and, in addition, corrected an error appearing in the published version of reference (4.116). Like Atkinson, Embley, and Sih considered a crack expanding at a uniform speed V having a collinear strip yield zone whose tip is simultaneously expanding at a speed β. The body is supposed to be infinite and to be acted on by a remote tensile applied stress σ. The crack face displacements for these conditions are given by

$$u_2(x) = \frac{\sigma_Y}{\pi\rho}\left[4C_2^3(C_2^2 - V^2)^{\frac{1}{2}} - C_1\frac{(V^2 - 2C_2^2)^2}{(C_1^2 - V^2)^{\frac{1}{2}}}\right]^{-1}$$
$$\cdot\left[Vt\log\left(\frac{t(\beta^2 - V^2)^{\frac{1}{2}} - (\beta^2t^2 - x^2)^{\frac{1}{2}}}{t(\beta^2 - V^2)^{\frac{1}{2}} + (\beta^2t^2 - x^2)^{\frac{1}{2}}}\right)\right. \tag{4.3-45}$$
$$\left. + x\log\left(\frac{x(\beta^2 - V^2)^{\frac{1}{2}} + V(\beta^2t^2 - x^2)^{\frac{1}{2}}}{x(\beta^2 - V^2)^{\frac{1}{2}} - V(\beta^2t^2 - x^2)^{\frac{1}{2}}}\right)\right]$$

As Embley and Sih pointed out, if $a = Vt$ and $c = \beta t$ are used to replace V and β in Equation (4.3-45), the result will be identical to Kanninen's result obtained for the Yoffe/Dugdale model. In particular, Equation (4.3-43) is the same in both models. However, because the singularity canceling equation differs, Equation (4.3-44) is not common to the two models. Indeed, as Atkinson found, the finiteness condition in the Broberg/Dugdale model is speed-dependent. As given by Embley and Sih, this relation can be written as

$$\frac{\sigma}{\sigma_Y} = \left[1 + \frac{2}{\pi}\frac{V^3(C_1^2 - V^2)^{\frac{1}{2}}(\beta^2 - V^2)^{\frac{1}{2}}}{C_1 C_2^4 D(V)}\right] \cdot \left[K(\lambda_1) + \frac{4C_2^4}{V^2\beta^2}E(\lambda_1) - E(\lambda_1) + \frac{4C_2^2}{V^2}\,II\left(\frac{\lambda^2}{\lambda_1^2}, \lambda_2\right) + \frac{(\beta^2 - 2C_2^2)^2}{V^2(C_1^2 - V^2)}\,II\left(\frac{\lambda_1^2}{\lambda_2^2}, \lambda_1\right)\right] \quad (4.3\text{-}46)$$

where $\lambda_1^2 = 1 - \beta^2/C_1^2, \lambda_2^2 = 1 - \beta^2/C_2^2$, and K, E, and II are elliptic functions of the first, second, and third kind, respectively.

Figure 4.37 presents a plot of Equation (4.3-46) showing the plastic zone tip velocity as a function of the applied stress and the crack speed. These results indicate that, for moderate applied stress levels (e.g., $\sigma/\sigma_Y < 0.5$) and for the crack speeds normally encountered, β is approximately equal to V. There will then be little difference between the speed-dependent finiteness condition and its static counterpart, Equation (1.4-9). Note that, for $\sigma = 0$, β and V are exactly the same whereupon Equation (4.3-46) reduces to Equation (1.4-9). It would therefore appear that the simpler results obtained by Kanninen could be used with reasonable accuracy.

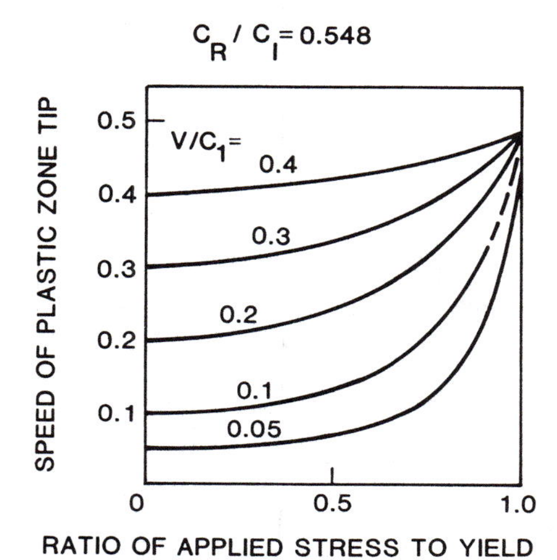

Figure 4.37 The finiteness conditions for an expanding crack with a strip yield zone model.

4.4 Applications of Dynamic Fracture Mechanics

In this section, emphasis is on the measurements and analyses of rapidly growing cracks, up to and including crack arrest. For this class of problems attention focuses on the K_{Ia} and K_{ID} parameters. The following sections describe some of the commonly used experimental methods and provide some typical results.

4.4.1 Crack Propagation Experimentation

Early experimental work in dynamic fracture mechanics was highly pragmatic. In essence, the objective was to develop reproducible test procedures that would correlate well with service failures. Such results would then be expected to produce "go-no go" material specifications—that is, either the material would be suitable for a given application or it would be rejected for those service conditions. The crack arrest temperature (CAT) test developed by Robertson (4.118) is one prominent example; see Figure 4.38. Here a crack is initiated at the cold side of a plate. The plate has a temperature gradient and is subjected to a uniform tensile stress. The crack is initiated by an impact from a bolt gun. The crack propagates rapidly, driven by the applied stress and assisted by the low toughness of the material at the colder side of the plate, but eventually encounters material at a higher temperature sufficiently tough to arrest it. The temperature at the point of crack arrest is the CAT for the given applied stress level; a value that presumably could be used in structural design.

Because Robertson-type tests tend to be difficult to conduct on a routine basis, a number of alternative procedures were developed. As described by Lange (4.119) and Pellini (4.120), researchers at the U.S. Naval Research Laboratory developed the explosion-bulge test and the Pellini drop weight tear test (DWTT). These are simpler to conduct than tests with a temperature gradient, but suffer in that the stress level in the test cannot be readily related to

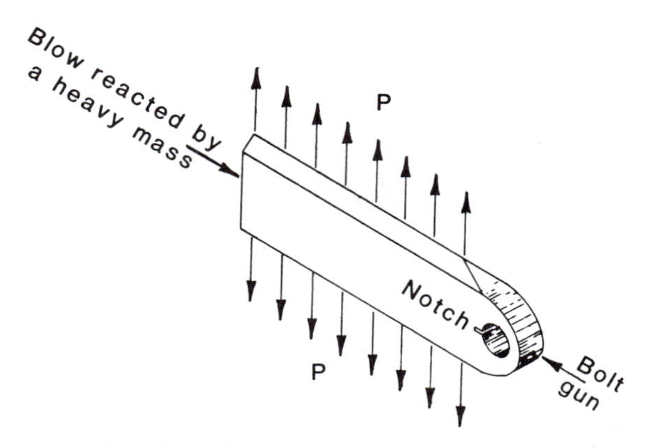

Figure 4.38 The Robertson crack arrest temperature specimen.

service conditions. Regardless, from the standpoint of the engineering analyst, they are of limited value as none of these tests is designed to produce values of a crack-tip characterizing parameter.

The various types of large size test specimens that have been used for fracture resistance determinations include: (1) the transversely welded wide plate tension specimen, (2) the ESSO (originally the SOD) specimen, (3) the deep notch specimen, and (4) the double tension test specimen. These are shown in Figures 4.39, 4.40, 4.41, and 4.42. For completeness, Figure 4.43 shows the various type of DCB test specimens that have been used.

While large specimen testing was once extensively performed—for example, see Nordell and Hall (4.121)—it was largely superceded by the much more economical testing that can be done with DCB and other small scale specimens. Lately, however, it has been recognized that crack arrest in very ductile conditions, which cannot be studied effectively in small-scale specimens, can be accommodated in wide plate testing. Accordingly, such experiments, properly instrumented and accompanied by dynamic finite element analyses, are now being used to probe the otherwise inaccessible

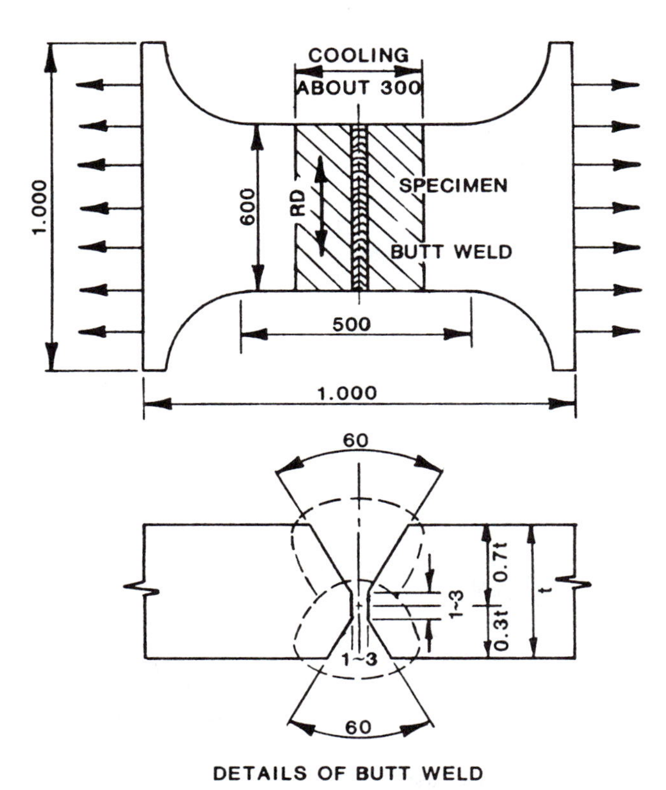

Figure 4.39 The transversely welded wide plate test specimen.

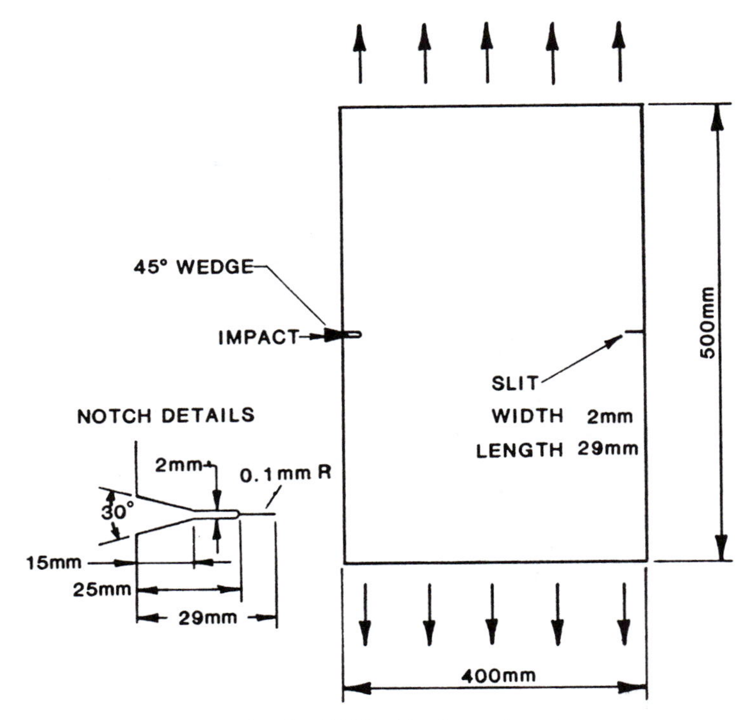

Figure 4.40 The ESSO wide plate test specimen.

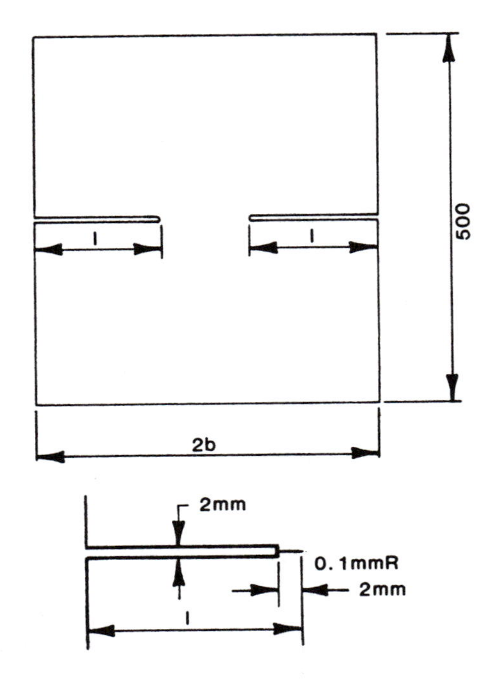

Figure 4.41 The deep notch test specimen.

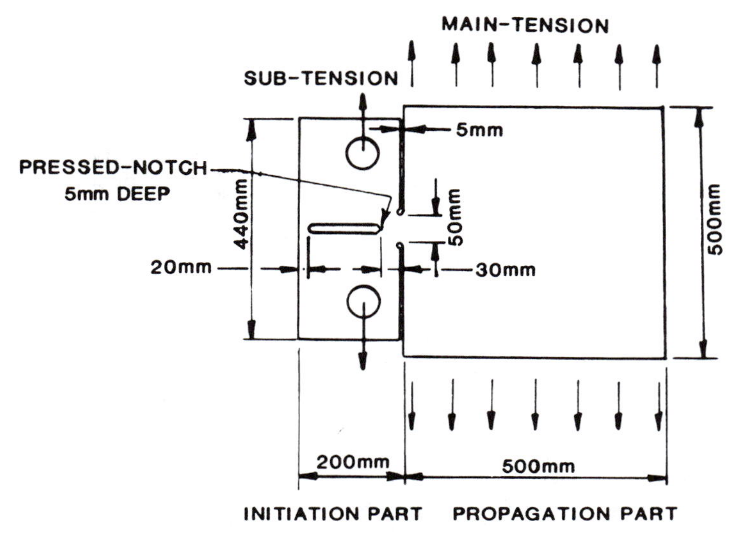

Figure 4.42 The double tension test specimen.

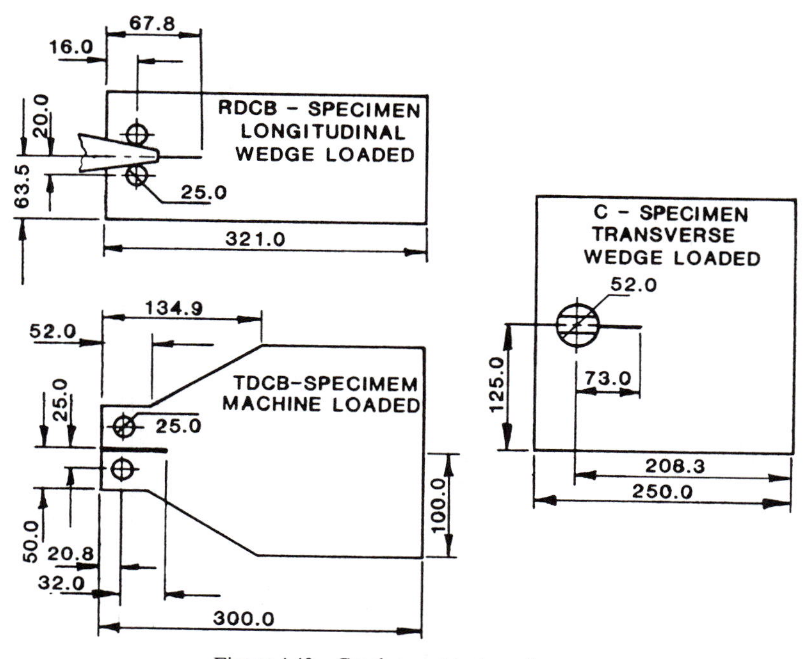

Figure 4.43 Crack arrest test specimens.

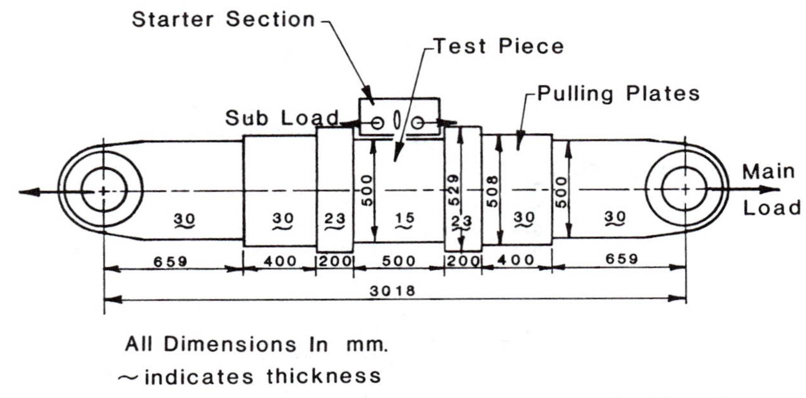

Figure 4.44 Typical geometry and loading arrangement for the double tension test.

temperature required for crack arrest data. Figure 4.44 shows the loading arrangement that would be typical of a wide plate test.

Rapid crack propagation/arrest experiments can be classified as either direct or indirect, depending upon whether crack-tip characterizing parameters are measured during the event or are inferred from a supplementary

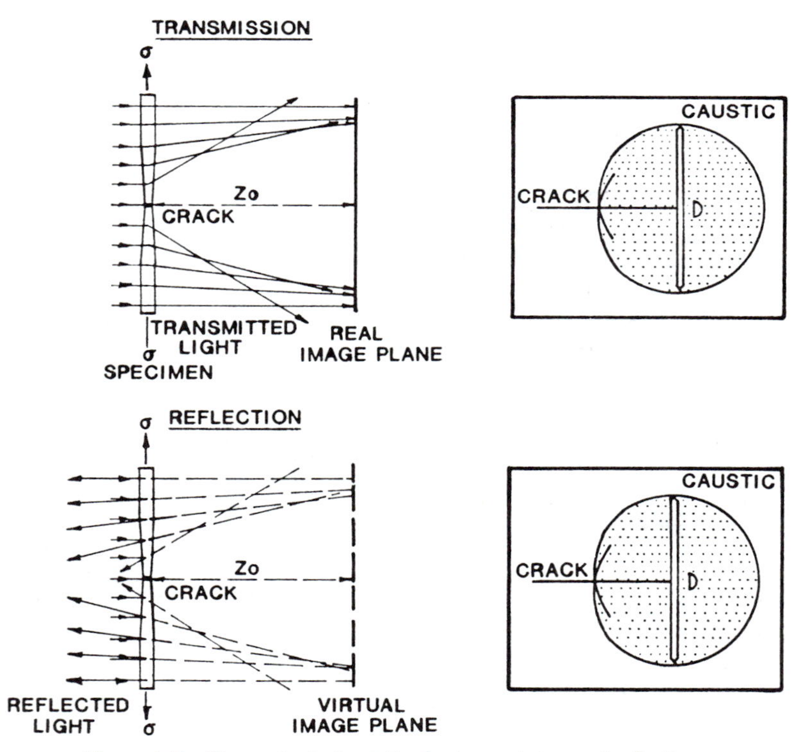

Figure 4.45 The method of caustics for transmission and reflection.

analysis. In the first category are the experiments on photoelastic and reflective materials where a shadow spot (caustic) or a fringe pattern is photographed by high speed cameras. Figure 4.45 shows the shadow optical method as used in transmission and in reflection. Figure 4.46 shows a typical set of results for a moving crack. Figure 4.47 shows corresponding results from a photoelastic study.

The second category contains experiments where only the crack growth history is measured—for example, by timing wires broken by the advancing crack. The resulting crack length versus time data can be used as input to, say, a finite element computation in which details (e.g., the dynamic stress intensity factor) that cannot be measured can then be calculated in what has been called a "generation-phase" calculation.

In both types of experiments some assumption about both the nature of the event and the constitutive behavior of the material during rapid crack propagation is required. While this is obvious in the indirect approach, it is equally true in the direct approach. The size of a reflected shadow spot may indeed correspond to the dimensions of the crack-tip plastic zone, but its relation to other features of the deformation will depend upon the material

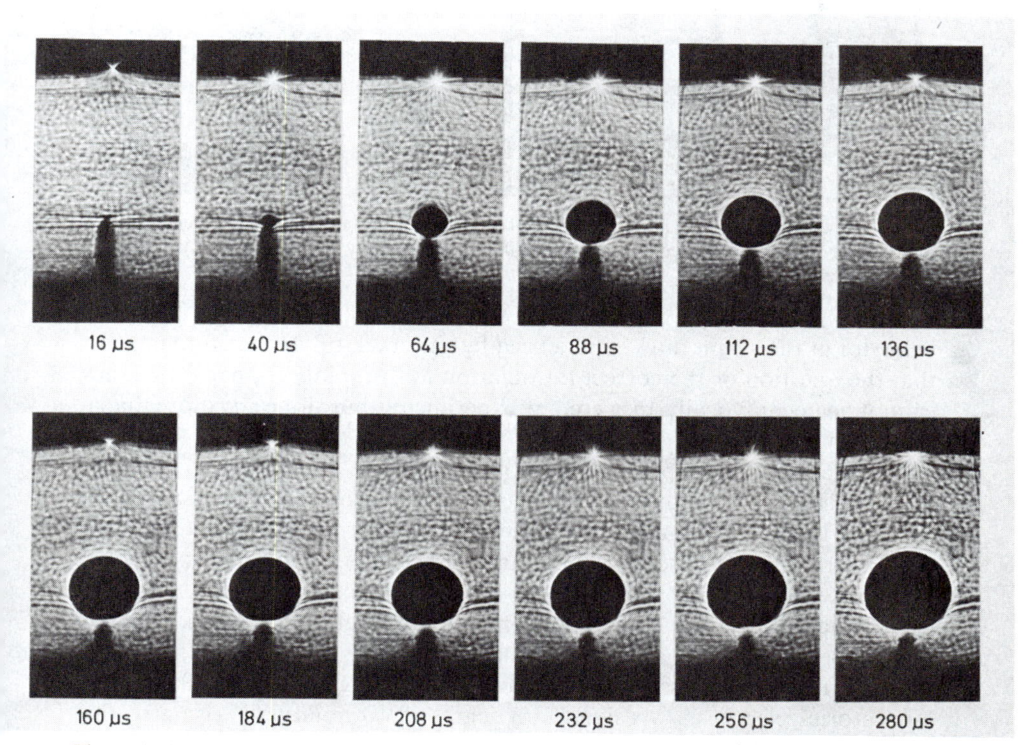

Figure 4.46 Use of the method of caustics with flash photography for measurements of stress intensity factors during dynamic crack propagation. (Provided by J. F. Kalthoff.)

Figure 4.47 Use of the photoelastic method with flash photography for measurements of stress intensity factors during dynamic crack propagation. (Provided by A. S. Kobayashi.)

behavior at the strain rates experienced by the crack tip. It therefore appears that the question of the correct formulation for a rapidly propagating crack cannot be unequivocally answered by experimentation alone any more than it can by analysis alone.

A cooperative test program has recently been conducted to assess a methodology for determining crack arrest values under the auspices of the ASTM. The program employed the compact crack arrest (CCA) test specimen (see Figure 4.43) and used two nuclear pressure vessel steels at two different test temperatures. The work focused on the determination of K_{Ia} and an average value of K_{ID} as a function of crack jump length. The results, based on the report of Crosley et al. (4.80), are provided in Figures 4.48, 4.49, and 4.50.

These results, representing the combined findings of some 30 different laboratories, clearly show a significant degree of variability. This is partly due to material differences but also to differences in the test procedure as executed at different laboratries. Regardless, a definite decrease in the K_{Ia} values can be

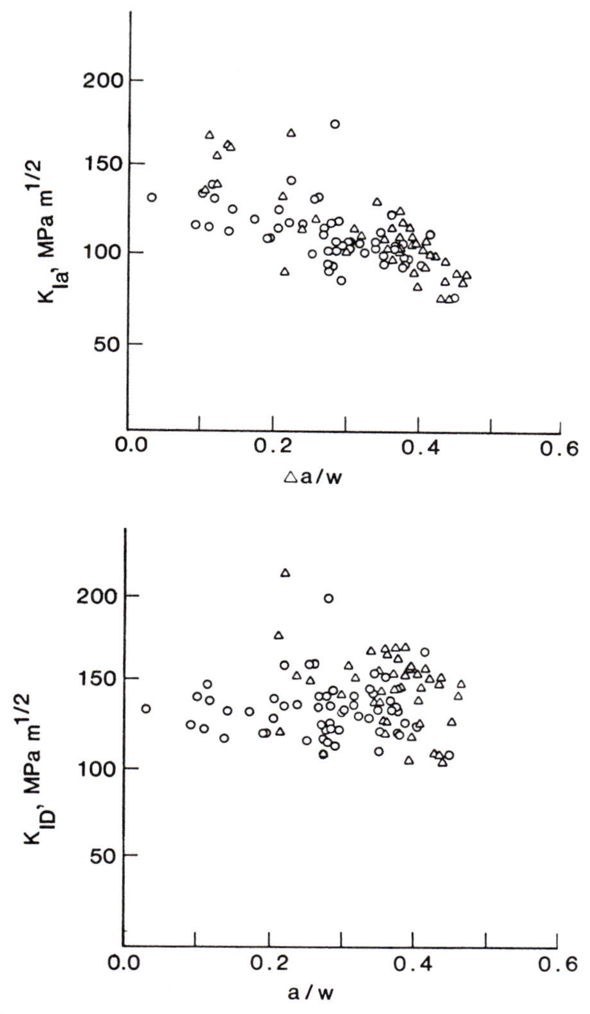

Figure 4.48 Cooperative crack arrest test program results for A533B steel at room temperature.

seen as the crack jump length increases. This result is consistent with the findings discussed in Section 4.1.7. The average K_{ID} values increase somewhat with crack jump length as would be expected from the nature of the K_{ID} relation (cf. Figures 4.15 to 4.17). However, caution should be exercised in using these K_{ID} values as they are not associated with any definite crack speed.

4.4.2 Dynamic Crack Propagation Analysis

Analyses of dynamic crack propagation and crack arrest for a specified initial flaw size, external geometry, and applied load can be conducted in either of two different ways. Given a material property relation such as $K_{ID} = K_{ID}(\dot{a})$,

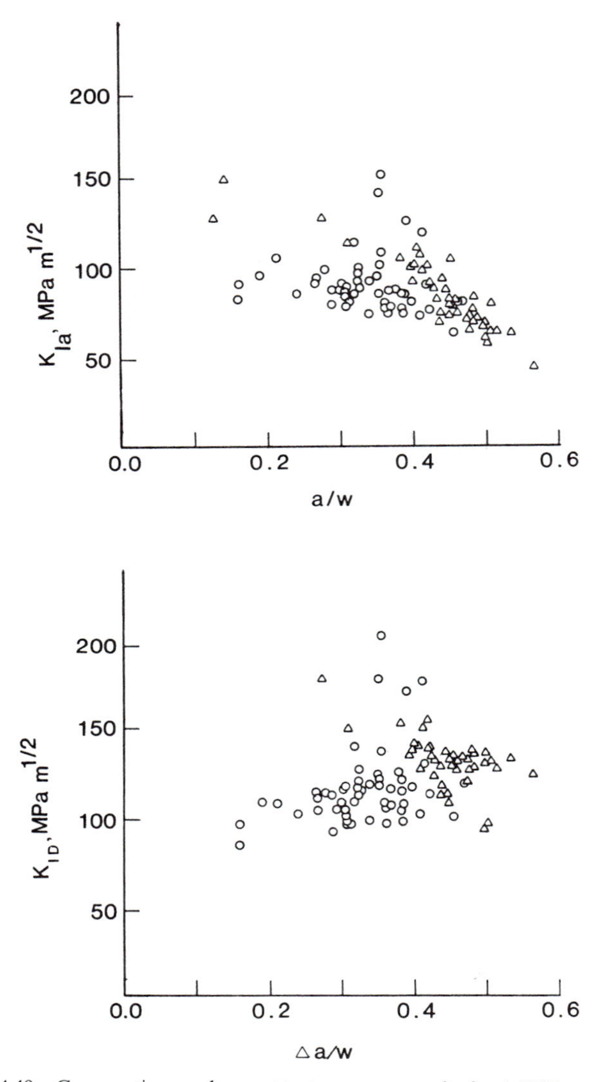

Figure 4.49 Cooperative crack arrest test program results for A533B steel at O°C.

one can compute the resulting crack growth history $a = a(t)$ and the crack arrest point. Alternatively, given the crack length-time behavior observed in an experiment, by forcing a computer model to respond in exactly the same manner, one can compute critical values of the fracture criterion. The former represents the procedure that would be followed in an "application phase" calculation to assess the safety of a structural component. The latter could be associated with a "generation phase" calculation to evaluate material properties from laboratory tests. The following discussion will use these names as defined here.

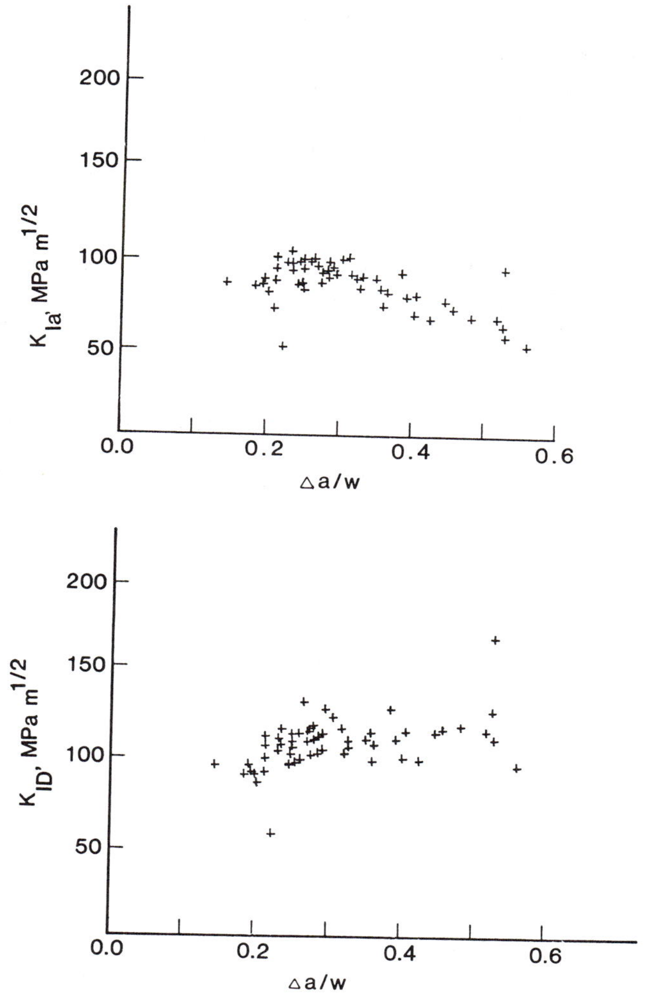

Figure 4.50 Cooperative crack arrest test program results for 1018 steel at room temperature.

One drawback to performing generation phase calculations based on experimental data is that a direct experimental determination of crack speeds does not appear to be possible. Experimentalists can at best supply crack length-time data from which crack speeds must be inferred by numerical differentiation.* Finite element models can be forced to follow the crack propagation/arrest behavior observed in photoelastic materials in order to compare the dynamic fracture toughness values determined from photoelastic

* While it is commonly assumed that the crack length history is a smooth function, van Elst (4.122) has shown that brittle fracture in steel can proceed by discrete steps of a few millimeters in length with pauses of 1 to 20 μsec between steps.

analysis with calculated values. Clearly, this can be done from point to point along the crack path without regard to the crack speed existing at each point. But, while this procedure has provided reasonable agreements with observed photoelastic results, such results can depend on the way in which the crack speeds are deduced. Consequently, whether or not material property data as a function of crack speed can be accurately extracted from experimental results in this way is still an open question.

Dynamic finite element codes were first based on an explicit time integration scheme and constant strain quadrilateral elements. Crack-tip motion was modeled by discontinuous jumps with the crack tip moving from one node to the next at discrete time intervals in accord with the crack speeds measured in the accompanying photoelastic experiments. The perturbations caused by this procedure are "filtered" in computing the dynamic energy release rate. This is done by using the time-averaged normal stress ahead of the advancing crack tip and the corresponding time-averaged crack opening displacement after crack advance. Specifically, the product of the two terms divided by the crack advance length is used as a measure of the dynamic energy release rate G.

Advancing a crack in the finite element method has raised some fundamental questions. In particular, if no work is done in the sudden release of the nodes connecting finite elements across the prospective crack plane, there can be no crack-tip energy dissipation in such a model. In this case, the energy contained in the model will be too great as crack growth progresses and an ever increasing error in the computational results expected. This artifact seems to have been contained in the analyses of Aberson et al. (4.123) to extend their work to propagating cracks. Keegstra et al. (4.124) and Yagawa et al. (4.125), it might be noted, overcame this difficulty by releasing nodes over a period of time.

Keegstra et al. developed a transient dynamic finite element model and have applied it in an attempt to obtain fundamental material properties from the Charpy test. Their predictions were "tuned" to experimental results by the adjustment of parameters representing the toughness of an initially blunt crack, the dynamic toughness of the running crack, and the striker/specimen contact stiffness. The crack growth criterion was based on the crack-tip node force, a quantity that is proportional to the stress intensity factor. To provide an energy sink, the crack-tip node was not suddenly released when a critical value was achieved. Instead, the force was reduced slowly so that work is done (and energy dissipated) at the crack tip. Encouraging agreement between the computer predictions and the experimental measurements was claimed, but, of course, much of the agreement was forced in view of the tuning process. It appears that what they have done can be considered as a "generation-phase" analysis (in the sense of the preceding discussion), albeit by a trial and error approach.

In addition to the errors introduced experimentally, significant inaccuracies in generation phase calculations can be introduced by the analysis model itself. Aside from the usual LEFM assumption that no inelastic deformation exists in the material, the most important source of inaccuracy is the perturbations caused by the manner of advancing the crack. Finite element models with crack growth simulated by sudden node release are particularly sensitive. Unless special measures are used—for example, to relax the node force over several time steps—quite large spurious oscillations will be introduced. Obviously, the oscillations will have a large effect on the crack speeds inferred from the model. These must be precluded if accurate crack-speed-dependent values are sought. However, by relaxing the node forces gradually, it becomes difficult to precisely locate the crack tip—a point of no small importance when calculations are to be made with a crack-speed-dependent fracture toughness.

The scheme employed by Kanninen et al. (4.126) to advance a crack in accord with a crack-speed-dependent fracture criterion presents an improvement. They considered that a crack advance increment has taken place at a time t_0. If the next increment of growth is to occur at any subsequent time t, then the crack speed would be $\Delta a/(t - t_0)$, where Δa is the increment length set by the model. The essence of the procedure was to compare the actual value of the crack driving force at each time step with the value of the crack growth resistance based on the hypothetical crack speed associated with that time. Clearly, this scheme depends upon having an exact crack-tip position in the model. An improvement on this approach was subsequently offered by Ahmad et al. (4.127) who were able to associate the partial relaxation of the crack-tip node forces with the advance of the crack part-way through an element in their finite element model.

4.4.3 Crack Growth Initiation Under Dynamic Loading

If a test specimen is loaded dynamically, it is well established that the apparent toughness can be different from that obtained under slow loading. To distinguish the toughness values obtained in rapid loading tests from those obtained in conventional slow testing, the dynamic values are designated as K_{Id}.* Typically, a monotonic diminution of K_{Id} with loading rate is found from slow loading rates (e.g., 1 ksi$\sqrt{\text{in.}}$/sec) to impact rates (e.g., 10^5 ksi$\sqrt{\text{in.}}$/sec) below the transition temperature while the reverse is true above the transition temperature. The values considered to be useful for design purposes in the lower toughness regime are those obtained under impact loading as these are felt to give a minimum.

For the most part, dynamic fracture initiation testing uses fracture specimens identical to those used for static testing. The test specimen most widely used to obtain K_{Id} values is a fatigue-cracked three-point bend specimen loaded by a striking tup mounted on a freely falling weight. The specimen and/or the tup is instrumented to record the applied load as a function of time. Usually, it is assumed that inertial effects do not significantly

* An alternative, which seems to be favored by ASTM, is to designate the rapidload plane strain fracture toughness as $K_{Ic}(t)$, where t is the loading time in the test.

change the state of deformation in the body from that which would exist under static loading. Strain gages are used to detect crack initiation and the static load at that time. Beam bending relations are then used to infer the value of the stress intensity factor at the time the crack initiates and, hence, determine K_{Id}.

To minimize inertial effects, the impact speed is typically limited to that obtained by dropping a 1600-pound weight a distance of about 9 inches. In addition, the impact is cushioned by placing a soft aluminum or lead pad where the tup strikes the test specimen. The pad is supposed to eliminate the elastic ringing waves in the specimen and to increase the loading time during the test. To check for possible inertial effects, the experimental parameters have been varied systematically. It is claimed that these do not cause any significant alternations in the measured K_{Id} values. Nevertheless, independent dynamic analyses indicate that the basis of these results is somewhat suspect.

Kalthoff et al. (4.61) have recently examined some of the assumptions in instrumented impact tests using an optical experimental procedure. They were able to measure dynamic stress intensity factors with the method of shadow patterns (caustics) for both propagating cracks and stationary cracks under dynamic loading. In applying their techniques to study notch bend specimens subjected to a dropped weight, they have come to a number of conclusions of interest to the numerical analysis of impact tests. First, the time at which crack growth begins does not generally coincide with the maximum load; typically, the stress intensity factor is still increasing when the load begins to decrease.

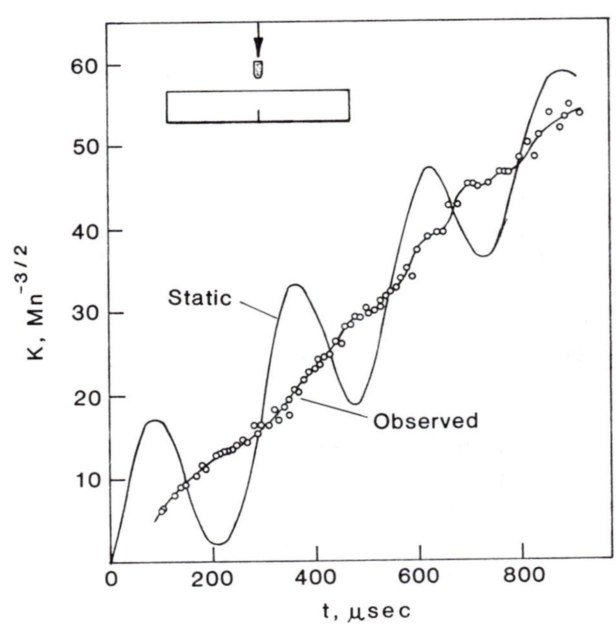

Figure 4.51 Comparison of static stress intensity factors with values observed by the method of caustics for an impact loaded high strength steel three-point bend specimen.

Second, when conditions for crack growth initiation are changed (e.g., by blunting the crack tip), the load as a function of time is essentially unchanged. Consequently, the K_{Id} value based on static relations and the maximum load can lead to erroneous results. It can be concluded that, for properly determined values, fully dynamic analyses must be performed that take inertia effects into account for the entire system (tup, specimen, and anvil). Figures 4.51 and 4.52 show a typical result. It might be noted that the short-time solution (i.e., infinite medium) gives $K \sim \sqrt{t}$ in accord with these results.

4.4.4 *Terminal Ballistics and Fragmentation*

There are two important areas in which dynamic fracture mechanics might be thought to be useful, but cannot be directly applied. The reason is that a pre-existing dominant crack that could provide the focal point for an analysis is not present. These are terminal ballistics—equivalently penetration and perforation processes—and fragmentation. In the latter area less work would appear to have been done even though the field can be dated back to the early contributions by Mott (4.128) and Taylor (4.129). More recent work has been given by Davison et al. (4.130) and Grady (4.131).

The hypervelocity impact processes that are involved in terminal ballistics produce a high concentration of energy in the vicinity of the impact point.

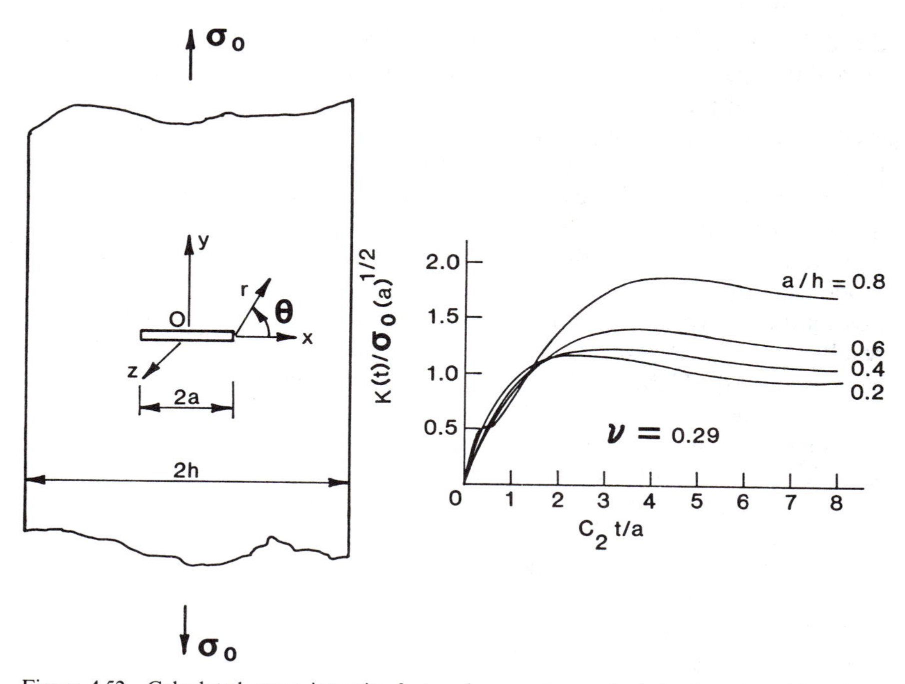

Figure 4.52 Calculated stress intensity factors in a center-cracked elastic strip subjected to suddenly applied remote tension.

This causes a violent interaction between the projectile and the target in which energy is dissipated through many different physical mechanisms. These include material vaporization, plastic flow, and fracture. Effective computer codes have been developed to model the resulting penetration/perforation events—for recent surveys, see Backman and Goldsmith (4.132) and Zukas et al. (4.133). However, all existing codes are handicapped by the relatively primitive failure models that are currently available. Indeed, it is generally agreed that the most serious limitation in the extensive use of current computational codes is not their cost or complexity, but the inadequacy of the material failure models that they employ. Considerable interest in dynamic fracture mechanics has been evidenced recently to alleviate these inadequate features.

Even though one failure mode tends to dominate, penetration/perforation processes usually occur by a combination of two or more mechanisms. But, as stated by Jonas and Zukas (4.134), with few exceptions, existing codes use simplistic criteria that assume instantaneous failure in a computational element once some critical condition is reached. Consequently, computations currently made are in accord with the philosophy best expressed by Wilkins (4.135); impact computations are not performed to predict results, but to enhance understanding of an experimental result. That is, the operative failure mechanisms must be specified in advance if the computation is to be successful.

While the use of computational codes to enhance physical understanding is certainly commendable, it would clearly be more desirable to have the failure modes arise in a natural way. What appears to be needed is a code that treats the individual failure modes as competing—but not mutually exclusive—events. According to Bodner (4.136), this can be done for the lower range of impact velocities (i.e., less than 1000 m/sec), but not for higher speeds. Models for the individual modes can be based on one or a combination of three distinct possibilities: (1) cumulative damage criteria based on elastic-plastic continuum concepts, (2) micromechanical hole growth and coalescence models, and (3) dynamic-plastic fracture mechanics. The first two of these possibilities have been vigorously pursued without entirely satisfactory results. The third, in contrast, appears to have not been adequately considered. Yet, as described above, there has been substantial progress in the field recently, which suggests that extensions of this type of approach to terminal ballistics problems could be useful. This progress, coupled with an intimate knowledge of currently available experimental and analysis results in terminal ballistics, could well lead to significant improvements in material failure modeling.

Phenomenologically, spalling, petalling, and plugging types of failure are observed in materials subjected to high-energy impact; see Figure 4.53. Spalling is common in thick targets composed of a material with relatively poor tensile properties. The mechanism of spall is usually explained as being due to the interaction of reflecting tensile waves from the free boundary and the secondary release wave at the point of contact; for example, see Gilman

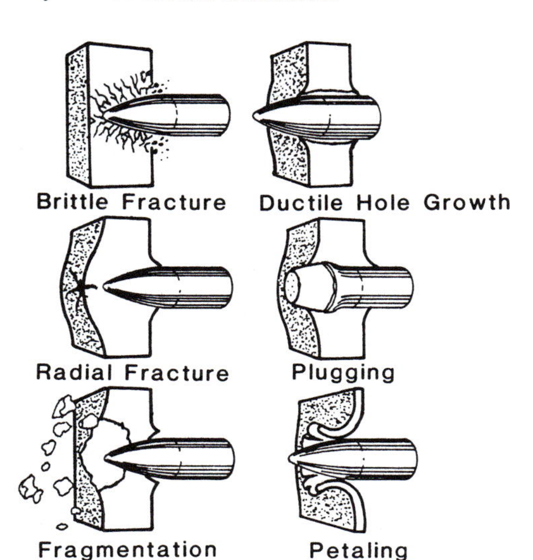

Figure 4.53 Schematic representation of failure events arising in terminal ballistics.

and Tuler (4.137). Petalling occurs in thin targets and the critical stress is perpendicular to the direction of impact. Plugging, often called adiabatic shear, is observed in thick targets made of ductile materials.

It has been hypothesized that plugging occurs when the transport of the heat generated due to the deformation of the target is not fast enough, leading to a concentrated band of material subjected to thermal softening. An example of alternating modes of failure is the point where a thick target becomes thin during the failure process whereupon plugging and petalling can be alternately induced. It is emphasized that, while the dominant mechanism that controls the particular failure mode can be identified in most cases, other mechanisms also participate. Hence, the great difficulty in modeling.

Most computer codes used for impact calculations contain some form of failure criterion. The most common of the criteria is based on the equivalent plastic strain reaching a critical value, generally the rupture strain for the material. The equivalent strain is a measure of the octahedral shear and, hence, is not connected to the hydrostatic stress state or to a preferred orientation. Consequently, if such a criterion is invoked, the material element or cell is isotropically precluded from transmitting tensile stresses. Another common criterion is the tensile strain in any direction reaching the critical rupture strain of the uniaxially tested material. In this case it is possible for the material to continue to transmit stresses in the perpendicular directions, thus allowing a directionality in the failure condition. But, this ignores the fact that the dilatational part of the deformation can also contribute to failure, and in some instances can achieve dominance in the material response.

An alternative approach resulted in the hole nucleation and growth models proposed by Shockey and Curran (4.138). In simple terms, their models (NAG/FRAG/SAG) are based upon the voids and shear bands that appear within metallic materials subjected to intense straining. A void growth-rate equation which contains scalar constants that depend on the material is embedded in the model to account for progression of damage due to wave reflections (cycles of stressing). Also degradation of the material due to damage accumulation is accounted for by reducing the stiffness of the particular cell.

Another damage model that has received attention is the Tuler-Butcher integral for calculating the time to initiate damage (4.139). It is representative of several models that calculate damage accumulation, generally, as a function of stress/strain, temperature, and the current damage fraction. In the Tuler-Butcher model the stress pulses that exceed a threshold stress value are assumed to cause increments of damage, the relation being through an exponent of the stress exceedances. When the sum of the increments reaches a critical value, the material is assumed to have failed. Some models including the NAG/FRAG/SAG models have sought to include progressive weakening of the material as the damage fraction grows.

4.5 References

(4.1) Kolsky, H., *Stress Waves in Solids*, Dover, New York (1963).

(4.2) Bland, D. R., *Nonlinear Dynamic Elasticity*, Blaisdell, Waltham, Mass. (1969).

(4.3) Achenbach, J. D., *Elastodynamic Wave Propagation*, North Holland, Amsterdam (1973).

(4.4) Erdogan, F., "Crack Propagation Theories," *Fracture*, Vol. II, H. Liebowitz (ed.), Academic, New York, pp. 497–590 (1968).

(4.5) Kolsky, H. and Rader, D., "Stress Waves and Fracture," *Fracture, Vol. I: Microscopic and Macroscopic Fundamentals*, H. Liebowitz (ed.), Academic, New York, pp. 533–569 (1968).

(4.6) Bluhm, J. I., "Fracture Arrest," Fracture, Vol. V, H. Liebowitz (ed.), Academic, New York, pp. 1–63 (1969).

(4.7) Sih, G. C. (ed.), *Dynamic Crack Propagation*, Noordhoff International, Leyden, The Netherlands (1973).

(4.8) Hahn, G. T. and Kanninen, M. F. (eds.), *Fast Fracture and Crack Arrest*, ASTM STP 627, American Society for Testing and Materials, Philadelphia (1972).

(4.9) Hahn, G. T. and Kanninen, M. F. (ed.), Crack Arrest *Methodology and Applications*, ASTM STP 711, American Society For Testing and Materials, Philadelphia (1980).

(4.10) Achenbach, J. D., "Dynamic Effects in Brittle Fracture," *Mechanics Today*, Vol. 1, S. Nemat-Nasser (ed.), Pergamon, Oxford, pp. 1–54 (1974).

(4.11) Freund, L. B., "Dynamic Crack Propagation," *The Mechanics of Fracture*, F. Erdogan (ed.), American Society of Mechanical Engineers Publication No. ASME AMD-19, pp. 105–134 (1976).

(4.12) Rose, L. R. F., "Recent Theoretical and Experimental Results on Fast Brittle Fracture," *International Journal of Fracture*, **12**, pp. 799–813 (1976).

(4.13) Francois, D., "Dynamic Crack Propagation and Arrest," *Developments in Pressure Vessel Technology—Flaw Analysis*, R. W. Nichols (ed.), Applied Science Pub., pp. 151–167 (1979).

(4.14) Kanninen, M. F., "A Critical Appraisal of Solution Techniques in Dynamic Fracture Mechanics," *Numerical Methods in Fracture Mechanics*, A. Luxmoore and R. Owen (eds.), U. Swansea, pp. 612–633 (1978); and "Whither Dynamic Fracture Mechanics?," *Numerical Methods in Fracture Mechanics*, R. D. G. Owen and A. R. Luxmoore (eds.), Pineridge, Swansea, U. K. (1980).

(4.15) Kamath, M. S., "Crack Propagation and Arrest," The Welding Institute Research Bulletin, Part 1—"Wide Plate Crack Arrest Studies and the Empirical Tests of Pellini and Others," pp. 14–25, Part 2—"Dynamic Aspects, Mechanisms and the Static LEFM Approach," pp. 193–203 (1980).

(4.16) Nilsson, F. and Brickstad, B., "Dynamic Fracture Mechanics—Rapid Crack Growth in

Linear and Non-Linear Materials," Report No. 49, The Royal Institute of Technology, Stockholm, Sweden (1983).
(4.17) Mott, N. F., "Fracture of Metals: Theoretical Considerations," *Engineering*, **165**, pp. 16–18 (1948).
(4.18) Roberts, D. K. and Wells, A. A., "The Velocity of Brittle Fracture," *Engineering*, **178**, pp. 820–821 (1954).
(4.19) Stroh, A. N., "A Theory of the Fracture of Metals," *Advances in Physics*, **6**, pp. 418–465 (1957).
(4.20) Berry, J. P., "Some Kinetic Considerations of the Griffith Criterion for Fracture," *Journal of the Mechanics of Physics and Solids*, **8**, pp. 194–216 (1960).
(4.21) Dulaney, E. N. and Brace, W. F., "Velocity Behavior of a Growing Crack," *Journal of Applied Physics*, **31**, pp. 2233–2236 (1960).
(4.22) Wells, A. A. and Post, D., "The Dynamic Stress Distribution Surrounding a Running Crack—A Photoelastic Analysis," *Proceedings of the Society for Experimental Stress Analysis*, **16**, pp. 69–92 (1958).
(4.23) Freund, L. B., "Crack Propagation in an Elastic Solid Subjected to General Loading—I. Constant Rate of Extension," *Journal of the Mechanics of Physics and Solids*, **20**, pp. 129–140 (1972).
(4.24) Freund, L. B., "Crack Propagation in an Elastic Solid Subjected to General Loading—II. Non-Uniform Rate of Extension," *Journal of the Mechanics of Physics and Solids*, **20**, pp. 141–152 (1972).
(4.25) Freund, L. B., "Crack Propagation in an Elastic Solid Subjected to General Loading—III: Stress Wave Loading," *Journal of the Mechanics of Physics and Solids*, **21**, pp. 47–61 (1973).
(4.26) Freund, L. B., "Crack Propagation in an Elastic Solid Subjected to General Loading—IV. Obliquely Incident Stress Pulse," *Journal of the Mechanics of Physics and Solids*, **22**, pp. 137–146 (1974).
(4.27) Winkler, S., Shockey, D. A., and Curran, D. R., "Crack Propagation at Supersonic Velocities," *International Journal of Fracture Mechanics*, **6**, pp. 151–158 and pp. 271–278 (1970).
(4.28) Hall, E. O., "The Brittle Fracture of Metals," *Journal of the Mechanics of Physics and Solids*, **1**, pp. 227–233 (1953).
(4.29) Stroh, A. N., "A Simple Model of a Propagating Crack," *Journal of the Mechanics and Physics of Solids*, **8**, pp. 119–122 (1960).
(4.30) Yoffe, E. H., "The Moving Griffith Crack," *Philosophical Magazine*, **42**, pp. 739–750 (1951).
(4.31) Broberg, K. B., "The Propagation of a Brittle Crack," *Arkiv for Fysik*, **18**, pp. 159–192 (1960).
(4.32) Craggs, J. W., "On the Propagation of a Crack in an Elastic-Brittle Material," *Journal of the Mechanics of Physics and Solids*, **8**, pp. 66–75 (1960).
(4.33) Baker, B. R., "Dynamics Stresses Created by a Moving Crack," *Journal of Applied Mechanics*, **29**, pp. 449–458 (1962).
(4.34) Nishioka, T. and Atluri, S. N., "Path-Independent Integrals, Energy Release Rates, and General Solutions of Near-Tip Fields in Mixed-Mode Dynamic Fracture Mechanics," *Engineering Fracture Mechanics*, **18**, pp. 1–22 (1983).
(4.35) Carlsson, J., "On the Mechanisms of Brittle Fracture Propagation," *Transactions of the Royal Institute of Stockholm*, No. 205, pp. 3–38 (1963).
(4.36) Cotterell, B., "On Brittle Fracture Paths," *International Journal of Fracture Mechanics*, **1**, pp. 96–103 (1965).
(4.37) Clark, A. B. J. and Irwin, G. R., "Crack-Propagation Behaviors," *Experimental Mechanics*, **6**, pp. 321–330 (1966).
(4.38) Saibel, E., "The Speed of Propagation of Fracture Cracks," *Fracturing of Metals*, American Society for Metals, Cleveland, pp. 275–281 (1948).
(4.39) Schardin, H., "Velocity Effects in Fracture," *Fracture*, M. I. T. Press, Cambridge, Mass., pp. 297–330 (1959).
(4.40) Congleton, J. and Petch, N. J., "Crack Branching," *Philosophical Magazine*, **16**, pp. 749–760 (1967).
(4.41) Anthony, S. R., Chubb, J. P., and Congleton, J., "The Crack Branching Velocity," *Philosophical Magazine*, **22**, p. 1201 (1970).
(4.42) Johnson, J. W. and Holloway, D. G., "On the Shape and Size of the Fracture Zones on Glass Fracture Surfaces," *Philosophical Magazine* **14**, pp. 731–743 (1966).

(4.43) Rabinovitch, A., "A Note on the Fracture Branching Criterion," *Philosophical Magazine*, **40**, pp. 873–874 (1979).
(4.44) Jacobson, A., "Bifurcation Velocity of Glassy Materials," *Israel Journal of Technology*, **3**, pp. 298–302 (1967).
(4.45) Andersson, H., "Stress Intensity Factors at the Tips of a Star-Shaped Contour in an Infinite Tensile Sheet," *Journal of the Mechanics and Physics of Solids*, **17**, pp. 405–417 (1969); see also errata, op. cit., **18**, 437 (1970).
(4.46) Kalthoff, J. F., "On the Characteristic Angle for Crack Branching in Ductile Materials," *International Journal of Fracture Mechanics*, **7**, pp. 478–480 (1971).
(4.47) Ramulu, M. and Kobayashi, A. S., "Dynamic Crack Curving—A Photoelastic Evaluation," *Experimental Mechanics* **23**, pp. 1–9 (1983).
(4.48) Wells, A. A., "Fracture Control: Past, Present and Future," *Experimental Mechanics*, **13**, pp. 401–410 (1973).
(4.49) Irwin, G. R. and Wells, A. A., "A Continuum-Mechanics View of Crack Propagation," *Metallurgical Reviews*, **10**, pp. 223–270 (1965).
(4.50) Crosley, P. B. and Ripling, E. J., "Dynamic Fracture Toughness of A533 Steel," *Journal of Basic Engineering*, **91**, pp. 525–534 (1969).
(4.51) Kanazawa, T., "Recent Studies on Brittle Crack Propagation in Japan," *Dynamic Crack Propagation*, G. C. Sih (ed.), Noordhoff, Leyden, The Netherlands, pp. 565–598 (1973).
(4.52) Hahn, G. T., Hoagland, R. G., Kanninen, M. F., and Rosenfield, A. R., "A Preliminary Study of Fast Fracture and Arrest in the DCB Test Specimen," *Dynamic Crack Propagation*, G. C Sih (ed.), Noordhoff, Leyden, pp. 649–662 (1973).
(4.53) Benbow, J. J. and Roesler, F. C., "Experiments on Controlled Fractures," *The Proceedings of the Physical Society*, Section B, **70**, pp. 201–211 (1957).
(4.54) Gilman, J. J., "Cleavage, Ductility, and Tenacity in Crystals," *Fracture*, M. I. T. Press, Cambridge, Mass., pp. 193–224 (1959).
(4.55) Kanninen, M. F., "A Dynamic Analysis of Unstable Crack Propagation and Arrest in the DCB Test Specimen," *International Journal of Fracture*, **10**, pp. 415–430 (1974).
(4.56) Srawley, J. E. and Gross, B., "Stress Intensity Factors for Crackline-Loaded Edge-Crack Specimens," *Materials Research and Standards*, **7**, pp. 155–162 (1967).
(4.57) Fichter, W. B., "The Stress Intensity Factor for the Double Cantilever Beam," *International Journal of Fracture*, **22**, pp. 133–143 (1983).
(4.58) Kanninen, M. F., "An Analysis of Dynamic Crack Propagation and Arrest for a Material Having a Crack Speed Dependent Fracture Toughness," *Prospects of Fracture Mechanics*, G. C. Sih et al. (eds.), Noordhoff, Leyden, The Netherlands, pp. 251–266 (1974).
(4.59) Kalthoff, J. F., Bienart, J., and Winkler, S., "Measurements of Dynamic Stress Intensity Factors for Fast Running and Arresting Cracks in Double-Cantilever-Beam Specimens," *Fast Fracture and Crack Arrest*, G. T. Hahn and M. F. Kanninen (eds.), ASTM STP 627, pp. 161–176 (1977).
(4.60) Kobayashi, A.S., Seo, K., K., Jou, J. Y., and Urabe, Y., "A Dynamic Analysis of Modified Compact-Tension Specimens Using Homolite -100 and Polycarbonate Plates," *Experimental Mechanics*, **20**, pp. 73–79 (1980).
(4.61) Kalthoff, J. F., Winkler, S., and Beinert, J., "The Influence of Dynamic Effects in Impact Testing," *International Journal of Fracture*, **13**, pp. 528–531 (1977).
(4.62) Kalthoff, J. F., Beinart, J., Winkler, S., and Klemm, W., "Experimental Analysis of Dynamic Effects in Different Crack Arrest Test Specimens," *Crack Arrest Methodology and Applications*, AST, STP 711, G. T. Hahn and M. F Kanninen, (eds.), American Society for Testing and Materials, Philadelphia, pp. 109–127 (1980).
(4.63) Hahn, G. T., Rosenfield, A. R., Marschall, C. W., Hoagland, R. G., Gehlen, P. C., and Kanninen, M. F., "Crack Arrest Concepts and Applications," *Fracture Mechanics*, N. Perrone et al. (eds.), University of Virginia Press, pp. 205–228 (1978).
(4.64) Kobayashi, T. and Dally, J. W., "Relation Between Crack Velocity and the Stress Intensity Factors in Birefringent Polymers," *Fast Fracture and Crack Arrest*, G. T Hahn and M. F. Kanninen (eds.), ASTM STP 627, American Society of Testing and Materials, Philadelphia, pp. 257–273 (1977).
(4.65) Kobayashi. A. S. and Mall, S., "Dynamic Fracture Toughness of Homalite-100," *Experimental Mechanics*, **18**, p. 11 (1978).
(4.66) Crosley, P. B. and Ripling, E. J., "Comparison of Crack Arrest Methodologies," *Crack Arrest Methodology and Applications*, G. T. Hahn and M. F. Kanninen (eds.), ASTM STP 711, pp. 211–209 (1980).

(4.67) Dahlberg, L., Nilsson, F., and Brickstad, B., "Influence of Specimen Geometry on Crack Propagation and Arrest Toughness," *Crack Arrest Methodology and Applications*, G. T. Hahn and M. F. Kanninen (eds.), ASTM STP 711, pp. 89–108 (1980).
(4.68) Kanazawa, T. and Machida, S., "Fracture Dynamics Analysis on Fast Fracture and Crack Arrest Experiments," *Fracture Tolerance Evaluation*, T. Kanazawa, A. S. Kobayashi, and K. Iido (eds.), Toyoprint, Japan (1982).
(4.69) Sih, G. C., "Dynamic Aspects of Crack Propagation," *Inelastic Behavior of Solids*, M. F. Kanninen et al. (eds.), McGraw-Hill, New York, pp. 607–633 (1970).
(4.70) Atkinson, C. and Eshelby, J. D., "The flow of Energy into the Tip of a Moving Crack," *International Journal of Fracture Mech.*, **4**, pp. 3–8 (1968).
(4-71) Rose, L. R. F., "An Approximate (Wiener-Hopf) Kernel for Dynamic Crack Problems in Linear Elasticity and Viscoelasticity," *Proceedings of the Royal Society of London, Series A*, **349**, pp. 497–521 (1976).
(4.72) Nilsson, F., "A Path-Independent Integral for Transient Crack Problems," *International Journal of Solids and Structures*, **9**, pp. 1107–1115 (1973).
(4.73) Shockey, D. A., Kalthoff, J. F., Klemm, W., and Winkler, S., "Simultaneous Measurements of Stress Intensity and Toughness for Fast Running Cracks in Steel," **15**, (1982).
(4.74) Rosakis, A. J. and Freund, L. B., "Optical Measurement of the Plastic Strain Concentration at a Crack Tip in a Ductile Steel Plate," *Journal of Engineering Materials Technology*, **104**, pp. 115–120 (1982).
(4.75) Popelar, C. H. and Kanninen, M. F., "A Dynamic Viscoelastic Analysis of Crack Propagation and Crack Arrest in a Double Cantilever Beam Test Specimen," *Crack Arrest Methodology and Applications*, G. T. Hahn and M. F. Kanninen (eds.), ASTM STP 711, pp. 3–21 (1980).
(4.76) Shukla, A., Fourney, W. L., and Dally, J. W., "Mechanisms of Energy Loss During a Fracture Process," *Proceedings of the Society for Experimental Stress Analysis* (1981).
(4.77) Eftis, J. and Krafft, J. M., "A Comparison of the Initiation With the Rapid Propagation of a Crack in a Mild Steel Plate," *Journal of Basic Engineering*, **87**, pp. 257–263 (1965).
(4.78) Paxon, T. L. and Lucas, R. A., "An Experimental Investigation of the Velocity Characteristics of a Fixed Boundary Fracture Model," *Dynamic Crack Propagation*, G. C. Sih (eds.), Noordhoff, Leiden, pp. 415–426 (1973).
(4.79) Jung, J. and Kanninen, M. F., "An Analysis of Dynamic Crack Propagation and Arrest in a Nuclear Pressure Vessel under Thermal Shock Conditions," *Aspects of Fracture Mechanics in Pressure Vessels and Piping*, S. G. Sampath and S. S. Palusamy (eds.), ASME PVP, Vol. 58 (1982).
(4.80) Crosley, P. B., Fourney, W. L., Hahn, G. T., Hoagland, R. G., Irwin, G. R., and Ripling, E. J., *Cooperative Test Program on Crack Arrest Toughness Measurements*, NUREG/CR-3261, Nuclear Regulatory Commision, Washington, D.C., April 1983.
(4.81) Kalthoff, J. F., "On Some Current Problems in Experimental Fracture Mechanics", *Workshop on Dynamic Fracture*, W. G. Knauss et al. (eds.), California Institute of Technology, pp. 11–35 (1983).
(4.82) Eshelby, J. D., "The Elastic Field of a Crack Extending Non-Uniformly Under General Anti-Plane Loading," *Journal of the Mechanics of Physics and Solids*, **17**, p. 177 (1969).
(4.83) Achenbach, J. D., Kanninen, M. F., and Popelar, C. H., "Crack-Tip Fields for Fast Fracture of an Elastic-Plastic Material," *Journal of the Mechanics and Physics of Solids*, **29**, pp. 211–225 (1981).
(4-84) Freund, L. B. and Douglas, A. S., "The Influence of Inertia on Elastic-Plastic Antiplane Shear Crack Growth," *Journal of the Mechanics of Physics and Solids*, **30**, pp. 59–74 (1982).
(4.85) Dantam, V. and Hahn, G. T., "Definition of Crack Arrest Performance of Tough Alloys," *Fracture Tolerance Evaluation*, T. Kanazawa, A. S. Kobayashi, and K. Iida (eds.), Toyoprint, Japan (1982).
(4.86) Ahmad, J., Jung, J., Barnes, C. R., and Kanninen, M. F., "Elastic-Plastic Finite Element Analysis of Dynamic Fracture," *Engineering Fracture Mechanics*, **17**, pp. 235–246 (1983).
(4.87) Tetelman, A. S., "The Plastic Deformation at the Tip of a Moving Crack," *Fracture of Solids*, D. C. Drucker and J. J. Gilman (eds.), Gordon and Breach, New York, pp. 461–501 (1963).
(4.88) Aboudi, J. and Achenbach, J. D., "Numerical Analysis of Fast Mode I Fracture of a Strip of Viscoplastic Work-Hardening Material," *International Journal of Fracture*, **21**, pp. 133–147 (1983).

(4.89) Aboudi, J. and Achenbach, J. D., "Arrest of Fast Mode I Fracture in an Elastic-Viscoplastic Transition Zone," *Engineering Fracture Mechanics*, **18**, pp. 109–119 (1983).

(4.90) Hoff, R., Rubin, C. A., and Hahn, G. T., "High Rate Deformation in the Field of a Crack," *Material Behavior Under High Stress and Ultrahigh Loading Rates*, J. Mescall and V. Weiss (eds.), Plenum, pp. 223–240, (1983).

(4.91) Lo, K. K., "Dynamic Crack-Tip Fields in Rate-Sensitive Solids," *Journal of the Mechanics and Physics of Solids*, **31**, pp. 287–305 (1983).

(4.92) Brickstad, B., "A Viscoplastic Analysis of Rapid Crack Propagation Experiments in Steel," *Journal of the Mechanics and Physics of Solids*, **31**, pp. 307–327 (1983).

(4.93) Bodner, S. D. and Partom, Y., "Constitutive Equations for Elastic-Viscoplastic Strain-Hardening Materials," *Journal of Applied Mechanics*, **42**, p. 305 (1975).

(4.94) Nilsson, F., "A Note on the Stress Singularity at a Non-uniformly Moving Crack Tip," *Journal of Elasticity*, 4, pp. 73–75 (1974).

(4.95) Freund, L. B. and Clifton, R. J., "On the Uniqueness of Plane Elastodynamic Solutions for Running Cracks," *Journal of Elasticity*, 4, pp. 293–399 (1974).

(4.96) Bergkvist, H., "The Motion of a Brittle Crack," *Journal of the Mechanics of Physics and Solids*, **21**, pp. 229–239 (1973).

(4.97) Popelar, C. H. and Atkinson, C., "Dynamic Crack Propagation in a Viscoelastic Strip," *Journal of Mechanics and Physics of Solids*, **28**, pp. 79–93 (1980).

(4.98) Nilsson, F., "Dynamic Stress Intensity Factors for Finite Strip Problems," *International Journal of Fracture Mechanics*, **8**, pp. 403–411 (1972).

(4.99) Aoki, S., Kishimoto, K., and Sakata, M., "Energy-Release Rate in Elastic-Plastic Fracture Problems," *Journal of Applied Mechanics*, **48**, pp. 825–928 (1981).

(4.100) Kishimoto, K., Aoki, S., and Sakata, M., "On the Path Independent Integral-$\hat{J}$," *Engineering Fracture Mechanics*, **13**, pp. 841–850, 1980.

(4.101) Gehlen, P. C., Popelar, C. H., and Kanninen, M. F., "Modeling of Dynamic Crack Propagation: I. Validation of One-Dimensional Analysis," *International Journal of Fracture*, **15**, pp. 281–294 (1979).

(4.102) Burns, S. J. and Chow, C. L., "Crack Propagation with Crack-Tip Critical Bending Moments in Double-Cantilever-Beam Spcimens," *Fast Fracture and Crack Arrest*, G. T. Hahn and M. F. Kanninen (eds.), ASTM STP 627, American Society of Testing and Materials, Philadephia, pp. 228–240 (1977).

(4.103) Malluck, J. F. and King, W. W., "Simulation of Fast Fracture in the DCB Specimen Using Kanninen's Model," *International Journal of Fracture*, **13**, pp. 656–665 (1977).

(4.104) Kanninen, M. F., Popelar, C., and Gehlen, P. C., "Dynamic Analysis of Crack Propagation and Crack Arrest in the Double-Cantilever-Beam Specimen," *Fast Fracture and Crack Arrest*, G. T. Hahn and M. F. Kanninen (eds.), ASTM STP 627, American Society for Testing and Materials, Philadelphia, pp. 19–38 (1977).

(4.105) Freund, L. B., "A Simple Model of the Double Cantilever Beam Crack Propagation Specimen," *Journal of the Mechanics of Physics and Solids*, **25**, pp. 69–79 (1977).

(4.106) Steverding, B. and Lehnigk, S. H., "The Propagation Law of Cleavage Fracture," *International Journal of Fracture*, **6**, p. 223 (1970).

(4.107) Bilek, Z. J. and Burns, S. J., "The Dynamics of Crack Propagation in Double Cantilever Beam Specimens," *Dynamic Crack Propagation*, G. S. Sih (ed.), Noordhoff, p. 371 (1973).

(4.108) Popelar, C. H., "A Model for Dynamic Crack Propagation in a Double-Torsion Fracture Specimen," *Crack Arrest Methodology and Applications*, ASTM STP 711, G. T. Hahn and M. F. Kanninen (eds.), American Society for Testing and Materials, Philadelphia, STP 711, pp. 24–37 (1980).

(4.109) Kanninen, M. F., "Research in Progress on Unstable Crack Propagation in Pressure Vessels and Pipelines," *International Journal of Fracture*, **6**, pp. 94–95 (1970).

(4.110) Maxey, W. A., Eiber, R. J., Podlasek, R. J., and Duffy, A. R., "Observations on Shear Fracture Propagation Behavior," *Crack Propagation in Pipelines*, Institute of Gas Engineers, London (1974).

(4.111) Kanninen, M. F., Sampath, S. G., and Popelar, C. H., "Steady-State Crack Propagation in Pressurized Pipelines Without Backfill," *Journal of Pressure Vessel Technology*, **98**, pp. 56–65 (1976).

(4.112) Emery, A. F., Love, W. J., and Kobayashi, A. S., "Dynamic Finite Difference Analysis of an Axially Cracked Pressurized Pipe Undergoing Large Deformations," *Fast Fracture and Crack Arrest*, G. T. Hahn and M. F. Kanninen (eds.), ASTM STP 627, American Society for Testing and Materials, Philadelphia, pp. 142–160 (1977).

(4.113) Freund, L. B., Parks, D. M., and Rice, J. R., "Running Ductile Fracture in a Pressurized Line Pipe," *Mechanics of Crack Growth*, ASTM STP 590, American Society for Testing and Materials, Philadelphia, pp. 243–260.

(4.114) Radok, J. R. M., "On the Solution of Problems of Dynamic Plane Elasticity," *Quarterly of Applied Mathematics*, **14**, pp. 289–298 (1956); and Sneddon, I. N., *Crack Problems in the Mathematical Theory of Elasticity*, North Carolina State College Report, Raleigh, N. C. (1961).

(4.115) Kanninen, M. F., "An Estimate of the Limiting Speed of a Propagating Ductile Crack," *Journal of the Mechanics of Physics and Solids*, **16**, pp. 215–228 (1968).

(4.116) Atkinson, C., "A Simple Model of a Relaxed Expanding Crack," *Arkiv for Fysik*, **26**, pp. 469–476 (1968).

(4.117) Embley, G. T. and Sih, G. C., "Plastic Flow Around an Expanding Crack," *Engineering Fracture Mechanics*, **4**, pp. 431–442 (1972).

(4.118) Robertson, T. S., "Propagation of Brittle Fracture in Steel," *Journal of the Iron and Steel Institute*, **175**, pp. 361–374 (1953).

(4.119) Lange, E. A., "Dynamic Fracture-Resistance Testing and Methods for Structural Analysis," Naval Research Laboratory Report 7979, Washington, D. C. (1976).

(4.120) Pellini, W. S., *Principles of Structural Integrity Technology*, Office of Naval Research, Arlington, Va. (1976).

(4.121) Nordell, W. J. and Hall, W. J., "Two Stage Fracturing in Welded Mild Steel Plates," *Welding Journal*, **44**, pp. 124s–134s. (1965).

(4.122) van Elst, H. C., "The Intermittent Propagation of Brittle Fracture," *Transactions of the Metallurgical Society of AIME*, **230**, pp. 460–469 (1964).

(4.123) Aberson, J. A., Anderson, J. M., and King, W. W., "Singularity-Element Simulation of Crack Propagation," *Fast Fracture and Crack Arrest*, ASTM STP 627, G. T. Hahn and M. F. Kanninen (eds.), American Society for Testing and Materials, Philadelphia, pp. 123–134 (1977).

(4.124) Keegstra, P. N. R., Head, J. L., and Turner, C. E., "A Transient Finite Element Analysis of Unstable Crack Propagation in Some Two- Dimensional Geometries," *Proceedings of the Fourth International Conference on Fracture*, University of Waterloo Press, Waterloo, Canada pp. 515–522 (1977).

(4.125) Yagawa, G., Sakai, Y., and Ando, Y., "Analysis of a Rapidly Propagating Crack Using Finite Elements," *Fast Fracture and Crack Arrest*, G. T. Hahn and M. F. Kanninen (eds.), ASTM STP 627, American Society for Testing and Materials, Philadelphia, pp. 109–122 (1977).

(4.126) Kanninen, M. F., Gehlen, P. C., Barnes, C. R., Hoagland, R. G., Hahn, G. T., and Popelar, C. H., "Dynamic Crack Propagation Under Impact Loading," *Nonlinear and Dynamic Fracture Mechanics*, N. Perrone and S. Atluri (eds.), ASME AMD Vol. 35, pp. 185–200 (1979).

(4.127) Ahmad, J., Barnes, C. R., and Kanninen, M. F., "An Elastoplastic Finite-Element Investigation of Crack Initiation Under Mixed-Mode Static and Dynamic Loading," *Elastic-Plastic Fracture: Second Symposium*, ASTM STP 803, C. F. Shih and J. P. Gudas (eds.), American Society for Testing and Materials, Philadelphia, Vol. I, pp. 214–239 (1983).

(4.128) Mott, N. F., "Fragmentation of Shell Cases," *Proceedings of the Royal Society of London, Series A*, **189**, pp. 300–308 (1947).

(4.129) Taylor, G. I., "The Fragmentation of Tubular Bombs," *Scientific Papers of G. I. Taylor*, Vol. III, No. 44, pp. 398–390, Cambridge University Press (1963).

(4.130) Davison, L., Stevens, A. L., and Kipp, M. E., "Theory of Spall Damage Accumulation in Ductile Metals," *Journal of the Mechanics and Physics of Solids*, **25**, pp. 11–28 (1977).

(4.131) Grady, D. E., "Fragmentation of Solids Under Impulsive Loading," *Journal of Geophysical Research*, **86**, pp. 1047–1054 (1981).

(4.132) Backman, M. E. and Goldsmith, W., "The Mechanics of Penetration of Projectiles into Targets," *International Journal of Engineering Science*, **16**, pp. 1–99 (1978).

(4.133) Zukas, J. A., Nicholas, T., Swift, H. F., Greszczuk, L. B., and Curran, D. R., *Impact Dynamics*, Wiley, New York (1982).

(4.134) Jonas, G. H. and Zukas, J. A., "Mechanics of Penetration: Analysis and Experiment," *International Journal of Engineering Science*, **16**, pp. 879–903 (1978).

(4.135) Wilkins, M. L., "Mechanics of Penetration and Perforation," *International Journal of Engineering Science*, **16**, pp. 793–807 (1978).

(4.136) Bodner, S. R., "Modelling Ballistic Perforation," *International Conference on Structural Impact and Crasshworthiness*, Imperial College, London, July (1984).
(4.137) Gilman, J. J. and Tuler , F. R., "Dynamic Fracture by Spallation in Metals," *International Journal of Fracture Mechanics*, **6**, pp. 169–182 (1970).
(4.138) Curran, D. R., Seaman, L., and Shockey, D. A., "Dynamic Failure in Solids," *Physics Today*, **30**, pp. 46–55 (1977).
(4.139) Tuler, F. R. and Butcher, B. M., "A Criterion for the Time Dependence of Dynamic Fracture," *International Journal of Fracture Mechanics*, **4**, pp. 431–437 (1968).

5

ELASTIC-PLASTIC FRACTURE MECHANICS

Linear elastic fracture mechanics is limited by the small-scale yielding condition that the plastic zone attending the crack tip be small compared to the size of the K-dominant region and any relevant geometric dimension. It is virtually impossible to satisfy this condition for high toughness, low strength materials which generally undergo extensive plastic deformation and crack-tip blunting prior to initiation of crack growth. Crack initiation in these materials is usually followed by stable crack growth or tearing. While LEFM can accommodate an increasing fracture resistance during stable growth, as demonstrated in Chapter 3, its prediction of the load carrying capacity of a degraded component may provide misleading estimates.

The need to include the influence of significant plastic deformation that may accompany crack initiation and the subsequent stable growth has been the main impetus for the development of the field of plastic fracture mechanics. While the development of this field is ongoing, sufficient advances have been made that plastic fracture mechanics is being used in design and assessment of the structural integrity of degraded components. The survey by Kanninen, Popelar, and Broek (5.1) summarizes the advances made through 1980. Recognizing that there will be further advances and refinements of existing approaches, we present in this chapter the developments of elastic-plastic fracture mechanics that will likely form the basis for these advances and refinements.

This chapter begins by considering the Dugdale model and its applications and limitations to plastic fracture. Because it is one of the few problems where closed-form solutions exist, the plastic antiplane strain problem is treated next. These solutions aid in developing an understanding and an appreciation of plastic fracture. After this the plastic crack-tip fields for power law hardening materials are investigated. The use of the J-integral as a fracture characterizing parameter follows naturally from this analysis. Simplified engineering methods for estimating the J-integral are presented. This is followed by a description of test procedures for determining the critical value of J at crack initiation. While J is, strictly speaking, only applicable for stationary cracks, conditions for J-controlled crack growth are considered. When these conditions are satisfied, a J-resistance curve analysis of stable crack growth is developed. Finally, extended crack growth beyond where J loses its significance is treated.

5.1 The Dugdale Model

Early attempts to model the plastic deformation attending the crack tip were based upon extensions of LEFM. An example is the Dugdale-Barenblatt yield strip or cohesive model for yielding in thin cracked sheets. The deformation in the plane stress plastic zone is due to slipping on planes at 45 degrees to the surface of the sheet as shown in Figure 3.14 and illustrated schematically in Figure 3.16(a). Experimental investigations by Hahn and Rosenfield (5.2) revealed that the height of the plane stress plastic zone is approximately equal to the thickness of the sheet.

Dugdale (5.3) assumed the length of the plastic zone to be much greater than the thickness of the sheet and modeled the plastic zone as a yielded strip ahead of the crack tip. The material is assumed to be elastic-perfectly plastic so that $\sigma_{22} = \sigma_y$ within the strip. Dugdale postulated that the effect of yielding is to increase the crack length by the extent of the plastic zone as depicted in Figure 5.1 for a finite length crack in an infinite medium subjected to a uniform remote stress $\sigma_{22} = \sigma$. Within the yielded strip, $a < |x_1| < c$, the opening of the crack faces is restrained by the stress $\sigma_{22} = \sigma_y$. The length d of the strip is established from the condition that the stress field be nonsingular.

The solution to this problem can be obtained by the superposition of the solutions for the uncracked sheet loaded by remote tension $\sigma_{22} = \sigma$ and for the cracked sheet with no remote loading and with pressure, $p_2(x_1) = \sigma$ for

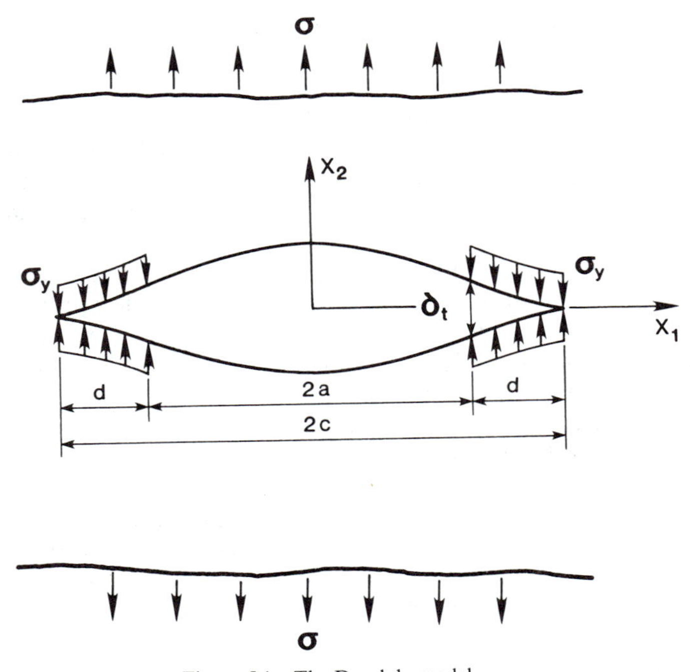

Figure 5.1 The Dugdale model.

$|x_1| < a$ and $p_2(x_1) = \sigma - \sigma_y$ for $a < |x_1| < c$, on the crack surfaces. For the former problem the potential functions are $\omega''(z) = 2\Omega'(z) = \sigma/2$ and for the latter $\psi'(z) = 0$ and $\Omega'(z)$ is given by Equation (3.2-17). The resultant expression for $\Omega'(z)$ is

$$\Omega'(z) = \frac{\sigma}{4} - \frac{(z^2 - c^2)^{-\frac{1}{2}}}{2\pi} \int_{-c}^{c} \frac{p_2(t)(c^2 - t^2)^{\frac{1}{2}}}{t - z} dt \tag{5.1-1}$$

The condition of bounded stresses [see Equation (3.4-7)] leads to

$$d = a\left[\sec\left(\frac{\pi\sigma}{2\sigma_y}\right) - 1\right] \tag{5.1-2}$$

for the length of the plastic zone. Dugdale found very good agreement between measured lengths of the plastic zones in steels and the predictions based upon Equation (5.1-2) for σ as large as $0.9\sigma_y$. Equally good agreement was reported by Mills (5.4) for cracked polycarbonate, polysulfane and polyvinylchloride sheets.

The evaluation of the integral in Equation (5.1-1) can be accomplished by replacing it with a closed contour integral shrunk onto the crack and then employing Cauchy's integral theorem. Effecting this integration and computing the displacements, one finds [e.g., see (5.5) and (5.6)] that the crack-tip opening displacement δ_t is

$$u_2(a, 0^+) - u_2(a, 0^-) \equiv \delta_t = \frac{8}{\pi}\frac{\sigma_y}{E} a \ln\left[\sec\left(\frac{\pi}{2}\frac{\sigma}{\sigma_y}\right)\right] \tag{5.1-3}$$

It is also possible to view δ_t as a measure of the stretching or deformation occurring in the plastic zone. By representing the crack and yielded zones by an inverted pile-up of dislocations, Bilby, Cottrell, and Swinden (5.7) obtained expressions analogous to Equation (5.1-3) for Modes II and III.

Within the context of the Dugdale model an elastic-plastic fracture criterion can be expressed in terms of a critical value of δ_t. At initiation of crack growth

$$\delta_t = \delta_{tc} \tag{5.1-4}$$

where δ_{tc} is the critical value that is considered to be a material property. Rather extensive reviews of this fracture criterion and of methods for measuring δ_{tc} are given in Broek (5.8) and Knott (5.9). Since by Equation (3.3-29) $J = \sigma_y \delta_t$, then Equation (5.1-3) can also be expressed as

$$J = \frac{8}{\pi}\frac{\sigma_y^2}{E} a \ln\left[\sec\left(\frac{\pi}{2}\frac{\sigma}{\sigma_y}\right)\right] \tag{5.1-5}$$

Consequently, any fracture criterion based upon δ_t attaining a critical value is equivalent to J reaching a critical value J_c; that is, at initiation of crack extension

$$J = J_c \tag{5.1-6}$$

where $J_c = \sigma_y \delta_{tc}$ is a material property.

Noting that $K_I = \sigma\sqrt{\pi a}$ and employing the small-scale yielding results of Equation (3.3-30) one can write

$$J_{ssy} \equiv \frac{K_I^2}{E} = \frac{\sigma^2 \pi a}{E} \tag{5.1-7}$$

Equations (5.1-5) and (5.1-7) provide that

$$\frac{J}{J_{ssy}} = \frac{8}{\pi^2}\left(\frac{\sigma_y}{\sigma}\right)^2 \ln\left[\sec\left(\frac{\pi}{2}\frac{\sigma}{\sigma_y}\right)\right] \tag{5.1-8}$$

The right-hand side of Equation (5.1-8) may be viewed as the plastic correction to the small-scale yielding prediction. For $\sigma/\sigma_y \ll 1$ this quantity approaches unity whereas it becomes unbounded as $\sigma \to \sigma_y$. To first order Equation (5.1-8) yields

$$\frac{J}{J_{ssy}} = 1 + \frac{\pi^2}{24}\left(\frac{\sigma}{\sigma_y}\right)^2$$

where the second term on the right-hand side is about 18 percent smaller than the Irwin plane stress plastic zone correction.

At fracture $\sigma = \sigma_f$, $J = J_c$ and Equation (5.1-8) gives

$$\left(\frac{J_{ssy}}{J_c}\right)_f = \left\{\frac{8}{\pi^2}\frac{\sigma_y^2}{\sigma_f^2} \ln\left[\sec\left(\frac{\pi}{2}\frac{\sigma_f}{\sigma_y}\right)\right]\right\}^{-1} \tag{5.1-9}$$

Within the context of small-scale yielding one can also write $J_c = K_c^2/E$, which when combined with Equations (5.1-7) and (5.1-9) leads to

$$\left(\frac{J_{ssy}}{J_c}\right)_f^{\frac{1}{2}} = \left(\frac{K_I}{K_c}\right)_f = \frac{\sigma_f}{\sigma_y}\left\{\frac{8}{\pi^2} \ln\left[\sec\left(\frac{\pi}{2}\frac{\sigma_f}{\sigma_y}\right)\right]\right\}^{-\frac{1}{2}} \tag{5.1-10}$$

Equation (5.1-10) can be used to establish K_c for intermediate scale yielding when σ_f is measured in a fracture test. Conversely, if J_c or, equivalently, K_c is known, then this equation can be used to predict the failure stress σ_f.

The latter concept was extended by Harrison, Loosemore, and Milne (5.10). They postulated that Equation (5.1-10) could be used to interpolate between linear elastic fracture ($\sigma_f/\sigma_y \ll 1$) for one extreme of failure and large-scale plastic yielding or collapse at the other extreme ($\sigma_f/\sigma_y \to 1$), if the yield stress were replaced by the plastic collapse stress σ_c in order to accommodate other flawed configurations. The resulting Failure Assessment Curve or R-6 Curve is

$$\left(\frac{K_I}{K_c}\right)_f = \frac{\sigma_f}{\sigma_c}\left\{\frac{8}{\pi^2} \ln\left[\sec\left(\frac{\pi}{2}\frac{\sigma_f}{\sigma_c}\right)\right]\right\}^{-\frac{1}{2}} \tag{5.1-11}$$

The plastic collapse stress is determined from a limit analysis. For example, in many instances it is possible to write

$$\sigma_c = \gamma\sigma_0(1 - a/W)^n \tag{5.1-12}$$

where σ_0 is the flow stress that attempts to accommodate work hardening and the through-thickness constraint. The width of the specimen per crack tip is denoted by W, γ is a dimensionless constant, $n = 1$ for center-cracked and double-edge cracked plates, and $n = 2$ for bend specimens (5.11, 5.12).

The failure assessment diagram is constructed in the $K_r - S_r$ plane where

$$S_r = \sigma/\sigma_c$$
$$K_r = K_I/K_c \tag{5.1-13}$$

are proportional to the applied load through the parameters σ and K_I. The failure assessment curve shown in Figure 5.2 is the loci of points $(K_r)_f$, $(S_r)_f$ satisfying Equation (5.1-11). Failure is associated with any combination of loading and crack size giving rise to a point (K_r, S_r) falling on or outside of this curve; and, conversely, the combination will be safe if the point lies inside the curve. Since K_r and S_r are proportional to the applied load, the distance from the origin to the point (K_r, S_r) is also proportional to the load. For a crack of fixed length, changing the applied load causes the point (K_r, S_r) to be displaced along the ray through the origin. The safety factor is the ratio of the distance from the origin to the point of intersection of this ray and the failure

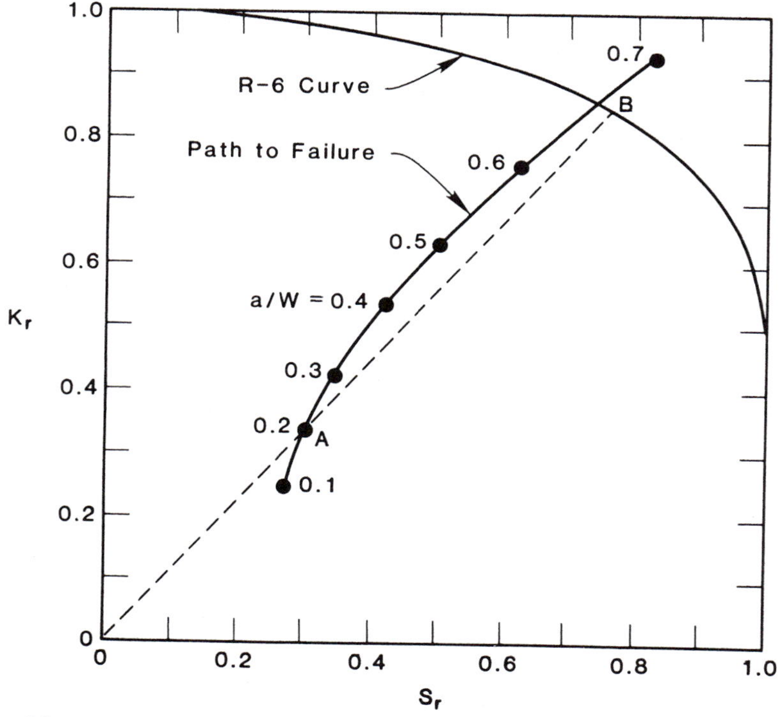

Figure 5.2 The Failure Assessment Diagram with the plane stress path to failure for a center-cracked panel having $\sigma/\sigma_0 = \frac{1}{4}$ and $\sigma\sqrt{\pi W}/K_c = \frac{3}{4}$.

assessment curve and the distance between the origin and the point (K_r, S_r). For a prescribed load intensity the loci of points (K_r, S_r) for different crack lengths is referred to as the path to failure. Once a path to failure has been established for one intensity of load, other paths can be constructed for other intensities by simple proportionality.

As an illustration of the use of the Failure Assessment Diagram consider a thin panel of width $2W$ containing a central crack of length $2a$ loaded by a uniform remote stress σ. In this case

$$K_{\mathrm{I}} = \sigma\sqrt{\pi W}\left[\frac{a}{W}\sec\left(\frac{\pi}{2}\frac{a}{W}\right)\right]^{\frac{1}{2}}$$

and

$$\sigma_c = \sigma_0(1 - a/W)$$

so that

$$K_r = \frac{\sigma\sqrt{\pi W}}{K_c}\left[\frac{a}{W}\sec\left(\frac{\pi}{2}\frac{a}{W}\right)\right]^{\frac{1}{2}}$$

$$S_r = \frac{\sigma}{\sigma_0}\left(1 - \frac{a}{W}\right)^{-1} \tag{5.1-14}$$

For a prescribed stress σ, Equation (5.1-14) is a parametric representation for the path to failure in the Failure Assessment Diagram. Such a path is shown in Figure 5.2 for $\sigma/\sigma_0 = \frac{1}{4}$ and $\sigma\sqrt{\pi W}/K_c = \frac{3}{4}$. From the point of intersection of this path with the failure assessment curve, failure can be expected for $a/W > 0.66$. By increasing σ/σ_0 by the product of the ratio of lengths OB to OA one finds that $\sigma/\sigma_0 = 0.63$ will produce failure for $a/W = 0.2$. Alternatively, for $\sigma/\sigma_0 = \frac{1}{4}$ and $a/W = 0.2$ the safety factor is 2.5.

In Figure 5.3 comparisons between the Failure Assessment Curve and similar curves based upon finite element computations for plane strain are made. In these computations J was computed and the linear elastic relationship, Equation (3.3-30), between J and K was used to cast these results into the required form. Chell (5.13) found that a universal failure curve does not exist. However, the differences between the curves are relatively small when compared to the uncertainties associated with the failure assessment of a real structure. With the obvious exceptions illustrated in Figure 5.3, the Failure Assessment Curve approximates a lower bound.

Chell and Milne (5.14) have extended this concept to include the influence of thermal and residual stresses and to investigate stable crack growth and tearing instability under load and displacement control. In the instability analysis the concept of a J-resistance curve is used. Chell (5.13) found that the failure assessment diagram can lead to nonconservative predictions when residual stresses exist.

The appealing aspect of the Failure Assessment Diagram is its simplicity. Only the LEFM stress intensity factor and the plastic limit or collapse load (stress) are required in the failure analysis. However, the question of what

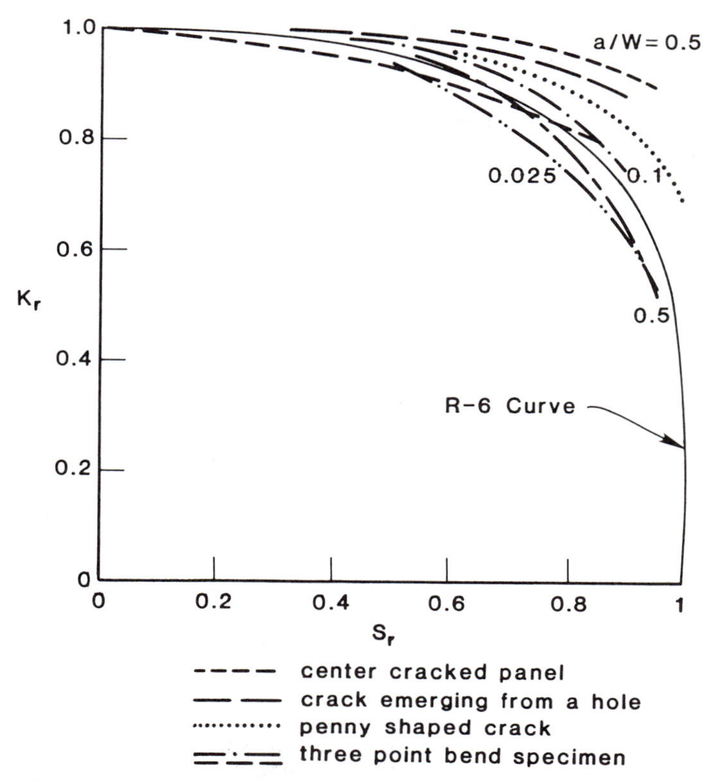

Figure 5.3 Comparison of Failure Assessment Curve and failure curves from plane strain finite element computations (5.13).

value should be used for the flow stress when computing the collapse load remains unanswered and becomes increasingly more difficult to answer for materials that exhibit appreciable strain hardening. The simplicity of the method associated with prescribed loading is diminished when conditions of controlled load-point displacement (e.g., fixed-grip loading) exist (5.13). For the latter loading condition the method may be considered in some applications as providing overly conservative estimates of failure.

For high toughness, low strength materials the failure can be expected to be due to plastic collapse of the remaining ligament or net section; that is, $K_r \simeq 0$ and $S_r \simeq 1$. The concept of net section collapse has been used by Feddersen (5.15) and Broek (5.8, 5.16–5.18) in the study of fracture of subsize aluminum panels. For a center cracked panel under plane stress loading the net section collapse criterion can be written as

$$\sigma = \sigma_c = \sigma_0(1 - a/W) \tag{5.1-15}$$

In the $\sigma - a/W$ plane Equation (5.1-15) is a straight line through the points $a/W = 0$, $\sigma = \sigma_0$, and $a/W = 1$, $\sigma = 0$.

Kanninen et al. (5.19, 5.20) found that this approach was also applicable to Type 304 stainless steel under noncompliant loading. The applied stress required for initiation of crack growth in 12-in. wide Type 304 stainless steel, center cracked panels was measured. Because the initial crack growth is stable, these panels could be further loaded before reaching a tearing instability at maximum load. The measured stress at crack initiation and at instability appears to fall on straight lines through the point, $2a = 12$ in. and $\sigma = 0$ in Figure 5.4. The flow stresses at crack initiation and maximum load are apparently 66 ksi and 73 ksi, respectively, compared to a yield stress of approximately 40 ksi. The disparity between the yield stress and the flow stress is indicative of the large degree of strain hardening that this material exhibits. Further tests also indicate that the type of loading (quasi-static, high strain rate, interrupted and cyclic) had only minor influences upon the applied stress at crack initiation and maximum load.

Kanninen et al. (5.19) used the net section collapse criterion to analyze circumferentially cracked Type 304 stainless steel pipes in four-point bending. In the limit load analysis of this configuration the previously determined flow

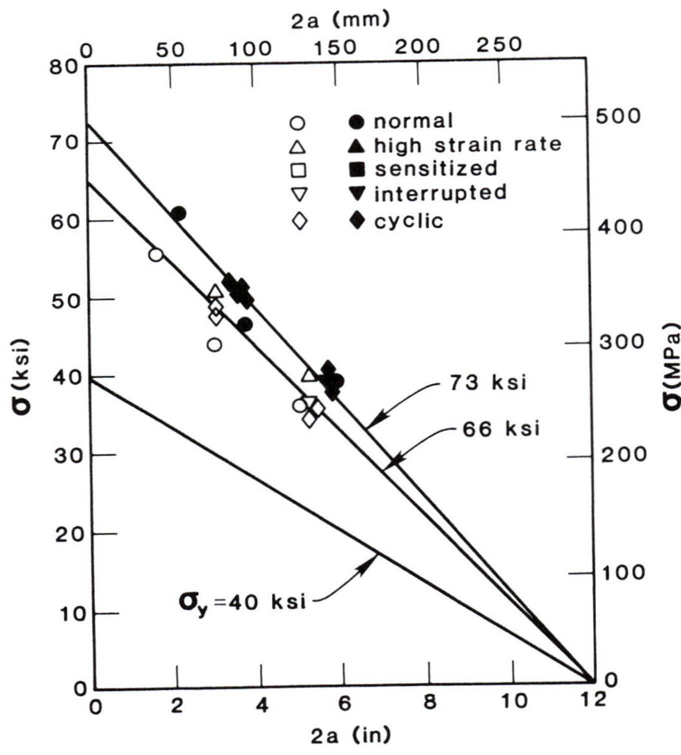

Figure 5.4 Residual strength as a function of crack size for Type 304 stainless steel center-cracked tensile panels (5.1).

stress obtained from the center cracked panel tests was used. The predicted load for crack initiation agreed very well with the observed load in quasi-static fracture tests of through-wall circumferentially cracked, 4-in. diameter pipes loaded primarily in bending. Because the load at crack initiation is nearly coincident with the maximum load, the net section collapse criterion can also be used to predict pipe fracture in this instance.

Another example of the use of the Dugdale model is the work of Hahn et al. (5.21) in the development of a criterion for crack extension in circular cylindrical pressure vessels. The driving force for a through-wall axial crack in the pressurized vessel is due to the hoop stress and to the bulging of the unsupported vessel wall in the flanks of the crack. The basic premise of this development is that a cracked, thin-walled pressure vessel can be treated as a flat panel, having the same thickness and crack length, in tension provided the nominal stress σ in the panel is written as

$$\sigma = M\sigma_h \tag{5.1-16}$$

where σ_h is the vessel's hoop stress. The factor M is a function of the crack length $2a$, radius R of the vessel, and wall thickness t. Folias (5.22) has established the basis of Equation (5.1-16) for small-scale yielding. The function M can be approximated by

$$M = \left(1 + 1.255\frac{a^2}{Rt} - 0.0135\frac{a^4}{R^2t^2}\right)^{\frac{1}{2}} \tag{5.1-17}$$

In the limit as $R \to \infty$, $M \to 1$, and the flat panel is recovered.

The introduction of Equation (5.1-16) into (5.1-5) yields

$$J = \frac{8\sigma_0^2}{\pi E} a \ln\left[\sec\left(\frac{\pi}{2}\frac{M\sigma_h}{\sigma_0}\right)\right] \tag{5.1-18}$$

where the yield stress has been replaced by the flow stress. The flow stress of line pipe materials has been found to average about 10 ksi (69 MPa) greater than the yield stress. At crack extension $J = J_c$ and $\sigma_h = \sigma_{hf}$ and Equation (5.1-18) becomes

$$J_c = \frac{8\sigma_0^2}{\pi E} a \ln\left[\sec\left(\frac{\pi}{2}\frac{M\sigma_{hf}}{\sigma_0}\right)\right] \tag{5.1-19}$$

A dimensionless plot of Equation (5.1-19) appears in Figure 5.5. For high toughness, low strength materials containing small cracks; that is, for large values of $J_cE/a\sigma_0^2$, $M\sigma_{hf}/\sigma_0$ approaches unity—a condition equivalent to large-scale yielding or plastic collapse.

Kiefner et al. (5.23) proposed that Equation (5.1-19) could also be used for part through wall axial cracks if M is modified appropriately. It is necessary to replace M by the empirical relation

$$M_p = \frac{t - d/M}{t - d} \tag{5.1-20}$$

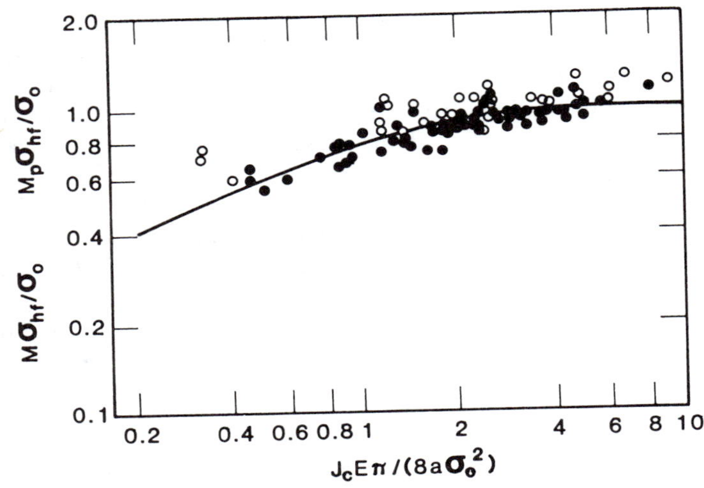

Figure 5.5 Comparison of predicted and measured failure stress for surface flaws (○) and through wall flaws (●) in pipes (5.25).

where d is the maximum depth of the surface flaw. Furthermore, when the depth of the surface flaw is not uniform, an effective crack length defined by

$$2a_e = A/d \tag{5.1-21}$$

where A is the surface area of the flaw, is to be used. As d/t tends to unity, M_p becomes unbounded. In light of the previous discussion, the implication is that only a small hoop stress is required in this case to produce plastic collapse in the remaining ligament of the wall.

For $M\sigma_{hf}/\sigma_0 < 0.8$, Maxey et al. (5.24) found a nearly one-to-one correlation between J_c given by Equation (5.1-19) and the Charpy-V-Notch upper plateau energy normalized with respect to the net cross-sectional area of the Charpy-V-Notch specimen. Using this value for J_c, Maxey (5.25) made the comparison in Figure 5.5 of the predictions based upon Equation (5.1-19) and measured results from fracture tests on pressurized pipes with axial through-wall and surface cracks. These tests involved pipes having radii from 3.3 in. to 21 in., wall thicknesses from $\frac{1}{4}$ in. to $1\frac{3}{4}$ in., and crack lengths from 1 in. to $24\frac{1}{2}$ in. The flow stress varied from 32 ksi to 120 ksi. In general the agreement is quite good with the experimental data scattered around the curve given by Equation (5.1-19). The large differences for the surface cracks are associated with either shallow ($d/t < 0.3$) or deep ($d/t > 0.7$) cracks. This may be due to tougher zones of materials lying near the surfaces of the pipe (5.23).

At failure the remaining ligament is severed and a through-wall crack is produced. If the stress level is less than that necessary to initiate axial growth of the resulting through-wall crack, then a leak occurs. On the other hand, a break (an axially propagating crack) will occur if the stress level is greater than the required initiating value. It follows from the present model that with everything else equal a leak-before-break is to be expected when $M < M_p$.

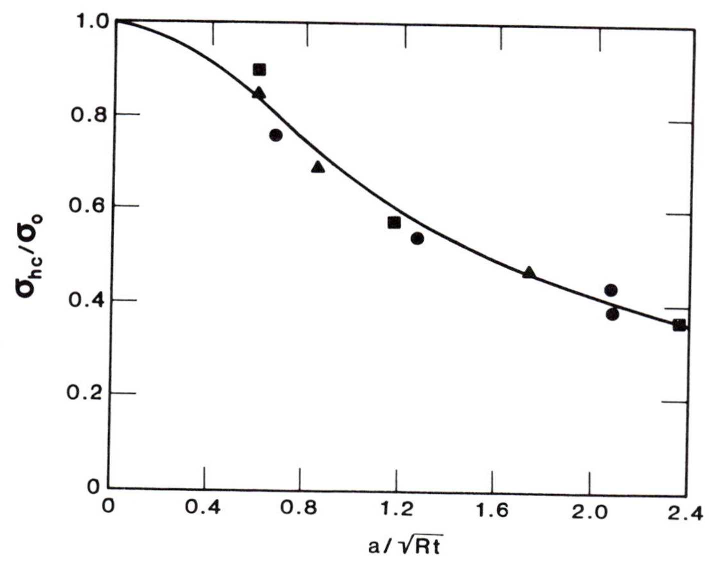

Figure 5.6 Comparison of predicted [Equation (5.1-22)] and measured failures of pressurized A106B steel pipes with axial through wall cracks.

As noted earlier, the Dugdale model predicts that large-scale yielding or plastic collapse will occur for small flaws in pipes made of high toughness, low strength materials. The condition for plastic collapse is

$$\sigma_{hc}/\sigma_0 = 1/M \tag{5.1-22}$$

for through-wall cracks and

$$\sigma_{hc}/\sigma_0 = 1/M_p \tag{5.1-23}$$

for surface flaws. In the latter equations σ_{hc} is the hoop stress for plastic collapse. Comparisons of predictions based upon Equations (5.1-22) and (5.1-23) and measured results are shown in Figures 5.6 and 5.7 for through-wall and surface flaws, respectively. In general, the agreement is quite good.

In closing this section we note that there are additional experiments in pipes with initial part through-wall cracks that produced failure loads somewhat in excess of those predicted by the net section or plastic collapse criterion. Situations also exist for which the net section collapse is simply inapplicable. For example, under nearly displacement controlled conditions extensive stable crack growth can occur beyond maximum load and only a plastic fracture mechanics approach is capable of predicting the failure load. Furthermore, there remains the question as to what value should be used for the flow stress in strain hardening materials. While the net section collapse criterion and the Failure Assessment Diagram based upon the Dugdale model have many appealing attributes, they do not represent a panacea for elastic-plastic fracture.

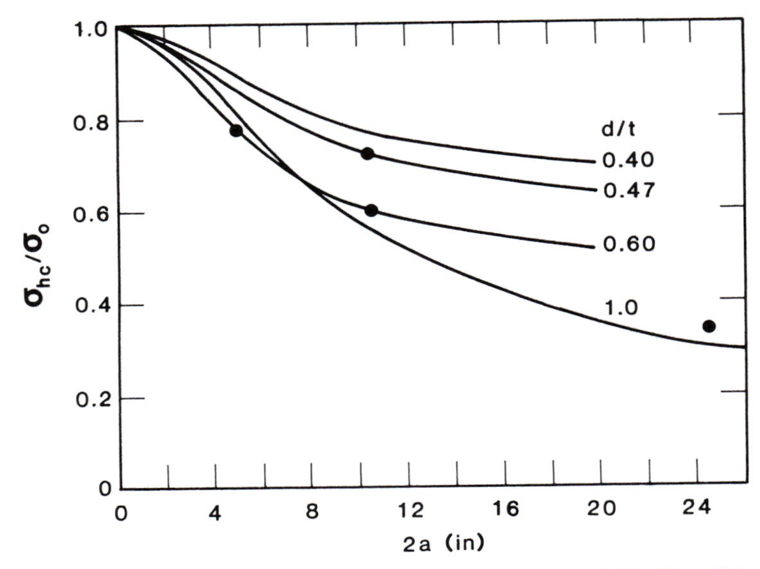

Figure 5.7 Comparison of measured and predicted failures in 24 × 1.50-in. Type 316 stainless steel pipes with surface flaws.

5.2 Antiplane Elastic-Plastic Solutions

In the Dugdale model plastic yielding is confined to a strip ahead of the crack tip. Without such a contrivance the mathematical difficulties for Mode I loading are such as to defy, even within the realm of small-scale yielding, an elastic-plastic solution in closed form. However, for Mode III loading these difficulties are sufficiently reduced that closed-form solutions are attainable for small-scale yielding. The solution to the antiplane problem is useful because it often provides a basis for a qualitative understanding of the opening mode behavior.

Consider the antiplane loading of a crack in an elastic-perfectly plastic material depicted in Figure 5.8. This problem was first addressed by Hult and McClintock (5.26). The nonzero displacement u_3 and stress components σ_{31} and σ_{32} are assumed to depend only upon x_1 and x_2. In the plastic zone ahead of the crack tip they must satisfy the equilibrium equation

$$\frac{\partial \sigma_{31}}{\partial x_1} + \frac{\partial \sigma_{32}}{\partial x_2} = 0 \tag{5.2-1}$$

the von Mises or Tresca yield condition

$$\sigma_{31}^2 + \sigma_{32}^2 = \tau_0^2 \tag{5.2-2}$$

and the Hencky deformation constitutive relations [cf. Equation (2.6-27)]

$$\frac{\partial u_3}{\partial x_1} = 2\Lambda\sigma_{31}, \qquad \frac{\partial u_3}{\partial x_2} = 2\Lambda\sigma_{32} \tag{5.2-3}$$

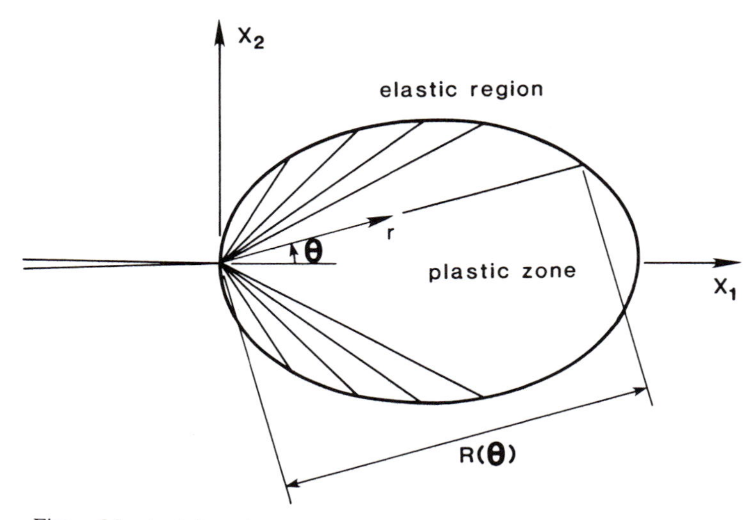

Figure 5.8 Antiplane shear deformation in an elastic-plastic cracked body.

where τ_0 is the yield stress in shear and Λ is a scalar function of the stress and strain invariants.

The form of Equation (5.2-2) suggests writing

$$\sigma_{31} = -\tau_0 \sin\theta, \qquad \sigma_{32} = \tau_0 \cos\theta \tag{5.2-4}$$

which will also satisfy Equation (5.2-1) provided

$$\theta = \tan^{-1}(x_2/x_1) \tag{5.2-5}$$

is the polar angle. Equations (5.2-4) and (5.2-5) imply that the shear stress $\sigma_{3\theta} = \tau_0$ on $\theta =$ constant. In addition Equations (5.2-3)–(5.2-5) provide that

$$du_3 = \frac{\partial u_3}{\partial x_1}\,dx_1 + \frac{\partial u_3}{\partial x_2}\,dx_2 = 0 \tag{5.2-6}$$

on $\theta =$ constant. Consequently, rays emanating from the crack tip are slip lines.

According to Equations (5.2-4) and (3.1-12)

$$\sigma_{31} + i\sigma_{32} = i\tau_0 e^{i\theta} \tag{5.2-7}$$

in the plastic zone and

$$\sigma_{31} + i\sigma_{32} = 2\overline{f'(z)} \tag{5.2-8}$$

within the elastic region. On the elastic-plastic boundary, $z = R(\theta)e^{i\theta}$,

$$2f'(z) = -i\tau_0 e^{-i\theta} \tag{5.2-9}$$

Furthermore, on the stress-free crack surfaces ($\theta = \pm\pi$) the analytic function $f(z)$ must satisfy

$$\operatorname{Im} f'(z) = 0 \tag{5.2-10}$$

For a plastic zone small compared to the K-dominant region (small-scale yielding) the stress field must approach asymptotically the singular elastic stress field at distances sufficiently removed from the plastic zone; that is, (see Section 3.1.),

$$f'(z) = -\frac{iK_{\text{III}}}{2\sqrt{(2\pi z)}}, \qquad |z| \to \infty \tag{5.2-11}$$

The solution [see (5.27) and (5.28)] to the boundary value problem described by Equations (5.2-10) and (5.2-11) is

$$f'(z) = -i\frac{K_{\text{III}}}{2}\left\{2\pi\left(z - \frac{K_{\text{III}}^2}{2\pi\tau_0^2}\right)\right\}^{-\frac{1}{2}} \tag{5.2-12}$$

From the introduction of Equation (5.2-12) into Equation (5.2-9) it follows that the elastic-plastic boundary is a circle of diameter $r_p = K_{\text{III}}^2/(\pi\tau_0^2)$ centered on the x_1-axis at $x_1 = K_{\text{III}}^2/(2\pi\tau_0^2)$. Consequently, the plastic zone does not engulf the crack tip for this material.

The substitution of Equation (5.2-12) into Equation (5.2-8) yields

$$\sigma_{32} + i\sigma_{31} = K_{\text{III}}\left\{2\pi\left(z - \frac{K_{\text{III}}^2}{2\pi\tau_0^2}\right)\right\}^{-\frac{1}{2}} \tag{5.2-13}$$

for the elastic stresses. The effect of yielding is to produce a stress field in the elastic region that is identical in character to the elastic singular field for a crack tip shifted to the center of the plastic zone. This is the basis of the Irwin correction introduced in Section 3.4.

Equation (5.2-6) implies that $u_3 = u_3(\theta)$ in the plastic region and, therefore, the engineering shear strain

$$\gamma_{3\theta} = \frac{1}{r}\frac{\partial u_3}{\partial \theta}$$

has a r^{-1} singularity at the crack tip. Furthermore, since $\gamma_{3\theta} = \tau_0/\mu$ on the elastic-plastic boundary, $r = R(\theta) = r_p \cos\theta$, then

$$\frac{\partial u_3}{\partial \theta} = \frac{K_{\text{III}}^2}{\pi\mu\tau_0}\cos\theta.$$

Consequently, in the plastic zone

$$\begin{aligned} \gamma_{3\theta} &= \frac{K_{\text{III}}^2}{\pi\mu\tau_0}\frac{\cos\theta}{r}, \qquad \gamma_{3r} = 0 \\ u_3 &= \frac{K_{\text{III}}^2}{\pi\mu\tau_0}\sin\theta \end{aligned} \tag{5.2-14}$$

Note that the plastic strains at each point increase proportionally with K_{III}^2 and, therefore, this solution also satisfies the equations of incremental plasticity theory.

The crack-tip opening displacement is

$$\delta_t = u_3\left(\frac{\pi}{2}\right) - u_3\left(\frac{-\pi}{2}\right) = \frac{2K_{\text{III}}^2}{\pi\mu\tau_0} \tag{5.2-15}$$

Since $J = K_{\text{III}}^2/2\mu$ for small-scale yielding, then Equation (5.2-15) yields

$$J = \frac{\pi}{4}\,\tau_0\delta_t \tag{5.2-16}$$

Equations (5.2-15) and (5.2-16) can be compared, respectively, with the small-scale yielding results,

$$\delta_t = \frac{1}{2}\frac{K_{\text{III}}^2}{\mu\tau_0} \qquad \text{and} \qquad J = \tfrac{1}{2}\tau_0\delta_t$$

from the Bilby, Cottrell, and Swinden (5.7) yield strip model.

Next, we consider the influence of isotropic strain hardening upon the shape of the plastic zone and the singularities of the Mode III stress and strain fields. Assume that the principal antiplane shear stress τ and engineering shear γ, where

$$\tau = (\sigma_{31}^2 + \sigma_{32}^2)^{\frac{1}{2}}, \qquad \gamma = (\gamma_{31}^2 + \gamma_{32}^2)^{\frac{1}{2}}, \tag{5.2-17}$$

obey a linear law to the yield point (τ_0, γ_0) and a hardening law thereafter; that is,

$$\begin{aligned} \tau &= \frac{\tau_0}{\gamma_0}\gamma = \mu\gamma, \qquad \gamma < \gamma_0 \\ \tau &= \tau(\gamma), \qquad \gamma > \gamma_0 \end{aligned} \tag{5.2-18}$$

The Hencky deformation relations are

$$\sigma_{31} = \frac{\tau(\gamma)}{\gamma}\gamma_{31}, \qquad \sigma_{32} = \frac{\tau(\gamma)}{\gamma}\gamma_{32} \tag{5.2-19}$$

These equations must be supplemented by the equation of equilibrium, Equation (5.2-1), and the compatibility equation

$$\frac{\partial\gamma_{31}}{\partial x_2} - \frac{\partial\gamma_{32}}{\partial x_1} = 0 \tag{5.2-20}$$

In much the same manner that was employed for perfect plasticity, Rice (5.27, 5.29) obtained the small-scale yielding solutions to these equations for a semi-infinite crack in an infinite medium. The contours of constant γ or τ in the plastic region are circles of radius $R(\gamma)$ centered on the x_1-axis at $x_1 = X(\gamma)$ as depicted in Figure 5.9. Here

$$\begin{aligned} R(\gamma) &= \frac{K_{\text{III}}^2}{2\pi\tau_0^2}\frac{\gamma_0\tau_0}{\gamma\tau(\gamma)} \\ X(\gamma) &= \frac{K_{\text{III}}^2}{2\pi\tau_0^2}\left\{2\gamma_0\tau_0\int_\gamma^\infty \frac{du}{u^2\tau(u)} - \frac{\gamma_0\tau_0}{\gamma\tau(\gamma)}\right\} \end{aligned} \tag{5.2-21}$$

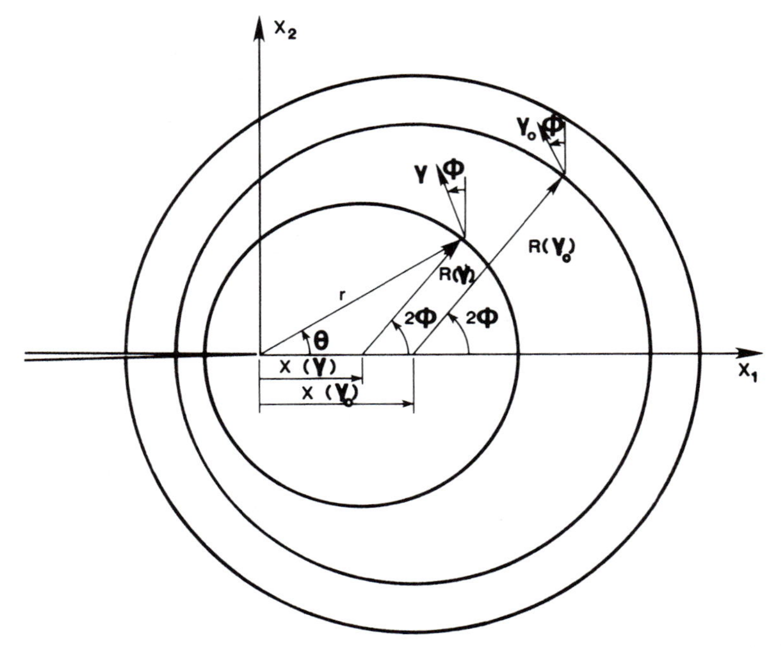

Figure 5.9 Contours of constant maximum antiplane shear strain for small-scale yielding in an elastic-strain hardening material.

The elastic-plastic boundary that now encompasses the crack tip is centered at $x_1 = X(\gamma_0)$ and has a radius $R(\gamma_0) = K_{\text{III}}^2/2\pi\tau_0^2$. In the elastic region the contours of constant τ or γ and the elastic-plastic boundary are concentric circles. The principal shear direction makes an angle ϕ, measured positive counterclockwise, from the x_2 direction. Consequently,

$$\tau e^{i\phi} = \sigma_{32} - i\sigma_{31} \tag{5.2-22}$$

and

$$\gamma e^{i\phi} = \gamma_{32} - i\gamma_{31} \tag{5.2-23}$$

It is clear from Figure 5.9 that on contours of constant γ

$$\begin{aligned} x_1 &= X(\gamma_0) + R(\gamma) \cos 2\phi \\ x_2 &= R(\gamma) \sin 2\phi, \qquad \gamma < \gamma_0 \end{aligned} \tag{5.2-24}$$

and

$$\begin{aligned} x_1 &= X(\gamma) + R(\gamma) \cos 2\phi \\ x_2 &= R(\gamma) \sin 2\phi, \qquad \gamma > \gamma_0 \end{aligned} \tag{5.2-25}$$

These equations can be used to determine the stress and strain fields. For example, in the elastic region Equation (5.2-24) can be rewritten as

$$x_1 + ix_2 = z = X(\gamma_0) + R(\gamma)e^{2i\phi} \tag{5.2-26}$$

where in this region

$$R(\gamma) = \frac{K_{\text{III}}^2}{2\pi\tau^2} \tag{5.2-27}$$

It follows from Equations (5.2-22), (5.2-26), and (5.2-27) that

$$\tau e^{-i\phi} = \sigma_{32} + i\sigma_{31} = \frac{K_{\text{III}}}{\{2\pi[z - X(\gamma_0)]\}^{\frac{1}{2}}} \tag{5.2-28}$$

Again, the effect of yielding is to produce a stress field within the elastic region identical to the elastic singular field for a crack tip shifted to the center of the plastic zone.

In the plastic zone Equation (5.2-25) permits writing

$$r = R(\gamma)\left\{\left[\frac{X(\gamma)}{R(\gamma)}\right]^2 + 2\frac{X(\gamma)}{R(\gamma)}\cos 2\phi + 1\right\}^{\frac{1}{2}} \tag{5.2-29}$$

It is apparent from Figure 5.9 that

$$\frac{X(\gamma)}{R(\gamma)} = \frac{\sin(2\phi - \theta)}{\sin\theta} \tag{5.2-30}$$

The latter can be used to simplify Equation (5.2-29) to

$$r = R(\gamma)\frac{\sin 2\phi}{\sin\theta} \tag{5.2-31}$$

Once the hardening law is specified, Equation (5.2-31) can be inverted to yield γ. The stress and strain fields are determined from Equations (5.2-22) and (5.2-23).

For example, consider a material exhibiting power law hardening according to

$$\begin{aligned} \gamma &= \frac{\gamma_0}{\tau_0}\tau, \qquad \tau < \tau_0 \\ \gamma &= \gamma_0\left(\frac{\tau}{\tau_0}\right)^n, \qquad \tau > \tau_0 \end{aligned} \tag{5.2-32}$$

In this case Equation (5.2-21) yields

$$\begin{aligned} R(\gamma) &= \frac{K_{\text{III}}^2}{2\pi\tau_0^2}\left(\frac{\gamma_0}{\gamma}\right)^{(n+1)/n} \\ X(\gamma) &= \frac{n-1}{n+1}R(\gamma) \end{aligned} \tag{5.2-33}$$

It follows from Equations (5.2-31)–(5.2-33) that

$$\gamma = \gamma_0\left[\frac{K_{\text{III}}^2\sin 2\phi}{2\pi\tau_0^2 r\sin\theta}\right]^{n/(n+1)}, \qquad \gamma > \gamma_0 \tag{5.2-34}$$

and

$$\tau = \tau_0 \left[\frac{K_{\mathrm{III}}^2 \sin 2\phi}{2\pi\tau_0^2 r \sin\theta} \right]^{1/(n+1)}, \qquad \tau > \tau_0 \tag{5.2-35}$$

where ϕ and θ are related by

$$\frac{\sin(2\phi - \theta)}{\sin\theta} = \frac{n-1}{n+1} \tag{5.2-36}$$

The stress and strain components follow directly upon the substitution of Equations (5.2-34) and (5.2-35) into Equations (5.2-22) and (5.2-23), respectively. Since the straining represented by Equation (5.2-34) is proportional, the use of the deformation theory is justified. The strain has a $r^{-n/(n+1)}$ singularity whereas the stress has a $r^{-1/(n+1)}$ singularity. The plastic zone extends a distance

$$r_p = X(\gamma_0) + R(\gamma_0) = \frac{2n}{n+1} \frac{1}{2\pi} \frac{K_{\mathrm{III}}^2}{\tau_0^2}$$

ahead of the crack tip and a distance

$$R(\gamma_0) - X(\gamma_0) = \frac{1}{n+1} \frac{1}{2\pi} \frac{K_{\mathrm{III}}^2}{\tau_0^2},$$

behind the tip. In the limit as $n \to 1$ and $n \to \infty$, the elastic and the elastic-perfectly plastic stress and strain fields are recovered.

Rice (5.27, 5.30) has also addressed the problem of an edge crack in an elastic-perfectly plastic, half-plane under uniform antiplane shear σ_{32}. In this case the length r_p over which the plastic zone extends ahead of the crack tip and the crack opening displacement are given by

$$\begin{aligned} r_p &= a\left[\frac{2}{\pi}\frac{1+s^2}{1-s^2} E_2\left(\frac{2s}{1+s^2}\right) - 1\right] \\ \delta_t &= 2\gamma_0 a\left[\frac{2}{\pi}(1+s^2)E_1(s^2) - 1\right] \end{aligned} \tag{5.2-37}$$

where $s = \sigma_{32}/\tau_0$, and E_1 and E_2 are the complete elliptic integrals of the first and second kind, respectively. For small loads such that s^2 can be neglected compared to unity, these expressions reduce to the small-scale yielding results. Significant departure between the plastic zone sizes for small- and large-scale yielding commence at 40–50 percent of the limit load ($s = 1$). Crack opening displacements begin deviating significantly at 60–70 percent of the limit load. The plastic zone elongates from the circular shape until it becomes unbounded in the x_1 direction at the limit load. Its height approaches asymptotically $4a/\pi$. Also, as can be seen from Equation (5.2-37), the crack-tip opening displacement becomes unbounded at the limit load.

For power hardening materials Rice (5.27, 5.29) also found that there is a transition from the circular plastic zone of small-scale yielding to a highly elongated zone as the applied stress is increased relative to the yield stress. As is to be expected, this transition is more gradual in a material exhibiting a larger degree of strain hardening.

5.3 Plastic Crack-Tip Fields

In this section the two-dimensional stress and strain fields described by the dominant singularity governing the plastic behavior at the tip of a line crack are considered. When the strain hardening can be characterized by a power law, the dominant singularity is referred to as the HRR singularity after the investigations of Hutchinson (5.31) and Rice and Rosengren (5.32). While the development is for small strains, nonlinearity enters into the study through the elastic-plastic constitutive relation.

The uniaxial stress-strain curve is modeled by the Ramberg-Osgood relation

$$\frac{\varepsilon}{\varepsilon_y} = \frac{\sigma}{\sigma_y} + \alpha\left(\frac{\sigma}{\sigma_y}\right)^n \tag{5.3-1}$$

where σ_y and ε_y are the yield stress and strain, respectively, $n > 1$ is the strain hardening exponent or index, and α is a dimensionless material constant. The first term on the right-hand side of Equation (5.3-1) describes the usual linear elastic behavior whereas the second term provides the nonlinear or plastic response. Typical curves are shown in Figure 5.10 for selected values of the hardening exponent n. In the limit as $n \to \infty$, an elastic-perfectly plastic material behavior is approached. Equation (5.3-1) is strictly applicable for a monotonically increasing stress and cannot accommodate elastic unloading. Alternatively, Equation (5.3-1) can be viewed as describing a nonlinear elastic material.

Because proportional loading in the plastic zone engulfing the crack tip is anticipated, a deformation theory of plasticity is used and Equation (5.3-1) is entirely appropriate. The plastic deformation is assumed to be incompressible and independent of the hydrostatic component of stress $\sigma_{kk}/3$. Under these conditions the generalized stress-strain relation [see Equation (2.6-32)] can be written as

$$\varepsilon_{ij} = \frac{1+\nu}{E} s_{ij} + \frac{1-2\nu}{3E} \sigma_{kk}\,\delta_{ij} + \frac{3}{2}\frac{\bar{\varepsilon}^p}{\bar{\sigma}} s_{ij} \tag{5.3-2}$$

where the deviatoric stress components are

$$s_{ij} = \sigma_{ij} - \tfrac{1}{3}\sigma_{kk}\,\delta_{ij} \tag{5.3-3}$$

The effective stress $\bar{\sigma}$ and plastic strain $\bar{\varepsilon}^p$ are given by Equations (2.6-18) and (2.6-30), respectively; that is,

$$\begin{aligned} \bar{\sigma} &= (3J_2)^{\frac{1}{2}} = \left(\tfrac{3}{2}s_{ij}s_{ij}\right)^{\frac{1}{2}} \\ \bar{\varepsilon}^p &= \left(\tfrac{2}{3}\varepsilon^p_{ij}\varepsilon^p_{ij}\right)^{\frac{1}{2}} \end{aligned} \tag{5.3-4}$$

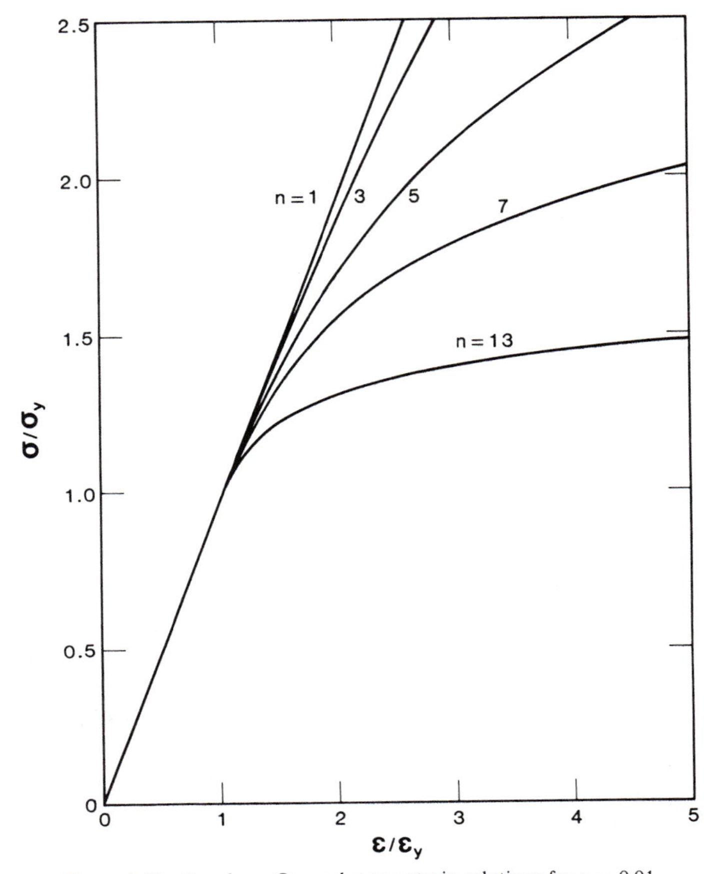

Figure 5.10 Ramberg-Osgood stress-strain relations for $\alpha = 0.01$.

where J_2 is the second deviatoric invariant and ε_{ij}^p denotes the plastic strain components. In accordance with Equation (5.3-1) the effective stress and plastic strain are assumed to be related by

$$\bar{\varepsilon}^p = \alpha\varepsilon_y\left(\frac{\bar{\sigma}}{\sigma_y}\right)^n \tag{5.3-5}$$

Consequently, the generalized stress-strain relation that reduces to the Ramberg-Osgood relation for a uniaxial stress state is

$$\varepsilon_{ij} = \frac{1+\nu}{E}s_{ij} + \frac{1-2\nu}{3E}\sigma_{kk}\delta_{ij} + \tfrac{3}{2}\alpha\varepsilon_y\left(\frac{\bar{\sigma}}{\sigma_y}\right)^{n-1}\frac{s_{ij}}{\sigma_y} \tag{5.3-6}$$

Since a singularity at the crack tip is expected, the plastic strain will be much greater than the elastic strain near the tip. Therefore, in a small region D (see Figure 5.11) encompassing the crack tip, the elastic strain will be negligible compared to the plastic strain. Within this region Equation (5.3-6) can be

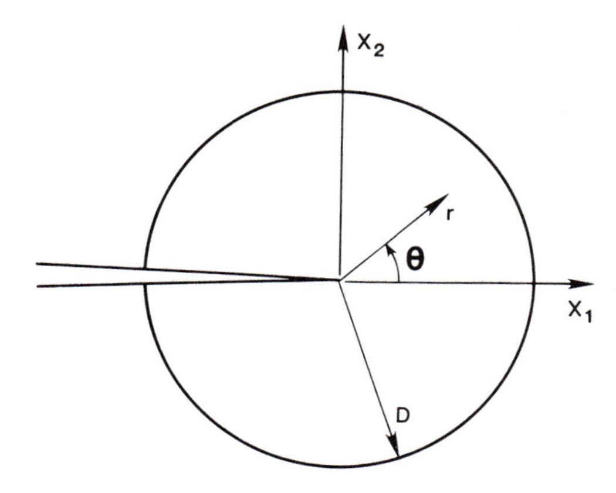

Figure 5.11 Crack-tip region.

approximated by

$$\varepsilon_{ij} = \tfrac{3}{2}\alpha\varepsilon_y\left(\frac{\bar{\sigma}}{\sigma_y}\right)^{n-1}\frac{s_{ij}}{\sigma_y} \tag{5.3-7}$$

Moreover, it is assumed that the only singularity contained within this region is associated with the crack tip. Hence, for a circular path of radius $r < D$ enclosing the crack tip one can write

$$J = r\int_{-\pi}^{\pi}\left(W[\varepsilon(r,\theta)]\cos\theta - T_i(r,\theta)\frac{\partial u_i(r,\theta)}{\partial x_1}\right)d\theta \tag{5.3-8}$$

Since the deformation theory of plasticity can be viewed as a nonlinear elasticity theory, then the J-integral of Equation (5.3-8) is also path-independent. Consequently, the integrand of Equation (5.3-8) must exhibit, at least in its angular average, an inverse r singularity to ensure this path independence. This was observed to be the case for linear elastic fracture and for elastic-plastic, antiplane strain problems. Since the terms of the integrand are essentially products of stress and strain-like components, then

$$\sigma_{ij}\varepsilon_{ij} \rightarrow \frac{f(\theta)}{r} \quad \text{as} \quad r \rightarrow 0 \tag{5.3-9}$$

In fact Hutchinson (5.31) has shown this to be the case for power law hardening materials if the stress and strain components can be written as separable forms in r and θ.

For power law hardening materials fulfilling Equation (5.3-7), Equation (5.3-9) implies that

$$\begin{aligned}
\sigma_{ij}(r,\theta) &= Kr^{-1/(n+1)}\tilde{\sigma}_{ij}(\theta)\\
\bar{\sigma}(r,\theta) &= Kr^{-1/(n+1)}\tilde{\sigma}(\theta)\\
\varepsilon_{ij}(r,\theta) &= \alpha\varepsilon_y K^n r^{-n/(n+1)}\tilde{\varepsilon}_{ij}(\theta)\\
u_i(r,\theta) &= \alpha\varepsilon_y K^n r^{1/(n+1)}\tilde{u}_i(\theta)
\end{aligned} \tag{5.3-10}$$

where K is a constant not to be confused with the elastic stress intensity factor. It is clear from Equation (5.3-10) that for other than linear elastic materials ($n = 1$) the orders of the stress and strain singularities are different. In the limit as $n \to \infty$, the stress field is nonsingular, whereas the strain field has a r^{-1} singularity for the stationary crack tip.

5.3.1 Mode I Fields

While Equation (5.3-10) provides the form of the dominant singularity at the crack tip, the θ-dependent functions must be determined in order to establish the structure of the crack-tip stress and strain fields. Toward this end introduce the Airy stress function Ψ, which identically satisfies the equations of equilibrium and which assumes the asymptotic form

$$\Psi = Kr^{(2n+1)/(n+1)}\tilde{\Psi}(\theta) \tag{5.3-11}$$

The near tip stresses in terms of this function are

$$\begin{aligned}
\sigma_{\theta\theta} &= \frac{\partial^2 \Psi}{\partial r^2} = Ks(s-1)r^{-1/(n+1)}\tilde{\Psi}(\theta) \\
\sigma_{r\theta} &= -\frac{\partial}{\partial r}\left(\frac{1}{r}\frac{\partial \Psi}{\partial \theta}\right) = K(1-s)r^{-1/(n+1)}\tilde{\Psi}'(\theta) \\
\sigma_{rr} &= \frac{1}{r}\frac{\partial \Psi}{\partial r} + \frac{1}{r^2}\frac{\partial^2 \Psi}{\partial \theta^2} \\
&= Kr^{-1/(n+1)}[s\tilde{\Psi}(\theta) + \tilde{\Psi}''(\theta)]
\end{aligned} \tag{5.3-12}$$

where the prime denotes differentiation with respect to θ and $s = (2n+1)/(n+1)$. The Airy stress function must satisfy the compatibility equation

$$\begin{aligned}
&\frac{1}{r}\frac{\partial^2}{\partial r^2}(r\varepsilon_{\theta\theta}) + \frac{1}{r^2}\frac{\partial^2 \varepsilon_{rr}}{\partial \theta^2} - \frac{1}{r}\frac{\partial \varepsilon_{rr}}{\partial r} \\
&\quad - \frac{2}{r^2}\frac{\partial}{\partial r}\left(r\frac{\partial \varepsilon_{r\theta}}{\partial \theta}\right) = 0
\end{aligned} \tag{5.3-13}$$

For plane strain where $\sigma_{33} = (\sigma_{rr} + \sigma_{\theta\theta})/2$, the crack-tip effective stress can be written as

$$\begin{aligned}
\bar{\sigma} &= [\tfrac{3}{4}(\sigma_{rr} - \sigma_{\theta\theta})^2 + 3\sigma_{r\theta}^2]^{\frac{1}{2}} \\
&= Kr^{-1(n+1)}\{\tfrac{3}{4}[\tilde{\Psi}'' - s^2\tilde{\Psi}]^2 + 3[(1-s)\tilde{\Psi}']^2\}^{\frac{1}{2}}
\end{aligned} \tag{5.3-14}$$

Equations (5.3-12) and (5.3-14) are substituted into Equation (5.3-7) to obtain the strain components, which, when introduced into Equation (5.3-13), yield

$$\begin{aligned}
&\left[\frac{d^2}{d\theta^2} + \frac{n}{n+1}\frac{n+2}{n+1}\right]\left[\tilde{\sigma}^{n-1}\left(\tilde{\Psi}'' + \frac{2n+1}{(n+1)^2}\tilde{\Psi}\right)\right] \\
&\quad + \frac{4n}{(n+1)^2}(\tilde{\sigma}^{n-1}\tilde{\Psi}')' = 0
\end{aligned} \tag{5.3-15}$$

A similar treatment for plane stress leads to

$$\begin{aligned}\left[\frac{d^2}{d\theta^2} + \frac{n}{n+1}\right]&\left[\tilde{\sigma}^{n-1}\left(2\tilde{\Psi}'' + \frac{n+2}{n+1}\frac{2n+1}{n+1}\tilde{\Psi}\right)\right] \\ &+ \frac{n}{(n+1)^2}\tilde{\sigma}^{n-1}\left(\tilde{\Psi}'' + \frac{n-1}{n+1}\frac{2n+1}{n+1}\tilde{\Psi}\right) \\ &+ \frac{6n}{(n+1)^2}(\tilde{\sigma}^{n-1}\tilde{\Psi}')' = 0\end{aligned} \tag{5.3-16}$$

where for plane stress

$$\begin{aligned}\tilde{\sigma}(\theta) &= [\tilde{\sigma}_{rr}^2 + \tilde{\sigma}_{\theta\theta}^2 - \tilde{\sigma}_{rr}\tilde{\sigma}_{\theta\theta} + 3\tilde{\sigma}_{r\theta}^2]^{\frac{1}{2}} \\ &= \{[\tilde{\Psi}'' + s\tilde{\Psi}]^2 + [s(s-1)\tilde{\Psi}]^2 \\ &\quad - s(s-1)[\tilde{\Psi}'' + s\tilde{\Psi}]\tilde{\Psi} + 3[(1-s)\tilde{\Psi}']^2\}^{\frac{1}{2}}\end{aligned} \tag{5.3-17}$$

It follows from Equation (5.3-12) that the conditions, $\sigma_{\theta\theta} = \sigma_{r\theta} = 0$ on $\theta = \pm\pi$, for traction-free crack surfaces require

$$\tilde{\Psi}(\pm\pi) = \tilde{\Psi}'(\pm\pi) = 0 \tag{5.3-18}$$

For the opening mode, Ψ must be symmetric with respect to $\theta = 0$ and the boundary conditions on $\theta = -\pi$ can be replaced by

$$\tilde{\Psi}'(0) = \tilde{\Psi}'''(0) = 0 \tag{5.3-19}$$

The homogeneous differential equations, Equations (5.3-15) and (5.3-16), together with the homogeneous conditions, Equations (5.3-18) and (5.3-19), define a two-point boundary value problem. Equivalently, the latter can be viewed as a nonlinear eigenvalue problem where the exponent of r in the solution for Ψ is the eigenvalue. However, the eigenvalue for the dominant singularity has been established through the arguments leading to Equation (5.3-10). Thus, there only remains to determine the eigenfunction $\tilde{\Psi}(\theta)$. Because it has been impossible to integrate Equations (5.3-15) and (5.3-16) in closed form, their integration has been performed numerically by treating the problem as an initial value one. Since the eigenfunction can only be determined to within a multiplicative constant, one can set $\tilde{\Psi}(0) = 1$ without loss in generality. Ultimately the eigenfunction is normalized such that $\tilde{\sigma}_{\max} = \sigma_y$. For an assumed value of $\tilde{\Psi}''(0)$ the numerical integration can be performed—for example, using a Runge-Kutta method. In general, the boundary conditions $\tilde{\Psi}(\pi) = \tilde{\Psi}'(\pi) = 0$ will not be satisfied. The value of $\tilde{\Psi}''(0)$ is adjusted systematically; say, by Newton's method, and the procedure repeated until these conditions are satisfied.

The θ-variations of the stress and strain components for $n = 3$ and 13 appear in Figure 5.12 for the plane strain opening mode. In these plots the stress and strain components have been normalized with respect to σ_y and ε_y, respectively. For purposes of comparison, companion plots of the stress and the slip-line fields for a perfectly plastic material are also presented. Due to the symmetry with respect to $\theta = 0$, only the upper half of the fields are illustrated.

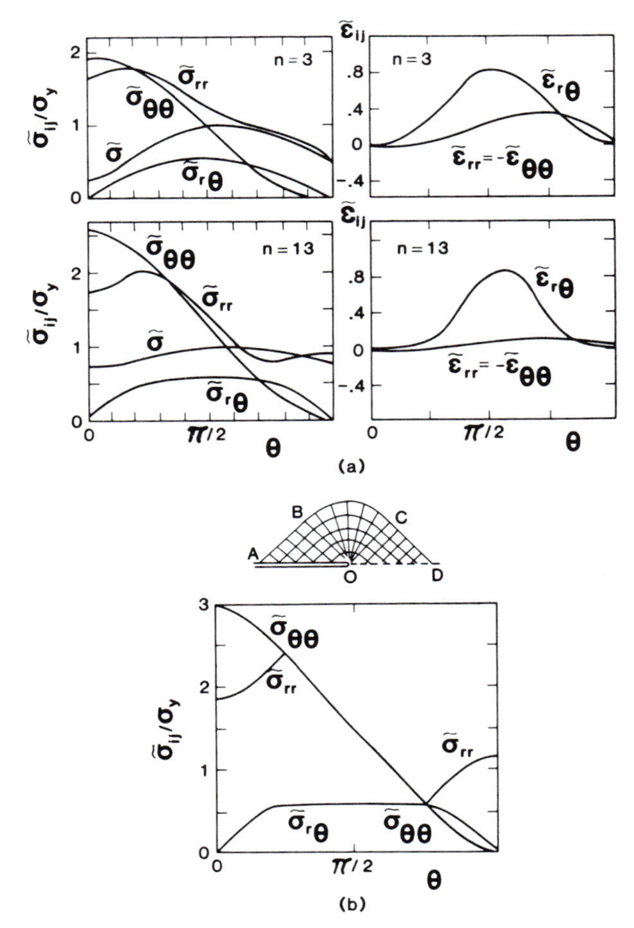

Figure 5.12 Plane strain structure of Mode I crack-tip fields for (a) a power law hardening material with $n = 3$ and 13, and (b) a perfectly plastic material (5.33).

For the perfectly plastic material the slip lines in the region AOB are straight and are indicative of a uniform stress state. Since $\sigma_{12} = \sigma_{22} = 0$ on the crack surfaces, then $\sigma_{12} = \sigma_{22} = 0$ and $\sigma_{11} = 2\sigma_y/\sqrt{3}$ throughout this region. Or, equivalently, for $\pi > \theta > 3\pi/4$

$$\begin{aligned}
\sigma_{rr} &= \sigma_y(1 + \cos 2\theta)/\sqrt{3} \\
\sigma_{\theta\theta} &= \sigma_y(1 - \cos 2\theta)/\sqrt{3} \qquad (5.3\text{-}20) \\
\sigma_{r\theta} &= \sigma_y(\sin 2\theta)/\sqrt{3}
\end{aligned}$$

On the β-line extending from the crack surface around the crack tip to the x_1-axis [see Equation (2.6-41)]

$$\frac{\sigma_{11} + \sigma_{22}}{2} + \frac{2\sigma_y \phi}{\sqrt{3}} = \frac{\sigma_y(1 + 3\pi/2)}{\sqrt{3}}$$

where ϕ is the angle that the α-line makes with the x_1-axis. On the plane of symmetry $x_1 > 0$, where $\sigma_{12} = 0$ and $\phi = \pi/4$, the yield condition [Equation (2.6-36)] implies

$$\sigma_{22} - \sigma_{11} = 2\sigma_y/\sqrt{3}$$

Consequently, for $0 < \theta < \pi/4$ these equations yield $\sigma_{11} = \pi\sigma_y/\sqrt{3}$, $\sigma_{22} = (2 + \pi)\sigma_y/\sqrt{3}$ and $\sigma_{12} = 0$ or

$$\begin{aligned} \sigma_{rr} &= \sigma_y(1 + \pi - \cos 2\theta)/\sqrt{3} \\ \sigma_{\theta\theta} &= \sigma_y(1 + \pi + \cos 2\theta)/\sqrt{3} \\ \sigma_{r\theta} &= \sigma_y(\sin 2\theta)/\sqrt{3} \end{aligned} \tag{5.3-21}$$

In the fan BOC the α-lines are the radial lines and $\phi = \theta$. Making use of Equations (2.6-38) and (2.6-41), one concludes that

$$\begin{aligned} \sigma_{rr} &= \sigma_{\theta\theta} = \sigma_y(1 + 3\pi/2 - 2\theta)/\sqrt{3} \\ \sigma_{r\theta} &= \sigma_y/\sqrt{3} \end{aligned} \tag{5.3-22}$$

for $\pi/4 < \theta < 3\pi/4$.

While numerical difficulties preclude treating the limiting case $n = \infty$, the near coincidence of the stress fields for perfect plasticity and for $n = 13$ in the hardening material suggests that the limiting stress field of the hardening solution is the perfectly plastic solution given by Equations (5.3-20)–(5.3-22). Because the governing equations for the perfectly plastic material are hyperbolic, it is not possible to establish the crack-tip strain field from a local analysis. Nevertheless, the plastic strains in the fan will have a $1/r$ singularity. Since $\sigma_{rr} = \sigma_{\theta\theta}$ in the fan BOC, then $\varepsilon_{rr} = \varepsilon_{\theta\theta} = 0$ and $\varepsilon_{r\theta}$ is the only nonzero strain component. The hardening solutions for large values of n have a similar characteristic.

The stress and strain fields for the plane stress opening mode determined by Hutchinson (5.33) are displayed in Figure 5.13. These distributions can be compared with the plane strain distributions. While the normal stresses are tensile in the latter, the former includes compressive stresses. Furthermore, a rather abrupt change in σ_{rr} exists near $\theta = 5\pi/6$ for $n = 13$. The origin of this change can be understood from examining the perfectly plastic solution where continuity of the tractions along a radial slip line implies that $\sigma_{\theta\theta}$ and $\sigma_{r\theta}$, but not necessarily σ_{rr}, are continuous. If the yield condition, represented by

$$\sigma_{rr}^2 + \sigma_{\theta\theta}^2 - \sigma_{rr}\sigma_{\theta\theta} + 3\sigma_{r\theta}^2 = \sigma_y^2 \tag{5.3-23}$$

is satisfied on either side of the radial line, then a discontinuity of the form

$$\sigma_{rr}^+ - \sigma_{rr}^- = (4\sigma_y^2 - 3\sigma_{\theta\theta}^2 - 12\sigma_{r\theta}^2)^{\frac{1}{2}} \tag{5.3-24}$$

is permissible. A discontinuity of this kind is characteristic of a perfectly plastic model. While the inclusion of elastic strains or strain hardening would preclude this discontinuity, rather large stress gradients would be anticipated.

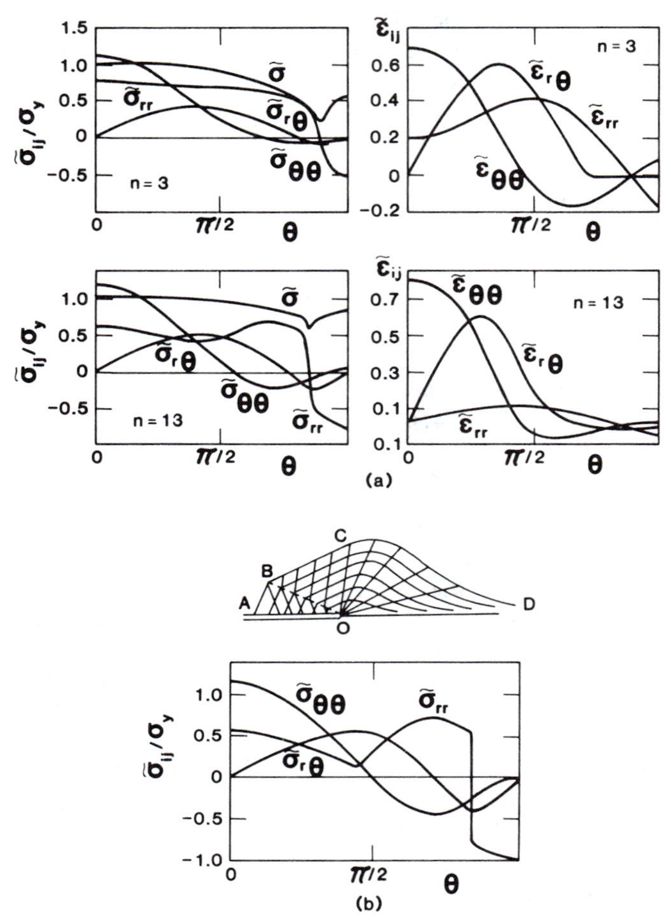

Figure 5.13 Plane stress structure of Mode I crack-tip fields for (a) a power law hardening material with $n = 3$ and 13, and (b) a perfectly plastic material (5.33).

In the constant stress region *AOB* of Figure 5.13, $\sigma_{11} = -\sigma_y$ and $\sigma_{22} = \sigma_{12} = 0$, or

$$\begin{aligned}\sigma_{rr} &= -\tfrac{1}{2}\sigma_y(1 + \cos 2\theta)\\ \sigma_{\theta\theta} &= -\tfrac{1}{2}\sigma_y(1 - \cos 2\theta)\\ \sigma_{r\theta} &= \tfrac{1}{2}\sigma_y \sin 2\theta\end{aligned} \tag{5.3-25}$$

The compressive stress behind the crack tip due to the discontinuity along *OB* is in marked contrast to the tensile stress for plane strain. This compressive stress can result in buckling of a cracked member. The stress field is also uniform in the region *BOC* except a jump, given by Equation (5.3-24), in σ_{rr}

occurs along the radial line OB defined by θ_{OB}. In this region

$$\begin{aligned}
\sigma_{rr} &= \sigma_y[\tfrac{1}{4}(-1 + 3\cos 2\theta_{OB}) + \tfrac{1}{4}(1 + \cos 2\theta_{OB})\cos 2(\theta - \theta_{OB}) \\
&\qquad + \tfrac{1}{2}\sin 2\theta_{OB}\sin 2(\theta - \theta_{OB})] \\
\sigma_{\theta\theta} &= -\sigma_{rr} + \sigma_y(-1 + 3\cos 2\theta_{OB})/2 \\
\sigma_{r\theta} &= \sigma_y[\tfrac{1}{2}\sin 2\theta_{OB}\cos 2(\theta - \theta_{OB}) - \tfrac{1}{4}(1 + \cos 2\theta_{OB})\sin 2(\theta - \theta_{OB})]
\end{aligned} \tag{5.3-26}$$

where θ_{OB} remains to be determined. Hutchinson (5.33) symmetrically centered the fan COD in order to satisfy $\sigma_{r\theta} = 0$ on $\theta = 0$. The stresses in this fan are

$$\begin{aligned}
\sigma_{\theta\theta} &= 2\sigma_{rr} = 2\sigma_y(\cos\theta)/\sqrt{3} \\
\sigma_{r\theta} &= \sigma_y(\sin\theta)/\sqrt{3}
\end{aligned} \tag{5.3-27}$$

Continuity of $\sigma_{\theta\theta}$ and $\sigma_{r\theta}$ on the radial line $OC(\theta = \theta_{OC})$ demands of Equations (5.3-26) and (5.3-27) that

$$\begin{aligned}
&\tfrac{1}{4}(-1 + 3\cos 2\theta_{OB}) - \tfrac{1}{4}(1 + \cos 2\theta_{OB})\cos 2(\theta_{OC} - \theta_{OB}) \\
&\qquad - \tfrac{1}{2}\sin 2\theta_{OB}\sin 2(\theta_{OC} - \theta_{OB}) = \frac{2}{\sqrt{3}}\cos 2\theta_{OC} \\
&-\tfrac{1}{4}(1 + \cos 2\theta_{OB})\sin 2(\theta_{OC} - \theta_{OB}) + \tfrac{1}{2}\sin 2\theta_{OB}\cos 2(\theta_{OC} - \theta_{OB}) \\
&\qquad = \frac{1}{\sqrt{3}}\sin 2\theta_{OC}
\end{aligned} \tag{5.3-28}$$

The simultaneous solution of these equations yields $\theta_{OC} = 79.7^\circ$ and $\theta_{OB} = 151.4^\circ$.

Again close similarity is observed between the stress distributions of the hardening solution for large values of n and those of the perfectly plastic solution. In the fan COD, $\sigma_{\theta\theta} = 2\sigma_{rr}$ and, hence, $s_{rr} = 0$. Consequently, for the perfectly plastic solution $\varepsilon_{rr} = 0$ for all θ. This characteristic is also reflected by the low strain hardening solution $n = 13$. Hutchinson (5.33) has also presented hardening solutions and a perfectly plastic slip-line solution for Mode II. Again many of the details of the perfectly plastic solution are reflected in the stress fields for low hardening materials.

Having established the stress and strain distributions, one can evaluate the J-integral. The introduction of Equation (5.3-7) into Equation (2.3-21) permits writing

$$W = \alpha\varepsilon_y\sigma_y\frac{n}{n+1}\left(\frac{\bar{\sigma}}{\sigma_y}\right)^{n+1} \tag{5.3-29}$$

in which Equation (5.3-4) has been used. The combination of Equations (5.3-8), (5.3-10), and (5.3-19) leads to

$$J = \alpha\varepsilon_y\sigma_y K^{n+1} I_n \tag{5.3-30}$$

where

$$I_n = \int_{-\pi}^{\pi} \left\{ \frac{n}{n+1} \left(\frac{\tilde{\sigma}}{\sigma_y} \right)^{n+1} \cos\theta - \left[\frac{\tilde{\sigma}_{rr}}{\sigma_y} (\tilde{u}_\theta - \tilde{u}_r') - \frac{\tilde{\sigma}_{r\theta}}{\sigma_y} (\tilde{u}_r + \tilde{u}_\theta') \right] \sin\theta \right.$$
$$\left. - \frac{1}{n+1} \left[\frac{\tilde{\sigma}_{rr}}{\sigma_y} \tilde{u}_r + \frac{\tilde{\sigma}_{r\theta}}{\sigma_y} \tilde{u}_\theta \right] \cos\theta \right\} d\theta \tag{5.3-31}$$

The quantities $\tilde{u}_r$ and $\tilde{u}_\theta'$ are obtainable from the strain displacement relations for ε_{rr} and $\varepsilon_{r\theta}$ as

$$\tilde{u}_r = (n+1) \left(\frac{\tilde{\sigma}}{\sigma_y} \right)^{n-1} [\tilde{\Psi}'' + s(3-s)\tilde{\Psi}/2]$$
$$\tilde{u}_\theta' = -\frac{1}{2} \left(\frac{\tilde{\sigma}}{\sigma_y} \right)^{n-1} [\tilde{\Psi}'' - s(2s-3)\tilde{\Psi}] - \tilde{u}_r \tag{5.3-32}$$

The variations of I_n with n for plane stress and plane strain are depicted in Figure 5.14.

When Equation (5.3-30) is used to replace K in favor of J in Equation (5.3-10), the HRR stress and strain singular fields and the displacement fields are

$$\sigma_{\alpha\beta} = \left(\frac{J}{\alpha\varepsilon_y \sigma_y I_n r} \right)^{1/(n+1)} \tilde{\sigma}_{\alpha\beta}(\theta)$$
$$\sigma_{\alpha 3} = 0$$
$$\varepsilon_{\alpha\beta} = \alpha\varepsilon_y \left(\frac{J}{\alpha\varepsilon_y \sigma_y I_n r} \right)^{n/(n+1)} \tilde{\varepsilon}_{\alpha\beta}(\theta)$$
$$\varepsilon_{\alpha 3} = 0 \tag{5.3-33}$$
$$u_\alpha = \alpha\varepsilon_y \left(\frac{J}{\alpha\varepsilon_y \sigma_y I_n} \right)^{n/(n+1)} r^{1/(n+1)} \tilde{u}_\alpha(\theta)$$
$$\sigma_{33} = \tfrac{1}{2}(\sigma_{11} + \sigma_{22}), \quad \varepsilon_{33} = 0 \quad \text{plane strain}$$
$$\sigma_{33} = 0, \quad \varepsilon_{33} = -(\varepsilon_{11} + \varepsilon_{22}) \quad \text{plane stress}$$

The parameter J is a function of the applied load, crack length, and the geometry of the body. Since the stress field increases monotonically with J everywhere, then the use of deformation plasticity theory is justifiable.

5.3.2 Fracture Criterion

It is clear from Equation (5.3-33) that the intensity of the crack tip fields in the J-dominant region D depends only upon the parameter J. Interpreting J not as an energy release rate but as a measure of the intensity of the HRR fields forms the basis of plastic fracture mechanics. Let R in Figure 5.15 denote the characteristic size of the fracture process zone where nonproportional loading, large strains, and other phenomena associated with fracture occur,

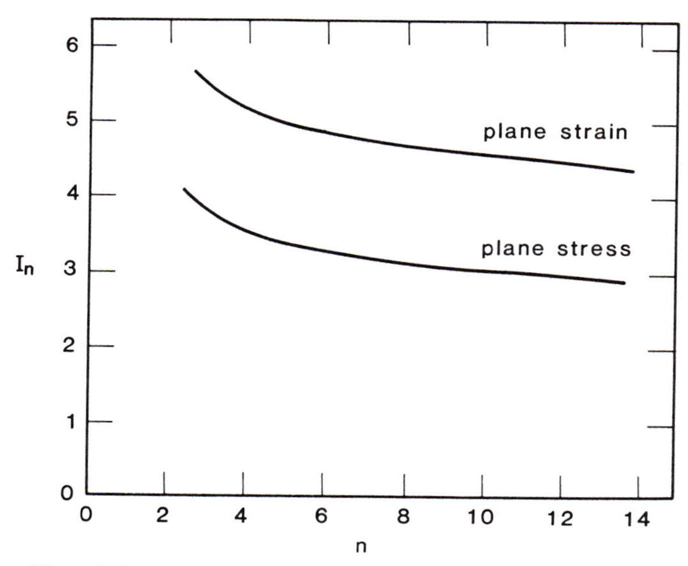

Figure 5.14 Variation of I_n with n for plane strain and plane stress.

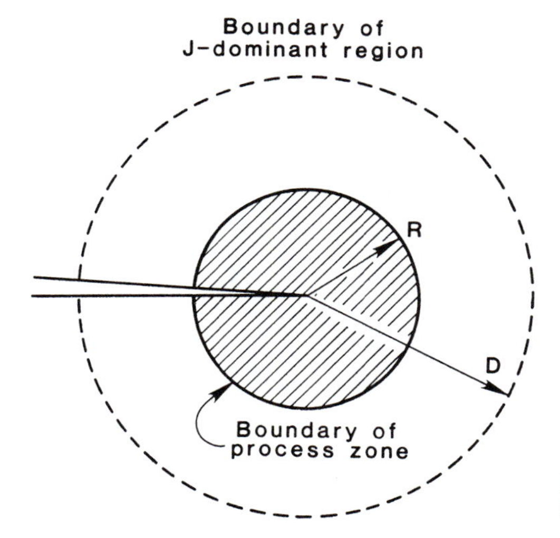

Figure 5.15 Basis for J-integral plastic fracture mechanics.

but are not properly accounted for in a small strain deformation theory of plasticity. If R is small compared to D, it can be argued that any event that occurs within this process zone must be controlled by the deformation in the surrounding "J-dominant" region. Therefore, where J-dominance exists, the initiation and growth of a crack can be expected to be governed by a critical value of J. Thus, the plastic fracture criterion can be expressed as

$$J(\sigma, a) = J_c \tag{5.3-34}$$

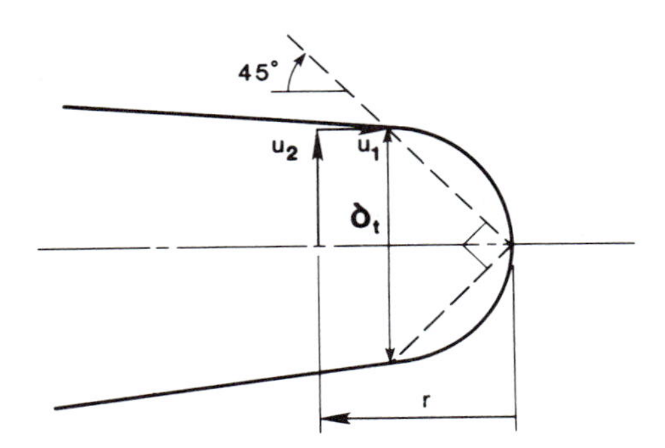

Figure 5.16 Definition of crack-tip opening displacement.

where J_c is a material property. Apparently, Begley and Landes (5.34) were the first to recognize the potential of J in plastic fracture mechanics.

From Equation (5.3-33) the displacement components of the upper crack face can be written as

$$\begin{Bmatrix} u_1 \\ u_2 \end{Bmatrix} = \alpha\varepsilon_y \left(\frac{J}{\alpha\varepsilon_y \sigma_y I_n} \right)^{n/(n+1)} r^{1/(n+1)} \begin{Bmatrix} \tilde{u}_1(\pi) \\ \tilde{u}_2(\pi) \end{Bmatrix} \tag{5.3-35}$$

The crack opening displacement is $\delta = 2u_2(r, \pi)$. Except for the limiting case of perfect plasticity ($n \to \infty$), these displacements tend to zero with r. Due to the latter property the definition of an effective crack-tip opening displacement is somewhat arbitrary. Tracy (5.35) used the definition, as illustrated in Figure 5.16, that the crack-tip opening displacement, δ_t, is the crack opening at the intercept of the two symmetric 45° lines from the deformed crack tip and the crack profile. Thus,

$$r - u_1(\pi) = \delta_t/2 = u_2(\pi) \tag{5.3-36}$$

The introduction of Equation (5.3-35) into Equation (5.3-36) leads to

$$r = (\alpha\varepsilon_y)^{1/n}[\tilde{u}_1(\pi) + \tilde{u}_2(\pi)]^{(n+1)/n} \frac{J}{\sigma_y I_n} \tag{5.3-37}$$

for the point of intersection. The substitution of Equation (5.3-37) into Equation (5.3-35) yields

$$\delta_t = d_n J/\sigma_y \tag{5.3-38}$$

where

$$d_n = 2(\alpha\varepsilon_y)^{1/n}[\tilde{u}_1(\pi) + \tilde{u}_2(\pi)]^{1/n}\tilde{u}_2(\pi)/I_n \tag{5.3-39}$$

The dependence of d_n upon n and $\varepsilon_y = \sigma_y/E$ in Figure 5.17 for plane stress and plane strain conditions was determined by Shih (5.36). For a perfectly plastic behavior ($n \to \infty$) under plane stress, $d_n = 1$ and Equation (5.3-38) coincides with the Dugdale model [cf. Equation (3.3-29)]. In general d_n depends strongly upon n and mildly upon σ_y/E.

Clearly, Equation (5.3-38) establishes an equivalence between J and δ_t. Therefore, any fracture criterion based upon a critical value of δ_t is equivalent

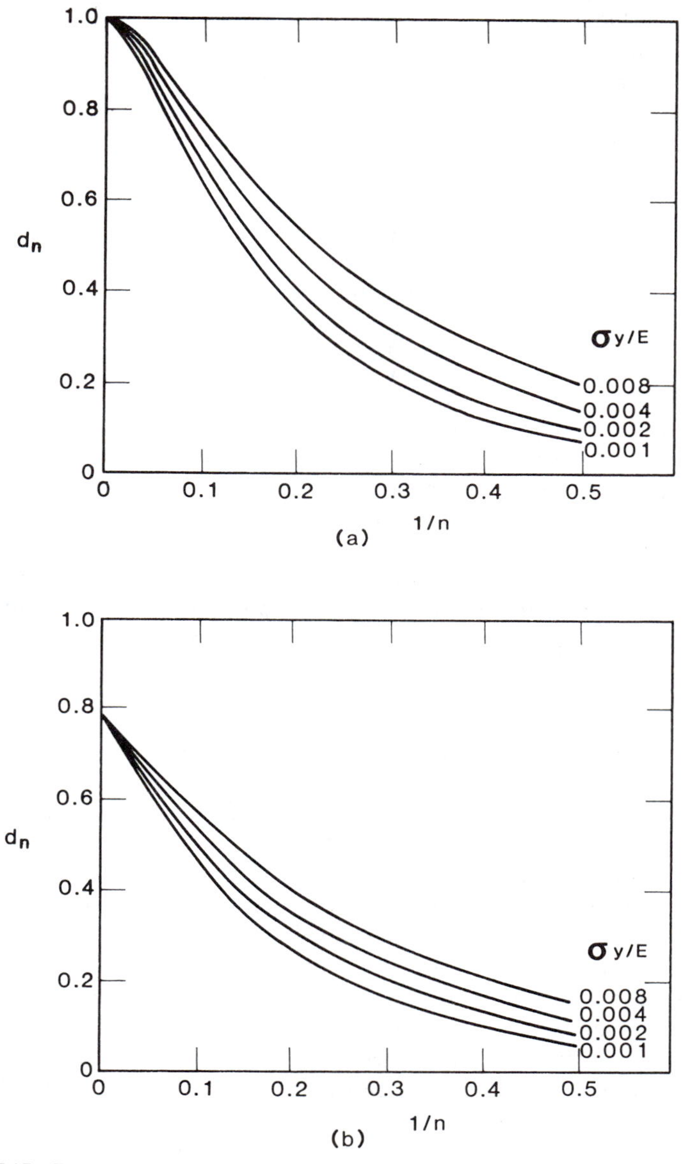

Figure 5.17 Dependence of d_n on n and σ_y/E for (a) plane stress and (b) plane strain (5.36).

to one based upon a critical value of J and vice versa. The J-integral has become the preferred fracture characterizing parameter because it is unambiguously defined, easier to compute and simpler to measure than δ_t. In order to ensure the validity of Equation (5.3-38), the HRR field must dominate the crack-tip deformation over a region at least as large as δ_t. The solution to the antiplane problem indicates the zone of dominance of the HRR singularity decreases with increasing n and vanishes in the limit of nonhardening materials where the plastic zone no longer engulfs the crack tip. This strongly suggests that there is no unique relationship between J and δ_t for nonhardening materials. Consequently, a one-parameter characterization of the intensity of the crack-tip fields, be it J or δ_t, assumes the existence of strain hardening.

For the same intensity of loading as measured by J, Equation (5.3-33) can be used to compare the stress, $\sigma_{22}(r, 0)$, ahead of the crack tip for conditions of plane strain and plane stress. In the limit as $n \to \infty$

$$\frac{\sigma_{22}(r, 0)_{\text{pl. strain}}}{\sigma_{22}(r, 0)_{\text{pl. stress}}} = 1 + \frac{\pi}{2} \tag{5.3-40}$$

which reflects the significantly higher tensile stress associated with plane strain. This is in contrast with the linear elastic solution for which this ratio is unity. The larger plane strain stresses ahead of the crack tip are more conducive to the process of formation, growth, and coalescence of voids that constitute plastic fracture. For $\theta = 0$ the ratio of the plane strain to plane stress triaxiality as measured by $\sigma_{kk}/3$ has the limiting value, $1 + \pi$, as $n \to \infty$. The increased triaxiality of plane strain is responsible for the development of elevated stresses ahead of the crack tip that are in excess of the yield stress. It is for these reasons that plane strain loading represents the more severe loading.

5.4 An Engineering Approach to Plastic Fracture

As reflected in the fracture criterion, Equation (5.3-34), the solution to a plastic fracture problem involves two parts: the determination of the driving force and the measurement of the fracture resistance. The computation of J usually involves a sophisticated analysis employing advanced finite element methods. Such capabilities are not universally available. Perhaps, to a lesser degree LEFM initially experienced a similar obstacle. However, the tabulation of LEFM solutions aided immensely in the application of LEFM to where its application is more or less commonplace now. In this section an engineering approach to plastic fracture is presented. As in LEFM, successful general implementation of this approach depends upon cataloging plastic solutions for a wide range of cracked configurations and loadings.

Within the confines of LEFM the fracture parameters, the J-integral, the crack opening displacement δ and the load point displacement Δ_c due to the presence of the crack, can be expressed as

$$\begin{aligned} J_e &= f_1(a/W)P^2/E' \\ \delta_e &= f_2(a/W)P/E' \\ \Delta_{ce} &= f_3(a/W)P/E' \end{aligned} \tag{5.4-1}$$

In Equation (5.4-1) P is a generalized load per unit thickness such that P acting through the generalized load-point displacement Δ produces work per unit thickness. Here and in the following the subscripts e and p will be used to designate elastic and plastic components, respectively. The load-point displacement can be decomposed as

$$\Delta = \Delta_{nc} + \Delta_c \tag{5.4-2}$$

where Δ_{nc} is the load-point displacement in the absence of a crack and is ordinarily obtainable from an elasticity solution. The functions f_1, f_2, and f_3 depend upon the crack length to width ratio, a/W, and possibly other geometric parameters. These functions or their equivalents can be found in linear elastic fracture handbooks; for example, see (5.37)–(5.39).

5.4.1 *Fully Plastic Solution*

Consider a fully plastic cracked body in which the elastic strain components are negligible compared to their plastic counterparts. Further assume that the plastic deformation can be described by J_2-deformation plasticity theory with power law hardening. The small strain constitutive relation can be written as

$$\frac{\varepsilon_{ij}}{\varepsilon_y} = \frac{3}{2}\alpha\left(\frac{\bar{\sigma}}{\sigma_y}\right)^{n-1}\frac{s_{ij}}{\sigma_y} \tag{5.4-3}$$

For such a material behavior Il'yushin (5.40) showed that a solution to a boundary value problem involving a single monotonically increasing load or displacement parameter assumes a relatively simple form. For example, if the tractions $T_i = PT'_i$, P denoting a loading parameter, are prescribed on the boundary S, then the solutions can be written as

$$\begin{aligned} \sigma_{ij} &= P\sigma'_{ij}(x_i, n) \\ \varepsilon_{ij} &= \alpha\varepsilon_y\left(\frac{P}{\sigma_y}\right)^n \varepsilon'_{ij}(x_i, n) \\ u_i &= \alpha\varepsilon_y\left(\frac{P}{\sigma_y}\right)^n u'_i(x_i, n) \end{aligned} \tag{5.4-4}$$

where the quantities σ'_{ij}, ε'_{ij}, u'_i are functions of x_i and n and are independent of P. This result follows trivially from the homogeneous nature of the equations of equilibrium and compatibility and the constitutive relation, Equation (5.4-3). According to Equation (5.4-4), if the fields are found for one value of the load parameter—for example, $P = 1$—then they can be determined immediately for any other value of P. Moreover, since the stress and strain fields increase proportionally at every point, the fully plastic solution based upon deformation plasticity theory coincides with the solution for incremental or flow theory.

Since the integrand of the J-integral involves products of stresses and displacement gradients, then the fully plastic J will be proportional to P^{n+1}

The fully plastic fracture parameters can be expressed as

$$\begin{aligned} J_p &= \alpha\varepsilon_y\sigma_y b g_1(a/W)h_1(a/W, n)(P/P_0)^{n+1} \\ \delta_p &= \alpha\varepsilon_y a g_2(a/W)h_2(a/W, n)(P/P_0)^n \\ \Delta_{cp} &= \alpha\varepsilon_y a g_3(a/W)h_3(a/W, n)(P/P_0)^n \\ \delta_{tp} &= \alpha\varepsilon_y b g_4(a/W)h_4(a/W, n)(P/P_0)^{n+1} \end{aligned} \tag{5.4-5}$$

where P_0 is the limit load based upon σ_y and $b = W - a$ is the remaining uncracked ligament (see Figure 5.18). The dimensionless functions, $h_1 - h_4$, depend upon a/W, n, and possibly other geometric parameters, but are independent of P. The known dimensionless functions, $g_1 - g_4$, are selected for convenience of tabulation. It follows from Equation (5.3-38) that

$$h_4 = d_n g_1 h_1 / g_4 \tag{5.4-6}$$

where d_n is given in Figure 5.17. The functional forms of Equation (5.4-5) can be compared to their linear elastic ($n = 1$) counterparts in Equation (5.4-1).

The functions, $h_1 - h_3$, can be computed using the finite element method. Computations of this kind were first performed by Goldman and Hutchinson (5.41). For the condition of plane stress, conventional finite element techniques suffice to establish the fully plastic solutions (5.42). The plane strain incompressible deformation introduces constraints on the displacements, and special techniques discussed in references (5.41), (5.43)–(5.45) were developed to treat this case. A compendium of plane stress and plane strain functions, $h_1 - h_3$, for a variety of fracture specimens and flawed cylinders under Mode I

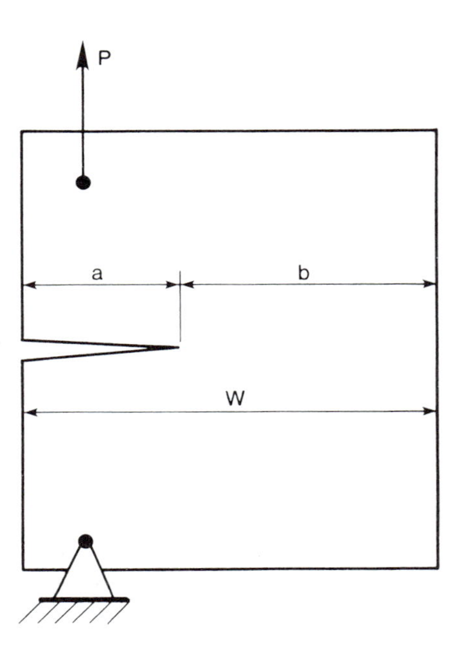

Figure 5.18 A flawed structural component.

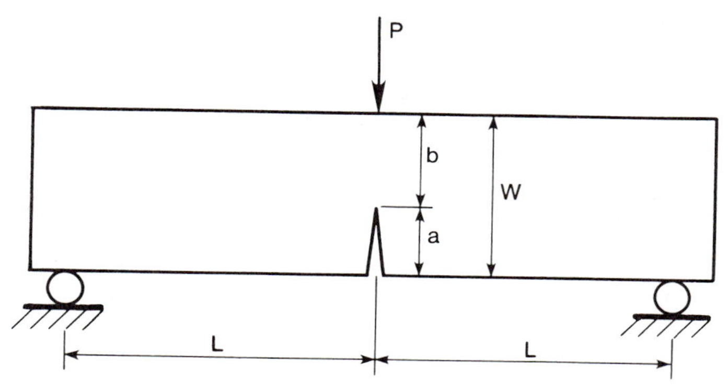

Figure 5.19 A three-point bend specimen.

loading can be found in reference (5.46). The differences between these numerical solutions and known solutions are typically less than 5 percent.

Tabulations of the functions for a three point bend specimen (Figure 5.19) in plane stress and plane strain are presented, respectively, in Tables 5.1 and 5.2 for $L/W = 2$.* For this case $g_i(a/W) = 1$. The plane strain limit load (5.10) and the plane stress limit load (5.11) are, respectively,

$$P_0 = 0.728\sigma_y b^2/L \quad \text{and} \quad P_0 = 0.536\sigma_y b^2/L \tag{5.4-7}$$

The difference in the coefficients is due to the difference between plane strain and plane stress constraints. In this example δ is the crack mouth opening displacement at the edge of the specimen. For the same load per unit thickness on the bend specimen, Tables 5.1 and 5.2 reveal that the plane stress J_p is greater than the plane strain J_p. However, for the same ratio of P/P_0, the plane stress J_p is generally the smaller.

5.4.2 Estimation Technique

Strictly speaking the fully plastic solution is only applicable when the cracked configuration has completely yielded and the elastic strains are negligible throughout the body. General yielding will occur when P is large compared to P_0. At the other extreme, small-scale yielding will occur for P small compared to P_0 and LEFM is applicable. For the Ramberg-Osgood stress-strain relation, Equation (5.3-1), Shih (5.47) and Shih and Hutchinson (5.48) proposed interpolating over the entire range of yielding by superposing the linear elastic and the fully plastic solutions according to

$$\begin{aligned} J &= J_e(a_e) + J_p(a, n) \\ \delta &= \delta_e(a_e) + \delta_p(a, n) \\ \Delta_c &= \Delta_{ce}(a_e) + \Delta_{cp}(a, n) \end{aligned} \tag{5.4-8}$$

* Tables for $h_1 - h_3$ for other specimens and loading can be found in Appendix A.

Table 5.1 h_1, h_2, and h_3 for Three-Point Bend Specimen in Plane Stress (5.46)

		$n = 1$	$n = 2$	$n = 3$	$n = 5$	$n = 7$	$n = 10$	$n = 13$	$n = 16$	$n = 20$
	h_1	0.676	0.600	0.548	0.459	0.383	0.297	0.238	0.192	0.148
$a/W = \frac{1}{8}$	h_2	6.84	6.30	5.66	4.53	3.64	2.72	2.12	1.67	1.26
	h_3	2.95	20.1	14.6	12.2	9.12	6.75	5.20	4.09	3.07
	h_1	0.869	0.731	0.629	0.479	0.370	0.246	0.174	0.117	0.0593
$a/W = \frac{1}{4}$	h_2	5.69	4.50	3.68	2.61	1.95	1.29	0.897	0.603	0.307
	h_3	4.01	8.81	7.19	4.73	3.39	2.20	1.52	1.01	0.508
	h_1	0.963	0.797	0.680	0.527	0.418	0.307	0.232	0.174	0.105
$a/W = \frac{3}{8}$	h_2	5.09	3.73	2.93	2.07	1.58	1.13	0.841	0.626	0.381
	h_3	4.42	5.53	4.48	3.17	2.41	1.73	1.28	0.948	0.575
	h_1	1.02	0.767	0.621	0.453	0.324	0.202	0.128	0.0813	0.0298
$a/W = \frac{1}{2}$	h_2	4.77	3.12	2.32	1.55	1.08	0.655	0.410	0.259	0.0974
	h_3	4.60	4.09	3.09	2.08	1.44	0.874	0.545	0.344	0.129
	h_1	1.05	0.786	0.649	0.494	0.357	0.235	0.173	0.105	0.0471
$a/W = \frac{5}{8}$	h_2	4.55	2.83	2.12	1.46	1.02	0.656	0.472	0.286	0.130
	h_3	4.62	3.43	2.60	1.79	1.26	0.803	0.577	0.349	0.158
	h_1	1.07	0.786	0.643	0.474	0.343	0.230	0.167	0.110	0.0442
$a/W = \frac{3}{4}$	h_2	4.39	2.66	1.97	1.33	0.928	0.601	0.427	0.280	0.114
	h_3	4.39	3.01	2.24	1.51	1.05	0.680	0.483	0.316	0.129
	h_1	1.086	0.928	0.810	0.646	0.538	0.423	0.332	0.242	0.205
$a/W = \frac{7}{8}$	h_2	4.28	2.76	2.16	1.56	1.23	0.922	0.702	0.561	0.428
	h_3	4.07	2.93	2.29	1.65	1.30	0.975	0.742	0.592	0.452

Such a procedure can be shown to be exact for the special case of an infinite strip with a semi-infinite crack under either antiplane strain (5.47) or plane stress loading.

To incorporate small-scale yielding effects the elastic contributions are based upon the effective crack length

$$a_e = a + \phi r_y \tag{5.4-9}$$

where

$$r_y = \frac{1}{\beta\pi}\left(\frac{n-1}{n+1}\right)\left(\frac{K}{\sigma_y}\right)^2 \tag{5.4-10}$$

is the Irwin correction modified for strain hardening. The form of this correction is based upon the antiplane strain hardening solution of Section 5.2 [cf. Equations (5.2-28) and (5.2-33)]. To simplify the calculation of r_y, K is based upon the crack length a. For plane stress and plane strain, respectively, $\beta = 2$ and $\beta = 6$. The coefficient

$$\phi = 1/[1 + (P/P_0)^2] \tag{5.4-11}$$

has been introduced in an attempt to reduce the correction under conditions of contained plasticity while retaining the plasticity correction r_y in the fully plastic regime where J, δ, and Δ_c are dominated by the second terms in

Table 5.2 h_1, h_2, and h_3 for Three-Point Bend Specimen in Plane Strain (5.46)

		$n = 1$	$n = 2$	$n = 3$	$n = 5$	$n = 7$	$n = 10$	$n = 13$	$n = 16$	$n = 20$
	h_1	0.936	0.869	0.805	0.687	0.580	0.437	0.329	0.245	0.165
$a/W = \frac{1}{8}$	h_2	6.97	6.77	6.29	5.29	4.38	3.24	2.40	1.78	1.19
	h_3	3.00	22.1	20.0	15.0	11.7	8.39	6.14	4.54	3.01
	h_1	1.20	1.034	0.930	0.762	0.633	0.523	0.396	0.303	0.215
$a/W = \frac{1}{4}$	h_2	5.80	4.67	4.01	3.08	2.45	1.93	1.45	1.09	0.758
	h_3	4.08	9.72	8.36	5.86	4.47	3.42	2.54	1.90	1.32
	h_1	1.33	1.15	1.02	0.084	0.695	0.556	0.442	0.360	0.265
$a/W = \frac{3}{8}$	h_2	5.18	3.93	3.20	2.38	1.93	1.47	1.15	0.928	0.684
	h_3	4.51	6.01	5.03	3.74	3.02	2.30	1.80	1.45	1.07
	h_1	1.41	1.09	0.922	0.675	0.495	0.331	0.211	0.135	0.0741
$a/W = \frac{1}{2}$	h_2	4.87	3.28	2.53	1.69	1.19	0.773	0.480	0.304	0.165
	h_3	4.69	4.33	3.49	2.35	1.66	1.08	0.669	0.424	0.230
	h_1	1.46	1.07	0.896	0.631	0.436	0.255	0.142	0.084	0.0411
$a/W = \frac{5}{8}$	h_2	4.64	2.86	2.16	1.37	0.907	0.518	0.287	0.166	0.0806
	h_3	4.71	3.49	2.70	1.72	1.14	0.652	0.361	0.209	0.102
	h_1	1.48	1.15	0.974	0.693	0.500	0.348	0.223	0.140	0.0745
$a/W = \frac{3}{4}$	h_2	4.47	2.75	2.10	1.36	0.936	0.618	0.388	0.239	0.127
	h_3	4.49	3.14	2.40	1.56	1.07	0.704	0.441	0.272	0.144
	h_1	1.50	1.35	1.20	1.02	0.855	0.690	0.551	0.440	0.321
$a/W = \frac{7}{8}$	h_2	4.36	2.90	2.31	1.70	1.33	1.00	0.782	0.613	0.459
	h_3	4.15	3.08	2.45	1.81	1.41	1.06	0.828	0.649	0.486

Equations (5.4-8)(5.49). Furthermore, this factor preserves the continuity of the partial derivatives of J and Δ_c with respect to P at $P = P_0$ (5.50). While this is not essential to the treatment here, it is important in the stability analysis of crack growth where these derivatives appear.

Figures 5.20 and 5.21 compare the estimation analysis based upon Equation (5.4-8) (solid curves) with the full finite element, deformation plasticity analyses (dashed curves) using Equation (5.3-1) for a center cracked panel and a bend specimen. The numerical results are for plane stress, $a/W = \frac{1}{2}$, $\alpha = \frac{3}{7}$, and a height to width ratio of three for the center cracked panel and $L/W = 2$ for the bend specimen. In these two specimens the remaining ligaments experience the extremes of either tension or bending. In both cases the overall agreement is quite good and improves with decreasing hardening. The comparison is better for the center cracked panel than for the bend specimen. In the limit as $n \to \infty$ both results agree with the predictions of reference (5.51). Equally good agreement between the estimation analysis, full finite element computations and experimental results has been found for other flawed configurations (5.46, 5.50). These limited investigations show that the estimation method describes adequately the elastic-plastic driving force for simple cracked configurations. As the plastic fracture handbook becomes more complete, this method offers a simple, attractive approach to design and safety analyses.

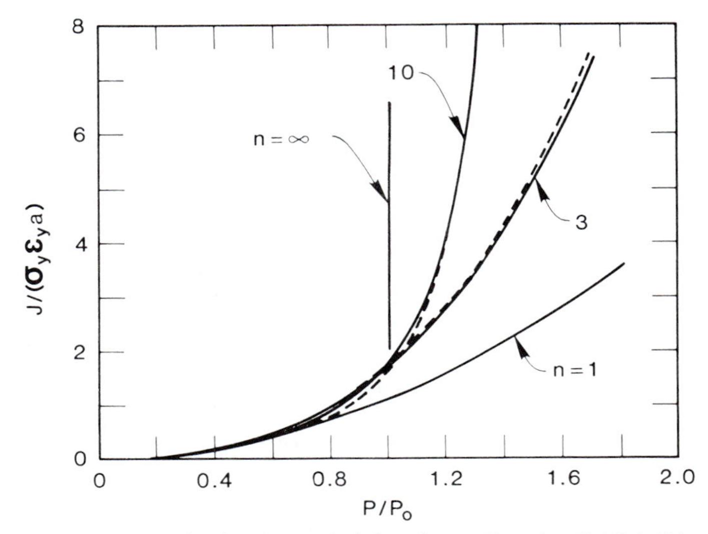

Figure 5.20 Comparison of estimation analysis based upon Equation (5.4-8) (solid curves) and full finite element analysis (dashed curves) of a center-cracked tensile panel (5.48).

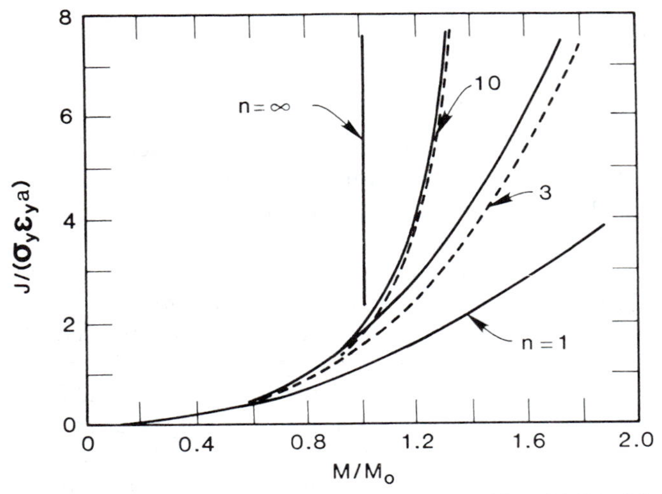

Figure 5.21 Comparison of estimation analysis based upon Equation (5.4-8) (solid curves) and full finite element analysis (dashed curves) of a bend specimen (5.48).

5.4.3 The Hardening Failure Assessment Diagram

The estimation procedure can be used to develop failure assessment curves analogous to the *R*-6 curve. It is possible to include in these the influence of strain hardening and the cracked configuration of which the *R*-6 curve is incapable. It is convenient to write

$$J_e(a_e) = \tilde{J}_e(a_e)(P/P_0)^2$$
$$J_p(a, n) = \tilde{J}_p(a, n)(P/P_0)^{n+1} \tag{5.4-12}$$

where $\tilde{J}_e$ and $\tilde{J}_p$ are independent of P and can be related to $f_1(a/W)$ and $h_1(a/W, n)$ via Equations (5.4-1) and (5.4-5), respectively. Within the estimation method, the fracture criterion can be written as

$$J = \tilde{J}_e(a_e)(P/P_0)^2 + \tilde{J}_p(a, n)(P/P_0)^{n+1} = J_c \tag{5.4-13}$$

The latter may also be expressed as

$$\frac{J_e}{\tilde{J}_e(a_e)(P/P_0)^2 + \tilde{J}_p(a, n)(P/P_0)^{n+1}} = \frac{J_e}{J_c} \tag{5.4-14}$$

where J_e is the elastic driving force having the form

$$J_e = \tilde{J}_e(a)(P/P_0)^2 \tag{5.4-15}$$

In addition J_e is also related to K by

$$J_e = K^2/E' \tag{5.4-16}$$

Hence, J_e and $\tilde{J}_e(a)$ are readily obtainable from solutions for K. Furthermore, the fracture resistance J_c and the fracture toughness K_c are connected by

$$J_c = K_c^2/E' \tag{5.4-17}$$

In a manner analogous to the development of the failure assessment diagram in Section 5.1, let the stress ratio S_r be defined as

$$S_r = P/P_0 \tag{5.4-18}$$

and the elastic driving force to fracture resistance ratios be

$$K_r = K/K_c \quad \text{and} \quad J_r = J_e/J_c \tag{5.4-19}$$

It follows from Equations (5.4-16) and (5.4-17) that

$$J_r = K_r^2 \tag{5.4-20}$$

The introduction of Equations (5.4-15) through (5.4-20) into Equation (5.4-14) permits writing

$$\frac{S_r^2}{H_e S_r^2 + H_n S_r^{n+1}} = J_r = K_r^2 \tag{5.4-21}$$

where

$$H_e = \tilde{J}_e(a_e)/\tilde{J}_e(a)$$
$$H_n = \tilde{J}_p(a, n)/\tilde{J}_e(\mathrm{a}) \tag{5.4-22}$$

The locus of points (S_r, K_r) in the $S_r - K_r$ space satisfying Equation (5.4-21) defines the failure assessment curve where the crack driving force J is in

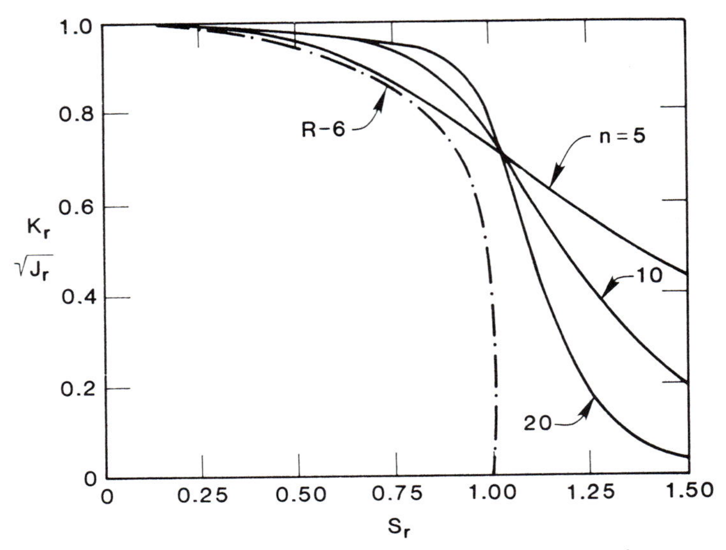

Figure 5.22 The influence of strain hardening on the failure assessment curve of a center-cracked panel in plane stress with $a/W = 0.5$ (5.46).

equilibrium with the fracture resistance of the material J_c. The influence of the cracked configuration, the hardening exponent, and other material properties occurs implicitly through the quantities H_e and H_n.

In Figure 5.22 the failure assessment curve is depicted for a center cracked panel (CCP) in plane stress for different values of the hardening exponent. The

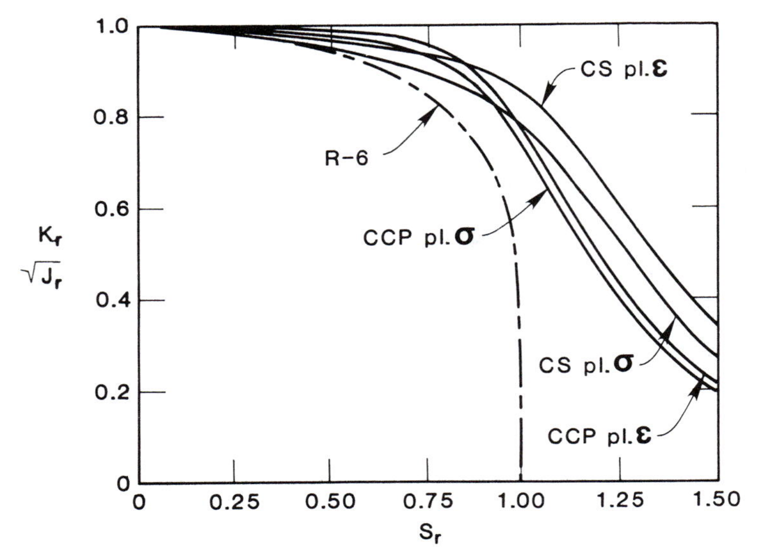

Figure 5.23 Comparison of plane strain and plane stress failure curves for a center-cracked panel (CCP) and a compact specimen (CS) with $a/W = 0.5$ and $n = 10$ (5.46).

effect of strain hardening is to permit S_r to exceed unity and still be safe which is impossible according to the R-6 curve. As the hardening exponent increases (strain hardening decreases) these failure curves move closer to the R-6 curve. The failure assessment curves for the center cracked panel and compact specimen (CS) are compared in Figure 5.23 for plane stress and plane strain. A modest dependence upon the configuration and the type of constraint is exhibited by these curves. Finally, Figure 5.24 indicates the relative dependence of the failure assessment curves on the crack length. In all these cases the R-6 curve is observed to be conservative.

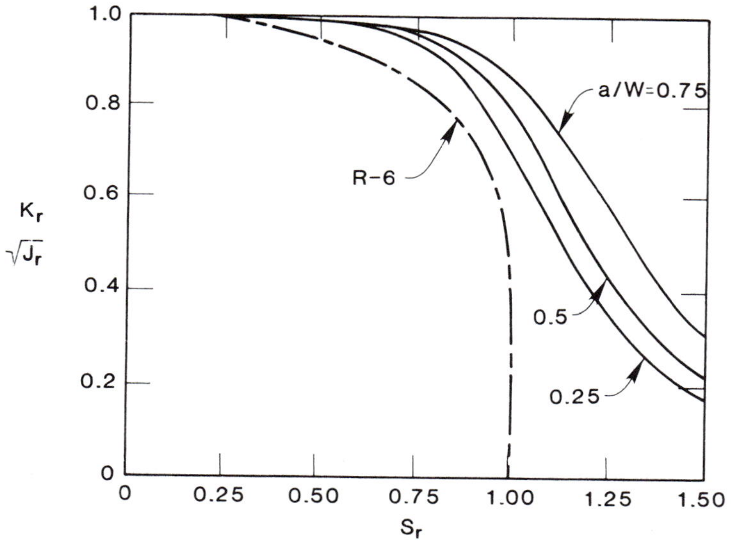

Figure 5.24 The influence of crack length on the failure assessment curve for a plane strain center-cracked panel with $n = 10$ (5.46).

Bloom (5.52) compared actual test data from GE/EPRI 4T-compact specimens of A533B, a reactor grade steel, with the respective failure assessment curve. The crack length to width ratio (a/W) for these specimens varied from 0.58 to 0.80. The plane strain failure assessment curves (Figure 5.25) for the compact specimen correspond to $a/W = 0.625$ and 0.75, $\alpha = 1.115$, and $n = 9.708$. In general the agreement between the predicted and measured values is very good with the predictions being conservative. Also shown in this figure is the failure assessment curve for an infinitely wide center cracked panel that is the hardening equivalent of the R-6 curve. Again this curve would provide conservative predictions.

5.4.4 Other Estimations

The estimation procedure can be generalized to other uniaxial stress-strain relations. Note the similarity between the forms of Equations (5.3-1) and

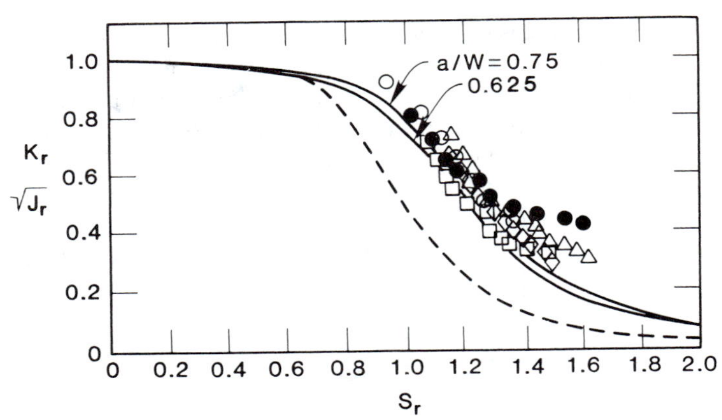

Figure 5.25 Failure assessment of GE/EPRI A533B 4T-compact specimens. Failure curves are shown for compact specimens (solid curves) with $a/W = 0.625$ and 0.75, and a center-cracked panel (dashed curve) (5.52).

(5.4-13). Hence, for the piecewise power law uniaxial stress-strain relation

$$\frac{\varepsilon}{\varepsilon_y} = \frac{\sigma}{\sigma_y}, \qquad \sigma \leqslant \sigma_y$$
$$\frac{\varepsilon}{\varepsilon_y} = \left(\frac{\sigma}{\sigma_y}\right)^n, \qquad \sigma > \sigma_y \tag{5.4-23}$$

the estimation for J is written as

$$J = J_e(a_e) = \tilde{J}_e(a_e)(P/P_0)^2, \qquad P \leqslant P_0$$
$$J = J_e(a_e) + \tilde{J}_p(a, n)(P/P_0)^{n+1}, \qquad P > P_0 \tag{5.4-24}$$

Analogous expressions can be written for δ and Δ_c. Equations (5.4-24) can be shown to be exact for the plane stress infinite strip with a semi-infinite crack.

The discontinuity of the slope of the piecewise power law stress-strain relation at the yield stress can present analytical difficulties when it comes to analyzing stable crack growth. The stress-strain relation

$$\frac{\varepsilon}{\varepsilon_y} = \frac{\sigma}{\sigma_y} \qquad \sigma < \sigma_y$$
$$\frac{\varepsilon}{\varepsilon_y} = \frac{1}{n}\left(\frac{\sigma}{\sigma_y}\right)^n + 1 - \frac{1}{n} \qquad \sigma > \sigma_y \tag{5.4-25}$$

has a continuous slope everywhere. In this case

$$J = J_e(a_e) = \tilde{J}_e(a_e)(P/P_0)^2, \qquad P \leq P_0$$
$$J = J_e(a_e) + \tilde{J}_p(a, n)[(P/P_0)^n - 1]/n, \qquad P > P_0 \tag{5.4-26}$$

Many materials do not strain harden indefinitely as a power law would indicate, but eventually saturate at a stress σ_s. A uniaxial stress-strain relation reflecting this latter phenomenon (5.46) is

$$\frac{\varepsilon}{\varepsilon_y} = \frac{\sigma}{\sigma_y} + \alpha\left(\frac{\sigma}{\sigma_y}\right)^n + \beta\left(\frac{\sigma}{\sigma_s}\right)^m \tag{5.4-27}$$

The saturation stress σ_s is chosen to ensure that $m/n \gg 1$. For a material of this kind there are three regimes of deformation: linear response for $\sigma < \sigma_y$, strain hardening for $\sigma_y < \sigma < \sigma_s$, and nearly perfect plasticity for $\sigma > \sigma_s$. The estimation for J is

$$J = \tilde{J}_e(a_e)(P/P_0)^2 + \tilde{J}_p(a, n)(P/P_0)^{n+1} + \tilde{J}_p(a, m)(P/P_s)^{m+1} \tag{5.4-28}$$

where

$$P_s = \lambda b \sigma_s \tag{5.4-29}$$

in which λ is a constraint factor. Again Equations (5.4-26) and (5.4-28) can be shown to be exact for the plane stress infinite strip with a semi-infinite crack.

5.5 *J*-Integral Testing

Begley and Landes (5.34) recognized that J and, hence, a critical value of J, could be evaluated experimentally from the interpretation of J as the energy release rate given by

$$J = -\frac{1}{B}\left(\frac{\partial U}{\partial a}\right)_\Delta \tag{5.5-1}$$

where U is the strain energy and B is the component thickness. Using multiple specimens with different crack lengths, they obtained the load-displacement records shown schematically in Figure 5.26(a). For a specified value of Δ the area under the load-displacement curve is the strain energy for the respective crack length. In this manner the strain energy per unit thickness versus crack length for fixed Δ can be established as illustrated in Figure 5.26(b). By Equation (5.5-1) the negative of the slope of these constant Δ-curves is J. Finally, J versus Δ for a fixed crack length can be plotted as in Figure 5.26(c).

Begley and Landes found that crack initiation in specimens having different crack lengths occurred at virtually the same value of J and, thereby, experimentally substantiated the use of a critical value of J as a fracture criterion. Furthermore, the plane strain critical value J_{Ic} agreed favorably with

$$J_{Ic} = K_{Ic}^2/E' \tag{5.5-2}$$

where K_{Ic} was obtained from independent K_{Ic}-fracture tests.

5.5.1 Single Specimen Testing

The major disadvantage of the preceding method for determining J_{Ic} is that five to ten specimens are necessary to develop the calibration of J versus displacement. Therefore, a technique for establishing J from a single specimen is very desirable. A method for estimating J from a single measured load-displacement record was proposed by Rice et al. (5.51).

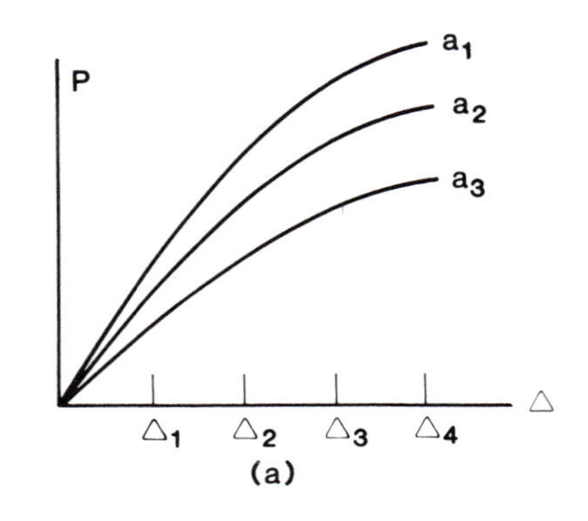

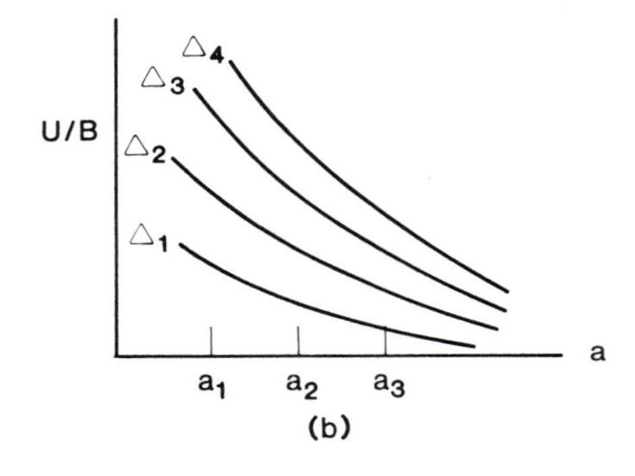

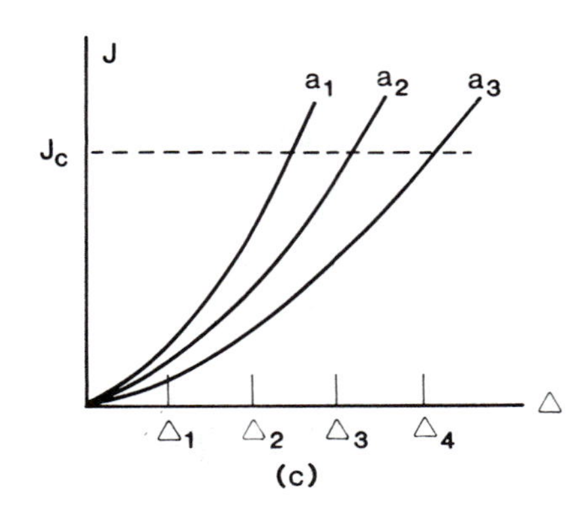

Figure 5.26 Schematic for extracting J versus Δ from a load-displacement record.

It is convenient to introduce alternative forms for J given by

$$J = -\int_0^{\Delta} \left(\frac{\partial P}{\partial a}\right)_{\Delta} d\Delta \tag{5.5-3}$$

or

$$J = \int_0^{P} \left(\frac{\partial \Delta}{\partial a}\right)_{P} dP \tag{5.5-4}$$

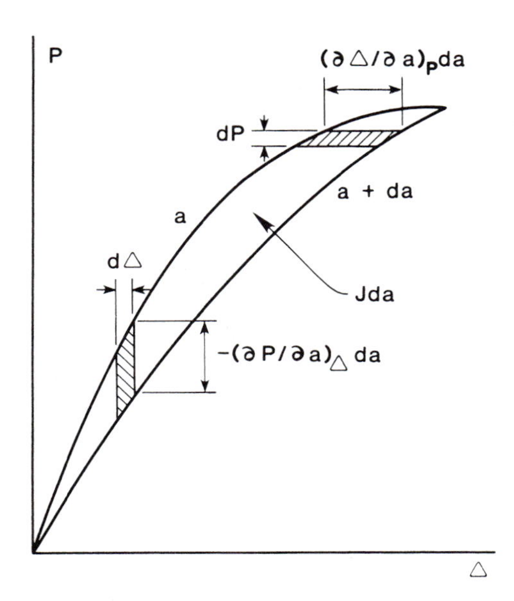

Figure 5.27 Equivalent representations for J.

where unless otherwise noted P is the generalized load per unit thickness of the component. When J is viewed as an energy release rate, it can be seen from Figure 5.27 that these are equivalent definitions of J. Equations (5.5-3) and (5.5-4) represent, respectively, the rates of decrease of the strain energy for fixed displacement and of increase of the complementary strain energy U^* for fixed load with crack extension.

Consider the deeply cracked bend specimen of Figure 5.28, where M is the applied moment per unit thickness and θ is the relative rotation of the ends. For this case Equation (5.5-4) becomes

$$J = \int_0^M \left(\frac{\partial \theta}{\partial a}\right)_M dM \tag{5.5-5}$$

The rotation θ can be decomposed according to

$$\theta = \theta_{nc} + \theta_c \tag{5.5-6}$$

where θ_{nc} is the relative rotation in the absence of a crack ($a = 0$) and θ_c is the remainder due to the crack. If the ligament b is small compared to W, then

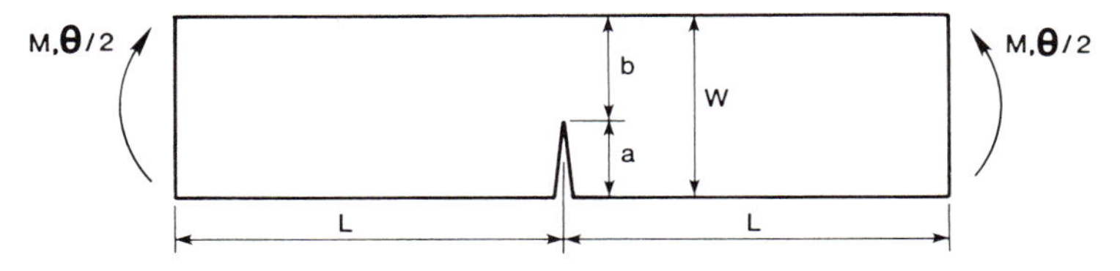

Figure 5.28 A deeply cracked bend specimen.

the rotation θ_c will be due largely to the deformation in the ligament and will be independent of L when L is large compared to W. From dimensional considerations

$$\theta_c = F(M/M_0, b/W) \tag{5.5-7}$$

where

$$M_0 = \sigma_y b^2/4 \tag{5.5-8}$$

is the fully plastic limit moment. While the function F may depend upon other dimensionless parameters, the present form is general enough for the purposes here. For sufficiently deep cracks, $b/W \ll 1$, F will depend weakly upon b/W; particularly, when the remaining ligament is fully yielded. For example, in the linear elastic range (5.38)

$$\theta_c = \frac{16M}{b^2E'} \quad \text{as } b/W \to 0 \tag{5.5-9}$$

The introduction of Equation (5.5-6) into Equation (5.5-5) yields

$$J = \int_0^M \left(\frac{\partial \theta_c}{\partial a}\right)_M dM \tag{5.5-10}$$

since θ_{nc} is independent of a. From Equation (5.5-7) it follows that

$$\left(\frac{\partial \theta_c}{\partial a}\right)_M = -\left(\frac{\partial \theta_c}{\partial b}\right)_M$$
$$= \frac{M}{M_0^2}\frac{dM_0}{db}\frac{\partial F}{\partial(M/M_0)} - \frac{1}{W}\frac{\partial F}{\partial(b/W)}$$

and

$$\left(\frac{\partial \theta_c}{\partial M}\right)_a = \frac{1}{M_0}\frac{\partial F}{\partial(M/M_0)}$$

which combine to give

$$\left(\frac{\partial \theta_c}{\partial a}\right)_M = \frac{M}{M_0}\frac{dM_0}{db}\left(\frac{\partial \theta_c}{\partial M}\right)_a - \frac{1}{W}\frac{\partial F}{\partial(b/W)} \tag{5.5-11}$$

The introduction of Equation (5.5-11) into Equation (5.5-10) leads to

$$J = \frac{1}{M_0}\frac{dM_0}{db}\int_0^{\theta_c} M\,d\theta_c - \frac{1}{W}\int_0^M \frac{\partial F}{\partial(b/W)}\,dM \tag{5.5-12}$$

where $d\theta_c = (\partial\theta_c/\partial M)\,dM$ for fixed a has been used in the first integral. Since $\partial F/\partial(b/W) = 0$ for a deeply cracked specimen, and since according to Equation (5.5-8)

$$\frac{1}{M_0}\frac{dM_0}{db} = \frac{2}{b}$$

Equation (5.5-12) reduces to

$$J = \frac{2}{b} \int_0^{\theta_c} M \, d\theta_c \tag{5.5-13}$$

The integral in Equation (5.5-13) is simply the area under the M versus θ_c curve. Hence, J is twice the work per ligament area done by the moment acting through the rotation due to the crack. Consequently, a single test record is sufficient to evaluate J.

If the remaining ligament supports primarily a bending moment due to an applied load P per unit thickness, then Equation (5.5-13) becomes

$$J = \frac{2}{b} \int_0^{\Delta_c} P \, d\Delta_c \tag{5.5-14}$$

Equation (5.5-14) is applicable to deeply cracked three-point bend and compact specimens. For the three-point bend specimen the displacement component Δ_{nc} can be appreciable and should be eliminated in evaluating J by Equation (5.5-14). On the other hand Δ_{nc} will be negligible compared to Δ_c for a typical compact specimen.

The second term of Equation (5.5-12) is frequently referred to as the Merkle-Corten correction (5.53). The remaining ligament of a compact specimen not only supports a bending moment but also a normal force. Depending upon the depth of the crack the latter can provide a significant contribution to J. Merkle and Corten (5.53) have examined the influence of this normal force on the J-integral for a compact specimen. Since Δ_{nc} can be neglected for a compact specimen, it is permissible and convenient to write

$$\Delta = \Delta_c = \Delta_e + \Delta_p \tag{5.5-15}$$

Consequently, Equation (5.5-4) can be written as

$$J = \int_0^P \left(\frac{\partial \Delta_e}{\partial a} \right)_P dP + \int_0^P \left(\frac{\partial \Delta_p}{\partial a} \right)_P dP \tag{5.5-16}$$

The first term is the linear elastic contribution and can be readily evaluated. The term

$$J_p = \int_0^P \left(\frac{\partial \Delta_p}{\partial a} \right)_P dP \tag{5.5-17}$$

is the plastic contribution and attention is now focused upon evaluating this quantity.

The stress distribution in the remaining ligament and the displacement diagram at plastic collapse are shown in Figure 5.29. Equilibrium demands that

$$P_0 = 2\sigma_y c\alpha = \sigma_y b\alpha \tag{5.5-18}$$

and

$$\alpha^2 + 2\alpha(a/c + 1) - 1 = 0 \tag{5.5-19}$$

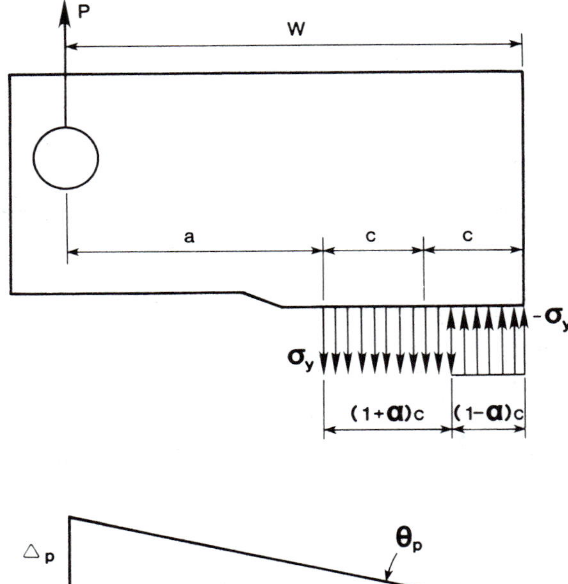

Figure 5.29 Stress distribution in the remaining ligament and deformation in a deeply cracked compact specimen at plastic collapse.

where $c = b/2$ is the half length of the ligament and αc is the distance from the center of the ligament to the point of stress reversal (neutral axis). For a specified value of a/c, Equation (5.5-19) can be used to establish α. Assuming that rotations occur about the neutral axis, one can write

$$\Delta_p = [a + (1 + \alpha)c]\theta_p = \frac{W}{2}\left[2 - (1 - \alpha)\frac{b}{W}\right]\theta_p \tag{5.5-20}$$

where θ_p is the plastic angle of rotation.

Analogous to the bend specimen the plastic rotation is assumed to be a function of the ratio of the applied load P to the fully plastic load P_0; that is,

$$\theta_p = f(P/P_0) \tag{5.5-21}$$

The substitution of Equation (5.5-21) into Equation (5.5-20) yields

$$\Delta_p = \frac{W}{2}\left[2 - (1 - \alpha)\frac{b}{W}\right] f\left(\frac{P}{P_0}\right) \equiv F\left(\frac{P}{P_0}, \frac{b}{W}\right) \tag{5.5-22}$$

When θ_c and Δ_p are viewed as generalized displacements, then it is clear from Equations (5.5-7) and (5.5-22) that these generalized displacements enjoy similar functional forms in b/W and the respective generalized forces M and P.

This similarity combined with that of Equations (5.5-9) and (5.5-17) permits writing

$$J_p = \frac{1}{P_0}\frac{dP_0}{db}\int_0^{\Delta_p} P\,d\Delta_p - \frac{1}{W}\int_0^P \frac{\partial F}{\partial(b/W)}\,dP \tag{5.5-23}$$

It follows from Equation (5.5-18) that

$$\frac{dP_0}{db} = \sigma_y\left(\alpha + b\frac{d\alpha}{db}\right)$$

Differentiating Equation (5.5-19) and using Equation (5.5-19) to eliminate a/c in the resulting expression, one obtains

$$\frac{d\alpha}{db} = \frac{1}{b}\frac{(1 + 2\alpha - \alpha^2)\alpha}{1 + \alpha^2}$$

Consequently,

$$\frac{1}{P_0}\frac{dP_0}{db} = \frac{2}{b}\frac{1 + \alpha}{1 + \alpha^2} \tag{5.5-24}$$

In a similar manner

$$-\frac{1}{W}\frac{\partial F}{\partial(b/W)} = \frac{f}{2}\frac{1 - 2\alpha - \alpha^2}{1 + \alpha^2} = \frac{2}{b}\frac{1 - 2\alpha - \alpha^2}{(1 + \alpha^2)^2}\alpha\Delta_p \tag{5.5-25}$$

where $a + (1 + \alpha)c = b(1 + \alpha^2)/4\alpha$, which follows from Equation (5.5-19), has been used. Finally, the combination of Equations (5.5-23)–(5.5-25) yields

$$J_p = \frac{2}{b}\frac{1 + \alpha}{1 + \alpha^2}\int_0^{\Delta_p} P\,d\Delta_p + 2\frac{\alpha}{b}\frac{1 - 2\alpha - \alpha^2}{(1 + \alpha^2)^2}\int_0^P \Delta_p\,dP \tag{5.5-26}$$

Once again the J-integral can be evaluated from a single compact specimen load-displacement record.

The first and second integrals of Equation (5.5-26) are, respectively, the strain energy and complementary energy per unit thickness. For a deeply cracked specimen ($a/c \gg 1$), $\alpha \approx 0$ and Equation (5.5-26) reduces to Equation (5.5-14) when Δ_p is replaced by Δ_c—that is, when the elastic contribution to Δ_c is negligible. For the rigid plastic case the complementary energy as well as the elastic deformations vanish and Equation (5.5-26) becomes

$$J = \frac{2}{b}\frac{1 + \alpha}{1 + \alpha^2}P_0\Delta \tag{5.5-27}$$

Merkle and Corten (5.53) found for compact specimens with $a/W > 0.5$ that the total J may be computed from Equation (5.5-26) using the total load-point displacement; that is,

$$J = \frac{2}{b}\frac{1 + \alpha}{1 + \alpha^2}\int_0^{\Delta} P\,d\Delta + 2\frac{\alpha}{b}\frac{1 - 2\alpha - \alpha^2}{(1 + \alpha^2)^2}\int_0^P \Delta\,dP \tag{5.5-28}$$

When the complementary energy is much smaller than the strain energy, Equation (5.5-28) reduces to

$$J = \frac{2}{b} \frac{1 + \alpha}{1 + \alpha^2} \int_0^{\Delta} P \, d\Delta \tag{5.5-29}$$

Using Equations (5.5-1) and (5.5-14) Landes et al. (5.54) evaluated J experimentally for three-point bend specimens having a span to height ratio of four. Equation (5.5-1) was used as the standard for comparison because it is simply the definition of J and involves no approximations. The best agreement between the two was obtained when Δ_c was replaced by Δ in Equation (5.5-14) that is, when the total strain energy was used in the evaluation rather than only that due to the crack. The comparison of the two approaches is shown in Figure 5.30. Slight differences are to be expected

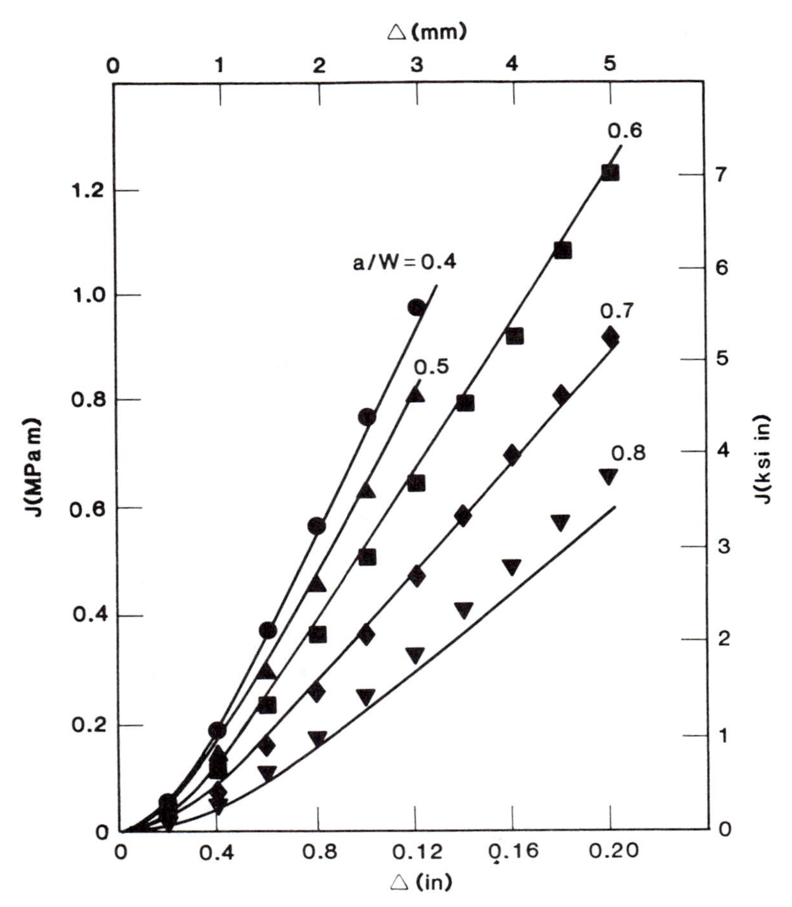

Figure 5.30 Comparison of J based upon Equation (5.5-1) (solid curves) and Equation (5.5-14) (solid points) using measurements from HY-130 steel, three-point bend specimens (5.54).

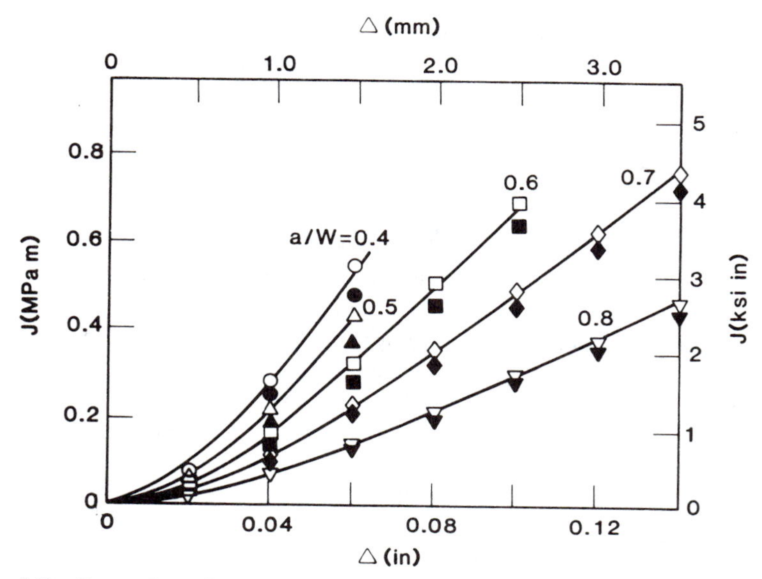

Figure 5.31 Comparison of J based upon Equation (5.5-1) (solid curves) and Equation (5.5-14) (solid points), and Equation (5.5-29) (open points) using measurements from HY-130 steel, 1T-compact specimens (5.54).

because of the inherent inaccuracies in effecting the numerical differentiation required in Equation (5.5-1). Similar comparisons were made for compact specimens based upon Equations (5.5-1), (5.5-14), (5.5-26), and (5.5-29). Figure 5.31 compares values of J for compact specimens using Equations (5.5-1), (5.5-14), and (5.5-29). The J values based upon Equation (5.5-26) tend in general to exceed the values predicted by the other equations. The best agreement occurs between Equations (5.5-1) and (5.5-29). These results support the use of J estimation methods in the experimental evaluation of the J-integral from a single specimen.

5.5.2 Standard J_{Ic} Test Method

A standard test method, E 813, has been issued by ASTM (5.55) for determining J_{Ic}, the plane strain value of J at initiation of crack growth. The value of J_{Ic} may be used to characterize the toughness of materials at or near the onset of crack extension from a pre-existing fatigue crack. It can also be used in

$$K_{Ic}^2 = J_{Ic}E \tag{5.5-30}$$

to obtain a conservative estimate of K_{Ic} when sufficient thickness precludes a valid K_{Ic} test according to the size requirements of the ASTM E 399 method. The method can be used to determine J_{Ic} for a wide range of ductile engineering materials. However, materials with extremely high resistances to tearing may not test satisfactorily because crack growth due to physical tearing may be indistinguishable from extensive crack-tip blunting. A pronounced nonlinear relationship between J and the amount of crack extension may present problems in determining J_{Ic}.

Preferred specimens are a three point bend specimen and a compact specimen. For a valid J_{Ic} value the remaining ligament b and thickness B* must satisfy

$$b, B > 25J_{Ic}/\sigma_y \tag{5.5-31}$$

where σ_y is the effective yield strength of the material at the test temperature. In this method σ_y is the average of the 0.2 percent offset yield strength and the ultimate strength. Since the crack-tip opening displacement is proportional to J_{Ic}/σ_y, Equation (5.5-31) requires that the ligament and the thickness be large compared to this displacement. Furthermore, the dimensions b and B must be greater than $15J/\sigma_y$ for all values of J calculated as data points. Since values of J and J_{Ic} are not known a priori, the selection of the specimen dimensions can only be based upon previous experience.

The initial crack length a_0 must be at least one-half the width W of the specimen but not greater than $0.75W$. Experience indicates that $a_0/W = 0.6$ is about optimum. The specimen thickness is nominally $0.5W$. The span to width ratio of the three-point bend specimen is four with an overall length of $4.5W$. The compact specimen is similar to the one recommended in Standard E 399 for K_{Ic} testing, but modified slightly to meet the needs of this method.

The procedure requires the measurement of applied load and load-point displacement to obtain the total work done on the specimen. Load versus displacement is either recorded autographically on an X-Y plotter or digitized for accumulation in a computer information storage facility. In the multiple specimen technique the crack extensions are marked after having deformed the specimens to selected values. The marking may be done either by heat tinting or by fatigue cycling the specimen. The specimens are then broken open and the crack extensions are measured. In the single specimen technique the specimen is periodically unloaded about 10 percent and the elastic unloading compliance is measured. From the unloading compliances the crack lengths and, hence, crack extensions are calculated. Upon completion of the test the total amount of crack extension predicted by the unloading compliance technique must agree within 15 percent of the average value determined by the heat tint method.

The area A under the measured load displacement curve (see Figure 5.32) is measured graphically or numerically integrated if the computer technique is used. For the three-point bend specimen J is computed from

$$J = \frac{2A}{bB} \tag{5.5-32}$$

in accordance with Equation (5.5-14) except that the total displacement is used

* If the specimen is side grooved, then B is the minimum thickness.

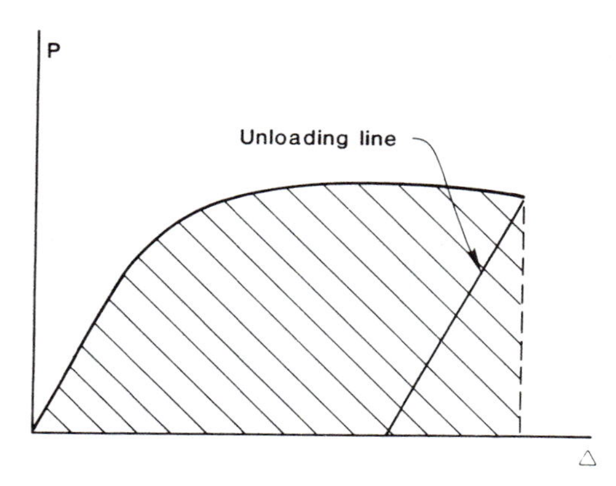

Figure 5.32 Illustration of area to be used in determining J.

as suggested by Landes et al. (5.54). The normal force in the compact specimen requires that J be computed from

$$J = \frac{2}{b}\frac{1+\alpha}{1+\alpha^2}\frac{A}{B} \tag{5.5-33}$$

Compare with Equation (5.5-29), where from Equation (5.5-19) α is given by

$$\alpha = \left[\left(\frac{2a_0}{b}\right)^2 + 2\left(\frac{2a_0}{b}\right) + 2\right]^{\frac{1}{2}} - \frac{2a_0}{b} - 1 \tag{5.5-34}$$

The values of J are plotted against the physical crack growth Δa_p as depicted in Figure 5.33. Superimposed on this same plot are three additional

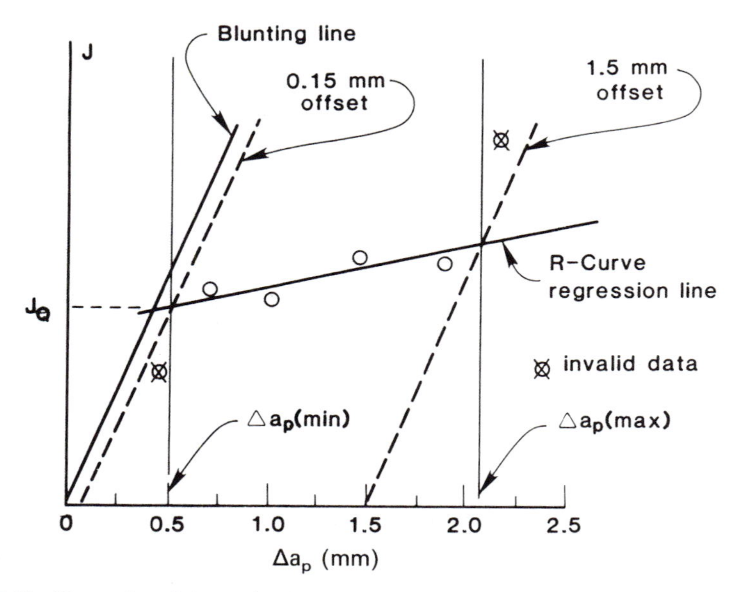

Figure 5.33 Illustration of data reduction required to establish J_{Ic} using ASTM E813 Standard.

lines. One is the blunting line defined by

$$J = 2\sigma_y \Delta a_p \tag{5.5-35}$$

This line approximates the apparent crack advance due to crack-tip blunting in the absence of tearing. It is based upon the assumption that prior to tearing the crack advance is one-half of the crack-tip opening displacement. The other two lines are parallel to the blunting line but offset 0.15 mm and 1.5 mm. For a valid test at least four data points must fall within the region bounded by the abscissa and the vertical lines through the points of interaction of a linear regression line and these offset lines. Points outside this region are considered invalid. The valid data points are used to establish a final linear regression line. The intersection of this line with the blunting line establishes J_Q. If B and b are greater than $25J_Q/\sigma_y$ and if the slope of the linear regression line is less than σ_y, then $J_Q = J_{Ic}$.

To demonstrate an advantage that J_{Ic} testing offers, consider the minimum thickness requirements. Assume that $J_{Ic} = 150\,\text{kJ/m}^2$, $\sigma_y = 280\,\text{MPa}$, and $E = 210 \times 10^3$ MPa, values that are typical of reactor grade steel. For a valid J_{Ic} test of this material

$$B > 25J_{Ic}/\sigma_y = 14 \text{ mm}$$

Equation (5.5-30) yields $K_{Ic} = 180 \text{ MPa m}^{\frac{1}{2}}$ and the minimum thickness for a valid K_{Ic} test is

$$B = 2.5\left(\frac{K_{Ic}}{\sigma_y}\right)^2 = 1 \text{ m!}$$

The main purpose of J_{Ic} testing is not, of course, to establish K_{Ic}, but to determine the fracture resistance that is an integral part of a J-based plastic fracture mechanics for high toughness, low strength materials.

5.6 *J*-Dominance and *J*-Controlled Crack Growth

For a single parameter characterization, be it J or δ_t, of the crack-tip fields to be valid, the region D over which the HRR singularity dominates must engulf the fracture process zone whose extent is typically on the order of δ_t for ductile rupture. As previously noted, the size of the region D for antiplane strain loading decreases with increasing n and vanishes in the nonhardening limit of $n \to \infty$. Furthermore, McClintock (5.56) found that the perfectly plastic slip-line fields for the cracked bend bar (CBB) in pure bending, the center crack panel (CCP) and the double-edge cracked panel (DECP) in tension depicted in Figure 5.34 are dramatically different. In addition there is no unique relationship between the crack-tip stress and strain fields—the latter field being dictated by the geometry of the body. Consequently, a one-parameter characterization requires the presence of strain hardening. Even in the presence of strain hardening the condition for J-dominance will likely depend upon the configuration. This is suggested by the rather drastic difference in the

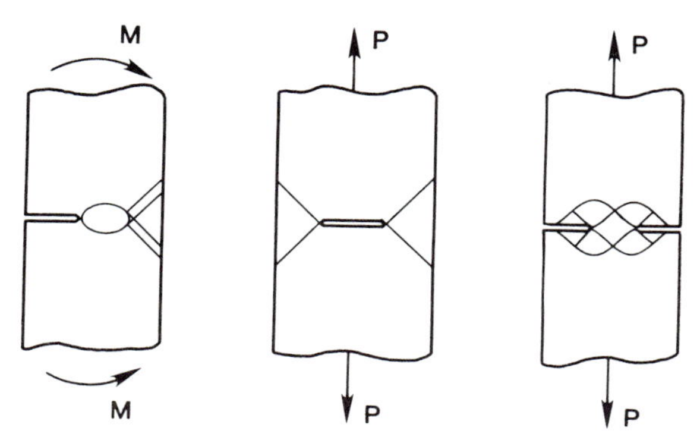

Figure 5.34 Perfectly plastic slip-line fields for cracked bend bar, center-cracked panel, and double edge notched specimens.

slip-line fields for the CBB and CCP specimens. Begley and Landes (5.57) also reported measuring a J-resistance curve for a CCP specimen that was quite different from that for a compact tension specimen.

McMeeking and Parks (5.58) performed finite-deformation, finite element analyses of plane strain CCP specimens subjected to uniform remote tension and CBB specimens in pure bending. A J_2-flow theory of plasticity was used in these analyses. The predicted normal stress distributions on the plane ahead of the crack tip for a CCP with $a/W = 0.5$, $n = 10$, and $\sigma_y/E = 1/300$ are shown in Figure 5.35 for various values of $b/(J/\sigma_y)$. Note that $\delta_t \approx J/2\sigma_y$ is a measure of

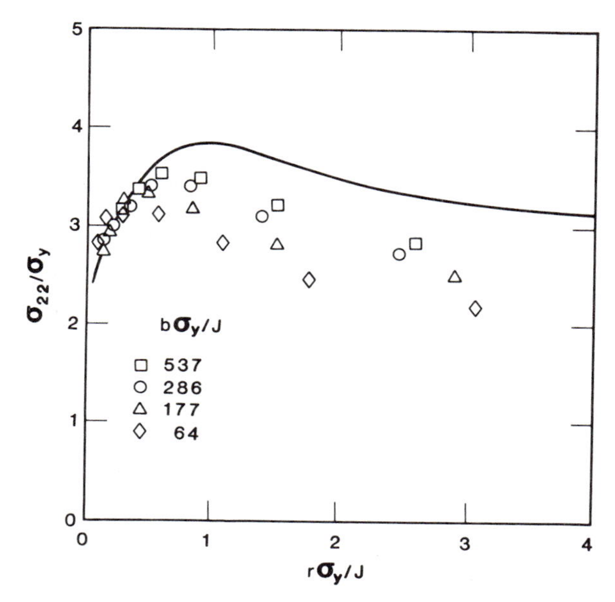

Figure 5.35 Normal stress distribution ahead of the crack tip in a CCP with $a/W = 0.5$, $n = 10$, $\sigma_y/E = \frac{1}{300}$ and selected values of $b\sigma_y/J$. The solid curve is the small-scale yielding prediction (5.58).

the size of the fracture process zone. When J-dominance holds, it follows from the HRR fields [see Equation (5.3-33)] that the stress field when plotted against $r/(J/\sigma_y)$ should be independent of J. These results as well as others in reference (5.58) suggest that

$$b > 200J/\sigma_y \tag{5.6-1}$$

is required for J-dominance in the center cracked panel.

Shih and German (5.59) have performed finite element computations for cracked bend bars, center cracked panels, and single-edge cracked panels. The analysis is based upon J_2-flow theory, a Ramberg-Osgood uniaxial stress-strain curve with $\alpha = 3/7$ and $\varepsilon_y = \sigma_y/E = 2 \times 10^{-3}$ and small strains. The small strain formulation permits direct comparison between the numerically calculated stress and strain fields and the HRR fields. The results of McMeeking and Parks indicate the effect of finite deformation occurs over a distance of approximately twice the crack-tip opening displacement or, equivalently, over a distance of about J/σ_y.

Figure 5.36 shows the normal stress σ_{22} ahead of the crack tip in CBB and CCP specimens based upon the finite element solutions and the HRR field for different levels of plastic deformation corresponding to $b\sigma_y/J$ equal to 600,

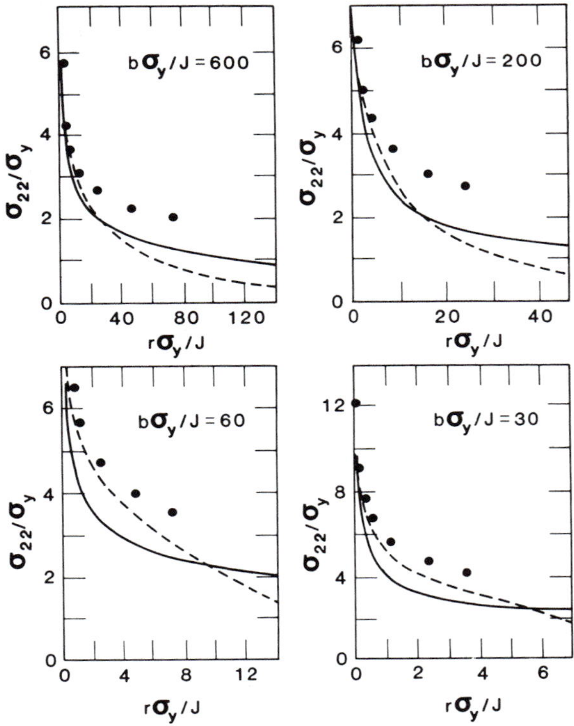

Figure 5.36 Variation of normal stress ahead of the crack tip in CBB (– – –) and CCP (——) from contained yielding to fully plastic behavior for $a/W = 0.75$, $n = 3$, and $\sigma_y/E = \frac{1}{500}$. Also shown is the HRR field (···) (5.59).

200, 60, and 30 and a relatively large strain hardening ($n = 3$). The larger values of $b\sigma_y/J$ correspond to contained plasticity whereas the smaller values are associated with nearly fully plastic conditions. The stress fields in the CBB agree favorably with the HRR singularity over a distance of about $3J/\sigma_y$ ($\approx 6\delta_t$) for the range of plastic deformation considered. By comparison the agreement for the CCP over the same distance is good for $b\sigma_y/J > 200$ and poor for $b\sigma_y/J \leqslant 60$.

Similar comparisons for a weakly hardening material ($n = 10$) are presented in Figure 5.37. Again good agreement between the calculated stress field and the HRR singular field is obtained for the CBB over the range of plastic deformation considered. The agreement for the CCP tends to deteriorate rapidly for $b\sigma_y/J < 200$.

Both the computations of McMeeking and Parks (5.58) and Shih and German (5.59) suggest that J-dominance will be preserved in cracked bend bars if $b\sigma_y/J > 30$. This condition compares favorably with experimentally observed behavior and the specifications in the E 813 Standard for J_{Ic} testing. On the other hand J-dominance can only be assured in the center crack panel when $b\sigma_y/J > 200$. The dependence of the condition for J-dominance upon specimen geometry is also reflected in the E 813 Standard by the caution that the use of specimen configurations other than those recommended in the method may involve different requirements for validity.

Crack growth is accompanied by elastic unloading and, hence, nonproportional plastic deformation in the neighborhood of the crack tip. The J-based plastic fracture mechanics that is founded on a deformation theory of

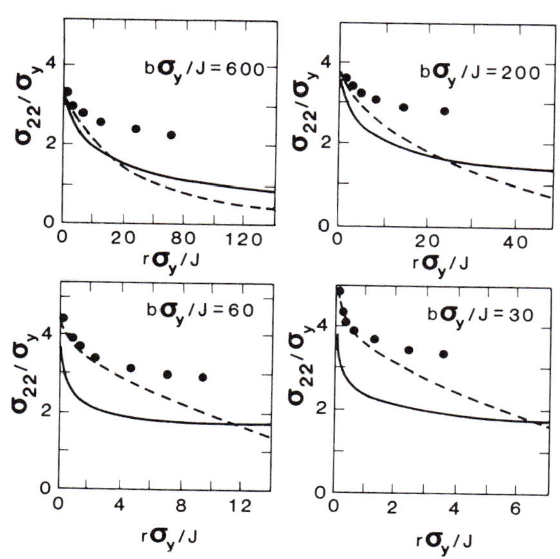

Figure 5.37 Variation of normal stress ahead of the crack tip in CBB (---) and CCP (——) from contained yielding to fully plastic behavior for $a/W = 0.75$, $n = 10$, and $\sigma_y/E = \frac{1}{500}$. Also shown is the HRR field (···) (5.59).

plasticity is incapable of rigorously modeling these characteristics of plastic crack extension. The implication is that J is strictly valid for analyzing stationary cracks. Nevertheless, if nearly proportional loading occurs everywhere except in a small neighborhood of the crack tip, then J can be used to analyze crack growth provided additional conditions for J-controlled crack growth are satisfied. For when nearly proportional loading occurs, the difference between a deformation theory and a corresponding flow theory of plasticity will be negligible. Such behavior is frequently observed in intermediate strength metals that can withstand substantial plastic deformation beyond crack initiation while exhibiting very limited amounts of crack growth.

When the conditions for J-controlled crack growth are satsified, then J is a meaningful fracture characterizing parameter and a unique, configuration-independent relationship between J and Δa exists. If conditions of plane strain are not satisfied, then the resistance curve relating J to Δa may depend upon the thickness of the body as the small-scale yielding resistance curve does. Under these conditions the small-scale yielding resistance curve analysis of Chapter 3 may be extended to form a J-resistance curve analysis for crack growth under large-scale yielding. The conditions for J-controlled crack growth have been examined by Hutchinson and Paris (5.60).

Consider a material with a J-resistance curve, $J_R(\Delta a)$, depicted in Figure 5.38 where the fracture resistance J_R increases with crack extension Δa. In particular, emphasis is placed upon materials for which a small amount of crack growth, say, a millimeter or two, is accompanied by a several-fold

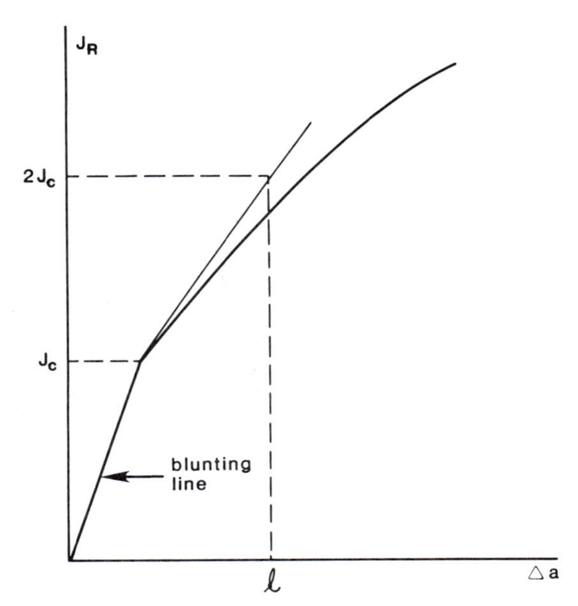

Figure 5.38 Typical J-resistance curve.

increase in J above the initiation value J_c. Since the discussion is also applicable when plane strain conditions do not exist, the plane strain designation for the initiation value of J has been dropped. According to the deformation theory the dominant strain field is

$$\varepsilon_{ij} = k_n(J/r)^{n/(n+1)}\tilde{\varepsilon}_{ij}(\theta) \tag{5.6-2}$$

where k_n is a constant. Again D in Figure 5.39 denotes the characteristic length of the region dominated by this HRR singular field. Since crack extension will produce an elastic unloading wake and a region of nonproportional loading on the order of Δa in length, then one condition for J-controlled crack growth is

$$\Delta a \ll D \tag{5.6-3}$$

The second condition follows from the requirement that predominantly proportional loading occurs within the annular region $l < r < D$ of Figure 5.39. An increment in the strain field due to increments of J and a is

$$\begin{aligned} d\varepsilon_{ij} = {} & k_n \frac{n}{n+1}\left(\frac{J}{r}\right)^{n/(n+1)} \frac{dJ}{J}\tilde{\varepsilon}_{ij}(\theta) \\ & - k_n J^{n/(n+1)}\, da \frac{\partial}{\partial x_1}\left[r^{-n/(n+1)}\tilde{\varepsilon}_{ij}(\theta)\right] \end{aligned} \tag{5.6-4}$$

where for a coordinate system attached to the crack tip the change due to a is $-(\partial/\partial x_1)\,da$. Since

$$\frac{\partial}{\partial x_1} = \cos\theta\frac{\partial}{\partial r} - \frac{\sin\theta}{r}\frac{\partial}{\partial\theta}$$

Equation (5.6-4) becomes

$$d\varepsilon_{ij} = k_n\left(\frac{J}{r}\right)^{n/(n+1)}\left[\frac{n}{n+1}\frac{dJ}{J}\tilde{\varepsilon}_{ij}(\theta) + \frac{da}{r}\tilde{\beta}_{ij}(\theta)\right] \tag{5.6-5}$$

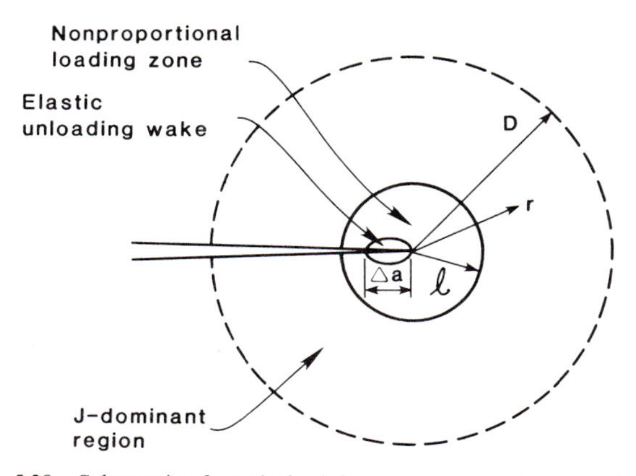

Figure 5.39 Schematic of crack tip deformation zones for a growing crack.

where

$$\tilde{\beta}_{ij}(\theta) = \frac{n}{n+1}\cos\theta\tilde{\varepsilon}_{ij} + \sin\theta\frac{\partial\tilde{\varepsilon}_{ij}}{\partial\theta} \tag{5.6-6}$$

The first term in the brackets of Equation (5.6-5) corresponds to proportional loading ($dJ > 0$) in that $d\varepsilon_{ij} \propto \varepsilon_{ij}$, whereas the second term is nonproportional. Since $\tilde{\varepsilon}_{ij}$ and $\tilde{\beta}_{ij}$ are of the same order of magnitude, then predominantly proportional loading will occur in the annular region of Figure 5.39 if

$$\frac{dJ}{J} \gg \frac{da}{r} \tag{5.6-7}$$

Let

$$\frac{1}{l} = \frac{1}{J}\frac{dJ}{da} \tag{5.6-8}$$

where according to Figure 5.38 l can be viewed as the crack growth just beyond initiation associated with a doubling of J above J_c. This definition of l reduces to the material-based length l introduced in the small-scale yield resistance-curve analysis of Section 3.5. If further

$$l \ll D \tag{5.6-9}$$

then there exists an annular region

$$l \ll r < D \tag{5.6-10}$$

in which the plastic loading is predominantly proportional and the HRR singular fields dominate. Consequently, if Equation (5.6-9) is satisfied, then a negligible difference can be expected between the strain fields predicted by a flow theory and a deformation theory for $r \gg l$. More importantly, J uniquely governs or controls the intensity of the fields in the region defined by Equation (5.6-10).

For a fully yielded configuration D will be some fraction of the smaller of the remaining ligament b or some other characteristic length from the crack tip to the boundary or load point. Thus, Equation (5.6-9) can be written as

$$l \ll b$$

or, equivalently,

$$\omega = \frac{b}{l} = \frac{b}{J}\frac{dJ}{da} \gg 1 \tag{5.6-11}$$

for J-controlled crack growth.

Shih et al. (5.61) performed a finite element analysis of crack growth in a A533B steel, $4T$-compact specimen using J_2-flow theory. The J-integral was computed for a wide variety of contours and was plotted against crack extension as illustrated in Figure 5.40. The larger the subscript on J in these plots the more remote is the contour from the crack tip; for example, J_2 is for

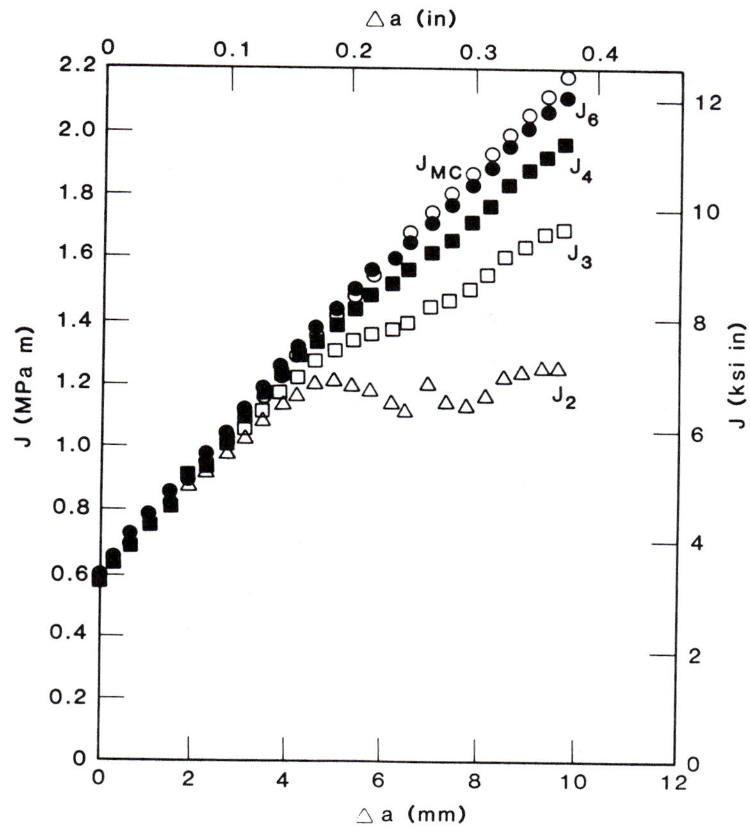

Figure 5.40 *J*-resistance curves for different contours on a A533B, 4T-compact tension specimen (5.61).

the contour closest to the crack tip. In this figure J_{MC} denotes the value of J computed from the experimental data using the Merkle-Corten method and can be viewed as a far field value for J. The path independence of J reflected by these computations for small amounts of crack growth ($\Delta a < 4$ mm) is consistent with the concept of J-controlled growth. Clearly, for large amounts of crack growth ($\Delta a > 4$ mm), J is no longer path-independent and J-controlled crack growth is lost. For the compact specimen in this analysis $\omega \approx 40$. Shih et al. (5.62) found for members subjected primarily to bending that the conditions expressed by Equations (5.6-3) and (5.6-11) become $\Delta a < 0.06b$ and $\omega > 10$, respectively.

The question of what is the smallest value of ω for which J-controlled crack growth is assured remains unanswered. Indications are that it will depend upon the configuration and the degree of strain hardening. It was demonstrated earlier in this section that the condition for J-dominance in a CCP specimen is more severe than for a CBB specimen. In order for J-controlled crack growth to exist, strain hardening must be present; otherwise, in the nonhardening limit the annular region defined by Equation (5.6-10) vanishes with D. Furthermore, as will be seen in Section 5.8, the $1/r$ strain singularity for a stationary crack in a perfectly plastic material becomes a $\ln r$ singularity for a propagating crack.

In summary, a consequence of J-controlled crack growth is that the J_R curve obtained from fully yielded specimens will coincide with the one obtained under small-scale yielding conditions if the plastic constraint remains unchanged. Except for perhaps a thickness dependence, the J_R curve will be independent of the configuration. When J-controlled growth exists, stable crack growth and the onset crack instability can be analyzed using a resistance curve approach based upon J or δ_t.

5.7 Stability of *J*-Controlled Crack Growth

As noted in the previous section, the J-resistance curve, $J_R(\Delta a)$, is a unique configuration-independent property of the material provided that the conditions ($\omega \gg 1$ and $\Delta a \ll D$) for J-controlled crack growth are satisfied. It is then feasible to extend the LEFM R-curve analysis and to perform a J-resistance curve analysis of ductile crack growth and tearing instability. For conditions under which the amount of crack growth is too great to be controlled only by J, a resistance curve must be determined for the geometry of interest or an alternate procedure found. Hutchinson and Paris (5.60) have developed a J-controlled crack growth and stability analysis. In the following a generalization (5.63) of this method is presented.

5.7.1 The Tearing Modulus

Consider a body with a through-thickness crack as illustrated in Figure 5.41. Let P be a generalized load per crack tip and per unit thickness of the body. When more than one crack tip exists, assume the flawed body and its loading are such that each tip experiences the same crack driving force. Take Δ to be the generalized load-point displacement through which P acts to do work on the body. The linear spring in Figure 5.41 can be veiwed as modeling the elastic compliance of the testing machine or any associated structure through which the body is loaded. The prescribed total displacement Δ_T can be written as

$$\Delta_T = C_M P + \Delta(P, a) \tag{5.7-1}$$

where $C_M = mBC_s$ in which m is the number of crack tips.

At impending or during crack extension equilibrium between the crack driving force and the material's resistance to ductile fracture and tearing requires that

$$J(P, a) = J_R(\Delta a) = J_R(a - a_0) \tag{5.7-2}$$

The crack extension is said to be stable if an arbitrarily small increase, $\delta a > 0$, in the current crack length with the total displacement Δ_T held fixed does not

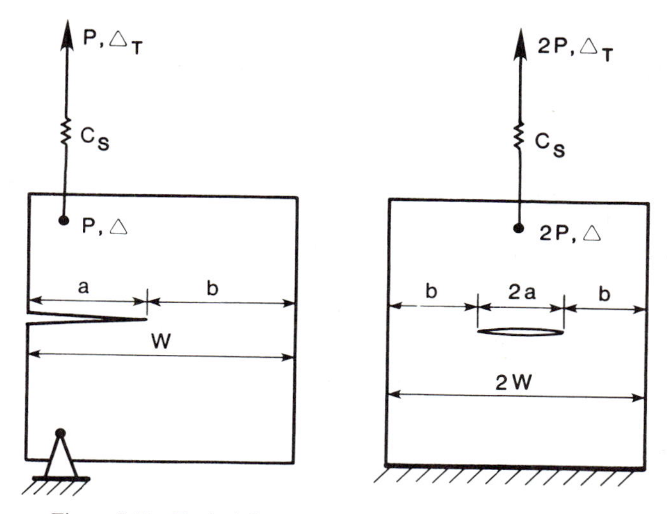

Figure 5.41 Typical flawed structures with one and two crack tips.

give rise to a driving force in excess of the material's fracture resistance. That is, the equilibrium state of Equation (5.7-2) will be stable if

$$J(P, a + \delta a) < J_R(a + \delta a - a_0) \tag{5.7-3}$$

for fixed Δ_T and $\delta a > 0$.

The expansion of this inequality about the current crack length yields

$$\left\{\left(\frac{dJ(P,a)}{da}\right)_{\Delta_T} - \frac{dJ_R(a-a_0)}{da}\right\}\delta a + \frac{1}{2}\left\{\left(\frac{d^2J(P,a)}{da^2}\right)_{\Delta_T} - \frac{d^2J_R(a-a_0)}{da^2}\right\}(\delta a)^2 + \cdots < 0 \tag{5.7-4}$$

in which Equation (5.7-2) has been used. Since this inequality must be satisfied for vanishingly small $\delta a > 0$, then the condition for the stability of the equilibrium state, Equation (5.7-2), reduces to

$$\left(\frac{dJ(P,a)}{da}\right)_{\Delta_T} < \frac{dJ_R(a-a_0)}{da} \tag{5.7-5}$$

The equilibrium is unstable if

$$\left(\frac{dJ(P,a)}{da}\right)_{\Delta_T} > \frac{dJ_R(a-a_0)}{da} \tag{5.7-6}$$

The demarcation between stable and unstable (neutral) equilibrium is expressed by

$$\left(\frac{dJ(P,a)}{da}\right)_{\Delta_T} = \frac{dJ_R(a-a_0)}{da} \tag{5.7-7}$$

Strictly speaking, the stability of the equilibrium defined by Equations (5.7-2) and (5.7-7) depends upon the sign of the next higher-order term in Equation (5.7-4). Equations (5.7-1) and (5.7-2) can be considered as parametric equations relating Δ_T as a function of J. When Equation (5.7-7) is satisfied, it can be shown (5.64) that $d\Delta_T/dJ = 0$. Depending upon the sign of the next higher-order term this condition is associated with either a maximum or an inflection point.

Paris et al. (5.65) introduced the dimensionless tearing moduli defined by

$$T = \frac{E}{\sigma_0^2}\left(\frac{dJ}{da}\right)_{\Delta_T} \quad \text{and} \quad T_R = \frac{E}{\sigma_0^2}\frac{dJ_R}{da} \tag{5.7-8}$$

where σ_0 is an appropriate flow stress. In terms of the tearing moduli, Equations (5.7-5) and (5.7-6) become

$$\begin{aligned} T &< T_R \text{ for stability} \\ T &> T_R \text{ for instability} \end{aligned} \tag{5.7-9}$$

Equations (5.7-2) and (5.7-7) represent two equations for determining the load and the crack length at the limit of stable crack growth. If, as is usually done in a resistance curve analysis for dead loading, J_R and J with P as a parameter are plotted against a as in Figure 5.42, then the limit of stable crack growth is identified with the point of tangency between the resistance curve, J_R, and the driving force curve, J.

A general expression for $(dJ/da)_{\Delta_T}$ can be developed. Consider the differential

$$dJ = \left(\frac{dJ}{\partial a}\right)_P da + \left(\frac{\partial J}{\partial P}\right)_a dP \tag{5.7-10}$$

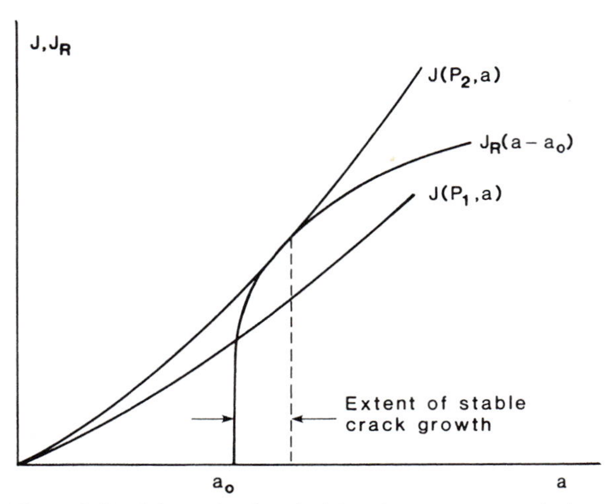

Figure 5.42 Schematic of typical J-resistance curve analysis.

With Δ_T held fixed it follows from (5.7-1) that

$$d\Delta_T = C_M\,dP + \left(\frac{\partial \Delta}{\partial P}\right)_a dP + \left(\frac{\partial \Delta}{\partial a}\right)_P da = 0$$

whence,

$$dP = -da\left(\frac{\partial \Delta}{\partial a}\right)_P \left[C_M + \left(\frac{\partial \Delta}{\partial P}\right)_a\right]^{-1} \tag{5.7-11}$$

The combination of Equations (5.7-10) and (5.7-11) yields

$$\left(\frac{dJ}{da}\right)_{\Delta_T} = \left(\frac{\partial J}{\partial a}\right)_P - \left(\frac{\partial J}{\partial P}\right)_a \left(\frac{\partial \Delta}{\partial a}\right)_P \left[C_M + \left(\frac{\partial \Delta}{\partial P}\right)_a\right]^{-1} \tag{5.7-12}$$

which can be compared with Equation (3.5-7) for LEFM. As in the elastic case the presence of the spring influences $(dJ/da)_{\Delta_T}$ but not J.

It is convenient to write Equations (5.5-3) and (5.5-4) as

$$J = -\left(\frac{\partial U}{\partial a}\right)_\Delta = \left(\frac{\partial U^*}{\partial a}\right)_P \tag{5.7-13}$$

where

$$U = \int_0^\Delta P\,d\Delta \qquad \text{and} \qquad U^* = \int_0^P \Delta\,dP \tag{5.7-14}$$

are, respectively, the strain energy and complementary energy per crack tip and per unit thickness of the body. According to Castigliano's theorem

$$\Delta = \left(\frac{\partial U^*}{\partial P}\right)_a$$

and, hence, by differentiating this expression

$$\left(\frac{\partial \Delta}{\partial a}\right)_P = \frac{\partial^2 U^*}{\partial a\,\partial P} = \frac{\partial}{\partial P}\left[\left(\frac{\partial U^*}{\partial a}\right)_P\right]_a = \left(\frac{\partial J}{\partial P}\right)_a \tag{5.7-15}$$

The introduction of Equation (5.7-15) into Equation (5.7-12) leads to

$$\left(\frac{dJ}{da}\right)_{\Delta_T} = \left(\frac{\partial J}{\partial a}\right)_P - \left(\frac{\partial J}{\partial P}\right)_a^2 \left[C_M + \left(\frac{\partial \Delta}{\partial P}\right)_a\right]^{-1} \tag{5.7-16}$$

For dead loading ($C_M \to \infty$) Equation (5.7-16) reduces to

$$\left(\frac{dJ}{da}\right)_{\Delta_T} = \left(\frac{\partial J}{\partial a}\right)_P \tag{5.7-17}$$

At the other extreme, corresponding to fixed grip loading, $(dJ/da)_{\Delta_T}$ is an absolute minimum when $C_M = 0$. Since in general $(dJ/da)_{\Delta_T} \leq (\partial J/\partial a)_P$, then clearly dead loading provides the most adverse condition for stable crack growth.

5.7.2 The η-Factor

To determine $(dJ/da)_{\Delta_T}$ it is necessary to evaluate the partial derivatives $(\partial J/\partial a)_P$ and $(\partial J/\partial P)_a$ in Equation (5.7-16). To accomplish this it is convenient to decompose the load-point displacement due to the crack into its linear elastic and plastic components so that

$$\Delta_c = \Delta_{ce} + \Delta_{cp} \tag{5.7-18}$$

Equation (5.7-18) permits rewriting Equation (5.5-4) as

$$J = \int_0^P \left(\frac{\partial \Delta_{ce}}{\partial a}\right)_P dP + \int_0^P \left(\frac{\partial \Delta_{cp}}{\partial a}\right)_P dP \tag{5.7-19}$$

since Δ_{nc} is independent of a. It is permissible to write

$$\Delta_{ce} = C_c(a)P, \qquad C_c(0) = 0 \tag{5.7-20}$$

where $C_c(a)/m$ is the contribution to the elastic compliance of the flawed body due to the presence of the crack. It follows from Equations (5.7-19) and (5.7-20) that

$$J_e = \frac{dC_c}{da}\int_0^P P\,dP = \frac{1}{C_c}\frac{dC_c}{da}\int_0^{\Delta_{ce}} P\,d\Delta_{ce} \tag{5.7-21}$$

Equivalently,

$$J_e = \frac{\eta_e}{b}\int_0^{\Delta_{ce}} P\,d\Delta_{ce} = \frac{\eta_e}{b}U_{ce} \tag{5.7-22}$$

where U_{ce} is the elastic contribution to the strain energy due to the crack, and

$$\eta_e = \frac{b}{C_c}\frac{dC_c}{da} = -\frac{b}{P}\left(\frac{\partial P}{\partial a}\right)_{\Delta_{ce}} \tag{5.7-23}$$

is a dimensionless geometric factor. The η_e-factor defined in Equation (5.7-23) is the reciprocal of the η-factor originally introduced by Turner (5.66).

Based upon the developments of Rice et al. (5.51) [cf. Equations (5.5-13) and (5.5-14)], Sumpter and Turner (5.67)* proposed writing

$$J_p = \frac{\eta_p}{b}\int_0^{\Delta_{cp}} P\,d\Delta_{cp} = \frac{\eta_p}{b}U_{cp} \tag{5.7-24}$$

where U_{cp} is the plastic (nonlinear elastic) contribution to the strain energy due to the crack. The dimensionless parameter η_p is similar to η_e in that it is assumed to be a function of the flawed configuration and independent of the deformation. It is necessary and sufficient for the existence of such an η_p that P and Δ_{cp} be related by the separable form

$$P = f(a)g(\Delta_{cp}) \tag{5.7-25}$$

in which $f(a)$ is a function of geometry only and $g(\Delta_{cp})$ is a function of Δ_{cp} but

* Sumpter and Turner used the total plastic strain energy rather than only the contribution due to the crack that is employed here.

independent of a. This form exists at limit load, for deeply cracked bodies in which the remaining ligament experiences primarily bending, and for a body exhibiting power law hardening that is subjected to a single monotonically increasing load parameter.

The use of the η_p-factor simplifies the task of determining J. It allows the stability of crack growth to be assessed rigorously when η_p exists and approximately when it does not. It also permits the stability of J-controlled crack growth to be formulated generally. The assumed existence of η_p does not appear to be any more severe than the assumptions regarding the form of the load-displacement function in alternative approaches. Paris et al. (5.68) argued that η_p does not rigorously exist when the plasticity in the remaining ligament changes substantially as it develops from small-scale yielding to the fully plastic state. In this case the separable form of Equation (5.7-25) does not exist. But, any other approach that relies on a relationship of this type will also suffer the same shortcoming.

The combination of Equation (5.7-24),

$$J_p = \int_0^P \left(\frac{\partial \Delta_{cp}}{\partial a}\right)_P dP \tag{5.7-26}$$

and

$$d\Delta_{cp} = \left(\frac{\partial \Delta_{cp}}{\partial P}\right)_a dP \tag{5.7-27}$$

for a fixed crack length yields

$$\eta_p = \frac{b}{P}\frac{(\partial \Delta_{cp}/\partial a)_P}{(\partial \Delta_{cp}/\partial P)_a} \tag{5.7-28}$$

Equivalently,

$$\eta_p = -\frac{b}{P}\left(\frac{\partial P}{\partial a}\right)_{\Delta_{cp}} \tag{5.7-29}$$

which is the plastic counterpart of Equation (5.7-23).

For a fixed crack length Equations (5.7-22) and (5.7-24) yield

$$dJ = \frac{\eta_e}{b} P\, d\Delta_{ce} + \frac{\eta_p}{b} P\, d\Delta_{cp}$$

and, whence,

$$\left(\frac{\partial J}{\partial P}\right)_a = \left(\frac{\partial J_e}{\partial P}\right)_a + \left(\frac{\partial J_p}{\partial P}\right)_a = P\left[\frac{\eta_e}{b}\left(\frac{\partial \Delta_{ce}}{\partial P}\right)_a + \frac{\eta_p}{b}\left(\frac{\partial \Delta_{cp}}{\partial P}\right)_a\right] \tag{5.7-30}$$

Since Δ_{nc} is independent of a, Equations (5.7-25) and (5.7-30) imply that

$$\left(\frac{\partial J_e}{\partial P}\right)_a = \left(\frac{\partial \Delta_{ce}}{\partial a}\right)_P = P\frac{\eta_e}{b}\left(\frac{\partial \Delta_{ce}}{\partial P}\right)_a \tag{5.7-31}$$

$$\left(\frac{\partial J_p}{\partial P}\right)_a = \left(\frac{\partial \Delta_{cp}}{\partial a}\right)_P = P\frac{\eta_p}{b}\left(\frac{\partial \Delta_{cp}}{\partial P}\right)_a \tag{5.7-32}$$

Noting that $U_{cp} = P\Delta_{cp} - U^*_{cp}$ (see Figure 5.43) one can write

$$J_p = \frac{\eta_p}{b}(P\Delta_{cp} - U^*_{cp})$$

which upon differentiation gives

$$\left(\frac{\partial J_p}{\partial a}\right)_P = \frac{\eta_p}{b}\left[P\left(\frac{\partial \Delta_{cp}}{\partial a}\right)_P - \left(\frac{\partial U^*_{cp}}{\partial a}\right)_P\right] + \frac{\eta_p}{b^2}\left[1 - \frac{b}{\eta_p}\frac{d\eta_p}{db}\right]U_{cp} \tag{5.7-33}$$
$$= \left(\frac{\eta_p P}{b}\right)^2\left(\frac{\partial \Delta_{cp}}{\partial P}\right)_a + \frac{\gamma_p}{b}J_p$$

where

$$\gamma_p = 1 - \eta_p - \frac{b}{\eta_p}\frac{d\eta_p}{db} \tag{5.7-34}$$

In arriving at Equation (5.7-33), $da = -db$ and Equations (5.7-12), (5.7-28), and (5.7-32) were used. In a similar way

$$\left(\frac{\partial J_e}{\partial a}\right)_P = \left(\frac{\eta_e P}{b}\right)^2\left(\frac{\partial \Delta_{ce}}{\partial P}\right)_a + \frac{\gamma_e}{b}J_e \tag{5.7-35}$$

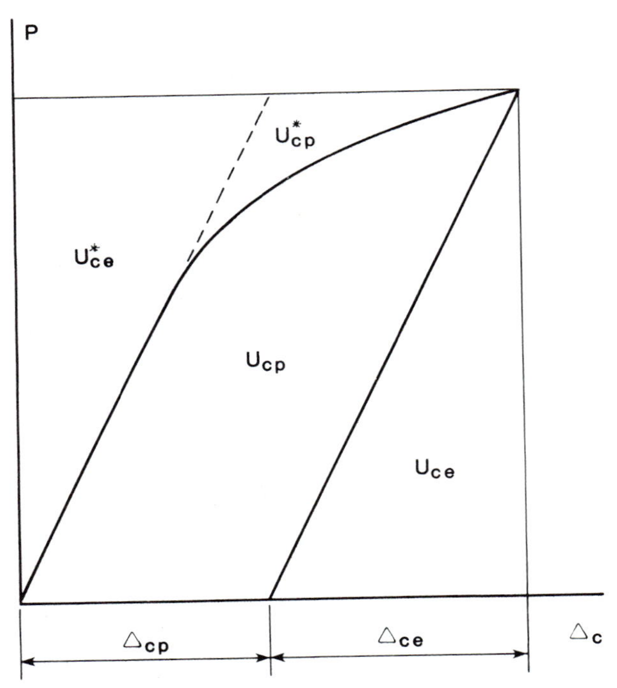

Figure 5.43 Load versus load-point displacement due to the crack.

where

$$\gamma_e = 1 - \eta_e - \frac{b}{\eta_e}\frac{d\eta_e}{db} \tag{5.7-36}$$

The combination of Equations (5.7-33) and (5.7-35) gives

$$\left(\frac{\partial J}{\partial a}\right)_P = \frac{\gamma_e}{b}J_e + \frac{\gamma_p}{b}J_p + \left(\frac{P}{b}\right)^2\left[\eta_e^2\left(\frac{\partial \Delta_{ce}}{\partial P}\right)_a + \eta_p^2\left(\frac{\partial \Delta_{cp}}{\partial P}\right)_a\right] \tag{5.7-37}$$

Finally, the introduction of Equations (5.7-30) and (5.7-37) into Equation (5.7-16) yields

$$\begin{aligned}\left(\frac{dJ}{da}\right)_{\Delta_T} = {} & \frac{\gamma_e}{b}J_e + \frac{\gamma_p}{b}J_p \\ & + \left(\frac{P}{b}\right)^2\left\{\left[\eta_e^2\left(\frac{\partial \Delta_{ce}}{\partial P}\right)_a + \eta_p^2\left(\frac{\partial \Delta_{cp}}{\partial P}\right)_a\right]\left[C_M + \left(\frac{\partial \Delta}{\partial P}\right)_a\right]\right. \\ & \left. - \left[\eta_e\left(\frac{\partial \Delta_{ce}}{\partial P}\right)_a + \eta_p\left(\frac{\partial \Delta_{cp}}{\partial P}\right)_a\right]^2\right\}\left\{C_M + \left(\frac{\partial \Delta}{\partial P}\right)_a\right\}^{-1}\end{aligned} \tag{5.7-38}$$

If the relationship between the load and the load-point displacement is known, either experimentally or analytically, then the tearing modulus can be computed from Equation (5.7-38). The stability of the crack growth can be examined by Equation (5.7-9).

Equations (5.7-22) and (5.7-24) are strictly valid only for a nonextending crack even though they are frequently used to determine J-resistance curves. Since J is based upon deformation theory, it is independent of the path leading to the current values of a and Δ_c provided that the conditions for J-controlled crack growth are satisfied. Thus, for arbitrary increments of a and Δ_{cp}, Equation (5.7-24) yields

$$dJ_p = \frac{\eta_p}{b}\left[\left(\frac{\partial U_{cp}}{\partial \Delta_{cp}}\right)_a d\Delta_{cp} + \left(\frac{\partial U_{cp}}{\partial a}\right)_{\Delta_{cp}} da\right] + \frac{\eta_p U_{cp}}{b^2}\left[1 - \frac{b}{\eta_p}\frac{d\eta_p}{db}\right]da$$

Since $J_p = -(\partial U_{cp}/\partial a)_{\Delta_{cp}}$ and $P = (\partial U_{cp}/\partial \Delta_{cp})_a$, then the preceding equation can be rewritten as

$$dJ_p = \frac{\eta_p}{b}P\,d\Delta_{cp} + \frac{\gamma_p}{b}J_p\,da$$

Because dJ_p is an exact differential, then

$$J_p = \int_0^{\Delta_{cp}} \frac{\eta_p}{b}P\,d\Delta_{cp} + \int_{a_0}^{a} \frac{\gamma_p}{b}J_p\,da \tag{5.7-39}$$

holds for any path leading to the current values of a and Δ_{cp}. The analogous expression for J_e is

$$J_e = \int_0^{\Delta_{ce}} \frac{\eta_e}{b}P\,d\Delta_{ce} + \int_{a_0}^{a} \frac{\gamma_e}{b}J_e\,da \tag{5.7-40}$$

Equation (5.7-40) has been presented to illustrate the symmetry between J_e and J_p. Rather than using Equation (5.7-40) it is simpler to compute J_e from $J_e = K^2/E'$.

The following method can be used to determine J_p for a growing crack from a P-Δ_{cp} curve depicted in Figure 5.44. Since within deformation plasticity theory Equation (5.7-39) for J_p is path-independent, then the path OA for a fixed crack length a_i may be followed to a load-point displacement Δ_{cp}^i. Because $da = 0$ on this path

$$J_A = J_p^i = \left(\frac{\eta_p}{b}\right)_i \int_0^{\Delta_{cp}^i} P\, d\Delta_{cp} \tag{5.7-41}$$

where the integral represents the area under the curve OA. Furthermore,

$$J_B = J_p^i + \left(\frac{\eta_p}{b}\right)_i \int_{\Delta_{cp}^i}^{\Delta_{cp}^{i+1}} P\, d\Delta_{cp} \tag{5.7-42}$$

To determine J_p^{i+1} for a crack length a_{i+1} and displacement Δ_{cp}^{i+1} integrate Equation (5.7-39) along the path OB and then along BC, where Δ_{cp} is constant to obtain

$$J_p^{i+1} = J_B + \int_{a_i}^{a_{i+1}} \frac{\gamma_p}{b} J_p\, da \tag{5.7-43}$$

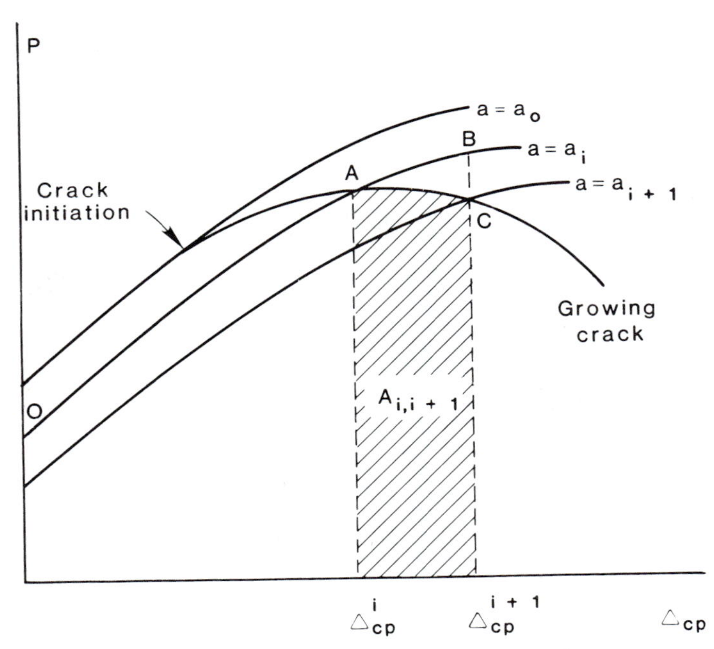

Figure 5.44 Typical P-Δ_{cp} curve for a growing crack.

To a first approximation

$$\int_{a_i}^{a_{i+1}} \frac{\gamma_p}{b} J_p \, da = \left(\frac{\gamma_p}{b}\right)_i J_B(a_{i+1} - a_i) \tag{5.7-44}$$

Hence,

$$J_p^{i+1} = [J_p^i + A_{i,i+1}(\eta_p/b)_i][1 + (a_{i+1} - a_i)(\gamma_p/b)_i] \tag{5.7-45}$$

where $A_{i,i+1}$ is the area under the P-Δ_{cp} curve between Δ_{cp}^i and Δ_{cp}^{i+1}.

In this manner the simultaneous measurement of the load, load-point displacement and crack extension permits until instability intercedes the determination of a J-resistance curve from a single test. The use of a mini-computer in automated data acquisition and reduction simplifies this task.

5.7.3 Illustrative Examples

For a compliant loading of a deeply cracked bend specimen [e.g., see Figure 5.45(a)] the generalized load and displacement are $P = M$ and $\Delta = \theta$, respectively. It follows from dimensional analysis that $M = b^2 F(\theta_c)$ [cf. Equations (5.5-7) and (5.5-8)]. Consequently, Equations (5.7-23) and (5.7-29) lead to $\eta_e = \eta_p = 2$ for this specimen. Equation (5.7-38) reduces to

$$\left(\frac{dJ}{da}\right)_{\theta_T} = -\frac{J}{b} + \frac{4M^2}{b^2} \frac{C}{1 + C(\partial M/\partial \theta_c)_a} \tag{5.7-46}$$

where

$$C = C_M + C_{nc} = C_M + d\theta_{nc}/dM \tag{5.7-47}$$

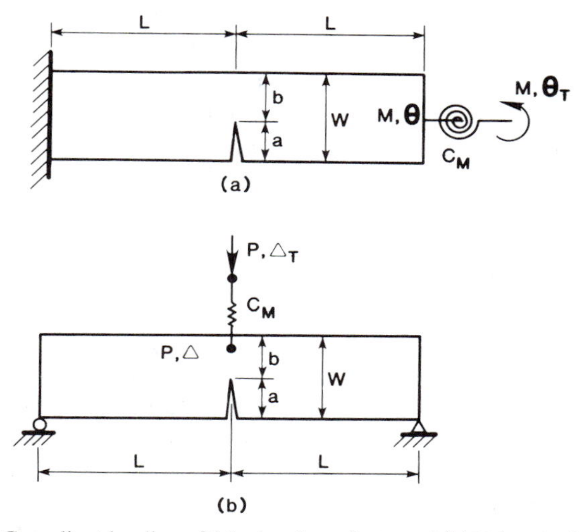

Figure 5.45 Compliant loading of (a) a bend specimen and (b) a three-point bend specimen.

is the combined compliance. Equation (5.7-46) agrees with the results of Hutchinson and Paris (5.60) who used a slightly different approach.

For fixed grip loading ($C_M = 0$) Equation (5.7-46) applies with $C = C_{nc}$. Under dead loading ($C_M = \infty$) Equation (5.7-46) reduces to

$$\left(\frac{dJ}{da}\right)_{\Delta_T} = \left(\frac{\partial J}{\partial a}\right)_M = -\frac{J}{b} + \frac{4M^2}{b^2}\left(\frac{\partial \theta_c}{\partial M}\right)_a \tag{5.7-48}$$

For a fully yielded, elastic-perfectly plastic specimen, M is the limit moment which is independent of θ_c so that Equation (5.7-46) becomes

$$\left(\frac{dJ}{da}\right)_{\theta_T} = -\frac{J}{b} + \frac{4M^2C}{b^2} \tag{5.7-49}$$

This analysis may also be applied to the compliant loading of the three-point bend specimen of Figure 5.45(b). In this case

$$\left(\frac{dJ}{da}\right)_{\Delta_T} = -\frac{J}{b} + \frac{4P^2}{b^2}\frac{C}{1 + C(\partial P/\partial \Delta_c)_a} \tag{5.7-50}$$

in which

$$C = C_M + C_{nc} = C_M + d\Delta_{nc}/dP \tag{5.7-51}$$

is the combined elastic compliance. With $\eta_e = \eta_p = 2$, Equations (5.7-39) and (5.7-40) combine to give

$$J = 2\int_0^{\Delta_c} \frac{P}{b}\, d\Delta_c - \int_{a_0}^{a} \frac{J}{b}\, da \tag{5.7-52}$$

for a growing crack.

Prior to the initiation of crack growth it is possible to evaluate all the quantities on the right-hand sides of Equations (5.7-46) and (5.7-50) from a single experimental record. Since these quantities are also continuous across the initiation point, then it is possible to use Equations (5.7-46) and (5.7-50) and a single experimental record to assess the stability of crack growth at initiation. When the remaining ligament is fully yielded and exhibits little strain hardening, it may be possible in some instances to neglect $C(\partial M/\partial \theta_c)_a$ in Equation (5.7-46) or $C(\partial P/\partial \Delta_c)_a$ in Equation (5.7-50). If these terms are not negligible, it will be necessary to determine them by some other means in order to assess the stability of crack growth beyond initiation.

Paris, Ernst, and Turner (5.68) found that

$$\eta_p = 2 - P\Delta_{cp} \bigg/ \int_0^{\Delta_{cp}} P\, d\Delta_{cp} \tag{5.7-53}$$

for a deeply center-cracked panel. According to Equation (5.7-53), η_p will be equal to or less than unity for this specimen. With this expression Equation (5.7-38) can be shown to yield the same tearing modulus obtained by Hutchinson and Paris (5.60). In general η_p given by Equation (5.7-53) is a function of the deformation. This is, of course, contrary to Turner's original assumption and the one assumed here that η_p be independent of the deformation. If the dependence of η_p upon the deformation is weak, then to a first approximation η_p can be taken outside the integral as in Equation (5.7-24). To the extent that the latter is an appropriate approximation, then Equation (5.7-38) is general since its development does not depend upon the restriction that η_p be independent of the deformation.

Equations (5.7-38)–(5.7-40) also contain the development of Ernst et al. (5.69) as a special case. The latter assumed that $\eta_e = \eta_p = \eta$, which holds in particular for deeply cracked bend specimens, but not in general.

5.7.4 *Tearing Instability for Power Law Hardening*

The forms of Equations (5.7-24) and (5.7-38) are particularly well suited for use with the GE/EPRI elastic-plastic fracture handbook (5.46). For a power law hardening material

$$U_{cp} = \frac{n}{n+1} P\Delta_{cp}$$

which upon the introduction into Equation (5.7-24) yields

$$J_p = \frac{n}{n+1} \frac{\eta_p}{b} P\Delta_{cp} \tag{5.7-54}$$

Substituting Equation (5.4-5) into Equation (5.7-54) one finds that

$$\eta_p = \frac{n+1}{n} \frac{b^2 \sigma_y}{aP_0} \frac{g_1(a/W)}{g_3(a/W)} \frac{h_1(a/W, n)}{h_3(a/W, n)} \tag{5.7-55}$$

which can be evaluated for the specimens and structures included in the handbook. Computations demonstrate that η_p can depend rather strongly upon the crack length and the hardening exponent for tension loading. A typical result is shown in Figure 5.46. However, as shown in Figure 5.47, η_p for a bend specimen is virtually independent of the hardening exponent for a range of crack lengths. For deeply cracked bodies η_p also appears to be independent of the hardening exponent and approaches a constant value that is specimen dependent.

Having established η_p, one can determine $d\eta_p/db$ numerically. Because it is the quantity,

$$\frac{b}{\eta_p} \frac{d\eta_p}{db} = \frac{d(\ln \eta_p)}{d(\ln b)} \tag{5.7-56}$$

that appears in Equation (5.7-34), it may be more convenient to form the numerical derivative of $d(\ln \eta_p)/d(\ln b)$. It also follows from Equation (5.4-5) that

$$\left(\frac{\partial \Delta_{cp}}{\partial P}\right)_a = \frac{n\alpha\sigma_y a}{EP_0} g_3\left(\frac{a}{W}\right) h_3\left(\frac{a}{W}, n\right)\left(\frac{P}{P_0}\right)^{n-1} \tag{5.7-57}$$

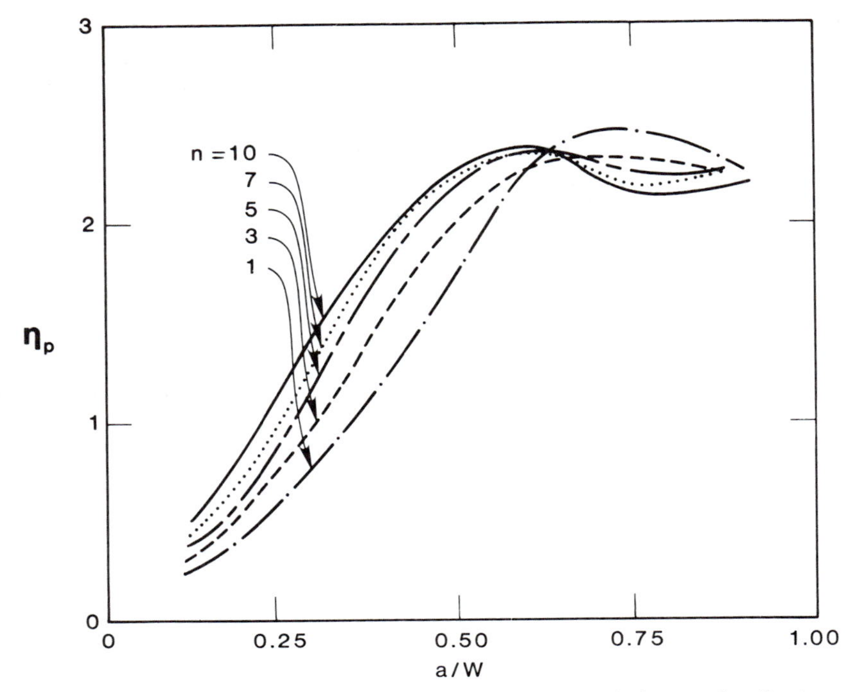

Figure 5.46 The η_p-factor determined from the GE/EPRI plastic fracture handbook as a function of crack length and hardening index for a single-edge-notched tensile specimen.

which completes the determination of the plastic contributions in Equation (5.7-38) for $(dJ/da)_{\Delta_T}$. The elastic contribution to this quantity can be readily evaluated using LEFM handbooks—for example, Tada et al. (5.38).

The principal advantage of this approach over that suggested in reference (5.46) is only a single numerical differentiation rather than four is required. Furthermore, as suggested by Parks et al. (5.70), a judicious choice of g_1 and g_3 can even simplify this computation. To the extent that the tabulated functions h_1 and h_3 exist in reference (5.46) for the configurations of interest, this approach can be used to assess the stability of crack growth in power law hardening materials.

For flawed configurations not included in the handbook, one can generate h_1 and h_3 for the configurations of interest or develop an alternative approach. A rather efficient approximate procedure is the following one. Suppose that an approximation for $\eta_p = \eta_p(b)$ has been established, say, through a combination of dimensional analysis and Equation (5.7-29) as was done in the previous examples. Since

$$J_p = \frac{\eta_p}{b} U_{cp} = -\left(\frac{\partial U_{cp}}{\partial a}\right)_{\Delta_{cp}} = \left(\frac{\partial U_{cp}}{\partial b}\right)_{\Delta_{cp}} \tag{5.7-58}$$

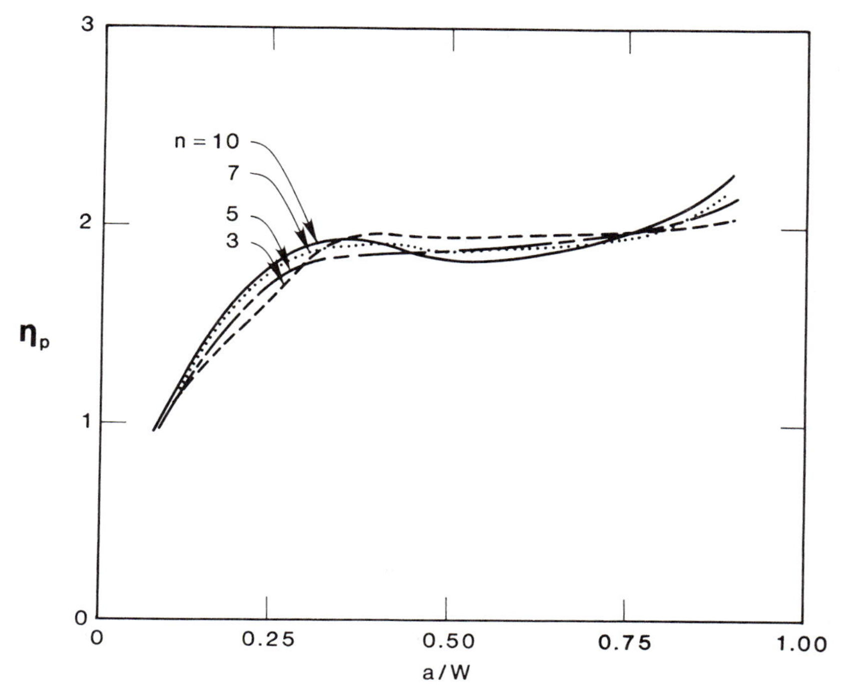

Figure 5.47 The η_p-factor determined from the GE/EPRI plastic fracture handbook as a function of crack length and hardening index for a bend specimen.

then

$$\frac{dU_{cp}}{U_{cp}} = \frac{\eta_p}{b}\,db \quad \text{for } \Delta_{cp} = \Delta_{cp}^0 = \text{constant} \tag{5.7-59}$$

Integrating Equation (5.7-59) yields

$$\frac{U_{cp}(b, \Delta_{cp}^0)}{U_{cp}(b_0, \Delta_{cp}^0)} = \left(\frac{b}{b_0}\right)^{\eta_p(b)} \Gamma\left(\frac{b}{b_0}\right) \tag{5.7-60}$$

where

$$\Gamma\left(\frac{b}{b_0}\right) = \exp\left\{-\int_{b_0}^{b} \ln\left(\frac{b}{b_0}\right)\frac{d\eta_p}{dp}\,db\right\} \tag{5.7-61}$$

A representative reference length of the remaining ligament of the flawed configuration is denoted by b_0 and Δ_{cp}^0 is an associated prescribed plastic load-point displacement. From Il'yushin's theorem it follows that

$$\frac{U_{cp}(b, \Delta_{cp})}{U_{cp}(b, \Delta_{cp}^0)} = \left(\frac{\Delta_{cp}}{\Delta_{cp}^0}\right)^{(n+1)/n} \tag{5.7-62}$$

The combination of Equations (5.7-60) and (5.7-62) leads to

$$U_{cp}(b, \Delta_{cp}) = U_{cp}(b_0, \Delta_{cp}^0)\left(\frac{b}{b_0}\right)^{\eta_p(b)} \Gamma\left(\frac{b}{b_0}\right)\left(\frac{\Delta_{cp}}{\Delta_{cp}^0}\right)^{(n+1)/n} \tag{5.7-63}$$

In addition

$$J_p = \frac{\eta_p}{b} U_{cp}(b_0, \Delta_{cp}^0)\left(\frac{b}{b_0}\right)^{\eta_p(b)} \Gamma\left(\frac{b}{b_0}\right)\left(\frac{\Delta_{cp}}{\Delta_{cp}^0}\right)^{(n+1)/n} \tag{5.7-64}$$

From Castigliano's first theorem

$$\left(\frac{\partial U_{cp}}{\partial \Delta_{cp}}\right)_a = P = \frac{n+1}{n}\frac{U_{cp}(b_0, \Delta_{cp}^0)}{\Delta_{cp}^0}\left(\frac{b}{b_0}\right)^{\eta_p(b)} \Gamma\left(\frac{b}{b_0}\right)\left(\frac{\Delta_{cp}}{\Delta_{cp}^0}\right)^{1/n} \tag{5.7-65}$$

from which it follows that

$$\left(\frac{\partial \Delta_{cp}}{\partial P}\right)_a = \left\{\frac{n+1}{n^2}\frac{U_{cp}(b_0, \Delta_{cp}^0)}{(\Delta_{cp}^0)^2}\left(\frac{b}{b_0}\right)^{\eta_p(b)} \Gamma\left(\frac{b}{b_0}\right)\left(\frac{\Delta_{cp}}{\Delta_{cp}^0}\right)^{(1-n)/n}\right\}^{-1} \tag{5.7-66}$$

For a given power law hardening material and a flawed configuration a single reference computation for a prescribed plastic load-point displacement Δ_{cp}^0 and remaining ligament length b_0 is sufficient to determine $U_{cp}(b_0, \Delta_{cp}^0)$. The finite element method can efficiently perform this type of computation. Having established $U_{cp}(b_0, \Delta_{cp}^0)$, it is clear from Equations (5.7-64)–(5.7-66) that P and $(\partial \Delta_{cp}/\partial P)_a$ can be determined for any other crack length and load-point displacement for this configuration. When these quantities are combined with their elastic counterparts, everything is in place for performing a tearing instability analysis.

A comparison of this approach and the known solution from the GE/EPRI handbook for bend specimens is made in Tables 5.3 and 5.4 for plane stress and plane strain. Based upon the previous dimensional analysis, $\eta_p = 2$ is used for the bend specimen. In these tables $b_0 = W/2$ is the reference length of the remaining ligament and J_0 is the corresponding value of J_p for a fixed Δ_{cp}. If there were perfect agreement, then the ratio of $(J_p/J_0)_h$ from the handbook and (J_p/J_0) determined from Equation (5.7-64) would be unity. For the most part the agreement is fairly good. Significant differences appear for $a/W < 0.25$ and

Table 5.3 Comparison of Handbook Solution and Approximate Solution for Plane Stress Three-Point Bend Specimen

$(J_p/J_0)_h/(J_p/J_0)$

a/W	$n = 1$	2	3	5	7	10	13	16	20
$\frac{1}{8}$	25.78	0.574	0.707	0.640	0.699	0.713	0.729	0.742	0.764
$\frac{1}{4}$	4.484	0.853	0.828	0.906	0.948	0.946	0.950	0.957	0.977
$\frac{3}{8}$	1.818	1.018	0.979	0.991	0.945	0.984	0.985	0.990	0.992
$\frac{1}{2}$	1.000	1.000	1.000	1.000	1.000	1.000	1.000	1.000	1.000
$\frac{5}{8}$	0.653	0.955	0.977	0.999	0.995	0.999	1.000	1.003	1.011
$\frac{3}{4}$	0.512	0.884	0.926	0.945	0.956	0.961	0.960	0.962	0.969
$\frac{7}{8}$	0.444	0.862	0.922	0.962	0.985	1.003	1.018	0.923	1.025

Table 5.4 Comparison of Handbook Solution and Approximate Solution for Plane Strain Three-Point Bend Specimen

$(J_p/J_0)_h/(J_p/J_0)$

a/W	$n = 1$	2	3	5	7	10	13	16	20
$\frac{1}{8}$	25.96	0.553	0.541	0.581	0.613	0.636	0.638	0.637	0.641
$\frac{1}{4}$	4.498	0.798	0.793	0.866	0.910	0.953	0.941	0.952	0.957
$\frac{3}{8}$	1.813	0.993	0.997	1.006	0.984	1.004	0.984	0.980	0.963
$\frac{1}{2}$	1.000	1.000	1.000	1.000	1.000	1.000	1.000	1.000	1.000
$\frac{5}{8}$	0.657	0.971	1.016	1.040	1.049	1.050	1.028	1.041	1.031
$\frac{3}{4}$	0.509	0.930	1.014	1.032	1.050	1.078	1.070	1.080	1.074
$\frac{7}{8}$	0.444	0.892	0.989	1.056	1.098	1.150	1.136	1.144	1.097

smaller departures for the larger values of n and $a/W > \frac{3}{4}$ exist. The large variances for $n = 1$ are of little practical consequence since linear elastic solutions are readily available. The range of a/W and n for which the differences are significant is, as expected, the same range where the plots in Figure 5.47 differ appreciably from $\eta_p = 2$.

It is also clear from Tables 5.3 and 5.4 that, except for $n = 1$, better agreement occurs for crack lengths close to the reference crack length. Thus, it is advisable to perform the reference computation for the anticipated flaw size. In this way the error introduced by the approximation inherent to this approach will be minimized when neighboring flaw sizes are considered. In addition, for reasons of precision it is better to work with Δ_{cp} than P. If Equation (5.7-65) is used to eliminate Δ_{cp} in Equation (5.7-64), then the term $(b/b_0)^{-\eta_p n}$ appears in J_p. Because of this term a small error in η_p, when multiplied by a large value of n, can produce a substantial error in J_p if b/b_0 differs appreciably from unity.

With this approach a tearing instability analysis can be performed in the following manner. For an assumed value of the crack extension Δa, $J_R(\Delta a)$ is determined from the resistance curve. A value of Δ_{cp} is selected and J_p and P are computed from Equations (5.7-64) and (5.7-65). The value of P can be used to compute J_e using LEFM methods. The sum, $J = J_e + J_p$, is compared with J_R. If $J \neq J_R$, then Δ_{cp} is adjusted appropriately and the procedure repeated until $J = J_R$. Next, $(dJ/da)_{\Delta_T}$ is computed using Equation (5.7-38), and the stability of the crack growth is assessed by means of Equation (5.7-9). Alternatively, the same procedure can be followed to determine $(dJ/da)_{\Delta_T}$ versus $J = J_R$. The value of J at instability is identified with the point of intersection (if it exists) of this curve and dJ_R/da versus J_R. The resistance curve can be used to determine the limit of stable crack growth. When J_e can be neglected compared to J_p, then $J_p = J_R$ so that Δ_{cp} can be determined without iteration from Equation (5.7-64); whereupon, P follows from Equation (5.7-65).

5.7.5 Applications

For application of a tearing instability analysis consider a long pressurized pipe of mean radius R and thickness t with a through wall axial crack of total

Table 5.5 Critical Crack Experiments on A106 Grade B Carbon Steel Pipe with Through-Wall Axial Flaws

Pipe	Experiment	Test Temperature (°F)	Total Axial Crack Length (in.)	σ_h, Nominal Hoop Stress at Failure (ksi)	Tensile Data		Outside Radius (in.)	Wall Thickness (in.)	Predicted Hoop Stress at Failure	
					Yield Stress (ksi)	Ultimate Stress (ksi)			J/T Analysis (ksi)	Plastic Collapse (ksi)
C1	3	575	24.5	13.49	33.0	75.4	12	1.735	13.6	14.3
C1	1	575	18.5	19.74	33.0	75.4	12	1.674	17.2	17.0
C1	2	587	18.5	18.03	32.8	75.0	12	1.593	17.2	17.3
C2	5	675	18.5	16.48	30.6	75.0	12	1.64	17.2	17.2
C2	7	670	18.5	17.05	30.6	75.0	12	1.635	17.2	17.2
C5	10	661	18.5	17.75	32.6	77.9	12	1.64	17.2	18.0
C5	15	639	18.5	19.85	32.6	77.9	12	1.64	17.2	18.0
C2	6	554	11.6	24.50	34.1	81.5	12	1.715	24.3	27.1
C2	17	642	6.0	33.94	33.6	82.3	12	1.65	35.5	38.1
				Averages	32.5	77.5				
					$\sigma_0 = 45.8$					
C8	13	555	14.5	17.3	36.5	74.7	12	0.700	15.5	15.9
C8	11	547	10.25	23.55	36.5	74.7	12	0.705	20.5	21.0
C8	12	561	5.25	33.0	36.5	74.7	12	0.710	32.1	32.4
C7	16	581	2.5	42.8	36.5	74.7	12	0.700	41.0	42.6
				Averages	36.5	74.7				
					$\sigma_0 = 46.3$					
C10	23	567	10.25	15.8	42.8	74.0	6.375	0.700	16.5	17.0
C10	22	538	5.25	24.8	42.8	74.0	6.375	0.707	27.0	28.1
C10	21	605	2.5	39.0	42.8	74.0	6.375	0.710	40.0	40.3
				Averages	42.8	74.0				
					$\sigma_0 = 48.7$					

$\sigma_0 = (\sigma_y + \sigma_u)/2.4$

length $2a$. The pressure loading in essence corresponds to a condition of load control. A tearing instability in this case usually results in a rapidly propagating axial crack accompanied by a depressurization of the pipe. A summary of failure test data for A106B carbon steel pressurized pipes is contained in Table 5.5 (5.71).

Typical J-resistance curves for compact specimens of ASTM A106 Class C steel obtained by Gudas and Anderson (5.72) are shown in Figure 5.48. These tests employed specimens with 20 percent side grooves to minimize the formation of shear lips. The test temperature of 550°F is in the temperature range of 538–625°F for the test data in Table 5.5. Within the scatter evident in Figure 5.48, the J-resistance curve after crack initiation can be approximated by a straight line. Average values of J_{Ic} and dJ_R/da from six tests are 1.8 ksi-in. and 18.6 ksi, respectively. In the following analysis the J-resistance curve is taken to be a straight line passing through $J_R = 1.8$ ksi-in. and $\Delta a = 0$ and having constant slope $dJ_R/da = 18.6$ ksi.

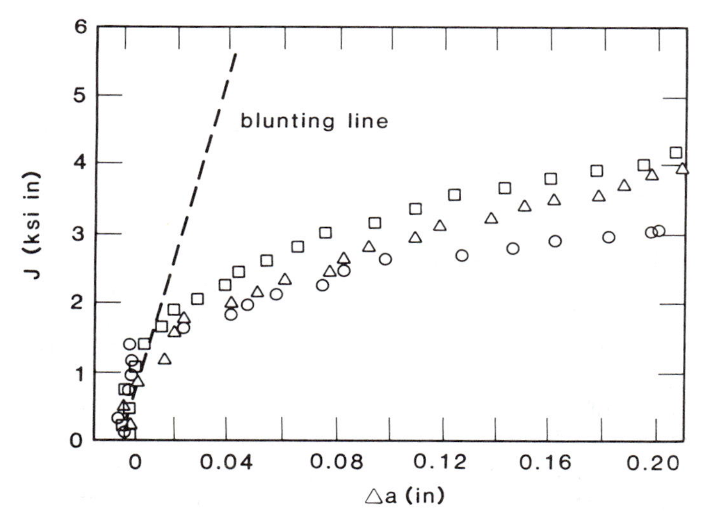

Figure 5.48 J-resistance curve for A106 carbon steel compact tension specimens (5.72).

Based upon a Dugdale model for the plastic zone ahead of the crack tip, the J-integral for the pressurized pipe is given by Equation (5.1-18). In this analysis the flow stress is taken to be

$$\sigma_0 = (\sigma_y + \sigma_u)/2.4 \tag{5.7-67}$$

where σ_u is the ultimate tensile stress. A family of curves of the J-integral versus the half-crack length, a, with the hoop stress as a parameter is shown in Figure 5.49 for the first data set of Table 5.5. Simflar curves for the remaining two data sets of Table 5.5 can be constructed.

Superimposed on Figure 5.49 is the resistance curve (shown dashed) of Figure 5.48 for Experiment 3 in Table 5.5 for which $a_0 = 12.25$ in. When

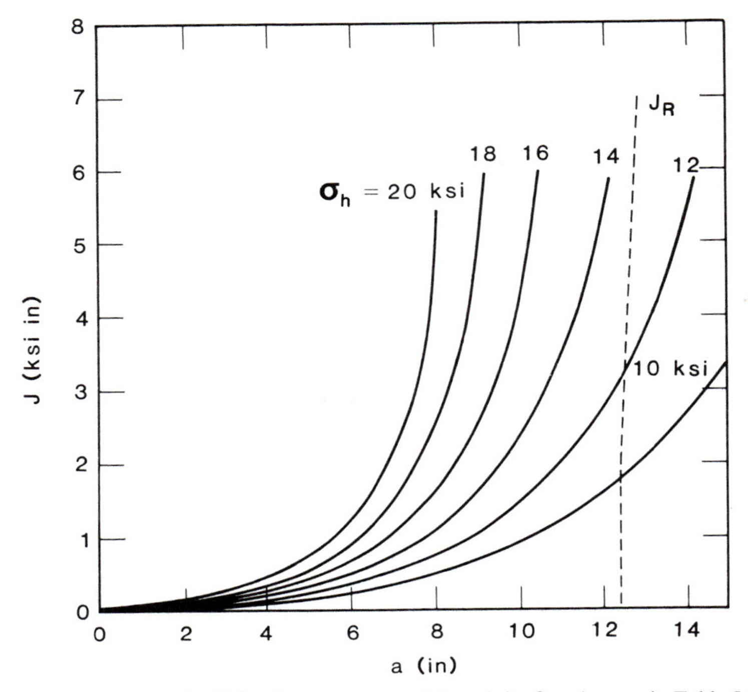

Figure 5.49 Crack driving force versus crack length for first data set in Table 5.5.

plotted on this scale, the relatively large value of dJ_R/da for this material becomes more vivid. The condition for tangency between the resistance curve and a crack driving force curve identifies the hoop stress at instability, which is predicted to be 13.6 ksi for this test. This prediction compares favorably with the measured value of 13.5 ksi. For a failure criterion based upon J attaining the critical value of $J_{Ic} = 1.8$ ksi-in., the predicted hoop stress would be 10 ksi. This underestimates the observed failure stress by 35 percent. With everything else constant the percentage difference between measured and predicted hoop stresses based upon a critical value of J will increase with increasing crack length.

The same procedure can be repeated for the other tests in the data set. In this manner the predictions of Table 5.5 were obtained. A graphical comparison of the measured and predicted hoop stresses at failure appears in Figure 5.50. If the agreement were perfect, all the solid points would fall on the straight line. When all things are considered, the agreement is very good. The successful use of the Dugdale model in this instability analysis depends upon using an appropriate value for the flow stress.

It is clear that dJ_R/da is relative large in this instance. If Equation (5.1-18) is differentiated, then

$$\frac{dJ}{da} = \frac{J}{a} + \frac{2\sigma_0\sigma_h}{ME}\left[2.51\,\frac{a^2}{Rt} - 0.054\,\frac{a^4}{R^2t^2}\right]\tan\left(\frac{\pi}{2}\,\frac{M\sigma_h}{\sigma_0}\right) \qquad (5.7\text{-}68)$$

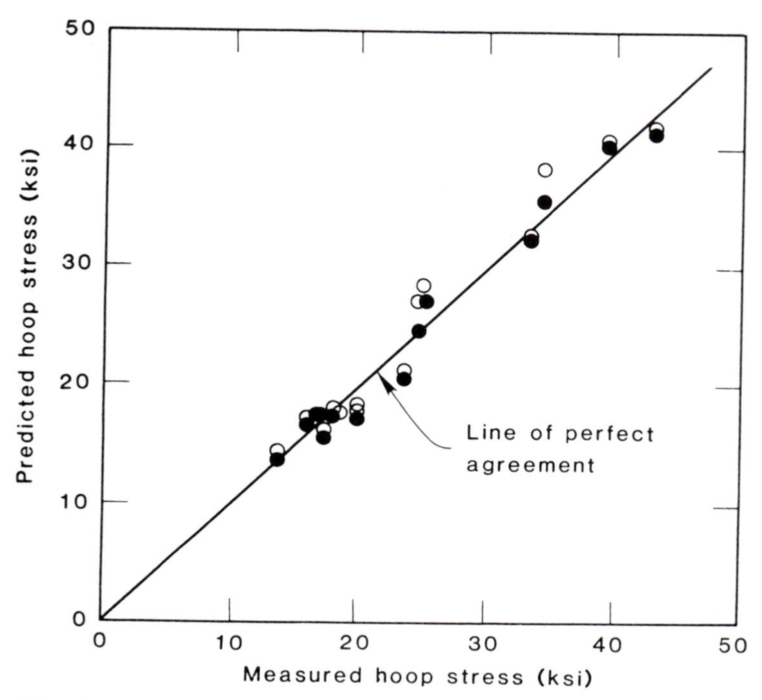

Figure 5.50 Comparison of measured and predicted hoop stress at failure of axially cracked A106B steel pipes based upon a J/T analysis (●) and plastic collapse (○).

As $M\sigma_h/\sigma_0 \to 1$, dJ/da becomes unbounded. Conversely, for very large (unbounded) values of dJ_R/da, instability can be expected to occur when

$$M\sigma_h/\sigma_0 = 1 \tag{5.7-69}$$

which in Section 5.1 was associated with plastic collapse. The hoop stress for plastic collapse in this case will bound from above the value for a tearing instability. Predicted hoop stresses for plastic collapse based upon Equation (5.7-69) are summarized in Table 5.5 and compared in Figure 5.50 (open circles) with the measured values at failure. Again very good agreement is observed. This is consistent with the observation that flawed structures made of high toughness, low strength materials under load control usually fail near limit load.

As an example of a compliant loading system consider the four-point bend test of a circumferentially cracked pipe depicted in Figure 5.51. This problem has been treated by Zahoor and Kanninen (5.73). The case of a noncompliant loading has been considered by Tada et al. (5.74).

The limit load for this configuration is

$$P_0 = \frac{8\sigma_0 R^2 t}{Z - L} h(\phi) \tag{5.7-70}$$

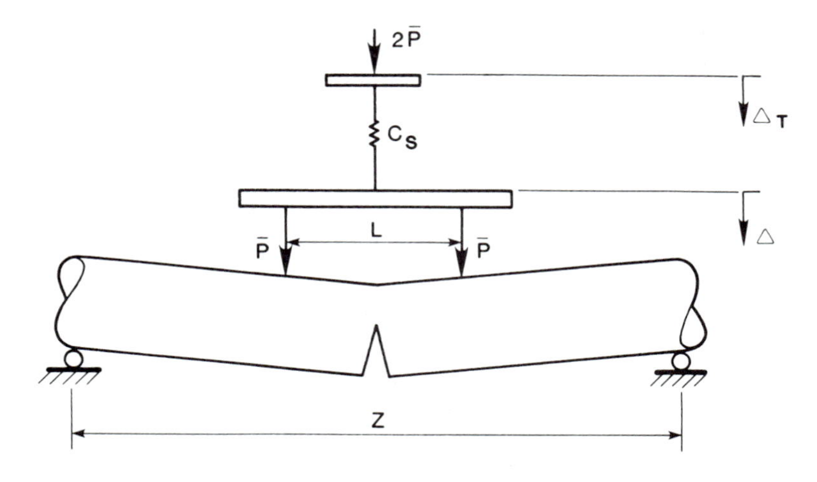

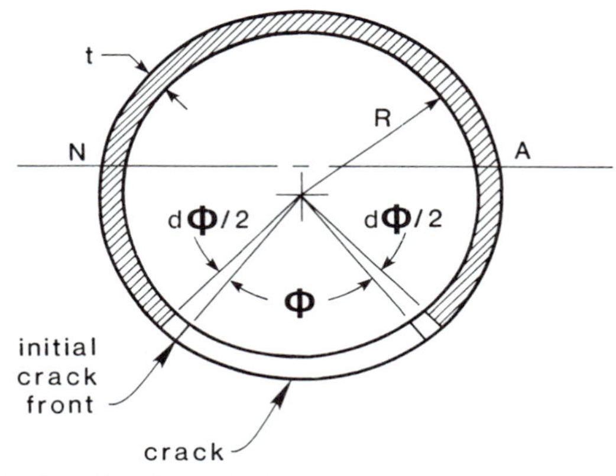

Figure 5.51 Compliant four-point bend loading of a circumferentially cracked pipe.

where

$$h(\phi) = \cos(\phi/4) - \tfrac{1}{2}\sin(\phi/2) \tag{5.7-71}$$

For this case

$$b = (2\pi - \phi)R/2 \quad \text{and} \quad P = \bar{P}/t \tag{5.7-72}$$

It can be argued from dimensional considerations that Δ_{cp} must be a function of $\bar{P}/P_0$ or, equivalently,

$$P = h(\phi)g(\Delta_{cp}) \tag{5.7-73}$$

Noting that $\partial/\partial a = (2/R)(\partial/\partial\phi)$ and introducing Equation (5.7-73) into Equation (5.7-29) one obtains

$$\eta_p = -(2\pi - \phi)h'/h \equiv (2\pi - \phi)Rt\beta \tag{5.7-74}$$

With the aid of Equations (5.7-72) and (5.7-74), Equation (5.7-24) yields*

$$J_p = 2\beta \int_0^{\Delta_{cp}} \bar{P}\, d\Delta_{cp} \tag{5.7-75}$$

for a nongrowing crack. For an extending crack Equations (5.7-39) and (5.7-74) combine to give

$$J_p = 2\beta \int_0^{\Delta_{cp}} \bar{P}\, d\Delta_{cp} + \int_{\phi_0}^{\phi} \gamma J_p \, d\phi \tag{5.7-76}$$

where $\gamma = h''/h'$ and β is evaluated for the initial crack angle ϕ_0.

When the elastic contributions to J and Δ can be neglected compared to their plastic counterparts, Equation (5.7-38) reduces to

$$\left(\frac{dJ}{da}\right)_{\Delta_T} = \left[1 - \eta_p - \frac{b}{\eta_p}\frac{d\eta_p}{db}\right]\frac{J}{b} + \left(\frac{\eta_p P}{b}\right)^2 \frac{C}{1 + C(\partial P/\partial \Delta_{cp})_a} \tag{5.7-77}$$

where $C = C_M + d\Delta_{nc}/dP$. Substitution of $C_M = 2C_s t$ and Equation (5.7-74) into Equation (5.7-77) leads to

$$\left(\frac{dJ}{da}\right)_{\Delta_T} = \frac{4t(\beta\bar{P})^2(2C_s + C_e)}{1 + (2C_s + C_e)(\partial \bar{P}/\partial \Delta_{cp})_a} + \frac{2\gamma}{R} J \tag{5.7-78}$$

where

$$C_e = \frac{d\Delta_{nc}}{d\bar{P}} = \frac{(Z - L)^2(Z + 2L)}{24EI} \tag{5.7-79}$$

is the elastic compliance of the uncracked pipe and EI is its flexural rigidity. The term $(\partial \bar{P}/\partial \Delta_{cp})_a$ can be evaluated from the load-displacement record up to the point of crack initiation. For many, but not all, ductile materials crack initiation occurs very near maximum load where $(\partial \bar{P}/\partial \Delta_{cp})_a \approx 0$. In this case Equation (5.7-78) reduces to

$$\left(\frac{dJ}{da}\right)_{\Delta_T} = 4t(\beta\bar{P})^2(2C_s + C_e) + \frac{2\gamma J}{R} \tag{5.7-80}$$

Figure 5.52 depicts J-resistance curves for Type 304 stainless steel obtained from a center crack panel, a three-point bend bar and two circumferentially cracked pipes in four-point bend (5.75). The pipes were tested in essentially the configuration shown in Figure 5.51 but without the spring. The load-displacement and the load-crack length records were used to establish the J-resistance curves for the pipes using Equation (5.7-76). It is clear from Figure 5.52 that these J-resistance curves exhibit significant geometry dependence. The variances in the initiation values of J are due in part to the lack of conditions of plane strain, to differences in material, and to estimation methods used to deduce J from measurements.

* Apparently, the factor of 2 in Equation (5.7-75) was inadvertently dropped in reference (5.73).

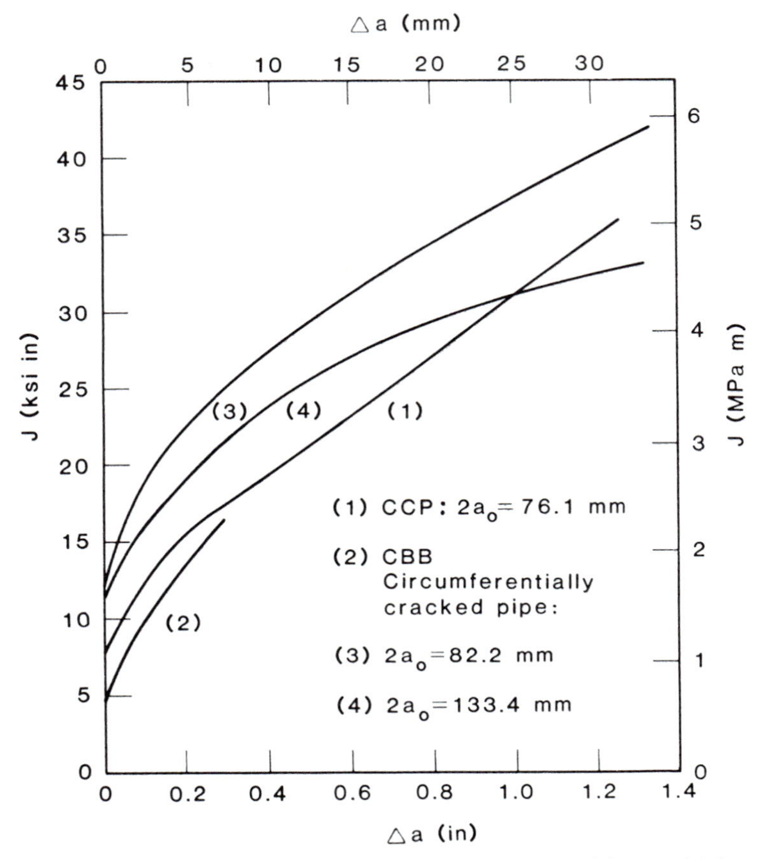

Figure 5.52 Comparison of J-resistance curves for Type 304 stainless steel inferred from experiments on four cracked configurations.

A pipe fracture instability experiment on a 102-mm (4-in.) diameter Schedule 80, Type 304 stainless steel pipe is reported by Wilkowski et al. (5.76). The flaw is a through wall crack having a total length of 104 mm ($\phi = 104°$). The loading configuration is the one depicted in Figure 5.51 with $Z = 1.35$ m and $L = 0.41$ m. The compliance of the spring is $C_s = 1.63$ m/MN and simulates the compliance of an approximately 9-m long pipe. This yields $C_s/C_e \approx 21$. The fracture instability was observed to occur after the load had decreased to 86 percent of its maximum value, which followed shortly after initiation of crack extension. The average stable crack growth at each tip was about 19 mm.

In this experiment the remaining ligament is fully yielded and near limit load conditions exist. For an assumed value of J the extent of crack growth can be determined from a J-resistance curve such as in Figure 5.52. When this crack extension is added to the initial crack length, the corresponding limit load can be computed from Equation (5.7-70). Finally, for the assumed value of J and with $\bar{P} = P_0$, $(dJ/da)_{\Delta_T}$ is determined from Equation (5.7-80). In this manner a plot of $(dJ/da)_{\Delta_T}$ versus J can be constructed. The solid curves in Figure 5.53 were established in this fashion for the J-resistance curves for pipes 3 and 4 in Figure 5.52. Superimposed on this plot is dJ_R/da versus J_R from the resistance curves of these pipes. These are the dashed curves in Figure 5.53.

Since $J = J_R$ has been employed, then at the point of intersection of the applied and material curves, $(dJ/da)_{\Delta_T} = dJ_R/da$, which defines the limit of stable crack extension. To the left of this point of intersection $(dJ/da)_{\Delta_T}$ is less than dJ_R/da and, hence, the crack growth is stable. Conversely, the crack growth will be unstable to the right of this point. The value of $J = J_R$ at the point of intersection can be used to determine from the J-resistance curve the amount of crack extension at instability. The extent of stable crack growth at each tip is predicted to be between 15.2 mm and 22.1 mm compared to the observed growth of 19 mm before instability. If the analysis is repeated using the J-resistance curve for the three-point bend bar, then the predicted crack growth at instability would be only 6.8 mm.

Further examples of tearing instability analyses for compliant loading systems can be found in reference (5.77) for three-point bend specimens and in

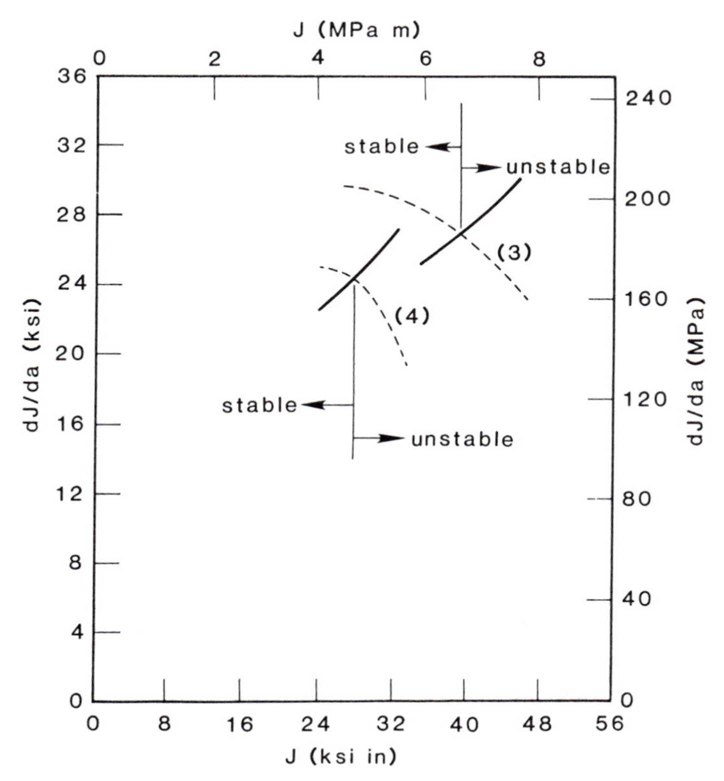

Figure 5.53 Fracture instability predictions for Type 304 stainless steel circumferentially cracked pipes using J-resistance curves for pipes.

reference (5.78) for compact specimens. The foregoing stability analysis for a circumferentially cracked pipe has been applied by Joyce (5.79). In these analyses strain hardening was accommodated only approximately through the use of a flow stress. Pan et al. (5.64) applied a simple method that accounts for material hardening to predict tearing instability in a circumferentially cracked pipe in bending.

The J-resistance curve will be unique only for limited amounts of stable crack growth; otherwise, as seen in Figure 5.52, the curves exhibit geometry dependence. The tearing instability analysis for the circumferentially cracked pipe demonstrates that to make reasonable predictions of extended stable crack growth and instability, the J-resistance curve that properly reflects the degree of plastic constraint at the crack tip must be used. The triaxiality that is known to determine the degree of plastic constraint varies significantly as the primary loading on the remaining ligament changes from tension to bending. Thus, for extended amounts of crack growth, at least two fracture parameters would be required to characterize the intensity of the deformation as well as the triaxiality. It has been suggested that a lower bound J-resistance curve, say, from a compact specimen would provide conservative estimates for the circumferentially cracked pipe in bending. This is borne out by the previous analysis.

In summary the J-integral, tearing modulus approach to the analysis of stable crack growth and tearing instability depends upon the uniqueness of the J-resistance curve. To the extent that the J-resistance curve is unique, then this approach, which comprises a resistance curve analysis, is well founded. From an analysis point of view it makes little difference whether or not the loading is compliant. When J-controlled crack growth exists any discrepancy between predicted and observed behavior must be due to approximations in the analysis and/or experimental scatter, but not due to the concept of a resistance curve analysis. The resistance curve analysis can still be performed when J-controlled crack growth is lost. It is only necessary to use a resistance curve that reflects the proper crack-tip constraint. When this necessitates developing a resistance curve for the specific flawed configuration of interest, than nearly all the appealing aspects of the approach are lost.

5.8 Extended Crack Growth

The J-integral is an appropriate fracture characterizing parameter governing the initiation of crack growth in ductile materials. Furthermore, its use in the analysis of quasi-statically extending cracks can be justified when the conditions for J-dominance and J-controlled growth ($\Delta a \ll D$ and $\omega \gg 1$) are satisfied. For certain flawed configurations the amount of crack extension permitted within these restrictions may be quite limited—for example, less than 10 percent of the remaining ligament. When the conditions for J-controlled growth are not fulfilled, the J-resistance curve is no longer a unique material property, but becomes a function of the flawed geometry. Under such conditions the application of the J-integral approach must be limited by the use of a J-resistance curve obtained for the specific geometry of concern. This severely diminishes the efficacy of this approach.

Because the triaxial constraint is usually greatest in bending, the J-resistance curve obtained from a bend specimen seems to have the lowest value. It can be argued that an analysis of a flawed configuration using a lower bound resistance curve will lead to conservative predictions of fracture instability that are still more realistic than those based upon a LEFM analysis. It should be recognized that not only must the J-resistance curve be a lower bound, but also its slope, dJ_R/da, must be smaller. Moreover, the stress analysis must not underestimate the actual values of J and dJ/da.

If these restrictions are satisfied and if the prediction is not inordinately conservative, then the J-based plastic fracture mechanics will suffice. If not, an alternate fracture criterion must be found. It is clear that the criterion should be independent of the geometry of the flawed configuration. Physical relevance and ease of application are other prime considerations.

An effective way of evaluating potential fracture characterizing parameters is by what is referred to as generation-application phase analyses. In this approach tests are conducted to gather data on initiation of crack extension and subsequent stable growth. In the generation phase, an analysis is performed in which the experimentally observed load-stable crack growth behavior is reproduced in a finite element model and each potential fracture characterizing parameter is evaluated. In the application phase an analysis is again executed using one of the candidate criteria to predict the load-crack growth phenomenon for an alternate specimen. The criterion is assessed by comparing the predicted and observed behavior. In this manner an appropriate fracture criterion can be identified. Figure 5.54 shows this method.

5.8.1 The Crack-Tip Opening Angle

A number of different criteria can be used for the basis of a plastic fracture methodology to predict crack growth and instability using the concept of a resistance curve. Two of the most appealing ones for extended crack growth are based upon the J-integral and the crack opening angle. There are two definitions for the crack opening angle. The crack-tip opening angle (CTOA) reflects the local slope of the crack faces near the crack tip. The average crack opening angle (COA) is the ratio of the crack opening displacement at the site of the initial crack tip to the current crack extension. While the value of the COA can be measured, its relationship to the events occurring at the crack tip is somewhat nebulous for extended stable growth. On the other hand, the CTOA reflects more closely the crack-tip behavior, but its measurement represents a formidable task. When J-dominance exists the intensity of the crack-tip displacement field is measured by J and, hence, J and CTOA are equivalent parameters. However, when J-dominance and J-controlled growth are lost after relatively small amounts of crack extension, J and CTOA are no longer equivalent.

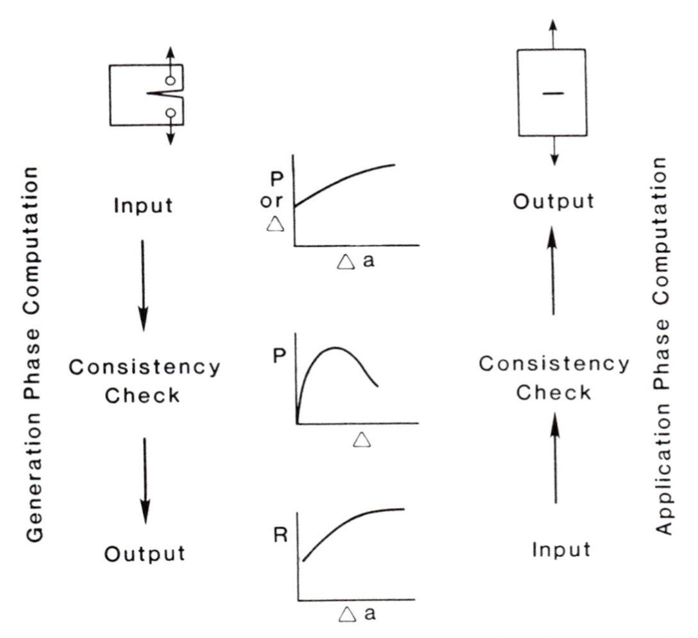

Figure 5.54 Schematic of generation/application phase analysis for developing a plastic fracture methodology.

Through a finite element simulation of experiments by Broek (5.80) on 2-mm thick, 2024-T3 aluminum, center cracked tension panels de Koning (5.81) demonstrated the constancy of CTOA during crack growth. It is clear from Figure 5.55 that while the J-resistance curve continually rises, the CTOA remains virtually constant until instability. The computed shape of the crack opening can be compared qualitatively with the observed profile for a crack in the Type 304 stainless steel center-cracked tension panel in Figure 5.56 (5.82).

Kanninen et al. (5.82) went one step beyond de Koning in their investigation. First, they performed a generation phase analysis for a 2219-T87 aluminum compact specimen for which the results are illustrated in Figure 5.57. Again the J-resistance curve increases with crack growth whereas the CTOA after an initial transient remains nearly invariant. Next they performed application phase analyses for a 2217-T3 aluminum center cracked panel. The results of four such analyses along with experimental measurements appear in Figure 5.58. In one computation the J-resistance curve from the compact specimen was used. In the second case a constant CTOA fracture criterion was employed. The third analysis used the J-resistance curve for about the first 10 mm of crack extension and then switched to the plateau value of 0.08 radians for the CTOA. The fourth analysis was similar but used the COA with a plateau value of 0.05 radians. It is seen that the combined J/CTOA criterion was still capable of handling the extended growth. Since the experiment was conducted under load-controlled conditions, instability occurred at maximum

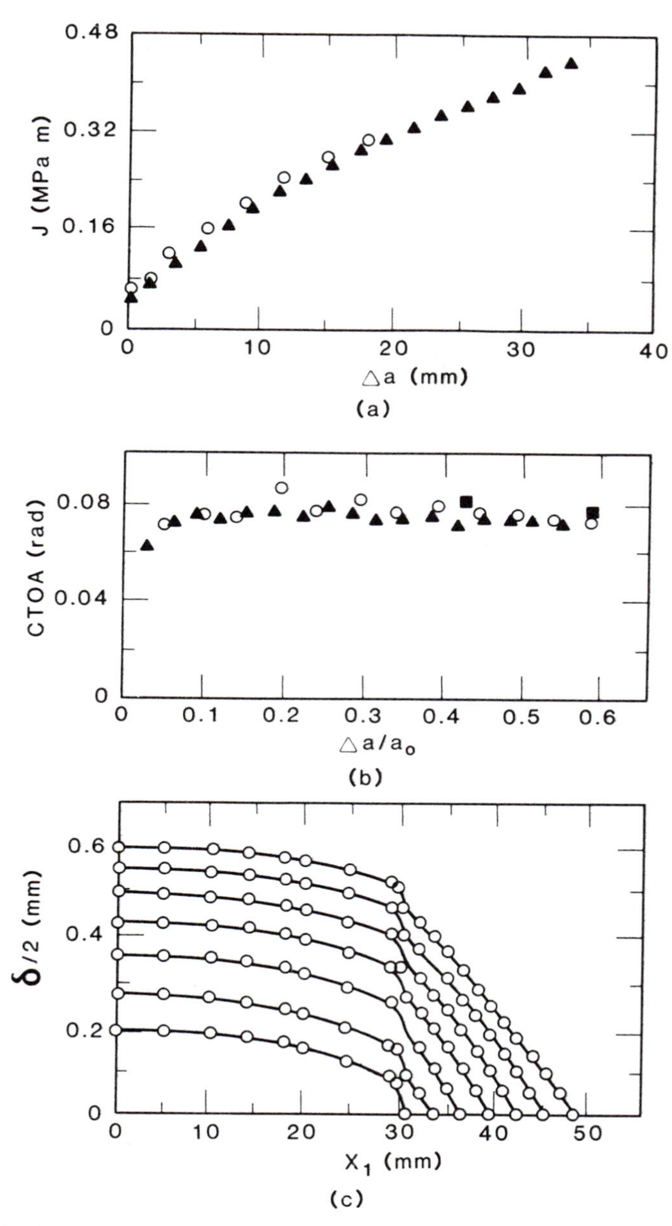

Figure 5.55 Generation phase analysis of stable crack growth in a 2024-T3 aluminum center-cracked panel: (a) J-resistance curve, (b) CTOA resistance curve, and (c) crack profile for $2a_0 = 61$ mm (5.81).

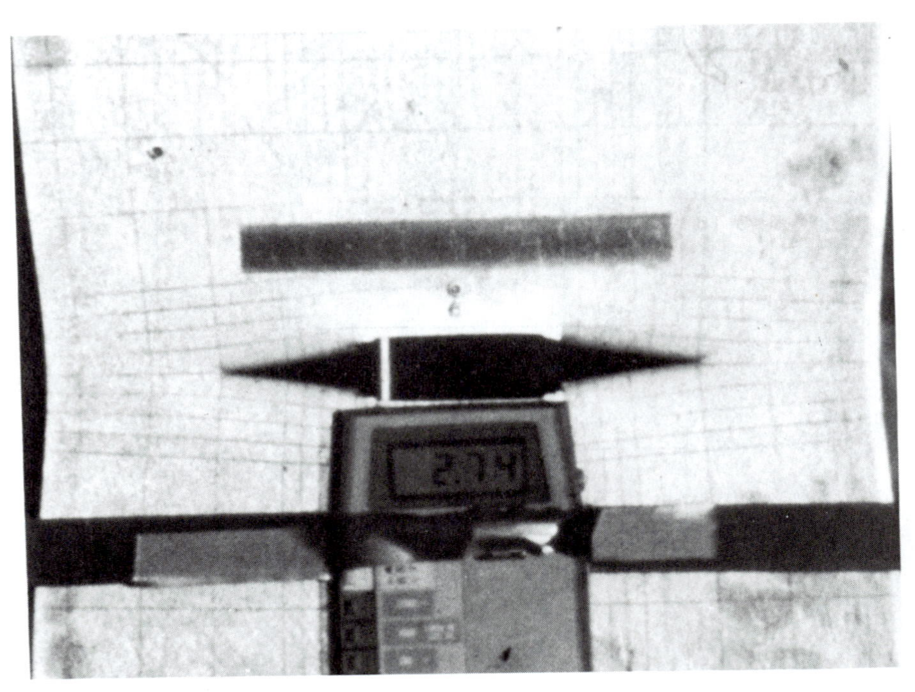

Figure 5.56 An illustration of stable crack growth profile in a Type 304 stainless steel center-cracked panel.

load. The J/CTOA approach predicted quite well not only this load, but also the crack extension at instability.

These results suggest a two-parameter criterion may be appropriate for crack extension beyond the validity of J-controlled growth. The J-resistance curve analysis would be used for crack initiation and a limited amount of stable growth. During this presumably J-controlled growth the CTOA is calculated. When the CTOA becomes constant, continued crack growth is permitted to occur at this constant value. In this regard the two-parameter approach offers the advantage that only a J-resistance curve is needed for the calculations; that is, no more information than in the usual J-resistance curve analysis is required. Because of the additional complexity in computing the CTOA, the price of this approach appears to be the relative simplicity frequently associated with the J-resistance curve analysis.

5.8.2 Asymptotic Fields for Growing Cracks

Consider the asymptotic stress and deformation fields at the tip of a steadily extending crack in an elastic-perfectly plastic solid under plane strain Mode I loading. The nature of the elastic-plastic strain singularity has been examined by Rice (5.27), Rice and Sorensen (5.83), and Cherepanov (5.28). In these analyses a full Prandtl field is assumed to exist at the crack tip. However, Rice et al. (5.84) subsequently found that the full Prandtl field is inappropriate for

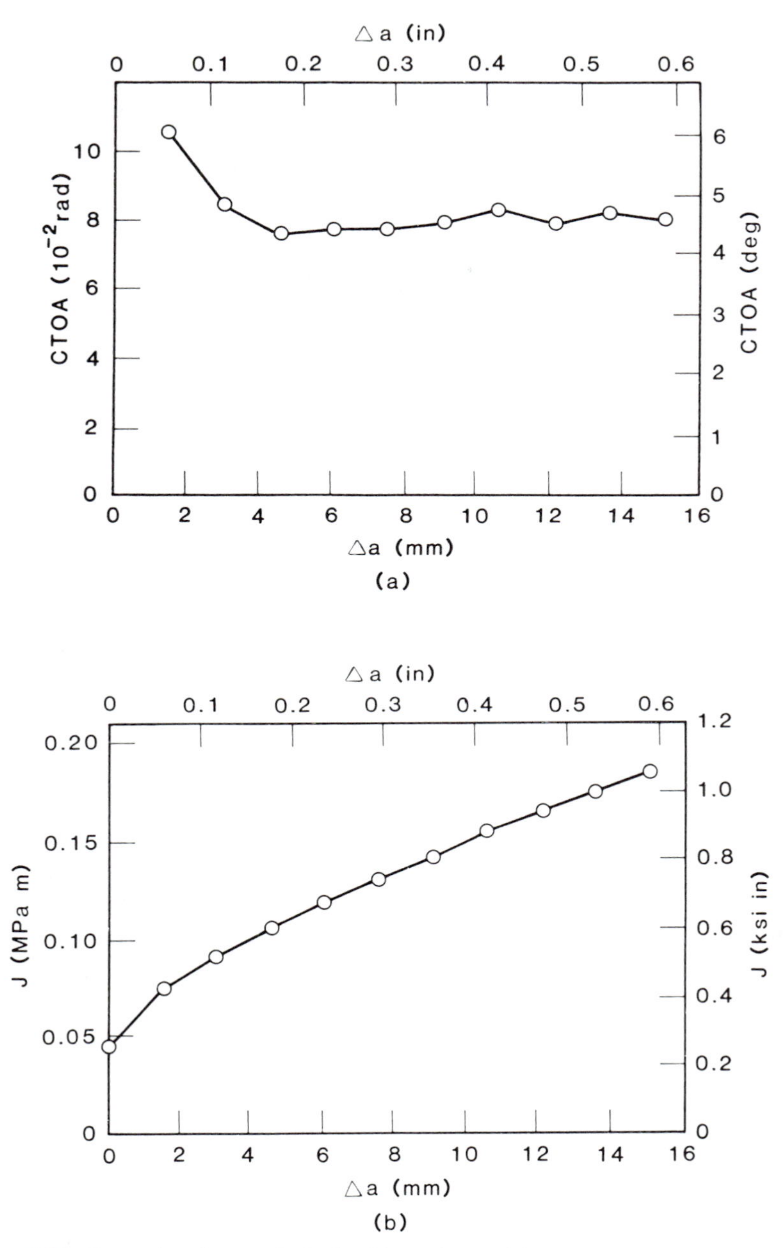

Figure 5.57 Generation phase analysis of stable crack growth in a 2219-T87 aluminum compact tension specimen; (a) CTOA resistance curve and (b) J-resistance curve.

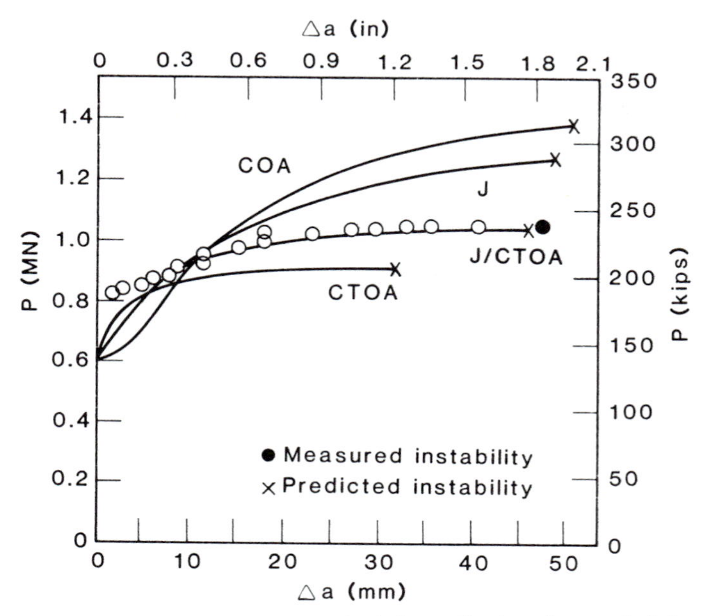

Figure 5.58 Comparison of plastic fracture criteria by application phase analysis for a center-cracked panel using resistance curves developed from tests on 2219-T87 aluminum compact tension specimens.

the extending crack because it fails to satisfy the requirement that the plastic work done be everywhere positive. Specifically, the stress component $\sigma_{r\theta}$ acting through the discontinuity of the radial velocity component v_r at the boundary between the centered fan C and the constant stress zone B in the Prandtl field of Figure 5.59(a) does negative work. This suggests the existence of an intervening elastic unloading zone.

Within the context of small strains, the stress field at the tip of a crack in an elastic-perfectly plastic solid is bounded and, consequently, $\sigma_{ij} = \sigma_{ij}(\theta)$ as $r \to 0$. Furthermore, Rice and Tracy (5.85) argue that $r\,\partial\sigma_{ij}/\partial r \to 0$ as $r \to 0$ and, hence, in the crack-tip region the equilibrium equations reduce to

$$\frac{\partial \sigma_{r\theta}}{\partial \theta} + \sigma_{rr} - \sigma_{\theta\theta} = 0$$
$$\frac{\partial \sigma_{\theta\theta}}{\partial \theta} + 2\sigma_{r\theta} = 0 \tag{5.8-1}$$

In addition,

$$\dot{\sigma}_{ij} = \sigma'_{ij}(\theta)\dot{\theta} = \sigma'_{ij}(\theta)\,\frac{\dot{a}}{r}\sin\theta \tag{5.8-2}$$

for a polar coordinate system attached to the steadily extending crack tip. The dot and prime are used to denote differentiation with respect to time and θ, respectively.

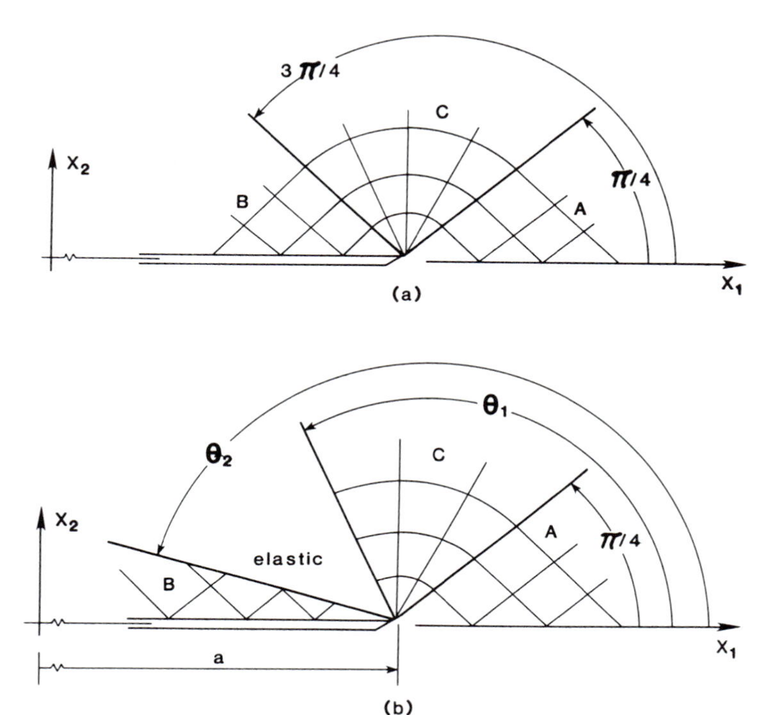

Figure 5.59 Slip-line fields for (a) a stationary crack and (b) a steadily growing crack.

The Prandtl-Reuss flow rule is expressed by

$$D_{ij} = D_{ij}^{e} + D_{ij}^{p} = \frac{1+v}{E}\,\dot{s}_{ij} + \frac{1-2v}{3E}\,\delta_{ij}\dot{\sigma}_{kk} + \dot{\Lambda} s_{ij} \qquad (5.8\text{-}3)$$

where the strain rate tensor D_{ij} defined in terms of the velocity components $v_i = \dot{u}_i$ is

$$D_{ij} = \tfrac{1}{2}(v_{i,j} + v_{j,i}) \qquad (5.8\text{-}4)$$

During elastic loading and unloading the plastic components D_{ij}^{p} of the strain rate tensor vanish while during plastic loading

$$D_{ij}^{p} = \dot{\Lambda} s_{ij} \qquad (5.8\text{-}5)$$

where

$$\dot{\Lambda} = (D_{ij}^{p} D_{ij}^{p} / s_{mn} s_{mn})^{\frac{1}{2}} \qquad (5.8\text{-}6)$$

The Mises yield condition is

$$s_{ij} s_{ij} = 2\sigma_y^2/3 \qquad (5.8\text{-}7)$$

Whenever

$$D_{33}^{p}(D_{ij}^{p} D_{ij}^{p})^{-\frac{1}{2}} = 0 \qquad (5.8\text{-}8)$$

it follows from Equation (5.8-5) that $s_{33} = 0$ and Equation (5.8-7) yields

$$(\sigma_{\theta\theta} - \sigma_{rr})^2/4 + \sigma_{r\theta}^2 = \sigma_y^2/3 \tag{5.8-9}$$

with $\sigma_{33} = (\sigma_{\theta\theta} + \sigma_{rr})/2$. Equation (5.8-9) can be expected to be valid asymptotically as $r \to 0$ when a plastic strain singularity exists. It should be noted that constant stress sectors do not produce unbounded plastic strains. In this case Equation (5.8-9) becomes an approximate criterion for plane strain yielding in these zones.

The differentiation of Equation (5.8-9) with the aid of Equation (5.8-1) leads to

$$\left[\frac{d\sigma_{r\theta}}{d\theta}\right]\left[\frac{d}{d\theta}(\sigma_{rr} + \sigma_{\theta\theta})\right] = 0 \tag{5.8-10}$$

In sectors for which $d\sigma_{r\theta}/d\theta = 0$, Equations (5.8-1) and (5.8-9) imply that

$$\sigma_{r\theta} = \pm\sigma_y/\sqrt{3}, \qquad \sigma_{rr} = \sigma_{\theta\theta} = \sigma_{33} = \text{constant} \pm (2\sigma_y/\sqrt{3})\theta \tag{5.8-11}$$

In slip-line theory these sectors correspond to centered fans. The regions in which $d(\sigma_{rr} + \sigma_{\theta\theta})/d\theta = 0$ are constant stress sectors since σ_{ij} is independent of θ there.

Let $e_i(\theta)$ and $h_i(\theta) = e_i'(\theta)$ be the direction cosines of unit vectors in the r and θ-directions, respectively. Under a transformation of coordinates it follows that

$$s_{rr} = s_{ij}e_ie_j, \qquad s_{r\theta} = s_{ij}e_ih_j, \qquad s_{\theta\theta} = s_{ij}h_ih_j \tag{5.8-12}$$

Since $s_{rr} = s_{\theta\theta} = 0$ and $s_{r\theta} = \sigma_y/\sqrt{3}$ in the centered fan, then

$$\begin{aligned} s_{ij}'e_ie_j &= (s_{ij}e_ie_j)' - 2e_i's_{ij}e_j \\ &= s_{rr}' - 2s_{r\theta} = -\sigma_y/\sqrt{3} \end{aligned}$$

While s_{rr} approaches zero as $r \to 0$, Rice (5.86), nevertheless, found that $\dot{\Lambda}s_{rr} = -(1 - 2\nu)(\sigma_y/\sqrt{3}E)(\dot{a}/r)\sin\theta$ as $r \to 0$. Failure to retain this term in earlier investigations (5.83, 5.84) produced minor inaccuracies. For example, this product vanishes for an elastic, incompressible material ($\nu = \frac{1}{2}$).

Multiplying Equation (5.8-3) by e_ie_j and noting from Equation (5.8-11) that $\sigma_{kk}' = -6\sigma_y/\sqrt{3}$, Rice (5.86) obtained

$$\begin{aligned} D_{ij}e_ie_j = D_{rr} = \frac{\partial v_r}{\partial r} &= \left[(1 + \nu)s_{ij}'e_ie_j + \frac{(1 - 2\nu)}{3E}\sigma_{kk}'\right]\frac{\dot{a}}{r}\sin\theta + \dot{\Lambda}s_{ij}e_ie_j \\ &= -\frac{5 - 4\nu}{\sqrt{3}}\frac{\sigma_y}{E}\frac{\dot{a}}{r}\sin\theta \end{aligned} \tag{5.8-13}$$

Therefore,

$$v_r = \frac{5 - 4\nu}{\sqrt{3}}\frac{\sigma_y}{E}\dot{a}\sin\theta\ln\left(\frac{\bar{R}}{r}\right) + f'(\theta) \tag{5.8-14}$$

where $f(\theta)$ and $\bar{R}$ remain undetermined in this asymptotic analysis. In a similar way

$$D_{11} + D_{22} \equiv \frac{\partial v_r}{\partial r} + \frac{v_r}{r} + \frac{1}{r}\frac{\partial v_\theta}{\partial \theta} = -\frac{6(1-2v)}{\sqrt{3}}\frac{\sigma_y}{E}\frac{\dot{a}}{r}\sin\theta \quad (5.8\text{-}15)$$

and

$$v_\theta = -\frac{5-4v}{\sqrt{6}}\frac{\sigma_y}{E}\dot{a}(1-\sqrt{2}\cos\theta)\left[\ln\left(\frac{\bar{R}}{r}\right) + \frac{1-8v}{5-4v}\right] - f(\theta) + g(r) \quad (5.8\text{-}16)$$

where $g(r)$ is also undetermined except $g(0) = 0$. Because the material is rate-independent, f and g will be homogeneous of degree one in $\dot{a}$ and in a loading parameter describing the intensity of the applied load.

Since D_{ij} for the centered fan can be computed from these velocity components and since D_{ij}^e can be determined from the known stress rates $\dot{\sigma}_{ij}$, then $D_{ij}^p = D_{ij} - D_{ij}^e$ can be ascertained. Because $s_{rr} = s_{\theta\theta} = 0$ in the centered fan, the only nonvanishing component of D_{ij}^p referred to the polar coordinate system is

$$D_{r\theta}^p = \frac{5-4v}{2\sqrt{6}}\frac{\sigma_y}{E}\frac{\dot{a}}{r}\ln\left(\frac{\bar{R}}{r}\right) + \frac{3(1-2v)}{\sqrt{6}}\frac{\sigma_y}{E}\frac{\dot{a}}{r}(1-\sqrt{2}\cos\theta) + \frac{f''+f}{2r} \quad (5.8\text{-}17)$$

When the latter is transformed into the Cartesian components, the resulting strain rate tensor may be integrated to yield (5.84)

$$\varepsilon_{ij}^p = \frac{5-4v}{2\sqrt{6}}\frac{\sigma_y}{E}G_{ij}(\theta)\ln\left(\frac{\bar{R}}{r}\right) + H_{ij}(\theta) \quad (5.8\text{-}18)$$

as $r \to 0$, where $H_{ij}(\theta)$ are undetermined and

$$\begin{aligned} G_{11}(\theta) &= G_{22}(\theta) = -2\sin\theta \\ G_{12}(\theta) &= G_{21}(\theta) = \ln[\tan(\theta/2)/\tan(\pi/8)] + 2(\cos\theta - 1/\sqrt{2}) \end{aligned} \quad (5.8\text{-}19)$$

Whereas the plastic strain singularity for the stationary crack in a non-hardening material is $1/r$, a logarithmic singularity exists for the steadily growing crack. When strain-hardening is present, the stress field also has a logarithmic singularity (5.87).

To the extent that Equation (5.8-9) is an adequate approximation for the yield condition, the crack-tip stress field in the inelastic region consists of either centered fans or constant stress sectors. Rice et al. (5.84) found it necessary to include an elastic unloading sector between the centered fan region C and the trailing constant stress sector B in the slip-line field shown in Figure 5.59(b). They found that for $v = 0.3$, $\theta_1 \approx 115°$, and $\theta_2 \approx 163°$ compared to $\theta_1 \approx 112°$ and $\theta_2 \approx 162°$ for $v = 0.5$. Surprisingly, the stress field ahead of the crack tip for this slip-line representation differs only about 1 percent from the Prandtl stress field. Even in the elastic region the largest difference is of the order of only 10 percent.

The form of the crack opening rate $\dot{\delta}$ near the tip is identical to that for the Prandtl field; that is,

$$\dot{\delta} = \beta\dot{a}\frac{\sigma_y}{E}\ln\left(\frac{\bar{R}}{r}\right) + \dot{A} \quad \text{as } r \to 0 \tag{5.8-20}$$

where $\beta = 5.08$ for $v = 0.3$ and $\beta = 4.39$ for $v = 0.5$. The quantity $\dot{A}$ remains undetermined in the asymptotic analysis but is a homogeneous function of degree one in $\dot{a}$ and the loading rate parameter.

When the applied loading does not significantly change the elastic-plastic boundary (e.g., by inducing elastic behavior in large portions of previously plastically deformed material), then $\dot{A}$ is expected to be linear in $\dot{a}$ and the loading rate parameter. The choice of a loading parameter is virtually arbitrary. When the plastic yielding is contained, the far field will be elastic and the far-field value of the J-integral is a convenient load parameter. However, in this case neither path independence nor any meaning in the plastic tip region can be attached to J. With α and μ undetermined, substitute

$$\dot{A} = \alpha\dot{J}/\sigma_y + \mu\dot{a} \tag{5.8-21}$$

into Equation (5.8-20) to obtain

$$\dot{\delta} = \alpha\frac{\dot{J}}{\sigma_y} + \beta\dot{a}\frac{\sigma_y}{E}\ln\left(\frac{R}{r}\right) \quad \text{as } r \to 0 \tag{5.8-22}$$

where μ has been combined with $\bar{R}$ to form a new length parameter R.

For monotonic loading of a stationary crack ($\dot{a} = 0$) Equation (5.8-22) yields the familiar result

$$\delta = \int \frac{\alpha}{\sigma_y}\,dJ \equiv \frac{\alpha^* J}{\sigma_y} \tag{5.8-23}$$

Finite element computations (5.88) indicate that the dimensionless parameter α^* varies from 0.65 for small-scale yielding to 0.51 for the fully plastic condition. Since α is constant for small-scale yielding, it is generally thought to be approximately constant up to general yielding in deeply cracked bend specimens.

When a increases monotonically with J, the asymptotic integration (replace da by dr) of Equation (5.8-22) for constant α yields

$$\delta = \frac{\alpha r}{\sigma_y}\frac{dJ}{da} + \beta r\frac{\sigma_y}{E}\ln\left(\frac{eR}{r}\right) \quad \text{as } r \to 0 \tag{5.8-24}$$

For the growing crack $\delta = 0$ and $d\delta/dr = \infty$ at the crack tip. The parameter R is expected to reflect the size of the plastic zone or at least the size of the region over which the slip-line field of Figure 5.59(b) prevails. Hence, for small-scale yielding one can write

$$R = \lambda EJ/\sigma_y^2 \tag{5.8-25}$$

where the parameter λ remains to be determined, say, from finite element computations.

Sham (5.89) has performed more refined finite element solutions for small-scale yielding than in reference (5.83). Depending upon how one interprets these solutions, values of α from 0.53 to 0.65 can be obtained. These results also suggest that, at least for small amounts of crack extension, α is nearly the same for stationary and growing cracks. For a growing crack under constant J, Equation (5.8-24) reduces to

$$\delta = \beta r \frac{\sigma_y}{E} \ln\left(\frac{eR}{r}\right) \tag{5.8-26}$$

Sham found that Equation (5.8-26) for the near-tip crack opening profile agreed favorably with the finite element solutions for $\beta = 5.4$ and $\lambda = 0.23$—the former value being somewhat greater than the theoretical value of 5.08 for $\nu = 0.3$. For this value of λ, the parameter R expressed by Equation (5.8-25) is found to be 15 to 30 percent greater than the maximum plastic zone radius predicted by the finite element solution.

Dean and Hutchinson (5.90) have also performed finite element computations for quasi-static crack growth under small-scale yielding. While they obtained near-tip stress fields in agreement with those determined for the slip-line field of Figure 5.59(b), they found no identifiable elastic unloading region near the crack tip. It may be that a further refinement of the mesh will be necessary to reveal it. The best least square fit of Equation (5.8-26) to the four computed values of δ closest to the tip yields $\beta = 4.28$ and $\lambda = 0.78$ for $\nu = 0.3$. When β is fixed at 5.08 the best least square fit for λ is 0.32. Clearly, further research is needed to determine definitive expressions for the dependence of α and R on the crack growth for both small-scale and large-scale yielding.

5.8.3 Comparison of Theory and Experiment

Since Equation (5.8-24) can be written as

$$\delta = (\beta r \sigma_y/E)\ln(\rho/r) \tag{5.8-27}$$

where

$$\begin{aligned} \rho &= eR\exp[(\alpha/\beta)(E/\sigma_y^2)(dJ/da)] \\ &= R\exp[1 + \alpha T/\beta] \end{aligned} \tag{5.8-28}$$

the form of the near-tip crack profile depends only upon the parameter ρ. Hermann and Rice (5.91) postulated that fracture proceeds such that the near-tip geometric profile of the steadily extending crack remains invariant. This is equivalent to constancy of the crack-tip opening angle. For the present development, this requires

$$\rho = \text{constant} \tag{5.8-29}$$

Consequently, Equation (5.8-28) yields the differential equation

$$\frac{dJ}{da} = \frac{\beta}{\alpha}\frac{\sigma_y^2}{E}\ln\left(\frac{\rho}{eR}\right) \tag{5.8-30}$$

relating J to Δa.

There are a number of implications associated with crack growth under an invariant near-tip geometrical profile. First, it is clear from Equation (5.8-30) that, unless $R = \rho/e$, J will in general vary with crack growth. This is strictly a plasticity effect. Crack growth with an invariant profile is not possible for a nonlinear elastic material (deformation plasticity) having the same stress-strain relation as the elastic-plastic material under monotonic loading. For such a nonlinear elastic material the near-tip crack profile and J are inherently connected; therefore, constancy of one implies constancy of the other.

Second, the crack growth criterion of Equations (5.8-28) and (5.8-29) implies that microcracking and void nucleation that are characteristic of the fracture process must be confined to the near crack-tip region. Obviously, this growth criterion cannot be expected to apply with much precision when these fracture processes occur in highly stressed regions removed from the immediate vicinity of the crack tip. Neither can it accommodate the transition from a stable ductile tearing mode of fracture to a cleavage fracture mode.

Apparently, other fracture criteria can give rise to a differential equation for crack growth having the same form as Equation (5.8-30). For example, Rice et al. (5.84) have shown that the equivalent plastic shear strain γ^p at points within the centered fan at a small distance r directly opposite the crack tip can be expressed as

$$\gamma^p = 0.94(5 - 4v)(\sigma_y/E)\ln(\zeta/r)$$

where

$$\zeta = L\exp\{[m/0.94(5 - 4v)](E/\sigma_y^2)(dJ/da)\}$$

with m and L undetermined in the asymptotic analysis. A crack growth criterion based upon a fixed strain state near the crack tip requires that ζ = constant and leads to a differential equation having the form of Equation (5.8-30). Wnuk (5.92) also obtained an equation of the same form using the Dugdale model and the final stretch criterion.

Hermann and Rice (5.91) conducted crack growth tests on four AISI 4140 sidegrooved compact specimens. The nominal tensile yield and ultimate strengths for this material are $\sigma_y = 1170$ MPa and $\sigma_u = 1330$ MPa. The remaining ligament b in these specimens is small compared to the overall dimensions and, hence, the specimens can be modeled as deeply cracked bend specimens. In these tests the unloading compliance method was used to determine the amount of crack extension. Typical experimental results are shown in Figure 5.60. The applied moment and fully plastic moment per unit

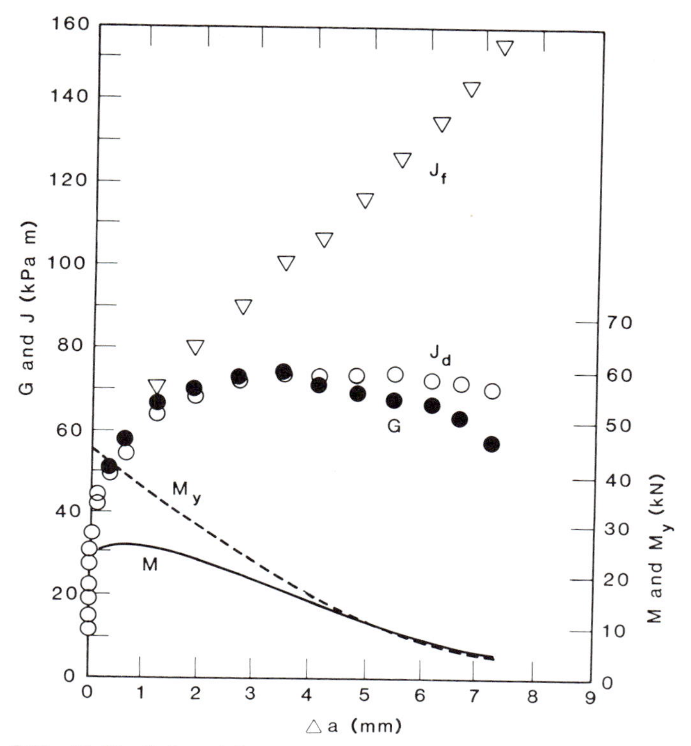

Figure 5.60 M, M_y, G, J_d, and J_f versus crack growth for specimen 2 of reference (5.91).

thickness are given, respectively, by $M = P(a + b/2)$ and $M_y = 0.364b^2\sigma_y$, where P is the applied load per unit thickness. This figure illustrates the change from contained yielding in the early stages of the test to general yielding in the later stages.

The elastic energy release rate is given by (5.93)

$$G = 16(1 - \nu^2)M^2/Eb^2 \tag{5.8-31}$$

The deformation theory value of J—that is, J_d—is obtained by integrating [see Equation (5.7-52)]

$$dJ_d = (2P/b)\, d\Delta - (J_d/b)\, da \tag{5.8-32}$$

If the material were nonlinear elastic or if the crack growth were J_d-controlled, then this integration would be path-independent. Moreover, J_d would agree with the definition of the J-integral and in the limit of small-scale yielding $J_d = G$. From Figure 5.60 it is clear that J_d and G agree until M almost equals M_y.

Finally, the far-field value of J—namely J_f—follows from integrating

$$\begin{aligned} dJ_f &= (2P/b)\, d\Delta \\ &= dJ_d + (J_d/b)\, da \end{aligned} \tag{5.8-33}$$

This value of J_f agrees with the J-integral for a contour coinciding with the outer boundary of a deeply cracked, rigid-plastic bend specimen. There is no evidence that a similar interpretation can be made for an elastic-plastic specimen or rigid-plastic specimens of other geometries. For relatively small amounts of crack extension there is little difference between J_f and J_d; however, the difference increases dramatically with increasing growth. It should be noted in passing that $\dot{J}_f$ has the property that at full plasticity it is independent of $\dot{a}$, whereas $\dot{J}_d$ depends upon $\dot{a}$. More will be said about this later.

Based upon small-scale yielding (taking $\alpha = 0.65$, $\beta = 5.08$, and $\lambda = 0.23$) and J replaced by J_d, Equation (5.8-30) becomes

$$\frac{dJ_d}{da} = 7.82\left(\frac{\sigma_u^2}{E}\right)\ln\left(\frac{\rho\sigma_y^2}{0.23eEJ_d}\right) \tag{5.8-34}$$

where in one place σ_y has been replaced by σ_u to accommodate strain hardening effects. As J_d approaches the limiting value

$$(J_d)_{ss} = \frac{\rho\sigma_y^2}{0.23eE} \tag{5.8-35}$$

for steady-state growth, Equation (5.8-34) implies that no further increase in J_d (i.e., $dJ_d/da = 0$) is required for continued growth.

Hermann and Rice integrated Equation (5.8-34) subject to $J_d = 35$ kN/m at crack initiation as suggested by the experimental data. Values of the remaining disposal parameter ρ were selected to best fit the experimental data for J_d prior to general yielding. The comparisons of theoretical predictions and experimental results for two tests are shown in Figure 5.61. The solid curves are identical and correspond to the best fit of the data of specimen 4 for which $\rho = 7.175$ mm and $(J_d)_{ss} = 80$ kN/m. The dashed curves conform to the value of ρ giving the best fit of the data for that specimen, even though the theory requires it to be specimen independent. The open circles are data points prior to general yielding ($M < M_y$) and the solid points identify post general yielding data. With the exception of small variations of ρ the predications of the small-scale yielding model agree reasonably well with the experimental data for crack extensions prior to general yielding.

5.8.4 *J for Extended Crack Growth*

Rice et al. (5.84) as well as Hermann and Rice (5.91) have speculated about crack growth under large-scale yielding. While the definition of the loading parameter is arbitrary, it is important to recognize that the definition of J influences through Equation (5.8-21) the values of μ and perhaps α. Since μ was ultimately absorbed by R, the latter will also depend upon the interpretation of J. Therefore, it is necessary to be more precise about the definition of J for extended crack growth and yielding. Some guidance in this direction can be provided by considering the limiting case of rigid-perfectly plastic material—that is, in the limit as $\sigma_y/E \rightarrow 0$. For a fully yielded specimen of this material the

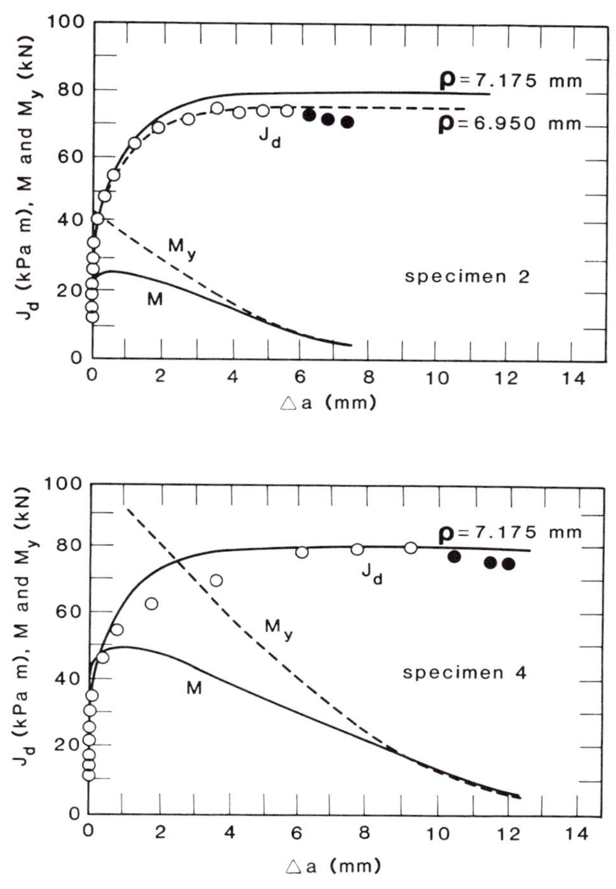

Figure 5.61 Comparison of theoretical predictions and experimental results (5.91).

crack opening rate $\dot{\delta}$ has the form $\dot{\delta} = \Omega\dot{\Delta}$, where Δ is a prescribed generalized displacement and Ω is a geometric factor. This expression is valid for both stationary and growing cracks; that is, it is independent of $\dot{a}$. For R to remain well defined in the rigid-plastic limit, whatever definition that is adopted for J must be such that $\dot{J}$ depends only upon $\dot{\Delta}$ and not on $\dot{a}$ in this limit. Otherwise R would have to contain a factor $\exp(E/\sigma_y)$ in order that the second term in Equation (5.8-22) annul the the $\dot{a}$ dependence of $\dot{J}$ in the first term as $\sigma_y/E \to 0$. This would result in an unsatisfactory behavior of R for this limiting case.

Rice et al. (5.84) have shown that the far field value, J_f, of the J-integral for the rigid-plastic bend-type specimen possesses this characteristic, but the deformation theory value of J, J_d, does not. On the other hand, J_d has the proper character for the rigid-plastic double-edge cracked tension specimen with sufficiently deep cracks to permit full development of the Prandtl field over the uncracked ligament. Due to the nonuniqueness of the stress field in the remaining rigid region, J_f is not uniquely defined. Both J_d and J_f have the proper limiting behavior for the rigid-plastic, center-cracked tension specimen. However, this specimen does not possess the Prandtl-like slip-line field at the crack tip.

Ernst (5.94) has proposed a definition of J which has the property that $\dot{J}$ is independent of $\dot{a}$ in the limit as $\sigma_y/E \to 0$. The deformation theory value of J can be decomposed into its linear elastic contribution and its plastic (nonlinear elastic) component; that is,

$$J_d = J_e + J_p = G + J_p \tag{5.8-36}$$

Assuming that $J_d = J_d(\Delta, a)$, then

$$\dot{J}_d = \left(\frac{\partial J_d}{\partial \Delta}\right)_a \dot{\Delta} + \left(\frac{\partial G}{\partial a}\right)_\Delta \dot{a} + \left(\frac{\partial J_p}{\partial a}\right)_\Delta \dot{a}$$

or upon rewriting

$$\dot{J}_d - \left(\frac{\partial J_p}{\partial a}\right)_\Delta \dot{a} = \left(\frac{\partial J_d}{\partial \Delta}\right)_a \dot{\Delta} + \left(\frac{\partial G}{\partial a}\right)_\Delta \dot{a}$$

Since $(\partial G/\partial a)_\Delta \to 0$ as $\sigma_y/E \to 0$, then

$$\dot{J} \equiv \dot{J}_d - \left(\frac{\partial J_p}{\partial a}\right)_{\Delta_p} \dot{a} \tag{5.8-37}$$

is independent of $\dot{a}$ and

$$J = J_d - \int_{a_0}^{a} \left.\frac{\partial(J_d - G)}{\partial a}\right|_{\Delta_p} da \tag{5.8-38}$$

This definition of J has the desired character and reduces, of course, to J_d in the absence of crack growth. Equation (5.8-37) can be rewritten as

$$\frac{dJ}{da} = \frac{dJ_d}{da} - \left.\frac{\partial(J_d - G)}{\partial a}\right|_{\Delta_p} \tag{5.8-39}$$

While J defined in Equation (5.8-38) yields the correct limiting behavior, there is no assurance that this is the only definition to do so.

Assuming the existence of the η_p factor, one can write

$$J_p = \frac{\eta_p}{b} U_p \tag{5.8-40}$$

and, whence,

$$\left.\frac{\partial(J_d - G)}{\partial a}\right|_{\Delta_p} = \left.\frac{\partial J_p}{\partial a}\right|_{\Delta_p} = \frac{\gamma_p}{b}(J_d - G) \tag{5.8-41}$$

where γ_p is given by Equation (5.7-34). Consequently, for known η_p and J_d Equation (5.8-41) can be introduced into Equation (5.8-38) to obtain

$$J = J_d - \int_{a_0}^{a} \frac{\gamma_p}{b}(J_d - G)\, da \tag{5.8-42}$$

Moreover, the substitution of Equation (5.8-41) into Equation (5.8-39) yields

$$\frac{dJ}{da} = \frac{dJ_d}{da} - \frac{\gamma_p}{b}(J_d - G) \tag{5.8-43}$$

For the deeply cracked bend specimen $\eta_p = 2$, $\gamma_p = -1$, and Equation (5.8-43) reduces to

$$\frac{dJ}{da} = \frac{dJ_d}{da} + \frac{(J_d - G)}{b} \tag{5.8-44}$$

in agreement with the interpretation of J given by Hermann and Rice.

It is apparent in Figure 5.61 that the J_d-resistance curve reached a plateau and then decreased ($dJ_d/da < 0$) with increasing crack growth. A negative slope of the resistance curve would indicate instability, whereas the test is quite stable in this case. However, with J defined by Equation (5.8-42) the slope of the associated resistance curve given by Equation (5.8-44) will be positive and vanish only if $d\Delta_p = 0$.

Ernst (5.94) has examined the resistance curves for geometrically similar compact specimens of A508 Class 2A steel at 400°F. Each specimen had a thickness one-half of its total width and $a/W = 0.6$. The sizes ranged from $1/2T$ ($W = 1$ in.) to $10T$ ($W = 20$ in.). The J_d-resistance curves for each specimen showed consistency during the early stages of crack growth, but significant departures occurred as the crack extension became an appreciable fraction of the remaining ligament. Such differences become more vivid when J_d is plotted against its respective tearing modulus T_d as in Figure 5.62. The large scatter for the smaller values of T_d is a reflection of the deviation of the resistance curves at the larger values of J_d.

Rather than using J_d as a fracture characterizing parameter and appealing to J_d-controlled crack growth, Ernst has suggested that it may be more appropriate to use J defined by Equation (5.8-38) and the normalized form of Equation (5.8-39) for the tearing modulus or, equivalently, Equations (5.8-42) and (5.8-43). When the previous data are interpreted in terms of these definitions of J and T, the result is shown in Figure 5.63. In this instance the scatter is significantly reduced and the implication is that J defined by Equations (5.8-38) and (5.8-42) correlates better than J_d for extended amounts of crack growth.

Not only should a potential fracture characterizing parameter correlate data for differing amounts of crack growth in the same kind of specimens, but also data from specimens of different configurations. Figure 5.64 compares the J_d-resistance curves for compact specimens and center cracked panels of 2024 aluminum alloy. The deviation of the resistance curves beyond the limit of J_d-controlled crack growth is apparent. When these data are replotted in terms of the newly defined J, Figure 5.65 shows the correlation to be enhanced significantly. These limited data suggest that J defined by Equations (5.8-38) and (5.8-42) improves the correlation of resistance curves obtained from specimens of different sizes and geometries.

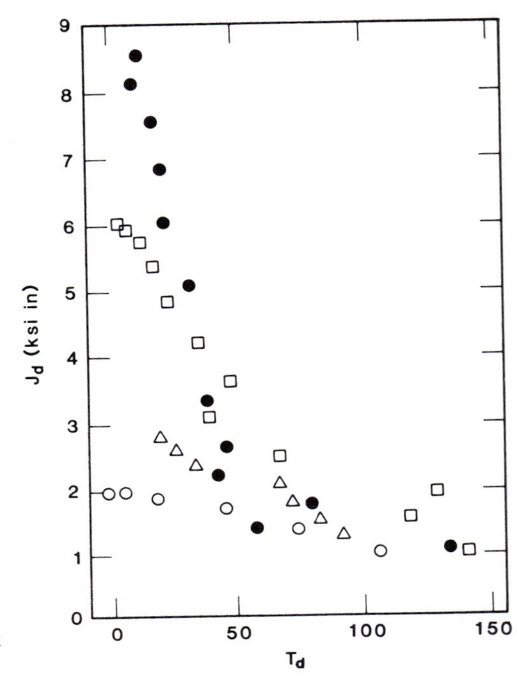

Figure 5.62 J_d versus T_d for compact specimens (5.94).

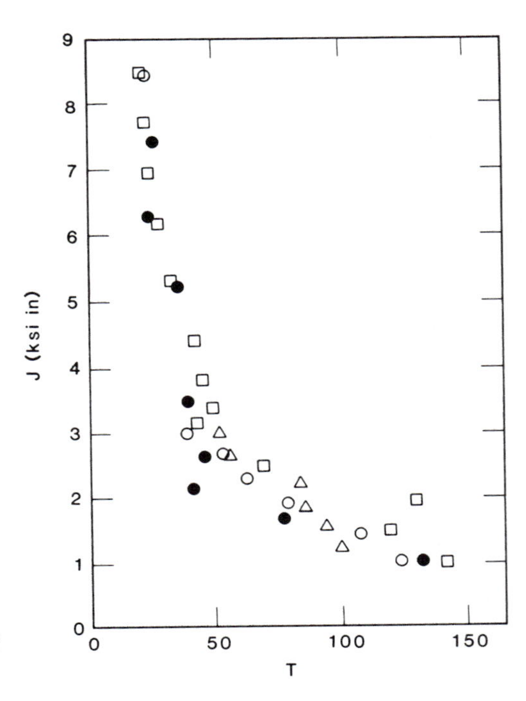

Figure 5.63 J versus T for data of Figure 5.62 (5.94).

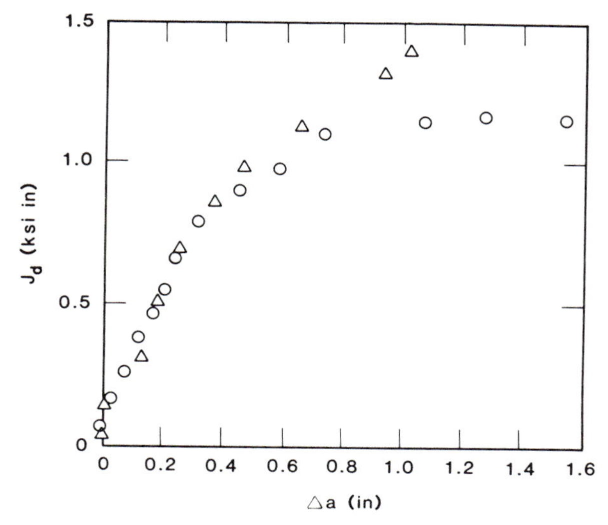

Figure 5.64 J_d-resistance curves for 2024-T351 aluminum 4T compact tension specimens (○) and center cracked panels (△) (5.94).

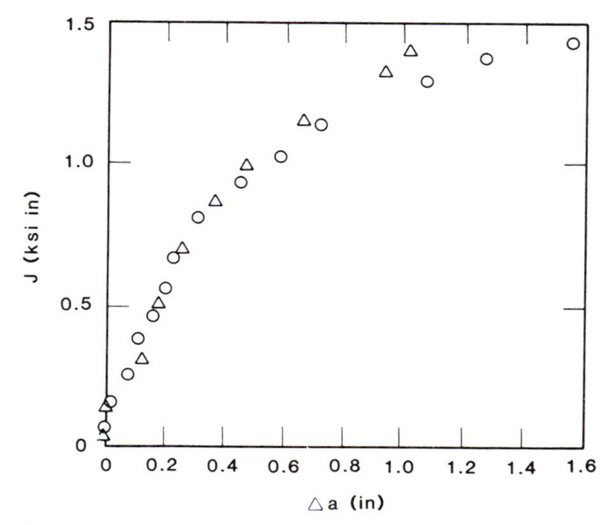

Figure 5.65 J-resistance curves for specimens and data of Figure 5.64 (5.94).

The concept of a resistance curve analysis and fracture instability is independent of the fracture characterizing parameter. Ideally, the resistance curve should be independent of the specimen's configuration. The tearing instability analysis of Section 5.7 is based upon a J_d-resistance curve. The disconcerting aspect of this analysis is that the J_d-resistance curve is specimen dependent once the ill-defined limit of J_d-controlled crack growth is exceeded.

If the proposed new definition of J in this section should prove to be applicable universally for even greater crack extensions, then Equations (5.8-38) and (5.8.39) or, equivalently, Equations (5.8-42) and (5.8-43) provide the connections between J_d/T_d and J/T. Consequently, the analysis of Section 5.7 can be carried over in terms of these new parameters in a straightforward manner and presumably would lead to more reliable predictions.

In summary, the J-integral in elastic-plastic fracture mechanics was originally employed as a measure of the intensity of the near-tip stress and deformation fields for stationary cracks subjected to monotonic loadings. Under such conditions J_d—the deformation theory value of J—suffices for characterizing crack initiation and limited amounts of crack extension. However, for extended amounts of crack growth accompanied by significant elastic unloading, J_d losses its significance as a crack-tip parameter and an alternative parameter must be used. The crack-tip opening angle (CTOA) represents one possibility, but it has the disadvantage that it is difficult to measure and to compute. The asymptotic solution for a growing crack in an elastic-perfectly plastic material has established a connection between the local crack profile (CTOA) and a loading parameter that is more readily measured and computed. This permitted the identification of certain limiting properties that the parameter must possess. A possible alternate definition of J [Equation (5.8-38)] that satisfies these properties has been proposed. Limited experimental evidence indicates that this definition permits correlating the fracture resistance for extended growth in specimens of different sizes and geometries. This has potentially strong implications with respect to tearing instability analyses that currently suffer from the dependence of J_d-resistance curves upon the specimen's geometry. While these results are certainly encouraging, further theoretical studies of the crack-tip mechanism involved in extended growth under general yielding and further experimental analyses of the resulting growth criterion are required.

5.9 Closure

While much has been accomplished in the development of a plastic fracture mechanics methodology, much remains to be done. The problem of extended crack growth in elastic-plastic materials is certainly worthy of more study. Efficient methods for analyzing large crack growth are needed. Currently, solution techniques are limited to two-dimensional problems. There is a distinct lack of research devoted to the three-dimensional problem involving cracks with curved fronts. It is even less clear here how growth proceeds and what the fracture criterion should be. The development of an effective plastic-fracture mechanics analyses for three-dimensional flawed structures and curved crack fronts will not likely be successful until an understanding of two-dimensional plastic fracture mechanics is achieved. Investigations to date have been limited primarily to monotonic Mode I loading involving a single loading parameter. The problem of multiple load parameters has many interesting facets that have yet to be explored. Work on mixed mode plastic fracture is nearly nonexistent. The transition from the ductile tearing mode to a cleavage-type fracture is not clearly understood. The problem of the interaction of the plastic crack-tip stress field and a residual stress field, say, in the neighborhood of the heat affected zones of welds deserves consideration. Most of the analyses have been limited to small strains that may be an oversimplification near the crack tip of high toughness–low strength materials. Clearly, the surface of plastic fracture mechanics has only been scratched.

5.10 References

(5.1) Kanninen, M. F., Popelar, C. H., and Broek, D., " A Critical Survey on the Application of Plastic Fracture Mechanics to Nuclear Pressure Vessels and Piping," *Nuclear Engineering and Design*, **67**, pp. 27–55 (1981).
(5.2) Hahn, G. T. and Rosenfield, A. R., "Local Yielding and Extension of a Crack under Plane Stress," *Acta Metallurgica*, **13**, pp. 293–306 (1965).
(5.3) Dugdale, D. S., "Yielding of Steel Sheets Containing Slits," *Journal of the Mechanics and Physics of Solids*, **8**, pp. 100–108 (1960).
(5.4) Mills, M. J., "Dugdale Yielded Zones in Cracked Sheets of Glassy Polymers," *Engineering Fracture Mechanics*, **6**, pp. 537–549 (1974).
(5.5) Rice, J. R., "Plastic Yielding at a Crack Tip, "*Proceedings of the First International Conference on Fracture*, Vol. 1, Sendai, Sept. 12–17, 1965, Japanese Society for Strength and Fracture of Materials, Tokyo, pp. 283–308 (1966).
(5.6) Burdekin, F. M. and Stone, D. E. W., "The Crack Opening Displacement Approach to Fracture Mechanics in Yielding," *Journal of Strain Analysis*, **1**, pp. 145–153 (1966).
(5.7) Bilby, B. A., Cottrell, A. H., and Swinden, K. H., "The Spread of Plastic Yield from a Notch," *Proceedings of the Royal Society, Series A*, **272**, pp. 304–314 (1963).
(5.8) Broek, D., *Elementary Engineering Fracture Mechanics*, 3rd ed., Martinus Nijhoff, The Hague (1982).
(5.9) Knott, J. F., *Fundamentals of Fracture Mechanics*, Butterworths, London (1979).
(5.10) Harrison, R. P., Loosemore, K., and Milne, I., "Assessment of the Integrity of Structures Containing Cracks," CEGB Report No. R/H/R6, Central Electricity Generating Board, United Kingdom (1976).
(5.11) Green, A. P. and Hundy, B. B., "Initial Plastic Yielding in Notch Bend Bars," *Journal of the Mechanics and Physics of Solids*, **4**, pp. 128–149 (1956).
(5.12) Ford, H. and Lianis, G., "Plastic Yielding of Notched Strips Under Conditions of Plane Stress," *ZAMP*, **8**, pp. 360–382 (1975).
(5.13) Chell, G. G., "A Procedure for Incorporating Thermal and Residual Stresses into the Concept of a Failure Assessment Diagram," *Elastic-Plastic Fracture*, ASTM STP 668, American Society for Testing and Materials, Philadelphia, pp. 581–605 (1979).
(5.14) Chell, G. G., and Milne, I., "A Simple Practical Method for Determining the Ductile Instability of Cracked Structures," CSNI Specialists Meeting on Plastic Tearing Instability, U.S. Nuclear Regulatory Commission, NUREG/CP-0010 (1980).
(5.15) Feddersen, C. E., "Evaluation and Prediction of the Residual Strength of Center Cracked Tension Panels," *Damage Tolerance in Aircraft Structures*, ASTM STP 486, American Society for Testing and Materials, Philadelphia, pp. 50–78 (1971).
(5.16) Broek, D., "Concepts in Fail Safe Design in Aircraft Structures," DMIC Memo 252 (1971).
(5.17) Broek, D., "Fail Safe Design Procedure," AGARDograph 176, pp. 121–166 (1974).
(5.18) Broek, D., "Particles and Crack Growth in Alumnum Alloys," *Prospects of Fracture Mechanics*, G. Sih et al. (ed.), Noordhoff, Groningen, pp. 19–34 (1974).
(5.19) Kanninen, M. F., et al., "Mechanical Fracture Predictions for Sensitized Stainless Steel Piping with Circumferential Cracks," EPRI NP-192, Sept. (1976).
(5.20) Kanninen, M. F., et al., "Instability Predictions for Circumferentially Cracked Large-Diameter Type 304 Stainless Steel Pipes under Dynamic Loading," Second Semi-Annual Report to EPRI on RP2554-2 June (1980).

(5.21) Hahn, G. T., Sarrate, M., and Rosenfield, A. R., "Criteria for Crack Extension in Cylindrical Pressure Vessels," *International Journal of Fracture Mechanics*, **5**, pp. 187–210 (1969).
(5.22) Folias, E. S., "A Finite Line Crack in a Pressurized Cylindrical Shell," *International Journal of Fracture Mechanics*, **1**, pp. 104–113 (1965).
(5.23) Kiefner, J. R., Maxey, W. A., Eiber, R. J., and Duffy, A. R., "Failure Stress Levels of Flaws in Pressurized Cylinders," *Progress in Flaw Growth and Fracture Toughness Testing*, ASTM STP 536, American Society for Testing and Materials, Philadelphia, pp. 461–481 (1973).
(5.24) Maxey, W. A., Kiefner, J. F., Eiber, R. J., and Duffy, A. R., "Ductile Fracture Initiation, Propagation and Arrest in Cylindrical Vessels," *Fracture Toughness, Part II*, ASTM STP 514, American Society for Testing and Materials, Philadelphia, pp. 70–81 (1972).
(5.25) Maxey, W. A., "Fracture Initiation, Propagation and Arrest," Fifth Symposium on Line Pipe Research, American Gas Association, Catalogue No. 230174, pp. J-1-J-13, Nov. (1974).
(5.26) Hult, J. A. H. and McClintock, F. A., "Elastic-Plastic Stress and Strain Distributions Around Sharp Notches Under Repeated Shear," *Proceedings of the 9th International Congress for Applied Mechanics*, Vol. 8, University of Brussels, pp. 51–58 (1957).
(5.27) Rice, J. R., "Mathematical Analysis in the Mechanics of Fracture," *Fracture—An Advanced Treatise*, Vol. II, H. Liebowitz (ed.), Academic, New York, pp. 191–308 (1968).
(5.28) Cherepanov, G. P., *Mechanics of Brittle Fracture*, McGraw-Hill, New York (1979).
(5.29) Rice, J. R., "Stresses Due to a Sharp Notch in a Work-Hardening Elastic-Plastic Material Loaded by Longitudinal Shear," *Journal of Applied Mechanics*, **34**, pp. 287–298 (1967).
(5.30) Rice, J. R., "Contained Plastic Deformation Near Cracks and Notches under Longitudinal Shear," *International Journal of Fracture Mechanics*, **2**, pp. 426–447 (1966).
(5.31) Hutchinson, J. W., "Singular Behavior at the End of a Tensile Crack in a Hardening Material," *Journal of the Mechanics and Physics of Solids*, **16**, pp. 13–31 (1968).
(5.32) Rice, J. R. and Rosengren, G. F., "Plane Strain Deformation near a Crack Tip in a Power-Law Hardening Material," *Journal of the Mechanics and Physics of Solids*, **16**, pp. 1–12 (1968).
(5.33) Hutchinson, J. W., "Plastic Stress and Strain Fields at a Crack Tip," *Journal of the Mechanics and Physics of Solids*, **16**, pp. 337–347 (1968).
(5.34) Begley, J. A. and Landes, J. D., "The *J*-Integral as a Fracture Criterion," *Fracture Toughness, Part II*, ASTM STP 514, American Society for Testing and Materials, Philadelphia, pp. 1–20 (1972).
(5.35) Tracy, D. M., "Finite Element Solutions for Crack-Tip Behavior in Small-Scale Yielding," *Journal of Engineering Materials and Technology*, **98**, pp. 146–151 (1976).
(5.36) Shih, C. F., "Relationship Between the *J*-Integral and the Crack Opening Displacement for Stationary and Extending Cracks," *Journal of the Mechanics and Physics of Solids*, **29**, pp. 305–326 (1981).
(5.37) Sih, G. C., *Handbook of Stress Intensity Factors*, Lehigh University, Bethlehem, Pa. (1973).
(5.38) Tada, H., Paris, P. C., and Irwin, G. R., *Stress Analysis of Cracks Handbook*, Del Research Corporation, Hellertown, Pa. (1973).
(5.39) Rooke, D. E. and Cartwright, D. J., *Compendium of Stress Intensity Factors*, Hillington, Uxbridge, England (1973).
(5.40) Il'yushin, A. A., "The Theory of Small Elastic-Plastic Deformations," *Prikadnaia Matematika i Mekhanika*, PMM, **10**, pp. 347–356 (1946).
(5.41) Goldman, N. L. and Hutchinson, J. W., "Fully Plastic Crack Problems: The Center-Cracked Strip Under Plane Strain," *International Journal of Solids and Structures*, **11**, pp. 575–591 (1975).
(5.42) Shih, C. F. and Kumar, V., "Estimation Technique for the Elastic-Plastic Fracture of Structural Components of Nuclear Systems," 1st Semiannual Report to EPRI, Contract No. RP1237-1, General Electric Company, Schenectady, N.Y., July 1, 1978–Jan. 31, 1979.
(5.43) Needleman, A. and Shih, C. F., "Finite Element Method for Plane Strain Deformations of Incompressible Solids," *Computer Methods in Applied Mechanics and Engineering*, **15**, pp. 223–240 (1978).
(5.44) Kumar, V., German, M. D., and Shih, C. F., "Estimation Techniques for the Prediction of Elastic-Plastic Fracture of Structural Components of Nuclear Systems," Combined 2nd and 3rd Semiannual Report to EPRI, Contract No. RP1237-1, General Electric Company, Schenectady, New York, Feb. 1, 1979–Jan. 31, 1980.
(5.45) Kumar, V., deLorenzi, H. G., Andrews, W. R., Shih, C. F., German, M. D., and Mowbray, D. F., "Estimation Technique for the Prediction of Elastic-Plastic Fracture of Structural

Components of Nuclear Systems," 4th Semiannual Report to EPRI, Contract No. RP1237-1, General Electric Company, Schenectady, New York, July 1, 1980–Jan. 31, 1981.
(5.46) Kumar, V., German, M. D., and Shih, C. F., "An Engineering Approach for Elastic-Plastic Fracture Analysis," EPRI NP-1931, Project 1287-1, Topical Report, July 1981.
(5.47) Shih, C. F., "J-Integral Estimates for Strain Hardening Materials in Antiplane Shear Using Fully Plastic Solution," *Mechanics of Crack Growth*, ASTM STP 590, American Society for Testing and Materials, Philadelphia, Pa., pp. 3–26 (1976).
(5.48) Shih, C. F. and Hutchinson, J. W., "Fully Plastic Solutions and Large Scale Yielding Estimates for Plane Stress Crack Problems," *Journal of Engineering Materials and Technology*, **98**, pp. 289–295 (1976).
(5.49) Kumar, V. and Shih, C. F., "Fully Plastic Crack Solutions, Estimation Scheme, and Stability Analyses for the Compact Specimen," *Fracture Mechanics: Twelfth Conference*, ASTM STP 700, American Society for Testing and Materials, Philadelphia, Pa., pp. 406–438 (1980).
(5.50) Shih, C. F., German, M. D., and Kumar, V., "An Engineering Approach for Examining Crack Growth and Stability in Flawed Structures," *International Journal of Pressure Vessels and Piping*, **9**, pp. 1–20 (1981).
(5.51) Rice, J. R., Paris, P. C., and Merkle, J. G., "Some Further Results of J-Integral Analysis and Estimates," *Progress in Flaw Growth and Fracture Toughness Testing*, ASTM STP 536, American Society for Testing and Materials, Philadelphia, Pa., pp. 231–245 (1973).
(5.52) Bloom, J. M., "Prediction of Ductile Tearing Using a Proposed Strain Hardening Failure Assessment Diagram," *International Journal of Fracture Mechanics*, **16**, pp. R163-R167 (1980).
(5.53) Merkle, J. G. and Corten, H. T., "A J-Integral Analysis for the Compact Specimen, Considering Axial Force as Well as Bending Effects," *Journal of Pressure Vessel Technology*, **96**, pp. 286–292 (1974).
(5.54) Landes, J. D., Walker, H., and Clarke, G. A., "Evaluation of Estimation Procedures Used in J-Integral Testing," *Elastic-Plastic Fracture*, ASTM STP 668, American Society for Testing and Materials, Philadelphia, Pa., pp. 266–287 (1979).
(5.55) "J_{IC}, A Measure of Fracture Toughness," *ASTM Annual Book of Standards*, Part 10, American Society for Testing and Materials, Philadelphia, Pa., E813, pp. 810–828 (1981).
(5.56) McClintock, F. A., "Plasticity Aspects of Fracture," *Fracture-An Advanced Treatise*, Vol. 3, H. Liebowitz (ed.), Academic, New York, pp. 47–225 (1971).
(5.57) Begley, J. A. and Landes, J. D., "Serendipity and the J-Integral," *International Journal of Fracture Mechanics*, **12**, pp. 764–766 (1976).
(5.58) McMeeking, R. M. and Parks, D. M., "On Criteria for J-Dominance of Crack-Tip Fields in Large-Scale Yielding," *Elastic-Plastic Fracture*, ASTM STP 668, American Society for Testing and Materials, Philadelphia, Pa., pp. 175–194 (1979).
(5.59) Shih, C. F. and German, M. D., "Requirements for a One Parameter Characterization of Crack-Tip Fields by the HRR Singularity," *International Journal of Fracture Mechanics*, **17**, pp. 27–43 (1981).
(5.60) Hutchinson, J. W. and Paris, P. C., "Stability Analysis of J-Controlled Crack Growth," *Elastic-Plastic Fracture*, ASTM STP 668, American Society for Testing and Materials, Philadelphia, Pa., pp. 37–64 (1979).
(5.61) Shih, C. F., deLorenzi, H. G., Andrews, W. R., Van Stone, R. H., and Wilkinson, J. P. D., "Methodology for Plastic Fracture," General Electric Cooperate Research and Development Reports to EPRI, on RP601-2, Schenectady, New York (1976–1979).
(5.62) Shih, C. F., Dean, R. H., and German, M. D., "On J-Controlled Crack Growth: Evidence, Requirements and Applications," General Electric Company, T13 Report, Schenectady, New York (1981).
(5.63) Popelar, C. H., Pan, J., and Kanninen, M. F., "A Tearing Instability Analysis for Strain Hardening Materials," *Fracture Mechanics: Fifteenth Symposium*, ASTM STP 833, American Society for Testing and Materials, Philadelphia, Pa., pp. 699–720 (1984).
(5.64) Pan, J., Ahmad, J., Kanninen, M. F., and Popelar, C. H., "Application of a Tearing Instability Analysis for Strain Hardening Materials to a Circumferentially Cracked Pipe in Bending," *Fracture Mechanics: Fifteenth Symposium*, ASTM STP 833, American Society for Testing and Materials, Philadelphia, Pa., pp. 721–745 (1984).
(5.65) Paris, P. C., Tada, H., Zahoor, A., and Ernst, H., "The Theory of Instability of the Tearing Mode of Elastic-Plastic Crack Growth," *Elastic-Plastic Fracture*, ASTM STP 668, American Society for Testing and Materials, Philadelphia, Pa., pp. 5–36 (1979).

(5.66) Turner, C. E., "Fracture Toughness and Specific Energy: A Re-analysis of Results," *Material Science and Engineering*, **11**, pp. 275-282 (1973).

(5.67) Sumpter, J. D. G. and Turner, C. E., "Method for Laboratory Determination of J_c," *Cracks and Fracture*, ASTM STP 601, American Society for Testing and Materials, Philadelphia, Pa., pp. 3–15 (1976).

(5.68) Paris, P. C., Ernst, H., and Turner, C. E., "A J-Integral Approach to the Development of η-Factors," *Fracture Mechanics: Twelfth Conference*, ASTM STP 700, American Society for Testing and Materials, Philadelphia, Pa., pp. 338–351 (1980).

(5.69) Ernst, H. A., Paris, P. C., and Landes, J. D., "Estimation on J-Integral and Tearing Modulus T from a Single Specimen Test Record," *Fracture Mechanics: Thirteenth Conference*, ASTM STP 743, American Society for Testing and Materials, Philadelphia, Pa., pp. 476–502 (1981).

(5.70) Parks, D. M., Kumar, V., and Shih, C. F., "Consistency Checks for Power Law Calibration Functions," *Elastic-Plastic Fracture*, Vol. I, ASTM STP 803, American Society for Testing and Materials, Philadelphia, Pa., pp. 370–383 (1983).

(5.71) Chang, C. I., Nakagaki, M., Griffis, C. A., and Masumura, R. A., "Piping Inelastic Fracture Mechanics Analysis," NUREG/CR 1119, U.S. Nuclear Regulatory Commission, June 30, 1980.

(5.72) Gudas, J. P. and Anderson, O. R., "J_I-R Curve Characteristics of Piping Material and Welds," U.S.N.R.C. 9th Water Reactor Safety Research Information Meeting, Washington, D.C., Oct. 29, 1981.

(5.73) Zahoor, A. and Kanninen, M. F., "A Plastic Fracture Mechanics Prediction of Fracture Instability in a Circumferentially Cracked Pipe in Bending—Part 1. J-Integral Analysis," *Journal of Pressure Vessel Technology*, **103**, pp. 352–358 (1981).

(5.74) Tada, H., Paris, P.C., and Gamble, R. M., "A Stability Analysis of Circumferential Cracks for Reactor Piping Systems," *Fracture Mechanics: Twelfth Conference*, ASTM STP 700, American Society for Testing and Materials, Philadelphia, Pa., pp. 296–313 (1980).

(5.75) Kanninen, M. F., et al., "Instability Predictions for Circumferentially Cracked Large Diameter Type 304 Stainless Steel Pipes Under Dynamic Loading," Battelle Columbus Laboratories Second Semi-Annual Report to ERPI on Project T118-2, June 1980.

(5.76) Wilkowski, G. M., Zahoor, A., and Kanninen, M. F., "A Plastic Fracture Mechanics Prediction of Fracture Instability in a Circumferentially Cracked Pipe in Bending—Part II. Experimental Verification on Type 304 Stainless Steel Pipe," *Journal of Pressure Vessel Technology*, **13**, pp. 359–365 (1981).

(5.77) Paris, P. C., Tada, H., Ernst, H., and Zahoor, A., "Initial Experimental Investigation of Tearing Instability Theory," *Elastic-Plastic Fracture*, ASTM STP 668, American Society for Testing and Materials, Philadelphia, Pa., pp. 251–265 (1979).

(5.78) Joyce, J. A. and Vassilaros, M. G., "An Experimental Evaluation of Tearing Instability Using the Compact Specimen," *Fracture Mechanics*: Thirteenth Conference, ASTM STP 743, American Society for Testing and Materials, Philadelphia, Pa., pp. 525–542 (1981).

(5.79) Joyce, J. A., "Instability Testing of Compact and Pipe Specimens Utilizing a Test System Made Compliant by Computer Control," *Elastic-Plastic Fracture Vol. II*, ASTM STP 803, American Society for Testing and Materials, Philadelphia, Pa., pp. 439–463 (1983).

(5.80) Broek, D., "The Residual Strength of Light Alloy Sheets Containing Fatigue Cracks," *Aerospace Proceedings*, Macmillan, New York, pp. 811–835 (1966).

(5.81) de Koning, A. U., "A Contribution to the Analysis of Quasi-Static Crack Growth," *Fracture 1977, Proceedings of the 4th International Conference on Fracture*, Vol. 3A, Pergamon, pp. 25–31 (1978).

(5.82) Kanninen, M. F., et al., "Development of a Plastic Fracture Methodology," EPRI NP-1734, Project 601-1, Final Report, Electric Power Research Institute, March 1981.

(5.83) Rice, J. R. and Sorensen, E. P., "Continuing Crack-Tip Deformation and Fracture for Plane-Strain Crack Growth in Elastic-Plastic Solids," *Journal of the Mechanics and Physics of Solids*, **26**, pp. 163–186 (1978).

(5.84) Rice, J. R., Drugan, W. J., and Sham, T.-L., "Elastic-Plastic Analysis of Growing Cracks," *Fracture Mechanics: Twelfth Conference*, ASTM STP 700, American Society for Testing and Materials, Philadelphia, Pa., pp. 189–221 (1980).

(5.85) Rice, J. R. and Tracey, D. M., "Computational Fracture Mechanics," *Numerical and Computer Methods in Structural Mechanics*, S. J. Fenves et al. (eds.), Academic, New York, pp. 525–623 (1973).

(5.86) Rice, J. R., "Elastic-Plastic Crack Growth," *Mechanics of Solids*, H. G. Hopkins and M. J. Sewell (ed.), Pergamon, Oxford, pp. 539–562 (1982).

(5.87) Gao, Y.-C., and Hwang, K.-C., "Elastic-Plastic Fields in Steady Crack Growth in a Strain-Hardening Material," *Advances in Fracture Mechanics, Proceedings of the Fifth International Conference on Fracture*, Cannes, D. Francois (ed.), Vol. 2, Pergamon, New York, pp. 669–682 (1981).

(5.88) Sorensen, E. P., "A Numerical Investigation of Plane Strain Stable Crack Growth Under Small-Scale Yielding Conditions," *Elastic-Plastic Fracture*, ASTM STP 668, American Society for Testing and Materials, Philadelphia, Pa., pp. 151–174 (1979).

(5.89) Sham, T.-L., "A Finite Element Analysis of Quasi-Static Crack Growth in an Elastic Perfectly Plastic Solid," Sc. M. Thesis, Brown University, Division of Engineering, Providence, R. I., March 1979.

(5.90) Dean, R. H. and Hutchinson, J. W., "Quasi-Static Steady Crack Growth in Small-Scale Yielding," *Fracture Mechanics: Twelfth Conference*, ASTM STP 700, American Society for Testing and Materials, Philadelphia, Pa., pp. 383–405 (1980).

(5.91) Hermann, L. and Rice, J. R., "Comparison of Theory and Experiment for Elastic-Plastic Plane Strain Crack Growth," *Metal Science*, **14**, pp. 285–291 (1980).

(5.92) Wnuk, M. P., "Occurence of Catastrophic Fracture in Fully Yielded Components. Stability Analysis," *International Journal of Fracture*, **15**, pp. 553–581 (1979).

(5.93) Clarke, G. A., Andrews, W. R., Paris, P. C., and Schmidt, D. W., "Single Specimen Tests for J_{Ic} Determination," *Mechanics of Crack Growth*, ASTM STP 590, American Society for Testing and Materials, Philadelphia, Pa., pp. 27–42 (1976).

(5.94) Ernst, H. A., "Material Resistance and Instability Beyond J-Controlled Crack Growth," Scientific Paper 81-107-JINF-P6, Westinghouse R & D Center, Pittsburgh, Pa., Dec. 3, 1981.

6

FRACTURE MECHANICS MODELS FOR FIBER REINFORCED COMPOSITES

Fiber composite materials are very strong for their weight and are generally fatigue and fracture tolerant. Consequently, they are highly attractive for use in aerospace, automotive, and other applications where weight is a primary concern. Reflecting this, a large number of textbooks have been written that provide the basic analysis approaches developed for the design of composite structures. The more recent of these are given as references (6.1)–(6.7). The more general treatment of Kelly (6.8) may also be profitably pursued. Tsai and Hahn (6.7) append a more complete listing—one that includes some 19 books on the subject that have appeared since 1966—but still omit several of the titles just given. There are currently eight journals devoted to composite materials with a steady stream of ASTM Special Technical Publications on the subject. This amount of publication is indicative of the importance of this class of materials—an importance that can only intensify in the years to come.

Despite their attractiveness, composites are not now being used as much as they could be. And, often when they are used, it is in low stress applications, or with such large factors of safety as to nullify much of their potential. The basic reason surely is the uncertainty that exists in determining their strength and safe-operating lifetime in service conditions—particularly when defects could be present. For example, it is well known that damage resulting from low velocity impact events that might readily occur in normal service can have an extremely detrimental effect on the subsequent performance of a fiber composite component. Hence, there has been a large amount of research addressed to determining the effect of a damaged condition on the strength of a fiber composite that have drawn upon fracture mechanics. The books of Vinson and Chou (6.2), Tewary (6.3), Agarwal and Broutman (6.5), and Piggott (6.6), each contain extensive treatments of linear elastic fracture mechanics while Jones (6.1) and Christensen (6.4) provide at least cursory treatments.

The importance of composite materials is not as well reflected by the fracture mechanics community per se; for example, of the fracture mechanics books that we have listed in Chapter 9, only the book of Jayatilaka (6.9) addresses this subject. However, there have been a number of review papers on the application of fracture mechanics to composites. These include Argon (6.10), Corten (6.11), Erdogan (6.12), Zweben (6.13), Beaumont (6.14), Smith (6.15), Kanninen et al. (6.16), Dharan (6.17), and Backlund (6.18). As reflected in these papers, because failure processes often emanate from crack-like defects in the material, it is natural to apply fracture mechanics techniques to obtain more precise strength and lifetime predictions. But, as stated by Potter (6.19):

> The great success of linear elastic fracture mechanics in predicting the behaviour of isotropic materials has led, almost hypnotically, to its direct application to composites with little or no modification.

Hence, despite the fact that LEFM techniques have been very successful for assessing defects in metals, limited success has so far been achieved for composites. What appear to be needed are more theoretically valid analysis models that recognize the inherent differences in the fracture processes in composites and metals. This chapter focuses on such fracture mechanics analysis models.

6.1 Preliminary Considerations

Composite structural materials present a fracture mechanics analyst with a very complex situation. First, because they are heterogeneous, crack growth does not generally progress in the relatively simple manner that usually occurs in metals. Second, the myriad of failure processes that can occur are asociated with mechanical properties that are time-dependent and sensitive to temperature, moisture, and other environmental parameters. Third, the constituents of the composite will generally have properties that can differ markedly from one to another and, in addition, from position to position in the material. The development of mathematical predictive models for fiber composites consequently presents a formidable challenge indeed.

Before addressing the various specific applications of fracture mechanics for fiber composites, it will be useful to highlight the key aspects that must underlie such applications. First, the conventional classifications and notation for fiber composite laminated structures are introduced. Next, some basic considerations in the application of fracture mechanics to composites are discussed. To illuminate the basic nature of the problem facing fracture mechanics analysts, the failure processes that are observed in fiber composites are briefly reviewed. Finally, the additional complications associated with laminates are described.

6.1.1 Classifications and Terminology

Recognizing that virtually all engineering materials are in a sense a composite of two or more constituents, the term "composite materials" may need a precise definition. It is commonly accepted that a composite material consists of at least two constituents that are chemically distinct on a macroscopic scale and have a clearly recognizable interface between them. Ordinarily, one constituent is a discontinuous phase that is bonded to, or embedded in, a continuous phase. The former is termed the reinforcement, the latter the matrix. The main classes of composites include, (1) embedded particle composites, (2) sandwich or layered composites, and (3) fiber reinforced composites. The latter are of most engineering interest and have received by far the most attention in fracture mechanics. They will be treated exclusively here.

Fiber reinforced composites (fiber composites for short) can be classified in various ways. One way distinguishes between glass fiber and so-called advanced composites—the latter designation referring to the use of boron, graphite, polyaramid, or other non-glass fiber materials. Another basis of classification is whether the matrix material is a polymer or a metal. In the former, a distinction exists between thermosets and thermoplastic resins—each having a distinct range of properties. Another classification basis is whether the fibers are (1) continuous and aligned or (2) chopped and distributed randomly in the matrix. Still another important distinction between the various types of advanced composites is in regard to the relative fiber size: boron fibers have a diameter in the order of 0.1 mm while glass and graphite fibers are roughly .01 mm in diameter. The latter are generally produced in tow form with 50 to 2000 fibers being contained in a single tow. Figure 6.1 illustrates this.

Regardless of the distinctions just described, the general behavior of a fiber composite is always governed by the same general principles. Specifically, a fiber composite material exploits the high strength of a material in the form of a fiber. As was shown long ago by Griffith—see Figure 1.20—the fiber form exhibits a high strength primarily because of the elimination of debilitating defects. In a fiber composite the fiber strength is retained by embedding them in a binding material that keeps the fibers in a desired location and orientation, transfers load between fibers, and protects them from environmental and handling damage.

Advanced composites are typically used in the form of laminates. A laminate is an assembly of "pre-pregs" or thin plies consisting of parallel fibers

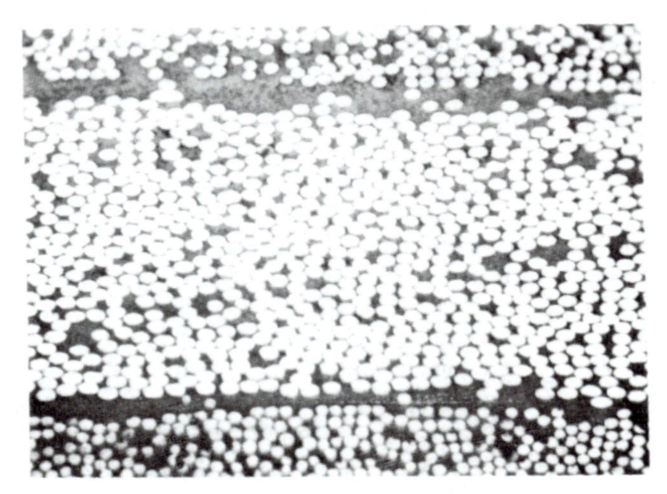

Figure 6.1 Enlarged view of a ply in a graphite/epoxy laminate showing individual fibers and the layers between two adjacent plies.

laid onto a resin and partially cured by heating. A number of pre-pregs normally are stacked (or layed-up) in some prescribed manner to obtain the properties needed for a particular application. The composite laminate is then final cured under heat and pressure. A commonly used notation for the lay up of a laminate is based upon the convention that designates the ply (or laminae) orientation by the angle between the fiber direction and primary load direction. That is, a 0° ply is one in which the load acts in the direction parallel to the fibers; a 90° ply is one in which the load acts in the direction transverse to the fiber length. Using this convention, a laminate designated as $[0/\pm 45/90]_4$ means that any four consecutive plies are in the 0°, 45°, −45°, and 90° directions and that there are four sets of these; 16 plies in all. An extension of this notation is to employ a subscript S to denote symmetry about the mid-plane. Thus, a $[0/\pm 45/90]_{2S}$ laminate is also one containing 16 plies, but arranged in a slightly different order then in the first example.

While many of the early analyses took a micromechanical approach (i.e., one in which the properties of the fiber and the matrix are specifically identified), it was gradually superceded by the macromechanical approach known as lamination theory. In lamination theory, the ply level is the smallest subdivision, with the mechanical properties of the fiber and the matrix being smeared into effective ply properties by the rule of mixtures. This superceded an approach known as netting analysis—an approach which assumes that the fibers carry all of the load with the matrix serving only to hold the fibers together. That the matrix plays an important role in a composite used in a loaded-bearing application is now clear. What is not so clear is its role in determining the strength of the composite and, in particular, the strength of the interface.

The terms "failure" and "strength" are not always used in a precise manner in engineering applications. But, because of the complexity of fiber composites, the distinction is of particular concern. There are many different ways that a structure made of a composite material can become unable to perform its primary function. In each such instance "failure" can be considered to have occurred. The term "strength" is conventionally associated with the load level at which failure occurs by some specific means—usually in a standard test specimen. The possible failure modes range from simple loss of structural stiffness due to gross inelastic deformation (e.g., yielding), through a premature warning of a potential loss of load-carrying capacity (e.g., first ply failure), to a catastrophic reduction in load-carrying capacity by gross macroscopic deformation and separation (e.g., fracture).

Failure can be gradual or rapid, and may or may not be catastrophic in nature. Clearly, the strength will depend upon the failure mode under consideration and will be a function of many different parameters in the test program. Hence, it may or may not be directly applicable to the material when used in a different form in service.

It is clear that an analysis procedure that provides a bridge between standard test procedures and engineering applications is absolutely necessary. Only in this way can reliable estimates of the failure loads expected in service be made using strengths determined in small-scale laboratory tests. While substantial failure analysis work has been performed on composite materials, a general theoretically valid capability does not exist. It is likely that the least progress has been made in the most critical problem area—when crack-like defects exist in the material. This, of course, is the motivation for the use of fracture mechanics for composite materials.

6.1.2 Basic Mechanical Behavior

The simplest point of view that can be taken in analyzing a composite laminate is to regard it as a homogeneous continuum and to make no distinction either between the fiber and matrix materials that make up the composite or between the individual plies. The laminate is then treated as a single material with suitably averaged properties of its constituents. As a result of such a sweeping generalization, much progress is possible. As already indicated, such an approach has been superceded by lamination theory. However, the "rule of mixtures" still forms the basis for many analyses. The manner in which the properties and geometrical arrangements of the individual materials in the composite contribute to the average mechanical properties of the material is therefore of interest.

A mathematical model of a composite can be constructed by applying combinations of elasticity, plasticity, and viscoelasticity to model the fiber and the matrix. The simplest approach is to envision the fiber and matrix as a simple series of parallel tensile elements. Then, if both the fiber and matrix are linear elastic materials and are firmly bonded to each other, the elastic moduli in axial tension, transverse tension, and shear, respectively, are given by

$$E_A = V_f E_f + (1 - V_f) E_m \tag{6.1-1}$$

$$E_T = \left(\frac{V_f}{E_f} + \frac{1 - V_f}{E_m} \right)^{-1} \tag{6.1-2}$$

$$G = \left(\frac{V_f}{G_f} + \frac{1 - V_f}{G_m} \right)^{-1} \tag{6.1-3}$$

where V_f denotes the volume fraction of fiber and the subscripts f and m denote properties of the fiber and the matrix, respectively. These results are known as the "rule of mixtures" and are, of course, only approximations. As an important example, the law of mixtures approach is completely invalid for predicting the toughness of a composite. Nevertheless, such relations can be useful for predicting the mechanical properties of a composite.

The introduction of an oriented family of fibers gives a definite directionality to the material. Hence, even if the constituents of the composite are isotropic, the composite itself will be anisotropic in its elastic properties. Moreover, unless the fibers are everywhere aligned in parallel straight lines (i.e., a unidirectional composite), the principle directions characterizing the elastic anistropy will vary from point to point. Finally, in a laminate where there is more than one family of fibers, the composite will have several preferred directions (possible varying from point to point) and the material exhibits an even more general form of anistropy. When the material contains a crack that is not aligned with a principle direction, even greater complications to a complete analysis are presented.

Other important combinations of continuum mechanics theories to fiber component materials include the viscoelastic fiber and matrix, elastic-plastic fiber and matrix, elastic fiber and viscoelastic matrix, and elastic fiber with an elastic-plastic matrix. Relations comparable to the rule of mixtures do not exist for these models. Nevertheless, some qualitative results can be stated. For example, if either or both constituents are viscoelastic, the composite is viscoelastic. Then, the five elastic constants of a directionally anisotropic (orthotropic) elastic material are replaced by five relaxation or creep functions. Similarly, if either or both constituents are elastic-plastic, the problem is one of an orthotropic elastic-plastic solid. In most cases, it can be assumed that the macroscopic response of the body in axial tension will be predominately due to the properties of the fibers, while its behavior in both transverse tension and in transverse or axial shear will be mainly governed by the properties of the matrix.

6.1.3 Anisotropic Fracture Mechanics

In their gross response to load, composite materials can be considered as being orthotropic in their elastic properties. (Orthotropic, meaning orthogonal anisotropy, is a special case of general anisotropy.) However, in a cracked orthotropic body where the crack is not associated with a plane of elastic symmetry, the analysis becomes a problem of a generally anisotropic body. In so far as the fracture mechanics of anisotropic bodies is concerned, the stress intensity factors are in most practical cases just the same as for isotropic bodies. In particular, except when unbalanced loads act on the crack faces, the stress intensity factors will be independent of the material constants and, therefore, will be identical to the K values derived in isotropic fracture mechanics. This holds for each of the three possible modes of crack extension. The analysis of the virtual work of crack extension, which relates the stress intensity factors to the energy release rates for each mode, is then much the same as in the isotropic case. In general, however, the relationship is nonlinear in the K values so that simple superposition of loads is not possible except under simple states of applied stress.

Following the derivation given by Sih et al. (6.20), let the generalized Hooke's law be written in index notation as

$$\varepsilon_i = \sum_{i=1}^{6} a_{ij}\sigma_j, \qquad i = 1, 2, \ldots, 6 \tag{6.1-4}$$

where $a_{ij} = a_{ji}$; $\varepsilon_1 = \varepsilon_x$, $\varepsilon_2 = \varepsilon_y, \ldots, \varepsilon_6 = \gamma_{xy}$; and $\sigma_1 = \sigma_x$, $\sigma_2 = \sigma_y, \ldots,$ $\sigma_6 = \tau_{xy}$. In general anisotropy there are 21 independent elastic constants. In plane problems this number is reduced to six. For plane stress these are: a_{11}, $a_{22}, a_{66}, a_{12}, a_{16}, a_{26}$. Notice that from a mathematical viewpoint the formulation of plane stress and plane strain problems are identical—a solution for the latter being obtained from the former by replacing the a_{ij}'s by b_{ij}'s, where

$$b_{ij} = a_{ij} - \frac{a_{i3}a_{j3}}{a_{33}}, \qquad i,j = 1, 2, \ldots, 6 \tag{6.1-5}$$

In either case (i.e., either using the a_{ij}'s or the b_{ij}'s) a solution by a complex variable technique leads to a result involving a parameter μ that is one of the roots of the characteristic equation

$$a_{11}\mu^4 - 2a_{16}\mu^3 + (2a_{12} + a_{66})\mu^2 - 2a_{26}\mu + a_{22} = 0 \tag{6.1-6}$$

Because the roots of Equation (6.1-6) are either complex or purely imaginary, μ can always be expressed in the conjugate pairs $(\mu_1, \bar{\mu}_1)$ and $(\mu_2, \bar{\mu}_2)$.

Just as in the isotropic case, a knowledge of the stress and displacement fields in the neighborhood of the crack tip is the key to expressing the fracture strength of a cracked body. The difficulty in the anisotropic case is that crack extension will not necessarily occur in a planar fashion. However, because the mathematical difficulties involved in treating angled cracks is prohibitive, this complication is usually ignored. Consequently, in the same manner as in the isotropic case, the energy release rate-stress intensity factor equivalence is determined from a virtual crack extension Δa. For Mode I conditions, this is

$$G_{\rm I} = \lim_{\Delta a \to 0} \int_0^{\Delta a} \frac{1}{\Delta a} \sigma_2(r, 0) u_2(\Delta a - r, \pi)\, dr \tag{6.1-7}$$

where the components of stress and displacement are considered to be given in terms of a polar coordinate system (r, θ) at the crack tip. Similar expressions for Mode II and Mode III deformation can also be written.

Omitting the details, if the stress and displacements corresponding to a crack in a rectilinearly anistropic body in plane stress are substituted into Equation (6.1-7), the results are found to be

$$G_{\rm I} = -\frac{K_{\rm I}^2}{2\pi} a_{22} \, {\rm Im}\left\{\frac{\mu_1 + \mu_2}{\mu_1 \mu_2}\right\} \tag{6.1-8}$$

Similarly,

$$G_{\rm II} = \frac{K_{\rm II}^2}{2\pi} a_{11} \, {\rm Im}\{\mu_1 \mu_2\} \tag{6.1-9}$$

where in Equations (6.1-8) and (6.1-9), μ_1 and μ_2 are the distinct roots of the characteristic equation (6.1-6). Notice that, provided there are no unbalanced loads acting on the crack faces, the stress intensity factors $K_{\rm I}$ and $K_{\rm II}$ are exactly the same functions of the applied loading and geometry as in the isotropic cases. For example, for a uniform tensile stress σ acting at an angle α to the crack plane in an infinite sheet, these are

$$\begin{aligned} K_{\rm I} &= \sigma\sqrt{\pi a} \sin^2 \alpha \\ K_{\rm II} &= \sigma\sqrt{\pi a} \sin \alpha \cos \alpha \end{aligned} \tag{6.1-10}$$

where $2a$ is the crack length.

Equations (6.1-8) and (6.1-9) simplify somewhat if the material is orthotropic with the crack on one plane of symmetry. Then, $a_{16} = a_{26} = 0$ so that there are only four independent elastic constants. More importantly, the roots of (6.1-6) can be extracted conveniently. Then (6.1-8) and (6.1-9) reduce to

$$G_{\mathrm{I}} = K_{\mathrm{I}}^2 \left(\frac{a_{11}a_{22}}{2}\right)^{\frac{1}{2}} \left[\left(\frac{a_{22}}{a_{11}}\right)^{\frac{1}{2}} + \frac{2a_{12} + a_{66}}{2a_{11}}\right]^{\frac{1}{2}} \tag{6.1-11}$$

and

$$G_{\mathrm{II}} = K_{\mathrm{II}}^2 \frac{a_{11}}{\sqrt{2}} \left[\left(\frac{a_{22}}{a_{11}}\right)^{\frac{1}{2}} + \frac{2a_{12} + a_{66}}{2a_{11}}\right]^{\frac{1}{2}} \tag{6.1-12}$$

For an isotropic material, $a_{11} = a_{22}$ and $a_{66} = 2(a_{11} - a_{12})$. Making these specializations then reduces (6.1-11) and (6.1-12) to the ordinary relations of linear elastic fracture mechanics given in Chapter 3.

The conclusion that can be drawn from the foregoing is that linear elastic fracture mechanics can be applied to generally anisotropic bodies with cracks in very much the same way as it is for isotropic bodies. In fact, if suitable values of K_c are known (or, when more than one mode is present, the appropriate functional form $(K_{\mathrm{I}}, K_{\mathrm{II}}, K_{\mathrm{III}}) = f_{cr}$ is known), in most cases of practical interest the analysis is identical to the isotropic case. Again, this is true only if unbalanced forces do not act on the crack faces and crack extension takes place in the place of the original crack. A modification of the isotropic fracture mechanics approach is necessary if the fracture criterion is derived from energy considerations—for example, work of fracture for a composite material. Then, Equations (6.1-8) and (6.1-9) must be used to cast the result in terms of a critical toughness level, a procedure that, in general, will require the determination of the roots of the characteristic equation.

6.1.4 Basic Considerations for Fracture Mechanics Applications

As repeatedly emphasized in this book, fracture mechanics should be applicable regardless of the constitutive behavior of a cracked body and of the origin, size, shape, and the direction of growth of the crack. However, most applications have called upon fracture mechanics relations that are valid only for rather specialized conditions—conditions that do not generally hold for fiber reinforced composites. A brief review of the basis of linear elastic fracture mechanics (LEFM) may be useful in seeing why this is so.

If a body everywhere obeys a linear elastic stress-strain relation, then the stresses at the tip of a crack in such a body will be given by

$$\sigma_{ij} = \frac{K_{\mathrm{I}}}{\sqrt{2\pi r}} f_{ij}(\theta) + \cdots \tag{6.1-13}$$

where r and θ are polar coordinates with their origin at the crack tip and K_{I} is the Mode I stress intensity factor. The omitted terms in Equation (6.1-13) are of higher order in r. Hence, very near the crack tip where r is small, only this term is significant. It follows that the remote stresses, the crack length and the external dimensions of the body will affect the stresses at the crack tip only as they affect K. Furthermore, there will be region surrounding the crack tip in which the $r^{-\frac{1}{2}}$ term given by Equation (6.1-13) will be a sufficiently good approximation to the actual stresses. This is called the "K-dominant" region.

Now, if the inelastic processes that occur in the vicinity of the crack tip are entirely contained within the K-dominant region, any event occurring in the inelastic region is controlled by the deformation that occurs in the annular elastic region surrounding it. Because this deformation is solely dependent upon the value of K_{I}, if crack growth occurs, it must do so at a critical K_{I} value. This leads directly to the basic equation of linear elastic fracture mechanics. Using notation common in work on composite materials, this is

$$K_{\mathrm{I}}(a, b, \sigma) = K_{1c}(T, \dot{\sigma}) \tag{6.1-14}$$

where K_{I} is a material-independent function of the crack size a, the component dimensions b, and the applied stress σ, while K_{1c} is a material property that can depend upon temperature T and loading rate $\dot{\sigma}$. Note that there is no restriction on the direction of crack growth in this argument.*

For metals, the above argument leads to the conclusion that, for Equation (6.1-14) to be valid, the crack length must be large in comparison to the value of $(K_{1c}/\sigma_Y)^2$, where σ_Y is the yield stress of the material. A comparable relation for composites is not so readily arrived at. Nevertheless, it is clear that the crack length in a composite must be large compared to the damage zone that would be experienced at a crack tip in the composite for this approach to be valid. It is therefore safe to say that most of the applications of fracture mechanics to composites have suffered as a consequence of violating this requirement.

In developing more appropriate fracture mechanics techniques for applications to fiber reinforced composite materials, several basic facts must be kept in mind. First, the initial defects that are apt to trigger a fracture are usually quite small. Sendeckyj (6.21) has recently cataloged the many different types of defects that occur and their effects on structural performance. These include preparation defects (e.g., resin-starved or fiber-starved areas), defects in laminates (e.g., fiber breaks, ply gaps, delaminations), and fabrication defects (edge delaminations caused by machinery, dents, and scratches). Of most significance to the present discussion is the inapplicability of Equation (6.1-14) to flaws such as these.

A second key fact involved in the application of fracture mechanics to FRP materials is the basic heterogeneous nature of fiber reinforced composites. This manifests itself in several different ways. Within a ply, cracking can be

* The notation used in mixed mode conditions must avoid use of K_{Ic} to denote fracture toughness as this term is generally reserved for plane strain conditions. Accordingly, the values K_{1c}, K_{2c}, and K_{3c} are used. Note that they will therefore depend upon the degree of triaxial constraint; for example, they will be thickness-dependent.

both discontinuous (e.g., fiber bridging) non-collinear crack growth (e.g., matrix splitting). On the laminate level, cracking can proceed in a distinctly different manner in different plys and, in addition, interply delamination can occur. Again, Equation (6.1-14) is ill-equipped to cope with these complexities. Consequently, many researchers have pursued an energy balance approach to the problem. This does not really present a significant improvement over the fundamental difficulties associated with the stress intensity factor point of view, however.

Until recently, composite fracture mechanics research generally fell into one of two broad categories: Either, (1) a continuum analysis for a homogeneous anisotropic linear elastic material containing an internal or external flaw of known length, or (2) a semi-empirical analysis of the micromechanical details of the crack-tip region in a unidirectional fiber composite. The continuum approach completely ignores the inherently heterogeneous nature of composite materials and the basic way that heterogeneity affects crack extension. In fact, this approach represents only a slight extension of ordinary linear elastic fracture mechanics to account for the anisotropic response of the material to load. It involves only an evaluation of the left-hand side of Equation (6.1-14) with the right-hand side tacitly being considered a material constant that can be obtained from experiments.

The micromechanical approach, the second of the two cited above, can also be related to Equation (6.1-14). It essentially represents a way to determine the right-hand side in terms of basic material properties by considering various mechanisms involved in composite fracture. For example, energy values have been deduced for debonding of the fiber from the matrix material, pull-out of the fiber from the matrix, and for inelastic deformation and fracture of the matrix material. Piggott (6.6) gives an extensive summary of these mechanisms and the relations that have been developed for them.

As stated by Wells (6.22), a composite material can suffer any of the modes of failure of its constituents together with a few more arising from their combination. It is for these reasons that approaches in which the two sides of Equation (6.1-14) are attacked independently will not ordinarily suffice for composites—a conclusion reached some time ago by Zweben (6.13). Nevertheless, it is useful to explore some of the results in the literature that have adopted such an approach. Understanding these may lead to more fundamentally sound approaches to the problem. Accordingly, consideration is given next to the micromechanical factors in crack growth.

6.1.5 Micromechanical Failure Processes

In the absence of a large cut, it can be assumed that failure in a fiber composite emanates from small inherent defects in the material. These defects may be broken fibers, flaws in the matrix, and/or debonded interfaces. The longer the fiber, the greater the probability that a critical defect exists that will cause individual fiber breakage at loads well below the average fiber strength of the composite. Regardless, after a single fiber break, more fiber breakage plus debonding and separation of the fiber and matrix will result. Defects in the matrix may also lead to yielding and fracture of the matrix between fibers. This will create further stress concentrations at the fibers and fiber/matrix interfaces. Thus, even at relatively low load levels, small cracks of a size on the order of several fiber diameters or fiber spacings are likely to exist.

Under higher load levels the microcracks can coalesce to form an identifiable crack-like flaw. Even so the size of the damage zone near the crack tip will still be small relative to the crack length and other dimensions of the body. Hence, qualitatively, the same failure mechanisms (e.g., fiber breakage, debonding, matrix yielding, cracking) will always be important in the fracture process. This will be true whether the composite is composed of uniaxially or multiaxially oriented fibers for all different fiber/matrix combinations. An illustration of various possible micromechanical damage processes for a macrocrack running transverse to the fibers is shown conceptually in Figure 6.2.

Figure 6.2 shows several possible local failure events occurring prior to the fracture of a fiber composite. At some distance ahead of the crack, the fibers are intact. In the high stress region near the tip they are broken, although not necessarily along the crack plane. Immediately behind the crack tip, fibers pull-out of the matrix. They absorb energy if the shear stress at the fiber/matrix interface is maintained while the fracture surfaces are separating. Theoretical treatments of the mechanics of the fiber/matrix interface have been given by several investigations; for examples, see Phillips and Tetelman (6.23), Cooper and Kelly (6.24 to 6.26), Gresaczuk (6.27), and Beaumont and Harris (6.28).

In some bonded composites, the stresses near the crack tip could cause the fibers to debond from the matrix before they break. When total debonding occurs, the strain energy in the debonded length of a fiber is lost to the material and is dissipated as heat. Fiber stress-relaxation, a variation of the debonding model, estimates the elastic energy that is lost from a broken fiber when the interfacial bond is not destroyed. It is also possible for a fiber to be left intact as

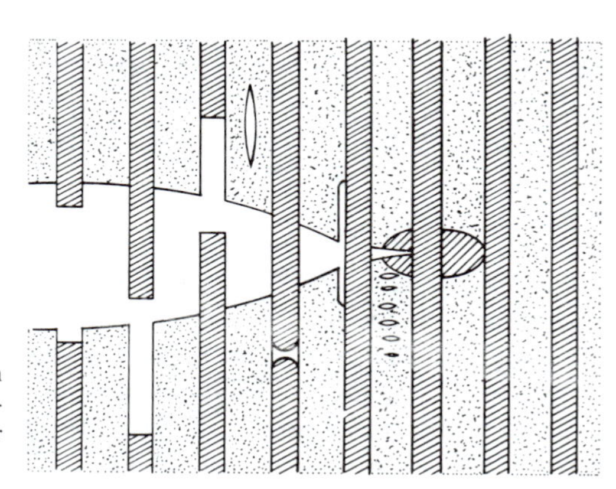

Figure 6.2 Schematic representation of micromechanical failure events accompanying crack extension in a fiber composite material.

the crack propagates and this process, known as crack bridging, can also contribute to the toughness of the material. An analysis of debonding and crack bridging for brittle fibers in a perfectly plastic material has been given by Piggott (6.29).

When brittle fibers are well bonded to a ductile metal matrix, the fibers tend to snap ahead of the crack tip leaving bridges of matrix material that neck down and fracture in a completely ductile manner. The fracture toughness in these circumstances is largely governed by the energy involved in the plastic deformation of the matrix to the point of failure. Of the energy absorption models, debonding and pull-out have been used most widely. Obviously, the same mechanisms will not be important in all combinations of matrix and fiber materials. For example, the fracture energies of carbon fiber composites have been more successfully correlated with the pull-out model; for boron and glass composites, debonding is more successful. Marston et al. (6.30) have shown that no single mechanism such as pull-out, debonding, or stress redistribution taken alone can account for the observed toughness of boron/epoxy composites.

Considerations of this kind can be contrasted with the simple "rule of mixtures" described above, which works quite well in determining the effective elastic moduli. In the predictions of effective moduli, it may be safely assumed that the generic stress/strain relations and geometries do not change under load. In contrast, the prediction of strength implies that process in the material have progressed to the extent that significant changes in material behavior and geometry have occurred—for example, yielding, damage formation, and crack growth. Thus, in order to perform a failure analysis properly, it is necessary to quantify these fundamental changes in the behavior of a structure.

While the micromechanical failure events that can take place may be the same for simple unidirectional composites as for multi-axial laminates, the macroscopic failure modes for each can be quite different. The reason is that the local stress states are different, even under the same applied loading. The situation is further complicated by the fact that the order in which a sequence of discrete failure modes occurs is important. In general, several processes are likely to occur more or less simultaneously with the amount of energy dissipated being dictated by the kind and rate of loading, the flaw size and orientation, external geometry, and the temperature. Hence, the appropriate value of K_{1c} to be used with a theoretically derived K_{I} for a given application—see Equation (6.1-14)—cannot be deduced by simply summing the effects of single mechanisms operating independently.

In the experimentation of Poe and Sova (6.31) on center-cracked sheet specimens of boron/aluminum, failure largely occurred by self-similar crack extension. Radiographs of the specimens indicated that the principal load carrying fibers began to break at the crack tip at about 80 percent of the eventual failure load. The breaks progressed from fiber to fiber, in effect producing resistance curve behavior and fracture instability after some stable crack growth, similar to that which occurs in elastic-plastic fracture

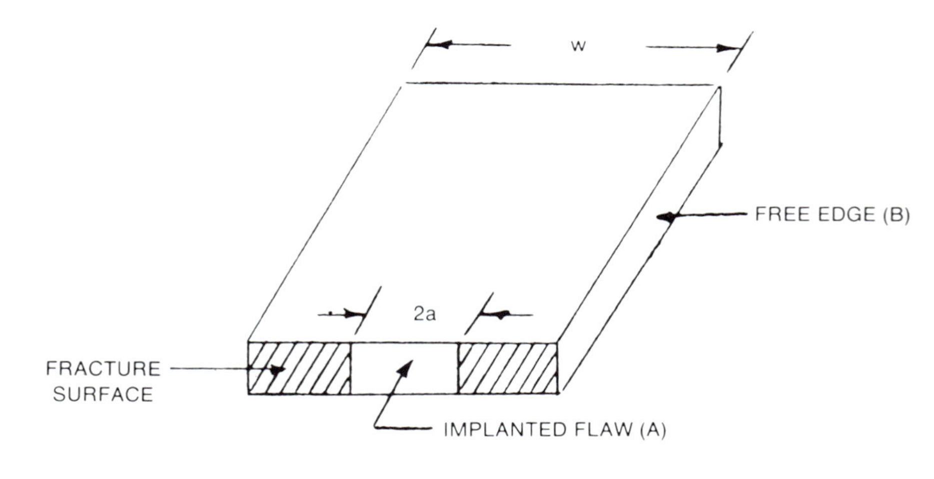

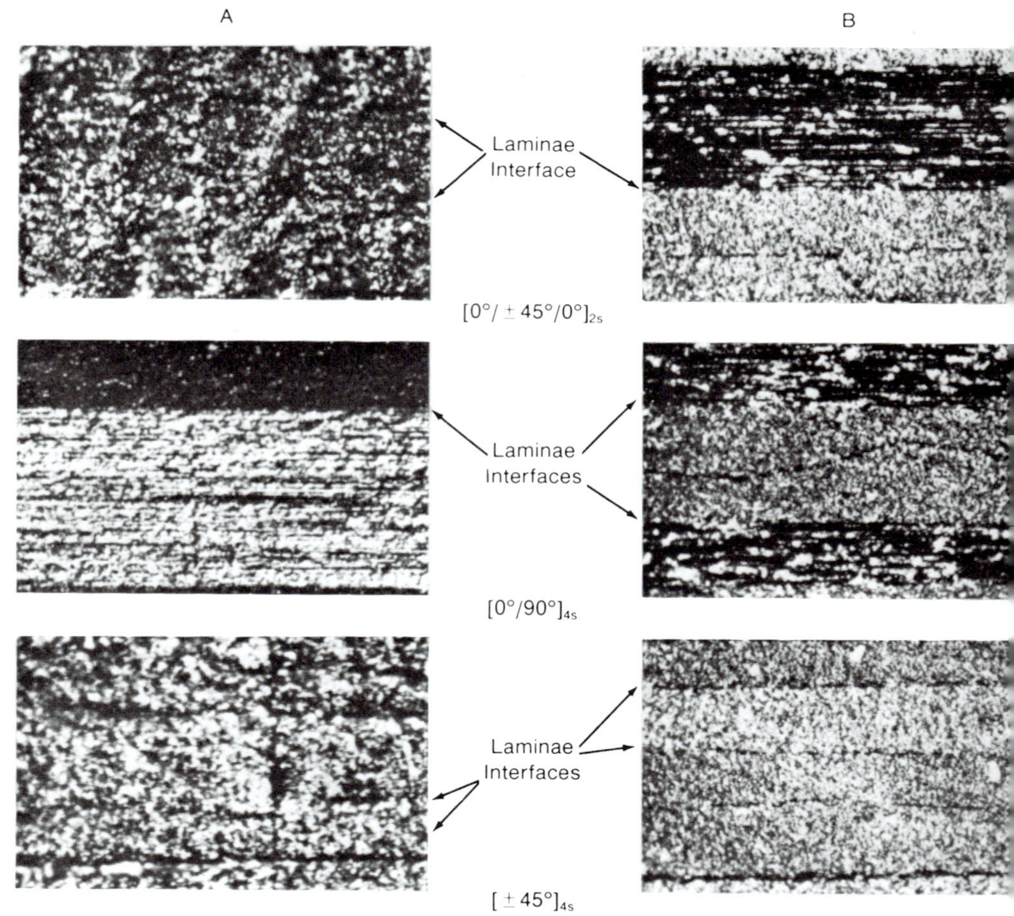

Figure 6.3 Edge view of graphite/epoxy laminae in three different laminates.

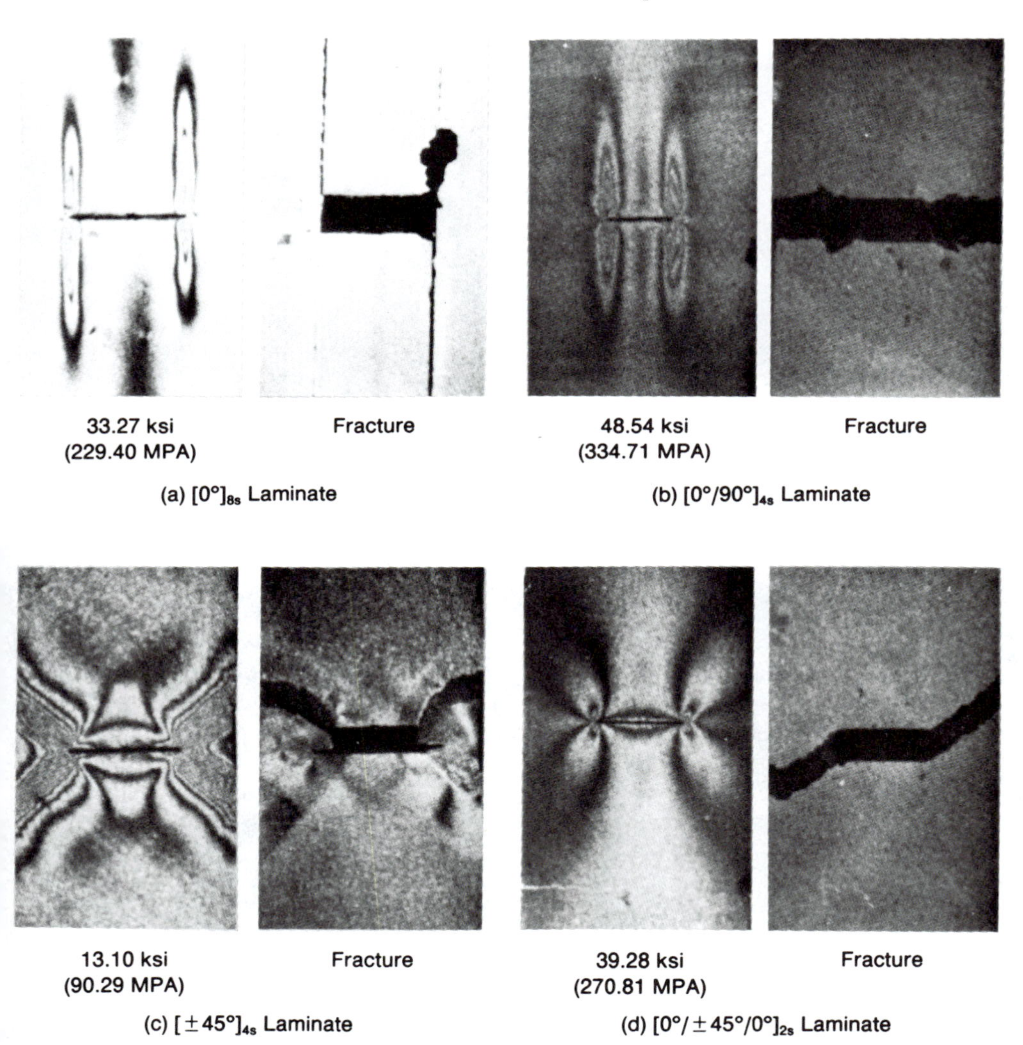

Figure 6.4 Fracture directions for center-cracked graphite/epoxy tension panels.

mechanics, in the manner first noted by Gaggar and Broutman (6.32) and by Mandell et al. (6.33).

In later work, Poe (6.34) found similar behavior in graphite/epoxy laminates. His radiographs show broken fibers and splits (matrix cracks) to a distance of 3 mm (about 20 percent of the crack length) ahead of the crack tip. It is of some interest to recognize that the damage in the $-45°$ ply, which consisted of matrix cracking, coincided with the fiber cracking damage in the $0°$ ply. Of further interest is his observation that damage progresses in a series of discrete jumps (accompanied by audible noise) that can start below one-half of the eventual failure load. Hence, it would appear, much as stated by Poe,

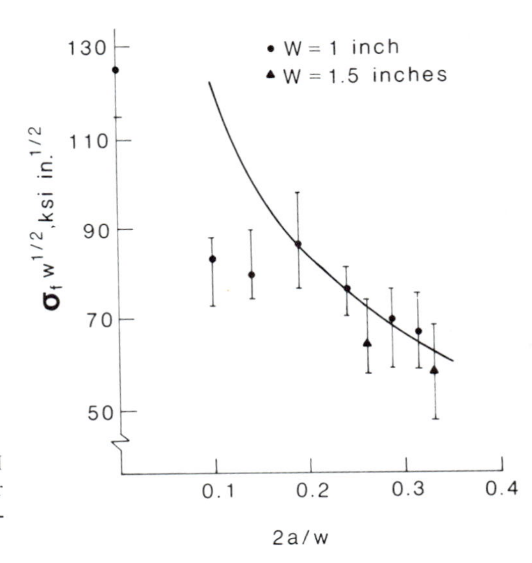

Figure 6.5 Comparison of LEFM prediction with fracture data on a fiber composite laminate—results of Zimmer.

that the fracture process cannot be properly analyzed without knowing the size and nature of the crack tip damage.

Following the work contributed by Wu (6.35), a number of investigators promulgated linear elastic fracture mechanics models for fiber composites; see references (6.36)–(6.45). Most of these can be summed up by noting that LEFM can be applied provided crack extension takes place prior to transverse splitting. The observations of Brinson and co-workers (6.46, 6.47) show that the effect is dependent upon the lay-up. Their results are shown in Figures 6.3 and 6.4. But, as shown for example by Zimmer (6.39), the agreement is not particularly satisfactory even when splitting does not occur. His results are shown in Figure 6.5.

6.2 Linear Fracture Mechanics Analysis Models

Despite the fact that composite materials are by their very nature heterogeneous materials, analyses developed for homogeneous materials are usually applied in treating composite fracture. A typical rationale has been given by Corten (6.11) who stated that, correct or not, such a point of view allows relationships between the variables to be developed for engineering purposes that could not otherwise be obtained. Of course, in composite materials, the picture is much more complex than in more conventional materials. Composite materials are inhomogeneous as well as anisotropic and, as described above, various fracture mechanisms with separate but perhaps coupled fracture energies must be accounted for. However, perhaps the single most complicating feature of composite fracture is that significant amounts of damage growth generally precede fracture in a fiber composite material. Self-similar crack growth is not likely to occur, even for undirectional or symmetric multi-directional laminates. Direct consideration (and implementation) or these key observations is one of the major delineations between the various approaches that have been offered.

6.2.1 Crack Length Adjustment and Other Simple Models

The early fracture theories of Waddoups et al. (6.36), Cruse and co-workers (6.40, 6.41), and Whitney and Nuisimer (6.48) are based entirely on linear elastic fracture mechanics considerations. Each of these investigators found that, in order to use various fracture solutions for either holes or cracks, the crack length had to be adjusted to include an "intense energy" region at each crack tip. The size of the intense region had to be found by experiment. Consider first the influential work of Waddoups et al. They have employed an empirical extension of linear elastic fracture mechanics for isotropic materials in which a damage zone of length l is taken as simply increasing the crack length, much as in Irwin's plastic zone correction factor (see Section 1.4.1).* For a crack of length $2a$ in an infinite body under tensile load normal to the crack, their result is expressed as

$$K_{\mathrm{I}} = \sigma \pi^{\frac{1}{2}}(a + l)^{\frac{1}{2}} \tag{6.2-1}$$

where l is taken to be the dimension of the characteristic intense energy region at the crack tip. The critical stress for crack extension is then

$$\sigma_c = \frac{K_{1c}}{\pi^{\frac{1}{2}}(l + a)^{\frac{1}{2}}} \tag{6.2-2}$$

where K_{1c} is the fracture toughness.

Waddoups et al. treated K_{1c} and l as disposable parameters that could be evaluated from experimental data. They then concluded that the agreement of this equation with the single body of data from which these two parameters were determined shows that linear elastic fracture mechanics can be applied to composite materials. In fact, they devised a two-parameter empirical correlation of certain test data that may or may not be applicable to other materials or to other crack orientations in the same material.

The actual crack-tip geometries (as opposed to the assumed ideal crack-tip geometries used in fracture analyses) are important for composite materials. Notch insensitivity is implicit in many fracture models as little distinction is made between a through circular flaw and a through crack. For this reason, the approach of Whitney and Nuisimer (6.48) has been of definite interest. They argued that, while different size holes in an infinite plate have the same stress concentration factor, the stress gradient is quite different for each. That is, large stresses are localized more closely to the edge of a small hole than a large hole. As a result, a critical defect is more likely to occur in a region of high stress for a large hole. Both a point stress and an average stress technique were

---

* There is a similarity between this approach and one described in Chapter 8 to elucidate the so-called short crack effect in fatigue.

used. The point stress criterion was given by

$$\frac{\sigma_c}{\sigma_0} = \frac{2}{2 + R^2 + 3R^4 + f(K_T^{\infty}, R)} \tag{6.2-3}$$

where $R = r/(r + d_0)$, $2r$ is the hole diameter, d_0 is the size of the damage zone, and σ_c and σ_0 are the critical notched and unnotched fracture stresses, respectively. The quantity $f(K_T^{\infty}, R)$ is a function of the hole size R and the orthotropic stress concentration factor for an infinite sheet K_T^{∞}.

6.2.2 Models Allowing Non-Self-Similar Crack Growth

Wu (6.35), though a combination of experiment and analysis on balsawood plates and unidirectional glass fiber composites, developed a crack initiation criterion for anisotropic materials having the form

$$\frac{K_{\text{I}}}{K_{1C}} + \left(\frac{K_{\text{II}}}{K_{2C}}\right)^2 = 1 \tag{6.2-4}$$

It was later suggested by Spencer and Barnby (6.49) that the fracture criterion for angle cracks in orthotropic materials can be obtained by generalizing Wu's criterion to

$$\left(\frac{K_{\text{I}}}{K_{1c}}\right)^m + \left(\frac{K_{\text{II}}}{K_{2c}}\right)^n = 1 \tag{6.2-5}$$

where K_{1c}, K_{2c}, m, and n are supposed to be properties of the composite. It can be seen that Equation (6.2-5) involves four disposable parameters. Mall et al. (6.50), for example, have developed specific values for wood using this type of relation and found that Wu's relation (i.e., $m = 1$, $n = 2$) best fit their data.

Laboratory results on balanced symmetric laminates containing a flaw perpendicular to the direction of loading indicate that the damage mode may either propagate in a self-similar manner or may change to another mode—for example, axial splitting. In the work given by Konish *et al.* (6.41), K_{1c} values in the laminates were found to depend upon the crack path. Not surprisingly, a high value of the fracture energy was obtained for tests in which fibers were broken while values approximately two orders of magnitude lower were obtained when the crack passed between fibers.

A further step towards a direct quantitative consideration of transverse crack growth in materials was taken by Harrison (6.51) and by Wright and Iannuzzi (6.42). To remove the restriction on self-similar crack growth, Harrison postulated different energy-release rates for crack growth in the plane of the crack and for growth normal to the crack. Denoting these as G_x and G_y, respectively, his approach leads to different relations for fiber breaking and splitting. These are, for splitting normal to the crack

$$G_y = R_y, \qquad G_x < R_x \tag{6.2-6}$$

while for fiber breakage in the plane of the crack

$$G_y < R_y, \qquad G_x = R_x \tag{6.2-7}$$

where R_x and R_y are the critical energy absorption requirements for crack growth in the two directions. The difficulty here, of course, lies in calculating energy release rate values for non-self-similar crack extension. In addition, in this form, the approach is strictly applicable only to unidirectional fiber composites.

A number of fracture theories subsequently generalized Harrison's approach. Wu (6.52) and Sih and Chen (6.53) each have approaches based on an anisotropic continuum interpretation of fracture. These techniques require that the intense energy region size be estimated either from the analysis or from an experiment. Both are designed to predict fracture and the direction of crack growth. In Wu's method this is accomplished by locating the intersection of the stress vector surface and the failure surface in the intense energy region ahead of the flaw. The failure surface must be obtained from experimental studies to determine remote properties. A disadvantage of the approach is that the intense energy region ahead of the crack must be obtained by experiment.

6.2.3 The Unifying Critical Strain Model

Poe (6.34), recognizing that experimentation to determine K_{1C} values for each combination of material and layup is prohibitive, has proposed a new fracture toughness parameter Q_c as an alternative to testing each laminate. This approach is based upon a failure strain criterion for fibers in the principal load-carrying piles. The Q_c parameter is supposed to be a material constant independent of the layup of the laminate, whereupon it can be determined from experiments on one single layup. Then, through the use of the ordinary elastic relations for a laminate, failure conditions can be predicted for other layups of the same material. Poe's data appear to suggest that the Q_c parameter is simply proportional to the ultimate tensile strain in the fibers measured on a unidirectional laminate whereupon K_{1C} can be obtained for any composite laminate from its unidirectional tensile strength. As Poe recognizes, this will not be true if either extensive delamination or splitting occurs. Consistent with the idea that matrix splitting effectively removes the stress concentration at the crack tip, Poe reports that splitting elevates Q_c whereupon his predictions provide lower bounds when splitting occurs. This would seem to confine the applicability of Poe's approach to through-wall cracked components where crack extension occurs in a self-similar manner, as in his work.

Poe's Q_c parameter was derived on the basis that the failure of a composite laminate is precipitated by breakage of the principal load-carrying fibers just ahead of the crack tip. Thus, damage growth should correspond to achieving critical strain levels in these fibers. If the damage region is small compared to the crack length, the LEFM regular stress field should be valid whereupon the strain field can be obtained from a laminate analysis. This gives

$$\varepsilon_{1c} = \frac{Q_c}{\sqrt{2\pi x}} \tag{6.2-8}$$

where ε_{1c} is the critical fiber strain and x is the distance from the crack tip. The critical value of the stress intensity factor can then be obtained as

$$K_c = \frac{E_y Q_c}{\xi} \tag{6.2-9}$$

where

$$\xi = \left[1 - \nu_{yx}\left(\frac{E_x}{E_y}\right)^{\frac{1}{2}}\right]\left[\cos^2\alpha + \left(\frac{E_y}{E_x}\right)^{\frac{1}{2}}\sin^2\alpha\right]$$

where α is the fiber orientation angle of the critical load-carrying plies while E_x, E_y, and ν_{xy} are the elastic constants of that ply (x is taken parallel to the crack, y is normal to the crack). It can be seen from Equation (6.2-9) that for an isotropic material where $E_x = E_y = E$ that Q_c reduces to

$$Q_c = \left(\frac{1-\nu}{1+\nu}\frac{G_c}{E}\right)^{\frac{1}{2}} \tag{6.2-10}$$

whereupon it would appear that Q_c is a variant of the critical strain energy release rate (n.b., the dimensions of Q_c are square root of length). The similarity with the point stress criterion of Whitney and Nuisimer described above can readily be seen from Equation (6.2-3).

By replacing ε_{Ic} by ε_{tuf} it follows that, if Q_c/ε_{tuf} is unique, there must be a characteristic distance

$$d_0 = (Q_c/\varepsilon_{tuf})^2/2\pi \tag{6.2-11}$$

Poe's data suggest that Q_c is proportional to the ultimate tensile strain of the fibers, ε_{tuf}, and that their ratio should therefore be unique for all fiber composites that fracture in a self-similar manner with limited crack-tip damage. The data indicate that Q_c/ε_{tuf} varies between 1.0 and 1.8 $mm^{\frac{1}{2}}$ for those layups that fail largely by self-similar crack extension and do not delaminate or split extensively. Using a median value of $Q_c/\varepsilon_{tuf} = 1.5\ mm^{\frac{1}{2}}$ allows a relation for the strength of the center cracked panels examined by Poe to be developed in terms of readily determinable parameters. This is

$$S_c = \frac{S_0}{\sec\left(\dfrac{\pi a}{2W}\right)}\left[1 + \frac{\pi a}{2}\left(\frac{\xi S_0}{1.5\varepsilon_{tuf} E_y}\right)^2\right]^{-\frac{1}{2}} \tag{6.2-12}$$

where S_0 is the tensile strength of an uncracked panel, a and W are the crack length and plate width, respectively, and ξ is the bracketed term defined in connection with Equation (6.2-9). Comparisons of the experimental results with the predictions of Equation (6.2-10) show reasonable agreement. However, no comparisons were made with other crack/structure geometries or for non-through wall cracks, assuming that the approach could be modified to treat such more complex conditions.

6.2.4 *Mixed Mode Fracture Models*

While a number of crack growth parameters have emerged in recent years, the four parameters that have so far been introduced—G, K, J, and δ—are adequate for the great majority of all fracture mechanics applications. Indeed, for linear elastic fracture mechanics, any one of the four would suffice. Obviously then, it is the necessity to confront nonlinear problems that has produced the proliferation of other parameters. We begin with the parameters devised to handle mixed-mode crack growth problems.

We have to this point concentrated on "opening mode" or "Mode I" crack growth problems on the basis that these adequately cover both the theoretical aspects and bulk of the applications in fracture mechanics. Nevertheless, there are conditions in which one must consider the two other possible modes. The distinction between the three modes was introduced by Irwin (see Chapter 3) who observed that there are three independent ways in which the two crack faces can move with respect to each other. These three modes describe all the possible modes of crack behavior in the most general elastic state. It might be noted that combinations of Modes I and II—the opening and the sliding modes—arise in problems involving a crack oriented at an acute angle to the applied stress or, more generally, when the applied stress is biaxial. Mode III—the antiplane tearing mode (not to be confused with the tearing instability theory)—can occur in very thin components. But, because the mathematical formulation of Mode III problems is considerably simplier than for in-plane crack growth problems, it also serves as a testing ground for new approaches to difficult problems.

The generalization of the relation connecting the stress intensity factor to the energy release rate is commonly written for plane strain conditions as

$$G = E'(K_{\mathrm{I}}^2 + K_{\mathrm{II}}^2) + \frac{1}{2\mu} K_{\mathrm{III}}^2 \tag{6.2-13}$$

where μ is the shear modulus. Note that the individual terms appearing in Equation (6.2-13) are designated as G_{I}, G_{II}, and G_{III}, respectively; Equation (6.2-13) therefore provides the special case where these are additive.

Sih and Liebowitz (6.20) provided a further generalization of Equation (6.2-13) to rectilinearly anisotropic bodies. As they point out, because cracks do not extend in a self-similar way in generally anisotropic materials, their general result based upon this assumption is of somewhat academic interest. However, in the more specialized conditions corresponding to an orthotropic material with the crack aligned with a line of symmetry, this result is meaningful. For these conditions, it can be written as

$$G = \left[\left(\frac{a_{22}}{a_{11}}\right)^{\frac{1}{2}} + \frac{2a_{12} + a_{66}}{2a_{11}}\right]^{\frac{1}{2}} \left[\left(\frac{a_{11}a_{22}}{2}\right)^{\frac{1}{2}} K_{\mathrm{I}}^2 + \frac{a_{11}}{\sqrt{2}} K_{\mathrm{II}}^2\right] + \frac{(a_{44}a_{55})^{\frac{1}{2}}}{2} K_{\mathrm{III}}^2 \tag{6.2-14}$$

where the a_{ii} are the coefficients in Hooke's law expressed in the form $\varepsilon_i = a_{ij}\sigma_j$: see Chapter 2 and Section 6.1.3.

The most prominent alternative to the above for the solution of mixed mode crack problems is the strain energy density theory introduced by Sih in the early 1970s to remove the restriction to self-similar crack extension. His strain energy density factor can be written as

$$S = \alpha_{11}K_{\mathrm{I}}^2 + 2\alpha_{12}K_{\mathrm{I}}K_{\mathrm{II}} + \alpha_{22}K_{\mathrm{II}}^2 \tag{6.2-15}$$

where

$$\alpha_{11} = \frac{1}{16\pi\mu}(1 + \cos\theta)(\kappa - \cos\theta)$$

$$\alpha_{12} = \frac{1}{16\pi\mu}(2\cos\theta + 1 - \kappa)$$

$$\alpha_{22} = \frac{1}{16\pi\mu}[(\kappa + 1)(1 - \cos\theta) + (1 + \cos\theta)(3\cos\theta - 1)]$$

and κ is defined as $(3 - \nu)/(1 + \nu)$ for plane stress and $(3 - 4\nu)$ for plane strain while θ denotes a generic angle from the crack line. The parameter S characterizes the local strain energy density on any radial plane that intersects the crack tip.

Sih's crack growth criterion involves two hypotheses: (1) that crack initiation occurs in the direction for which S is a minimum (i.e., where $dS/d\theta = 0$) and (2) at a point when S achieves a material-dependent critical value, S_c. Clearly, S_c is related to the conventional toughness measure in Mode I conditions via

$$S_c = \frac{\kappa - 1}{8\mu\pi}K_{\mathrm{Ic}}^2 \tag{6.2-16}$$

In applying the S-theory to determine the initial direction of crack growth in an angled crack problem, note that the appropriate stress intensity factors are

$$K_{\mathrm{I}} = \sigma(\pi a)^{\frac{1}{2}}\sin^2\beta$$

and

$$K_{\mathrm{II}} = \sigma(\pi a)^{\frac{1}{2}}\sin\beta\cos\beta \tag{6.2-17}$$

where β denotes the angle between the applied stress and the crack line. The relation for determining θ_0, the angle of crack extension, is then given by ($\beta \neq 0$)

$$(\kappa - 1)\sin(\theta_0 - 2\beta) - 2\sin(2\theta_0 - 2\beta) - \sin 2\theta_0 = 0 \tag{6.2-18}$$

More details of this approach, together with comparisons with experimental results, are given by Sih and Chen (6.54). A variation on the strain energy density approach has recently been given by Partizjar et al. (6.55).

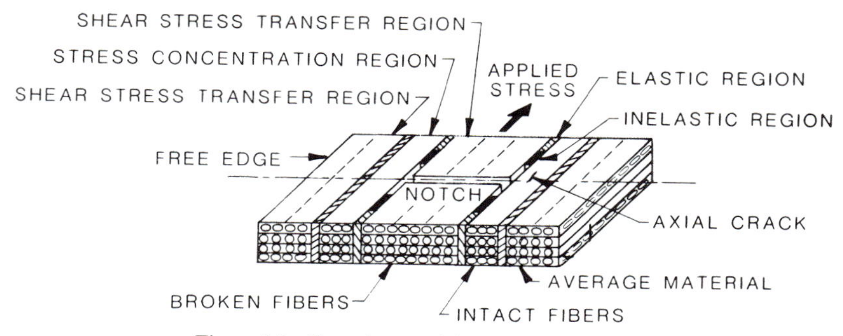

Figure 6.6 Rosen's materials engineering model.

The approach of Rosen and his associates (6.56) is shown in Figure 6-6. They limit their model to consider crack growth normal to the crack plane, as observed for both unidirectional and general laminates that contain 90 degree plies. In their model, the region adjacent to the notch is taken as a shear stress transfer region. The regions adjacent to the crack are subdivided into a region of local shear stress transfer, a region of stress concentration, and a region of shear stress transfer in the average material. It is necessary to identify experimentally the size of the region of intense energy adjacent to the flaw and the demarcation between mode I and mode II behavior. If axial fracture (normal to the original crack surface) is assumed, their technique requires two arbitrary parameters. If greater freedom is allowed, additional arbitrary parameters must be specified.

Recently, Zhen (6.57) has developed a variant of the Whitney-Nuismer average stress criterion approach. This approach is based upon a damage parameter that represents the aggregate of the inelastic micromechanical processes that occur at a notch or other stress concentrators. Zhen defines the damage parameter D in terms of σ_0, the unnotched tensile strength of the material, such that

$$\frac{1}{D}\int_0^D \sigma_y \, dx = \sigma_0 \tag{6.2-19}$$

where σ_y is the normal stress acting ahead of the stress concentration. Failure occurs when D becomes equal to the ultimate damage, D_c, which is supposed to be a material property independent of the load and geometry. Zhen was able to show that unique values of D_c exist for a variety of composites used in a number of laminate configurations and specimen geometries. He was also able to show that D_c, which has the dimensions of length, compares well with measured damage zones.

Zhen's damage parameter has a clear physical basis and appears to provide an impressive correspondence with existing experimental data. However, it has the obvious limitations of any two-parameter linear elastic approach: it can only be applied when a major stress concentrator is present and it cannot discern individual ply behavior. In fact, it can be shown that this approach is only modestly different from a normal LEFM approach. For example, for a center-cracked tension panel, Equation (6.2-19) leads to

$$D = \frac{2}{\pi}\left(\frac{K_{\mathrm{I}}}{\sigma_0 - \sigma}\right)^2 \tag{6.2-20}$$

where σ is the applied stress. Hence, when $\sigma \ll \sigma_0$, a reasonable assumption in most cases, it can readily be found from Equation (6.2-20) that $D_c = (2/\pi)(K_{1c}/\sigma_0)^2$. A similar result would be obtained for other geometries.

6.2.5 Perspective

In concluding this section on linear elastic approaches to fracture of fiber composites, we remark that one trend in developing analytical representations clearly is to introduce more and more disposable parameters into a linear elastic continuum mechanics formulation. With enough parameters, of course, any model can be made to match a body of experimental results. If the number of disposable parameters is restricted but comparisons are limited to those data used to set the parameter values, a similar conclusion can of course be reached. Experimental verification is not really at issue and therefore is not of much concern to us.

A more appropriate basis for judgment is whether or not the parameters contained in the model have a clear-cut physical significance. If not, the model is nothing more than a curve fitting device that can be used only for interpolation of a specific set of experimental results. If the parameters do have physical significance, then experiments can be designed to obtain accurate values and the model can be expected to be valid over a wide range of conditions. That is, it will have an "extrapolative" function. It can then be used with some degree of confidence to predict behavior beyond the regime in which the parameters contained in them were experimentally established. Models that bindly employ LEFM concepts will inevitably fall somewhat short of this goal.

This is not to say that "interpolative" models—those that cannot be relied upon beyond the regime that contains their experimental support—are not useful. There are applications that can be effectively simulated by characterization experiments whereupon simple models are indispensible for evaluating various matrix (fiber/interface properties). Nicholls and Gallagher (6.58) have recently employed cantilever beam specimens in this way. Wells and Beaumont (6.59, 6.60) have devised a fracture map approach that provides a systematic way to achieve this purpose. Kunz-Douglass et al. (6.61) have provided more specific models for investigating the toughness of epoxy-rubber particular composites.

An area of particular concern currently is inter-ply delamination. The first use of fracture mechanics to model this aspect was that given by Rybicki et al. (6.62) who employed an energy release rate approach. More recent models have been offered by Wang and co-workers (6.63), Jurf and Pipes (6.64), and O'Brien (6.65). The approaches that have been developed in this still evolving area are similar to those that have been developed for the fracture of adhesive joints discussed below. Another currently active research area is that of compression strength; see, for example, Kwiashige (6.66).

6.3 Nonlinear Fracture Mechanics Analysis Models

In the many other applications discussed in this book, the well-demonstrated applicability of fracture mechanics to metals is being steadily advanced to cope effectively with nonlinear and dynamic conditions. In contrast, fracture mechanics applications to fiber-reinforced composites have been less successful. In essence, these have mainly been generally empirical extensions of linear elastic fracture mechanics that are limited in their ability to treat the complexity of the crack extension process as seen from the micromechanical point of view. The key to a more appropriate approach was suggested by Potter (6.19):

> Clearly, a simple, reliable fracture criterion may be developed only by the identification of those particular factors which govern the propagation of the crack (or the cracks) which will ultimately cause failure. The identification and isolation of these factors is the only means by which an acceptable degree of material idealization may be determined and from which such reliable fracture criteria may be developed.

In accordance with this sentiment, some of the more adventurous possibilities that have recently been suggested to improve upon this situation will be discussed in this section.

Because of the pronounced effects of subcritical damage growth on laminate performance due to load redistributions and stress concentrations, incorporation of micromechanical failure processes into a fracture model of a fiber composite laminate would seem to be necessary for realistic fracture predictions. A convenient catagorization of such events would include fiber breakage, matrix cracking, fiber/matrix debonding, and interply delamination. Clearly, however, the direct consideration of the sequence of such events requires an extremely localized focus that cannot be included in continuum treatments of actual structural components. Nor is the consideration of such events independent of the load and crack/structure geometry of such components likely to be particularly fruitful. Approaches that integrate micromechanical damage growth into an overall continuum point of view are therefore required. In this section the work of representative investigators that take this point of view will be described.

6.3.1 Continuum Models

A great number of analysis approaches have been advanced that attempt to model the micromechanical events that precede fracture at a stress riser in a fiber composite material. Some of those are given as references (6.67–6.77). In this section, attention will be focused on the representative approach developed by Potter (6.72). Potter has developed a fracture model based upon three observations: (1) tensile failure of a laminate is governed by the failure of the plies having fibers parallel to the load direction, (2) these fibers fail sequentially by a process in which the failure of the fiber causes its neighbor to fail, and (3) other laminate failure processes (e.g., longitudinal splitting of the axial plies, transverse tensile or shear failure of the angle plies, and delamination) are important only insofar as they influence the fiber failure process in the axial plies.

These observations have led Potter to develop a model for a notched laminate based upon two conditions. First, the *initiation* of failure requires that the laminate unnotched tensile strength σ_0 must be reached at the notch tip. This criterion can be expressed as

$$C\sigma = \sigma_0 \tag{6.3-1}$$

where σ is the remote applied stress and C is the notch tip elastic stress concentration factor. The second condition involves failure *propagation*. It is based on the idea that sustained propagation of a crack across axial load-bearing fibers requires a critical load to be transferred from the broken fiber to its neighbor. This condition involves the stress gradient ahead of the notch tip, σ'_x $(=\partial\sigma_x/\partial y)$, and can be expressed as

$$s|\sigma'_x| = \delta D \tag{6.3-2}$$

where s is the fiber spacing, δ is the increase in the fiber stress due to the failure of an adjacent fiber, and D $(=\sigma_x/\sigma_f)$ is the ratio of the laminate stress in the fiber direction to the fiber stress. When Equation (6.3-2) is satisfied, the fibers will fail in sequence.

A useful feature of Potter's model is that it distinguishes between large blunt notches and small inherent defects. In the former case, laminates generally behave in a brittle manner whereupon the initiation criterion, Equation (6.3-1), governs. Potter found it convenient to introduce the parameter R $(=\sigma/\sigma_0)$ as a dimensionless failure stress. Here, the failure criterion for a laminate containing a blunt notch is simply

$$R = \frac{1}{C} \tag{6.3-3}$$

Conversely, for small sharp notches where the initiation criterion is readily satisfied, the propagation criterion governs. However, in this regime, failure will be preceded by the development of a crack-tip damage zone that expands as the load is increased. This presents a difficult analysis problem. But, by modeling the damage zone as an increase in notch tip radius, Potter deduced a relation for failure due to small defects of length $2a$ in the form of the polynominal

$$R^3 + [(K_4 - 2) - (K_4 + 2)b\Delta]R^2 + (1 - 2K_4)R + K_4 = 0 \tag{6.3-4}$$

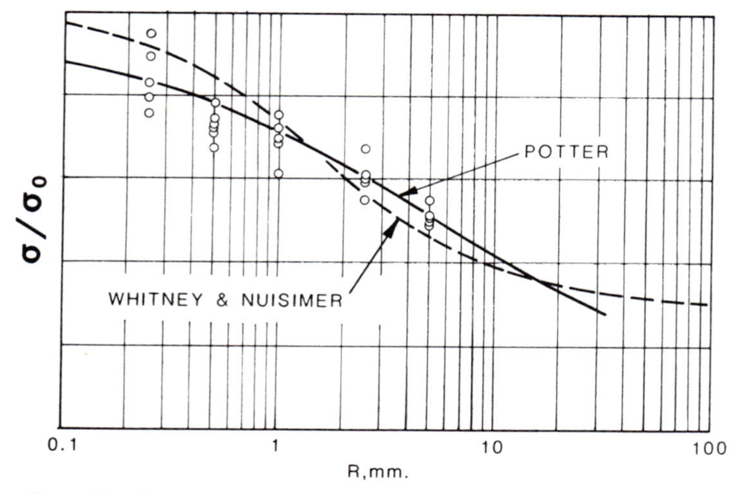

Figure 6.7 Comparison of Whitney-Nuisimer and Potter analysis models.

Here, b is the notch tip radius and

$$K_4 = 2\alpha\sqrt{E_x E_y} \tag{6.3-5}$$

$$\alpha = \left(\frac{1}{2G_{xy}} - \frac{\nu_{xy}}{E_x}\right) \tag{6.3-6}$$

$$\Delta = \frac{\delta D}{s\sigma_0} \tag{6.3-7}$$

where E_x and E_y are the longitudinal and transverse elastic moduli, G_{xy} is the in-plane shear modulus, and ν_{xy} is Poisson's ratio.

In Potter's model, failure is predicted at the lesser of the values given by Equations (6.3-3) and (6.3-4). As shown in Figure 6.7, the model predictions are similar to those obtained using the average stress criterion of Whitney and Nuismer (6.48). This is not too surprising in that both approaches utilize maximum and minimum strength values based upon the unnotched strength and the elastic stress intensity factor, respectively. The importance of Potter's approach is rather that, perhaps for the first time, the failure mechanisms that are involved in the fracture of a composite were directly confronted. And, in so doing, it was possible to discriminate between small and large flaws.

6.3.2 Hybrid Models

To achieve an improved predictive capability for composite fracture, a mathematical model directly incorporating the various micromechanical failure processes must be developed. A model initiated by Kanninen and his associates (6.78) and (6.79) seeks this end by merging a micromechanical failure analysis with a macromechanical analysis. Their hybrid approach treats the material as homogeneous where it is acceptable to do so and as discrete only where it is necessary. In this way, the manner in which the sequence of microstructural failure events that accompany the stable growth of a flaw under an increasing load up to the point of catastrophic fracture can be directly determined. Moreover, because micromechanical damage events are modeled only locally, actual structural configurations and loads can be treated.

The approach suggested by Kanninen et al. can be likened conceptually to the well-established singular perturbation and matched asymptotic expansion techniques of fluid mechanics. That is, the problem of a composite material containing a flaw is divided into distinct "inner" and "outer" regions. In each of these regions, the material is modeled in different ways. The inner region, which contains the tip of a macroscopic crack or any other type of stress riser and can also include an entire microscopic crack, is considered on the microscopic level and treats the material as being heterogeneous. This region was called a local heterogeneous region (LHR). The outer region surrounds the crack-tip region and treats the material as a homogeneous orthotropic continuum. A typical LHR model is shown in Figure 6.8.

Figure 6.8 depicts a unidirectional fiber composite containing three distinct components: the fibers, the matrix, and the fiber/matrix interface zones. For precise quantitative results, the constitutive relations of these elements, up to and including their rupture points, must be known. In the model, any element of a fiber composite ruptures when an intrinsic critical energy dissipation rate can be provided at some point of that element. (Alternatively, the somewhat simpler approach based on a critical effective stress can be used.) These critical values are assumed to be independent of local stress field environment at the

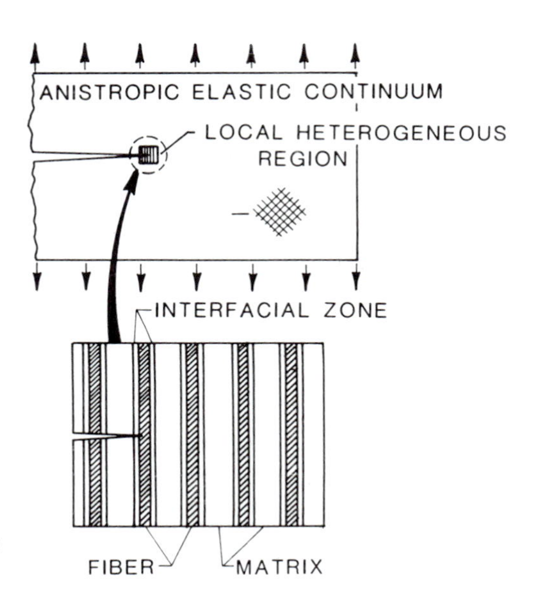

Figure 6.8 Hybrid fracture mechanics model.

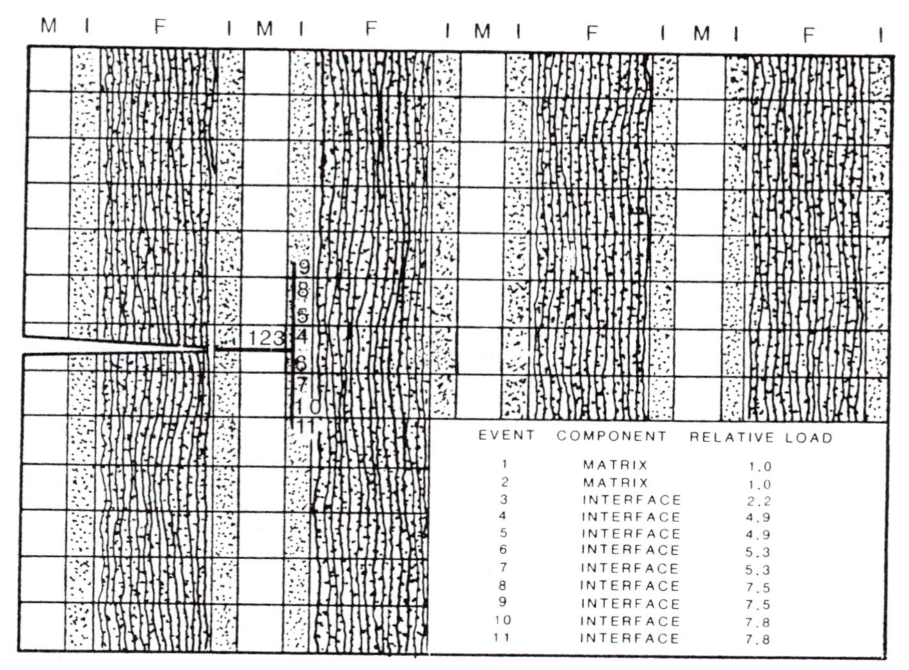

Figure 6.9 Computational result of a hybrid fracture mechanics model showing damage growth by splitting.

point of incipient rupture. Data from fracture tests on isolated fibers and on neat matrix materials must therefore be available to be inserted into the model, or they can be inferred from the computational results.

Computations for arbitrary flaw size and orientation were first performed for unidirectional composites with linear elastic/brittle constituent behavior. The mechanical properties were nominally those of graphite/epoxy. With the rupture properties arbitrarily varied to test the capability of the model to reflect real fracture modes in fiber composites, it was shown that fiber breakage, matrix crazing, crack bridging, matrix/fiber debonding, and axial splitting can all occur during a period of (gradually) increasing load prior to catastrophic fracture. Qualitative comparisons with experimental results of Brinson and Yeow (6.46) on edge-notched unidirectional graphite/epoxy specimens have also been made. Figures 6.9 and 6.10 show some typical results for damage growth at the tip of a sharp crack in a unidirectional composite.

The example computational result presented in Figure 6.9 shows that the propagation of damage (in this instance, fiber/matrix splitting) requires an increasing load. Figure 6.10 similarly shows fiber bridging. These results were found to be in reasonable qualitative agreement with the experiments of Yeow et al. (6.46). However, the absence of realistic *in situ* matrix/fiber/interface properties, along with the restriction of the work of reference (6.78) to a single

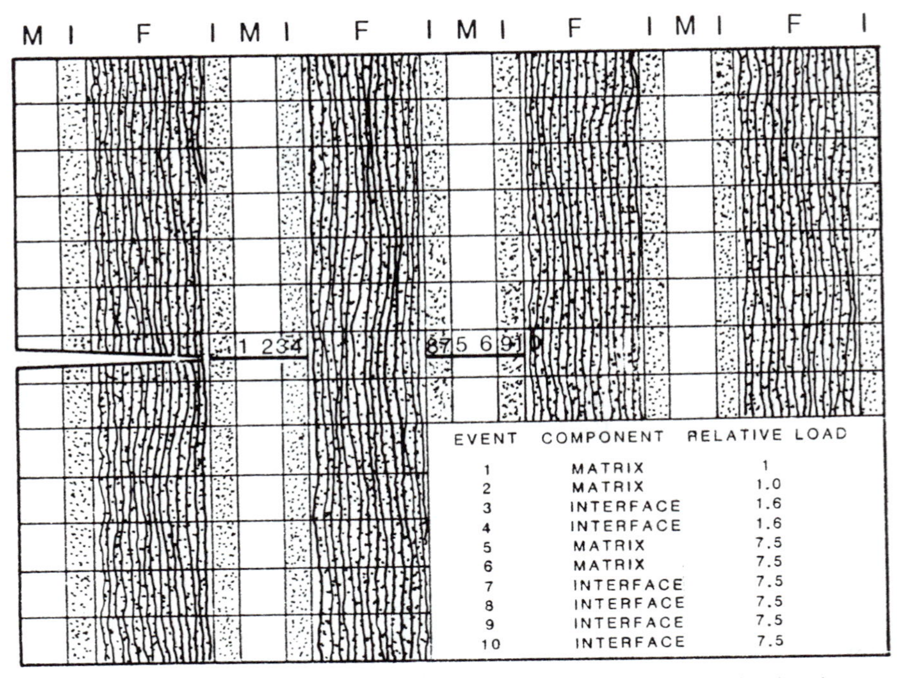

Figure 6.10 Computational result of a hybrid fracture mechanics model showing damage growth by fiber bridging.

ply, precluded quantitative comparisons. Nevertheless, while the work reported in reference (6.78) was focused on a sharp crack normal to the fiber, direction, any type and direction of stress riser can be accommodated in this approach. Indeed, any type of loading, including impact, can be accommodated.

The work reported in reference (6.79) recognized that, while events in the fiber/matrix/interface scale at a stress riser are crucial to a quantitative understanding of damage growth and fracture, these can be significantly affected by events occurring elsewhere in the composite. In particular, the load transfer that occurs in a laminate when damage progresses at different rates in the individual plies must considered (and/or when delamination occurs). To address this, a parallel-spring continuum model was developed. Computations were performed for center-cracked graphite/epoxy tension panels using fiber fracture and matrix splitting toughness values inferred from the experiments of Yeow et al. (6.47). Crack growth was allowed to occur within each ply either parallel to or normal to the fiber direction—in essence, using Harrison's approach via the crack closure method suggested by Rybicki and Kanninen (6.80). Typical results are shown in Figures 6.11 and 6.12.

The LHR concept was developed by taking the through-thickness fiber/matrix representation in the form of a series of plates. This is shown in Figure 6.13(b) [c.f. the actual configuration shown in Figure 6.13(a)]. Thus,

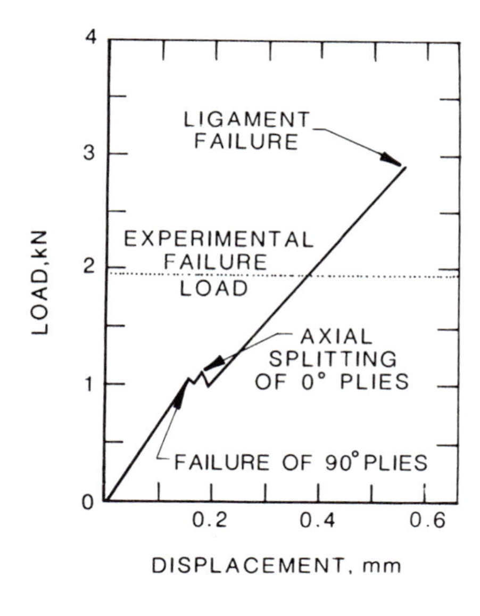

Figure 6.11 Comparison of calculated and observed behavior in a $[0/90]_{4S}$ graphite/epoxy center—cracked tension panel.

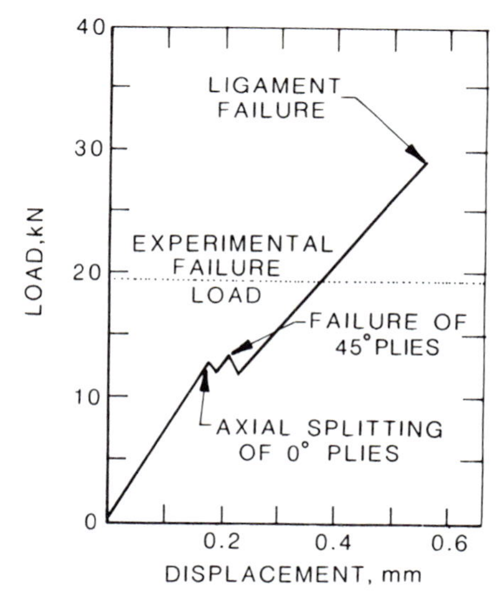

Figure 6.12 Comparison of calculated and observed behavior in a $[0/\pm 45/0]_{2S}$ graphite/epoxy center—cracked tension panel.

the quasi-three-dimensional representation shown in Figure 6.13(c) was developed. This approach assumes that the fibers are square in cross section and are distributed in a regular array. The analysis model then takes as its focal point the strip of material far enough from a ply interface for periodicity conditions to be appropriate. Comparisons between some heuristic computational results and experimental results as graphite/epoxy tension panels show reasonable agreement (6.114).

Ouyang and Lu (6.81) have adopted a point of view that coincides with Kanninen and coworkers. They maintain that, even for small-scale yielding

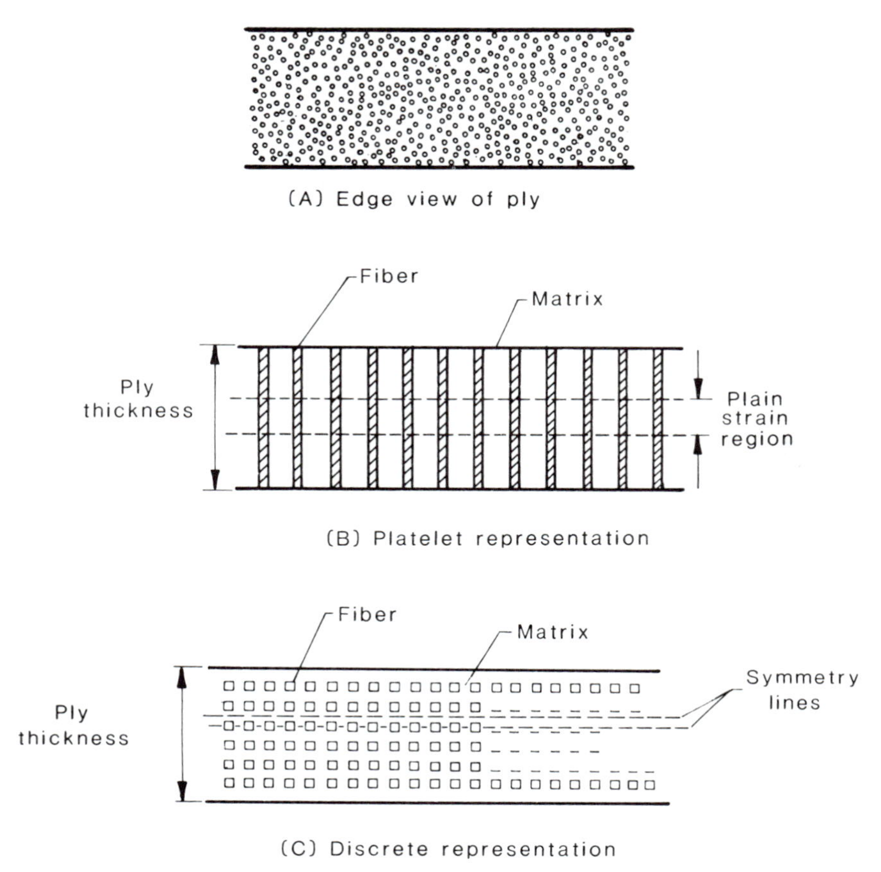

Figure 6.13 Two-dimensional and quasi-three-dimensional idealizations of a fiber composite ply. (a) Edge view of ply .(b) Platelet representation. (c) Discrete representation.

conditions, the crack-tip region must be addressed in terms of elastic-plastic, finite deformations, and heterogeneity. Because an elastic anisotropic continuum description is used outside the crack-tip region, their formulation of the problem is just as shown in Figure 6.8.

Ouyang and Lu assume that the effect of the inner heterogeneous region on the outer continuum region can be neglected. This effectively decouples the problem and gives boundary conditions for the inner region that depend only upon the applied stress and the overall crack/structure geometry. The inner problem is that formulated with an elastic-plastic large deformation finite element model in which individual fiber, matrix, and interface elements are represented by four-noded elements. When the effective stress of an element reaches a critical value, the nodes were relaxed to simulate local failure. However, no computational results were given, nor have they indicated how the laminate configuration would enter into their approach.

Other noteworthy approaches to the laminate fracture problem include the analyses of Nuismer and his associates (6.82, 6.83). As these investigators have recognized, while simple approaches such as the average stress theory of Whitney and Nuismer seem to suffice for notched laminates under uniaxial loading conditions, they cannot account for the progressive damage and consequent stress redistribution that occur from small defects and nonproportional biaxial loadings. Nuismer and Brown therefore developed an approach that allows for the damage to the matrix and fibers in individual plys that precedes the ultimate failure of the laminate. This is accomplished with an incremental finite element model that traces the damage and invokes appropriate stiffness changes in the laminate via laminated plate theory.

In the Nuismer-Brown approach, failure of a ply in the fiber direction is assumed when the strain in that direction reaches an experimentally determined critical value. Similarly, matrix failure occurs when either the transverse or the shear strain reaches a critical value. Recognizing that the ply does not completely lose its stiffness after failure, Nuismer and Brown developed post-failure constitutive relations, as follows. First, when a ply element fails in the fiber direction, it loses its stiffness in the fiber direction and in shear, but not its transverse stiffness; that is, $E_y = \nu_{xy} = G_{xy} = 0$, $E_x \neq 0$. Second, when matrix failure occurs, there is a loss of stiffness in the transverse direction and in shear, but not in the fiber direction; that is, $E_x = \nu_{xy} = G_{xy} = 0$, $E_y \neq 0$. Because the laminate stiffness is determined using the ply stiffness (damaged or undamaged) in conjunction with laminate theory, the approach is justifiable unless extensive interply delamination occurs.

It might be noted that the Nuismer-Brown approach differs in one important respect from recent work of Kanninen et al. The latter have employed the "death" option, as it is known from work on concrete, to represent local failure. In this procedure, failure of an element is enforced by simply deleting the appropriate terms in the global stiffness matrix. Of more importance, the Nuismer-Brown model is a completely continuum representation that does not attempt to include the actual micromechanical failure events. However, more recent work by Nuismer and co-workers has been addressed to this end—for example, by incorporating a micromechanical model to describe the initiation and growth of matrix cracking in a ply subjected to transverse tension. Their preliminary computations have revealed that, after damage begins, the constitutive behavior of a ply is laminate dependent. An example of their computational modeling is shown in Figure 6.14.

6.3.3 Finite Element Models

The first attempts to develop finite element fracture models for fiber composites seem to be those of Adams (6.84) and Wang et al. (6.85). Adams and his co-workers have progressed to consider the details of the fracture process in composites; for example, see reference (6.86). Other models are those of Reedy (6.87), who has concentrated on boron aluminum laminates,

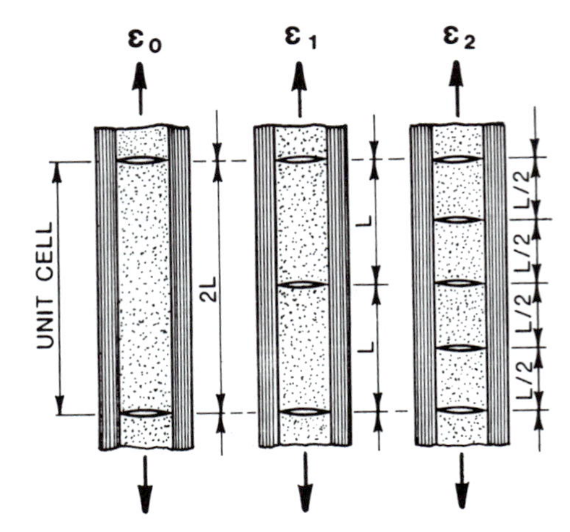

Figure 6.14 Nuisimer's model for progressive damage growth in fiber composites.

and Wang and Crossman (6.88) to (6.89), who have studied cracking and edge delamination in graphite/epoxy laminates. Reedy's nonlinear finite element approach is focused on a unidirectional, notched boron fiber composite. As noted in Section 6.1, the fiber size is then comparable to the ply thickness, necessitating a treatment that is somewhat different in detail from those developed for graphite/epoxy. Reedy employs a shear-lag assumption that is somewhat analogous to Rosen's approach described above. His finite element formulation allows elastic work hardening constitutive behavior to be used for both the fibers and the matrix materials. Stable crack growth is modeled using a critical stress criterion. Reedy's model is shown in Figure 6.15.

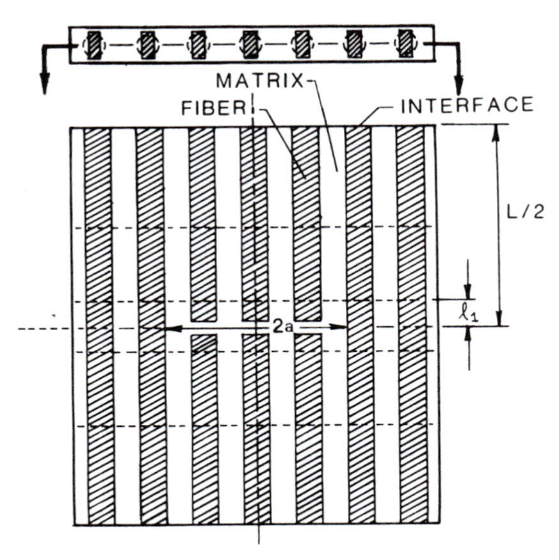

Figure 6.15 Reedy's model for fracture of a boron/aluminum composite.

6.4 Related Topics

This chapter has concentrated on analysis models for damage growth and fracture under a monotonically increasing load. The loading conditions were tacitly assumed to be slow enough that dynamic effects were unimportant, but to be rapid enough that environmental and other time-dependent (viscoelastic) effects need not be of concern. As complex as the fracture mechanics problem is under those conditions, it must be recognized that such a loading is highly idealized in comparison to actual design and service conditions. Fatigue and environmental effects—especially those due to temperature and moisture—are a primary concern in most practical situations. Because the polymeric materials commonly used as the matrix material in a fiber composite exhibit viscoelastic behavior, time-dependent alterations in the matrix-dominated mechanical and failure properties of a polymer-based composite due to temperature and state of stress can indeed be important. Some of these complications will be briefly touched upon in this section.

6.4.1 Fracture of Adhesive Joints

The difficulties in the fracture analysis of fiber composite materials that preclude the direct application of the linear elastic fracture mechanics (LEFM) techniques developed for metals stem from two main sources. First, the scale of the damage that accumulates in the form of inelastic deformation around a crack is not usually negligible. The necessity for this condition for the validity of LEFM was outlined in Section 6.1.4. The second source of difficulty arises because fiber composites in structural applications are generally used in the form of laminates. The variation in the damage that occurs from one ply to another in a laminate can be significant. Compounding these effects is the possibility of interply delamination. Because these effects cannot be directly accounted for within LEFM, many current efforts are therefore attempting to develop nonlinear fracture mechanics approaches based upon hybrid micromechanical/continuum analysis models. Thus, at first sight, the analysis of fracture in an adhesive joint would appear to be much less difficult.

An adhesive is a homogeneous material, and in the form of a joint, is subjected to constraints that force crack growth to proceed in a reasonably predictable manner under conditions that closely approximate plane strain. However, while the same difficulties that attend fiber composites do not arise for adhesives, there are difficulties nontheless. Cracks that lie on an interface between two dissimilar materials (adhesive fracture) give rise to an oscillatory form of singular behavior that not only differs from that associated with cracks embedded in a continuous material, it gives rise to physically unacceptable behavior. Thus, the fracture mechanics techniques described in Section 6.1.3 do not apply. At the same time, referring to the argument of Section 6.1.4, because of the narrowness of adhesive joints, cohesive cracking will not likely occur with a damage zone that is small compared to this dimension.

Despite the possibility that an LEFM approach is as unlikely to be successful for an adhesive joint as it has proven to be for fiber composite laminates, nonlinear fracture mechanics approaches are not much in evidence for adhesives. One reason may be that, unlike most structural applications, adhesive joints that closely simulate service conditions can be tested readily in the laboratory. Thus, one can rely upon the concept of "similitude" that is often invoked for fatigure crack growth in metals; see Chapter 8. In essence, even though the assumptions of LEFM are violated, provided the conditions in the laboratory test and the intended application are similar, an LEFM approach can be used. Tacitly, this argument supports the universal use of LEFM parameters and concepts in adhesive fracture. As described in this section, most current work in the assessment of adhesive joints with flaws is based upon this approach.

While not unique to adhesives, mixed mode crack growth is much more prominent in adhesive fracture than in other situations. This most often involves combined modes I and II, but can also include a mode III component—for example, in a scarf joint. The well-accepted yield criteria of either von Mises or Tresca make it possible to predict the onset of yielding under a multiaxial state of stress knowing only the uniaxial yield stress value. However, there currently is no counterpart for fracture. The nearest analog is probably that given by Equation (6.2-5).

One reason for the lack of a fracture surface is that, since most engineering materials are most susceptible to fracture in mode I conditions, by far the bulk of the experimental data have been collected for this condition alone. Another reason is the difficulty of obtaining valid experimental results under either pure mode II or mode III conditions. For fiber composites, and especially for adhesive joints, the general lack of a fracture counterpart of the yield condition is probably the single most serious obstacle to the analysis of flawed adhesive joints. For this reason, it is therefore appropriate to review briefly the experimental results before considering the analyses that have been developed for adhesives.

Recognizing the difficulties involved in determing K_{I}, K_{II}, and K_{III} values for adhesive layers, the early work of Ripling et al. (6.90, 6.91) was focused on G, the strain energy release rate. Because adhesives are usually brittle, and because the constraint imposed by the near proximity of high modulus adherends precludes a volume change in the adhesive, cohesive fracture tends to occur under plane strain conditions. If this is valid, there should be unique fracture toughness values for each mode—that is, the plane strain values, $K_{\mathrm{I}c}$, $K_{\mathrm{II}c}$, and $K_{\mathrm{III}c}$. Nonetheless, for the sake of generality, the designations K_{1c}, K_{2c}, and K_{3c} will be used to preserve the possibility that a constraint effect could arise in some applications.

Ripling et al. (6.92) have recently completed a preliminary study on the behavior of flawed joints subjected to mixed mode I and III loadings. They used graphite/epoxy plates bonded with an epoxy adhesive. Their test specimen was modified tapered double cantilever beam specimen with a scarf joint. By increasing the joint angle θ, the ratio of the mode III to the Mode I contribution can be systematically increased in this configuration. Ripling et al. found that, as K_{III} was increased, G_c initially increased, then decreased. This reversal could be correlated with an observed change in the fracture character. That is, the cracking was at the center of the bond for $\theta = 0$ (no mode III contribution), but tended towards the adherend/adhesive interface as θ was increased (with increased amounts of fiber pull-out). At the point at which G_c began to decrease with increasing θ, the fracture changed from cohesive (within the adhesive) to adhesive (on the interface).

Everett (6.93) has conducted a combined experimental and analysis investigation on mode I failure of an adhesive that focused on the role of "peel stresses"—that is, stresses resulting from tensile loading normal to the bond line. His experiments were conducted on graphite/epoxy laminates bonded by an epoxy adhesive in the form of a cracked lap shear specimen. Debonding was monitored by a photoelastic method. Tests were run under constant amplitude cyclic loading with the peel stresses systematically altered by the imposition of a compressive load applied through a spring-loaded clamp. The failure mode in all tests was cohesive.

A finite element model focused on the strain energy release rate parameters was developed to provide an analysis of the experimental results. It was noted that crack growth normally proceeds in this specimen by a mixed mode process with $G_{II} \simeq 2G_I$. However, G_{II}/G_I increases when the clamping device is used. It was found that, for the clamping force that was required to arrest the debond process, G_{II} increased while G_I became almost equal to zero. From this finding, Everett concluded that G_I governs debond crack growth with G_{II} being irrelevant.

Ikegami and Kamiya (6.94) have conducted an experimental investigation on the tensile and shear strength of joints with interfacial flaws. Butt joint specimens consisting of thin wall tubes were subjected to axial tension and to torsion loadings. Their results were analyzed using the stress intensity factor for layered materials developed by Erdogan and Gupta (6.95). The strength of joints without a flaw were correlated by the use of an effective flaw size. Experiments were conducted with carbon steel and other metals as the adherends and an epoxy resin as the adhesive. The stress-strain curves reported by Ikegami and Kamiya show that this resin behaves essentially as an elastic-brittle material.

The experimental results obtained by Ikegami and Kamiya were similar for all the adherends they tested and can be summarized as follows. First, the strength in shear (torsion) is always greater than in tension. Second, the fracture stress decreases with increasing thickness of the adhesive layer. Third, the strength of a joint with a large crack is roughly one-half that of a joint with a small or no crack for thin adhesive layers, but the presence of a crack does not have a large effect for thick adhesive layers.

Ikegami and Kamiya observed that fracture was initiated from the artificial crack and, because of the elastic-brittle nature of the adhesive that they used, concluded that a linear elastic fracture mechanics approach could be used. They also noted that, while the problem involved mixed mode fracture, K_{II} is small compared to K_I in tension while K_I is small in comparison to K_{II} in shear. Thus, values of K_{1c} and K_{2c} could be readily obtained. Their results indicate that the fracture toughness values so inferred depend upon the adherend.

Other experimental work of note on the fracture of adhesives would include that of Saxena (6.96) who suggests that an "effective" stress intensity factor given by $(K_I^2 + K_{II}^2)^{\frac{1}{2}}$ can be used to predict fracture. Crack growth measurements were made by Gledhill and Kinloch (6.97) who found that fracture could be uniquely described by a critical plastic zone size developed at the crack tip. Finally, Bitner et al. (6.98) have studied the time-dependent behavior of structural adhesives and have concluded that elastomer-modified adhesives achieve their very good fracture properties through viscoelastic processes that vary substantially with temperature and loading history.

Turning to the analysis approaches that have been developed for adhesives, one of earliest efforts is that of Bikerman (6.99). Subsequently, important contributions were made by Cook and Gordon (6.100), Williams (6.101), Gent and Kinlock (6.102), Knauss (6.103), and Trantina (6.104). Much of this work was based upon the use of simple beam models for which a more rigorous approach was later provided by Chang et al. (6.105). Here, the more recent work of Wang and Yau (6.106) using finite element methods will be highlighted.

As Wang and Yau point out, the interfacial cracking (or debonding) of an adhesively bonded joint can occur both at geometric boundaries (e.g., edges, re-entrant corners) due to local stress concentrations and from internal flaws from faulty joining (e.g., from incomplete wetting between adherend and adhesive). Because it is one of the most widely used structural joint configurations, they focused on the lap-shear joint. Recognizing that the adhesive layer thickness is a crucial characteristic dimension in the problem, they have advanced a conservation integral approach via the J- and M-integrals; see Chapter 3. The finite element mesh and the integration path for the M-integral are shown in Figure 6.16.

The procedure developed by Wang and Yau reduces the problem to a pair of linear algebraic equations whereupon stress intensity factors can be evaluated directly from far-field information, once the conservation integrals have been evaluated. For this purpose, Wang and Yau have used a conventional finite element method with eight-noded isoparametric elements. Example results showing K_I and K_{II} as a function of the system elastic moduli, the adhesive thickness, and the crack length are given in reference (6.106).

6.4.2 Fiber Pull-Out Models

Following the basic picture developed by Cottrell (6.107), consider a bundle of parallel elastic fibers, each of length l, embedded in a ductile metal rod that is then strained longitudinally. Let τ be the tangential stress exerted on a fiber by the matrix material as it flows plastically. The stress $\sigma(x)$ in the fiber at a distance x from one end, assumed to be free, for a fiber of radius r is given by

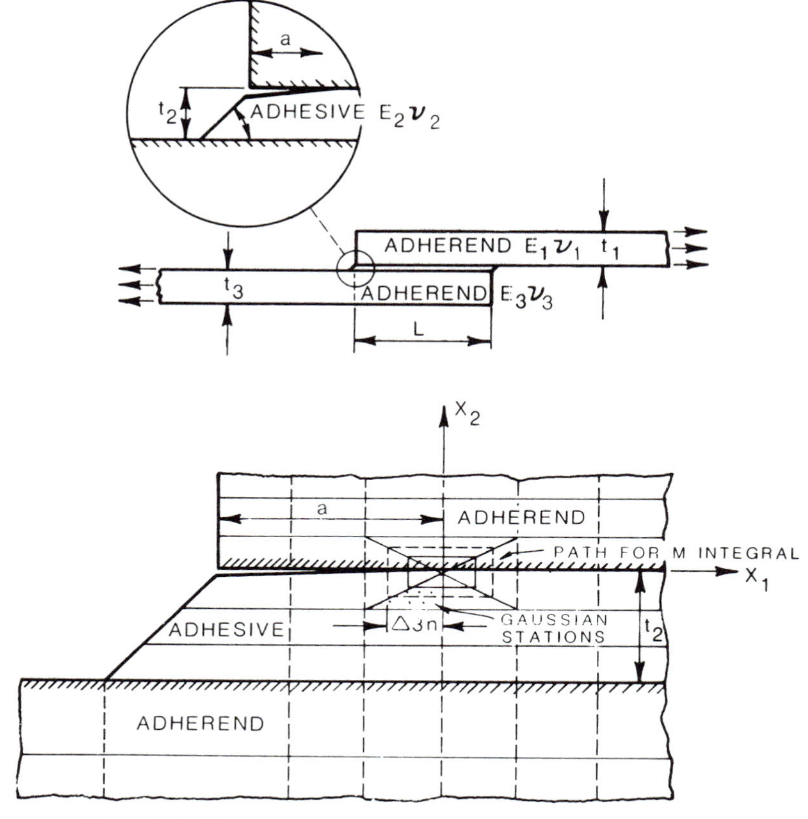

Figure 6.16 Finite element model for crack growth in an adhesive layer.

the force balance

$$2\pi r \tau x = \pi r^2 \sigma(x)$$

or

$$\sigma(x) = \frac{2\tau x}{r} \tag{6.4-1}$$

If τ is a constant, the fiber stress $\sigma(x)$ will increase linearly to a maximum value and will be constant thereafter. This maximum value is approximately equal to $E\varepsilon$, ε denoting the overall tensile strain of the rod, because there will be no relative slip between the matrix and the fiber beyond the point where the elastic strain in the fiber becomes equal to ε.

Let x_0 denote the distance from the free end at which the maximum stress is attained. This load transfer length reaches its highest value, $l_c/2$, when $\sigma(x_0) = \sigma_f$, where σ_f is the fracture stress of the fiber. Hence

$$l_c = \frac{r\sigma_f}{\tau} \tag{6.4-2}$$

To exploit the intrinsic strength of the fibers, the composite must be designed such that $l \gg l_c$ whereupon the overall strength of the composite will be

$$\sigma_c = \sigma_f V_f + Y_m(1 - V_f) \tag{6.4-3}$$

Here V_f is the volume fraction of the fibers and Y_m is the yield strength of the matrix. This is the highest strength that the composite can have. However, if the matrix has a limited ability to carry additional load by work hardening, because for $l \gg l_c$ the fibers are stressed to their limit, if one fiber breaks a fracture can easily spread. Hence, this is not an optimum situation.

By contrast, if $l < l_c$, the fibers do not break but are pulled out of their holes. Assuming that the tangential stess τ continues to be exerted during the pulling out, the work done on each fiber is

$$\text{work done/fiber} = \int_0^l (2\pi r\tau)x\, dx = \pi r\tau l^2$$

The number of fibers per unit surface area normal to the fiber axes is $A_f/\pi r^2$. Hence, the fracture energy per unit surface is

$$G_c = V_f \frac{\tau l^2}{r} \tag{6.4-4}$$

(Note that Cottrell finds the same result but with the factor $\frac{1}{12}$ appearing on the right-hand side.) For fibers of length l_c, this can be rewritten to eliminate τ in favor of σ_f. The result is

$$G_c = A_f \sigma_f l_c \tag{6.4-5}$$

which shows that the fracture toughness increases with l_c. It follows that for greatest toughness l should be slightly smaller than l_c while l_c should be as large as possible. The low τ necessary for a large l_c can be obtained either from a weak matrix or a weak interface.

More recent work on fiber pull-out has been contributed by Gent and Yeoh (6.108), Phan-Thien et al. (6.109), and Atkinson et al. (6.110).

6.4.3 Accelerated Characterization

Fiber composites can readily be designed to achieve high in-plane strength and stiffness characteristics. However, for applications involving loadings such as flexure and torsion, the out-of-plane behavior is dominated by the matrix properties alone. Thus, elevated temperature and stress levels can lead to significant property changes in time and can also induce basic material phase changes. Moisture diffusion can have similar effects. Increased water content can cause swelling of organic materials, which, in turn, induces internal stresses, causes basic material property alterations, and degrades the integrity of the matrix/fiber bond. This section will briefly review the treatments of these effects from the point of veiw of damage accumulation and fracture.

Ordinarily, a fiber composite structure is designed so that the fibers are in the high stress directions while the relatively weak and compliant polymeric matrix operates under low stress levels. While this might suggest that the matrix is relatively unimportant as a load carrying agent, this is not so. The matrix serves to transfer the applied stress from fiber to fiber and from ply to ply. It is for this reason that the matrix, and hence matrix-dominated properties, play an important role in structural design and/or behavior of a composite material and cannot be ignored.

Matrix dominated moduli and strength properties of fiber reinforced plastic laminates are time-dependent or viscoelastic and as such are sensitive to environmental conditions—for example, temperature and moisture. Consequently, the long-term integrity of a composite structural component is an important consideration in the initial design process. Therefore, how viscoelastic matrix dominated modulus (compliance) and strength properties vary with time over the design lifetime is necessary to the initial design process. As many structural components are designed for years of service, property variations over years are often needed. Obviously, long-term testing equivalent to the lifetime of a structure is impractical and undesirable. The alternative is to develop analytical and experimental methods that can be reliably used for extrapolation.

The accelerated characterization procedure developed for graphite/epoxy composite laminates by Brinson and co-workers (6.111, 6.112) is based upon the well-known time-temperature superposition principle (TTSP) for polymers and the widely used lamination theory for composite materials. The objective was to develop a method by which the time-dependent deterioration of laminate moduli (compliances) and strength could be calculated from the results of a minimum number of tests. Advanced composite laminates are most frequently designed using lamination theory. This theory allows the calculation of the properties of a general laminate from the knowledge of the behavior of a single lamina or ply. The stress-strain properties of a single ply may be found from constant strain-rate tests on unidirectional laminates and are normally routinely obtained when a general laminate is made. Thus, Brinson's accelerated characterization plan assumes that lamina stress-strain properties from zero load to failure are known.

The transformation equation for the moduli of orthotropic materials has been shown to be valid for unidirectional laminates. Also, various orthotropic failure theories have been shown to be valid for unidirectional laminates. Therefore, modulus and strength properties as a function of fiber angle are known. But, before the time-dependent properties of a general laminate can be predicted, knowledge of the time-dependent behavior of a single ply is required. For this reason, constant strain-rate behavior is insufficient for viscoelastic predictions. Long-term creep or relaxation tests to determine the necessary lifetime information are impractical since the objective is to make long-term predictions from a minimum number of tests conducted in a short time.

The fundamental concept employed by Brinson and co-workers to overcome the above obstacle was to use the well-known time-temperature-superposition principle to produce a modulus (compliance) master curve for a single fiber orientation. The TTSP principle applied to composite laminates requires the short-term creep tests to be conducted on a unidirectional laminate at various temperatures. (These can likely be performed in a single day or a few days at most.) Next, either an Arrhenius or WLF-type equation could be modified to predict the variation of the shift function with fiber angle for a single lamina without further testing. If this can be done, then the results can be combined to produce the modulus (compliance) master curves by simple scaling procedures without additional testing.

Delayed failure predictions, of course, require knowledge of time-dependent strength properties. The determination of such properties often requires large amounts of testing over a prolonged period of time. To avoid an extensive creep rupture testing program, the assumption was made that strength master curves were of the same shape as modulus (compliance) master curves for any particular fiber angle. From this assumption lamina strength master curves as a function of fiber angle and temperature were determined again by simple scaling procedures. Important work in this area has also been contributed by Christensen (6.113).

6.5 References

(6.1) Jones, R. M., *Mechanics of Composite Materials*, Scripta, Washington, D.C. (1975).

(6.2) Vinson, J. R. and Chou, T-W., *Composite Materials and Their Uses in Structures*, Wiley, New York (1975).

(6.3) Tewary, V. K., *Mechanics of Fibre Composites*, Wiley, New York (1978).

(6.4) Christensen, R. M., *Mechanics of Composite Materials*, Wiley, New York (1979).

(6.5) Agarwal, B. D. and Broutman, L. J., *Analysis and Performance of Fiber Composites*, Wiley, New York (1980).

(6.6) Piggott, M. R., *Load Bearing Fibre Composites*, Pergamon, New York (1980).

(6.7) Tsai, S. W. and Hahn, H. T., *Introduction to Composite Materials*, Technomic, Westport, Ct. (1980).

(6.8) Kelly, A., *Strong Solids*, Oxford, London (1973).

(6.9) Jayatilaka, A. de S., *Fracture of Engineering Brittle Materials*, Applied Science, London (1979).

(6.10) Argon, A. S., "Fracture of Composites," *Treatise on Materials Science and Technology*, Vol. 1, H. Herman (ed.), Academic, New York, pp. 79–114 (1972).

(6.11) Corten, H. T., "Fracture Mechanics of Composites," *Fracture*, Vol. VII, H. Liebowitz (ed.), Academic, New York, pp. 676–769 (1972).

(6.12) Erdogan, F., "Fracture Problems in Composite Materials." *Engineering Fracture Mechanics*, **4**, pp. 811–840 (1972).

(6.13) Zweben, C., "Fracture Mechanics and Composite Materials: A Critical Analysis," ASTM STP 521, American Society for Testing and Materials, Philadelphia, pp. 65–97 (1973).

(6.14) Beaumont, P. W. R., "A Fracture Mechanics Approach to Failure in Fibrous Composites," *Journal of Adhesion*, **6**, pp. 107–137 (1974).

(6.15) Smith, C. W., "Limitations of Fracture Mechanics as Applied to Composites," *Inelastic Behavior of Composite Materials*, C. T. Herakovich (ed.), ASME AMD, **13**, pp. 157–176 (1975).

(6.16) Kanninen, M. F., Rybicki, E. F., and Brinson, H. F., "A Critical Look at Current Applications of Fracture Mechanics to the Failure of Fibre-Reinforced Composites," *Composites*, **8**, pp. 17–22 (1977).

(6.17) Dharan, C. K. H., "Fracture Mechanics of Composite Materials," *Journal of Engineering Materials and Technology*, **100**, pp. 233–247 (1978).

(6.18) Backlund, J., "Fracture Analysis of Notched Composites," *Computers and Structures*, **13**, pp. 145–154 (1981).

(6.19) Potter, R. T., "On the Mechanism of Tensile Fracture in Notched Fibre Reinforced Plastics," *Proceedings of the Royal Society of London*, **A361**, pp. 325–341 (1978).
(6.20) Sih, G. C. and Liebowitz, H., "Mathematical Theories of Brittle Fracture," *Fracture*, Vol. II, H. Liebowitz (ed.), Academic, New York, pp. 67–190 (1968).
(6.21) Sendeckyj, G., "Effect of Defects in Composite Structures," *Seventh Annual Mechanics of Composites Review*. Dayton, Ohio, October 28–30, 1981.
(6.22) Wells, A. A., "Composite Materials and the Designer," *Composites*, **3**, pp. 112–118 (1972).
(6.23) Phillips, D. C. and Tetelman, A. S., "The Fracture Toughness of Fibre Composites," *Composites*, **3**, pp. 216–223 (1972).
(6.24) Cooper, G. A. and Kelly, A., "Role of the Interface in the Fracture of Fibre-Composite Materials," ASTM STP 452, pp. 90–105, (1969).
(6.25) Kelly, A., "Interface Effects and the Work of Fracture of a Fiberous Composite," *Proceedings of the Royal Society*, **A319**, pp. 95–116 (1970).
(6.26) Cooper, G. A., "The Fracture Toughness of Composites Reinforced with Weakened Fibres," Journal of Materials Science, **5**, pp. 645–654 (1970).
(6.27) Gresaczuk, L. B., "Theoretical Studies of the Mechanisms of the Fibre-Matrix Interface in Composites," ASTM STP 452, American Society for Testing and Materials, Philadelphia, pp. 42–54 (1969).
(6.28) Beaumont, P. W. R. and Phillips, D. C., "Tensile Strength of Notched Composites," *Journal of Composite Materials*, **6**, pp. 32–46 (1972).
(6.29) Piggott, M. R., "Theoretical Estimation of Fracture Toughness of Fiberous Composites," *Journal of Materials Science*, **5**, pp. 669–675 (1970).
(6.30) Marston, T. U., Atkins, A. G., and Felbeck, D. K., "Interfacial Fracture Energy and the Toughness of Composites," *Journal of Materials Science*, **9**, pp. 447–455 (1974).
(6.31) Poe, C. C., Jr. and Sova, J. A., "Fracture Toughness of Boron/Aluminum Laminates with Various Proportions of 0 and ± 45 Piles," NASA Technical Paper 1707, Nov. 1980.
(6.32) Gaggar, S. and Broutman, L. J., "Crack Growth Resistance of Random Fiber Composites," *Journal of Composite Materials*, **9**, pp. 216–227 (1975).
(6.33) Mandell, J. F., Wang, S. S., and McGarry, F. J., "The Extension of Crack Tip Damage Zones in Fiber Reinforced Plastic Laminates," *Journal of Composite Materials*, **9**, pp. 266–287 (1975).
(6.34) Poe, C. C., Jr., "A Unifying Strain Criterion for Fracture of Fibrous Composite Laminates," *Engineering Fracture Mechanics*, **17**, pp. 153–171 (1983).
(6.35) Wu, E. M., "Application of Fracture Mechanics to Anisotropic Plates," *Journal of Applied Mechanics*, **34**, pp. 967–974 (1967).
(6.36) Waddoups, M. E., Eisenmann, J. R., and Kaminski, B. E., "Macroscopic Fracture Mechanics of Advanced Composite Materials," *Journal of Composite Materials*, **5**, pp. 446–454 (1971).
(6.37) Zweben, C., "On the Strength of Notched Composites," *Journal of the Mechanics and* Physics of Solids, **19**, pp. 103–116 (1971).
(6.38) Gerberich, W. W., "Fracture Mechanics of a Composite with Ductile Fibers," *Journal of the Mechanics and Physics of Solids*, **19**, pp. 71–87 (1971).
(6.39) Zimmer, J. E., "Fracture Mechanics of a Fiber Composite," *Journal of Composite Materials*, **6**, pp. 312–315 (1972).
(6.40) Cruse, T. A., "Tensile Strength of Notched Composites," *Journal of Composite Materials*, **7**, pp. 218–229 (1973).
(6.41) Konish, J. J., Swedlow, J. L., and Cruse, T. A., "Fracture Phenomena in Advanced Fibre Composite Materials," *AIAA Journal*, **11**, pp. 40–43 (1973).
(6.42) Wright, M. A. and Iannuzzi, F. A., "The Application of the Principles of Linear Elastic Fracture Mechanics to Unidirectional Fiber Reinforced Composite Materials," *Journal of Composite Materials*, **7**, pp. 430–447 (1973).
(6.43) Owen, M. J. and Bishop, P. T., "Critical Stress Intensity Factors Applied to Glass Reinforced Polyester Resin," *Journal of Composite Materials*, **7**, pp. 146–159 (1973).
(6.44) Scott, J. M. and Phillips, D. C., "Carbon Fibre Composites with Rubber Toughened Matrices," *Journal of Materials Science*, **10**, pp. 551–562 (1975).
(6.45) Sendeckyj, G. P., "On Fracture Behavior of Composite Materials," *Strength and Structure of Solid Materials*, Miyamoto et al. (eds.), Noordhoff, Leyden, pp. 373–388 (1976).
(6.46) Brinson, H. F. and Yeow, Y. T., "An Experimental Study of the Fracture Behavior of Laminated Graphite/Epoxy Laminates", *Composite Materials: Testing and Design (Fourth Conference)*, ASTM STP 617, American Society for Testing and Materials, Philadelphia, pp. 18–38 (1977).

(6.47) Yeow, Y. T., Morris, D. H., and Brinson, H. F., "The Fracture Behavior of Graphite/Epoxy Laminates," *Experimental Mechanics*, **19**, pp. 1–8 (1979).
(6.48) Whitney, J. M. and Nuismer, R. J., "Stress Fracture Criteria for Laminated Composites Containing Stress Concentrations," *Journal of Composite Materials*, **8**, pp. 253–265 (1974).
(6.49) Spencer, B. and Barnby, J. T., "The Effects of Notch and Fibre Angles on Crack Propagation in Fibre-Reinforced Polymers," *Journal of Materials Science*, **11**, pp. 83–88 (1976).
(6.50) Mall, S., Murphy, J. F., and Shottafer, J. E., "Criterion for Mixed Mode Fracture in Wood," *Journal of Engineering Mechanics*, **109**, pp. 680–690 (1983).
(6.51) Harrison, N. L., "Strain Energy Release Rates for Turning Cracks," *Fibre Science and Technology*, Vol. 5, pp. 197–212 (1972); and "Splitting of Fibre-Reinforced Materials," *Fibre Science Technology*, **6**, pp. 25–38 (1973).
(6.52) Wu, E. M., "Optimal Experimental Measurements of Anisotropic Failure Tensors," *Journal of Composite Materials*, **6**, pp. 472–489 (1972).
(6.53) Sih, G. C. and Chen, E. P., "Fracture Analysis of Unidirectional Composites," *Journal of Composite Materials*, **7**, pp. 230–244 (1973).
(6.54) Sih, G. C. and Chen, E. P., *Cracks in Composite Materials*, Martinus Nijhoff, The Hague (1981).
(6.55) Parkizjar, S., Zachary, L. E., and Sun, C. T., "Application of the Principles of Linear Fracture Mechanics to the Composite Materials," *International Journal of Fracture*, **20**, pp. 3–15 (1982).
(6.56) Rosen, B. W., Kulkarni, S. V., and McLaughlin, P. V., Jr., "Failure and Fatigue Mechanisms in Composite Materials," *Inelastic behavior of Composite Materials*, C. T. Herakovich (ed.), ASME AMD, Vol. 13, American Society of Mechanical Engineers, New York, pp. 17–72 (1975).
(6.57) Zhen, S., "The *D* Criterion in Notched Composite Materials," *Journal of Reinforced Plastics and Composites*, **2**, pp. 98–110 (1983).
(6.58) Nicholls, D. J. and Gallagher, J. P., "Determination of G_{Ic} in Angle Ply Composites Using a Cantilever Beam Test Method," *Journal of Reinforced Plastics and Composites*, **2**, pp. 2–17 (1983).
(6.59) Wells, J. K. and Beaumont, P. W. R., "Correlations for the Fracture of Composite Materials," *Scripta Metallurgica*, **16**, pp. 99–103 (1982).
(6.60) Wells, J. K. and Beaumont, P. W. R., "Prediction of Notch-Tip Energy Absorption in Composite Laminates," *Testing, Evaluation, and Quality Control of Composites*, Butterworths Scientific, London (1983).
(6.61) Kunz-Douglass, S., Beaumont, P. W. R., and Ashby, M. F., "A Model for the Toughness of Epoxy-Rubber Particulate Composites," *Journal of Materials Science*, **15**, pp. 1109–1123 (1980).
(6.62) Rybicki, E. F., Schmueser, D. W., and Fox, J., "An Energy Release Rate Approach for Stable Crack Growth in the Free-Edge Delamination Problem," *Journal of Composite Materials*, **11**, pp. 470–487 (1977).
(6.63) Wang, S. S., "Fracture Mechanics for Delamination Problems in Composite Materials," *Journal of Composite Materials*, **17**, pp. 210–223 (1983).
(6.64) Jurf, R. A. and Pipes, R. B., "Interlaminar Fracture of Composite Materials," *Journal of Composite Materials*, **16**, pp. 386–394 (1982).
(6.65) O'Brien, T. K., "Characterization of Delamination Onset and Growth in a Composite Laminate," *Damage in Composite Materials*, ASTM STP 775, K. L. Reifsnyder (ed), American Society for Testing and Materials, Philadelphia, pp. 140–167 (1982).
(6.66) Kurashige, M., "Compressive Strength of Fiber-Reinforced Materials," *Acta Mechanica*, **49**, pp. 49–56 (1983).
(6.67) Ko, W. L., Nagy, A., Francis, P. H., and Lindholm, U. S., "Crack Extension in Filamentary Materials," *Engineering Fracture Mechanics*, **8**, pp. 411–424 (1976).
(6.68) Pipes, R. B., Wetherhold, R. C., and Gillespie, J. W., Jr., "Notched Strength of Composite Materials," *Journal of Composite Materials*, **13**, pp. 148–160 (1979).
(6.69) Mandel, J. A. and Pack, S. C., "Crack Growth in Fiber-Reinforced Materials," *Journal of the Engineering Mechanics Division of ASCE*, **108**, pp. 509–526 (1982).
(6.70) Konishi, D. Y., Lo, K. H., and Wu, E. M., "Progressive-Failure Model for Advanced Composite Laminates Containing a Circular Hole," *Composite Structures*, I. H. Marshall (ed.), Applied Science, New York, pp. 646–655 (1981).
(6.71) Altus, E. and Rotem, A., "A 3-D Fracture Mechanics Approach to the Strength of Composite Materials," *Engineering Fracture Mechanics*, **14**, pp. 637–644 (1981).

(6.72) Potter, R. T., "The Notch Size Effect in Carbon Fibre, Glass Fibre and Kevlar Reinforced Plastic Laminates," Technical Report 81027, Royal Aircraft Establishment, Farnborough, Feb. 1981.

(6.73) Mandell, J. F., Huang, D. D., and McGarry, F. J., "Tensile Fatigue Performance of Glass Fiber Dominated Composites," *Composites Technology Review*, **3**, No. 3, pp. 96–102 (1981).

(6.74) Lifshitz, J. M., "Nonlinear Matrix Failure Criteria for Fiber-Reinforced Composite Materials," *Composites Technology Review*, **4**, No. 3. pp. 78–83 (1982).

(6.75) Morley, J. G., "Some Aspects of Crack Growth and Failure in Fibre Reinforced Composites," *Journal of Materials Science*, **18**, pp. 1564–1576 (1983).

(6.76) Dharani, L. R., Jones, W. F., and Goree, J. G., "Mathematical Modeling of Damage in Unidirectional Composites," *Engineering Fracture Mechanics*, **17**, pp. 555–573 (1983).

(6.77) Lo, K. H., Wu, E. M., and Konishi, D. Y., "Failure Strength of Notched Composite Laminates," *Journal of Composite Materials*, **17**, 00. 384–398 (1983).

(6.78) Kanninen, M. F., Rybicki, E. F., and Griffith, W. I., "Preliminary Development of a Fundamental Analysis Model for Crack Growth in a Fiber Reinforced Composite Material," *Composite Materials: Testings and Design* (Fourth Conference), ASTM STP 617, pp. 53–69 (1977).

(6.79) Griffith, W. I., Kanninen, M. F., and Rybicki, E. F., "A Fracture Mechanics Approach to the Analysis of Graphite/Epoxy Laminated Precracked Tension Panels," *Nondestructive Evaluation and Flaw Criticality for Composite Materials*, ASTM STP 696, pp. 185–201 (1979).

(6.80) Rybicki, E. F. and Kanninen, M. F., "A Finite Element Calculation of Stress Intensity Factors by a Modified Crack Closure Integral," *Engineering Fracture Mechanics*, **9**, pp. 931–938 (1977).

(6.81) Ouyang, C. and Lu, M. Z., "On a Micromechanical Fracture Model for Cracked Reinforced Composites," *International Journal of Non-Linear Mechanics*, **18**, pp. 71–77 (1983).

(6.82) Nuismer, R. J. and Brown, G. E., "Progressive Failure of Notched Composite Laminates Using Finite Elements," *Proceedings of the 13th Annual Meeting of the Society of Engineering Science*, Vol. 1, NASA CP 2001, pp. 183–192, 1976.

(6.83) Nuismer, R. J. and Tan, S. C., "The Role of Matrix Cracking in the Continuum Constitutive Behavior of a Damaged Composite Ply," *Proceedings of the IUTAM Symposium on Mechanics of Composite Materials*, Blacksburg, Va., 1982.

(6.84) Adams, D. F., "Elastoplastic Crack Propagation in a Transversely Loaded Unidirectional Composite," *Journal of Composite Materials*, **8**, pp. 38–54 (1974).

(6.85) Wang, S. S., Mandell, J. F., and McGarry, F. J., "Three-Dimensional Solution for a Through-Thickness Crack With Crack Tip Damage in a Cross-Piled Laminate," *Fracture Mechanics of Composites*, ASTM STP 593, American Society for Testing and Materials, Philadelphia, pp. 61–85 (1975).

(6.86) Mahishi, J. M. and Adams, D. F., "Fracture Behavior of a Single-Fibre Graphite/Epoxy Model Composite Containing a Broken Fibre or Cracked Matrix," *Journal of Materials Science*, **18**, pp. 447–456 (1983).

(6.87) Reedy, E. D., Jr., "Analysis of Center-Notched Monolayers with Application to Boron/Aluminum Composites," *Journal of the Mechanics of Physics and Solids*, **28**, pp. 265–286 (1980).

(6.88) Wang, A. S. D. and Crossman, F. W., "Initiation and Growth of Transverse Cracks and Edge Delamination in Composite Laminates Part 1. An Energy Method," *Journal of Composite Materials Supplement*, **14**, pp. 71–87, (1980); and Crossman, F. W. et al., "Part 2. Experimental Correlation," *Journal of Composite Materials Supplement*, **14**, pp. 88–108 (1980).

(6.89) Crossman, F. W. and Wang, A. S. D., "The Dependence of Transverse Cracking and Delamination on Ply Thickness in Graphite/Epoxy Laminates," *Damage in Composite* Materials, ASTM STP 775, K. L. Reifsnider (ed.), American Society for Testing and Materials, pp. 118–139 (1982).

(6.90) Ripling, E. J., Mostovoy, S., and Patrick, R. L., "Measuring Fracture Toughness of Adhesive Joints," *Materials Research and Standards*, **4**, pp. 129–134 (1964).

(6.91) Ripling, E. J., Mostovoy, S., and Corten, H. T., "Fracture Mechanics: A Tool for Evaluating Structural Adhesives," *Journal of Adhesion*, **3**, pp. 107–123 and pp. 125–144 (1971).

(6.92) Ripling, E. J., Santner, J. S., and Crosley, P. B., "Fracture Toughness of Composite Adherend Adhesive Joints Under Mixed Mode I and III Loading," *Journal of Materials Science*, **18**, pp. 2274–2282 (1983).

(6.93) Everett, R. A., Jr., "The Role of Peel Stresses in Cyclic Debonding," *Adhesive Age*, **26**, pp. 24–29 (1983).

(6.94) Ikegami, K. and Kamiya, K., "The Effect of Flaws in the Adhering Interface on the Strength of Adhesive-Bonded Butt Joints," *Journal of Adhesion*, **14**, pp. 1–17 (1982).

(6.95) Erdogan, F. and Gupta, G., "The Stress Analysis of Multi-Layered Composites With a Flaw," *International Journal of Solids and Structures*, **7**, pp. 39–61 (1971).

(6.96) Saxena, A., "Applications of Linear Elastic Fracture Mechanics to the Evaluation of Aluminum-Epoxy Bonds," *Fibre Science and Technology*, **12**, pp. 111–128 (1979).

(6.97) Gledhill, R. A. and Kinlock A. J., "Mechanics of Crack Growth in Epoxide Resins," *Polymer Engineering and Science* , **19**, pp. 82–88 (1979).

(6.98) Bitner, J. L., Rushford, J. L., Rose, W. S., Hunston, D. L., and Riew, C. K., "Viscoelastic Fracture of Structural Adhesives," *Adhesion*, **13**, pp. 3–28 (1981).

(6.99) Bikerman, J. J., "Theory of Peeling Through a Hookean Solid," *Journal of Applied Physics*, **28**, pp. 1484–1485 (1957).

(6.100) Cook, J. and Gordon, J. E., "A Mechanism for the Control of Crack Propagation in All-Brittle Systems," *Proceedings of the Royal Society of London*, **A282**, pp. 508–519 (1964).

(6.101) Williams, M. L., "The Relation of Continuum Mechanics to Adhesive Fracture," *Journal of Adhesion*, **4**, pp. 307–332 (1972).

(6.102) Gent, A. N. and Kinlock, A. J., "Adhesion of Viscoelastic Materials to Rigid Substrates, III. Energy Criterion for Failure," *Journal of Polymer Science*, Part A-2, **9**, pp. 659–668 (1971).

(6.103) Knauss, W. G., "Fracture Mechanics and the Time-Dependent Strength of Adhesive Joints," *Journal of Composite Materials*, **5**, pp. 176–192 (1971).

(6.104) Trantina, G. G., "Fracture Mechanics Approach to Adhesive Joints," *Journal of Composite Materials*, **6**, pp. 192–207 (1972).

(6.105) Chang, D. J., Muki, R., and Westmann, R. A., "Double Cantilever Beam Models in Adhesive Fracture Mechanics," *International Journal of Solids and Structures*, **12**, pp. 13–26 (1976).

(6.106) Wang, S. S. and Yau, J. F., "Interface Cracks in Adhesively Bonded Lap-Shear Joints," *International Journal of Fracture*, **19**, pp. 295–309 (1982).

(6.107) Cottrell, A. H., "Strong Solids," *Proceedings of the Royal Society*, **A282**, pp. 2–9 (1964).

(6.108) Gent, A. N. and Yeoh, O. H., "Failure Loads for Model Adhesive Joints Subjected to Tension, Compression or Torsion," *Journal of Materials Science*, **17**, pp. 1713–1722 (1982).

(6.109) Phan-Thien, N., Pantelis, G., and Bush, M. B., "On the Elastic Fibre Pull-Out From a Fixed and Flat Surface: Asymptotic and Numerical Results," *Applied Mathematics Modeling*, **6**, pp. 257–261 (1982).

(6.110) Atkinson, C., Avila, J., Betz, E., and Smelsar, R.E., "The Rod Pull Out Problem, Theory and Experiment," *Journal of the Mechanics of Physics and Solids*, **30**, pp. 97–120 (1982).

(6.111) Brinson, H. F., "The Viscoelastic Constitutive Modeling of Adhesives." *Composites*, **8**, pp. 377–382 (1977).

(6.112) Brinson, H. F., Griffith, W. I., and Morrison, D. H., "Creep Rupture of Polymer-Matrix Composites," *Experimental Mechanics*, pp. 329–335 (Sept. 1981).

(6.113) Christensen, R. M., "Lifetime Predictions for Polymers and Composites Under Constant Load," *Society for Rheology*, **25**, pp. 517–528 (1981).

(6.114) Papaspyropoulos, V., Ahmad, J., and Kanninen, M. F., "A Micromechanical Fracture Mechanics Analysis of a Fiber Composite Laminate Containing a Defect," *Effects of Defects in Composite Materials*, ASTM STP 836, American Society for Testing and Materials, Philadelphia, Pa., pp. 237–249 (1984).

7

TIME-DEPENDENT FRACTURE

Materials exist for which a suddenly applied constant stress produces an instantaneous elastic strain followed by a slow continuous straining or creep. The creep strain is usually characterized by three different stages: primary, secondary (steady-state), and tertiary. Straining in the primary stage proceeds at a decreasing rate until a constant strain rate associated with the secondary stage is attained. The latter stage is followed by the tertiary stage where straining occurs at an increasing rate and terminates in rupture of the material.

The failure of structural components can occur by crack growth in the presence of creep—that is, creep crack growth. Metals will exhibit creep at temperatures greater than about thirty percent of their absolute melting temperatures. Many components of gas turbines, fossil and nuclear power plants and aerospace structures are required to perform at service temperatures in excess of the creep threshold temperature. On the other hand, substantial creep deformation in polymeric structures; for example, plastic piping components, can be observed at room temperature. In any case the service life of a structural component can be dictated by time-dependent crack growth. While the micromechanisms for creep crack growth in a polycrystalline steel and an amorphous polymer may be quite different, there are similarities in the macroscopic crack growth behavior in each.

There are two competing mechanisms involved in creep crack growth. The creep deformation is characterized by crack-tip blunting in the material ahead of the crack tip. This relaxes the crack-tip stress field and tends to retard crack growth. The other mechanism results in an accumulation of creep damage in the form of microcracks and voids that enhance crack growth as they coalesce. Whichever phenomenon dominates determines whether or not creep crack growth will take place. Steady-state crack growth will occur when an equilibrium between these two effects is attained.

The time-dependent fracture considered in this chapter is due to creep crack growth. While environmentally assisted crack growth is important and at times is difficult to separate from creep crack growth, time-dependent fracture due solely to environmental effects is not treated here.

This chapter begins with a study of the near stress and strain fields for a stationary crack tip under the influence of the different creep regimes. Appropriate load parameters are identified and estimates for the time intervals over which they are applicable are presented. This is then followed by a study of creep crack growth. Unlike other fields the crack-tip field for creep crack growth does not contain an undetermined parameter. The implication of this characteristic is discussed. Steady-state and transient creep crack growth within the confines of small-scale yielding are investigated. In Section 7.3 creep crack growth correlations are presented. Viscoelastic crack growth is treated in Section 7.4. The chapter closes with a discussion of further research and problems that remain to be addressed.

7.1 Stationary Crack-Tip Fields

Macroscopic crack growth in a creeping material occurs by local failure resulting from nucleation and coalescence of microcavities in the highly strained region near the crack tip. These and other noncontinuum processes may be aided by environmental effects. When the fracture process zone can be considered small compared to the region over which the singular terms of the stress and strain fields dominate, then a detailed accounting of the fracture process is not essential. For whatever the fracture mechanism is within the zone, it must be dictated by the surrounding singular fields. Under such conditions the creep crack growth is governed by a time-dependent loading parameter that characterizes the geometry of the flawed body and its loading. The stress intensity factor, the crack-tip opening displacement rate, and the path-independent energy rate integral among others have been proposed as relevant loading parameters for creep crack growth. The appropriate parameter can be determined from an analysis for the crack-tip stress and strain fields. Such analyses have been performed by Riedel (7.1–7.3) and Riedel and Rice (7.4) for stationary cracks.

Consider the crack-tip fields for a stationary crack in a time-dependent material undergoing either plane stress or plane strain Mode I loading. The singular character of these fields can also be expected to be the same for plane problems subjected to either Mode II or Mode III loadings. Let the x_1-x_3 plane contain the crack plane with the x_3-axis parallel to the crack front. The dependent variables are independent of x_3. The stress components $\sigma_{\alpha 3}$ and the strain components $\varepsilon_{\alpha 3}$ vanish as well as σ_{33} and ε_{33} for the plane stress and plane strain problems, respectively.

7.1.1 Elastic-Secondary Creep

We begin by considering the crack-tip fields in an elastic-secondary creeping material because of the close similarity of these fields to the ones for an elastic-plastic material. The uniaxial material law

$$\dot{\varepsilon} = \dot{\sigma}/E + B\sigma^n \tag{7.1-1}$$

is assumed. The total strain rate $\dot{\varepsilon}$ is composed of the sum of an elastic component, $\dot{\sigma}/E$, and a nonlinear secondary creep component, $B\sigma^n$. In Equation (7.1-1), E is the modulus of elasticity and the creep exponent n and temperature-dependent coefficient B are material parameters for the power law creep. Material aging can be modeled by Equation (7.1-1) if B is further allowed to be a function of time t. The superposed dot is used to denote a time derivative. The material described by Equation (7.1-1) is variously referred to as an elastic-power-law creep material, a Maxwell elastic-nonlinear viscous material, or simply an elastic-viscoplastic solid. It can also be considered as describing a creeping material with negligible primary and tertiary creep regimes.

A generalization of Equation (7.1-1) to a multiaxial stress state is

$$\dot{\varepsilon}_{ij} = \frac{1+\nu}{E}\dot{s}_{ij} + \frac{1-2\nu}{3E}\dot{\sigma}_{kk}\delta_{ij} + \tfrac{3}{2}B\bar{\sigma}^{n-1}s_{ij} \tag{7.1-2}$$

where

$$\begin{aligned} s_{ij} &= \sigma_{ij} - \sigma_{kk}\delta_{ij}/3 \\ \bar{\sigma} &= (\tfrac{3}{2}s_{ij}s_{ij})^{\frac{1}{2}} \end{aligned} \tag{7.1-3}$$

are, respectively, the deviatoric stress components and effective stress. According to Equation (7.1-2) the creep component of the deformation is incompressible. For a small strain formulation Equation (7.1-2) must be supplemented by the equilibrium condition

$$\sigma_{\alpha\beta,\beta} = 0 \tag{7.1-4}$$

and the compatibility relation

$$\dot{\varepsilon}_{\alpha\beta,\alpha\beta} - \dot{\varepsilon}_{\alpha\alpha,\beta\beta} = 0 \tag{7.1-5}$$

for the plane problems.

The equilibrium condition will be automatically satisfied when the stress components are expressed in terms of the Airy stress function Ψ defined by

$$\sigma_{\alpha\beta} = -\Psi_{,\alpha\beta} + \Psi_{,\gamma\gamma}\delta_{\alpha\beta} \tag{7.1-6}$$

The combination of Equations (7.1-2), (7.1-5), and (7.1-6) provides the governing equation for the Airy stress function. For the plane stress problem it is

$$\frac{2}{E}\nabla^4\dot{\Psi} - B[(\Psi_{,\gamma\gamma}\delta_{\alpha\beta} - 3\Psi_{,\alpha\beta})\bar{\sigma}^{n-1}]_{,\alpha\beta} = 0 \tag{7.1-7}$$

where

$$\bar{\sigma} = [\tfrac{3}{2}\Psi_{,\delta\rho}\Psi_{,\delta\rho} - \tfrac{1}{2}\Psi_{,\delta\delta}\Psi_{,\rho\rho}]^{\frac{1}{2}} \tag{7.1-8}$$

In Equation (7.1-7), $\nabla^4(\ ) \equiv (\ )_{,\alpha\alpha\beta\beta}$ is the biharmonic operator.

The deviatoric stress component s_{33} can not be expressed simply in terms of Ψ for the plane strain problem. Nevertheless, it follows directly from the condition $\dot{\varepsilon}_{33} = 0$. Instead of a single equation we have the two coupled equations

$$2\frac{1-\nu}{E}[\nabla^4\dot{\Psi} + \dot{s}_{33,\alpha\alpha}] - B\{[(\Psi_{,\gamma\gamma} - s_{33})\delta_{\alpha\beta} - 2\Psi_{,\alpha\beta}]\bar{\sigma}^{n-1}\}_{,\alpha\beta} = 0 \tag{7.1-9}$$

and

$$\frac{1-2\nu}{3E}\dot{\Psi}_{,\alpha\alpha} + \frac{1}{E}\dot{s}_{33} + B\bar{\sigma}^{n-1}s_{33} = 0 \tag{7.1-10}$$

where now

$$\bar{\sigma} = \frac{\sqrt{3}}{2}[2\Psi_{,\delta\rho}\Psi_{,\delta\rho} - \Psi_{,\delta\delta}\Psi_{,\rho\rho} + 3s_{33}^2]^{\frac{1}{2}} \tag{7.1-11}$$

For an incompressible material ($\nu \to \frac{1}{2}$), Equation (7.1-10) implies that $s_{33} = 0$, which simplifies Equations (7.1-9) and (7.1-11).

The crack-tip fields are anticipated to be singular at the crack tip for the assumed material law of Equation (7.1-2). If it is further assumed that the creep exponent is greater than unity (i.e., $n > 1$), then the creep strain rates will dominate the elastic rates near the crack tip. Consequently, sufficiently close to the crack tip the linear (elastic) terms in Equations (7.1-7) and (7.1-9) can be neglected. As Goldman and Hutchinson (7.5) noted the resulting equations have the same form as the equations governing the asymptotic behavior in a rate-insensitive, power law strain hardening material. Hence, the stress and strain singularities are of the HRR type (see Section 5.3).

For a local polar coordinate system at the crack tip with $\theta = 0$ corresponding to the positive x_1 direction, the asymptotic fields can be written as

$$\sigma_{ij} = \left[\frac{C(t)}{BI_n r}\right]^{1/(n+1)} \tilde{\sigma}_{ij}(\theta)$$
$$\dot{\varepsilon}_{ij} = B\left[\frac{C(t)}{BI_n r}\right]^{n/(n+1)} \tilde{\varepsilon}_{ij}(\theta) \tag{7.1-12}$$

The time-dependent loading parameter $C(t)$, which depends upon the applied loading and the geometry of the crack body, is defined as

$$C(t) = \lim_{\Gamma_\varepsilon \to 0} \int_{\Gamma_\varepsilon} \frac{n}{n+1}\sigma_{ij}\dot{\varepsilon}_{ij}\,dx_2 - \sigma_{ij}n_j\frac{\partial \dot{u}_i}{\partial x_1}\,ds \tag{7.1-13}$$

The contour Γ_ε is a vanishingly small loop enclosing the crack tip and is traversed in a counterclockwise direction. The unit normal n_i to Γ_ε is directed away from the crack tip. It is impossible to establish $C(t)$ from only an asymptotic analysis. The angular dependence of σ_{ij} and $\dot{\varepsilon}_{ij}$ is the same as that for the HRR fields (see Figures 5.12 and 5.13). The factor I_n, which depends upon the creep exponent, appears in Figure 5.14. It is clear from Equation (7.1-12) that $C(t)$ is the loading parameter that determines the strength of the crack-tip singular fields.

If the applied load remains fixed in time, then the constitutive relation, Equation (7.1-2), implies that the stress field becomes time-independent ($\dot{\sigma}_{ij} \to 0$) as $t \to \infty$ for a stationary crack tip. The elastic strain rates vanish in this limit and secondary (steady-state) creep extends throughout the body. Under this condition of steady-state creep the material exhibits nonlinear viscous flow. In this case $C(t) \to C^*$ as $t \to \infty$, where C^* is the path-independent integral

$$C^* = \int_\Gamma W(\dot{\varepsilon}_{ij})\, dx_2 - \sigma_{ij} n_j \frac{\partial \dot{u}_i}{\partial x_1}\, ds \qquad (7.1\text{-}14)$$

in which

$$W(\dot{\varepsilon}_{ij}) = \int_0^{\dot{\varepsilon}_{ij}} \sigma_{kl}\, d\dot{\varepsilon}_{kl} \qquad (7.1\text{-}15)$$

is the strain energy rate density and the contour Γ is an arbitrary loop enclosing the crack tip and no other defect. Again, the integration is to be performed in a counterclockwise direction. The proof of the path-independence of C^* follows along the same lines as that for path-independence of the J-integral.

Alternatively, C^* can be expressed as

$$C^* = -\left(\frac{\partial \dot{U}}{\partial a}\right)_{\dot{\Delta}} \qquad (7.1\text{-}16)$$

where

$$\dot{U} = \int_0^{\dot{\Delta}} P\, d\dot{\Delta} \qquad (7.1\text{-}17)$$

is the rate of work done by the load P per unit thickness acting through the conjugate displacement rate $\dot{\Delta}$. In this respect the C^*-integral can be viewed as a path-independent energy rate integral.

The near-tip fields for steady-state creep are

$$\begin{aligned} \sigma_{ij} &= \left[\frac{C^*}{BI_n r}\right]^{1/(n+1)} \tilde{\sigma}_{ij}(\theta) \\ \dot{\varepsilon}_{ij} &= B\left[\frac{C^*}{BI_n r}\right]^{n/(n+1)} \tilde{\varepsilon}_{ij}(\theta) \end{aligned} \qquad (7.1\text{-}18)$$

It follows that the C^*-integral is the loading parameter that determines the strength of the crack-tip fields in a body undergoing steady-state creep.

Under conditions of steady-state creep, the material law of Equation (7.1-2) reduces to a nonlinear elastic stress-strain relationship [cf. Equation (5.3-7)] when the strain rates are replaced by strains. In a similar way the C^*-integral becomes the J-integral if the rates $\dot{u}_i$ and $\dot{\varepsilon}_{ij}$ are replaced by u_i and ε_{ij}, respectively. Conversely, the C^*-integral can be obtained from expressions for the J-integral for power law hardening plastic (nonlinear elastic) materials by replacing $\alpha\varepsilon_y/\sigma_y^n$ by B. For example, Equation (5.4-5) becomes

$$C^* = Bbg_1\left(\frac{a}{W}\right) h_1\left(\frac{a}{W}, n\right)\left(\frac{P\sigma_y}{P_0}\right)^{n+1} \qquad (7.1\text{-}19)$$

where P_0 is the limit load for an equivalent perfectly plastic body having a yield stress σ_y. Therefore, the elastic-plastic handbook (7.6) (see Appendix A) can also be used to determine C^*. In addition, the J-estimation techniques of the Rice-Paris-Merkle type (7.7) (see Section 5.5) can be used to approximate C^* for large values of n by replacing the load point displacement Δ with its rate. While the crack opening rate is related to C^*, it is easier to compute and to measure C^* (7.8, 7.9).

Steady-state creep can be viewed as the limiting process that occurs under constant load as $t \to \infty$. Also of interest is the character of the crack-tip fields at the other end of the spectrum—that is, for short times after loading. According to Equation (7.1-2) the instantaneous response of the material to a suddenly applied load at time $t = 0$ is elastic. Subsequently, the initial elastic stress concentration at the crack tip is relaxed by creep deformation as creep strains grow in the crack-tip region. If the creep zone is viewed as a time-dependent plastic zone, then small-scale yielding is said to occur as long as the creep zone is small compared to the crack length and other relevant dimensions of the cracked body. Solutions for the time during which this condition exists are referred to as small-scale yielding solutions or short-time solutions since they describe the stress and strain fields for the period shortly after application of the load. Small-scale yielding solutions have been investigated by Riedel (7.2) for Mode III loading and by Riedel and Rice (7.4) for Mode I.

For short times after application of the load, the stress field is a function of the independent variables r, θ, and t and the parameters K, E, B, ν, and n. The Mode I stress intensity factor K is used for the load parameter in small-scale yielding. It is apparent from an examination of Equations (7.1-7), (7.1-9), and (7.1-10) that E, B, and t must occur as the product EBt. Furthermore, $(EBt)^{-1/(n-1)}$ has the dimensions of stress and, hence, $K^2/(EBt)^{-2/(n-1)}$ has the units of length. Dimensional consistency requires that the stress field for plane strain, small-scale yielding have the form

$$\sigma_{ij} = \frac{E}{1-\nu} T^{-1/(n-1)} \Sigma_{ij}(R, \theta) \tag{7.1-20}$$

where $\Sigma_{ij}(R, \theta)$ is a dimensionless function. The dimensionless radial coordinate R and dimensionless time T are defined by

$$R = \frac{2\pi r T^{-2/(n-1)}}{[(1-\nu)K/E]^2} \tag{7.1-21}$$

$$T = \frac{n-1}{2}\left(\frac{E}{1-\nu}\right)^n Bt \tag{7.1-22}$$

The plane stress field σ_{ij} is independent of ν since it appears neither in Equation (7.1-7) nor in the prescribed traction boundary conditions. Hence, the factor $1 - \nu$ must be replaced by unity for plane stress. For aging materials where B is time-dependent, Bt is replaced everywhere by $\int_0^t B(\tau)\,d\tau$.

Due to the complexity of the nonlinear Equations (7.1-7), (7.1-9), and (7.1-10), a closed-form solution is not to be anticipated. In lieu of a numerical solution an approximate description of the small-scale yielding stress and strain fields will be developed. It was previously shown that HRR singular fields exist near the crack tip for $t > 0$. Accordingly $\Sigma_{ij}(R, \theta)$ must exhibit a $R^{-1/(n+1)}$ singularity as $R \to 0$. Therefore, Equation (7.1-20) must have the asymptotic form

$$\sigma_{ij} = \alpha_n \left[\frac{(1 - v^2)K^2}{(n+1)EBI_n rt} \right]^{1/(n+1)} \tilde{\sigma}_{ij}(\theta) \tag{7.1-23}$$

where α_n is an unspecified amplitude factor that depends upon the creep exponent. For plane stress the factor $1 - v^2$ is replaced by unity. The introduction of Equation (7.1-23) into Equation (7.1-2) and an integration with respect to time yield

$$\varepsilon_{ij} = \tfrac{3}{2} Bt(n+1)\alpha_n^n \left[\frac{(1 - v^2)K^2}{(n+1)EBI_n rt} \right]^{n/(n+1)} \times [\tilde{\sigma}(\theta)]^{n-1} \tilde{s}_{ij}(\theta) \tag{7.1-24}$$

In arriving at Equation (7.1-24) the elastic strain rates have been neglected compared to their creep counterparts. The combination of Equation (7.1-23) and (7.1-24) leads to

$$\varepsilon_{ij} = \tfrac{3}{2} Bt(n+1)\bar{\sigma}^{n-1} s_{ij} \tag{7.1-25}$$

The rigorous determination of α_n would require a numerical solution to the nonlinear partial differential Equations (7.1-7), (7.1-9), and (7.1-10). An estimate of α_n can be made by assuming the J-integral, which is generally path-dependent for creep problems, to be path-independent. Significant creep straining occurs mainly in the creep zone. Within this zone where the HRR fields are dominant, the stress-strain relation, Equation (7.1-25), is independent of r and θ. The existence of such a unique relationship implies that J is path-independent in the HRR region. On this basis Riedel and Rice (7.4) argue that the approximation of a path-independent J is reasonable.

The form of Equation (7.1-25) is the same as that for power law hardening plasticity [cf. Equation (5.3-7)]. The J-integral for such a material law has been calculated in Section 5.3. For the present problem the near tip value of J is

$$J_0 = (\alpha_n)^{n+1}[(1 - v^2)K^2/E] \tag{7.1-26}$$

Sufficiently removed from the crack tip the fields are elastic and the J-integral is

$$J_\infty = (1 - v^2)K^2/E \tag{7.1-27}$$

Again, the plane stress equivalents of Equations (7.1-26) and (7.1-27) are obtained by replacing $1 - v^2$ by unity. The approximation that the J-integral is path-independent implies that $J_0 = J_\infty$ and, hence, $\alpha_n = 1$ for plane strain and plane stress. A comparison of the HRR fields expressed alternately by

Equations (7.1-12) and (7.1-23) with $\alpha_n = 1$ reveals that

$$C(t) = \frac{J_\infty}{(n+1)t} \tag{7.1-28}$$

for small-scale yielding.

It is apparent from Equations (7.1-23) and (7.1-24) that the strengths of the crack-tip fields are governed by the elastic stress intensity factor for short times under small-scale yielding. The asymptotic analysis indicates that K and C^* are loading parameters for the extreme cases of small and large creep regions, respectively. Which case governs depends upon the size of the creep zone relative to the size of the body and the crack length.

The creep zone boundary has been somewhat arbitrarily defined by Riedel (7.2) as the locus of points where the effective creep strain $\bar{\varepsilon}_{cr}$ equals the effective elastic strain $\bar{\varepsilon}_e$.* The creep zone boundary $r_{cr}(\theta, t)$ can be approximated by using the HRR-creep strain field, Equation (7.1-24), and the remote elastic singular strain field. This leads to

$$r_{cr}(\theta, t) = \frac{1}{2\pi}\left(\frac{K}{E}\right)^2\left[\frac{(n+1)I_n E^n Bt}{2\pi(1-\nu^2)\alpha_n}\right]^{2/(n-1)} F_{cr}(\theta) \tag{7.1-29}$$

The angular function $F_{cr}(\theta)$ for plane strain and plane stress is shown in Figure 7.1. Alternatively, a boundary $r_1(\theta, t)$ can be defined by equating the effective stresses of the crack-tip HRR field [Equation [7.1-23)] and the remote elastic singular field. This yields

$$r_1(\theta, t) = \frac{1}{2\pi}\left(\frac{K}{E}\right)^2\left[\frac{(n+1)I_n E^n Bt}{2\pi(1-\nu^2)\alpha_n}\right]^{2/(n-1)} F_1(\theta) \tag{7.1-30}$$

where

$$F_1(\theta) = \left\{\frac{[(1-2\nu)^2 + 3\sin^2(\theta/2)]\cos^2(\theta/2)}{[\tilde{\sigma}(\theta)]^2}\right\}^{(n+1)/(n-1)} \tag{7.1-31}$$

for plane strain. Replace $1 - 2\nu$ and $1 - \nu^2$ by unity for plane stress. The function $F_1(\theta)$ is also shown in Figure 7.1. The difference between r_{cr} and r_1 is only in the angular functions. The plane stress boundary extends beyond the plane strain boundary. A third measure of the creep-zone boundary is given by setting $R = 1$, which yields a circular zone. In any case all three measures indicate that the creep zone has a self-similar shape that expands as $t^{2/(n-1)}$.

The characteristic time for transition from small-scale yielding to extensive creep can be estimated by equating the intensities of the HRR fields for short times [Equation (7.1-23)] and long times [Equation (7.1-18)]. For $\alpha_n = 1$ this transition time t_T is

$$t_T = \frac{J_\infty}{(n+1)C^*} \tag{7.1-32}$$

* The effective strain $\bar{\varepsilon} = (2e_{ij}e_{ij}/3)^{\frac{1}{2}}$, where e_{ij} are the deviatoric strain components.

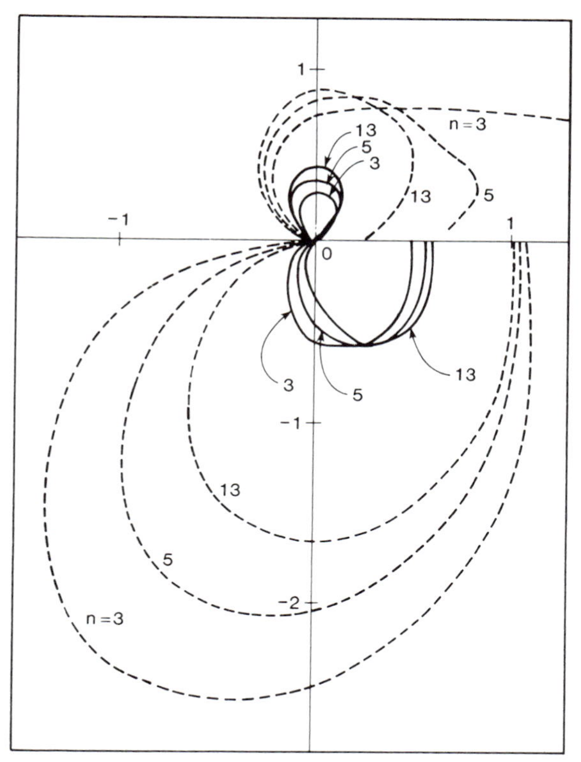

Figure 7.1 Polar diagrams for angular variations of $F_1(\theta)$ (dashed curves) and $F_{cr}(\theta)$ (solid curves) under conditions of plane strain (upper half) and plane stress (lower half) for $\nu = 0.3$ and $n = 3, 5, 13$ (7.4).

where J_∞ is given by Equation (7.1-27). For times less than this transition time, small-scale yielding can be expected whereas extensive creep is to be anticipated for greater times. With Equations (5.4-5) and (7.1-19) the transition time can be computed from

$$t_T = \frac{h_1(a/W, 1)}{EB(n+1)h_1(a/W, n)(P\sigma_y/P_0)^{n-1}} \tag{7.1-33}$$

when $h_1(a/W, n)$ can be found in Appendix A or estimated. Equations (7.1-30) and (7.1-32) or (7.1-33) can be used to estimate the size of the creep zone at the transition time. For a center cracked panel Riedel and Rice (7.4) found the creep zone size at the transition time to be approximately a tenth of the half crack length.

The approximate small-scale yielding solution can be compared with the plane strain finite element computations performed by Bassani and McClintock (7.10) for an edge crack in the elastic power law creep specimen illustrated in Figure 7.2. This configuration is meant to simulate an edge crack in a remotely loaded half space. At $t = 0$ a constant nominal stress $\sigma_{22} = \sigma_N$ is

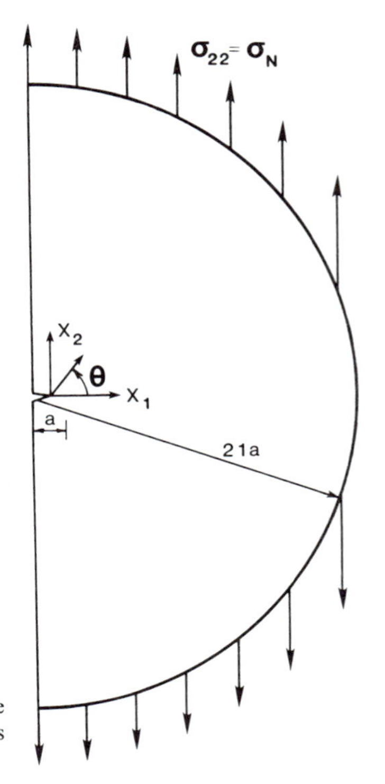

Figure 7.2 Model of plane strain edge crack under constant nominal stress $\sigma_{22} = \sigma_N$.

applied to the circular boundary. The material law specified by Equation (7.1-2) is used in these computations. In the comparisons $\dot{\varepsilon}_N^c = B\sigma_N^n$ represents the nominal creep strain rate and $t_N = \sigma_N/(EB\sigma_N^n)$ is the time for the creep strain to equal the elastic strain for a uniaxial stress σ_N. For the edge crack configuration $t_N \approx (n + 1)t_T$.

Figure 7.3(a) compares the finite element computations for $C(t)$ with the approximate small-scale yielding prediction of Equation (7.1-28) for $n = 3$. The transition time for this example corresponds to $t_T/t_N \approx 0.23$. The agreement is very good for times less than approximately one-half of the transition time t_T and the numerical results seem to reproduce the $1/t$ decay of $C(t)$ in Equation (7.1-28). For greater times the two solutions depart as the numerically computed $C(t)$ approaches the nonzero asymptote C^*. The numerical results are in agreement with the short- and long-term asymptotic analyses in that K is the relevant loading parameter for times small compared to the transition time; whereas, C^* is the appropriate loading parameter for greater times. For intermediate times there is a smooth transition from K- to C^*-controlled behavior or, equivalently, from small-scale yielding to extensive creep. Except near the transition time, an interpolation between the small-scale yielding and the extensive creep predictions provides an effective estimation. This figure also illustrates the rapid relaxation of the maximum effective stress $\bar{\sigma}_{max}$.

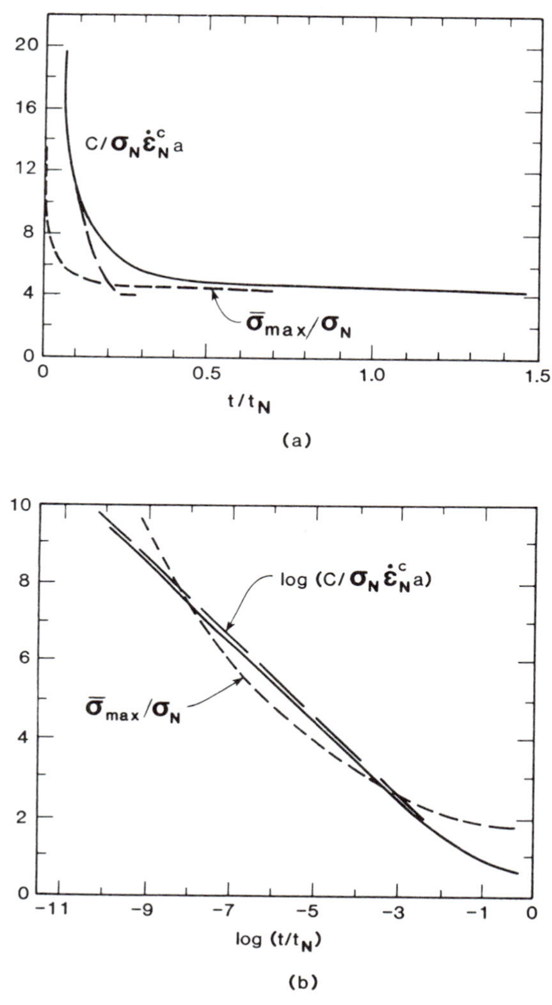

Figure 7.3 Comparison of $C(t)$ computed from finite element analysis (——) and Equation (7.1-28) (– –) for $\sigma_N/E = \frac{1}{2000}$, $\nu = 0.3$, (a) $n = 3$ and (b) $n = 10$ (7.10).

Figure 7.3(b) is a similar comparison for $n = 10$. Again very good agreement is observed between the finite element and the small-scale yielding predictions. Morjaria and Mukherjee (7.11) found that the asymptotic results correlated well with the numerical computations for the relaxation of the stress concentration in a center cracked panel.

The short-time creep zone boundaries for $n = 3$ and $n = 10$ are illustrated in Figure 7.4. These boundaries, which are reminiscent of the small-scale yielding zones for time-independent plasticity [e.g., see Shih (7.12)], expand approximately with a self-similar shape. They extend further ahead of the crack tip than the small-scale estimates depicted in Figure 7.1. The extent of

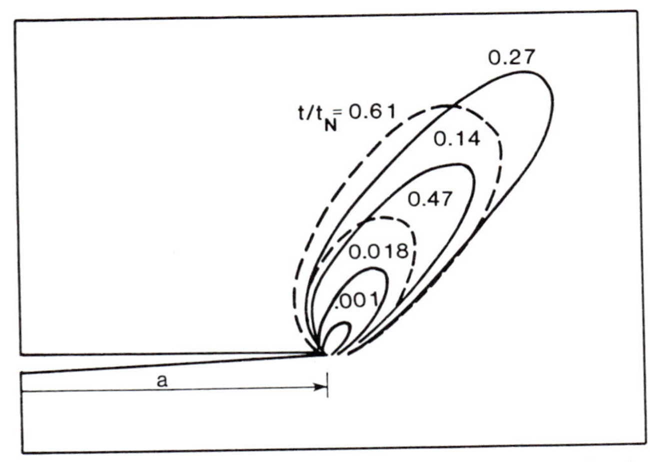

Figure 7.4 Short-time creep zones for $\sigma_N/E = \frac{1}{2000}$, $\nu = 0.3$ and $n = 3$ (--) and $n = 10$ (——) (7.10).

maximum creep occurs approximately along a 60 degree ray. For the same normalized time, t/t_N, the maximum extent of the zone increases with increasing n. The maximum extent of the creep zone at the transition time is still less than the crack length.

Ehlers and Riedel (7.13) have performed a finite element analysis of the transient creep in a compact tension specimen. Their numerical findings are very similar to those of Bassani and McClintock. The introduction of $C(t) = C^*(1 + t_T/t)$—which has the correct short- and long-time limits—into Equation (7.1-12) yields to a good degree of precision the near-tip stress field for all time. In general, the numerical results support the findings—particularly, with respect to the relevance of K and C^* as loading parameters—of the approximate small-scale yielding analysis based upon the assumed path-independence of the J-integral. Riedel (7.3) has performed similar small-scale yielding analyses to establish the appropriate loading parameters for other time-dependent material laws.

7.1.2 Elastic-Primary Creep

First consider the crack-tip fields for a body undergoing primary or tertiary creep. The material law for the creep component of the strain is assumed to have the Bailey-Norton form

$$\dot{\varepsilon}_{ij} = \tfrac{3}{2} B \bar{\sigma}^{n-1} \bar{\varepsilon}^{-p} s_{ij} \qquad (7.1\text{-}34)$$

where B, n, and p are material parameters and $\bar{\sigma}$ and $\bar{\varepsilon}$ are the effective stress and creep strain, respectively. According to Equation (7.1-34) the creep deformation is incompressible. Depending upon the value of the exponent p, Equation (7.1-34) describes strain hardening primary creep ($p > 0$), secondary creep ($p = 0$), or tertiary creep ($p < 0$). When $p = 2$, Andrade's uniaxial primary creep law ($\varepsilon \propto t^{\frac{1}{3}}$) is recovered.

For time-dependent proportional loading the homogeneous nature of the material law [Equation (7.1-34)], the equilibrium equations [Equation (7.1-4)] and the compatibility condition [Equation (7.1-5)] permits writing the stress field as

$$\sigma_{ij}(r, \theta, t) = P(t)\bar{\sigma}_{ij}(r, \theta) \tag{7.1-35}$$

The time-dependent loading parameter is $P(t)$ and $\bar{\sigma}_{ij}$ is a time-independent function of the coordinates r and θ only. For proportional loading the time integration of Equation (7.1-34) yields the creep stress-strain relation

$$\varepsilon_{ij} = \frac{3}{2}\left\{\frac{B(1+p)\bar{\sigma}^n}{[P(t)]^n}\int_0^t [P(\tau)]^n\, d\tau\right\}^{1/(1+p)} \frac{s_{ij}}{\bar{\sigma}}, \qquad p > -1 \tag{7.1-36}$$

When a unique stress-strain relationship, such as Equation (7.1-36), is independent of the spatial coordinates, then the J-integral is path-independent. The time-dependence of J and $\varepsilon_{ij}\sigma_{ij}$ are identical. For example, $J \propto t^{1/(1+p)}$ for a constant applied load. It is convenient to introduce a path-independent integral that is also time-independent for a constant loading. Since the J-integral multiplied by an arbitrary function of time remains path-independent, then

$$C_h^* = J(t)\left\{[P(t)]^n \middle/ \int_0^t [P(\tau)]^n\, d\tau\right\}^{1/(1+p)} \tag{7.1-37}$$

is also path-independent. The C_h^*-integral depends upon the current load as $P^{1+n/(1+p)}$ and is independent of the proportional loading history. In the limiting case of $p = 0$, C_h^* and C^* are identical. However, there is a fundamental difference between C^* and C_h^* for $p \neq 0$. The path independence of C_h^* demands that the loading be proportional. This is not the case for the C^*-integral, which is independent of the loading path followed to the current state. There may be other effects—for example, elastic unloading—that would also limit its applicability.

The C_h^*-integral offers the potential for measurement by using Equation (7.1-37) and

$$J = -\left(\frac{\partial U}{\partial a}\right)_{\Delta, t} \tag{7.1-38}$$

where

$$U = \int_0^{\Delta} P\, d\Delta \tag{7.1-39}$$

is the work done by the applied load P per unit thickness acting through the conjugate load-point displacement Δ. The derivative, $(\partial U/\partial a)_{\Delta,t}$, can be formed by measuring the work done by the same loading history for two slightly different crack lengths in otherwise identical specimens.

The stress-strain law of Equation (7.1-36) has the form of that for a power law hardening plastic material (nonlinear elastic material) [cf. Equation

(5.3-7)] but with a time-dependent coefficient. Consequently, Equation (7.1-37) and the fully plastic analyses for J can be used to determine C_h^*. For a constant load it is only necessary to replace $\alpha\varepsilon_y/\sigma_y^n$ by $[B(1+p)t]^{1/(1+p)}$ in Equation (5.4-5) for J and Equation (7.1-37) becomes

$$C_h^* = [B(1+p)]^{1/(1+p)} b g_1\left(\frac{a}{W}\right) h_1\left(\frac{a}{W}, m\right)\left(\frac{P\sigma_y}{P_0}\right)^{m+1} \tag{7.1-40}$$

where $m = n/(1+p)$ replaces the plastic strain hardening index n. Tabulated values of $g_1(a/W)$ and $h_1(a/W, m)$ are available for a number of specimens in Appendix A.

Extending this analogy an additional step one can write

$$\sigma_{ij} = \left\{\frac{C_h^*(t)}{[B(1+p)]^{1/(1+p)} I_m r}\right\}^{1/(m+1)} \tilde{\sigma}_{ij}(\theta, m), \qquad \mathrm{r} \to 0 \tag{7.1-41}$$

for the HRR stress field under extensive primary creep and proportional loading. The angular functions $\tilde{\sigma}_{ij}(\theta, m)$ are normalized such that the maximum value of $\tilde{\sigma}(\theta, m)$ for a fixed m is unity. The primary creep stress field has a $r^{-1/(m+1)}$ singularity at the crack tip. It is also apparent from Equation (7.1-41) that C_h^* is an appropriate loading parameter for extensive primary creep.

Because creep deformation takes time to develop, elastic deformation can be significant for short times after a sudden application of the load. When the elastic strain rates are added to the creep rates of Equation (7.1-34), the isotropic material law becomes

$$\dot{\varepsilon}_{ij} = \frac{1+\nu}{E}\dot{s}_{ij} + \frac{1-2\nu}{3E}\dot{\sigma}_{kk}\delta_{ij} + \tfrac{3}{2}B\bar{\sigma}^{n-1}\bar{\varepsilon}_{cr}^{-p}s_{ij} \tag{7.1-42}$$

where $\bar{\varepsilon}_{cr}$ is the effective creep strain.

For a suddenly applied constant load at $t = 0$, an approximate small-scale yielding anlysis in the spirit of that performed for secondary creep will be considered. The remote boundary condition for small-scale yielding or, equivalently, for short times is the asymptotic approach of the stress field to the elastic singular field. Mode I self-similar solutions that satisfy this condition, the constitutive relation Equation (7.1-42), equilibrium and compatibility equations are

$$\sigma_{ij} = \frac{E}{1-\nu} T^{-1/(n-p-1)} \Sigma_{ij}(R, \theta) \tag{7.1-43}$$

The dimensionless radial coordinate R and dimensionless time T are (7.3)

$$R = \frac{2\pi r T^{-2/(n-p-1)}}{[(1-\nu)K/E]^2} \tag{7.1-44}$$

and

$$T = \frac{n-p-1}{2}\left(\frac{E}{1-\nu}\right)^n Bt \tag{7.1-45}$$

for plane strain. For plane stress replace $1 - \nu$ by unity. Equations (7.1-43)–(7.1-45) can be compared to Equations (7.1-20)–(7.1-22) for secondary creep.

Near the crack tip the creep strain rates will dominate the elastic rates if $m = n/(1 + p) > 1$. According to Equation (7.1-41) the dominant stress field has a $r^{-1/(m+1)}$ singularity and, therefore, Σ_{ij} must have a singularity of the same order in R. This deduction and Equations (7.1-43)–(7.1-45) permit writing

$$\sigma_{ij} = \alpha(n,p)\left\{\frac{(1-\nu^2)K^2/E}{[B(n+p+1)t]^{1/(p+1)}I_m r}\right\}^{1/(m+1)} \tilde{\sigma}_{ij}(\theta, m) \quad (7.1\text{-}46)$$

as $r \to 0$ for plane strain and $1 - \nu^2$ replaced by unity for plane stress. For $p = 0$ this equation reduces to Equation (7.1-23). The amplitude $\alpha(n, p)$ can be determined approximately by assuming as for secondary creep that the J-integral is path-independent. This approximation leads to $\alpha(n, p) = 1$. It is clear from Equation (7.1-46) that the strength of the near tip fields under conditions of small-scale yielding is governed by the stress intensity factor. Hence, for short times K is the appropriate crack-tip loading parameter. With $R = 1$ as a measure of the characteristic size of the creep zone, Equation (7.1-44) predicts a self-similar growth of the zone with time as $t^{2/(n-p-1)}$.

A time will come when the size of the creep zone will no longer be small compared to the relevant dimensions of specimen and small-scale yielding is inapplicable. After a longer period of time the creep strains will extend throughout the specimen. The near-tip stress field will be described by Equation (7.1-41) and C_h^* is the appropriate loading parameter. A characteristic time for transition from small-scale yielding to extensive creeping under constant load is determined by equating the stress fields of Equations (7.1-41) and (7.1-46). This transition time is

$$t_1 = \frac{1}{m+1}\left(\frac{J_\infty}{C_h^*}\right)^{p+1} \quad (7.1\text{-}47)$$

It is supposed that J_∞ is related to K through Equation (7.1-27) and that C_h^* is attainable from a fully plastic analysis via Equation (7.1-40). In this case Equation (7.1-47) assumes the form

$$t_1 = \frac{1}{m+1}\left\{\frac{h_1(a/W, 1)}{E[B(1+p)]^{1/(p+1)}h_1(a/W, m)(P\sigma_y/P_0)^{m-1}}\right\}^{p+1} \quad (7.1\text{-}48)$$

which is convenient to use with fracture handbook values for $h_1(a/W, n)$.

7.1.3 *Primary-Secondary Creep*

Following the development of extensive primary creep in the body, a small secondary creep zone develops near the crack tip and grows. Typically, the elastic strain rates will be smaller than the creep rates and may be neglected. Under such conditions the material may be modeled by the constitutive relation

$$\dot{\varepsilon}_{ij} = \tfrac{3}{2}B_1\bar{\sigma}^{n_1-1}\bar{\varepsilon}^{-p}s_{ij} + \tfrac{3}{2}B\bar{\sigma}^{n-1}s_{ij} \quad (7.1\text{-}49)$$

where the right-hand side is the sum of the primary and secondary strain rates. In this case the primary creep stress field given by Equation (7.1-41) sets the remote boundary condition for the small-scale yielding solution.

Provided $p > 0$ and $n > n_1/(1 + p)$ Riedel (7.3) found the stress field has the self-similar form

$$\sigma_{ij} = \left(\frac{B_1}{B}\right)^{1/(n-n_1)} T^{-p/[n(1+p)-n_1]} \Sigma_{ij}(R, \theta) \tag{7.1-50}$$

where now

$$R = \frac{rI_m[B_1(1 + p)]^{1/(p+1)}}{C_h^* T^{p(m+1)/[n(1+p)-n_1]}} \left(\frac{B_1}{B}\right)^{(m+1)/(n-n_1)} \tag{7.1-51}$$

$$T = \frac{n(1 + p) - n_1}{p(m + 1)} \left(\frac{B_1}{B}\right)^{n_1/(n-n_1)} B_1 t \tag{7.1-52}$$

and $m = n_1/(1 + p)$. In the secondary creep zone near the crack tip the stress field is known to have a $r^{-1/(n+1)}$ singularity. Hence, for $r \to 0$ Equation (7.1-50) assumes the asymptotic form

$$\sigma_{ij} = \alpha(n, n_1, p) \left\{\frac{n + p + 1}{(1 + p)(1 + n)} \frac{C_h^*}{BI_n r t^{p/(1+p)}}\right\}^{1/(n+1)} \tilde{\sigma}_{ij}(\theta, n) \tag{7.1-53}$$

Once again $\alpha(n, n_1, p) = 1$ if path-independence of J is assumed.

As long as the secondary creep zone is small compared to the crack length and relevant dimensions of the specimen, the intensity of the primary creep field, C_h^*, is the appropriate loading parameter. Setting $R = 1$ as a measure of the characteristic size of the secondary creep zone, one concludes that this zone grows as $t^{p(m+1)/[n(1+p)-n_1]}$. For times greater than the transition time

$$t_2 = \left\{\frac{n + p + 1}{(1 + p)(n + 1)} \frac{C_h^*}{C^*}\right\}^{1+1/p} \tag{7.1-54}$$

obtained by equating Equations (7.1-18) and (7.1-53), extensive secondary creep of the body occurs and C^* becomes the relevant loading parameter.

It is now possible to describe the sequence of events when a constant load is suddenly applied to a body with a stationary crack. If the instantaneous response of the cracked body is essentially elastic except for possibly a small plastic zone at the crack tip, the stress intensity factor K governs the growth of the ensuing creep zone and the strength of the HRR field while the creep zone is small compared to the crack length and other relevant dimensions. If the material has a significant primary creep regime, extensive primary creep will occur for times greater than the transition time t_1 [Equation (7.1-47)]. The C_h^*-integral then controls the growth of the secondary creep zone and the amplitude of the near-tip fields. This secondary creep zone continues growing until after the transition time t_2 [Equation (7.1-54)] when it has virtually superceded the primary creep throughout the body. The C^*-integral then becomes the appropriate loading parameter. If the primary creep regime of

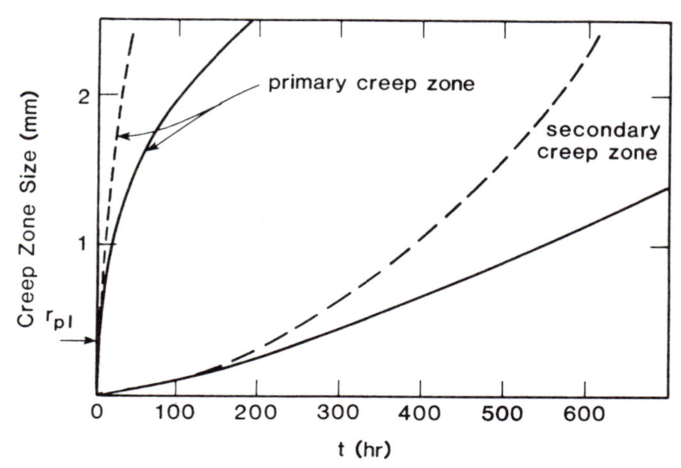

Figure 7.5 Growth of primary and secondary creep zones in a center-cracked panel (7.3). ($\sigma_\infty = 80$ MPa, $E = 140$ GPa, $\sigma_y = 320$ MPa, $B_1 = 2.5 \times 10^{-29}$ hr^{-1} MPa^{-9}, $B = 5 \times 10^{-16}$ hr^{-1} MPa^{-5}, $n_1 = 9$, $p = 2$, $n = 5$, $t_1 = 0.17$ hr, $t_2 = 102$ hr)

the material is insignificant or if the secondary creep zone overtakes the primary creep zone, then the C_h^*-controlled period is inconsequential. Instead, a transition from K- to C^*-dominated behavior will occur at the transition time t_T [Equation (7.1-32)].

Figure 7.5 illustrates for a hypothetical material the development of the plane strain primary and secondary creep zones in a center-cracked panel subjected to a remotely applied stress $\sigma_{22} = \sigma_\infty$. The size r_{p1} of the instantaneous plastic zone is small compared to the half crack length $a = 10$ mm and ligament length $b = 10$ mm. The solid curves represent the small-scale yielding estimates for the creep zone size obtained from setting $R = 1$ in Equation (7.1-44) and (7.1-51). The dashed curves represent rough estimates presented by Riedel (7.3) for the creep zone size outside the small-scale yielding range. It is apparent that after the transition time t_1 the accuracy of the small-scale yielding approximation for the size of the primary creep zone diminishes rapidly. Shortly after the time t_1 the primary creep zone has engulfed the remaining ligament and C_h^* commences to govern the near-tip fields. For times greater than t_2 the secondary creep zone extends over the entire ligament and C^* is the controlling load parameter.

7.1.4 Plastic-Primary Creep

It may happen that a suddenly applied load to a cracked body is sufficient to produce an instantaneous fully plastic state rather than an elastic one. In this case the creep zone will begin growing not in an elastic region but in a plastic one. As long as the creep zone is small compared to the relevant dimensions, the J-integral can be expected to control the near tip fields and the development of the creep zone. The material law

$$\dot{\varepsilon}_{ij} = \tfrac{3}{2} B_0 \bar{\sigma}^{1/N - 2} \dot{\bar{\sigma}} s_{ij} + \tfrac{3}{2} B \bar{\sigma}^{n-1} \bar{\varepsilon}_{cr}^{-p} s_{ij} \tag{7.1-55}$$

models instantaneous power law hardening plasticity with superimposed creep. The coefficient B_0 can be expressed in terms of the yield stress σ_y as $\alpha\sigma_y^{1-1/N}/E$.

Riedel (7.3) found that sufficiently strong strain-hardening, $N > (1 + p)/n_1$, is required for the existence of a small-scale creep zone in a fully plastic specimen. When such a condition exists, the self-similar stress field has the form

$$\sigma_{ij} = B_0^{-N} T^{-N(nN-1-p)} \Sigma_{ij}(R, \theta) \tag{7.1-56}$$

where

$$R = \frac{I_{1/N} r}{B_0^N J T^{(N+1)/(nN-1-p)}} \tag{7.1-57}$$

$$T = \frac{nN - 1 - p}{N + 1} \frac{Bt}{B_0^{nN}} \tag{7.1-58}$$

Since the primary creep stress field has a $r^{-1/(m+1)}$ singularity, the combination of Equations (7.1-56)–(7.1-58) provides the creep dominated crack-tip field

$$\sigma_{ij} = \alpha(n, p, N) \left\{ \frac{J}{[(n + p + 1)Bt]^{1/(p+1)} I_m r} \right\}^{1/(m+1)} \tilde{\sigma}_{ij}(\theta, m) \tag{7.1-59}$$

where $m = n/(1 + p)$. The assumption of approximate path-independence of J leads to $\alpha(n, p, N) = 1$.

Equating Equations (7.1-41) and (7.1-59) one obtains the transition time

$$t_3 = \frac{1 + p}{n + p + 1} \left(\frac{J}{C_h^*} \right)^{p+1} \tag{7.1-60}$$

at which extensive creep replaces the initial plastic state.

When the step loading of the cracked body produces a fully plastic state, the J-integral determines the strength of the near-tip fields and the initial development of the creep zone for $0 < t < t_3$. During this period the creep zone grows as $t^{(N+1)/(nN-1-p)}$ [set $R = 1$ in Equation (7.1-57)]. When $t > t_3$, primary creep displaces the fully plastic state and C_h^* becomes the appropriate loading parameter. For even greater times $(t > t_2)$ secondary creep predominates and C^* is the relevant loading parameter. In materials with a weak primary creep regime, a transition from J- to C^*-controlled crack-tip fields occurs for $t > t_3$ with $p = 0$ and $C_h^* = C^*$.

A criterion for crack extension necessary to predict the initiation time for crack growth is lacking. Riedel and Rice (7.4) have proposed that crack extension will occur when the effective creep strain at a small structural distance, r_c, from the crack tip attains its critical value. Both the critical value of the effective strain and the structural distance are considered material properties to be determined from experiments. Direct measurements of these quantities present formidable problems. While such a criterion may yield the proper qualitative behavior, it will likely yield unreliable quantitative predictions because the fracture process is more complicated than this simple criterion implies. Further research on the initiation of creep crack growth is necessary.

7.1.5 Elastic-Exponential Law Creep

The asymptotic fields for a stationary crack under Mode III loading have been investigated by Bassani (7.14, 7.15) for the uniaxial material law

$$\dot{\varepsilon} = \frac{\dot{\sigma}}{E} + \dot{\varepsilon}_0\left[\sinh\left(\frac{\sigma}{\sigma_0}\right)\right]^n \tag{7.1-61}$$

where $\dot{\varepsilon}_0$ and σ_0 are normalizing parameters. At low stress levels the creep strain rate according to Equation (7.1-61) obeys a power law, $\dot{\varepsilon}_c = \dot{\varepsilon}(\sigma/\sigma_0)^n$, whereas at high stress levels it follows the exponential law

$$\dot{\varepsilon}_c = \dot{\varepsilon}_0 \exp\left(\frac{\sigma}{\sigma_0}\right) \tag{7.1-62}$$

where 2^{-n} and n have been absorbed by $\dot{\varepsilon}_0$ and σ_0, respectively. For high stress levels Bassani used the technique, employed by Hult and McClintock (7.16), and Rice (7.17), for the Mode III analysis of a crack in an elastic-plastic material (see Section 5.2), whereby the nonlinear equations are reduced to linear equations by taking the coordinates x_1 and x_2 to be functions of the stress components.

The elastic strain rates can be neglected compared to the creep rates near the crack tip. As $r \to 0$, the asymptotic stress field is expressed parametrically by

$$\begin{aligned} x_1 &= \frac{A(\sigma_{32}/\bar{\sigma})^2}{(\bar{\sigma}/\sigma_0)\exp(\bar{\sigma}/\sigma_0)} \\ x_2 &= \frac{A\sigma_{31}\sigma_{32}/\bar{\sigma}^2}{(\bar{\sigma}/\sigma_0)\exp(\bar{\sigma}/\sigma_0)} \end{aligned} \tag{7.1-63}$$

where A is an undetermined amplitude. The latter equations can be combined to yield

$$\left(\frac{\bar{\sigma}}{\sigma_0}\right)\exp\left(\frac{\bar{\sigma}}{\sigma_0}\right) = \frac{A\cos\theta}{r} \tag{7.1-64}$$

Because the left-hand side of Equation (7.1-64) is positive, the asymptotic behavior is limited to $-\pi/2 < \theta < \pi/2$. The loci of constant effective stress are circles centered on $x_2 = 0$ ahead of the crack tip at distances equal to their radii. Note that these contours do not encompass the crack tip. The contours of constant effective stress resemble the small-scale yielding elastic-perfectly plastic boundary for a stationary crack tip (cf. Figure 5.9 for $n \to \infty$).

For the creep law, Equation (7.1-62), it follows that

$$\left(\frac{\bar{\sigma}}{\sigma_0}\right)\left(\frac{\dot{\bar{\varepsilon}}_c}{\dot{\varepsilon}_0}\right) = \frac{A\cos\theta}{r} \tag{7.1-65}$$

Letting $\xi = (A\cos\theta)/r$ and taking the logarithm of Equation (7.1-64), one can write

$$\frac{\bar{\sigma}}{\sigma_0} = \ln\xi - \ln\left(\frac{\bar{\sigma}}{\sigma_0}\right) \tag{7.1-66}$$

Successive replacement of $\bar{\sigma}/\sigma_0$ on the right-hand side of Equation (7.1-66) leads to

$$\frac{\bar{\sigma}}{\sigma_0} = \ln\{\xi\ln[\xi\ln(\ldots)]\}$$

from which the leading term as $r \to 0$ is

$$\frac{\bar{\sigma}}{\sigma_0} = \ln\left[\frac{A\cos\theta}{r}\right] \tag{7.1-67}$$

The introduction of Equation (7.1-67) into Equation (7.1-65) yields as $r \to 0$

$$\frac{\dot{\bar{\varepsilon}}_c}{\dot{\varepsilon}_0} = \frac{A\cos\theta}{r\ln[(A\cos\theta)/r]} \tag{7.1-68}$$

The stress singularity of Equation (7.1-67) is weaker than the $r^{-1/(n+1)}$ singularity for the power law creeping material. On the other hand, the strain rate singularity of Equation (7.1-68) is stronger than the $r^{-n/(n+1)}$ singularity of power law creep. Qualitatively, these singularities fall between the large n limit of power law creep and the nonhardening limit of time-independent plasticity (7.18). While analyses have not been performed yet for Mode I and Mode II loadings, the same kind of singularity is expected. This is certainly known to be the case for power law creep. Simiarly, the elastic stress intensity factor and the C^*-integral are the relevant loading parameters for short-time (small-scale yielding) and long-time (extensive creep), respectively.

In the foregoing treatments the fracture process zone has been assumed to be negligibly small compared to the crack-tip creep zone. However, the size of the process zone in a small specimen of a ductile material can be comparable to the length of the remaining ligament. Under such circumstances, the asymptotic analysis has no physical relevance. The stress and strain distributions are likely to be more uniform in the ligament than predicted by such an analysis. In this case the net section stress or the reference stress would likely be the loading parameter that governs the life of the specimen. More will be said about the reference stress in Section 7.3.

7.1.6 The ΔT_k-Integral

Atluri (7.19) has developed a path-independent integral ΔT_k that is amenable to numerical solution using the finite element method. Because it includes elasto-viscoplastic material behavior as a special case, it is informative to examine its relationship to other integrals introduced in this section.

In its most general form ΔT_k includes finite deformation, material acceleration, and arbitrary traction and displacement conditions on the crack faces. In the absence of these complicating factors, ΔT_k under quasi-static loading reduces for small strains and deformations to

$$\Delta T_k = \lim_{\Gamma_\varepsilon \to 0} \int_{\Gamma_\varepsilon} [\Delta W n_k - (\sigma_{ij} + \Delta\sigma_{ij}) n_j \Delta u_{i,k}]\, ds \tag{7.1-69}$$

where the incremental internal work density ΔW in the time increment Δt is given by

$$\Delta W = \left(\sigma_{ij} + \tfrac{1}{2}\Delta\sigma_{ij}\right)\Delta u_{i,j} \tag{7.1-70}$$

The incremental components of the stress tensor and displacement vector in this interval are $\Delta\sigma_{ij}$ and Δu_i, respectively. The vanishingly small contour Γ_ε encloses the crack tip. The retention of the second-order terms in Equations (7.1-69) and (7.1-70) can offer computational advantages.

The following physical interpretation of ΔT_k has been given by Atluri (7.19). The increment ΔE of the total energy in the interval Δt is the sum of the incremental internal work and the incremental kinetic energy—the latter being zero for the problem under consideration—less the incremental work done by the external loading. Consider two bodies that are otherwise identical except that the crack in the second body has been extended infinitesimally by da_i. The loading history of both bodies is the same. It can be shown that

$$\Delta T_i da_i = -(\Delta E_2 - \Delta E_1) \tag{7.1-71}$$

where the subscripts 1 and 2 are used to identify energy increments belonging to the first and second body, respectively. Hence, $\Delta T_i da_i$ represents the decrease in the incremental total energy of two bodies that are identical except for an infinitesimal crack length difference da_i. Therefore, ΔT_k can be given an incremental energy release rate interpretation.

For self-similar crack extension the relevant component of ΔT_k is

$$\begin{aligned}\Delta T_1 &= \lim_{\Gamma_\varepsilon \to 0} \int_{\Gamma_\varepsilon} \Delta W\, dx_2 - (\sigma_{ij} + \Delta\sigma_{ij}) n_j \frac{\partial \Delta u_i}{\partial x_1}\, ds \\ &= \int_{\Gamma} \Delta W\, dx_2 - (\sigma_{ij} + \Delta\sigma_{ij}) n_j \frac{\partial \Delta u_i}{\partial x_1}\, ds - \int_A \Delta u_{i,j} \frac{\partial \sigma_{ij}}{\partial x_1}\, dA \end{aligned} \tag{7.1-72}$$

where A is the area enclosed by a contour Γ surrounding the crack tip and no other singularity. The latter representation for ΔT_1 follows from the former after the application of the divergence theorem for the region bounded by Γ_ε, Γ and the crack faces. Equations (7.1-69) and (7.1-72) are valid under steady as well as nonsteady creep conditions and, also, in the presence of elastic strains.

Dividing Equation (7.1-72) by Δt and proceeding to the limit as $\Delta t \to 0$, one finds that

$$\dot{T}_1 = \lim_{\Gamma_\varepsilon \to 0} \int_{\Gamma_\varepsilon} \dot{W}\,dx_2 - \sigma_{ij} n_j \frac{\partial \dot{u}_i}{\partial x_1}\,ds$$

$$= \int_{\Gamma} \dot{W}\,dx_2 - \sigma_{ij} n_j \frac{\partial \dot{u}_i}{\partial x_1}\,ds - \int_A \dot{u}_{i,j} \frac{\partial \sigma_{ij}}{\partial x_1}\,dA \qquad (7.1\text{-}73)$$

where

$$\dot{W} = \sigma_{ij} \dot{\varepsilon}_{ij} \qquad (7.1\text{-}74)$$

is the rate of internal-work (stress power) density.

When the first representation for $\dot{T}_1$ is compared with Equation (7.1-13) for $C(t)$ for a power law creeping material, it follows that

$$\dot{T}_1 = C(t) + \frac{1}{n+1} \lim_{\Gamma_\varepsilon \to 0} \int_{\Gamma_\varepsilon} \sigma_{ij} \dot{\varepsilon}_{ij}\,dx_2 \qquad (7.1\text{-}75)$$

Hence, $\dot{T}_1$ characterizes the strength of the crack-tip fields.

Under conditions of steady secondary creep, Equation (7.1-75) reduces to

$$\dot{T}_1 = C^* + \frac{B}{n+1} \lim_{\Gamma_\varepsilon \to 0} \int_{\Gamma_\varepsilon} \bar{\sigma}^{n+1}\,dx_2 \qquad (7.1\text{-}76)$$

Equivalently,

$$\frac{\dot{T}_1}{C^*} = 1 + \frac{1}{(n+1)I_n} \int_{-\pi}^{\pi} \left[\frac{\tilde{\sigma}(\theta)}{\sigma_y} \right]^{n+1} \cos\theta\,d\theta \qquad (7.1\text{-}77)$$

where $\tilde{\sigma}(\theta)/\sigma_y$ of the HRR field is given in Figures 5.12 and 5.13 for plane strain and plane stress, respectively. Stonesifer and Atluri (7.20) found that $\dot{T}_1$ and C^* differ by less than 2 percent for plane strain and less than 15 percent for plane stress.

It is clear that $C(t)$ and C^* are included in $\dot{T}_1$ as special cases. Furthermore, since ΔT_k also includes elastic behavior, it has the attractive feature that it can characterize the crack-tip fields from very localized creep behavior associated with short times after loading to steady-state creep attained after very long times. Hence, ΔT_k has the potential for rather wide applications as a crack-tip characterizing parameter. As Stonesifer and Atluri have demonstrated, it can be readily incorporated into a finite element method of analysis. Because ΔT_k has an energy release rate interpretation, there exists the potential to extract it from measurements similar to those originally performed by Landes and Begley (7.8) for experimental determination of C^*.

7.2 Creep Crack Growth

Creep fracture involves an incubation period followed by a period of crack growth that left unchecked can result in a loss of structural integrity. During the incubation period creep deformation develops in the creep zone emanating from the crack tip until sufficient accumulative damage has occurred to produce crack growth. Crack-tip stress and strain fields and approximate small-scale yielding solutions for this period were presented in Section 7.1 for a number of material laws. The far more physically important and complicated problem of a quasi-statically growing crack in a creeping material will be addressed in this section. The asymptotic stress and strain fields for a crack growing in an elastic-power-law creeping material have been determined by Hui and Riedel (7.21). A small-scale yielding analysis of steady-state crack growth in this material has been performed by Hui (7.22). Approximate analyses of nonsteady growth have been conducted by Riedel and Wagner (7.23).

7.2.1 Elastic-Secondary Creep Crack Fields

The plane problem of symmetric (Mode I) loading of a crack extending in an elastic-power law creeping material with the constitutive relation of Equation (7.1-2) will be investigated. Again, it is expedient to use the Airy stress function Ψ to express the in-plane stress components. The Cartesian components are given by Equation (7.1-16), whereas in polar coordinates

$$\begin{aligned}\sigma_{rr} &= \frac{1}{r^2}\frac{\partial^2\Psi}{\partial\theta^2} + \frac{1}{r}\frac{\partial\Psi}{\partial r}\\ \sigma_{\theta\theta} &= \frac{\partial^2\Psi}{\partial r^2}\\ \sigma_{r\theta} &= -\frac{\partial^2(\Psi/r)}{\partial r\,\partial\theta}\end{aligned} \tag{7.2-1}$$

The plane stress and plane strain nonlinear equations governing the stress functions are given by Equations (7.1-7)–(7.1-11).

For a Cartesian coordinate system attached to the crack tip the total time derivative of the stress function is

$$\dot{\Psi} = -\dot{a}(t)\frac{\partial\Psi}{\partial x_1} + \frac{\partial\Psi}{\partial t} \tag{7.2-2}$$

Under steady-state crack growth the deformation is time-independent when viewed by an observer fixed to the moving crack tip. Hence, Equation (7.2-2) reduces to

$$\dot{\Psi} = -\dot{a}\frac{\partial\Psi}{\partial x_1} \tag{7.2-3}$$

Since

$$\frac{\partial}{\partial x_1} = \cos\theta\frac{\partial}{\partial r} - \sin\theta\frac{1}{r}\frac{\partial}{\partial\theta} \tag{7.2-4}$$

the first term on the right-hand side of Equation (7.2-2) will give rise for $\Psi \propto r^s$

to a singularity as $r \to 0$ that is one order greater than that due to the second term. As a consequence, the asymptotic stress and strain fields as $r \to 0$ are identical for steady and nonsteady crack growth. While the following asymptotic analysis is for steady crack growth, it is also applicable to nonsteady growth unless explicitly stated otherwise.

For plane stress Equations (7.1-7), (7.1-8), and (7.2-3) combine to yield

$$\frac{2\dot{a}}{E}\nabla^4\left(\frac{\partial \Psi}{\partial x_1}\right) + B[(\Psi_{,\gamma\gamma}\delta_{\alpha\beta} - 3\Psi_{,\alpha\beta})(\tfrac{3}{2}\Psi_{,\delta\rho}\Psi_{,\delta\rho} - \tfrac{1}{2}\Psi_{,\delta\delta}\Psi_{,\rho\rho})^{(n-1)/2}]_{,\alpha\beta} = 0 \tag{7.2-5}$$

The solution to Equation (7.2-5) must satisfy the boundary condition that the crack faces be traction-free. It follows with Equation (7.2-1) that this condition in polar coordinates requires

$$\frac{\partial^2 \Psi(r, \pm\pi)}{\partial r^2} = 0, \qquad \frac{\partial \Psi(r, \pm\pi)}{\partial \theta} = 0 \tag{7.2-6}$$

For a stationary crack tip the elastic strain is negligible compared to the creep strain near the tip. Hence, as $r \to 0$ the nonlinear creep terms not only dominate the linear (elastic) terms in Equation (7.1-7), but also the contribution to the crack-tip stress and strain fields. The stress singularity is of the HRR type; that is, $\sigma \propto r^{-1/(n+1)}$ or, equivalently, $\Psi \propto r^{(2n+1)/(n+1)}$ as $r \to 0$. On the other hand, the elastic strain rate can not be neglected for a growing crack. To do so would imply that a HRR singularity must exist at the crack tip. The introduction of $\Psi \propto r^{(2n+1)/(n+1)}$ into Equation (7.2-5) leads to a linear elastic contribution of the order $(\dot{a}/E)r^{-(3n+4)/(n+1)}$ compared to a nonlinear creep contribution of the order $Br^{-(3n+2)/(n+1)}$. The higher-order singularity of the elastic term contradicts the original supposition and, therefore, the asymptotic field of a growing crack cannot be of the HRR type.

Suppose that the linear term of Equation (7.2-5) dominates the asymptotic behavior. In this case a linear elastic stress singularity would occur with $\Psi \propto r^{\frac{3}{2}}$. The linear contribution to Equation (7.2-5) would be of the order $(\dot{a}/E)r^{-\frac{7}{2}}$ compared to the nonlinear contribution of the order $Br^{-(n+4)/2}$ as $r \to 0$. If $n < 3$, then the assumption that the elastic strain rate dominates the creep rate at the crack tip is valid. For $n > 3$, the nonlinear term has a stronger singularity and the assumption is contradicted. Consequently, the linear (elastic) and the nonlinear (creep) terms in Equation (7.2-5) play equally important roles in determining the asymptotic character of the near tip fields when $n > 3$.

For $n < 3$ the asymptotic solution of Equation (7.2-5)—which satisfies the boundary conditions, Equation (7.2-6)—is

$$\Psi = Ar^{\frac{3}{2}}\left[\cos\left(\frac{\theta}{2}\right) + \tfrac{1}{3}\cos\left(\frac{3\theta}{2}\right)\right] \tag{7.2-7}$$

The factor A cannot be determined from the asymptotic analysis but is determined from the complete solution to the problem. It depends upon the applied loading and the crack growth history. When small-scale yielding conditions exist, the elastic strain will be dominant in the far field. However, for $n < 3$ the creep strain rate varies as $\dot{\varepsilon}_{cr} \propto r^{-n/2}$ and the creep strain for steady-state crack growth varies as $\varepsilon_{cr} \propto r^{-(n-2)/2}$. In the remote field ($r \to \infty$) the creep strain field will dominate the elastic strain field which varies as $\varepsilon_e \propto r^{-1/2}$. But, this is contradictory to the assumed small-scale yielding behavior. Therefore, steady-state crack growth cannot exist in an elastic-power-law creeping material when $n < 3$. Hart (7.24) has shown that steady-state crack extension is possible if the creep exponent depends upon the effective stress such that $n > 3$ as $\bar{\sigma} \to 0$, but $n < 3$ as $\bar{\sigma} \to \infty$.

For $n > 3$ the stress function can be written as

$$\Psi = \left[\frac{\dot{a}(t)}{EB}\right]^{1/(n-1)} r^s f(\theta, n) \tag{7.2-8}$$

near the crack tip. Equality of the order of the linear and nonlinear singularities in Equation (7.2-5) requires that

$$s = (2n - 3)/(n - 1) \tag{7.2-9}$$

The introduction of Equation (7.2-8) into Equation (7.2-5) leads to the nonlinear differential equation (7.10)

$$2L(f) + n(2 - s)[q_2 + (ns - 2n + 1)q_1] + q_2'' + 6(ns - 2n + 1)q_3' = 0 \tag{7.2-10}$$

for the dimensionless function $f(\theta, n)$. In Equation (7.2-10) a prime is used to denote a differentiation with respect to θ and

$$\begin{aligned} L(f) &= (4 - s)k(\theta)\cos\theta + k'(\theta)\sin\theta \\ k(\theta) &= f^{iv} + 2(s^2 - 2s + 2)f'' + s^2(s - 2)^2 f \\ p(\theta) &= \{[f'' + s(3 - s)f/2]^2 + 3(s - 1)^2[f'^2 + (sf/2)^2]\}^{(n-1)/2} \\ q_1(\theta) &= p(\theta)[f'' + s(3 - 2s)f] \\ q_2(\theta) &= p(\theta)[2f'' + s(3 - s)f] \\ q_3(\theta) &= (s - 1)p(\theta)f' \end{aligned} \tag{7.2-11}$$

The boundary conditions on $\theta = \pi$ and the symmetry conditions for Mode I on $\theta = 0$ require that

$$\begin{aligned} f(\pi, n) &= 0, \qquad f'(\pi, n) = 0 \\ f'(0, n) &= 0, \qquad f'''(0, n) = 0 \end{aligned} \tag{7.2-12}$$

The highest-(fifth-) order derivative of the linear operator $L(\ )$ is multiplied by $\sin\theta$. In order to avoid solutions that are singular at $\theta = 0$, the sum of the remaining terms in Equation (7.2-10) must vanish at $\theta = 0$. This regularity condition leads to the fifth boundary condition

$$\begin{aligned} 2(4 - s)k(0) + n(2 - s)[q_2(0) + (ns - 2n + 1)q_1(0)] \\ + q_2''(0) + 6(ns - 2n + 1)q_3'(0) = 0 \end{aligned} \tag{7.2-13}$$

When Equations (7.2-11) and (7.2-12) are substituted into Equation (7.2-13), $f^{iv}(0,n)$ can be expressed in terms of $f(0,n)$ and $f''(0,n)$.

Equations (7.2-10)–(7.2-13) define a nonlinear two-point boundary value problem that can be solved by the shooting method. When trial values for $f(0,n)$ and $f''(0,n)$ are assumed, the numerical integration can be performed by treating the problem as an initial value one. These trial values are adjusted systematically until the boundary conditions at $\theta = \pi$ are satisfied.

Having determined $f(\theta, n)$ one can use Equations (7.2-8) and (7.1-6) or (7.2-1) to write the stress components as

$$\sigma_{\alpha\beta} = \alpha_n \left[\frac{\dot{a}(t)}{BEr}\right]^{1/(n-1)} \hat{\sigma}_{\alpha\beta}(\theta)$$
$$\bar{\sigma} = \alpha_n \left[\frac{\dot{a}(t)}{BEr}\right]^{1/(n-1)} \hat{\sigma}(\theta) \tag{7.2-14}$$

where α_n is a numerical factor that depends upon n. The angular distributions of the stress components are shown in Figure 7.6 for $n = 4$ and $n = 6$. In these representations the components $\hat{\sigma}_{\alpha\beta}(\theta)$ have been normalized such that the maximum value of $\hat{\sigma}(\theta)$ is unity. The values of α_n are $\alpha_4 = 0.815$ and $\alpha_6 = 1.064$.

The strain rate field can be determined by introducing the stress field, Equation (7.2-14), into the material law, Equation (7.1-2). This rate field has a $r^{-n/(n-1)}$ singularity at the crack tip. For steady-state crack growth $\dot{\varepsilon}_{\alpha\beta}$ can be replaced by $-\dot{a}\partial\varepsilon_{\alpha\beta}/\partial x_1$. An x_1-integration of the strain rate field leads to

$$\varepsilon_{\alpha\beta} = \frac{\alpha_n}{E}\left[\frac{\dot{a}}{BEr}\right]^{1/(n-1)} \hat{\varepsilon}_{\alpha\beta}(\theta) \tag{7.2-15}$$

where

$$\hat{\varepsilon}_{\alpha\beta}(\theta) = \hat{\varepsilon}^{e}_{\alpha\beta}(\theta) + \hat{\varepsilon}^{cr}_{\alpha\beta}(\theta) \tag{7.2-16}$$

is a dimensionless function of θ and n. The elastic contributions to the angular distribution of the strain components are

$$\hat{\varepsilon}^{e}_{\alpha\beta} = (1 + \nu)\hat{s}_{\alpha\beta}(\theta) + (1 - 2\nu)\hat{\sigma}_{\gamma\gamma}(\theta)\,\delta_{\alpha\beta}/3 \tag{7.2-17}$$

where $\hat{s}_{\alpha\beta}(\theta)$ is the deviatoric component of $\hat{\sigma}_{\alpha\beta}(\theta)$. The creep contributions are

$$\hat{\varepsilon}^{cr}_{\alpha\beta}(\theta) = \frac{3}{2}\frac{\alpha_n^{n-1}}{(\sin\theta)^{1/(n-1)}} \int_0^{\theta} \frac{\hat{s}_{\alpha\beta}(\phi)\hat{\sigma}^{n-1}(\phi)}{(\sin\phi)^{(n-2)/(n-1)}}\, d\phi \tag{7.2-18}$$

To obtain Equation (7.2-18), Equation (7.2-4) has been used to replace the x_1-integration with a θ-integration. The angular variation of the creep strain field is depicted in Figure 7.7 for $n = 4$ and $n = 6$.

Unlike the plane stress problem, the stress function for plane strain depends upon Poisson's ratio. This dependence complicates the analysis by producing the coupled set of Equations (7.1-9)–(7.1-11). However, these equations simplify for incompressible materials ($\nu = \frac{1}{2}$) for which $s_{33} = 0$. In this instance

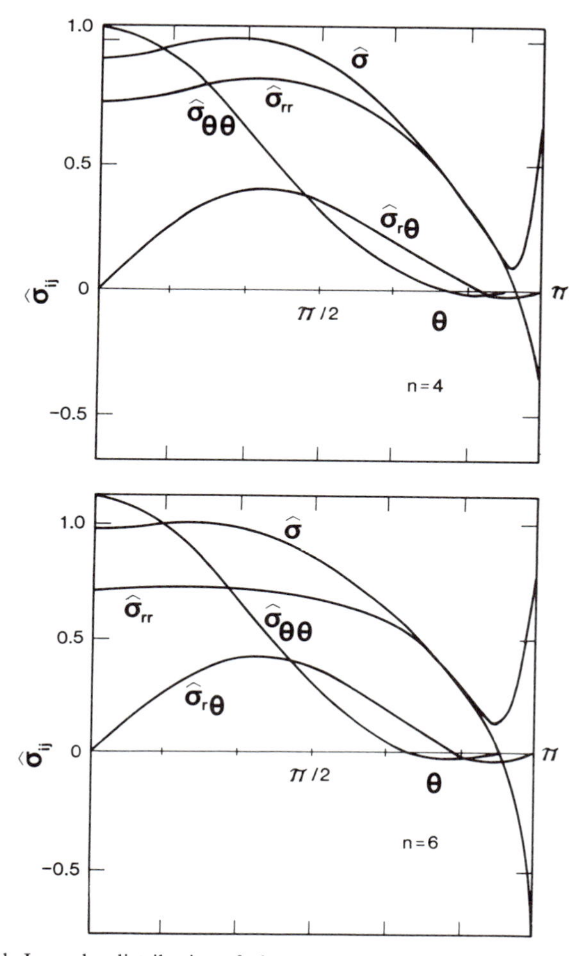

Figure 7.6 Mode I angular distribution of plane stress components for $n = 4$ and $n = 6$ (7.21).

the governing equation for the stress function reduces to

$$\frac{\dot{a}}{E}\nabla^4\left(\frac{\partial\Psi}{\partial x_1}\right) + B[(\Psi_{,\gamma\gamma}\delta_{\alpha\beta} - 2\Psi_{,\alpha\beta})(\tfrac{3}{2}\Psi_{,\delta\rho}\Psi_{,\delta\rho} - \tfrac{3}{4}\Psi_{,\delta\delta}\Psi_{,\rho\rho})^{(n-1)/2}]_{,\alpha\beta} = 0 \qquad (7.2\text{-}19)$$

The stress function must also satisfy Equation (7.2-6) for traction-free crack faces.

Again the stress function for $n > 3$ can be taken in the form of Equations (7.2-8) and (7.2-9). The nonlinear differential equation for the angular distribution of the stress function is (7.19)

$$L(f) + n(2 - s)(ns - 2n + 2)q_1(\theta) + q_1''(\theta) + 4(ns - 2n + 1)q_3(\theta) = 0 \qquad (7.2\text{-}20)$$

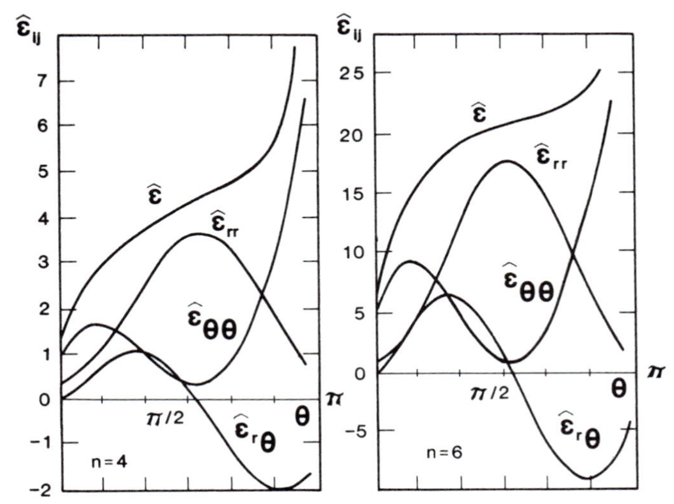

Figure 7.7 Mode I plane stress angular variation of creep strain components for $n = 4$ and $n = 6$ (7.21).

where $L(f)$ is defined in Equation (7.2-11) and now

$$p(\theta) = \left(\frac{\sqrt{3}}{2}\right)^{n-1} \{[f'' + s(2-s)f]^2 + 4(1-s)^2 f'^2\}^{(n-1)/2}$$

$$q_1(\theta) = p(\theta)[f'' + s(2-s)f] \tag{7.2-21}$$

$$q_3(\theta) = (s-1)p(\theta)f'$$

In addition to the boundary conditions, Equation (7.2-12), $f(\theta, n)$ must also satisfy the regularity condition

$$(4-s)k(0) + n(2-s)(ns - 2n + 2)q_1(0) + q_1''(0) + 4(ns - 2n + 1)q_3(0) = 0 \tag{7.2-22}$$

The same numerical technique that was used to solve the plane stress problem may be used here.

With the exception that

$$\hat{\sigma}(\theta) = \frac{\sqrt{3}}{2}[(\hat{\sigma}_{rr} - \hat{\sigma}_{\theta\theta})^2 + 4\hat{\sigma}_{r\theta}^2]^{\frac{1}{2}} \tag{7.2-23}$$

the stress and strain fields are given by Equations (7.2-14)–(7.2-18) with $\nu = \frac{1}{2}$. When the maximum value of $\hat{\sigma}(\theta)$ is normalized to unity, α_n has the values $\alpha_4 = 1.042$ and $\alpha_6 = 1.237$. The angular variations of the stress and creep strain components are illustrated in Figures 7.8 and 7.9 for $n = 4$ and $n = 6$. The dashed curve in Figure 7.8 is the angular dependence of the effective stress for steady crack extension in an elastic-ideally plastic material (7.25). The incompressibility condition implies that $\varepsilon_{\alpha\alpha} = 0$ for plane strain (i.e., $\varepsilon_{\theta\theta} = -\varepsilon_{rr}$).

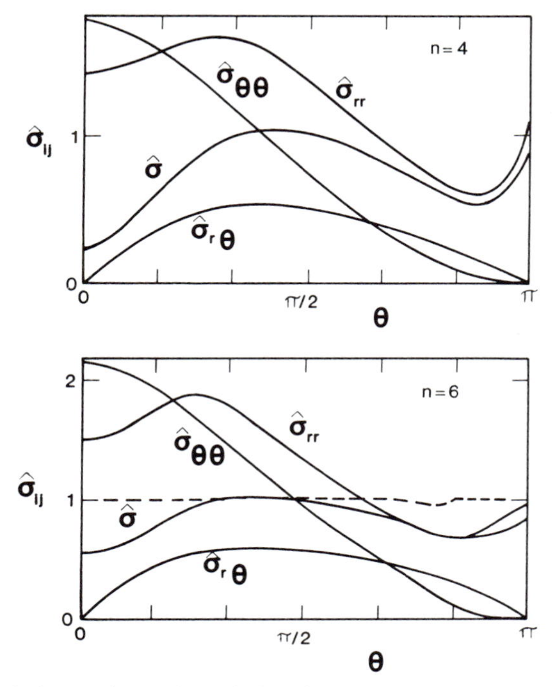

Figure 7.8 Mode I plane strain angular variation of stress components for $n = 4$ and $n = 6$ (7.21).

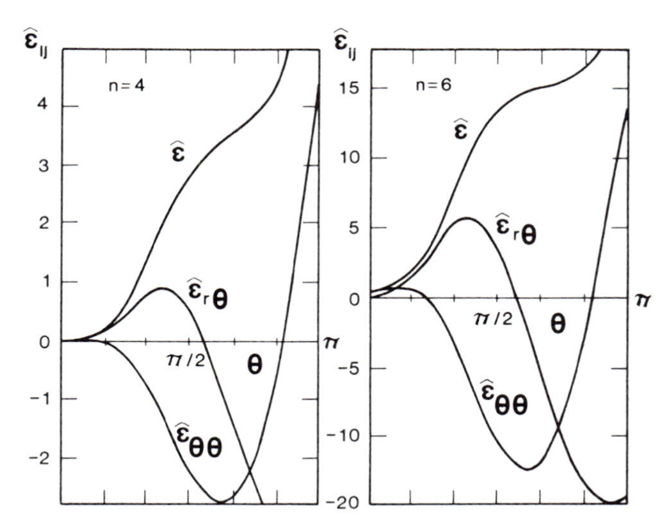

Figure 7.9 Mode I plane strain angular distribution of creep strain components for $n = 4$ and $n = 6$ (7.21).

Unlike the HRR fields for a stationary crack, the stress and strain singularities for the growing crack are of the same order. In contrast to other asymptotic fields, the present ones contain no free parameter and are independent of the applied loading and the geometry of the body. Moreover, the strengths of these fields are independent of the prior crack history and depend only upon the material properties and the current crack speed. Consequently, any fracture criterion that involves only a combination of these fields will lead to a crack growth rate that is independent of the applied loading and that is universal for all flawed configurations of the same material. This is a physically untenable result that is contrary to observed behavior. The implication is that the load- and geometry-dependent nonsingular terms of the stress and strain fields play an important role in creep crack growth.

7.2.2 Steady-State Crack Growth

Steady-state crack growth under the conditions of small-scale yielding will be considered for Mode I. The equivalent problem for Mode III has been treated by Hui and Riedel (7.21). Small-scale yielding implies that the elastic field

$$\sigma_{\alpha\beta} = K(2\pi r)^{-\frac{1}{2}} f_{\alpha\beta}(\theta) \tag{7.2-24}$$

establishes the remote boundary conditions for the solution to Equations (7.2-5) and (7.2-19) for plane stress and incompressible plane strain, respectively.

Dimensional analysis indicates that the stress field for small-scale yielding has the form

$$\begin{aligned} \sigma_{\alpha\beta} &= \left[\frac{\dot{a}}{EBK^2}\right]^{1/(n-3)} \Sigma_{\alpha\beta}(R,\theta) \\ \bar{\sigma} &= \left[\frac{\dot{a}}{\mathrm{EBK}^2}\right]^{1/(n-3)} \bar{\Sigma}(R,\theta) \end{aligned} \tag{7.2-25}$$

where the dimensionless radial coordinate R is

$$R = r\left(\frac{EBK^{n-1}}{\dot{a}}\right)^{-2/(n-3)} \tag{7.2-26}$$

A measure of the characteristic length over which the asymptotic field, Equation (7.2-14), dominates is obtained by setting R equal to unity. For small-scale yielding this length must be small compared to the crack length, the remaining ligament and other relevant dimensions of the body. A comparison of Equation (7.2-25) with Equations (7.2-14) and (7.2-24) reveals that the asymptotic behavior of the dimensionless quantity $\Sigma_{\alpha\beta}(R,\theta)$ is

$$\begin{aligned} \Sigma_{\alpha\beta}(R \to 0, \theta) &= \alpha_n R^{-1/(n-1)} \hat{\sigma}_{\alpha\beta}(\theta) \\ \Sigma_{\alpha\beta}(R \to \infty, \theta) &= (2\pi R)^{-\frac{1}{2}} f_{\alpha\beta}(\theta) \end{aligned} \tag{7.2-27}.$$

In order to relate the crack growth rate to the applied load a fracture criterion is required. To date no generally accepted criterion exists and is the subject of ongoing research. A criterion that has received consideration requires the crack to extend in such a manner that the effective creep strain $\bar{\varepsilon}_{cr}$ at a distance r_c from the crack tip is at its critical value ε_c. For plane stress r_c is defined ahead of the crack tip ($\theta = 0$), whereas for plane strain it is measured in the direction for which $\bar{\Sigma}(R, \theta)$ is a maximum. The critical strain criterion for crack growth may be an oversimplification. Hui (7.22) has introduced a more sophisticated criterion that attempts to model the void growth ahead of the crack tip. Qualitatively, the relationship between $\dot{a}$ and K is very similar regardless of which criterion is used. This is not surprising since both are deformation-based criteria. For the purpose of discussion here the plane stress problem with a critical strain criterion will be investigated.

The combination of the stress field of Equation (7.2-25) and the material law of Equation (7.1-2) permits writing the effective creep strain rate as

$$\dot{\bar{\varepsilon}}_{cr} = B\bar{\sigma}^n = B\left[\frac{\dot{a}}{EBK^2}\right]^{n/(n-3)} \bar{\Sigma}^n(R, \theta) \tag{7.2-28}$$

With $\dot{\bar{\varepsilon}}_{cr} = -\dot{a}\, d\bar{\varepsilon}_{cr}/dr$ on $\theta = 0$, an integration of Equation (7.2-28) in from infinity where $\bar{\varepsilon}_{cr} = 0$ in the small-scale yielding limit leads to

$$\frac{E\varepsilon_c\sqrt{r_c}}{K} = F(R_c) \tag{7.2-29}$$

where

$$F(R) = \sqrt{R}\int_R^{\infty} \bar{\Sigma}^n(\rho, 0)\, d\rho \tag{7.2-30}$$

and

$$R_c = r_c\left(\frac{EBK^{n-1}}{\dot{a}}\right)^{-2/(n-3)} \tag{7.2-31}$$

To evaluate $F(R)$ it is necessary to know the dimensionless effective stress $\bar{\Sigma}(R, 0)$. The asymptotic character of $F(R)$ can be determined using the limiting stress fields in Equation (7.2-27). This yields

$$F(R \to 0) = \alpha_n^n(n-1)\,\hat{\sigma}^n(0)\mathrm{R}^{(n-3)/2(n-1)} \tag{7.2-32}$$

$$F(R \to \infty) = \frac{2(2\pi)^{-n/2}}{(n-2)R^{(n-3)/2}} \tag{7.2-33}$$

Riedel and Wagner (7.23) proposed using

$$\bar{\Sigma}(R, 0) = \left[R(\alpha_n\hat{\sigma}(0))^{1-n} + (2\pi R)^{(n-1)/2}\right]^{-1/(n-1)} \tag{7.2-34}$$

to interpolate between the near field stress represented by the first term within the brackets and the far field stress associated with the second term. Hui (7.22) found that a similar interpolation for $\Sigma_{22}(R, 0)$ agreed with plane strain finite element computations to within 10 percent. Figure 7.10(a) shows for $n = 5$ the asymptotic behavior [Equations (7.2-32) and (7.2-33)] of $F(R)$ (solid curves) along with $F(R)$ based upon the interpolated stress (dashed curve). It is clear

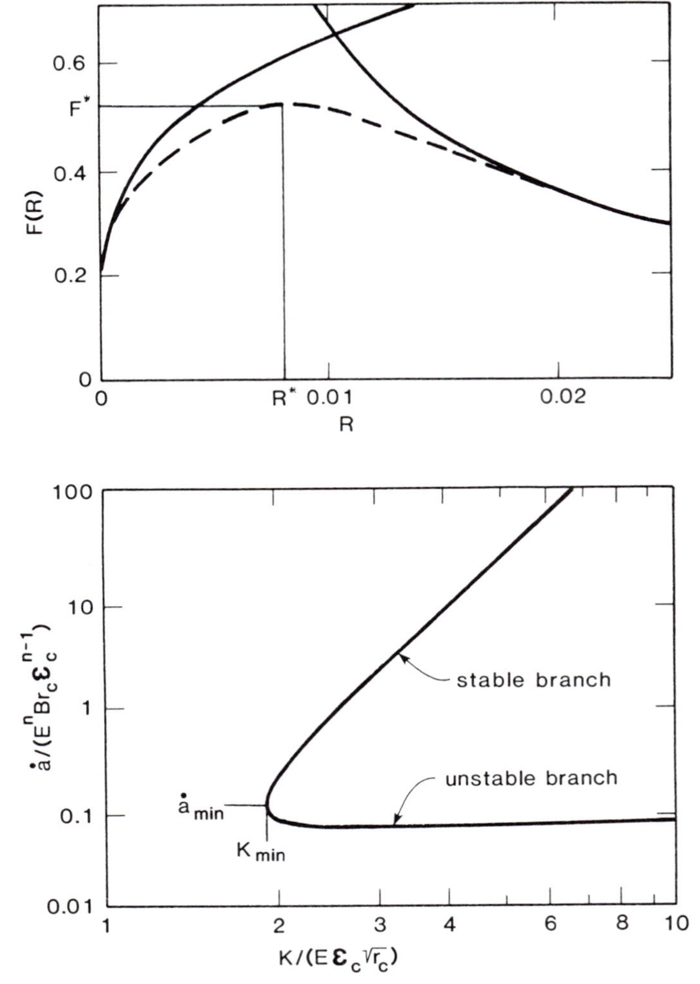

Figure 7.10 Asymptotic (solid curve) and interpolated (dashed curve) representations for $F(\mathrm{R})$ and the variation of $\dot{a}$ with K for small-scale yielding (7.21).

from this figure that $F(R)$ will have a maximum F^* at $R = R^*$. Associated with this maximum is a minimum value of K,

$$K_{\min} = E\varepsilon_c\sqrt{r_c}/F^* \tag{7.2-35}$$

below which no solution to Equation (7.2-29) exists. Therefore, within the confines of small-scale yielding steady-state crack growth is not possible for $K < K_{\min}$.

With $F(R)$ known, Equations (7.2-29) and (7.2-31) can be viewed as a pair of parametric equations relating $\dot{a}$ to K. For $K > K_{\min}$ or, equivalently, $F(R_c) < F^*$, two values of R_c and, hence, $\dot{a}$ satisfying Equation (7.2-29) exist. The existence of two values of $\dot{a}$ for the same K produces the two branches depicted in Figure 7.10(b). The lower branch associated with $R_c < R^*$ is unstable in the sense that a decrease of K gives rise to an increase in the crack speed; that is, $d\dot{a}/dK < 0$. The upper branch ($d\dot{a}/dK > 0$) for which $R_c > R^*$ is stable. From Equation (7.2-31) the mimimum steady crack speed possible under small-scale yielding conditions is found to be

$$\dot{a}_{\min} = \left(\frac{R^*}{r_c}\right)^{(n-3)/2} EBK_{\min}^{n-1} = R^{*(n-3)/2} E^n r_c \left(\frac{\varepsilon_c}{F^*}\right)^{(n-1)} \qquad (7.2\text{-}36)$$

It follows from setting $R = 1$ in Equation (7.2-26) that the size of the creep zone depends upon $K^{n-1}/\dot{a}$. On the upper branch ($R_c > R^*$), $\dot{a}$ must increase faster than K^{n-1} if Equation (7.2-29) is to be satisfied. Therefore, with increasing K the creep zone decreases from its maximum extent when $K = K_{\min}$. For $K \gg K_{\min}$ the elastic far field dominates the creep strain at $x_1 = r_c$. It follows from Equations (7.2-29) and (7.2-33) that

$$\dot{a} = \frac{2}{n-2} \frac{Br_c}{\varepsilon_c} \left[\frac{K}{\sqrt{2\pi r_c}}\right]^n \qquad (7.2\text{-}37)$$

That is, $\dot{a}$ and K obey a power law relationship under conditions of small-scale yielding. The void fracture model of Hui (7.22) also predicts $\dot{a} \propto K^n$ for high crack growth rates.

7.2.3 Transient Crack Growth

The problem of transient crack growth is considerably more difficult and only limited results exist. For this reason only an approximate analysis will be considered. Such an analysis is useful because it can be expected to describe qualitatively the behavior. The steady-state stress field, Equation (7.2-25), can be used to approximate a transient field under small-scale yielding provided that $\dot{K}$ and $\ddot{a}$ are, respectively, small compared to K and $\dot{a}$ while the crack tip traverses the region where the near-tip field dominates. For the approximation to be appropriate

$$|\ddot{a}| < \left[\frac{\dot{a}^{n-2}}{EBK^{n-1}}\right]^{2/(n-3)} \qquad (7.2\text{-}38)$$

$$|\dot{K}| < \left[\frac{\dot{a}^{n-1}}{EB^2K^{n+1}}\right]^{1/(n-3)} \qquad (7.2\text{-}39)$$

The critical strain criterion can be approximated by

$$\frac{B\bar{\sigma}^n}{\dot{a}} \frac{r_c}{\varepsilon_c} = -\frac{\partial \bar{\varepsilon}_{cr}}{\partial x_1} \frac{r_c}{\varepsilon_c} \qquad (7.2\text{-}40)$$

where $\bar{\sigma}$ and $\partial\bar{\varepsilon}_{cr}/\partial x_1$ are understood to be evaluated at $x_1 = r_c$. Equation (7.2-40) is exact for steady-state crack growth.

The total creep strain is the sum of the strain accumulated while the crack is stationary and the strain while the crack is growing. The far field stress can be used to approximate the creep strain while the crack is stationary. The creep strain, $(\bar{\varepsilon}_{cr})_1$, due to crack growth is given by

$$(\bar{\varepsilon}_{cr})_1 = B \int_{a_0}^{a} \frac{\bar{\sigma}^n(a + r_c - \zeta)}{\dot{a}(\zeta)} \, d\zeta \tag{7.2-41}$$

where the integral extends over all previous crack-tip positions and a_0 is the initial crack length. It is clear that $(\bar{\varepsilon}_{cr})_1$ depends upon the crack-tip history. The effective stress in Equations (7.2-40) and (7.2-41) is approximated by the steady-state field of Equation (7.2-25) with $\bar{\Sigma}(R, 0)$ interpolated by Equation (7.2-34). As a consequence the left-hand side of Equation (7.2-40) depends only upon the current crack growth rate whereas the right-hand side depends upon the prior crack history.

Equation (7.2-40) has been solved numerically for $\dot{a}$ by Riedel and Wagner (7.23). Analogous to steady-state crack growth this equation, depending upon the value of K, yields two values of $\dot{a}$ or none. Two values exist if

$$K > K_{\min} = \frac{n}{2(n-1)} \frac{n^{n/(n-1)}}{(\alpha_n \hat{\sigma}(0))^{n-1}} E\varepsilon_c \sqrt{2\pi r_c} \tag{7.2-42}$$

Again the smaller crack speed is associated with the unstable branch whereas the larger crack speed belongs to the stable branch. If the stress intensity factor $(K > K_{\min})$ is held fixed and if the remaining ligament is sufficiently large, the crack growth rate increases and approaches asymptotically the steady-state rate. If after initiation of crack growth the stress intensity factor decreases continuously with the crack growth but within the restriction of Equation (7.2-39), the condition arises where Equation (7.2-40) can no longer be satisfied. When this occurs, the present model predicts an instantaneous drop to zero of the crack growth rate and crack arrest. This rapid drop in the crack speed would violate the restriction of Equation (7.2-38) and invalidate the model. In reality further nonsteady crack growth would likely occur before crack arrest.

If the duration of the loading is sufficiently long that extensive creep engulfs the whole specimen, then the elastic strains can be neglected except in the zone near the extending tip where the fields of Equations (7.2-14) and (7.2-15) dominate. When this zone is small compared to the crack length and the remaining ligament, the bulk of the material according to the constitutive relation, Equation (7.1-2), will exhibit nonlinear viscous flow. In this case the HRR field of Equation (7.1-18) sets the remote boundary condition for the crack-tip zone. Under conditions of extensive creep the appropriate loading parameter is the C^*-integral.

Dimensional consistency requires that the steady-state stress field have the form

$$\begin{aligned} \sigma_{\alpha\beta} &= \left[\frac{EC^*}{\dot{a}}\right]^{\frac{1}{2}} \Sigma_{\alpha\beta}(R, \theta) \\ \bar{\sigma} &= \left[\frac{EC^*}{\dot{a}}\right]^{\frac{1}{2}} \bar{\Sigma}(R, \theta) \end{aligned} \tag{7.2-43}$$

where $\Sigma_{\alpha\beta}(R, \theta)$ and $\bar{\Sigma}(R, \theta)$ are dimensionless functions of θ and the dimensionless coordinate

$$R = r \frac{B}{C^*} \left[\frac{EC^*}{\dot{a}} \right]^{(n+1)/2} \tag{7.2-44}$$

The characteristic length of the crack-tip region in which the singular crack-tip field of Equation (7.2-14) can be expected to dominate is defined by setting R equal to unity. The present analysis requires that this length be small compared to the crack length and the remaining ligament.

The steady-state stress field, Equation (7.2-43), is approximately valid if variations of $\dot{a}$ and C^* are sufficiently small while the crack extends through the crack-tip region. Specifically,

$$\begin{aligned} |\ddot{a}| &\lesssim \frac{B\dot{a}^2}{C^*} \left[\frac{EC^*}{\dot{a}} \right]^{(n+1)/2} \\ |\dot{C}^*| &\lesssim B\dot{a} \left[\frac{EC^*}{\dot{a}} \right]^{(n+1)/2} \end{aligned} \tag{7.2-45}$$

A plausible interpolation between the near and far fields ahead of the crack tip is

$$\begin{aligned} \bar{\Sigma}(R, 0) = [(I_n R)^{(1-n)/(n+1)} (\tilde{\sigma}(0))^{n-1} \\ + (\alpha_n \hat{\sigma}(0))^{n-1}/R]^{1/(n-1)} \end{aligned} \tag{7.2-46}$$

From this point on the analysis proceeds as for small-scale yielding except that the HRR field, Equation (7.1-18), is used to determine the creep strain while the crack tip is stationary. However, contrary to the small-scale yielding analysis two solutions to Equation (7.2-40) for $\dot{a}$ exist only if (7.23)

$$C^* < \left[\frac{(n-1)n^{1/(1-n)} E\varepsilon_c}{(n+1)(\alpha_n \hat{\sigma}(0))^{n-1} \tilde{\sigma}(0)} \right]^{n+1} I_n B r_c \tag{7.2-47}$$

otherwise, none exists. The smaller value of $\dot{a}$ is associated with the stable branch while the larger value belongs to the unstable branch. Under constant C^*-loading the crack tip accelerates after initiation and never attains a constant speed according to the present model. When the crack speed exceeds

$$\begin{aligned} \dot{a}_{\max} = (n-1) \left[\frac{\tilde{\sigma}(0)}{\alpha_n \hat{\sigma}(0) I_n^{1/(n+1)}} \right]^{n-1} \\ \times E(Br_c)^{2/(n+1)} C^{*(n-1)/(n+1)} \end{aligned} \tag{7.2-48}$$

solutions to Equation (7.2-40) no longer exist and unstable crack growth follows. This instability is a consequence of the increased contribution of the near-tip stress field, Equation (7.2-14), to the creep strain at $x_1 = r_c$ for the larger crack speeds. Not only the strength of this field, but also the size of the near-tip zone increases with the crack growth rate. At sufficiently large crack speeds the zone engulfs the point $x_1 = r_c$. A condition develops whereby a further increase in $\dot{a}$ produces a creep strain at $x_1 = r_c$ that exceeds ε_c. When this happens, unstable crack growth occurs. It should be noted that the large crack-tip accelerations that develop as the point of instability is approached may invalidate the present theory due to the failure to satisfy Equation (7.2-45).

In the preceding development the crack growth is assumed to occur continuously. There is some evidence (7.26) to indicate that creep crack growth in some alloys does not proceed continuously but rather by a series of discrete steps. If the time t_G between increments Δa of discrete crack extension is large relative to the transient time t_T for the development of the secondary or steady-state creep, then most of the creep damage will occur in the steady-state crack-tip region while the tip is stationary. This means that the sequence of discrete crack advances will be controlled by C^* if the secondary (steady-state) creep field can develop around the crack tip long before the next increment of crack growth. Consequently, the macroscopic crack growth rate for this kind of crack extension should correlate with C^*. When the increment of crack growth carries the tip through and well beyond the crack-tip plastic zone, McMeeking and Leckie (7.27) estimate that

$$\frac{t_T}{t_G} = \frac{3\dot{a}[1 - (2C^*/3B\Delta a)^{1/(n+1)}]}{4EB\Delta a\sigma_y^{n-1}} \tag{7.2-49}$$

When this ratio is small compared to unity (i.e., $t_T \ll t_G$), the C^*-integral can be expected to be the loading parameter governing the rate of intermittent crack growth in a plastic-power-law creeping material. On the other hand if the ratio is much greater than unity ($t_T \gg t_G$), the crack growth can be considered to be nearly continuous.

A complete analysis of the stress field around a crack tip that starts from a stationary position and begins to grow after an incubation period has not been performed. Due to the inherent mathematical complexities associated with the different regimes of load and crack velocities, numerical methods will undoubtedly be required to carry out such an analysis.

7.2.4 Elastic-Primary Creep Crack Fields

It can be shown (7.28) that the crack-tip stress field belonging to an elastic-primary creep law of Equation (7.1-42) is

$$\begin{aligned}
\sigma_{\alpha\beta} &= \alpha(n,p)\left[\frac{\dot{a}(t)}{BE^{1+p}r}\right]^{1/(n-1-p)} \hat{\sigma}_{\alpha\beta}(\theta,n,p) \\
\bar{\sigma} &= \alpha(n,p)\left[\frac{\dot{a}(t)}{BE^{1+p}r}\right]^{1/(n-1-p)} \hat{\sigma}(\theta,n,p)
\end{aligned} \tag{7.2-50}$$

provided that $n - p > 3$. Otherwise, the fields have an elastic $r^{-\frac{1}{2}}$ singularity. The numerical coefficient $\alpha(n,p)$ depends upon the values of n and p as do the θ-variations $\hat{\sigma}_{\alpha\beta}(\theta,n,p)$ and $\hat{\sigma}(\theta,n,p)$. When $p = 0$, this stress field reduces to that of Equation (7.2-14). Like the latter field the strength of the present one is independent of the loading and prior crack history and depends only upon the material properties and the current crack growth rate. Again, the implication is that the load- and geometry-dependent nonsingular fields have a vital role in creep crack growth.

For steady-state crack growth the crack-tip strain field is

$$\varepsilon_{\alpha\beta} = \alpha(n, p)\left[\frac{\dot{a}}{BE^n r}\right]^{1/(n-1-p)} \hat{\varepsilon}_{\alpha\beta}(\theta, n, p) \qquad (7.2\text{-}51)$$

where $\hat{\varepsilon}_{\alpha\beta}(\theta, n, p)$ is a dimensionless function of θ, n, and p. The elastic component of $\hat{\varepsilon}_{\alpha\beta}(\theta, n, p)$ is given by Equation (7.2-17) while the creep component has a form similar to that of Equation (7.2-18). The same method used to analyze crack growth in the secondary creep regime can also be used here.

7.3 Creep Crack Growth Correlations

Numerous experimental studies on creep crack growth at elevated temperatures have been conducted for several structural alloys. Time-dependent crack growth can result from creep effects as well as environmental effects or from a combination of both. An oxidizing environment can accelerate the creep crack growth rate in superalloys by an order of magnitude or more. In many tests no attempt was made to separate the effects. This has made the interpretation and comparison of data difficult. Efforts have concentrated on trying to identify the loading parameter with which the crack growth rate correlates. The most commonly employed loading parameters are the elastic stress intensity factor K, the energy rate integral C^*, and the reference stress σ_{ref}. If a particular loading parameter is applicable, the crack growth rate should correlate with it regardless of the specimen's geometry. The number of correlation studies using several specimen geometries is limited. While more test results for different geometries are becoming available, it is not always possible to make comparisons because the influence of the environment is unknown. Some results of these correlation studies are presented in this section in order to attempt to establish the condition(s) under which a specific loading parameter is applicable. Rather extensive reviews of crack growth studies at elevated temperatures under static, cyclic, and combined loading have been prepared by Sadananda and Shahinian (7.29, 7.30).

Perhaps the most appealing aspect of the stress intensity factor as a correlating parameter is the ease with which it can be calculated. This is particularly advantageous when conducting creep crack growth tests at elevated temperatures since only the load and the crack length are required for its determination. The stress intensity factor has previously been shown to be the relevant loading parameter under conditions of small-scale yielding—that is, when the size of the plastic and/or creep zone attending the crack tip is small compared to the elastic K-dominant region. However, increasing the temperature of most materials is more conducive to plastic flow and creep, which can ultimately limit the range of applicability of the stress intensity factor. If the plastic deformation is negligible compared to the creep deformation, the stress intensity factor can be expected to be applicable if the test duration over which the crack growth is observed is less than the transition time t_1 given by Equation (7.1-47). When the primary creep is also insignificant, the creep crack growth rate can be expected to correlate with K if the test period is less than the transition time t_T of Equation (7.1-32). These times are only estimates since the transition times t_1 and t_T were established for a stationary crack. Analogous transition times have not been determined for a growing crack.

When the plastic deformation is negligible and the secondary creep is extensive, then the C^*-integral is expected to be the relevant loading parameter. Landes and Begley (7.8) have developed an experimental technique for the determination of C^* based upon the definition of Equation (7.1-16). In this method, shown schematically in Figure 7.11, multiple specimens are subjected to different constant displacement rates. The load P per unit crack plane thickness and the crack length are measured as a function of time as

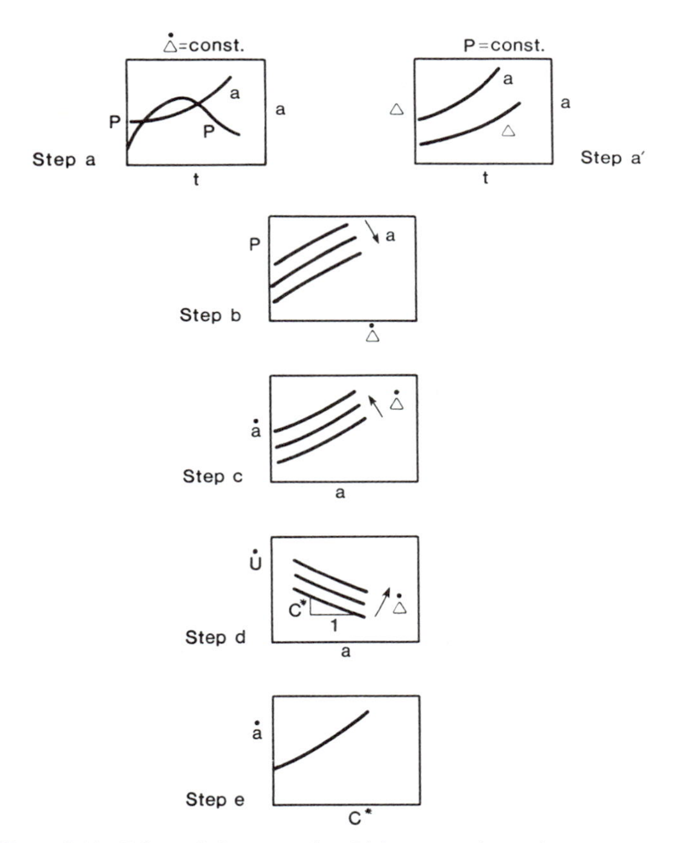

Figure 7.11 Schematic for extracting C^* from experimental measurements.

depicted in Step a. These data are cross plotted to yield the load as a function of the displacement rate for fixed crack lengths, Step b, and the crack growth rate versus crack length, Step c. The area under the curves in Step b is the rate of work done $\dot{U}$ per unit of crack plane thickness. The latter is plotted against the crack length in Step d. According to Equation (7.1-16) the slope of the curve in Step d is C^*. Finally, the curves of Step c permit plotting the crack growth rate as a function of C^*, Step e.

Normally creep tests are conducted under constant load. Sadananda and Shahinian (7.9) have modified the above procedure to accommodate the latter type of loading. In this approach the load-point displacement and the crack length as a function of time are recorded as in Step a′ in Figure 7.11. For a fixed crack length the load, displacement rate, and crack growth rate are determined from these plots. The data are used to prepare the plots in Step b and Step c. From here on the two procedures are identical. The crack length can be determined from measurements of the drop of the electrical potential across the crack when a current is applied to the specimen.

Clearly, either approach to analyzing the data is more involved than simply computing the stress intensity factor from the applied load and crack length. On the other hand, when handbook values for $h_1(a/W, n)$ are available, Equation (7.1-19) can be used to compute C^* for extensive creep in a power law creeping material. Alternatively, the estimation approaches used to determine J for elastic-plastic materials can be extended to establish C^* from measured P-$\dot{\Delta}$ curves. For example, assume, analogous to elastic-plastic fracture, that the generalized load P per crack tip and per unit crack plane thickness and the conjugate displacement rate $\dot{\Delta}$ are related by the separable form

$$P = f(a)g(\dot{\Delta}) \tag{7.3-1}$$

where $f(a)$ is a function of geometry only, whereas, $g(\dot{\Delta})$ is a function of $\dot{\Delta}$ and independent of a. The combination of Equations (7.1-16), (7.1-17), and (7.3-1) yields

$$C^* = -\int_0^{\dot{\Delta}} \left(\frac{\partial P}{\partial a}\right)_{\dot{\Delta}} d\dot{\Delta} = \frac{\eta}{b}\dot{U} \tag{7.3-2}$$

where b is the remaining ligament and

$$\eta = -\frac{bf'}{f} = -\frac{b}{P}\left(\frac{\partial P}{\partial a}\right)_{\dot{\Delta}} \tag{7.3-3}$$

is the creep analog of Turner's η-factor [cf. Equation (5.7-29)].

The problem of determining C^* from a P-$\dot{\Delta}$ curve reduces to establishing η. For a deeply cracked bend specimen $\eta = 2$. For a power law creeping material it is possible to write

$$\dot{U} = \frac{n}{n+1} P\dot{\Delta} \tag{7.3-4}$$

It also follows from a generalization of Il'yushin's theorem that $P \propto \dot{\Delta}^{1/n}$. For a deeply cracked plane tension specimen (e.g., a center crack panel or a double-edge cracked plate) and for a large value of the creep exponent, dimensional analysis requires that

$$P = Cb(\dot{\Delta}/b)^{1/n} \tag{7.3-5}$$

where C is a constant independent of b. The introduction of Equation (7.3-5) into Equation (7.3-3) yields

$$\eta = \frac{n-1}{n} \tag{7.3-6}$$

and, hence,

$$C^* = \frac{n-1}{n+1}\frac{P\dot{\Delta}}{b} \tag{7.3-7}$$

Harper and Ellison (7.31) found for Discaloy at 650° C very good agreement between C^* determined by the Landes-Begley method (Figure 7.11) and this estimation technique. On the other hand, Saxena (7.32) reported that the two methods yield values for C^* that differ from 8 to 40 percent for Type 304 stainless steel.

Many materials when tested under conditions conducive to high ductile creep deformation are notch insensitive. Under these conditions no special significance can be attached to the singular crack-tip fields because of the large degree of crack-tip blunting. The reference stress method of analysis for creeping components is often used under these circumstances. The idea behind this method is to correlate the creep deformation in a body with the deformation in a simple creep specimen (7.33). The method is an approximate technique that has been employed successfully to predict creep rupture (7.34). The reference stress is defined as the stress in a component that when applied to a uniaxial specimen will result in the same deformation rate. For a tension specimen without bending—for example, a center cracked panel—the reference stress reduces to the net section stress.

Williams and Price (7.35) have established reference stresses for a number of fracture specimens. They found that the reference stress is virtually independent of the creep exponent and is equivalent to the stress at the sketal point—the location within the body where the stress is nearly independent of n (7.36). Since the limiting solution for an infinite creep exponent corresponds to the solution for perfect plasticity, then the reference stress can be approximated by

$$\sigma_{\text{ref}} = \frac{P\sigma_y}{P_0} \tag{7.3-8}$$

where again P_0 is the limit load for the equivalent perfectly plastic structure and σ_y is the yield stress. Ponter and Leckie (7.37) have shown that this approximation for the reference stress leads to an upper bound to the deformation. When Equation (7.3-8) is introduced into Equation (7.1-19), the connection,

$$C^* = \sigma_{\text{ref}}\dot{\varepsilon}(\sigma_{\text{ref}})bg_1\left(\frac{a}{W}\right)h_1\left(\frac{a}{W}, n\right) \equiv \sigma_{\text{ref}}\dot{\varepsilon}(\sigma_{\text{ref}})R \tag{7.3-9}$$

between C^* and σ_{ref} is obtained.

Failure due to creep rupture is to be expected when sufficient crack-tip blunting relaxes the stress concentrations there. Assuming the crack tip blunts into a semi-circular arc, Ainsworth (7.38) obtained an estimate for the initiation time t_i for crack growth in terms of the crack opening displacement δ_i at initiation. When the crack opening immediately after loading is negligible compared to δ_i, then

$$t_i = \frac{2(n+1)}{n\sqrt{3}}\left[\frac{2n\,\delta_i}{(3n+4)B^{1/n}C^*}\right]^{n/(n+1)} \tag{7.3-10}$$

If the simple creep rupture criterion is adopted that failure occurs when a critical accumulated strain ε_c is attained, then for a constant stress the creep rupture time $t_r(\sigma)$ for a power law creeping material is expressed by

$$\dot{\varepsilon}(\sigma)t_r(\sigma) = \varepsilon_c \tag{7.3-11}$$

The combination of Equations (7.3-9)–(7.3-11) leads to

$$\frac{t_i}{t_r(\sigma_{\text{ref}})} = \frac{2(n+1)}{n\sqrt{3}\,\varepsilon_c}\left[\frac{2n}{3n+4}\frac{\delta_i}{R}\right]^{n/(n+1)} \tag{7.3-12}$$

The time $t_r(\sigma_{\text{ref}})$ represents the life expectancy of a structure when crack-tip processes are unimportant and failure is governed by overall creep rupture mechanisms.

Equation (7.3-12) forms the basis for deciding whether it is necessary to consider crack initiation and crack growth in determining the life of a structure or to take the creep rupture time $t_r(\sigma_{\text{ref}})$ as the lifetime. When the right-hand side of Equation (7.3-12) is greater than or only slightly less than unity—that is, when no initiation of crack growth occurs or it takes place only shortly before creep rupture—then failure is governed by creep rupture mechanisms and not by macroscopic crack growth. Under such conditions the reference stress will be the relevant loading parameter. At the other end of the time spectrum, $t_i/t_r(\sigma_{\text{ref}}) \ll 1$, crack growth will play an important role in the failure. On the basis of Equation (7.3-12) this type of failure can be expected in materials that exhibit a small value for the crack opening displacement at initiation and/or large critical accumulated creep strain.

Koterazawa and Mori (7.39) report results of a rather extensive study of creep crack growth in three heats of Type 304 stainless steel at 650° C. Center notched (CN), two different types of single edge notched (SEN), and three different sizes of double edge notched (DEN) specimens were used. Fractographic examinations reveal the fracture process to be intergranular. Measured crack growth rates in the DEN specimens of heat treatment A and C and

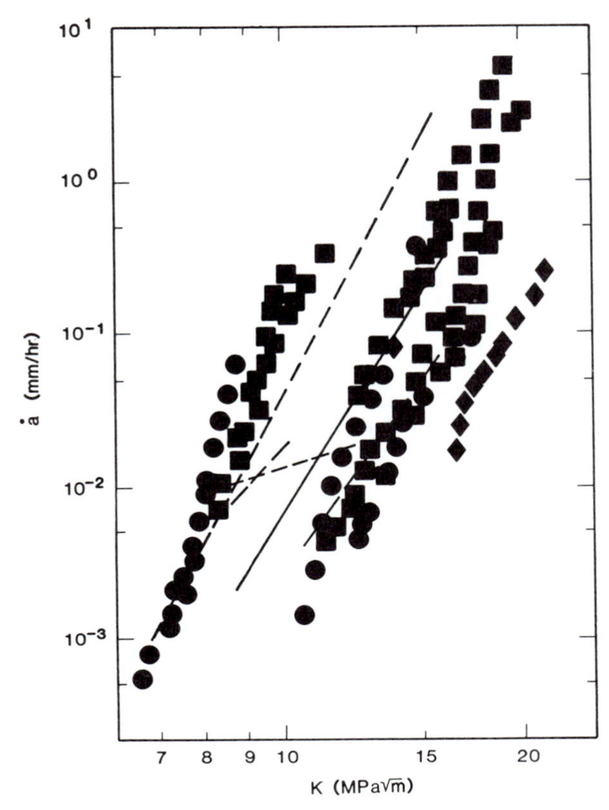

Figure 7.12 Crack growth rate in Type 304 stainless steel CN (◆), DEN (dashed curves, ■ heat A, ● heat C), and NRB (solid curves) specimens versus the stress intensity factor (7.39).

in the CN specimens with heat treatment C are plotted as a function of the stress intensity factor in Figure 7.12. Also shown are the results (dashed curves) from similar tests by Yokobori et al. (7.40) and Ohji et al. (7.41) on DEN and notched round bar (NRB) specimens together with the NRB results (solid curve) of Koterazawa and Iwata (7.42). For a fracture mechanics parameter to be applicable its correlation with crack growth rate must be independent of the specimen's geometry. Clearly, the stress intensity factor does not satisfy this requirement for the loading levels used in these tests. However, it appears as though the results might be converging at the lower levels of the stress intensities. This is to be expected because at even lower stresses the creep deformation would be confined to the vicinity of the crack tip and its influence should be minimal. Therefore, the stress intensity factor could conceivably describe creep crack propagation if in addition the conditions for small-scale yielding are satisfied. That is, the stress intensity factor would be expected to be applicable for high strength-low ductility (creep-brittle) materials.

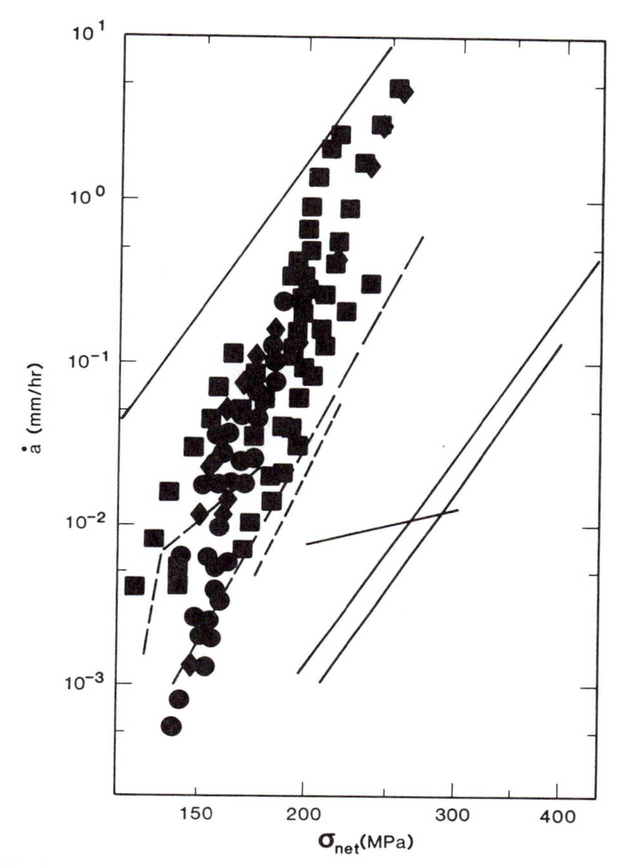

Figure 7.13 Crack growth rate in Type 304 stainless steel CN (◆), DEN (dashed curves, ■ heat A, ● heat C), and NRB (solid curves) specimens versus the net section stress (7.39).

In Figure 7.13 the crack growth rate in DEN, CN, and NRB specimens is plotted against the net section stress σ_{net}. For these types of tension specimens the net section stress represents an approximation to the reference stress. Here the correlation between the results from DEN and CN specimens is much better than that in Figure 7.12. The growth rates tend to be greater in the wider specimens. The results for the NRB specimens are also much closer together but the crack growth rate in the latter for the same net section stress is generally several orders of magnitude smaller than in the CN and DEN specimens. The net section (reference) stress exhibits geometry dependence, but to a smaller degree than the stress intensity factor.

The crack growth rate in CN specimens and in three different sizes of DEN specimens as a function of C^* is shown in Figure 7.14. The scatter in the growth rates for a specific value of C^* is less than four. This is considerably better than when either the stress intensity factor or the net section stress is

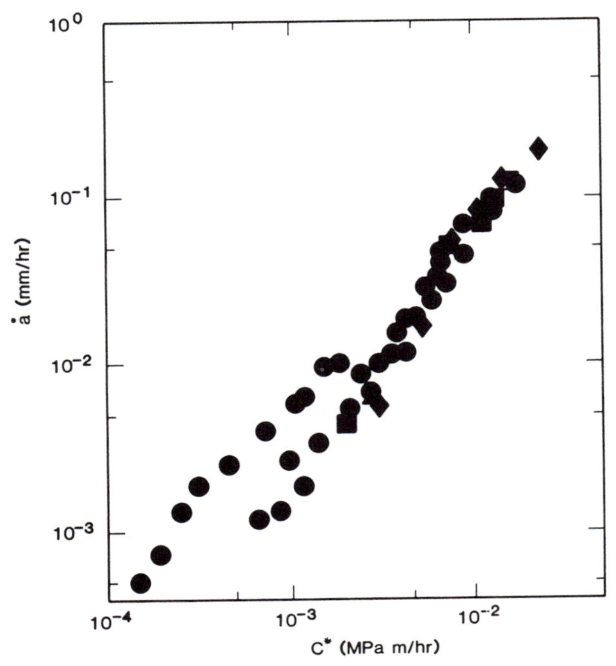

Figure 7.14 Crack propagation rates in Type 304 stainless steel CN (◆) and DEN (●) specimens versus the C^*-integral (7.39).

used. Also, the measurements indicate that a linear relationship between the crack growth rate and C^* exists. Of the three fracture mechanics parameters considered the crack propagation rate correlates best with C^* for $\sigma_{net} >$ 118 MPa. While the scatter band in Figure 7.15 is wider, the correlation of crack growth rates with C^* for CN, CT, DEN, and NRB specimens is still judged quite good by present standards of comparison. Taira et al. (7.43) found an equally good correlation between the crack growth rate and C^* for center notched cylinders (CNC) and NRB specimens of 0.16 percent carbon steel at 400° C. Again a linear relationship between $\dot{a}$ and C^* was observed.

When Equation (7.3-9) is rewritten as

$$C^* = \sigma_{ref}\,\dot{\varepsilon}(\sigma_{ref})WH\left(\frac{a}{W}, n\right) \tag{7.3-13}$$

where

$$H\left(\frac{a}{W}, n\right) = \left(1 - \frac{a}{W}\right)g_1\left(\frac{a}{W}\right)h_1\left(\frac{a}{W}, n\right) \tag{7.3-14}$$

it is clear that C^* depends upon the width W of the specimen, the reference stress, and the geometry of the specimen through $H(a/W, n)$. Since $\dot{a}$ appears to be proportional to C^*, then the crack growth rate in geometrically similar

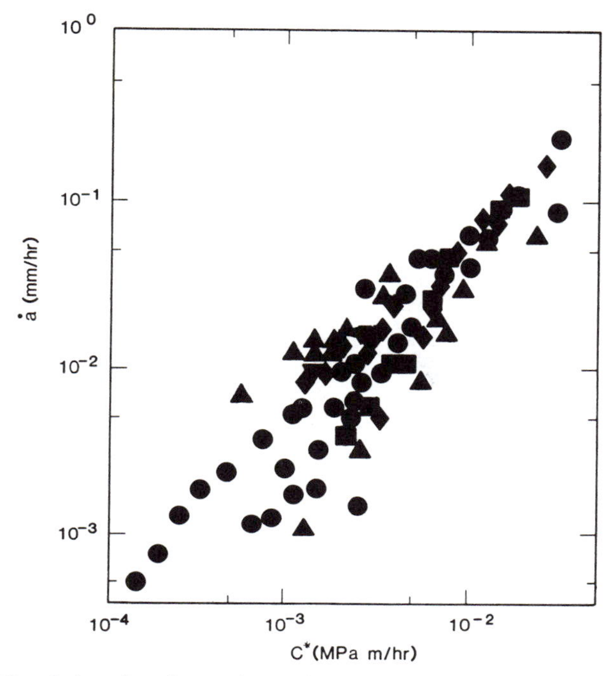

Figure 7.15 Correlation of crack growth rates in Type 304 stainless steel CN (◆), DEN (■), CT (▲), and NRB (◀) specimens with the C^*-integral (7.39).

specimens of a power law creeping material should increase proportionally with the size of the specimen for the same reference stress. This prediction is consistent with the data in Figure 7.14. Furthermore, $\dot{a}/W$ ought to correlate with the reference stress whenever $\dot{a}$ varies linearly with C^* and changes in $H(a/W, n)$ during crack growth in a power law creeping material are negligible. For example, $H(a/W, n) = 1.21 \pm .07$ for a DEN specimen in the range $\frac{1}{8} \leqslant a/W < \frac{5}{8}$ and for $n = 7$, a representative value of the creep exponent for Type 304 stainless steel at 650° C. A similar argument for the correlation of $\dot{a}/W$ with σ_{ref} can be made for deeply cracked specimens on the basis of the estimation of Equation (7.3-7) for C^* and Equations (7.3-5) and (7.3-8). Figure 7.16 demonstrates that the correlation of $\dot{a}/W$ with the net section stress, which is used to approximate the reference stress, for Type 304 stainless steel CN, DEN, and NRB specimens of different sizes is very good. The plot for the NRB specimens does not coincide with one for the DEN specimens because the respective $H(a/W, n)$ is different—that is, because of the aforementioned geometry dependence of the net section (reference) stress.

To assess the influence of environment and temperature on creep crack growth in 0.16 percent carbon steel, Taira et al. (7.43) conducted tests with CNC specimens at 400° C and 500° C in air and in a vacuum. When the crack growth rate is plotted against the net section stress as in Figure 7.17(a), a rather strong dependence of crack growth rate on temperature and environment

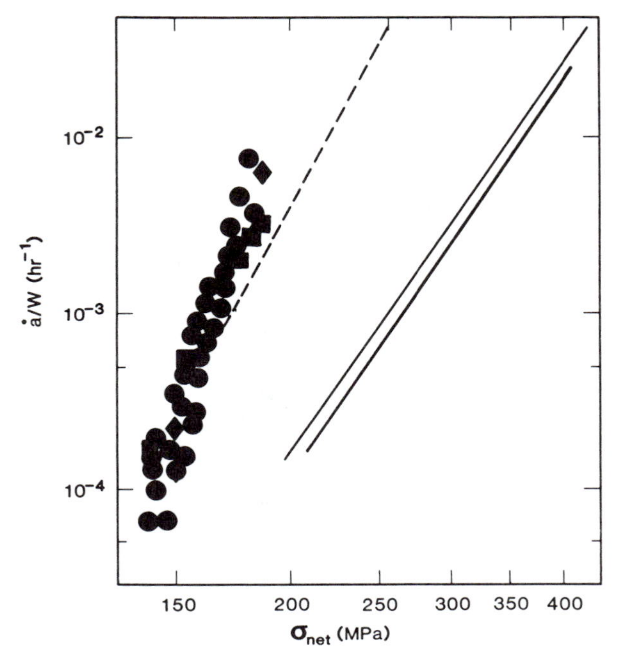

Figure 7.16 Crack growth rate normalized with respect to specimen width versus the net section stress in Type 304 stainless steel CN (◆), DEN (● and dashed curves), and NRB (solid curves) specimens (7.39).

appears. However, when C^* is used as the correlating load parameter in Figure 7.17b, the same data fall into a narrow band with slightly larger crack propagation rates occurring at 500° C. Again the crack growth rate is seen to be nearly proportional to C^*. Similar behavior has been observed for Type 304 stainless steel.

While the preceding correlations have been for creep crack growth in metals, such growth in polymers is becoming a concern as these materials are being increasingly used in load carrying members—for example, natural gas pipelines. Figure 7.18 shows the results of a creep crack growth study in Phillips M 8000, a high density polyethylene, at room temperature (7.44). The stress intensity factor is seen to correlate with the creep crack growth rate in three-point bend and compact tension specimens.

Because existing data are somewhat limited, it is difficult to quantify the conditions under which a loading parameter can be expected to be applicable. While conditions have been established for a stationary crack, analogous conditions for a growing crack remain to be developed. In the interim the conditions for a stationary crack may be used only as an approximate guide. Existing data indicate that for relatively ductile materials the C^*-integral is an appropriate loading parameter when the crack growth in the secondary creep regime is predominantly deformation controlled. The C^*-integral appears to

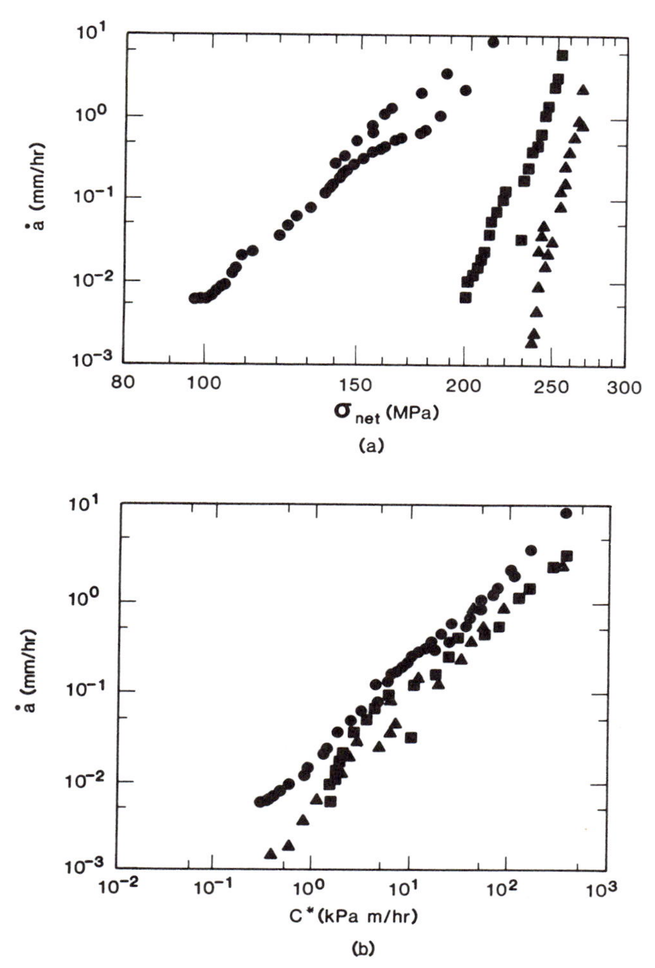

Figure 7.17 Correlation of crack growth rate in 0.16 percent carbon steel CNC specimens at 400°C (■) and 500°C (●) in air and 400°C (▲) in vacuum with (a) the net section stress and (b) the C^*-integral (7.39).

be applicable for large values of the creep exponent ($n > 5$–6) and relatively small crack growth rates (7.45). The stress intensity factor may be more appropriate for creep-brittle materials and environmentally assisted creep crack growth wherein the creep zone is confined to the crack tip region. A confined creep zone can be expected for small values of the creep exponent ($n < 3$) and relatively large growth rates.

It is clear from Equation (7.1-77) that $\dot{T}_1$ will be an appropriate loading parameter for crack growth in the secondary creep regime whenever C^* is. Moreover, because $\dot{T}_1$ and in general not C^* can be given an energy interpretation, it appears that the experimentalist may be "measuring" $\dot{T}_1$ rather thant C^* when an energy approach is used (7.46).

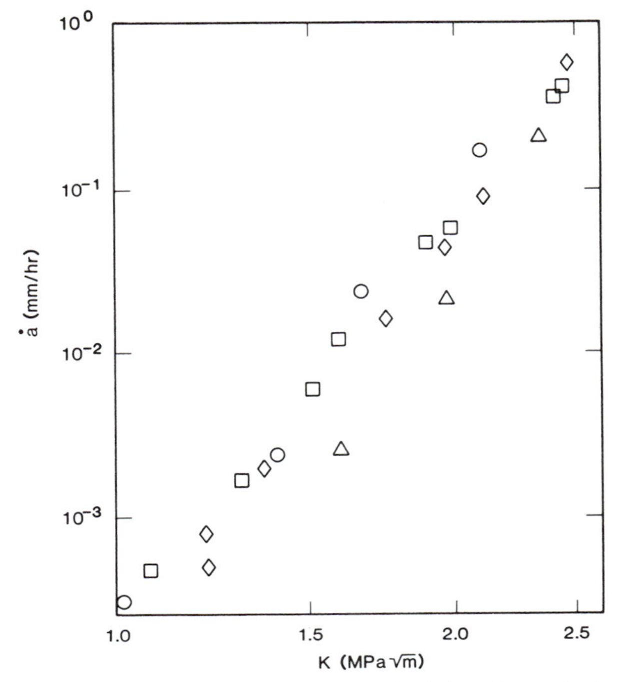

Figure 7.18 Creep crack propagation rate in M 8000 polyethylene three-point bend (◇, △) and compact tension (○, □) specimens versus the stress intensity factor.

7.4 Viscoelastic Crack Growth

While all solids under appropriate conditions will exhibit to various degrees viscoelastic behavior or rate sensitivity in their response to mechanical loadings, polymers appear particularly susceptible in even the most benign environment. Furthermore, polymers are finding more applications in which their load carrying capacity becomes an integral part of the performance of the structure—for example, plastic pipelines, binders in advanced composites, and adhesives to mention only a few. Consequently, time-dependent fracture plays an important role in determining the service life of polymeric as well as viscoelastic components.

To date most of the fracture mechanics approaches to crack propagation in polymeric structures have modeled the material as being linear viscoelastic. Such behavior can be reasonably expected in a crosslinked polymer above the glass transition temperature. In the absence of a generally accepted nonlinear viscoelastic theory, the linear analysis may be viewed as a first-order approximation to be used qualitatively, if not quantitatively, to gain an appreciation and understanding of the importance of parameters in the time-dependent fracture phenomenon. The literature on crack growth in linear viscoelastic solids is extensive; for example, see references (7.47)–(7.50).

When an element of a linear viscoelastic material is subjected to a plane strain tension, as depicted in Figure 7.19, the strain due to a stepped stress σ_0 can be written as

$$\varepsilon(t) = C(t)\sigma_0 \tag{7.4-1}$$

where $C(t)$ is the plane strain creep compliance. The short- and long-time limits of the strain, when they exist, correspond to the instantaneous and long time elastic response with

$$C(0) = (1 - \nu_0^2)/E_0, \qquad C(\infty) = (1 - \nu_\infty^2)/E_\infty \tag{7.4-2}$$

where ν_0, ν_∞ and E_0, E_∞ are the corresponding values of Poisson's ratio and modulus of elasticity of the isotropic material. For plane stress loading replace $1 - \nu_0^2$ and $1 - \nu_\infty^2$ by unity.

The stress field in a plane viscoelastic body subjected only to prescribed tractions is the same as that in an elastic body of the same geometry and loading provided the resultant force on any closed boundary vanishes. It follows as a corollary that the stress intensity factors for similarly loaded linear elastic and viscoelastic bodies satisfying this condition are equal.

Consider a linear viscoelastic material whose work of fracture $2\bar{\gamma}$ required to produce a unit of surface area is independent of the crack growth rate $\dot{a}$. For a very slowly growing crack ($\dot{a} \to 0^+$) the material is expected to respond nearly elastically according to the long time compliance $C(\infty)$. Based upon the energy balance criterion of fracture, the stress intensity factor K_0 required for this growth rate is given by

$$G = C(\infty)K_0^2 = 2\bar{\gamma}, \qquad \dot{a} \to 0^+ \tag{7.4-3}$$

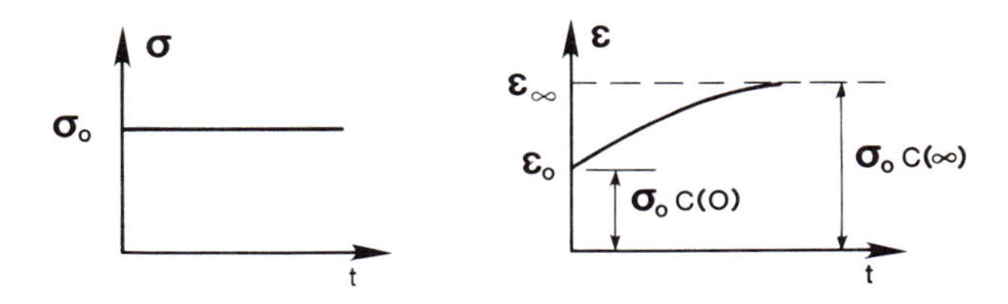

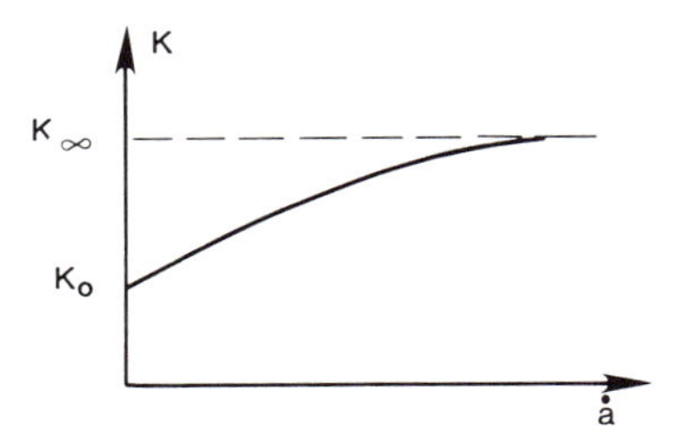

Figure 7.19 Typical viscoelastic behavior.

At the other extreme of a very rapidly propagating crack ($\dot{a} \to \infty$) the material is expected to respond nearly elastically according to its instantaneous properties. With the neglect of inertia effects the required stress intensity factor K_∞ satisfies

$$G = C(0)K_\infty^2 = 2\bar{\gamma}, \qquad \dot{a} \to \infty \tag{7.4-4}$$

The relationship between K and $\dot{a}$ for intermediate crack growth rates is depicted qualitatively in Figure 7.19.

If the crack tip is modeled as a mathematically sharp one, then an entirely different behavior is predicted. Due to the unloading of the crack-tip elements produced by the advancing tip, the local crack opening displacement is governed by the instantaneous elastic properties. Based upon a local work argument or, equivalently, the crack closure integral [Equation (3.3-15)], the energy balance criterion yields

$$G = C(0)K^2 = 2\bar{\gamma} \tag{7.4-5}$$

for all growth rates (7.51). This is in opposition to the anticipated behavior in Figure 7.19. The source of this apparent paradox is the neglect of the finite size crack-tip fracture process zone in the latter model (7.51, 7.52).

7.4.1 Cohesive Fracture Model

Recognizing the need to incorporate a fracture process zone in a time-dependent fracture model, Knauss et al. (7.53-7.55) and Schapery (7.56–7.58) extended the Dugdale model to include viscoelastic materials. The cohesive fracture model is depicted in Figure 7.20 for symmetric Mode I loading. The crack plane is defined by $x_2 = 0$ and the crack front is parallel to the x_3-axis and is located at $x_1 = a(t)$ relative to this fixed coordinate system. The coordinate $r = a(t) - x_1$ is attached to the moving crack tip as is the coordinate r_1. In the spirit of Dugdale the cohesive fracture process zone in which the material may exhibit nonlinearity and viscosity is confined to the strip $0 < r < \omega$. Outside of this strip the material is assumed to exhibit linear, isotropic viscoelastic behavior.

When the solution for the equivalent elastic problem is known, the solution to the viscoelastic problem can be produced by means of the viscoelastic correspondence principle (see Section 2.5). This principle, which is generally valid for stationary cracks, is also applicable for growing cracks provided that in the equivalent elastic problem the stress $\sigma_{\alpha 2}$ on $x_2 = 0$ is independent of E and ν and that the crack plane displacements have the separable form $v = f(E, \nu)g(r)$ (7.59, 7.60). The first provision is met if the cohesive stress $\sigma_c(r)$ is prescribed a priori as in the Dugdale model. The second condition will be satisfied if the resultant force on any closed boundary vanishes. For other boundary conditions it may not be possible to express v in the separable form. However, if Poisson's ratio is a constant, a condition approximately satisfied by many polymers, then the correspondence principle is still applicable even though ν is not separable in the elastic solution.

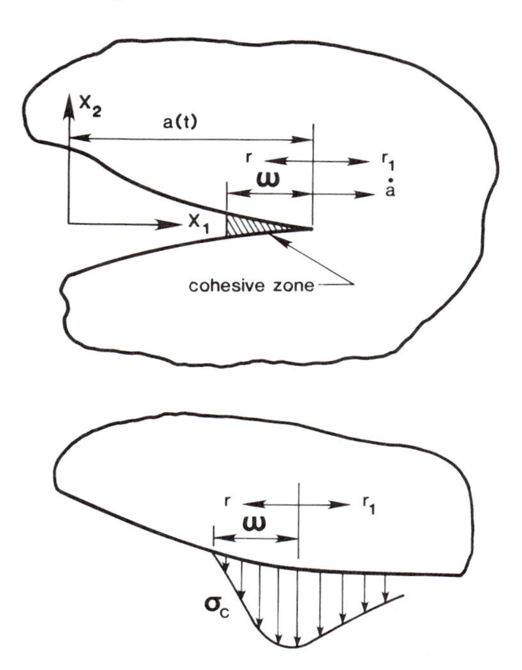

Figure 7.20 Cohesive viscoelastic fracture model depicting cohesive zone and nonsingular crack plane stress distribution.

If v_i is the elastic displacement field for a unit value of the elastic material compliance that is, $1/E' = 1$—the corresponding viscoelastic displacement field is

$$u_i = \int_0^t C(t-\tau)\frac{\partial v_i}{\partial \tau}\,d\tau \tag{7.4-6}$$

The dependence of v_i upon time stems from crack growth. If the material is thermorheologically simple, the influence of temperature on the viscoelastic compliance can be included through the time-temperature superposition principle. In this case $C(t)$ is replaced by $C\big(t/a_T(T)\big)$, where $a_T(T)$ is the shift function that depends upon the temperature T.

The elastic displacement can be established by superposing: (1) the solution for a cracked body with traction-free crack faces subjected to the applied remote loading and (2) the solution for the same cracked configuration with no remote loading but with a cohesive stress $\sigma_{22}^{(2)} = \sigma_c(r)$ acting over $0 < r < \omega$ on $x_2 = 0$. The crack plane stress $\sigma_{22}^{(1)}$ and the crack face displacement $v_2^{(1)}$ on $x_2 = 0^+$ from the first solution for Mode I loading [see Equations (3.1-38) and (3.1-39)] are

$$\sigma_{22}^{(1)} = \frac{KH(r_1)}{(2\pi r_1)^{\frac{1}{2}}}, \qquad r_1 \to 0 \tag{7.4-7}$$

and

$$v_2^{(1)} = 4KH(r)(r/2\pi)^{\frac{1}{2}}, \qquad r \to 0 \tag{7.4-8}$$

where $H(r)$ is the Heaviside step function. Within the confines of small-scale yielding, $\omega \ll a$, these quantities for the second solution are (7.61)

$$\sigma_{22}^{(2)} = -\frac{H(r_1)}{\pi(r_1)^{\frac{1}{2}}} \int_0^\omega \frac{\sigma_c(\zeta)}{\zeta^{\frac{1}{2}}} d\zeta, \quad r_1 \to 0 \tag{7.4-9}$$

and

$$v_2^{(2)} = -\frac{2}{\pi} H(r) \int_0^\omega \sigma_c(\zeta) \ln \left| \frac{\sqrt{\zeta} + \sqrt{r}}{\sqrt{\zeta} - \sqrt{r}} \right| d\zeta \tag{7.4-10}$$

The condition that the total stress σ_{22}, the sum of Equations (7.4-7) and (7.4-9), be nonsingular at $r_1 = 0$ requires that

$$K = \left(\frac{2}{\pi}\right)^{\frac{1}{2}} \int_0^\omega \frac{\sigma_c(\zeta)}{\zeta^{\frac{1}{2}}} d\zeta \tag{7.4-11}$$

For a prescribed cohesive stress Equation (7.4-11) determines the length ω of the cohesive zone. When the cohesive stress is written as $\sigma_c = \sigma_m f(r)$, where σ_m is the maximum value of σ_c and when the change of variable $\eta^2 = \zeta/\omega$ is made, Equation (7.4-11) can be rewritten as

$$\omega = \frac{\pi}{8} \frac{K^2}{\sigma_m^2 D^2} \tag{7.4-12}$$

where

$$D = \int_0^1 f(\eta^2)\, d\eta \leqslant 1 \tag{7.4-13}$$

For a constant cohesive stress, $\sigma_c = \sigma_y$, $D = 1$, and

$$\omega = \frac{\pi}{8} \left(\frac{K}{\sigma_y}\right)^2 \tag{7.4-14}$$

which is identical with the small-scale yielding result of Dugdale. Measured yielded zones in cracked sheets of the glassy polymers, polycarbonate, polysulfane, and polyvinylchloride, have been found to be in good agreement with predictions based upon the Dugdale model (7.62).

The total crack face displacement v_2 on $x_2 = 0^+$, the sum of Equations (7.4-8) and (7.4-10), can be written with the aid of Equation (7.4-11) as

$$v_2 = \frac{H(r)}{\pi} \int_0^\omega \sigma_c(s) \left[2\left(\frac{r}{s}\right)^{\frac{1}{2}} - \ln \left| \frac{\sqrt{s} + \sqrt{r}}{\sqrt{s} - \sqrt{r}} \right| \right] ds \tag{7.4-15}$$

The correspondence principle permits writing the crack opening displacement δ in the viscoelastic body as

$$\delta = \frac{8}{\pi} \sigma_y \int_{t_1}^t C(t - \tau) \frac{\partial}{\partial \tau} \left\{ \sqrt{\omega[a(\tau) - x_1]} \right.$$
$$\left. - \frac{\omega + x_1 - a(\tau)}{2} \ln \left| \frac{\sqrt{\omega} + \sqrt{a(\tau) - x_1}}{\sqrt{\omega} - \sqrt{a(\tau) - x_1}} \right| \right\} d\tau \tag{7.4-16}$$

when the cohesive stress is uniform. The lower limit of integration has been changed to t_1 since there is no crack opening displacement at the position $x_1 = a(t_1)$ prior to the arrival of the crack tip.

The work of fracture of an element in the cohesive zone from the time t_1 when it begins to deform until time t_2 after having attained its maximum extension δ_c at rupture is

$$2\bar{\gamma} = \int_{t_1}^{t_2} \sigma_c \frac{\partial \delta}{\partial t} dt \tag{7.4-17}$$

A more convenient form of Equation (7.4-17) is obtainable for a growing crack if δ is considered to be a function of x_1 and $r(x_1, t) = a(t) - x_1$, where $r(x_1, t_1) = 0$. The left end ($r = \omega$) of the cohesive zone will arrive at the point $x_1 = a(t_1)$ at $t = t_2$ when the crack-tip opening displacement δ_t is at its critical value δ_c. Hence, Equation (7.4-17) becomes

$$2\bar{\gamma} = \int_0^{\omega} \sigma_c \frac{\partial \delta}{\partial r} dr, \qquad \dot{a} > 0 \tag{7.4-18}$$

which reduces to

$$2\bar{\gamma} = \sigma_y \delta_c = \sigma_y \delta_t \tag{7.4-19}$$

for a constant cohesive strength.

During the time interval $t_2 - t_1$ for the crack to extend an increment ω, $\dot{a}$ and ω are assumed to be constants; i.e., $\omega = \dot{a}(t_2 - t_1)$. The combination of Equations (7.4-16) and (7.4-19) and the change of variables $s = \dot{a}(\tau - t_1)/\omega$ lead to

$$2\bar{\gamma} = \sigma_y \delta_c = \frac{K^2}{2} \int_1^0 C\left(\frac{\omega s}{\dot{a}}\right) \ln\left[\frac{1 - \sqrt{1 - s}}{1 + \sqrt{1 - s}}\right] ds \tag{7.4-20}$$

for the equation of motion of the crack tip. Equation (7.4-20) reduces to the anticipated limiting behavior of Equations (7.4-3) and (7.4-4). Furthermore, this equation of motion is universal in the sense that given two different cracked configurations in the same material the rate of crack growth in each will be identical if their stress intensity factors are equal. It is also apparent that the energy balance and the critical crack opening displacement fracture criteria are equivalent for a constant cohesive strength. For a given value of the work of fracture $2\bar{\gamma}$ or, equivalently, the cohesive strength and the critical crack-tip opening displacement and the stress intensity factor of the cracked configuration, Equation (7.4-20) can be used to determine the crack growth rate. This rate can be further integrated to yield the history of crack growth. In the limit as $\sigma_y \to \infty$ and, consequently, $\omega \to 0$, Equation (7.4-20) reduces to Equation (7.4-5) for a structureless crack-tip region.

Until the time t_i for initiation of crack growth the crack-tip opening displacment at $x_1 = a_0$, where a_0 is the initial crack length, is given by Equation (7.4-16) with $a(\tau) - x_1 = \omega(\tau)$ and $t_1 = 0$. The combination of

Equations (7.4-14), (7.4-16) and (7.4-19) yields

$$2\bar{\gamma} = \sigma_y \delta_c = \int_0^{t_i} C(t_i - \tau) \frac{\partial K^2(\tau)}{\partial \tau} d\tau \tag{7.4-21}$$

for the initiation time. For a step loading with $K(t) = KH(t)$, Equation (7.4-21) reduces to the transcendental equation

$$2\bar{\gamma} = \sigma_y \delta_c = K^2 C(t_i) \tag{7.4-22}$$

for t_i.

As an illustration of the use of this theory assume that over the time period of interest the creep compliance can be represented by

$$C(t) = C_0 + C_2 t^m \tag{7.4-23}$$

where $C_0 = C(0)$, C_2 and m are positive quantitites. When this compliance is introduced into Equation (7.4-20), the crack growth rate is

$$\dot{a} = \frac{\pi}{8}\left[\frac{C_2 I}{2\bar{\gamma}(1 - K^2/K_\infty^2)}\right]^{1/m} \frac{K^{2+2/m}}{\sigma_y^2} \tag{7.4-24}$$

where

$$I = \frac{1}{2}\int_1^0 s^m \ln\left[\frac{1-\sqrt{1-s}}{1+\sqrt{1-s}}\right] ds \tag{7.4-25}$$

and K_∞ is given by Equation (7.4-4). Note that the growth rate becomes unbounded as K approaches K_∞ from below. Within the confines of small-scale yielding, nonuniformity of the cohesive stress does not change the essential character of the kinetics of crack growth (7.58). One needs to only view σ_y as an effective uniform cohesive stress; that is, $\sigma_y = D\sigma_m$.

Based upon the observation that the second derivative of the logarithm of the creep compliance with respect to the logarithm of time for many viscoelastic materials is small, Schapery (7.58) has developed approximations to Equation (7.4-20) and its equivalent for a nonuniform cohesive stress. McCartney (7.63–7.65) has also developed approximate crack growth laws based upon the character of the second derivative of the creep compliance. The cohesive model has been analyzed by Wnuk (7.66) using a final stretch fracture criterion. The final stretch criterion represents an adaptation of the critical strain criterion. It is postulated that just prior to decohesion the amount of deformation within the cohesive zone is invariant.

7.4.2 Experimental Comparison

Consider a remotely tensioned center cracked panel with an instantaneous crack length $2a$ much less than its width. The stress intensity factor can be approximated by

$$K = \sigma\sqrt{\pi a} \tag{7.4-26}$$

where σ is the remotely applied step stress. When Equation (7.4-26) is introduced into Equation (7.4-24) and the integration performed, the crack history is given by

$$t = \frac{8}{\pi^2}\left(\frac{\sigma_y}{\sigma}\right)^2\left(\frac{C_0}{C_2 I}\right)^{1/m}\int_{a_0}^{a}(a_\infty/\zeta - 1)^{1/m}\,\frac{d\zeta}{\zeta} \qquad (7.4\text{-}27)$$

where $2a_0$ is the initial crack length. The crack length for which the crack propagation rate becomes unbounded is

$$a_\infty = \frac{K_\infty^2}{\pi\sigma^2} = \frac{2\bar{\gamma}}{\pi C_0 \sigma^2} \qquad (7.4\text{-}28)$$

Failure occurs at time t_f when the crack has grown to $a = a_\infty$. After a change of integration variable Equation (7.4-27) yields

$$t_f = \frac{8}{\pi^2}\left(\frac{\sigma_y}{\sigma}\right)^2\left(\frac{C_0}{C_2 I}\right)^{1/m}\int_{a_0/a_\infty}^{1}(1-\eta)^{1/m}\,\eta^{-(1+1/m)}\,d\eta \qquad (7.4\text{-}29)$$

Figure 7.21 shows measured failure times of a center cracked panel of Solithane 50/50, a crosslinked amorphous polyurethane rubber. The applied stress is normalized with respect to the minimum stress

$$\sigma_\infty = [2\bar{\gamma}/\pi a_0 C(\infty)]^{\frac{1}{2}} \qquad (7.4\text{-}30)$$

required to produce crack growth. The data for different temperatures have been reduced to the reference temperature of 0°C through the time–temperature superposition principle; that is, t is replaced by $t/a_T(T)$. The reported values of the material parameters for Solithane are $C(\infty) = 10^{-2.51}$ psi^{-1} and $\bar{\gamma} = 2.41 \times 10^{-2}$ lb/in. when adjusted for conditions of plane strain with $\nu = \frac{1}{2}$ (7.53, 7.54). With these values and $a_0 = 0.125$ in., Equation (7.4-30) yields $\sigma_\infty = 8.40$ psi for the critical stress. For the range of failure

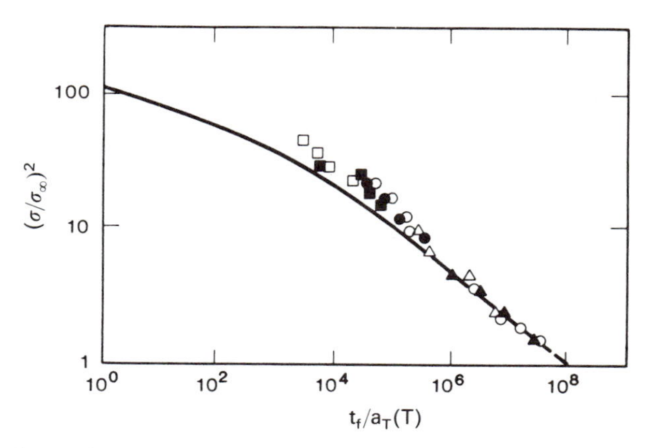

Figure 7.21 Comparison of measured and predicted [Equation (7.4-31)] failure times of Solithane 50/50 center-cracked panels at different temperatures as a function of the applied remote stress (7.54).

times of interest the creep compliance of Solithane can be fitted with the generalized power law of Equation (7.4-23) with $C_0 = 10^{-4.9}$ psi^{-1}, $C_2 = 10^{-1.8}$ psi^{-1} min$^{-\frac{1}{2}}$, and $m = \frac{1}{2}$ at a reference temperature of 0°C.

With $m = \frac{1}{2}$, Equations (7.4-25) and (7.4-29) yields, respectively, $I = 0.524$ and

$$t_f = \frac{8}{\pi^2}\left(\frac{\sigma_y}{\sigma}\right)^2\left(\frac{C_0}{C_2 I}\right)^2\left[\frac{3}{2}+\frac{1}{2}\left(\frac{a_\infty}{a_0}\right)^2 - 2\frac{a_\infty}{a_0} - \ln\left(\frac{a_\infty}{a_0}\right)\right] \quad (7.4\text{-}31)$$

When $a_\infty/a_0 \gg 1$, this equation simplifies to

$$t_f = \left[\frac{4}{\pi^2}\frac{\bar{\gamma}\sigma_y}{a_0 C_2 I}\right]^2 \sigma^{-6} \quad (7.4\text{-}32)$$

The data in Figure 7.21 indicates that $t_f = 10^8$ sec when $\sigma = \sigma_\infty$. This result along with Equation (7.4-31) can be used to deduce a cohesive strength of $\sigma_y = 7.7 \times 10^3$ psi. Now with all the material parameters established Equation (7.4-31) can be used to determine the failure time as a function of the applied stress, which is depicted by the solid curve in Figure 7.21. The agreement between the measured and predicted failure times is very good for the lower values of the applied stress. At the higher stress levels the theory underpredicts the failure time. This may be due to nonlinearities not included in the analysis that produce further blunting of the crack tip, and thereby contribute to an increased life by reducing the stress concentration and the damage ahead of the crack tip.

Figure 7.22 compares predicted crack growth rates based upon Equation (7.4-24) and measured rates in Solithane strips at various temperatures. The clamped lateral boundaries are displaced normal to the crack plane to produce

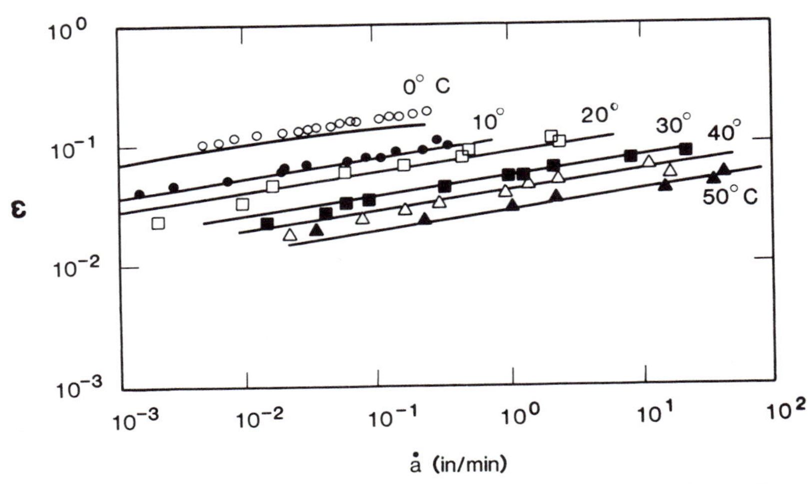

Figure 7.22 Comparison of measured and predicted [Equation (7.4-24)] crack growth rates in Solithane 50/50 cracked strips at different temperatures (7.53).

a uniform strain, $\varepsilon_{22} = \varepsilon$, in the remote regions of the specimen. Again very good agreement between the predicted and measured crack propagation rates exists for the small levels of strain. For the larger strains where nonlinear effects may be significant the predicted rate underestimates the measured rate.

Implicit in this development is the hypothesis that the intensity of the cohesive strength remains unchanged during crack growth. The implication of Equation (7.4-14) is that the length of the cohesive zone will change with K as the crack extends. The favorable comparisons of predicted and measured results are in agreement with this hypothesis. Kaminskii (7.49), on the other hand, observes that experiments using PMMA indicate that the crack-tip region for a growing crack does not depend upon the crack length; that is, the length of the cohesive zone remains invariant. Consequently, according to Equation (7.4-14) the intensity of the cohesive stress must change with crack growth. Kaminskii presents comparisons of measured times to failure in Solithane panels and of predicted times for both a constant cohesive stress and an invariant length of the cohesive zone. In general, the predictions based upon the latter premise agree better with the measured values. However, a number of approximations have been made to obtain these predictions. An essential one is equivalent to modeling the materials as a standard linear viscoelastic solid that is known to be an oversimplified model for the behavior of Solithane. To date the preponderance of experimental evidence seems to support a constant cohesive stress.

7.5 Closure

The analytical investigation of nonlinear creep crack growth is in its infancy and much remains to be learned and done. Because of the mathematical complexities, most of the studies have been confined to small-scale yielding for power law material behavior. Aside from those cases for which the latter conditions are satisfied, the understanding that these solutions afford is necessary in the pursuit of the study of creep crack growth under extensive creep and more general material laws. Progress is being made in modeling creep crack growth based upon grain-boundary diffusional cavity growth and coalescence in the highly strained crack-tip region (7.22, 7.67–7.69). The cavities are assumed to be aligned ahead of the crack tip only and off-axis cavities remain to be included. For the purpose of mathematical simplicity the stress field ahead of the crack tip is determined independently of the cavitation process even though the stress analysis and the cavity growth are not separable problems. Rather strong coupling is to be expected during the latter stages of the coalescence of the cavity and the crack tip when the remaining ligament between the two is smaller than the cavity size. This coupled problem has yet to be addressed. Unfortunately, these studies do not indicate any simple correlation between the overall loading and the crack growth rate. There are indications that the mechanisms controlling cavity nucleation may influence the crack growth rate.

While the C^*-integral has shown promise as a loading parameter to correlate crack growth rates, the question of what is the appropriate loading parameter is still unanswered. It is likely that there is no single all inclusive parameter, but instead the appropriate parameter probably depends, as for the stationary crack, upon the relative size of the creep zone. While the time periods for applicability of a parameter developed for a stationary crack can serve as a guide, there is a need to develop similar criteria, if they exist, for growing cracks. Because of its generality, the ΔT_k-integral represents a potential correlating parameter that deserves further study.

While analytical developments are being made, it is equally important that they be supplemented by experimental studies to aid the development and assessment of emerging mathematical models. Innovative experiments that are capable of furthering the understanding of diffusional cavity growth and coalescence associated with the fracture process are essential. There is also a need for better controlled creep crack growth tests that are capable of distinguishing the influence of environmental effects on crack growth.

Models of crack growth in viscoelastic materials have been primarily limited to linear material behavior. These models have employed almost exclusively an extension of the Dugdale yielded strip model. Even though they have been used to treat plane strain problems, they are in reality limited to plane stress. A rigorous plane strain model is lacking. Because many viscoelastic materials exhibit rather pronounced nonlinearities, there is a growing need for a nonlinear viscoelastic crack growth theory, particularly, for the new generation of ductile, tough polymers and composites that are being developed and used. In this vein efforts (7.70, 7.71) are underway to combine the correspondence principle and a generalized J-integral to form a nonlinear fracture theory.

Finally, as progress is made in relaxing the many existing restrictions—for example, small-scale yielding, steady-state crack growth, simplified material laws, and so on—efficient numerical algorithms will have to be developed if time-dependent fracture mechanics is to have any practical application.

7.6 References

(7.1) Riedel, H., "A Dugdale Model for Crack Opening and Crack Growth under Creep Conditions," *Materials Science and Engineering*, **30**, pp. 187–196 (1977).

(7.2) Riedel, H., "Cracks Loaded in Anti-Plane Shear under Creep Conditions," *Zeitschrift fur Metallkunde*, **69**, pp. 755–760 (1978).

(7.3) Riedel, H., "Creep Deformation at Crack Tips in Elastic-Viscoplastic Solids," *Journal of the Mechanics and Physics of Solids*, **29**, pp. 35–49 (1981).

(7.4) Riedel, H. and Rice, J. R., "Tensile Cracks in Creeping Solids," *Fracture Mechanics: Twelfth Conference*, ASTM STP 700, American Society for Testing and Materials, Philadelphia, Pa., pp. 112–130 (1980).

(7.5) Goldman, N. L. and Hutchinson, J. W., "Fully Plastic Crack Problems: The Center-Cracked Strip Under Plane Strain," *International Journal of Solids and Structures*, **11**, pp. 575-591 (1975).

(7.6) Kumar, V., German, M. D., and Shih, C. F., "An Engineering Approach for Elastic-Plastic Fracture Analysis," EPRI NP-1931, Project 1287–1, Topical Report, July 1981.

(7.7) Rice, J. R., Paris, P. C., and Merkle, J. G., "Some Further Results of J-Integral Analysis and Estimates," *Progress in Flaw Growth and Fracture Toughness Testing*, ASTM STP 536, American Society for Testing and Materials, Philadelphia, Pa., pp. 231–245 (1973).
(7.8) Landes, J. D. and Begley, J. A., "A Fracture Mechanics Approach to Creep Crack Growth," *Mechanics of Crack Growth*, ASTM STP 590, American Society for Testing and Materials, Philadelphia, Pa., pp. 128–148 (1976).
(7.9) Sadananda, K. and Shahinian, P., "Creep Crack Growth in Alloy 718," *Metallurgical Transactions A*, **8A**, pp. 439–449 (1977).
(7.10) Bassani, J. L. and McClintock, F. A., "Creep Relaxation of Stress Around a Crack Tip," *International Journal of Solids and Structures*, **17**, pp. 479–492 (1981).
(7.11) Morjaria, M. and Mukherjee, S., "Numerical Solutions for Stresses near Crack Tips in Time-Dependent Inelastic Fracture Mechanics," *International Journal of Fracture*, **18**, pp. 293–310 (1982).
(7.12) Shih, C. F., "Small-Scale Yielding Analysis of Mixed Mode Plane-Strain Crack Problems," *Fracture Analysis*, ASTM STP 560, American Society for Testing and Materials, Philadelphia, Pa., pp. 187–210 (1974).
(7.13) Ehler, R. and Riedel, H., "A Finite Element Analysis of Creep Deformation in a Specimen Containing a Macroscopic Crack," *Advances in Fracture Research*, Proceedings of the Fifth International Conference on Fracture, Cannes, D. Francois (ed.), Vol. 2, Pergamon, New York, pp. 691–698 (1981).
(7.14) Bassani, J. L., "Cracks in Materials with Hyperbolic-Sine Law Creep Behavior," *Elastic-Plastic Fracture Vol. I*, ASTM STP 803, American Society for Testing and Materials, Philadelphia, Pa., pp. 532–550 (1983).
(7.15) Bassani, J. L., "Macro and Micro-Mechanical Aspects of Creep Fracture," *Advances in Aerospace Structures and Materials-I*, S. S. Wang and W. J. Penton (eds.), ASME, New York, pp. 1–8 (1981).
(7.16) Hult, J. A. H. and McClintock, F. A., "Elastic-Plastic Stress Strain Distributions Around Sharp Notches Under Repeated Shear," *Proceedings of the 9th International Congress for Applied Mechanics*, Vol. 8, University of Brussels, pp. 51–58 (1957).
(7.17) Rice, J. R., "Mathematical Analysis in the Mechanics of Fracture," *Fracture-An Advanced Treatise*, Vol. II, H. Liebowitz (ed.), Academic, New York, pp. 191–308 (1968).
(7.18) Leckie, F. A. and McMeeking, R. M., "Stress and Strain Fields at the Tip of a Stationary Tensile Crack in a Creeping Material," *International Journal of Fracture*, **17**, pp. 467–476 (1981).
(7.19) Atluri, S. N., "Path-Independent Integrals in Finite Elasticity and Inelasticity, with Body Forces, Inertia, and Arbitrary Crack-Face Conditions," *Engineering Fracture Mechanics*, **16**, pp. 341–369 (1982).
(7.20) Stonesifer, R. B. and Atluri, S. N., "On a Study of the $(\Delta T)_c$ and C^* Integrals for Fracture Analysis Under Non-Steady Creep," *Engineering Fracture Mechanics*, **16**, pp. 625–643 (1982).
(7.21) Hui, C. Y. and Riedel, H., "The Asymptotic Stress and Strain Field Near the Tip of a Growing Crack under Creep Conditions," *International Journal of Fracture*, **17**, pp. 409–425 (1981).
(7.22) Hui, C. Y., "Steady-State Crack Growth in Elastic Power Law Creeping Materials," *Elastic-Plastic Fracture Vol. I*, ASTM STP 803, American Society for Testing and Materials, Philadelphia, Pa., pp. 573–593 (1983).
(7.23) Riedel, H. and Wagner, W., "The Growth of Macroscopic Cracks in Creeping Materials," *Advances in Fracture Research*, Proceedings of the Fifth International Conference on Fracture, Cannes, D. Francois (ed.), Vol. 2, Pergamon, New York, pp. 683–690 (1981).
(7.24) Hart, E. W., "A Theory for Stable Crack Extension Rates in Ductile Materials," *International Journal of Solids and Structures*, **16**, pp. 807–823 (1980).
(7.25) Rice, J. R., Drugan, W. S., and Sham, T.-L., "Elastic-Plastic Analysis of Growing Cracks," *Fracture Mechanics: Twelfth Conference*, ASTM STP 700, American Society for Testing and Materials, Philadelphia, Pa., pp. 189–221 (1980).
(7.26) Pilkington, R., "Creep Crack Growth in Low-Alloy Steels," *Metal Science Journal*, **13**, pp. 555–564 (1979).
(7.27) McMeeking, R. M. and Leckie, F. A., "An Incremental Crack Growth Model for High Temperature Rupture in Metals," *Advances in Fracture Research*, Proceedings of the Fifth International Conference on Fracture, Cannes, D. Francois (ed.), Vol. 2, Pergamon, New York, pp. 699–704 (1981).

(7.28) Popelar, C. H., Staab, G. H., and Chang, T. C., "Crack Growth in Elastic-Primary Creep Regime," The Ohio State University (1983).

(7.29) Sadananda, K. and Shahinan, P., "Review of the Fracture Mechanics Approach to Creep Crack Growth in Structural Alloys," *Engineering Fracture Mechanics*, **15**, pp. 327–342 (1981).

(7.30) Sadananda, K. and Shahinan, P., "Creep-Fatigue Crack Growth," *Cavities and Cracks in Creep and Fatigue*, J. Gittus (ed.), Applied Science, London, pp. 109–195 (1981).

(7.31) Harper, M. P. and Ellison, E. G., "The Use of the C^* Parameter in Predicting Creep Crack Propagation Rates," *Journal of Strain Analysis*, **12**, pp. 167–179 (1977).

(7.32) Saxena, A., "Evaluation of C^* for the Characterization of Creep-Crack-Growth Behavior in 304 Stainless Steel," *Fracture Mechanics: Twelfth Conference*, ASTM STP 700, American Society for Testing and Materials, Philadelphia, Pa. pp. 131–157 (1980).

(7.33) Kraus, H., *Creep Analysis* Wiley, New York (1981).

(7.34) Leckie, F. A. and Hayhurst, D. R., "Creep Rupture of Structures," *Proceedings of the Royal Society of London*, **340**, pp. 323–347 (1974).

(7.35) Williams, J. A. and Price, A. T., "A Description of Crack Growth From Defects Under Creep Conditions," *Journal of Engineering Materials and Technology*, **97**, pp. 214–222 (1975).

(7.36) MacKenzie, A. C., "On the Use of a Single Uniaxial Test to Estimate Deformation Rates in Some Structures Undergoing Creep," *International Journal of Mechanical Sciences*, **10**, pp. 441–453 (1968).

(7.37) Ponter, A. R. S. and Leckie, F. A., "The Application of Energy Theorems to Bodies Which Creep in the Plastic Range," *Journal of Applied Mechanics*, **37**, pp. 753–758 (1970).

(7.38) Ainsworth, R. A., "The Initiation of Creep Crack Growth," *International Journal of Solids and Structures*, **18**, pp. 873–881 (1982).

(7.39) Koterazawa, R. and Mori, T., "Applicability of Fracture Mechanics Parameters of Crack Propagation Under Creep Condition," *Journal of Engineering Materials and Technology*, **99**, pp. 298–305 (1977).

(7.40) Yokobori, T., Kawasaki, T., and Horiguchi, M., "Creep Crack Propagation in Austenitic Stainless Steel at Elevated Temperatures," Third National Conference on Fracture at Law Tatry, Slovakia, May 11, 1976.

(7.41) Ohji, K., Ogura, L., and Katada, Y., "Creep Crack Propagation Rate in SUS 304 Stainless Steel and Interpretation in Terms of Modified J-Integral," 25th Annual Meeting of the Society of Materials Sciences, Japan, pp. 19–20 (1976).

(7.42) Koterazawa, R. and Iwata, Y., "Fracture Mechanics and Fractography of Creep and Fatigue Creep Propagation at Elevated Temperature," *Journal of Engineering Materials and Technology*, **98**, pp. 296–304 (1976).

(7.43) Taira, S., Ohtani, R., and Kitamura, T., "Application of J-Integral to High-Temperature Crack Propagation Part I—Creep Crack Propagation," *Journal of Engineering Materials and Science*, **101**, pp. 154–161 (1979).

(7.44) Popelar, C. H. and Staab, G. H., "An Evaluation of the Critical Nature of Flaws in Plastic Pipeline Gas Distribution Systems," Proceedings Eighth Plastic Fuel Gas Pipe Symposium, New Orleans, pp. 62–68, Nov. 29–30, Dec. 1, 1983.

(7.45) Ainsworth, R. A., "Some Observations on Creep Crack Growth," *International Journal of Fracture*, **20**, pp. 147–159 (1982).

(7.46) Stonesifer, R. B. and Atluri, S. N., "Moving Singularity Creep Crack Growth Analysis with the $(\dot{\mathrm{T}})_c$ and C^*-Integrals," *Engineering Fracture Mechanics*, **16**, pp. 769–782 (1982).

(7.47) Knauss, W. G., "The Mechanics of Polymer Fracture," *Applied Mechanics Reviews*, **26**, pp. 1–17 (1973).

(7.48) Knauss, W. G., "Fracture of Solids Possessing Deformation Rate Sensitive Material Properties," *The Mechanics of Fracture*, F. Erdogan (ed.), AMD-Vol. 19, ASME, New York, pp. 69–103 (1976).

(7.49) Kaminskii, A. A., "Investigation in the Field of the Mechanics of the Fracture of Viscoelastic Bodies," *Soviet Applied Mechanics*, **16**, pp. 741–759 (1980).

(7.50) Williams, J. G., "Application of Linear Fracture Mechanics," *Advances in Polymer Science*, **27**, pp. 67–120 (1978).

(7.51) Kostrov, B. and Nikitin, L. V., "Some General Problems of Mechanics of Brittle Fracture," *Archiwum Mechaniki Stosowanej*, **22**, pp. 749–775 (1970).

(7.52) Barenblatt, G. I., Entov, V. M., and Salganik, R. L., "Some Problems of the Kinetics of Crack Propagation," *Inelastic Behavior of Solids*, M. F. Kanninen et al. (eds.), McGraw-Hill, New York, pp. 559–584 (1970).

(7.53) Knauss, W. G., "Delayed Failure-The Griffith Problem for Linearly Viscoelastic Materials," *International Journal of Fracture*, **6**, pp. 7–20 (1970).

(7.54) Mueller, H. K. and Knauss, W. G., "Crack Propagation in a Linearly Viscoelastic Strip," *Journal of Applied Mechanics*, **38**, pp. 483–488 (1971).

(7.55) Knauss, W. G., "On the Steady Propagation of a Crack in a Viscoelastic Sheet: Experiments and Analysis," *Deformation and Fracture of High Polymers*, H. H. Kausch et al. (eds.), Plenum, New York, pp. 501–541 (1974).

(7.56) Schapery, R. A., "A Theory of Crack Initiation and Growth in Viscoelastic Media I. Theoretical Development," *International Journal of Fracture*, **11**, pp. 141–159 (1975).

(7.57) Schapery, R. A., "A Theory of Crack Initiation and Growth in Viscoelastic Media II. Approximate Methods of Analysis," *International Journal of Fracture*, **11**, pp. 369–388 (1975).

(7.58) Schapery, R. A., "A Theory of Crack Initiation and Growth in Viscoelastic Media III. Analysis of Continuous Growth," *International Journal of Fracture*, **11**, pp. 549–562 (1975).

(7.59) Graham, G. A. C., "The Correspondence Principle of Linear Viscoelastic Theory for Mixed Boundary Value Problems Involving Time-Dependent Boundary Regions," *Quarterly of Applied Mathematics*, **26**, pp. 167–174 (1968).

(7.60) Graham, G. A. C., "The Solution of Mixed Boundary Value Problems that Involve Time-Dependent Boundary Regions for Viscoelastic Materials with One Relaxation Function," *Acta Mechanica*, **8**, pp. 188–204 (1969).

(7.61) Barenblett, G. J., "The Mathematical Theory of Equilibrium Cracks in Brittle Fracture," *Advances in Applied Mechanics*, Vol. VII, Academic, New York, pp. 55–129 (1962).

(7.62) Mills, N. J., "Dugdale Yielded Zones in Cracked Sheets of Glassy Polymers," *Engineering Fracture Mechanics*, **6**, pp. 537–549 (1974).

(7.63) McCartney, L. N., "Crack Propagation, Resulting from a Monotonic Increasing Applied Stress, in a Linear Viscoelastic Material," *International Journal of Fracture*, **13**, pp. 641–654 (1977).

(7.64) McCartney, L. N., "Crack Propagation in Linear Viscoelastic Solids: Some New Results," *International Journal of Fracture*, **14**, pp. 547–554 (1978).

(7.65) McCartney, L. N., "On the Energy Balance Approach to Fracture in Creeping Materials," *International Journal of Fracture*, **19**, pp. 99–113 (1982).

(7.66) Wnuk, M. P., "Quasi-Static Extension of a Tensile Crack Contained in a Viscoelastic-Plastic Solid," *Journal of Applied Mechanics*, **41**, pp. 234–242 (1974).

(7.67) Bassani, J. L., "Creep Crack Extension by Grain-Boundary Cavitation," *Creep and Fracture of Engineering Materials and Structures*, B. Wilshire and D. R. J. Owens (eds.), Pineridge, Swansea, U.K., pp. 329–344 (1981).

(7.68) Bassani, J. L. and Vitek, V., "Propagation of Cracks under Creep Conditions," Symposium on Nonlinear Fracture Mechanics, L. B. Freund and C. F. Shih (eds.), Proceedings of the Ninth U.S. National Congress of Applied Mechanics, 1982.

(7.69) Wilkinson, D. S. and Vitek, C., "The Propagation of Cracks by Cavitation: A General Theory," *Acta Metallurgica*, **30**, pp. 1723–1732 (1982).

(7.70) Schapery, R. A., "Nonlinear Fracture Analysis of Viscoelastic Composite Materials Based on a Generalized *J*-Integral Theory," Proceedings of the Japan-U.S. Conference on Composites, Tokyo, January 1981.

(7.71) Schapery, R. A., "Continuum Aspects of Crack Growth in Time-Dependent Materials," Texas A&M University Report MM4665-83-2 (prepared for *Encyclopedia of Materials Science and Engineering*, Pergamon, New York), February 1983.

8

SOME NONLINEAR ASPECTS OF FATIGUE CRACK PROPAGATION

Fatigue crack propagation is perhaps the most thoroughly studied area in all of fracture mechanics. A vast amount of data has been collected and the basic mechanisms have been identified. Yet, the theoretical relations that have been so far developed, which largely depend upon linear elastic fracture mechanics considerations, are not fully capable of treating the crack growth processes that occur in service. Specifically, while fatigue growth data are conventionally collected and correlated under constant amplitude cyclic loading, service conditions are generally characterized by nonuniform (spectrum) loading. As such, nonlinear effects play a key role.

Most applications to predict fatigue crack growth are made by appealing to the concept of "similitude." In essence, predictions are made from linear elastic fracture mechanics-based correlations of data obtained under conditions similar to those existing in the application area. In particular, similitude is achieved by matching the plastically deformed regions generated by the prior loading history. Because it is not possible to proceed in this way for the vast number of service conditions that exist, the development of reliable predictive relations for fatigue crack growth currently represents one of the most significant research opportunities in fracture mechanics.

Reflecting the importance of fatigue crack propagation behavior to structural integrity, substantial information already exists in the literature. In fracture mechanics terms, the books of Barsom and Rolfe, Broek, and Hertzberg (see Chapter 9 for references) all contain comprehensive accounts of fatigue experimentation and applications. In addition, the book of Kocanda (8.1) and that edited by Fong (8.2) are of interest for their comprehensive reviews of fatigue mechanisms. Among many noteworthy review articles are those of Plumbridge (8.3), Nelson (8.4), and Weertman (8.5). Because a plethora of further information also exists, a complete treatment of the subject will not be attempted here. The focus instead will be placed on the nonlinear analysis methods that are called for in the resolution of three specific aspects of the subject: (1) crack growth following an overload, (2) the short crack problem, and (3) crack growth in welds. We begin by providing some perspective for these applications by reviewing the conventional analytical characterizations of fatigue crack growth.

8.1 Basic Considerations in the Prediction of Fatigue Crack Propagation

The research that first decisively revealed the key role of the stress intensity factor in fatigue crack growth is attributable to work performed at the Boeing Company around 1960—see Section 1.3.1 and Paris' reminiscent article (8.6). General acceptance of the idea followed from the 1963 publication of Paris and Erdogan (8.7). They analyzed crack growth data from center-cracked panels of a high strength aluminum alloy under two different types of loading conditions. One condition was a remote tension, the other a concentrated force acting on the crack surface. When the load is cycled between constant values, the first of these produces K values that increase with crack length, while the second gives K values that decrease with crack length. Thus, because the fatigue crack growth data reported by Paris and Erdogan could be consolidated by ΔK, the use of this parameter as the driving force for fatigue cracking was well on its way to being established.*

8.1.1 Constant Amplitude Fatigue Crack Growth Relations

It is significant that virtually all verifications of the applicability of LEFM in fatigue have been of a generally empirical nature. For example, it is widely recognized that fatigue crack growth rates can often be represented by a simple relation of the form

$$\frac{da}{dN} = C(\Delta K)^m \tag{8.1-1}$$

where $\Delta K = K_{\max} - K_{\min}$, these referring to the maximum and minimum values of the stress intensity factors in any given load cycle. But, effective data correlations over the complete range of ΔK values require a modification of this relation. Figure 8.1 illustrates data collected by Paris et al. (8.8) for A533B steel (room temperature, $R = K_{\min}/K_{\max} = 0.1$), which shows the three distinct types of behavior that occur in fatigue crack propagation.

At one extreme, Figure 8.1 clearly shows a marked increase in growth at high ΔK values as $K_{\max}$ approaches K_c. This has led to the relation proposed by Forman et al. (8.9)

$$\frac{da}{dN} = \frac{C(\Delta K)^n}{(1 - R)K_c - \Delta K} \tag{8.1-2}$$

where K_c is the fracture toughness of the material. At the opposite extreme, Figure 8.1 shows that Equation (8.1-1) is similarly violated due to the threshold behavior. In this regime Donahue et al. (8.10) have suggested a

* It may be interesting to note that a key finding in the use of the K_{Ia} parameter for the arrest of rapid crack propagation was that crack arrest with K increasing with crack length could be predicted by using data taken from experiments with K decreasing; see Chapter 4.

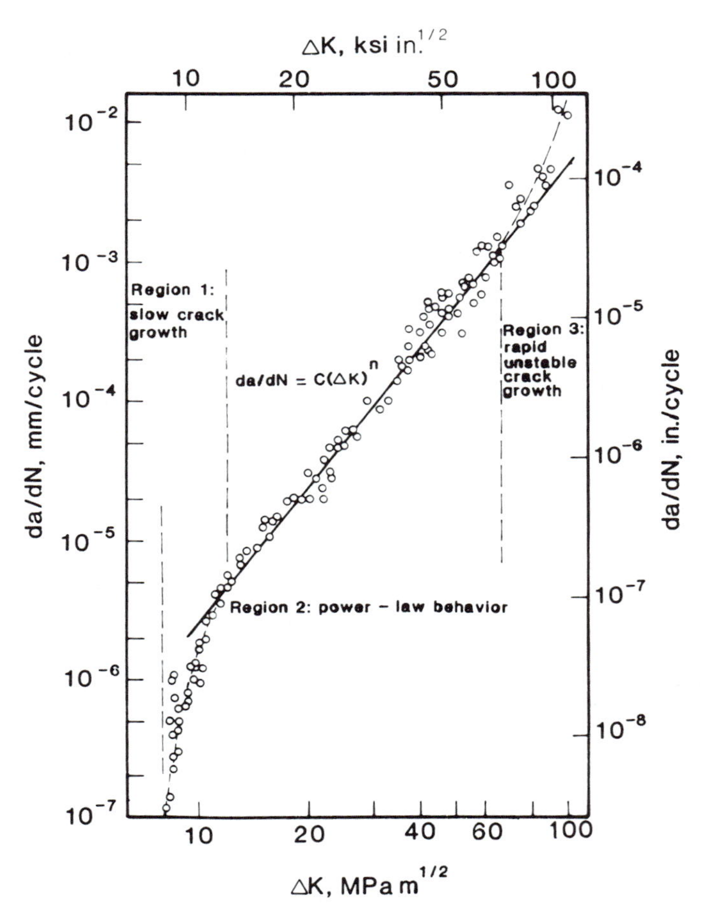

Figure 8.1 Fatigue crack growth behavior in A533 steel at ambient temperature and $R = 0.1$.

relation that can be generalized to

$$\frac{da}{dN} = C\{\Delta K - (\Delta K)_{\text{th}}\}^m \tag{8.1-3}$$

where $(\Delta K)_{\text{th}}$ denotes the limiting or threshold value of the stress intensity factor range. Relations combining the departures from power law behavior at high and low ΔK values also exist; for example, the relation developed by Priddle (8.11) is

$$\frac{da}{dN} = C\left(\frac{\Delta K - (\Delta K)_{\text{th}}}{K_c - K_{\max}}\right)^m \tag{8.1-4}$$

where the threshold stress intensity range is not a material constant but can depend upon R. For example, as reported by Schijve (8.12), for steel $(\Delta K)_{\text{th}} = A(1 - R)^\gamma$ in which γ is a constant that varies between 0.5 and 1.0 depending upon the type of steel.

As another illustrative example of the diversity of relations that have been developed, the relation suggested by Walker (8.13) can be written as

$$\frac{da}{dN} = C\left(\frac{\Delta K}{(1 - R)^n}\right)^m \tag{8.1-5}$$

This exhibits a particularly strong dependence upon R. Typical values that have been used are $m = 4$ and $n = 0.5$. An extensive review of the effect of mean stress upon fatigue crack propagation has been given by Maddox (8.14).

It is significant that none of the relations cited so far in this account (together with the many more that exist) has been developed from basic mechanics considerations. Each formula contains two or more disposable parameters that are evaluated from observed crack growth data. But, theoretical analyses to reduce the arbitrariness have also been made. One of the more notable efforts is that of McEvily and Groeger (8.15). Their relation can be written as

$$\frac{da}{dN} = \frac{A}{E\sigma_Y}(\Delta K - (\Delta K)_{\text{th}})^2\left[1 + \frac{\Delta K}{K_{Ic} - K_{\text{max}}}\right] \tag{8.1-6}$$

where A is an environment-sensitive material property with other symbols as commonly used. McEvily and his co-workers have introduced a further refinement by recognizing that the threshold stress intensity factor range is not really constant, but will generally depend upon the mean stress and the environment. They suggest that

$$(\Delta K)_{\text{th}} = \left(\frac{1 - R}{1 + R}\right)^{\frac{1}{2}} (\Delta K)_0 \tag{8.1-7}$$

where $R = K_{\text{min}}/K_{\text{max}}$ and $(\Delta K)_0$ is an environment-dependent material constant.

It is worth noting here that the inverse dependence of the crack growth rate upon the yield stress, exemplified by Equation (8.1-6) and typical of many other theoretical formulations, appears to be at odds with the observed behavior. Figure 8.2 shows results of Imhoff and Barsom (8.16) for 4340 steel indicating that the trend can actually be the opposite. Petrak and Gallagher (8.17) similarly found that constant amplitude crack growth rates are affected by the yield strength of steel with the higher strength steel exhibiting the faster cracking rates. Their results show that the magnitude of retardation following an overload cycle is also influenced by the yield strength: the lower strength material displaying significantly more retardation. However, it should be recognized that such results can sometimes be misleading. By altering one metallurgical property such as the yield stress, other properties will generally also be changed at the same time.

As a final example of the fatigue crack growth relations that exist, consider the general form suggested by Erdogan (8.18). This is

$$\frac{da}{dN} = \frac{C(1 + \beta)^m(\Delta K - (\Delta K)_{\text{th}})^n}{K_c - (1 + \beta)\,\Delta K} \tag{8.1-8}$$

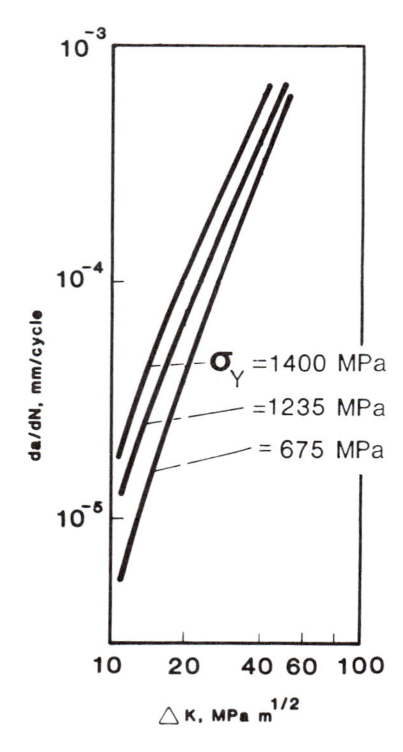

Figure 8.2 Effect of yield stress on fatigue crack growth rates in 4340 steel.

where

$$\beta = \frac{K_{\max} + K_{\min}}{K_{\max} - K_{\min}}$$

This gives the sigmoidal response exhibited by the data shown in Figure 8.1. According to Erdogan, this relation can be made to fit data over a range from 10^{-8} to 10^{-2} in./cycle by proper choice of the constants. Note that this relation contains five such constants: m, n, C, K_c and $(\Delta K)_{\text{th}}$. There is no theoretical basis for this relation. In fact, as Erdogan points out, despite the efforts that have been made (mostly by materials scientists) to prove the validity of any power law relation for fatigue, these relations remain mostly empirical.

It has been suggested by a number of investigators that $(\Delta K)_{\text{th}}$ must be related to the fatigue endurance limit of the material, σ_e, through the size of a dominant inherent crack. It follows from this that $(\Delta K)_{\text{th}} \propto \sigma_e a^{\frac{1}{2}}$, where $2a$ denotes the size of a characteristic natural flaw. This argument produces reasonable correlations. Nevertheless, the extent to which the threshold stress intensity range is a material property is currently an open issue. For example, see the recent article of Paris (8.6). This question is also intimately connected with the so-called "short crack" problem discussed in Section 8.2.3.

8.1.2 Load Interaction Effects

All of the above relations are based upon fatigue crack growth under constant amplitude loading cycles. With all of these results, there is a problem of predicting crack length for a load spectrum that may not be uniform. The simplest approach is that originated by Miner (8.19), which ignores the effect of load interactions. This can be written most succinctly as

$$\sum_{i=1}^{p} \frac{n_i}{N_i} = 1 \tag{8.1-9}$$

where n_i denotes the number of cyclic loads of a given intensity while N_i is the maximum number of load levels that could be experienced at that load level. In Equation (8.1-9), p is the number of distinct load levels in the load spectrum of interest. Unfortunately, this simple approach is not generally valid. The nonlinear effect of load intereactions is most clearly seen from the crack growth retardation that follows the imposition of a peak overload in a constant amplitude cyclic loading series. A particularly striking example, taken from von Euw et al. (8.20), is shown in Figure 8.3. Similar findings for other metals were reported by Jones (8.21) and by Matsuska and Tanaka (8.22).

Two mechanical explanations have been suggested to explain retardation: residual compressive plastic stresses and crack closure. The semiempirical models of Wheeler (8.23) and of Willenborg et al. (8.24) exemplify those based upon the former mechanism while that of Elber (8.25) exemplifies the latter. It is more likely that both effects occur simultaneously and, as discussed later in this section, some models have been offered that recognize this. However, it has not yet been possible to obtain closed-form relations such as those of Section 8.1.1 for conditions in which arbitrary load sequences occur. Nevertheless, as shown by Elber (8.26), Wood et al. (8.27), Broek and Smith (8.28), Schijve (8.29), and Johnson (8.30), acceptable crack growth predictions for service conditions can be made, albeit with some degree of empiricism.

Other than Elber's crack closure model, analyses of crack growth retardation following an overload have focused on the plastic deformation ahead of the crack tip and, in essence, the consequent difficulty experienced by the crack in forcing its way through it. One of the two most prominent models is that of Wheeler (8.23). His basic relations can be written as

$$a_n = a_0 + \sum_{j=1}^{n} C_{pi} f(\Delta K) \tag{8.1-10}$$

where

$$C_p = \begin{cases} \left(\dfrac{r_y}{a_p - a_n} \right)^m, & a_n + r_y < a_p \\ 1, & a_n + r_y \geqslant a_p \end{cases} \tag{8.1-11}$$

Here, a_n is the crack length at the time that the overload is applied, a_p is the current crack length, r_y is the size of the plastic zone created by the

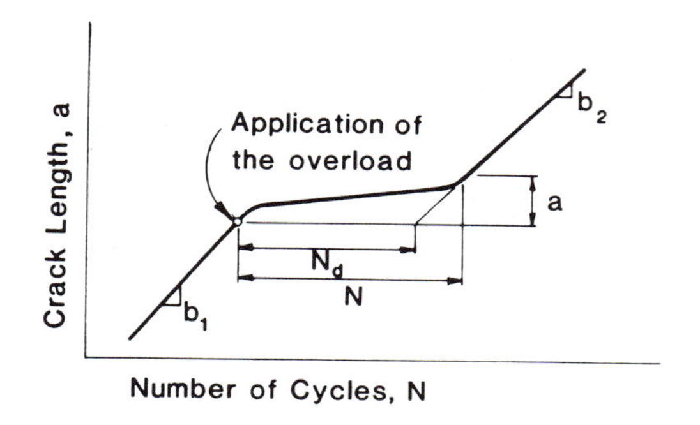

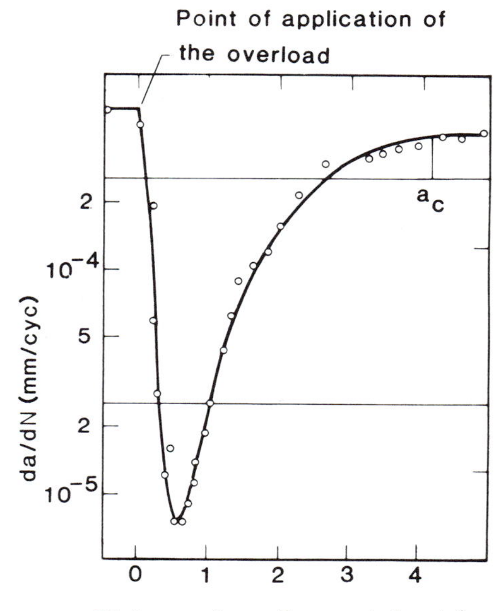

Figure 8.3 Crack growth retardation following a peak overload.

overload—for example, as estimated from Equation (1.4-2)—and m is an empirically determined "shape-fitting" constant. It is significant that m will generally depend upon the material and the nature of the load spectrum that is under consideration. Consequently, accurate results can only be expected from Wheeler's model if crack growth results are already available for load spectra similar to that for which a prediction is required.

Unlike Wheeler's model, the model advanced by Willenborg et al. (8.24) does not require an empirical parameter. It instead uses the material yield stress and is then used to calculate crack growth retardation according to the distance traveled into the overload plastic zone. Their basic equation can be expressed in terms of stress intensity factor in the form

$$K_R = (K_{\max})_{OL}\left(1 - \frac{\Delta a}{Z_{OL}}\right)^{\frac{1}{2}} - K_{\max} \tag{8.1-12}$$

where $(K_{\max})_{OL}$ is the maximum stress intensity factor achieved during the overload cycle, Δa is the amount of crack growth since the overload cycle, and Z_{OL} is the plastic zone radius associated with $(K_{\max})_{OL}$. It is of course assumed that when $\Delta a \geqslant Z_{OL}$, Equation (8.1-12) no longer applies. As later developed by Johnson (8.30), K_R is implemented via an effective R value defined as

$$R^{\text{eff}} = \frac{K_{\min} - K_R}{K_{\max} - K_R} \tag{8.1-13}$$

whereupon R^{eff} is used in place of R, for example, in the Foreman equation. Johnson further defines

$$Z_{OL} = \frac{1}{\beta\pi}\left[\frac{(K_{\max})_{OL}}{\sigma_Y}\right]^2 \tag{8.1-14}$$

where β, the plastic zone constraint factor, has a value of either one for plane stress or of three for plane strain.

A detailed study of the plastic zones attending fatigue crack growth was conducted by Lankford et al. (8.31). Using the parameter α to relate the plastic zone size to the characteristic dimension $(K/\sigma_Y)^2$, they were able to establish values of α for a variety of materials and loading conditions. Of most significance, they found that while α in the range from .06 to .10 adequately characterizes the maximum plastic zone dimension under monotonic loading, a value of $\alpha = .014$ is needed for cyclic loading. Both the Wheeler and the Willenborg models just described depend upon an explicit dimension of the plastic zone. Commonly, these use the monotonic values that Lankford et al. show can be a substantial overestimate. Hence, their predictions must be somewhat suspect.

8.1.3 The Crack Closure Concept

Although crack closure had long previously been recognized in fracture testing, use of the concept in fatigue awaited the work of Elber (8.25) in 1970. His was an empirical study based on observations revealing that the faces of fatigue cracks produced under zero-to-tension loading close during unloading whereupon large residual compressive stresses act on the crack surfaces at zero load. Consequently, the load at which the crack closes to preclude further propagation is tensile rather than zero or compressive. Figure 8.4 shows Elber's conceptualization, which connects crack closure to the existence of a zone of plastically defomed material behind the crack tip having residual tensile strains. Figure 8.5 shows the measured results by which Elber was able to make the crack closure concept quantitative.

Figure 8.5(a) shows the crack configuration and the gage location for the experiments conducted by Elber. Figure 8.5(b) shows a typical result for the applied stress as a function of the gage displacement. Of most significance, Elber observed that (1) the slope of the line between points A and B is linear and equal to that of the uncracked body, and (2) the slope of the line between points C and D is also linear but equal to that of the body having an open crack

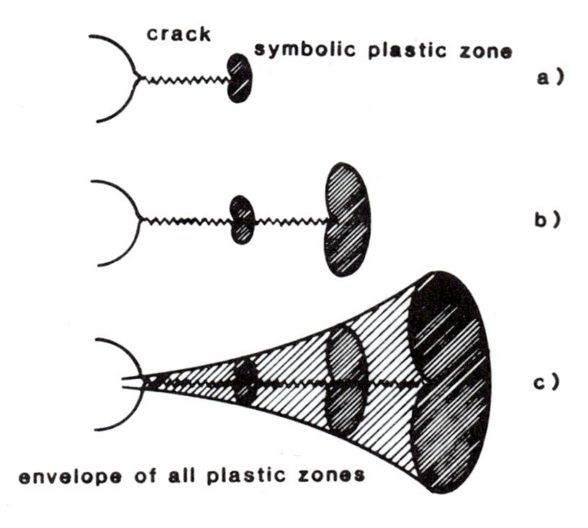

Figure 8.4 Conceptual illustration of plastic zones attending a fatigue crack.

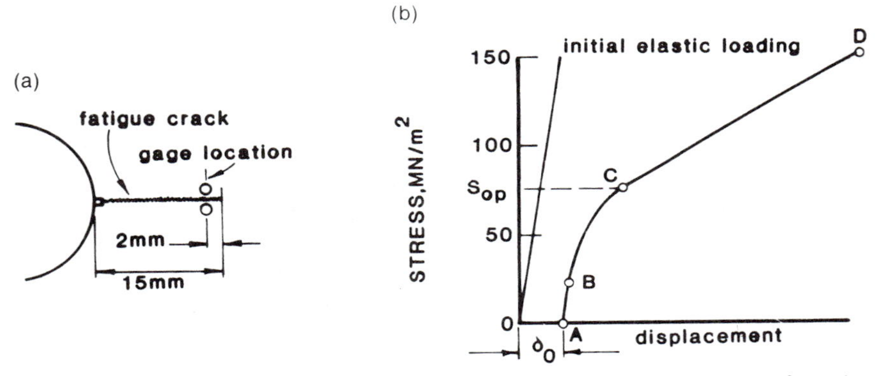

Figure 8.5 Crack closure results for fatigue crack propagation: (a) experimental configuration, (b) typical load-displacement record.

equal to the length of the fatigue crack. From this he concluded that the nonlinear line between points B and C corresponds to a transition between a fully open and a fully closed crack. Arguing that crack propagation can only occur during the portion of the loading cycle in which the crack is fully open, Elber defined an effective stress intensity factor range as

$$(\Delta K)_{\text{eff}} = K_{\max} - K_{\text{op}} \tag{8.1-15}$$

where K_{op} corresponds to the point at which the crack is fully open.

Elber suggested that this concept could be most effectively utilized through the use of an effective stress intensity factor range ratio taken in the form

$$U = \frac{K_{\max} - K_{\text{op}}}{K_{\max} - K_{\min}} \tag{8.1-16}$$

whereupon existing fatigue crack growth relations could be utilized by simply replacing ΔK by $U\Delta K$. For example, using the effective stress range concept,

Equation (8.1-1) would become

$$\frac{da}{dN} = C(U\Delta K)^m \tag{8.1-17}$$

The difficulty here, of course, is to determine appropriate K_{op} values to obtain U for the materials and conditions of concern. To accomplish this, Elber (8.25) performed a series of constant amplitude loading tests for aluminum 2024-T3. The results are shown in Figure 8.6.

Of some significance, the data of Figure 8.6 led Elber to conclude that U is independent of the crack length and of K_{max}. The results show an approximate linear behavior that he expressed as

$$U = 0.5 + 0.4R \tag{8.1-18}$$

in the range $-0.1 \leqslant R \leqslant 0.7$ covered by these experiments. Elber has also applied this concept to explain crack growth retardation and, while more complicated relations than those developed for constant amplitude loading would seem to be required, it is clear that crack closure is playing a key role in this process. Elber (8.26) subsequently applied an equivalent effective stress concept to predict crack growth under spectrum loading—that is, by choosing the maximum load and the crack opening load in the constant amplitude loading to be equal to those for the spectrum.

The following subsection addresses some of the mathematical modeling techniques that have been developed to predict fatigue crack propagation under variable amplitude loading. But, before turning to these, it will be useful to report upon the empirical efforts to expand upon Equation (8.1-18). First, it should be recognized that this has been a source of considerable controversy. For example, Shih and Wei (8.32) have taken direct issue with Elber's findings

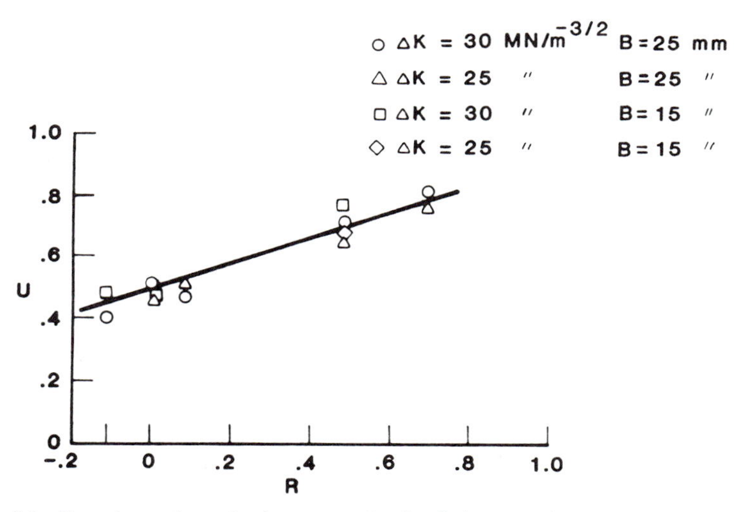

Figure 8.6 Experimental crack closure results for fatigue crack propagation in 2023-T3 aluminum alloy.

on the basis that many of Elber's specimens were under net section yielding conditions and, in addition, that his major conclusions were based upon a very limited set of data. Their experiments on Ti-6A1-4V titanium alloy indicated that, while crack closure does occur, the effective stress intensity range as developed by Elber may be over simplified. Hence, in claiming to account for delayed retardation and other aspects of fatigue crack growth, according to Shih and Wei, it may be over optimistic.

Gomez et al. (8.33) came to Elber's defense, presenting data reputed to support Equation (8.1-18) except near the fatigue crack growth threshold. For their troubles they were rebuked by Shih and Wei (8.34) who replotted the data of reference (8.33) to show a definite K_{max} dependence. These results are shown in Figure 8.7. Results demonstrating a lack of correlation with Equation (8.2-9) for 2219-T851 aluminum have also been given by Unangst et al. (8.35). In contrast, Schijve (8.36) reported test results on 2024-T3 aluminum that not only confirm Elber's relation, but extend it to negative R ratios. His suggested relation is

$$U = 0.55 + 0.33R + 0.12R^2 \tag{8.1-19}$$

which is supposed to be valid for the range $-1.0 < R < 0.54$. However,

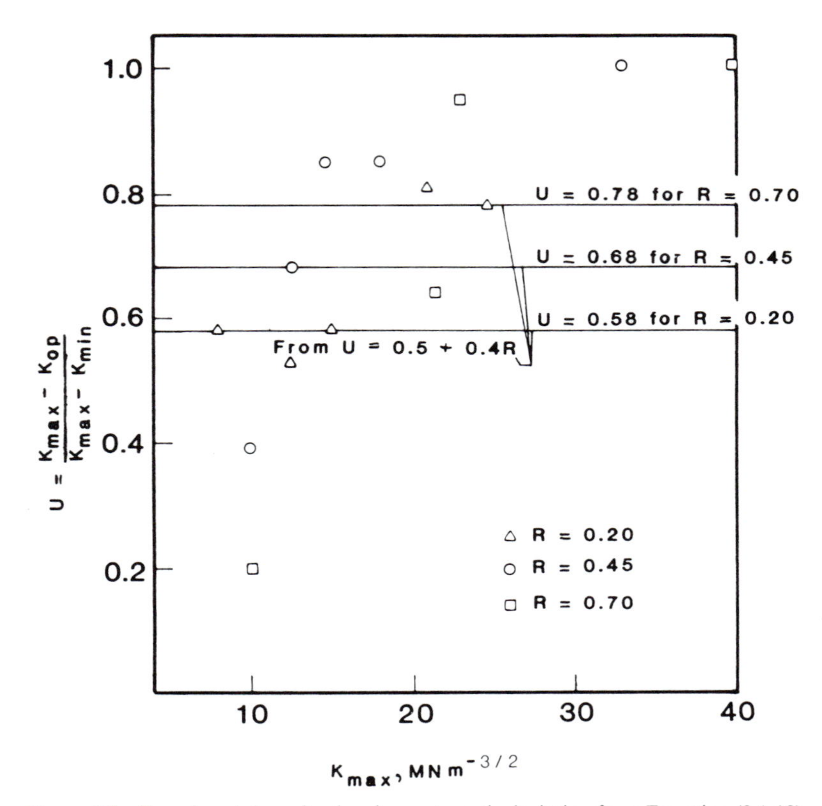

Figure 8.7 Experimental results showing systematic deviation from Equation (8.1-18).

Schijve did not address the possibility of the K_{max} dependence raised by Shih and Wei.

More recently, Chand and Garg (8.37) presented an empirical correlation for 6063-T6 aluminum alloy in the form

$$U = K_{max}(0.0088R + 0.006) + 1.3R + 0.2 \tag{8.1-20}$$

where K_{max} is in kg/mm$^{\frac{3}{2}}$. Such a result, albeit one that describes the data, is objectionable on esthetic grounds as U, by definition, must be dimensionless. Nonetheless, it does point up a conclusion reached earlier by Unangst et al.—that U is a function of K_{max}, R, and specimen thickness (or state of stress). However, they were unable to arrive at any sensible correlation and, on this basis, concluded that crack closure is a three-dimensional phenomenon that is not amenable to a simple treatment like that leading to Equation (8.1-18) et seq.

To conclude this discussion, the somewhat contradictory work of Probst and Hillberry (8.38) and of Lindley and Richards (8.39) might be briefly touched upon. The former performed fatigue crack growth experiments in 2024-T3 aluminum alloy and found fair agreement with Elber's equation—that is, the combination of Equations (8.1-17) and (8.1-18). Then, by extending this concept in a semiempirical manner to describe the effects of single peak overloads, a good correlation with the number of delay cycles was obtained. This would appear to add substantiation to the crack closure concept. However, the experimental observations of Lindley and Richards did not support this view, at least in predominantly plane strain conditions.

8.1.4 Closing Remarks

Further discussion on the predictive ability of LEFM-based fatigue crack growth models is beyond the scope of this book. It will suffice at this point to remark that, just as the foregoing suggests, such models do not perfectly mirror even constant amplitude fatigue crack behavior. Conventional fatigue crack growth models work because, being semiempirical, experimental results can be predicted when similitude exists. This is true despite the fact that the basic assumptions of LEFM are violated for fatigue. To understand this, consider the modern view of linear elastic fracture mechanics given in Chapter 3. As shown there, K-dominance exists only when the inelastic region can be contained within an annular region surrounding the crack tip. Obviously, for a growing crack that leaves a wake of residual plasticity behind it, this condition cannot be satisfied. Thus, similitude is important in that fatigue crack growth predictions can at least be made from correlations of data observed under conditions similar to the application. Needless to say, when similitude does not exist, nonlinear methods are required.

Most analysis work in fatigue crack propagation is empirical with the fatigue "laws" generally being relations containing enough disposable parameters to give a reasonable representation of any given body of data. To further illustrate this, we note that Kocanda (8.1) lists nearly 100 different fatigue laws; see also Mogford (8.40), for example. As might be expected, considerable controversy swirls about the research activity in the field. One key controversial area—crack closure—was briefly outlined in the above. Another, the so-called short crack problem, will be taken up in Section 8.2.3. What is important to recognize is not that such differences exist. All active developing research areas (e.g., dynamic fracture mechanics) will inevitably have their share of controversy. Rather, especially from the applied mechanics point of view, it is important to differentiate between what is established and what is at issue.

In the area of crack closure, the fact that the crack faces can impinge upon each other prior to reaching the minimum load in the unloading portion of a fatigue load cycle is not in dispute. That contact between the crack faces alters the driving forces for crack advance in some way is also definitely agreed upon. Questions on the validity of the "crack closure concept" instead center on the particular characterization introduced by Elber via his effective stress intensity factor range parameter, $(\Delta K)_{\text{eff}}$, and, more specifically, upon the generality of the simple form given by Equation (8.1-18).

Perhaps the most vociferous critics of this approach have been R. P. Wei and his associates. Their statement, even though written in 1977, would likely still be representative of the criticism that has been expressed. On the basis of their work with two different metals, they concluded that (8.32):

> No sensible correlation could be made between the fatigue crack growth kinetics and $(\Delta K)_{\text{eff}}$ obtained from the crack closure studies. Hence, the effective stress intensity concept, based on crack closure, is not able to account for the various aspects of fatigue crack growth under constant amplitude loading. Its extension, in its present form, to the more complex problems of fatigue crack growth and fatigue life prediction under variable amplitude loading does not appear to be warranted.

The final sentence is most damning. That is, it is precisely to cope with the variable amplitude loading that occurs in service that any new fatigue crack propagation approach is conceived and developed. Consequently, there would appear to be a need for more detailed analysis models. Some of the more prominent of those that have been offered are described next.

8.2 Theoretical Models for Fatigue Crack Propagation

Before describing the specific approaches in which nonlinear methods are needed for fatigue crack growth problems, one possibly useful finding of elastic-plastic fracture mechanics might be noted. This is the observation that stable crack propagation in ductile materials, after some initial transient, occurs with a virtually constant crack opening profile; see Chapter 5. It follows from these observations that the instantaneous crack-tip opening displacement (CTOD) must be very nearly a constant. While this fact is not directly related to subcritical crack growth, it does lend credence to the possibility that the CTOD can be an effective measure of the crack driving force for fatigue. Indeed, since under LEFM conditions K and CTOD are directly related, no loss of generality arises from the use of CTOD. At the same time the intriguing possibility exists that it may offer the basis for a more broadly applicable elastic-plastic fatigue relation.

8.2.1 The Model of Budiansky and Hutchinson

Budiansky and Hutchinson (8.41) have presented a theoretical model for the explicit purpose of examining Elber's crack closure concept. They employed an ideally plastic Dugdale model to consider the steady-state growth of a long crack under small-scale yielding conditions. Their aim was to estimate closure contact stresses and the consequent reduced effective stress intensity factors. Consistent with the use of the Dugdale model, they considered plane stress conditions. The unique feature of their analysis was that, not only does plastic deformation occur when the stress on the crack line becomes equal to σ_Y, but compressive crushing is allowed at $-\sigma_Y$. Hence, through the use of residual plastic stretches left in the wake of a growing fatigue crack, crack closure can occur. An illustration of their exercise of this concept on a cycle of unloading from K_{max} to zero with reloading to K_{max} is shown in Figure 8.8. It can be seen that contact occurs well before the applied stress has reached its minimum value.

The essence of the Budiansky-Hutchinson steady-state fatigue crack growth model is contained in two sets of assumptions. First, it is assumed that the crack line displacements at K_{max} are the same as in the ordinary Dugdale model except that a wake of plastically stretched material remains appended

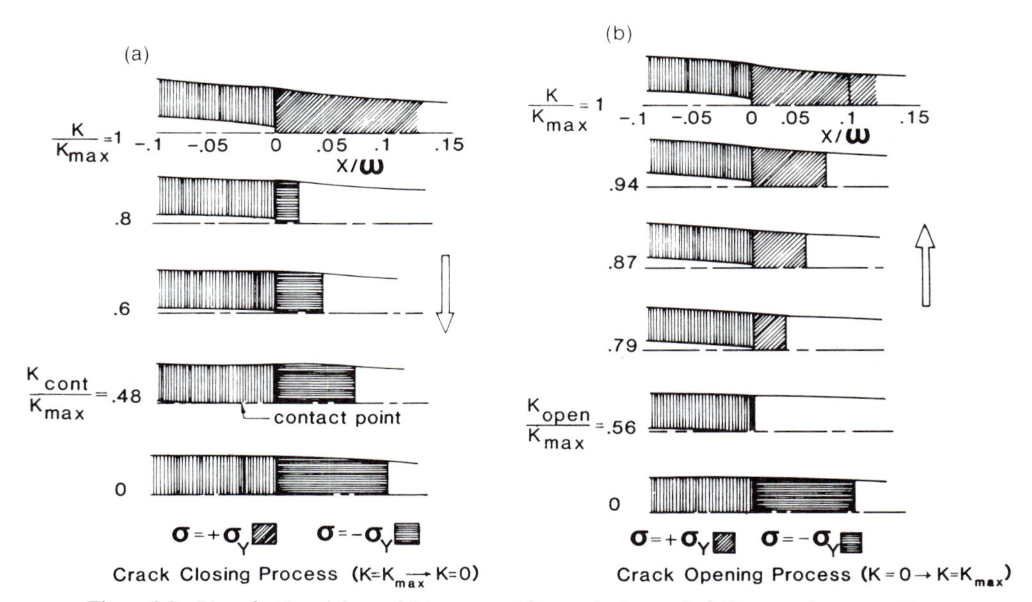

Figure 8.8 Use of a Dugdale model to account for crack closure in fatigue crack propagation: (a) crack closing process, (b) crack opening process.

to the upper and lower faces of the crack. The size of the stretched material denoted as $\delta_R/2$. The second set of assumptions is that for the state at K_{min}. These assumptions are that (1) the plastically stretched crack surfaces are in contact all along their lengths, (2) there is a region ahead of the crack tip that has gone into reverse plastic flow, and (3) beyond this region the plastic stretch that existed at K_{max} remains unchanged. A consequence of this assumption, it might be noted, is that the crack-tip strain at K_{min} is δ_R.

The analysis procedure used by Budiansky and Hutchinson is one employing the complex variable formulation of the theory of elasticity and parameters made nondimensional through the use of results from the stationary crack problem. For the convenience of the reader, the latter parameters can be expressed as follows. First, the plastic zone size is

$$l = \frac{\pi}{8}\left(\frac{K}{\sigma_Y}\right)^2 \tag{8.2-1}$$

while the crack opening displacement is

$$\delta_0 = \frac{K^2}{E\sigma_Y} \tag{8.2-2}$$

Second, the plastic zone displacements are given by

$$\delta(x) = \delta_0 g(x/l), \qquad 0 \leqslant x \leqslant l \tag{8.2-3}$$

where

$$g(\xi) = \sqrt{1-\xi} - \tfrac{1}{2}\xi \log\left(\frac{1+\sqrt{1-\xi}}{1-\sqrt{1-\xi}}\right) \tag{8.2-4}$$

We also note that, as shown by Rice (8.42), if the stationary crack is unloaded to $K_{min} = 0$, reverse plastic flow under compressive stresses $-\sigma_Y$ will occur in the interval $(0, l/4)$ while the plastic stretch in the remainder of the zone is unchanged. Hence, the plastic zone displacements become

$$\delta = \delta_0 \begin{cases} g(x/l) - \tfrac{1}{2}g(4x/l), & 0 < x < l/4 \\ g(x/l), & l/4 < x < l \end{cases} \tag{8.2-5}$$

It can readily be seen that the crack-tip stretch at $K_{min} = 0$ is reduced to half the value that it had at K_{max}.

Returning to the steady-state crack growth problem of Budiansky and Hutchinson, their modifications of the stationary crack problem result in the pair of integral integrations given by

$$\frac{\delta_R}{\delta_0} = \frac{\pi^2\alpha}{4} - \int_\alpha^1 \left(\frac{\xi-\alpha}{\xi}\right)^{\frac{1}{2}} g'(\xi)\,d\xi$$

and (8.2-6)

$$\frac{\delta_R}{\delta_0} = -\frac{\pi^2\alpha}{4} - \int_\alpha^1 \left(\frac{\xi}{\xi-\alpha}\right)^{\frac{1}{2}} g'(\xi)\,d\xi$$

where

$$g'(\xi) = \tfrac{1}{2}\log\left(\frac{1 + \sqrt{1-\xi}}{1 - \sqrt{1-\xi}}\right) \tag{8.2-7}$$

and α denotes the ratio of the reverse plastic zone for the growing crack to that of the stationary crack. A numerical solution of Equations (8.2-6) gives $\alpha = 0.09286$ whereupon Budiansky and Hutchinson obtained $\delta_R = 0.8562\delta_0$. Thus, in contrast to the stationary crack where reverse plastic yielding occurs in a quarter of the plastic zone, the growing crack experiences reverse yielding in less than 10 percent of the zone. In contrast, the residual stretch is a somewhat surprising 86 percent of the maximum value while in the stationary crack it is only one-half the original value.

Having determined the magnitude of the residual stretch, Budiansky and Hutchinson were then able to study the unloading and reloading process in detail. If there were no residual stretches, the crack opening displacement as a function of K, $0 \leqslant K \leqslant K_{\max}$, would be given by

$$\frac{\delta}{\delta_0} = g\left(\frac{x}{l}\right) - \frac{2\lambda}{l} g\left(\frac{x}{\lambda}\right) \tag{8.2-8}$$

where

$$\lambda = \frac{l}{4}\left(1 - \frac{K}{K_{\max}}\right)^2 \tag{8.2-9}$$

Equation (8.2-8) has a local minimum very near to the crack tip. When the stretch $\delta_R/2$ is attached to the crack faces, the first contact upon unloading will occur there. Denoting the applied K value at this instant as K_{cont}, it was found that

$$\frac{K_{\text{cont}}}{K_{\max}} = 0.48 \tag{8.2-10}$$

As the load is reduced further from K_{cont} the size of the contact zone will increase until complete closure occurs at $K = 0$.

In reloading from $K = 0$, the contact region begins to open from the center of the crack. The edge of the contact region moves progressively towards the crack tip as K increases until it vanishes at a level that can be denoted as K_{op}. It was found that

$$\frac{K_{\text{op}}}{K_{\max}} = 0.56 \tag{8.2-11}$$

As K is increased above K_{op}, the zone of plastic reloading spreads into the region ahead of the crack tip until the initial configuration is reasserted at $K = K_{\max}$. This process was illustrated in Figure 8.8. Figures 8.9 and 8.10 provide further details of this calculational procedure.

Figure 8.9 shows how the crack opening stretch varies during one loading cycle for the case of $K_{\min} = 0$. We note here that the actual process of crack

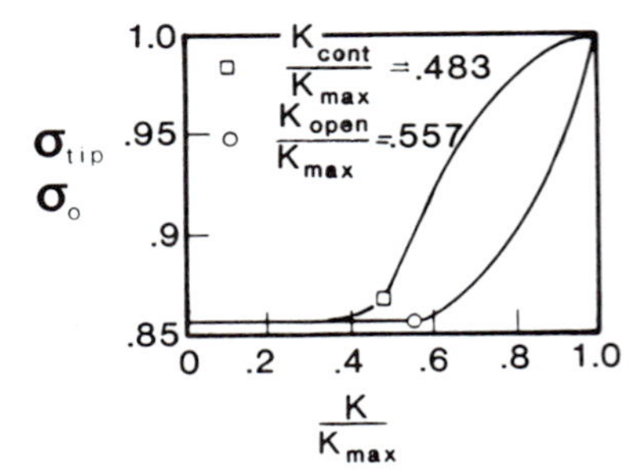

Figure 8.9 Calculated results for the crack-tip stretch during fatigue crack propagation.

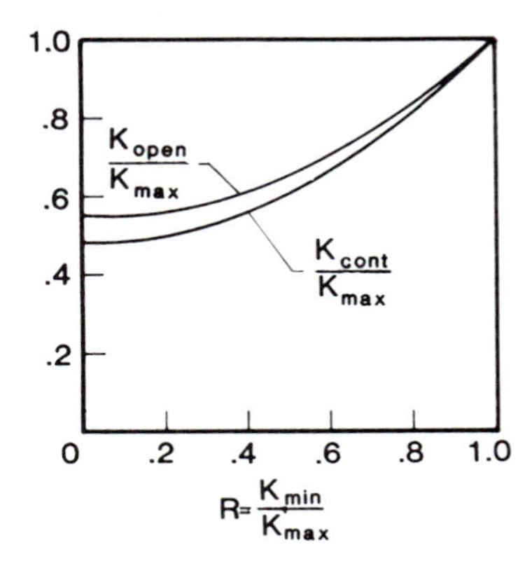

Figure 8.10 Calculated results for the opening and contact load ratios during fatigue crack propagation.

growth, which presumably could have been inferred from the change in δ_{tip} during a loading cycle, was ignored in this model. However, Budiansky and Hutchinson did use their results to provide indirect support for Elber's crack closure effective stress intensity factor range. Figure 8.10 shows the calculated values of K_{cont} and K_{op} as a function of R, results that compare favorably to Equation (8.1-18), which, when combined with Equation (8.1-16), can be written as (Elber's result):

$$\frac{K_{op}}{K_{max}} = 0.5 + 0.1R + 0.4R^2 \tag{8.2-12}$$

Next, they expressed the cyclic stretch at the crack tip in the approximate forms that can be expressed as

$$\delta_0 - \delta_R = 0.73\frac{(K_{max} - K_{op})^2}{E\sigma_Y} \tag{8.2-13}$$

and

$$\delta_0 - \delta_R = 0.54\frac{(K_{max} - K_{cont})^2}{E\sigma_Y} \tag{8.2-13}$$

Budiansky and Hutchinson, asserted that the crack growth rate for a particular material ought to be a function of the cyclic crack-tip stretch—that

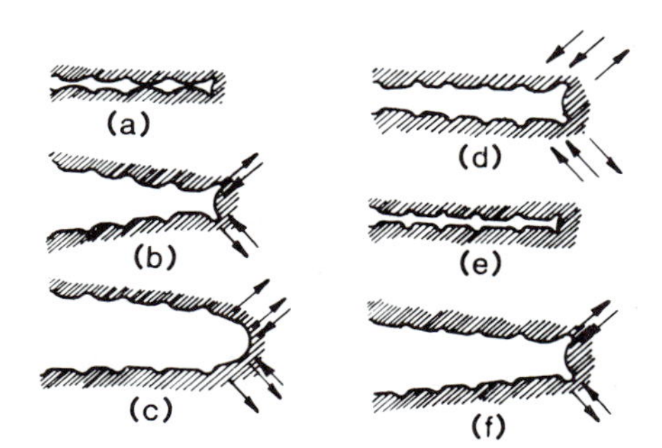

Figure 8.11 Schematic representation of fatigue crack growth by dislocation slip at the crack tip: (a) to (e) represent one load cycle, (f) represents a stage in the following load cycle.

is, of $\delta_0 - \delta_R$. Consequently, on the basis of Equations (8.2-13), they would also expect it to be a function of $(\Delta K)_{\mathrm{eff}} = K_{\max} - K_{\mathrm{op}}$ or $(\Delta K)_{\mathrm{eff}} = K_{\max} - K_{\mathrm{cont}}$. As they point out, their model, like Elber's, indicates that $(\Delta K)_{\mathrm{eff}}$ is independent of R.

The contribution of Budiansky and Hutchinson is particularly noteworthy in that it provided for the first time a theoretical approach to evaluating the effect of crack closure in fatigue. But, as they recognized, their model suffers from three major shortcomings. First, as already noted, they do not include cyclic crack growth in their model. Second, use of the Dugdale model is not representative of the plane strain conditions that better characterize fatigue crack growth. Third, their work is confined to constant amplitude cyclic loading. However, Lo (8.43) has extended this approach to admit a step change in the load. He finds that the effect is significant only so long as the extent of crack growth is within about one plastic zone size.

8.2.2 The Inclined Strip Yield Model

As illustrated in Figures 8.11 and 8.12, the generally representative models of Laird and Smith (8.44) and Neumann (8.45) associate crack growth with inelastic sliding off processes that occur off the crack plane. These are not well accommodated by yielding that is confined to the crack plane as in the

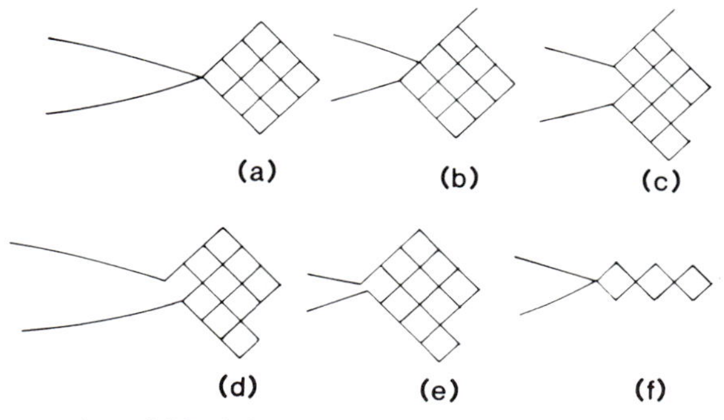

Figure 8.12 Geometric model for fatigue crack growth by shear sliding at the crack tip: (a) to (e) represent one load cycle, (f) represents a number of repetitions of the mechanism.

Dugdale model used by Budiansky and Hutchinson. The effect of residual plastic deformation also cannot adequately be given by the Dugdale model. Accordingly, a plane strain analysis would appear to be needed. Such a model was developed by Kanninen and co-workers (8.46, 8.47). Their approach contains a number of highly desirable features. These include the ability of the model to predict the crack advance increment from micromechanical considerations and to automatically account for residual plasticity and crack closure in plane strain conditions with appropriate plastic yielding.

The key element in the model is the simulation of the (symmetric) plastic field surrounding the crack tip by dislocation slip. This was originally developed by Bilby and Swinden (8.48) who introduced dislocation pileups on slip planes inclined at angles of $\pm\theta$ to the crack plane. However, the mathematics proved to be intractable. This led Atkinson and Kay (8.49) to replace each pileup by an aggregate "superdislocation." This idea was carried to fatigue by Kanninen et al. (8.46) who considered multiple load cycles with a superdislocation pair generated and retained for each load cycle. Because of the complexity that this entailed, the basis for a simpler version was later developed by Atkinson and Kanninen (8.50). This led to the steady-state crack propagation model advanced by Kanninen and Atkinson (8.47) that is discussed next.

Consider that a crack of length $2a$ exists in an infinite plane. A remote tensile loading σ is applied in the direction normal to the crack line with a remote stress $\lambda\sigma$ applied parallel to the crack line. Because force equilibrium of the superdislocations is specified to obtain a solution, a key relation is one for the shear stress τ_n acting along the slip plane at the position occupied by the nth superdislocation. For plane stress conditions this relation is

$$\tau_n = \sigma h_n + (1 - \lambda)\sigma \sin\theta\cos\theta + \frac{E}{8\pi}\sum_{j=1}^{M} b_j g_{jn}, \qquad n = 1, 2, \ldots, M \qquad ((8.2\text{-}15)$$

where the b_j's denote the superdislocation strengths with E and ν, respectively, denoting the elastic modulus and Poisson's ratio. The two remaining undefined quantities in Equation (8.2-15), h_n and g_{jn}, are rather involved functions of the complex variable representation of the superdislocation position given by

$$Z_j = a_j + l_j e^{i\theta}, \qquad j = 1, 2, \ldots, M$$

where θ is the angle between the crack plane and slip plane (a constant), a_j is the crack length at the time that the jth dislocation was emitted, and l_j is the distance from the crack tip along the slip plane to the jth dislocation. The complete expressions for these functions can be found in reference (8-50). They will not be repeated here as only the linearized versions given below are needed.

It is supposed that the material provides an intrinsic friction stress that resists the movement of the superdislocation. This parameter can be taken as $\sigma_Y \sin\theta\cos\theta$, where σ_Y is the tensile yield stress, as this will give rise to general yielding in the model when $\sigma = \sigma_Y$. The equations of force equilibrium that are the basis for the mathematical model are then formed by setting the right-hand side of Equation (8.2-15) equal to $\sigma_Y \sin\theta\cos\theta$ for each of the M superdislocations in the problem.

The second key relation involved in the Kanninen-Atkinson work is one for the crack-tip stress intensity factor. For the infinite medium under consideration here, this is

$$K = \sigma(\pi a)^{\frac{1}{2}} - \frac{E}{8(\pi a)^{\frac{1}{2}}} \sum_{j=1}^{M} b_j f_j \tag{8.2-16}$$

where f_j is a known function of position that represents the contribution of a dislocation of unit strength to the singular term. An expression for the crack-opening displacement is also needed. Again omitting the details, this result is

$$v = \frac{4\sigma}{E}(a^2 - x^2)^{\frac{1}{2}} + \sin\theta \sum_{j=1}^{M} b_j d_j \tag{8.2-17}$$

where $v = v(x)$ is one half of the crack-opening displacement at a position on the crack face $x < a$. In Equation (8.2-17), x is measured from the center of the crack and d_j is another known function of position given in reference (8.50). It gives the contribution of a dislocation of unit strength to the crack-tip displacement.

The final step in the formulation of the fatigue problem is to formulate the crack growth rate. The length of new crack surface projected along the original crack plane from the nascent superdislocation is $b_M \cos\theta$, where b_M is the strength of the dislocation emanating from the crack tip. By associating the crack advance increment with this length, and also taking account of environmental influences on the permanency of the new crack surface, this can be expressed as

$$\frac{da}{dN} = kb_M \cos\theta \tag{8.2-18}$$

where k is taken to be an environment-material constant that could account for the "rewelding" of the crack surfaces. Because the extent to which rewelding occurs depends upon the environment, in benign environments, k would be expected to be small compared to unity while in aggressive environments it would approach unity.

In the work reported in reference (8.46), fatigue crack growth computations were carried out using a "cycle-by-cycle" model—that is, one superdislocation pair to represent the crack-tip plasticity in each and every load cycle. In contrast to this complex approach, the calculations of reference (8.47) were performed using a much simpler two superdislocation model. The basic idea, partly supported by results obtained using the cycle-by-cycle model, is that one of the superdislocations left behind by the growing crack invariably dominates. Hence, a suitable model for steady-state crack growth under constant amplitude load cycles is one composed of one "residual" superdislocation pair and a "nascent" superdislocation pair. This model is illustrated in Figure 8.13.

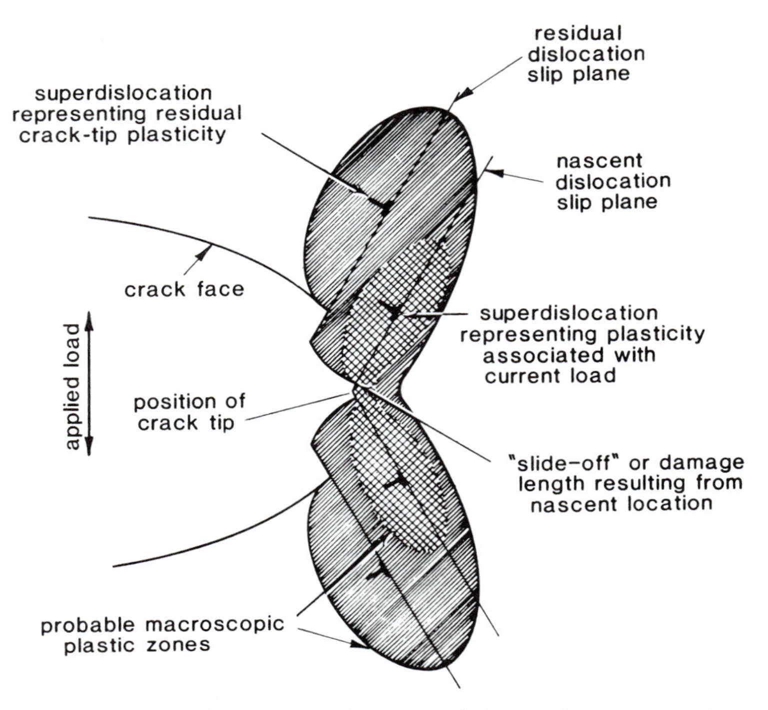

Figure 8.13 Superdislocation representation of crack-tip plastic zones in fatigue crack propagation.

The equations governing the conditions at maximum load are for the force equilibrium of the nascent dislocation and for the cancellation of the singularity. The latter condition is based on the assumption that crack growth occurs on the increasing part of the load cycle and ceases when the crack driving force is just zero. The residual superdislocation is positioned to reflect the wake of plasticity resulting from a large number of load cycles between two fixed K values.

Because only two superdislocation pairs will henceforth be considered, the subscript notation used in the general problem is unnecessary. Instead, the parameters B and b will be used to denote the strengths of the residual and the nascent superdislocations, respectively, while L and l will denote their distances from the crack tip along the slip plane. Consider for the moment that B and L are fixed by the prior load history and are known in some way. The variables to be determined then are b and l. This is done by satisfying the equations for the equilibrium of the nascent superdislocation and for the cancellation of the singularity at the maximum load in the cycle. Thus, from Equations (8.2-15) and (8.2-16)

$$\sigma_Y \sin\theta\cos\theta = \sigma_{\max}[h(l) + (1-\lambda)\sin\theta\cos\theta] + \frac{E}{8\pi}[bg(l) + Bg(L, l)] \tag{8.2-19}$$

and

$$a\sigma_{\max} = \frac{E}{8\pi}\,[bf(l) + Bf(L)] \tag{8.2-20}$$

In small-scale yielding where $l \ll a$, the functions appearing in Equations (8.2-19) and (8.2-20) are simply

$$f(l) = 6\left(\frac{2a}{l}\right)^{\frac{1}{2}} \sin\theta\cos\frac{\theta}{2} \tag{8.2-21}$$

$$g(l) = -\frac{1}{l} \tag{8.2-22}$$

$$h(l) = \frac{1}{4}\left(\frac{2a}{l}\right)^{\frac{1}{2}} \sin\theta\cos\frac{\theta}{2} \tag{8.2-23}$$

An expression for the function $g(L, l)$ can be deduced by assuming that the distance between the two slip planes can be neglected and, further, that the superdislocations occupy positions such that $a \gg L \gg l$. This is

$$g(L, l) = -\frac{3}{(lL)^{\frac{1}{2}}}\sin^2\theta\cos^2\frac{\theta}{2} \tag{8.2-24}$$

Notice that $g(l, L)$, the function that would appear in the equilibrium equation for the residual superdislocation, is given by $-lg(L, l)/L$ for the same assumptions. Thus, the effect of the nascent superdislocation on the residual superdislocation is negligible, consistent with the assumption that the parameters B and L are independent of b and l.

For the conditions under consideration (i.e., an infinite medium under remote tensile loading normal to the crack plane), it is appropriate to set

$$K_{\max} = \sigma_{\max}(\pi a)^{\frac{1}{2}} \tag{8.2-25}$$

Also, for convenience, a "residual plasticity stress intensity factor," K_R, can be defined as

$$K_R = \left(\frac{9}{8\pi}\right)^{\frac{1}{2}} \frac{EB}{L^{\frac{1}{2}}}\sin\theta\cos\frac{\theta}{2} \tag{8.2-26}$$

Then, Kanninen and Atkinson found that

$$\frac{da}{dN} = \frac{(3 - \sin^{-2}\theta\cos^{-2}\theta/2)}{9\sin\theta}\,\frac{k}{EY}\,\frac{(K_{\max} - K_R)^2}{(1 - (1 - \lambda)\sigma_{\max}/\sigma_Y)} \tag{8.2-27}$$

which expresses the steady-state crack growth rate for biaxial loading conditions.

Further simplification is possible by replacing B and L as parameters governing the residual plasticity in favor of two dimensionless parameters. This can be done by setting $B = \beta B_0$ and $L = \gamma L_0$, where B_0 and L_0 refer to the values of B and L that would exist under the monotonic loading of a virgin specimen to a given K_{max} level. These can be evaluated by solving the equilibrium and singularity canceling equations corresponding to this situation. Equation (8.2-27) then becomes

$$\frac{da}{dN} = \frac{(3 - \sin^{-2}\theta \cos^{-2}\theta/2)}{9\sin\theta}\frac{k}{EY}\left(1 - \frac{\beta}{\gamma^{\frac{1}{2}}}\right)\frac{K_{max}^2}{(1-(1-\lambda)\sigma_{max}/\sigma_Y)} \quad (8.2\text{-}28)$$

where $\beta = \beta(R)$ and $\gamma = \gamma(R)$ are dimensionless functions of $R = K_{min}/K_{max}$. It is to be expected that $\beta \leqslant 1$ while $\gamma \geqslant 1$ to take proper account of residual plasticity.

The parameters β and γ can be evaluated by employing the crack closure prediction capability of the model. The appropriate expression can be obtained by using a linearized version of the d_j function appearing in Equation (8.2-17). Assuming that $c \ll L$ as well as $L \ll a$, this is

$$d(L, a - c) = -\frac{6}{\pi}\left(\frac{c}{L}\right)^{\frac{1}{2}}\cos\frac{\theta}{2} \quad (8.2\text{-}29)$$

where c denotes the distance between the crack tip and the point of intersection of the superdislocation slip plane and the crack face; that is, the point $x = a - c$.

Substituting Equation (8.2-29) into Equation (8.2-17) gives an expression for the crack opening displacement at the potential point of contact between the two crack surfaces. Neglecting the influence of the nascent superdislocation gives

$$v = \frac{4\sigma}{E}(2ac)^{\frac{1}{2}} - \frac{6}{\pi}B\left(\frac{c}{L}\right)^{\frac{1}{2}}\sin\theta\cos\frac{\theta}{2} \quad (8.2\text{-}30)$$

where the load-dependent contribution has also been linearized for consistency. For a given B and L, the applied stress at which contact of the crack face will occur is simply the applied stress level at which the right-hand side of (8.2-30) vanishes. Call this σ_c. Then, defining $K_{cont} = \sigma_c(\pi a)^{\frac{1}{2}}$, an expression for the stress intensity at crack closure can be obtained by setting $v = 0$. This is

$$K_{cont} = \frac{\beta}{\gamma^{\frac{1}{2}}}K_{max} \quad (8.2\text{-}31)$$

It can be shown that K_{cont} is just the same as K_R. Thus, it would seem that whether one takes the point of view that residual stress or crack closure effects govern fatigue crack growth rates is irrelevant if, as these results suggest, both effects occur simultaneously.

As an expedient to test the results of the model against experimental results, Kanninen and Atkinson introduced a degree of empiricism, via the experimental result of Elber, to set values of β and γ. Using Equations (8.1-16) and (8.1-18) and ignoring any difference between an "opening" and a "closing" level, it can be seen that if

$$\frac{\beta}{\gamma^{\frac{1}{2}}} = 0.5 + 0.1R + 0.4R^2 \quad (8.2\text{-}32)$$

then the model will give a prediction of crack closure that is in exact agreement with Elber's observation. Taking this is to be appropriate, Equation (8.2-32) serves to define β and γ (n.b., individual values are not needed).

To specialize the above results, the particular angle $\theta = \cos^{-1}(\frac{1}{3})$, the angle that maximizes the extent of plastic yielding, can be chosen. Substituting in Equation (8.1-28) then gives

$$\frac{da}{dN} = \frac{7}{64\sqrt{2}} \frac{k}{E\sigma_Y} (1 - 0.2R - 0.8R^2) \frac{K_{\max}^2}{1 - (1 - \lambda)\sigma_{\max}/\sigma_Y} \tag{8.2-33}$$

which, apart from the environment-dependent constant k, provides a crack growth prediction in terms only of load-history parameters and mechanical-material constants.

The data used to test Equation (8.2-33) must include either explicit $\sigma_{\max}$ or crack length values in addition to $K_{\max}$ and R. Unfortunately, these are not often reported. Thus, a particularly useful set of experimental data is that of von Euw et al. (8.20) who obtained data on uniaxial loaded ($\lambda = 0$) 76-mm-wide SEN specimens at two different crack length to specimen width ratios; $a/w = 0.25$ and 0.45. As they point out, the conventional view is that, for the same values of $K_{\max}$ and $K_{\min}$, the same crack growth rate should be obtained for all a/w values. However, this was not the case. To determine if Equation (8.2-33) could account for their observed differences, the net section values in their experiments were calculated. Specifically, letting $f(a/w)$ denote the finite width correction for the SEN specimen used by von Euw et al., then $f(0.25) = 2.665$ and $f(0.45) = 4.294$. For $w = 76$ mm, this procedure gives $\sigma_{\max} = K_{\max}/1.731$ and $K_{\max}/2.744$ for $a/w = 0.25$ and 0.45, respectively. Other values used in the calculations are $E = 7 \times 10^4$ MN/m^2 and $\sigma_Y = 362$ MN/m^2.

Comparisons between the prediction of Equation (8.2-33) and the results of von Euw et al. for two different kinds of tests are shown in Figures 8.14 and 8.15. These results were obtained by setting $\lambda = 0$ in Equation (8.2-33) and using the experimental results to establish an empirical value of $k = 0.7$. In both figures it can be seen that the model provides an excellent prediction of the growth rates. Also, the difference between the results for the different a/w ratios is accounted for. The conventional Paris law formulations will not, of course, differentiate between the different crack lengths.

The biaxial loading case gives a result that is conveniently written in the form

$$\frac{da}{dN} = \left(\frac{da}{dN}\right)_{\lambda=0} \left(1 + \lambda \frac{\sigma_{\max}}{\sigma_Y - \sigma_{\max}}\right)^{-1} \tag{8.2-34}$$

where $(da/dN)_{\lambda=0}$ denotes the result obtained under uniaxial loading. This form indicates that a tensile stress applied parallel to the crack line decreases the crack growth rate while a compressive stress increases it. Further, the change induced by a parallel applied stress will depend on its magnitude relative to the yield stress. The results are in quantitative agreement with the

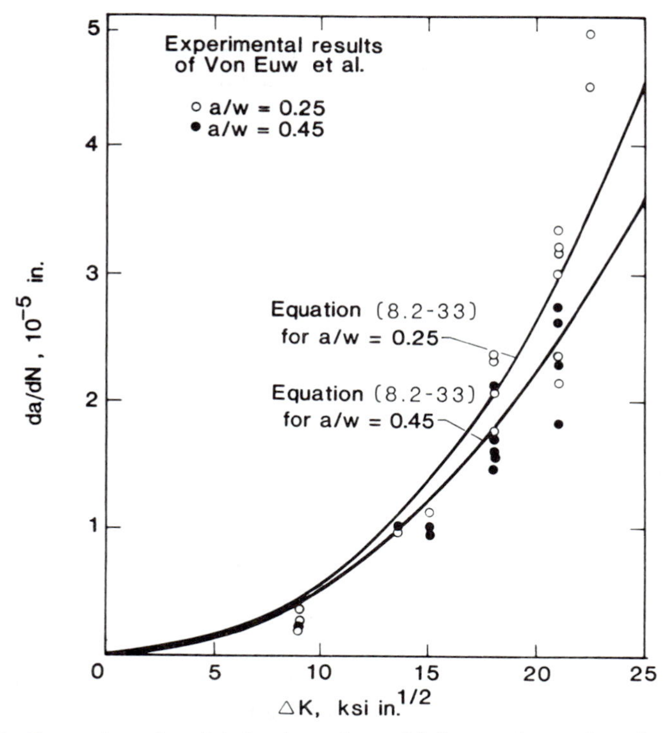

Figure 8.14 Comparison of predicted and experimental fatigue crack growth results in 2024-T3 aluminum alloy for $R = 0$.

observations of the effect of biaxial loading reported in the literature. That is, compressive loadings acting parallel to the crack plane act to increase the growth rate while tensile loadings decrease it. Quantitatively, however, the predicted effect appears to be too large. The differences that some investigators have reported between cycling the parallel stress and holding it constant are also not reflected by Equation (8.2-34).

The exhaustive data of Liu and Dittmer (8.51) on aluminum alloys 2024-T351 and 7075-T351 under multiaxial loading reveals that a constant amplitude fatigue crack will grow in a straight line so long as the stress component parallel to the crack does not exceed that normal to the crack (it will curve at higher values). Liu and Dittmer claim that the effect of the parallel stress component was negligible. While seemingly contradicting the predictions of the model developed in this paper, a close examination of their data does indicate a slight, but discernable, effect of λ that is in qualitative agreement with Equation (8.2-34). Nevertheless, while the model predictions are again consistent with these observations, at the current level of development, the effect is exaggerated; see also Adams (8.52).

A number of other research efforts have employed dislocation models to develop a general fatigue crack growth model. Included are the efforts of

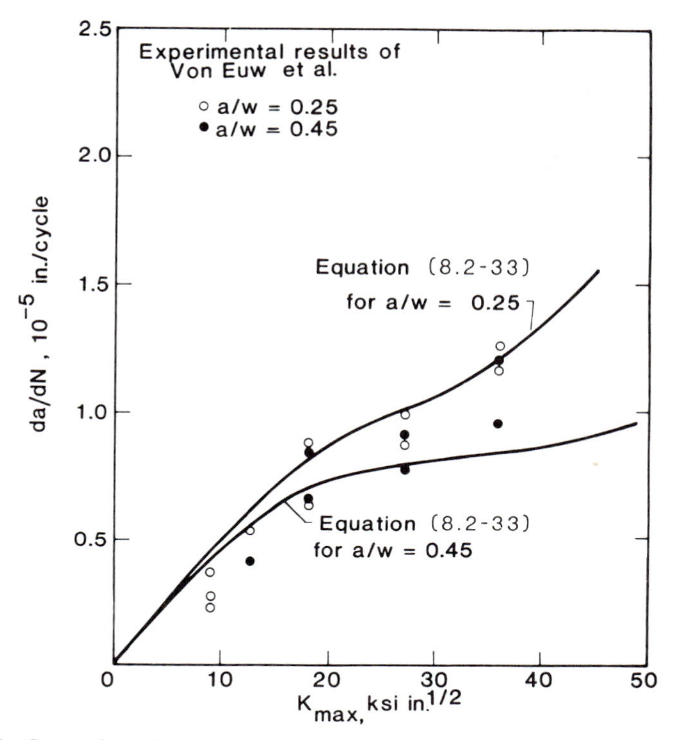

Figure 8.15 Comparison of predicted and experimental fatigue crack growth results in 2024-T3 aluminum alloy for $\Delta K = 9$ ksi-in.$^{\frac{1}{2}}$

Sadananda and Shahinian (8.53), Swenson (8.54), Pook and Frost (8.55). Weertman (8.56), and Yokobori et al. (8.57). These all utilize either a simplified representation of the crack-tip plasticity, or incorporate a mechanism for the crack closure condition, or both. The essential differences between the work described in the foregoing and the work of others is that the latter are apparently not able to incorporate a crack growth criterion directly into their models nor is it possible to realistically model (and distinguish) the plastic deformation occurring in individual load cycles. The Kanninen-Atkinson model is capable of providing this key feature and, as a consequence, reliance on a postulated controlling mechanism (e.g., crack closure, residual plasticity) is avoided. These mechanisms are automatically included in this model and make their contribution in a natural way.

It is clear that the parameters governing the residual plasticity in the wake of the crack must be determined without the expedient of using experimental observations as in the Kanninen-Atkinson model. One way to accomplish this is by calibration of the model with elastic-plastic finite element solutions, as Newman (8.58 to 8.61) has done. This avoids the uncertainty associated with crack closure observations. Figures 8.16 to 8.19 show his key results.

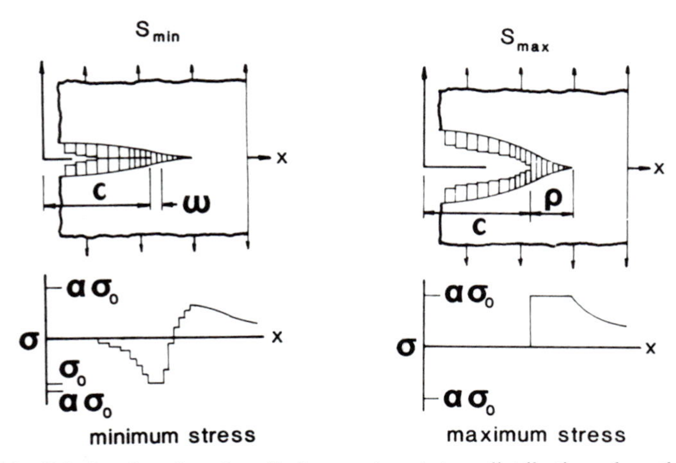

Figure 8.16 Calculated crack surface displacements and stress distributions along the crack line in fatigue crack propagation.

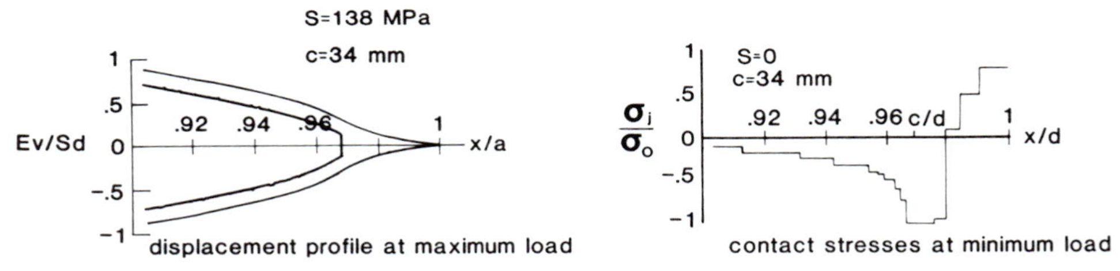

Figure 8.17 Calculated crack surface displacements and contact stresses in constant amplitude fatigue crack propgation.

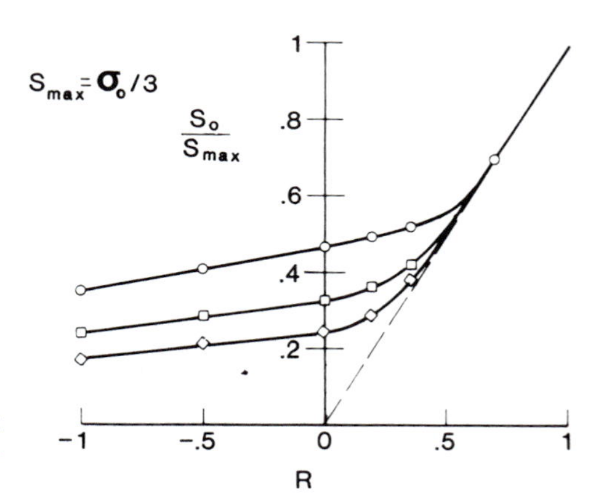

Figure 8.18 Calculated crack opening stresses in fatigue crack propagation as a function of R.

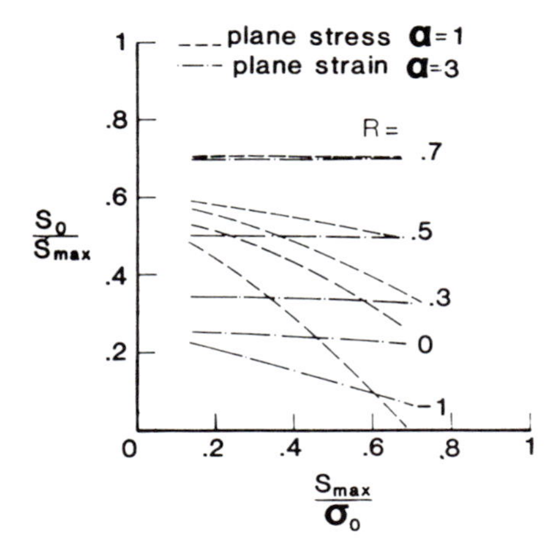

Figure 8.19 Calculated crack opening stresses in fatigue crack propagation as a function of applied stress.

8.2.3 The Short Crack Problem in Fatigue

The short crack effect in fatigue results from the inability of linear elastic fracture mechanics procedures to predict the growth rate of physically small cracks. Several different possible reasons exist for this effect. It is entirely conceivable, for example, that very small cracks are influenced by microstructural features of the material that are not addressed by the conventional continuum-based LEFM techniques. In addition, three-dimensional effects (as arise in conjunction with corner cracks) exist to compound the problem.

In view of these difficulties, the development of predictive techniques to address the full range of crack lengths down to very small cracks is a formidable undertaking. Nonetheless, it is likely that an intermediate regime exists where progress can be made. The crack sizes for this regime lie just below the limit of validity for LEFM and above that where heterogeneity on the micro-mechanical scale strongly influences the crack growth process. In this regime an elastic-plastic fracture mechanics approach to fatigue can be effective. By pursuing such an approach, most investigators are not expecting to completely erase the short crack effect. Rather, they expect to shift the short crack demarcation point downwards to the limit of a continuum mechanics approach.

Analysis approaches addressed to short cracks can be divided into two main categories. One includes those attempts that have incorporated some modification or correction into a conventional fracture mechanics formulation. These are generally semiempirical treatments. The second category includes the more rigorous approaches that attempt to treat short crack fatigue from first principles of continuum mechanics. More particularly, because the essence of the problem is the presence of plastic deformation, these lie within the domain of elastic-plastic fracture mechanics.

El Haddad et al. (8.62, 8.63) have recognized that a generalization of conventional fatigue crack growth rate predictive techniques is necessary and have chosen to provide one within the confines of deformation plasticity. Specifically, they have suggested the addition of a small pseudo crack length, l_0, to the actual crack length. Because the material constant l_0 is small, long crack behavior is unchanged, as is essential in any such approach. However, for small crack lengths, the use of l_0 increases K. This escalates the crack growth rate, as required to match the short crack effect.*

The approach of El Haddad et al. can be criticized on both fundamental and pragmatic grounds. They argue that l_0 can be related to the threshold stress intensity and the fatigue limit. If this interpretation is correct, the physical meaning of l_0 must be that of an inherent defect length in the material that dominates the specimen response when the artificial crack length is of the order of l_0. But, it surely is disingenuous to suppose that such defects are both noninteracting and always located so as to increase the artificial crack length. While it would be permissible to limit the smallest crack length to be l_0, it cannot then logically be incorporated as an additive term.

On a practical basis, the approach of El Haddad et al. similarly appears to be wanting. For example, their approach fails to consolidate the short crack data for mild steel developed by Leis and Forte (8.64). This is shown in Figure 8.20. This version of the El Haddad et al. approach is based upon the use of the J-integral in an effort to formulate a plastic fracture mechanics formalism. But, the pseudo-crack-length concept, even when enhanced by the use of J, is not in correspondence with these data. Similar discouraging results for other J-based approaches have been reviewed by Leis et al. (8.65). On a theoretical basis, the argument given in Chapter 5 for the basis of J in elastic-plastic fracture mechanics clearly shows why the substitution of J for K as the crack driving force parameter in fatigue is not necessarily an improvement.

Worthy of note are the approaches using an extended version of the Dugdale (collinear) strip yield model and those using elastic-plastic finite elements models. In the former group are Kanninen and co-workers (8.66) and Fuhring and Seeger (8.67), while in the latter are Newman (8.61) and Trantina et al. (8.68). It might be noted that, while the use of an inclined strip yield model has yet to be applied to the short crack problem per se, as the work of Kanninen and Atkinson (8.50) shows, this approach offers an intriguing compromise between a realistic representation of plane strain plastic deformation and computational convenience that may warrant serious consideration in future attempts.

Work based upon the use of the Dugdale crack-tip plasticity model can be used to investigate the effectiveness of using a CTOD criterion for the growth of short fatigue cracks. Ordinarily, some key effects—e.g., crack closure—are thereby omitted. That this is not precluded through the adoption of a Dugdale model is clearly shown in the work performed by Newman (8.61) who has devised a "ligament model" generalization of the Dugdale model, via a finite element calibration. This work has recently been focused on the short crack

* The similarity with the crack length adjustment approach in LEFM (see Section 1.4.1) and with that used by Waddoups et al. for fiber composite materials (see Section 6.2.1) is evident.

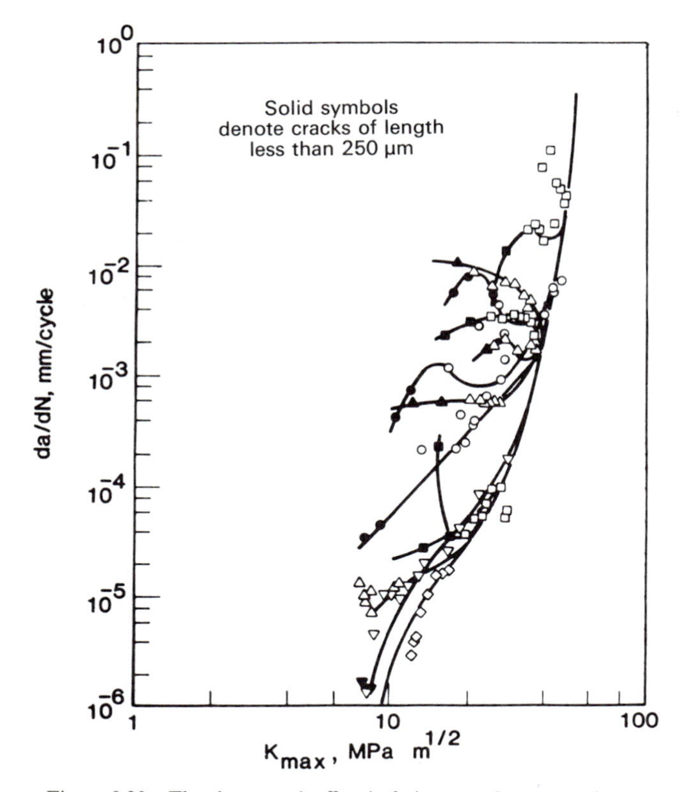

Figure 8.20 The short crack effect in fatigue crack propagation.

problem by modifying the Dugdale model to leave plastically deformed material along the crack faces as the crack grows. His purpose was to study the related effects of small crack growth rates and of large cracks under the load reduction schemes used to determine threshold stress intensity values.

Newman cites as the primary advantage of the Dugdale model that linear superposition is valid even when such obstensibly nonlinear effects as crack closure are included. This is true because the crack closure effects take place only from residual plasticity in the line of the crack. By leaving plastically deformed material behind the crack in this way, the crack surface displacements used to calculate contact (closure) stresses under cyclic loading are influenced by the plastic yielding both ahead and behind the crack tip. Specifically, bar elements, assumed to behave like rigid-perfectly plastic materials, are used. At any applied stress level, these elements are intact ahead of the crack tip or are broken to represent the residual plasticity behind the crack tip.

It is important to recognize that the broken elements carry compressive loads only, and then only if they are in contact. Those elements that are not in contact apparently do not affect the calculation in any way. They are used simply to calculate crack-opening stresses (both crack-tip closure and closure elsewhere on the crack edges can be handled) for use in determining an "effective" ΔK value in the manner suggested by Elber—that is, to determine ΔK_{eff}, where

$$\Delta K_{eff} = \frac{\sigma_{max} - \sigma_0}{\sigma_{max} - \sigma_{min}} \Delta K \qquad (8.2\text{-}35)$$

where σ_0 denotes the crack opening stress. Note that, although not specifically indicated here, Newman used plasticity corrected K values in his calculations. Newman's calculations were made for a center-cracked tension panel and for cracks emanating from a hole. From these results, Newman concluded that the short crack effect is a result of the differences in the crack closure effect between long and short cracks. In particular, at equal K values, the applied stress needed to open a small crack is less than that required to open a large crack. Consequently, the effective stress range is greater for small cracks. This, in turn, gives rise to the higher crack growth rates that exemplify the short crack effect. Experimental results on the short crack effect are given by Hudak (8.69).

8.2.4 Fatigue Crack Growth in Welds

A significant proportion of all structural failures can be traced to cracks emanating in and around welds. Crack growth in welded regions would appear to involve three major effects: (1) the presence of thermoplastically deformed material in the crack path, (2) the attendant residual stresses, and (3) the possible change in microstructure. Fracture mechanics analysis procedures based on linear elastic conditions do not usually treat these complications. The conventional rationale is that the K_{max} values in fatigue are small enough that the plastic zones will be negligible. In addition, the presence of residual stresses will simply change the mean stress. Hence, particularly for those materials that exhibit a limited amount of R-dependence, LEFM fatigue crack growth relations established for the material condition of concern can be applied without modification to predict crack growth in or around a weld. However, recent work has been focused on eliminating these assumptions.

Perhaps the leading contributor to the analysis of residual stresses in welds is Masubuchi; see reference (8.70). Early work in this area was also contributed by Jahsman and Field (8.71). The use of linear elastic fracture mechanics techniques with estimates of residual stress fields to obtain subcritical crack growth predictions has been pursued by many investigators. Recent work of note would include Glinka (8.72), Parker (8.73), Berge and Eide (8.74), Nelson (8.75), and de Koning (8.76). Generally, these approaches introduce the normal component of a residual stress state as a surface traction on the prospective crack plane with the component otherwise remaining linear elastic. Linear superposition is then used to obtain a stress intensity factor due to the combined effect of the applied stresses and the residual stresses. Note that this can be done for a range of selected crack lengths, whereupon Equation (8.1-1) or any of the alternatives can be used to compute the crack growth rate.

Elastic-plastic fracture mechanics analysis procedures have previously been applied to account only for crack-tip plasticity itself. Recently, a further step has been taken by Kanninen and co-workers (8.77–8-79) through the use of postulated elastic-plastic crack growth relations for crack growth in weld-induced plastic deformation fields. The objective of their work was to critically examine the assumptions of linear elastic material behavior commonly made in analyzing weld cracking problems. Fatigue crack growth in welded regions will be affected by the presence of the plastically deformed material indigenous to the welding process. Yet, the present-day analyses, based on linear elastic conditions, do not directly treat such complications.

The analysis procedure followed in references (8.77)–(8.79) consisted of three main steps. First, the residual stress field induced in a welding process was computed using an incremental thermoplastic finite element analysis procedure. Second, crack growth was simulated by sequential node release along a pre-set crack plane with values of δ, the crack-tip opening displacement, being obtained. Third, a postulated elastic-plastic crack growth relation based on δ was used to infer crack length as a function of loading history. A comparison with a commonly accepted linear approach was then made. This comparison allowed the significance of the linear elastic assumption and the essential neglect of residual plasticity inherent in such an approach to be assessed.

The residual stress analysis procedure requires a thermal analysis to be made to obtain the time-temperature history for each point in the body for each individual welding pass. These histories provide the input to an incremental elastic-plastic finite element model. This determines the stress and deformation state of the weld and the base material as the weld is deposited. Because each welding pass is considered on an individual basis, the residual stress and strain state that exists at the completion of one weld pass constitutes the initial condition for the next. The final residual stress state is that which exists at the completion of all of the weld passes.

Consider the butt-welded plate typical of the ship structure steel HY-80 shown in Figure 8.21. Typical heat input values were assumed. Together with the weld pass and structure geometry, these sufficed to determine the residual stress distribution. The computed normal stresses acting on the prospective crack plane for the welded configuration are then shown in Figure 8.22. It can be seen that a high tensile stress exists near both surfaces. Thus, an edge crack

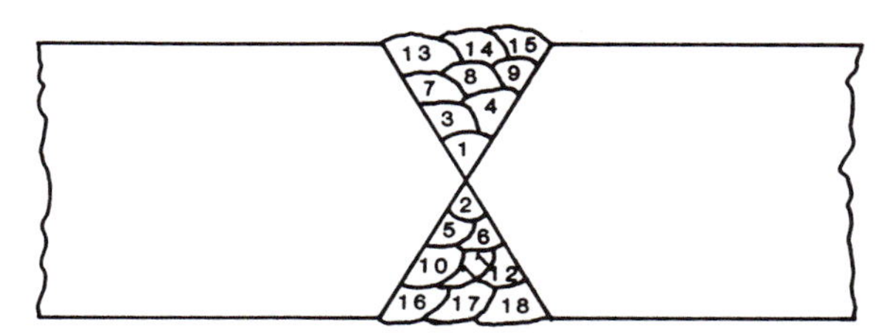

Figure 8.21 Cross section of 18-pass weld typical of a HY-80 steel ship structure.

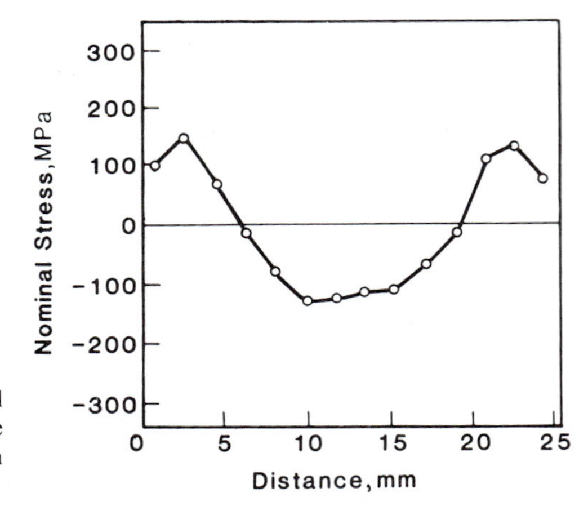

Figure 8.22 Calculated residual stresses (normal component) on the potential crack line for the weld shown in Figure 8.21.

would be likely to grow, particularly when the residual stresses are abetted by a tensile applied stress.

Starting from an assumed initial crack, crack growth was simulated in the finite element model by sequential node release along a line of double-noded elements. Each node pair was released by gradually diminishing the initial force that exists between them to zero. This was typically done over from five to ten load increments. The value of δ for a given crack length was then the value of the CTOD that existed when the load vanished. The results obtained from the finite element model for an assumed maximum load of $0.67\sigma_Y$ and a corresponding minimum load of zero are shown in Figure 8.23.

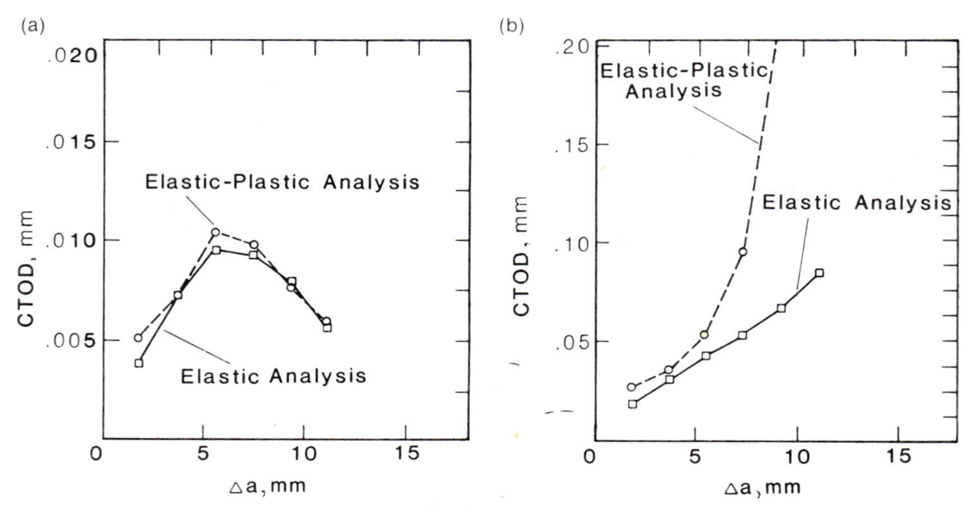

Figure 8.23 Calculated crack-tip opening displacement for monotonic crack growth through the weld shown in Figure 8.21: (a) zero applied load, (b) applied load ($= 0.67\sigma_Y$).

Figure 8.23 contrasts the δ values obtained by advancing the crack through the finite element model under two different conditions. First, the computed weld-induced plastic deformation was left unaltered and an incremental plasticity computation made. This is the elastic-plastic analysis. Second, a simplified approach was followed wherein, (1) only the normal component of the residual stress acting on the potential crack plane was retained, and (2) linear elastic behavior is assumed. This is denoted as the simple elastic analysis. It typifies that commonly used by others for this kind of problem.

The equivalence between the crack-tip opening displacement and the stress intensity factor in small-scale yielding was described in Section 1.4. Recent progress in elastic-plastic fracture mechanics has further revealed the distinctive role played by the crack-tip opening displacement in crack initiation and stable growth in large-scale yielding conditions; see Chapter 5. Specifically, for the initiation of crack growth, the CTOD can be expressed as

$$\delta = \begin{cases} \alpha \dfrac{K^2}{\sigma_Y} & \text{small-scale yielding} \\ d_n \dfrac{J}{\sigma_Y} & \text{deformation plasticity} \end{cases} \qquad (8.2\text{-}36)$$

where α and d_n are numerical constants on the order of unity. In addition, for extended stable crack growth, the CTOD appears to take on a constant value. While certainly not conclusive evidence that the CTOD is the controlling parameter for subcritical crack growth as well, for lack of an alternative, it can be so taken. Note that, because of the equivalence represented by Equation (1.4-23), this choice is not inferior to one based on either K or J in any event.

Fatigue crack growth under a uniform cyclic loading can very often be adequately characterized in the form of Equation (8.1-1). That is,

$$\frac{da}{dN} = C(K_{\max} - K_{\min})^m \qquad (8.2\text{-}37)$$

where C and m are material constants. Introducing the CTOD from Equation (8.2-36) gives

$$\frac{da}{dN} = C'(\delta_{\max}^{\frac{1}{2}} - \delta_{\min}^{\frac{1}{2}})^m \qquad (8.2\text{-}38)$$

where $\delta_{\max}$ and $\delta_{\min}$ are the CTOD values that would be attained under the maximum and minimum load levels, respectively. Consequently, the number of cycles required to achieve a given crack length can be found by integrating Equation (8.2-38). The results, using the CTOD values given in Figure 8.23, are shown in Figure 8.24.

The computational results presented in Figure 8.24 show a wide disparity between the fatigue crack growth obtained using an elastic-plastic approach and that obtained in the more usual way. Perhaps surprisingly, this result indicates that the simple elastic analysis may be highly anticonservative. This

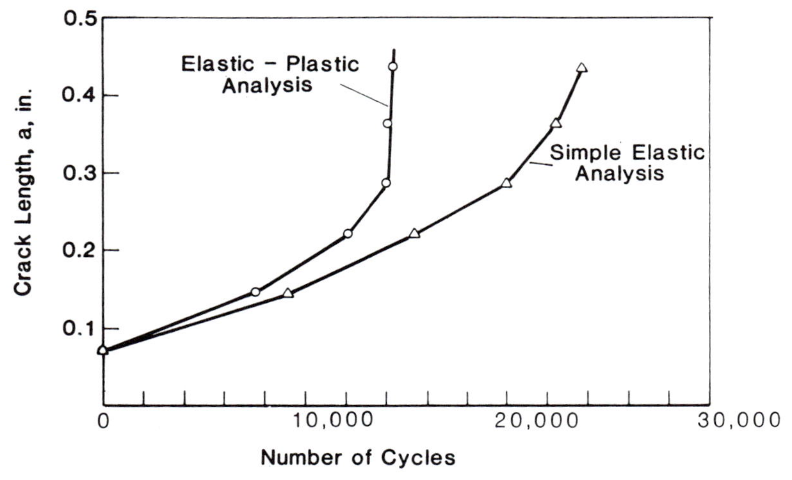

Figure 8.24 Calculated fatigue crack growth results for the weld shown in Figure 8.21 when subjected to uniform amplitude loading from $0.67\sigma_Y$ to zero.

would appear to be a very significant finding in view of the great practical importance of such problems.

The basic assumption that has been called into question here is the applicability of linear elastic fracture mechanics in the presence of weld-induced residual stress fields. This has been addressed by performing two parallel computations where this assumption has and has not been made. There are undeniably many aspects of the calculations that can be improved upon. But, because the two computations were otherwise performed on exactly the same basis, these cannot be of critical importance. Indeed, the comparison has revealed such wide disparities that neglect of the inelastic deformation accompanying welding would appear to be of serious practical concern.

8.3 References

(8.1) Kocanda, S., *Fatigue Failures of Metals*, Sijthoff and Noordhoff, Alphen aan den Rijn, The Netherlands (1978).

(8.2) Fong, J. T. (ed.), *Fatigue Mechanisms*, ASTM STP 675, American Society for Testing and Materials, Philadelphia, Pa. (1979).

(8.3) Plumbridge, W. J., "Review: Fatigue-Crack Propagation in Metallic and Polymeric Materials," *Journal of Materials Science*, **7**, pp. 939–962 (1972).

(8.4) Nelson, D. V., "Review of Fatigue-Crack-Growth Prediction Methods," *Experimental Mechanics*, **17**, pp. 41–49 (1977); see also discussion by J. C. Newman, Jr. and Author's Closure, *Experimental Mechanics*, **17**, pp. 399–400 (1977).

(8.5) Weertman, J., "Fatigue Crack Propagation Theories," *Fatigue and Microstructure*, American Society for Metals, Metals Park, Ohio, pp. 279–306 (1979).

(8.6) Paris, P. C., "Twenty Years of Reflection on Questions Involving Fatigue Crack Growth," *Fatigue Thresholds*, J. Backlund, A. F. Blom, and C. J. Beevers (eds.), Chamelem, London, pp. 3–10 (1982).

(8.7) Paris, P. and Erdogan, F., "A Critical Analysis of Crack Propagation Laws," *Journal of Basic Engineering*, **85**, pp. 528–534 (1963).
(8.8) Paris, P. C. Bucci, R. J., Wessel, E. T., Clark, W. G., Jr., and Mager, T. R., "An Extensive Study on Low Fatigue Crack Growth Rates in A533 and A508 Steels," *Stress Analysis and Growth of Cracks*, Part I, ASTM STP 513, American Society for Testing and Materials, Philadelphia, Pa., pp. 141–176 (1972).
(8.9) Foreman, R. G., Kearney, V. E., and Engle, R. M., "Numerical Analysis of Crack Propagation in Cyclic-Loaded Structures," *Journal of Basic Engineering*, **89**, pp. 459–464 (1967).
(8.10) Donahue, R. J., Clark, H. M., Atanmo, P., Kumble, R., and McEvily, A. J., "Crack Opening Displacement and the Rate of Fatigue Crack Growth," *International Journal of Fracture Mechanics*, **8**, pp. 209–219 (1972).
(8.11) Priddle, E. K., "High Cycle Fatigue Crack Propagation Under Random and Constant Amplitude Loadings," *International Journal of Pressure Vessels and Piping*, **4**, pp. 89–117 (1976).
(8.12) Schijve, J., "Four Lectures on Fatigue Crack Growth," *Engineering Fracture Mechanics*, **11**, pp. 167–221 (1979).
(8.13) Walker, K., "The Effect of Stress Ratio During Crack Propagation and Fatigue for 2024-T3 and 7075-T6 Aluminum," *Effects of Environment and Complex Load History on Fatigue Life*, ASTM STP 462, American Society for Testing and Materials, Philadelphia, Pa., pp. 1–14 (1970).
(8.14) Maddox, S. J., "The Effect of Mean Stress on Fatigue Crack Propagation. A Literature Review," *International Journal of Fracture*, **11**, pp. 389–408 (1975).
(8.15) McEvily, A. J. and Groeger, J., "On the Threshold for Fatigue-Crack Growth," *Fourth International Conference on Fracture*, Vol. 2, University of Waterloo Press, Waterloo, Canada, pp. 1293–1298 (1977).
(8.16) Imhoff, E. J. and Barsom, J. M., "Fatigue and Corrosion-Fatigue Crack Growth of 4340 Steel at Various Yield Strengths," *Progress in Flaw Growth and Fracture Toughness Testing*, ASTM STP 536, American Society for Testing and Materials, Philadelphia, pp. 182–205 (1973).
(8.17) Petrak, G. J. and Gallagher, J. P., "Predictions of the Effect of Yield Strength on Fatigue Crack Growth Retardation in HP-9Ni-4Co-30C Steel," *Journal of Engineering Materials and Technology*, **97**, pp. 206–213 (1975).
(8.18) Erdogan, F., "Stress Intensity Factors," *Journal of Applied Mechanics*, **50**, pp. 992–1002 (1963).
(8.19) Miner, M. A., "Cumulative Damage in Fatigue," *Journal of Applied Mechanics*, **12**, pp. A159–A164 (1945).
(8.20) von Euw, E. F. J., Hertzberg, R. W., and Roberts, R., "Delay Effects in Fatigue-Crack Propagation," *Stress Analysis and Growth of Cracks*, ASTM STP 513, American Society for Testing and Materials, Philadelphia, Pa., pp. 230–259 (1972).
(8.21) Jones, R. E., "Fatigue Crack Growth Retardation After Single-Cycle Peak Overload in Ti-6Al-4V Titanium Alloy," *Engineering Fracture Mechanics*, **5**, pp. 585–604 (1973).
(8.22) Matsuoka, S. and Tanaka, K., "Delayed Retardation Phenomena of Fatigue Crack Growth in Various Steels and Alloys," *Journal of Materials Science*, **13**, pp. 1335–1353 (1978).
(8.23) Wheeler, O. E., "Spectrum Loading and Crack Growth," *Journal of Basic Engineering*, **94**, pp. 181–186 (1972).
(8.24) Willenborg, J., Engle, R. M., Jr., and Wood, R. A., "A Crack Growth Retardation Model Using an Effective Stress Concept," Air Force Flight Dynamics Laboratory Report AFFDL-TM-71-1-FBR, January 1971.
(8.25) Elber, W., "Fatigue Crack Closure Under Cyclic Tension," *Engineering Fracture Mechanics*, **2**, pp. 37–45 (1970).
(8.26) Elber, W., "Equivalent Constant-Amplitude Concept for Crack Growth Under Spectrum Loading," *Fatigue Crack Growth Under Spectrum Loads*, ASTM STP 595, American Society for Testing and Materials, Philadelphia, Pa., pp. 236–250 (1976).
(8.27) Wood, H. A., Gallagher, J. P., and Engle, R. M., Jr., "Application of Fracture Mechanics to the Prediction of Crack Growth Damage Accumulation in Structures," *The Mechanics of Fracture*, F. Erdogan (ed.), ASME AMD-Vol. 19, New York, pp. 171–190 (1976).
(8.28) Broek, D. and Smith, S. H., "The Prediction of Fatigue Crack Growth Under Flight-by-Flight Loading," *Engineering Fracture Mechanics*, **11**, pp. 123–141 (1979).
(8.29) Schijve, J., "Prediction Methods for Fatigue Crack Growth in Aircraft Material," *Fracture*

Mechanics: Twelfth Conference, ASTM STP 700, American Society for Testing and Materials, Philadelphia, Pa., pp. 3–34 (1980).
(8.30) Johnson, W. S., "Multi-Parameter Yield Zone Model for Predicting Spectrum Crack Growth," *Methods and Models for Predicting Fatigue Crack Growth Under Random Loading*, ASTM STP 748, J. B. Chang and C. M. Hudson (eds.), American Society for Testing and Materials, Philadelphia, Pa., pp. 85–102 (1981).
(8.31) Lankford, J., Davidson, D. L., and Cook, T. S., "Fatigue Crack-Tip Plasticity," *Cyclic Stress-Strain and Plastic Deformation Aspects of Fatigue Crack Growth*, ASTM STP 637, American Society for Testing and Materials, Philadelphia, Pa., pp. 36–55 (1977).
(8.32) Shih, T. T. and Wei, R. P., "A Study of Crack Closure in Fatigue," *Engineering Fracture Mechanics*, **6**, pp. 19–32 (1974).
(8.33) Gomez, M. P., Ernst, H., and Vazquez, J., "On the Validity of Elber's Results on Fatigue Crack Closure for 2024-T3 Aluminum," *International Journal of Fracture*, **12**, pp. 178–180 (1976).
(8.34) Shih, T. T. and Wei, R. P., "Discussion," *International Journal of Fracture*, **13**, pp. 105–106 (1977).
(8.35) Unangst, K. D., Shih, T. T., and Wei, R. P., "Crack Closure in 2219-T851 Aluminum Alloy," *Engineering Fracture Mechanics*, **9**, pp. 725–734 (1977).
(8.36) Schijve, J., "Some Formulas for the Crack Opening Stress Level," *Engineering Fracture Mechanics*, **14**, pp. 461–465 (1981).
(8.37) Chand, S. and Garg, S. B. L., "Crack Closure Studies Under Constant Amplitude Loading," *Engineering Fracture Mechanics*, **18**, pp. 333–347 (1983).
(8.38) Probst, E. P. and Hillberry, B. M., "Fatigue Crack Delay and Arrest Due to Single Peak Tensile Overloads," *AIAA Journal*, **12**, pp. 330–335 (1974).
(8.39) Lindley, T. C. and Richards, C. E., "The Relevance of Crack Closure to Fatigue Crack Propagation," *Materials Science and Engineering*, **14**, pp. 281–293 (1974).
(8.40) Mogford, I. L., "Recent Developments in Fatigue Crack Growth Assessment," *Developments in Pressure Vessel Technology—1. Flaw Analysis*, R. W. Nichols (ed.), Applied Science, London, pp. 169–202 (1979).
(8.41) Budiansky, B. and Hutchinson, J. W., "Analysis of Closure in Fatigue Crack Growth," *Journal of Applied Mechanics*, **45**, pp. 267–276 (1978).
(8.42) Rice, J. R., "Mechanics of Crack-Tip Deformation and Extension by Fatigue," *Fatigue Crack Propagation*, ASTM STP 415, American Society of Testing and Materials, Philadelphia, Pa., pp. 247–309 (1967).
(8.43) Lo, K. K., "Fatigue Crack Closure Following a Step-Increase Load," *Journal of Applied Mechanics*, **47**, pp. 811–814 (1980).
(8.44) Laird, C. and Smith, G. C., "Crack Propagation in High Stress Fatigue," *Philosophical Magazine*, **7**, pp. 847–857 (1962); see also Laird, C., "Mechanisms and Theories of Fatigue," *Fatigue and Microstructure*, American Society for Metals, Metals Park, Ohio, pp. 149–203 (1979).
(8.45) Neumann, P., "The Geometry of Slip Processes at a Propagating Fatigue Crack," *Acta Metallurgica*, **22**, pp. 1155–1178 (1974).
(8.46) Kanninen, M. F., Atkinson, C., and Feddersen, C. E., "A Fatigue Crack-Growth Analysis Method Based on a Simple Representation of Crack-Tip Plasticity," Cyclic Stress-Strain and Plastic Deformation Aspects of Fatigue Crack Growth, ASTM-STP 637, pp. 122–140 (1977).
(8.47) Kanninen, M. F. and Atkinson, C., "Application of an Inclined-Strip-Yield Crack-Tip Plasticity Model to Predict Constant Amplitude Fatigue Crack Growth," *International Journal of Fracture*, **16**, pp. 53–69 (1980).
(8.48) Bilby, B. A. and Swinden, K. H., "Representation of Plasticity at Notches by Linear Dislocation Arrays," *Proceedings of the Royal Society*, **A285**, pp. 22–33 (1965).
(8.49) Atkinson, C. and Kay, T. R., "A Simple Model of Relaxation at a Crack Tip," *Acta Metallurgica*, **19**, pp. 679–683 (1971).
(8.50) Atkinson, C. and Kanninen, M. F., "A Simple Representation of Crack Tip Plasticity: The Inclined Strip-Yield Superdislocation Model," *International Journal of Fracture*, **13**, pp. 151–163 (1977).
(8.51) Liu, A. F. and Dittmer, D. F., *Effect of Multiaxial Loading on Crack Growth*, Northrop Corporation, Report to the Air Force Flight Dynamics Laboratory AFFDL-TR-78-175, Dec. 1978.
(8.52) Adams, N. J. I., "Some Comments on the Effect of Biaxial Stress on Fatigue Crack Growth and Fracture," *Engineering Fracture Mechanism*, **5**, pp. 983–991 (1973).

(8.53) Sadananda, K. and Shahinian, P., "Prediction of Threshold Stress Intensity for Fatigue Crack Growth Using a Dislocation Model," *International Journal of Fracture*, **13**, pp. 585–594 (1977).
(8.54) Swenson, D. O., "Discrete Dislocation Model of a Fatigue Crack Under Shear Loading—Part I," *Journal of Applied Mechanics*, **36**, pp. 723–730 (1969); "Mean Stress Effect on the Shear Fatigue Crack Model-Part II," *Journal of Applied Mechanics*, **36**, pp. 731–735 (1969).
(8.55) Pook, L. P. and Frost, N. E., "A Fatigue Crack Growth Theory," *International Journal of Fracture*, **9**, pp. 53–61 (1973).
(8.56) Weertman, J., "Theory of Fatigue Crack Growth Based on a BCS Crack Theory With Work Hardening," *International Journal of Fracture*, **9**, pp. 125–131 (1973).
(8.57) Yokobori, T., Yokobori, A. T., and Kamei, A., "Dislocation Dynamics Theory for Fatigue Crack Growth," *International Journal of Fracture*, **11**, pp. 781–788 (1975).
(8.58) Newman, J. C., Jr., "A Finite Element Analysis of Crack Closure," *Mechanics of Crack Growth*, ASTM STP 590, American Society for Testing and Materials, Philadelphia, Pa., pp. 281–301 (1976).
(8.59) Newman, J. C., Jr., "Finite-Element Analysis of Crack Growth Under Monotonic and Cyclic Loading," *Cyclic Stress-Strain and Plastic Deformation Aspects of Fatigue Crack Growth*, ASTM STP 637, American Society for Testing and Materials, Philadelphia, Pa., pp. 56–80 (1977).
(8.60) Newman, J. C., Jr., "A Crack-Closure Model for Predicting Fatigue Crack Growth Under Aircraft Spectrum Loading," *Methods and Models for Predicting Fatigue Crack Growth Under Random Loading*, ASTM STP 748, American Society for Testing and Materials, Philadelphia, Pa., pp. 53–84 (1981).
(8.61) Newman, J. C., Jr., "A Nonlinear Fracture Mechanics Approach to the Growth of Small Cracks," AGARD Specialists Meeting on Behavior of Short Cracks in Airframe Components, Toronto, Canada, Sept. 1982.
(8.62) El Haddad, M. H., Topper, T. H., and Smith, K. N., "Prediction of Non-Propagating Cracks," *Engineering Fracture Mechanics*, **11**, pp. 573–584 (1979).
(8.63) El Haddad, M. H., Dowling, N. E., Topper, T. H., and Smith, K. N., "*J*-Integral Applications for Short Fatigue Cracks at Notches," *International Journal of Fracture*, **16**, pp. 15–30 (1980).
(8.64) Leis, B. N. and Forte, T. P., "Fatigue Growth of Initially Physically Short Cracks in Notched Aluminum and Steel Plates," *Fracture Mechanics: Thirteenth Conference*, ASTM STP 743, American Society for Testing and Materials, Philadelphia, Pa., pp. 100–124 (1981).
(8.65) Leis, B. N., Kanninen, M. F., Hopper, A. T., Ahmad, J., and Broek, D., "A Critical Review of the Short Crack Problem in Fatigue," Air Force Aeronautical Laboratories Report AFWAL-TR-83-4019, May 1983.
(8.66) Kanninen, M. F., Ahmad, J., and Leis, B. N., "A CTOD-Based Fracture Mechanics Approach to the Short-Crack Problem in Fatigue," *Mechanics of Fatigue*, ASME AMD-Vol. 47, American Society of Mechanical Engineers, New York, pp. 81–90 (1981).
(8.67) Fuhring, H. and Seeger, T., "Dugdale Crack Closure Analysis of Fatigue Cracks Under Constant Amplitude Loading," *Engineering Fracture Mechanics*, **11**, pp. 99–122 (1979).
(8.68) Trantina, G. G., deLorenzi, H. G., and Wilkening, W. W., "Three-Dimensional Elastic-Plastic Finite Element Analysis of Small Surface Cracks," *Engineering Fracture Mechanics*, in press, 1984.
(8.69) Hudak, S. J., Jr., "Small Crack Behavior and the Prediction of Fatigue Life," *Journal of Engineering Materials and Technology*, **103**, pp. 26–35 (1981).
(8.70) Masubuchi, K., *Analysis of Welded Structures*, Pergamon, New York (1980).
(8.71) Jahsman, W. E. and Field, F. A., "The Effect of Residual Stresses on the Critical Crack Length Predicted by the Griffith Theory," *Journal of Applied Mechanics*, **30**, pp. 613–616 (1963).
(8.72) Glinka, G., "Effect of Residual Stresses on Fatigue Crack Growth in Steel Weldments Under Constant and Variable Amplitude Loads," *Fracture Mechanics*, ASTM STP 677, C. W. Smith (ed.), American Society for Testing and Materials, Philadelphia, Pa., pp. 198–214 (1979).
(8.73) Parker, A. P., "Stress Intensity Factors, Crack Profiles, and Fatigue Crack Growth Rates in Residual Stress Fields," *Residual Stress Effects in Fatigue*, ASTM STP 776, American Society for Testing and Materials, Philadelphia, Pa., pp. 13–31 (1982).
(8.74) Berge, S. and Eide, O. I., "Residual Stress and Stress Interaction in Fatigue Testing of

Welded Joints," *Residual Stress Effects in Fatigue*, ASTM STP 776, American Society for Testing and Materials, Philadelphia, Pa., pp. 115–131 (1982).

(8.75) Nelson, D. V., "Effects of Residual Stress on Fatigue Crack Propagation," *Residual Stress Effects in Fatigue*, ASTM STP 776, American Society for Testing and Materials, Philadelphia, Pa., pp. 172–194 (1982).

(8.76) de Koning, A. U., "A Simple Crack Closure Model for Prediction of Fatigue Crack Growth Rates Under Variable-Amplitude Loading," *Fracture Mechanics: Thirteenth Conference*, ASTM STP 743, R. Roberts, editor, American Society for Testing and Materials, Philadelphia, Pa., pp. 63–85 (1981).

(8.77) Kanninen, M. F., Brust, F. W., Ahmad, J., and Abou-Sayed, I. S., "The Numerical Simulation of Crack Growth in Weld-Induced Residual Stress Fields," *Residual Stress and Stress Relaxation*, E. Kula and V. Wiess (eds.), Plenum, New York, pp. 227–247 (1982).

(8.78) Kanninen, M. F., Brust, F. W., Ahmad, J., and Papaspyropoulos, V., "An Elastic-Plastic Fracture Mechanics Prediction of Fatigue Crack Growth in the Heat-Affected Zone of a Butt-Welded Plate," *Fracture Tolerance Evaluation*, T. Kanazawa et al. (eds.), Toyoprint, Tokyo, pp. 113–128 (1982).

(8.79) Abou-Sayed, I. S., Brust, F. W., Ahmad, J., and Kanninen, M. F., "A Plastic Fracture Mechanics Prediction of Crack Growth in the Presence of Weld-Induced Residual Stresses With Application to Stress Corrosion Cracking of BWR Piping," *Fracture Mechanics: Fourteenth Symposium*, Vol. I, J. C. Lewis and G. Sines (eds.), ASTM STP 791, American Society for Testing and Materials, Philadelphia, Pa., pp. 482–496 (1983).

9

SOURCES OF INFORMATION IN FRACTURE MECHANICS

According to Grogan (9.1), the information seeking habits of scientists and engineers are typified by the principle of least effort: information will be sought only when it is less troublesome and painful to obtain it than to not have it. He quotes a medical scientist as categorizing consumers of the technical literature as "generally speaking, arrogant, conservative, lazy and ignorant." As this is clearly not true of engineers and others interested in fracture mechanics, we believe that it will be appropriate to provide a guide to sources of further information on fracture mechanics to facilitate further study and applications.

We will adopt Grogan's categorization to present a list of sources to the fracture mechanics literature. In descending order of freshness, these are: journals, conference proceedings, standards, dissertations, abstracting periodicals, progress reviews, handbooks, treatises, and textbooks. The best sources of fracture mechanics information in each category are summarized in this chapter. We hope that these will enable the reader both to find the more detailed information on fracture mechanics needed for applying the concepts given in this book, and to keep abreast of further developments in the continuing evolution of the subject.

9.1 Technical Journals

As should be evident from scanning the reference lists given at the end of each chapter of this book, a great many journals contain articles on fracture mechanics. Of this number, three are devoted exclusively to the subject. These are the *International Journal of Fracture* (originally, the *International Journal of Fracture Mechanics*), published since 1965; *Engineering Fracture Mechanics*, published since 1968; and the *International Journal of Fatigue*, published since 1978. A reader generally looks to the first of these for the more theoretical articles and to the second for more applications-oriented articles. There are many exceptions, of course. The *International Journal of Fracture* also contains a special section on reports of work in progress that strives for quick publication of short articles highlighting current work. A fourth journal, *Fracture Mechanics Technology*, has begun to appear at the time of this writing. It will apparently be focused on very much more applied work than the three existing journals.

For the convenience of the reader, a list of the journals that will likely be of most interest in the future is given as Table 9.1. These are grouped as being of

Table 9.1 Journals Containing Fracture Mechanics Analyses and Results

Title	Publisher	Issues per Year
Primary Interest		
International Journal of Fracture	Martinus Nijhoff	12
Engineering Fracture Mechanics	Pergamon	12
International Journal of Fatigue	Butterworths	12
Theoretical and Applied Fracture Mechanics	North Holland	4
Secondary Interest		
Experimental Mechanics	Society of Experimental Stress Analysis	12
Journal of Engineering Materials and Technology	American Society of Mechanical Engineers	4
Journal of Pressure Vessel Technology	American Society of Mechanical Engineers	4
International Journal of Pressure Vessels and Piping	Applied Science	12
International Journal of Engineering Science	Pergamon	12
International Journal of Mechanical Sciences	Pergamon	12
International Journal of Solids and Structures	Pergamon	12
Journal of Materials Science	Chapman and Hall	12
Journal of Composite Materials	Technomic	6
Journal of Mechanics and Physics of Solids	Pergamon	6
Nuclear Engineering and Design	North Holland	4
Occasional Interest		
Journal of Applied Mechanics	American Society of Mechanical Engineers	4
Journal of Applied Physics	American Institute of Physics	12
Philosophical Magazine	Taylor and Francis	12
AIAA Journal	American Institute of Aeronautics and Astronautics	12
Journal of Biomechanics	Pergamon	12
Standardization News	American Society of Testing and Materials	12
Acta Metallurgica	Pergamon	12
Mechanics of Materials	North Holland	4
Journal of Computational and Applied Mathematics	North Holland	6

primary interest—those being exclusively devoted to the subject; secondary interest—those in which one might look to one or more articles on the subject in each issue; and occasional interest—those in which articles of interest to fracture mechanics do appear, but with no regularity.

Any of the journals listed in Table 9.1 will from time to time devote a substantial portion of an issue to collections of articles that bear on fracture mechanics in some way. For example, the *International Journal of Fracture* has published special issues of review articles on fatigue of nonmetals (9.2) and fatigue of metals (9.3), both edited by H. W. Liu. This journal has also published the compendium of sources to fracture toughness and fatigue crack growth data compiled by Hudson and Seward (9.4). *The ASTM Standardization News* has presented a set of comprehensive review articles on nondestructive testing techniques (9.5) that could be read with profit for the insight it brings to the precision that flaws can be detected.

As a further example, a special issue of *Wear*, a journal on the science and technology of friction, lubrication and wear—a journal not ordinarily concerned with fracture mechanics—was devoted to delamination wear and ferrography (9.6). The articles, largely contributed by Nam P. Suh and his associates, made direct use of fracture mechanics concepts in an area that one might not consider to have any connection with fracture mechanics. However, because the appearance of special volumes such as these is not predictable, a researcher wishing to keep abreast of developments in fracture mechanics should make a habit of visiting a technical library. The journals listed in Table 9.1 will cover most articles of interest. But, as this example should indicate, it will not include them all.

The listing given in Table 9.1 is consistent with the general viewpoint taken throughout this book in that it has been compiled from the viewpoint of applied mechanics. A listing from a metallurgical point of view might well include some additional titles while omitting others. We also note that several journals well-represented in the reference lists of previous chapters are not listed; for example, *Proceedings of the Royal Society* and the *Journal of Applied Physics*. While articles in such journals have greatly contributed to the development of fracture mechanics, based upon trends in evidence over the past several years, these would not appear to be fruitful repositories for future progress in this area. Conversely, journals such as the *Journal of Pressure Vessel Technology* and *Nuclear Engineering and Design*, which were not even in existence during the formative years of fracture mechanics, are rich in fracture mechanics applications articles. This shift most probably reflects the evolution of the subject from one of mainly scientific curiosity to one of intense practical interest.

9.2 Conference Proceedings

The second freshest source of information is conference proceedings. Of these the special technical publications of the American Society for Testing and Materials (ASTM, 1916 Race Street, Philadelphia, Pa. 19103) and the conference volumes of the American Society of Mechanical Engineers (ASME, 345 East 47th Street, New York, N.Y. 10017) are probably the most useful. While there is not likely to be a great deal of difference in the technical level between these and the proceedings of other conferences, the ASTM and ASME publications appear both frequently and with some degree of regularity. The proceedings of the annual ASTM National Symposium on

Fracture Mechanics, the most recent of which at the time of this writing was the Sixteenth Symposium held in Columbus, Ohio in August, 1983, are of particular interest. Unfortunately, the appearances of the published volumes have lately been delayed by 2 or more years. Specific references to articles appearing in the ASTM and ASME conference volumes are contained in every chapter of this book.

Other conferences that produce proceedings containing significant amounts of fracture mechanics analyses and data would include the biennial *Structural Mechanics in Reactor Technology* (SMIRT) conference, which meets in odd-numbered years. The published proceedings are generally available in advance of the conference. The eighth of these conferences was held in Chicago in August 1983. The post-SMIRT conferences are also of interest, as are the associated CSNI workshops. They generally provide volumes that are focused on a single theme. An example is the CSNI specialist meeting on Leak-Before-Break in nuclear plant piping held following the 8th SMIRT conference. The proceedings, to be published by the U.S. Nuclear Regulatory Commission, will be edited by J. Strosnider of the NRC. Of a similar nature are the AGARD and NATO conferences, which usually also provide a proceedings volume. Finally, the quadrennial *International Conference on Fracture* produces a multi-volume set from each of its gatherings. These are available at the time of the meeting. The next of these from the time of this writing is scheduled to be held in New Delhi, India, in December 1984.

The disadvantage of conference proceedings as a source of up-to-date information stems from their proliferation and their generally sporadic appearance. Because of the number and generally high prices of such books, few libraries (not to say individuals) will acquire them all. This makes perusal of them a sometime thing. These factors also tend to attract papers of somewhat lower quality—authors wishing to put their work before a larger audience will tend to favor a periodical. On the other hand, because conference proceedings tend to focus on specialized topics, merely keeping track of the titles can provide a good indication of just where the frontiers of research and application are at any time. An example is the multi-volume set on fracture mechanics of ceramics (9.7). We have not provided a compilation of conference proceedings to parallel Table 9.1 as the reader could well do so on his own from the references that we have given, many of which have been taken from conference proceedings.

9.3 Standards

Whether or not one feels that standards reflect fresh information, no one will deny their importance to the technical community. Indeed, one could argue that no research in a subject like fracture mechanics has any value if it does not in some way eventually help a more realistic and reliable standard to be written. Having said this, it should also be said that no current standard is based upon anything much more than linear elastic fracture mechanics. While advances in nonlinear and dynamic fracture mechanics will surely be reflected eventually, changes in regulatory standards are (and should be) made cautiously.

The standards containing fracture mechanics contributions that are of most significance are those of the American Society of Mechanical Engineers, particularly Section III, Appendix G, of the Boiler and Pressure Vessel Code, and Section XI, for nuclear components. The Code of Federal Regulations is also vital for applications to nuclear systems. In the United Kingdom, the British Standards Institution has provided rules on acceptance of defects in welds and on fracture toughness testing.

For testing purposes, the American Society for Testing and Materials *Annual Book of Standards* should be consulted. Of most interest to fracture mechanics is Part 10: *Metals-Mechanical, Fracture, and Corrosion Testing; Fatigue: Erosion and Wear; Effects of Temperature*, which contains a number of standards bearing on the subject. Of most importance are E338, for testing high strength sheet materials; E399, for determining the plane-strain fracture toughness of a metal; E561 for R-curve determinations; E604, for determining dynamic tearing energy; E647, for conducting fatigue crack growth rate studies; and E813, for determining J_{Ic} values. The reader should also be aware that a standard exists in regard to the definitions, symbols, and abbreviations that are used in fracture testing. These are given in E616, standard terminology relating to testing. Standards for crack arrest testing and for the determination of J resistance curves are currently being developed and should appear by 1985.

9.4 Dissertations

University research has on many occasions provided new and vital information in fracture mechanics. The work of Professor J. R. Rice with several of his Ph. D. students at Brown University comes quickly to mind in this regard. These and other dissertations written at Universities in the United States can be obtained at a nominal cost from Xerox University Microfilms, Ann Arbor, Michigan. However, much the same reasons that make conference proceedings a difficult source of current information also apply to dissertations. It is also likely today that Ph. D. research will be quickly written up for publication in a technical journal or presented at a conference whereupon the dissertation itself will be of interest only for particular details. Hence, keeping track of dissertations is not likely to be an effective way to keep abreast of progress in fracture mechanics.

9.5 Abstracting Periodicals

In the category of abstracting periodicals, *Applied Mechanics Reviews*, a monthly publication issued under the auspices of the American Society of Mechanical Engineers, is clearly the most significant source of information.

While too expensive for most individuals to obtain personal copies, virtually all technical libraries subscribe to it. It offers critical capsulations of key articles and books in many areas of fracture mechanics. Unfortunately, probably because the reviews are signed, they tend to be bland. They also tend to lag behind the actual publication, sometimes by a year or more. Nevertheless, one could probably obtain a comprehensive look at an unfamiliar area in fracture mechanics expeditiously in no better way than to inspect AMR. Periodic perusal to reduce the chance of missing an important contribution is also recommended. Occasional review articles on fracture mechanics also appear in AMR—for example, that of Atkinson (9.8).

9.6 Progress Reviews

Review articles appear from time to time in most technical journals. The appearance of such articles is generally unpredictable. But, there are several book series that appear at regular intervals that occasionally contain an article on some aspect of fracture mechanics. These include the *Annual Reviews of Materials Science*, *Advances in Applied Mechanics*, and *Mechanics Today*. Other volumes also appear that have been commissioned by a technical society—for example, the book of Campbell, Gerberich, and Underwood published by the American Society for Metals (9.9). These reviews are usually much more lengthy than journal articles usually are and, thereby, a much more in-depth treatment is provided. A disadvantage is that writers are understandably most enamoured of their own work and often suffer from a considerable degree of myopia with regard to the subject as a whole.

9.7 Handbooks

Three fairly complete collections of linear elastic stress intensity factors exist in handbook form. These are the collections compiled by Tada, Paris, and Irwin (9.10), by Sih (9.11), and by Rooke and Cartwright (9.12). In the elastic-plastic regime, the handbook developed by the General Electric Company under the auspices of the Electric Power Research Institute exists. This is described in Chapter 5 and Appendix A of this book. Corresponding compilations of materials fracture property data are more rare. One that is available is that put together by Battelle's Metals and Ceramics Information Center, Columbus, Ohio (9.13); see also reference (9.4).

9.8 Treatises

A landmark effort in fracture mechanics was the seven volume treatise on fracture assembled by Professor Harold Liebowitz that appeared from 1968 to 1972 (9.14). While some of the more applied articles are by now out of date, many more are timeless. In particular, Vol. II, *Mathematical Fundamentals*, will likely always remain of value to an applied mechanics readership. No textbook could hope to reproduce the wealth of detail that is contained in these articles. Any serious analyst is advised to secure a copy, if one can be found. Also of some interest to fracture mechanics is the seven-volume treatise on composite materials edited by Broutman and Krock that appeared in 1974 (9.15). It contains several chapters on the fracture of composites.

9.9 Textbooks

Thirteen individual textbooks have appeared prior to the writing of this book. The honor of producing the first book on the subject probably belongs to Yokobori (9.16) with the book of Tetelman and McEvily (9.17) a close second. These two books appeared in 1964 (English translation) and 1967, respectively. Next came the books of Knott (9.18), Broek (9.19), Lawn and Wilshaw (9.20), Hertzberg (9.12), and Rolfe and Barsom (9.22). It might be noted that all of these generally excellent offerings were written by materials scientists and, while the analysis side of the subject is represented, they generally reflect the materials point of view. Broek's book, which now appears in its third edition, and Hertzberg's, now in its second edition, probably best complement the present book.

Subsequently, new books have appeared that more directly reflect the point of view of applied mechanics. The first of these are the books by Parton and Morozov (9.23) and Cherepanov (9.24)—both of these by Soviet authors—and by Jayatilaka (9.25) and Parker (9.26). Recently appearing are the books written by Owen and Fawkes (9.27) and Hellan (9.28). These books do focus on the analysis aspects of the subject. However, much like those already listed, these newer offerings are almost exclusively concerned with linear elastic fracture mechanics. While this subject matter does indeed cover the bulk of current applications, extensions of the subject now rapidly coming to fruition are not treated completely in any of these books. Accordingly, the present text was undertaken to provide a treatment encompassing current and advanced nonlinear and dynamic fracture mechanics techniques from the applied mechanics viewpoint.

Readers interested in the early (predominately metallurgical) views of the subject could consult the books of Gensamer et al. (9.29), Parker (9.30), or Tipper (9.31). Textbooks that address a specific area (e.g., composite materials) are given in the reference lists of other chapters. Textbooks covering areas of fracture mechanics that are not addressed in this book would include the important topic of failure analysis. In this area are the book of Thielsch (9.32), which addresses the origin of defects, and those of Engle and Klingele (9.33) and Boyer (9.34), which are concerned with post-mortem examination of failure surfaces. The collection of case studies compiled by Hutchings and Unterweiser (9.35) may be interest for the practical side of fracture. Finally, there have been a number of specialized texts written for specific audiences. An example is one on fracture mechanics for bridge design prepared by Roberts et al. (9.36). Another is that of Fisher (9.37).

9.10 References

(9.1) Grogan, D., *Science and Technology—An Introduction to the Literature*, 4th ed., Clive Bingley, London (1982).

(9.2) Multiple authors, "Fatigue of Non-Metals," *International Journal of Fracture*, **16**, pp. 481–616 (1980).

(9.3) Multiple authors, "Fatigue of Metals," *International Journal of Fracture*, **17**, pp. 101–247 (1981).

(9.4) Hudson, C. M. and Seward, S. K., "A Compendium of Sources of Fracture Toughness and Fatigue-Crack Growth Data for Metallic Alloys," *International Journal of Fracture*, Part I **14**, pp. R151–R184 (1978); Part II **20**, pp. R59–R117 (1982).

(9.5) Multiple authors, *ASTM Standardization News*, Nov. 1982.

(9.6) Multiple authors, *Wear*, **44**, pp. 1–202 (1977).

(9.7) Bradt, R. C., Evans, A. G., Hasselman, D. P. H., and Lange, F. F. (eds.), *Fracture Mechanics of Ceramics*, Plenum, New Y ork (1983).

(9.8) Atkinson, C., "Stress Singularities and Fracture Mechanics," *Applied Mechanics Reviews*, **32**, pp. 123–135 (1979).

(9.9) Campbell, J. E., Gerberich, W. W., and Underwood, J. H. (ed.), *Application of Fracture Mechanics for Selection of Metallic Structural Materials*, American Society for Metals, Metals Park, Ohio (1982).

(9.10) Tada, H., Paris, P. C., and Irwin, G. R., *The Stress Aanlysis of Cracks Handbook*, Del Research Corporation, Hellertown, Pa. (1973).

(9.11) Sih, G. C., *Handbook of Stress Intensity Factors*, Institute of Fracture and Solid Mechanics, Lehigh University, Bethlehem, Pa. (1973).

(9.12) Rooke, D. P. and Cartwright, D. J., *Compendium of Stress Intensity Factors*, Her Majesty's Stationery Office, London (1976).

(9.13) Anon. *Damage Tolerant Design Handbook*, two volumes, Metals and Ceramics Information Center, Battelle, Columbus, Ohio.

(9.14) Liebowitz, H. (ed.), *Treatise on Fracture*, in seven volumes, Academic, New York (1968).

(9.15) Broutman, L. J. and Krock, R. H. (ed.), *Composite Materials*, in seven volumes, Academic, New York (1974).

(9.16) Yokobori, T., *The Strength, Fracture and Fatigue of Materials*, Noordhoff, Groningen, The Netherlands (1964).

(9.17) Tetelman, A. and McEvily, A. J., *Fracture of Structural Materials*, Wiley, New York (1967).

(9.18) Knott, J. F., *Fundamentals of Fracture Mechanics*, Wiley, New York (1973).

(9.19) Broek, D., *Elementary Engineering Fracture Mechanics*, Noordhoff, Leyden, The Netherlands (1974); Third Revised Edition (1982).

(9.20) Lawn, B. R. and Wilshaw, T. R., *Fracture of Brittle Solids*, Cambridge University Press (1975).

(9.21) Hertzberg, R. W., *Deformation and Fracture Mechanics of Engineering Materials*, Wiley, New York (1976); second Edition (1983).

(9.22) Rolfe, S. T. and Barsom, J. M., *Fracture and Fatigue Control in Structures; Applications of Fracture Mechanics*, Prentice-Hall, Englewood Cliffs, N.J. (1977).

(9.23) Parton, V. Z. and Morozov, E. M., *Elastic-Plastic Fracture Mechanics*, Mir, Moscow (1978).

(9.24) Cherepanov, G. P., *Mechanics of Brittle Fracture*, McGraw-Hill, New York (1979).

(9.25) Jayatilaka, A. de S., *Fracture of Engineering Brittle Materials*, Applied Science, London (1978).

(9.26) Parker, A. P., *The Mechanics of Fracture and Fatigue, An Introduction*. E. and F. N. Spin, London (1981).

(9.27) Owen, D. R. J. and Fawkes, A. J., *Engineering Fracture Mechanics: Numerical Methods and Applications*, Pineridge Press Ltd., Swansea, U.K. (1983).

(9.28) Hellan, K., *Introduction to Fracture Mechanics*, McGraw-Hill, New York (1984).

(9.29) Gensamer, M., Saibel, E., Ransom, J. T., and Laurie, R. E., *The Fracture of Metals*, American Welding Society, New York (1947).

(9.30) Parker, E. R., *Brittle Behavior of Engineering Structures*, Wiley, New York (1957).

(9.31) Tipper, C. F., *The Brittle Fracture Story*, Cambridge University Press (1962).

(9.32) Thielsch, K., *Defects and Failures in Pressure Vessels and Piping*, Reinhold, New York (1965).

(9.33) Engle, L. and Klingele, H. *An Atlas of Metal Damage*, Prentice-Hall, Englewood Cliffs, N.J. (1981).
(9.34) Boyer, H. E. (ed.), *Metals Handbook*, Vol. 10, *Failure Analysis and Prevention*, American Society for Metals, Metals Park, Ohio, 8th ed. (1975).
(9.35) Hutchings, F. R. and Unterweiser, P. M., *Failure Analysis: The British Engine Technical Reports*, American Society for Metals, Metals Park, Ohio (1981).
(9.36) Roberts, R., Barsom, J. M., Rolfe, S. T., and Fisher, J. W., *Fracture Mechanics for Bridge Design*, Report No. FHWA-RD-78-68, Federal Highway Administration, Washington, D.C. (1977).
(9.37) Fisher, J. W., *Fatigue and Fracture in Steel Bridges*, Wiley, New York (1984).

Appendix A

COMPILATION OF FULLY PLASTIC SOLUTIONS

In the following, tabulated values of the functions $h_1(a/W, n)$, $h_2(a/W, n)$, and $h_3(a/W, n)$ for the fully plastic solutions in the engineering approach to plastic fracture of Section 5.4 are presented for a variety of laboratory fracture specimens and flawed structural components. The corresponding expressions for $g_i(a/W)$ are also given. The development of these functions was sponsored by the Electric Power Research Institute.*

A.1 Compact Tension Specimen

For the standard ASTM compact tension specimen of Figure A.1, the limit load P_0 per unit thickness is

$$P_0 = 1.455\beta b\sigma_y \tag{A.1-1}$$

for plane strain and

$$P_0 = 1.071\beta b\sigma_y \tag{A.1-2}$$

for plane stress wherein

$$\beta = [(2a/b)^2 + 4a/b + 2]^{\frac{1}{2}} - 2a/b - 1 \tag{A.1-3}$$

Note that β is the positive root of Equation (5.5-19). For this specimen Δ is the crack opening displacement at the load line and δ is the crack mouth opening displacement. For the tabulated functions h_1, h_2, and h_3 in Table A.1 for plane strain and Table A.2 for plane stress, $g_i = 1$.

A.2 Center Cracked Panel

The center cracked panel in Figure A.2 is loaded in tension by a uniform remote stress $\sigma = P/2W$. The limit load P_0 per unit thickness is

$$P_0 = 4b\sigma_y/\sqrt{3} \tag{A.2-1}$$

for plane strain and

$$P_0 = 2b\sigma_y \tag{A.2-2}$$

* Copyright © 1981, Electric Power Research Institute. EPRI NP-1931, "An Engineering Approach for Elastic-Plastic Fracture Analysis." Reprinted with permission.

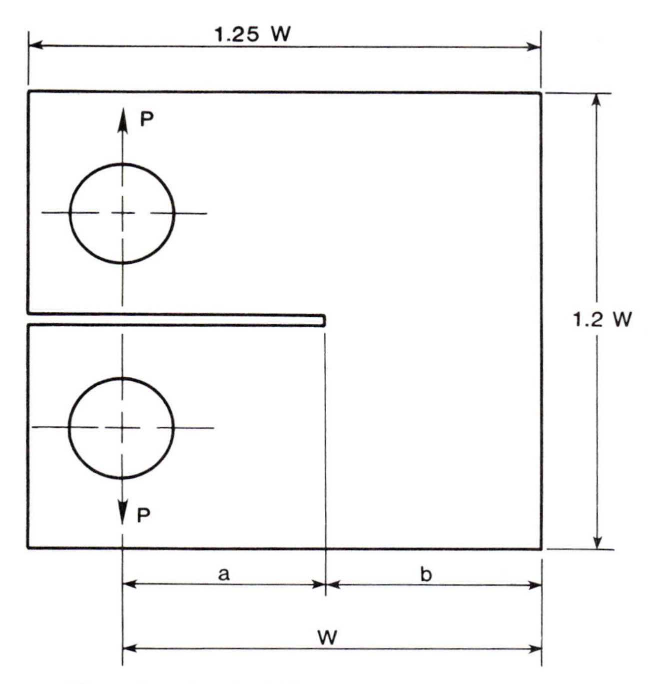

Figure A.1 Standard ASTM compact tension specimen.

Table A.1 Plane Strain h-Functions for Standard ASTM Compact Tension Specimen

		$n = 1$	$n = 2$	$n = 3$	$n = 5$	$n = 7$	$n = 10$	$n = 13$	$n = 16$	$n = 20$
	h_1	2.23	2.05	1.78	1.48	1.33	1.26	1.25	1.32	1.57
$a/W = \frac{1}{4}$	h_2	17.9	12.5	11.7	10.8	10.5	10.7	11.5	12.6	14.6
	h_3	9.85	8.51	8.17	7.77	7.71	7.92	8.52	9.31	10.9
	h_1	2.15	1.72	1.39	0.970	0.693	0.443	0.276	0.176	0.098
$a/W = \frac{3}{8}$	h_2	12.6	8.18	6.52	4.32	2.97	1.79	1.10	0.686	0.370
	h_3	7.94	5.76	4.64	3.10	2.14	1.29	0.793	0.494	0.266
	h_1	1.94	1.51	1.24	0.919	0.685	0.461	0.314	0.216	0.132
$a/W = \frac{1}{2}$	h_2	9.33	5.85	4.30	2.75	1.91	1.20	0.788	0.530	0.317
	h_3	6.41	4.27	3.16	2.02	1.41	0.888	0.585	0.393	0.236
	h_1	1.76	1.45	1.24	0.974	0.752	0.602	0.459	0.347	0.248
$a/W = \frac{5}{8}$	h_2	7.61	4.57	3.42	2.36	1.81	1.32	0.983	0.749	0.485
	h_3	5.52	3.43	2.58	1.79	1.37	1.00	0.746	0.568	0.368
	h_1	1.71	1.42	1.26	1.033	0.864	0.717	0.575	0.448	0.345
$a/W = \frac{3}{4}$	h_2	6.37	3.95	3.18	2.34	1.88	1.44	1.12	0.887	0.665
	h_3	4.86	3.05	2.46	1.81	1.45	1.11	0.869	0.686	0.514
	h_1	1.57	1.45	1.35	1.18	1.08	0.950	0.850	0.730	0.630
$a/W \to 1$	h_2	5.39	3.74	3.09	2.43	2.12	1.80	1.57	1.33	1.14
	h_3	4.31	2.99	2.47	1.95	1.79	1.44	1.26	1.07	0.909

Table A.2 Plane Stress h-Functions for Standard ASTM Compact Tension Specimen

		$n=1$	$n=2$	$n=3$	$n=5$	$n=7$	$n=10$	$n=13$	$n=16$	$n=20$
	h_1	1.61	1.46	1.28	1.06	0.903	0.729	0.601	0.511	0.395
$a/W=\frac{1}{4}$	h_2	17.6	12.0	10.7	8.74	7.32	5.74	4.63	3.75	2.92
	h_3	9.67	8.00	7.21	5.94	5.00	3.95	3.19	2.59	2.023
	h_1	1.55	1.25	1.05	0.801	0.647	0.484	0.377	0.284	0.220
$a/W=\frac{3}{8}$	h_2	12.4	8.20	6.54	4.56	3.45	2.44	1.83	1.36	1.02
	h_3	7.80	5.73	4.62	3.25	2.48	1.77	1.33	0.990	0.746
	h_1	1.40	1.08	0.901	0.686	0.558	0.436	0.356	0.298	0.238
$a/W=\frac{1}{2}$	h_2	9.16	5.67	4.21	2.80	2.12	1.57	1.25	1.03	0.814
	h_3	6.29	4.15	3.11	2.09	1.59	1.18	0.938	0.774	0.614
	h_1	1.27	1.03	0.875	0.695	0.593	0.494	0.423	0.370	0.310
$a/W=\frac{5}{8}$	h_2	7.47	4.48	3.35	2.37	1.92	1.54	1.29	1.12	0.928
	h_3	5.42	3.38	2.54	1.80	1.47	1.18	0.988	0.853	0.710
	h_1	1.23	0.977	0.833	0.683	0.598	0.506	0.431	0.373	0.314
$a/W=\frac{3}{4}$	h_2	6.25	3.78	2.89	2.14	1.78	1.44	1.20	1.03	0.857
	h_3	4.77	2.92	2.24	1.66	1.38	1.12	0.936	0.800	0.666
	h_1	1.13	1.01	0.775	0.680	0.650	0.620	0.490	0.470	0.420
$a/W\to 1$	h_2	5.29	3.54	2.41	1.91	1.73	1.59	1.23	1.17	1.03
	h_3	4.23	2.83	1.93	1.52	1.39	1.27	0.985	0.933	0.824

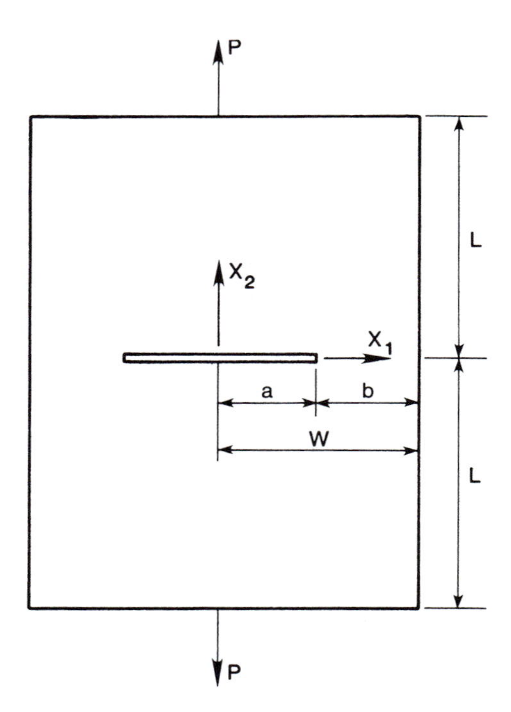

Figure A.2 Center cracked panel in tension.

Table A.3 Plane Strain h-Functions for a Center-Cracked Panel

		$n=1$	$n=2$	$n=3$	$n=5$	$n=7$	$n=10$	$n=13$	$n=15$	$n=20$
	h_1	2.80	3.61	4.06	4.35	4.33	4.02	3.56	3.06	2.46
$a/W=\frac{1}{8}$	h_2	3.05	3.62	3.91	4.06	3.93	3.54	3.07	2.60	2.06
	h_3	0.303	0.574	0.840	1.30	1.63	1.95	2.03	1.96	1.77
	h_1	2.54	3.01	3.21	3.29	3.18	2.92	2.63	2.34	2.03
$a/W=\frac{1}{4}$	h_2	2.68	2.99	3.01	2.85	2.61	2.30	1.97	1.71	1.45
	h_3	0.536	0.911	1.22	1.64	1.84	1.85	1.80	1.64	1.43
	h_1	2.34	2.62	2.65	2.51	2.28	1.97	1.71	1.46	1.19
$a/W=\frac{3}{8}$	h_2	2.35	2.39	2.23	1.88	1.58	1.28	1.07	0.890	0.715
	h_3	0.699	1.06	1.28	1.44	1.40	1.23	1.05	0.888	0.719
	h_1	2.21	2.29	2.20	1.97	1.76	1.52	1.32	1.16	0.978
$a/W=\frac{1}{2}$	h_2	2.03	1.86	1.60	1.23	1.00	0.799	0.664	0.564	0.466
	h_3	0.803	1.07	1.16	1.10	0.968	0.796	0.665	0.565	0.469
	h_1	2.12	1.96	1.76	1.43	1.17	0.863	0.628	0.458	0.300
$a/W=\frac{5}{8}$	h_2	1.71	1.32	1.04	0.707	0.524	0.358	0.250	0.178	0.114
	h_3	0.844	0.937	0.879	0.701	0.522	0.361	0.251	0.178	0.115
	h_1	2.07	1.73	1.47	1.11	0.895	0.642	0.461	0.337	0.216
$a/W=\frac{3}{4}$	h_2	1.35	0.857	0.596	0.361	0.254	0.167	0.114	0.0810	0.0511
	h_3	0.805	0.700	0.555	0.359	0.254	0.168	0.114	0.0813	0.0516
	h_1	2.08	1.64	1.40	1.14	0.987	0.814	0.688	0.573	0.461
$a/W=\frac{7}{8}$	h_2	0.889	0.428	0.287	0.181	0.139	0.105	0.0837	0.0682	0.0533
	h_3	0.632	0.400	0.291	0.182	0.140	0.106	0.0839	0.0683	0.0535

Table A.4 Plane Stress h-Functions for a Center-Cracked Panel

		$n=1$	$n=2$	$n=3$	$n=5$	$n=7$	$n=10$	$n=13$	$n=16$	$n=20$
	h_1	2.80	3.57	4.01	4.47	4.65	4.62	4.41	4.13	3.72
$a/W=\frac{1}{8}$	h_2	3.53	4.09	4.43	4.74	4.79	4.63	4.33	4.00	3.55
	h_3	0.350	0.661	0.997	1.55	2.05	2.56	2.83	2.95	2.92
	h_1	2.54	2.97	3.14	3.20	3.11	2.86	2.65	2.47	2.20
$a/W=\frac{1}{4}$	h_2	3.10	3.29	3.30	3.15	2.93	2.56	2.29	2.08	1.81
	h_3	0.619	1.01	1.35	1.83	2.08	2.19	2.12	2.01	1.79
	h_1	2.34	2.53	2.52	2.35	2.17	1.95	1.77	1.61	1.43
$a/W=\frac{3}{8}$	h_2	2.71	2.62	2.41	2.03	1.75	1.47	1.28	1.13	0.988
	h_3	0.807	1.20	1.43	1.59	1.57	1.43	1.27	1.13	0.994
	h_1	2.21	2.20	2.06	1.81	1.63	1.43	1.30	1.17	1.00
$a/W=\frac{1}{2}$	h_2	2.34	2.01	1.70	1.30	1.07	0.871	0.757	0.666	0.557
	h_3	0.927	1.19	1.26	1.18	1.04	0.867	0.758	0.668	0.560
	h_1	2.12	1.91	1.69	1.41	1.22	1.01	0.853	0.712	0.573
$a/W=\frac{5}{8}$	h_2	1.97	1.46	1.13	0.785	0.617	0.474	0.383	0.313	0.256
	h_3	0.975	1.05	0.970	0.763	0.620	0.478	0.386	0.318	0.273
	h_1	2.07	1.71	1.46	1.21	1.08	0.867	0.745	0.646	0.532
$a/W=\frac{3}{4}$	h_2	1.55	0.970	0.685	0.452	0.361	0.262	0.216	0.183	0.148
	h_3	0.929	0.802	0.642	0.450	0.361	0.263	0.216	0.183	0.149
	h_1	2.08	1.57	1.31	1.08	0.972	0.862	0.778	0.715	0.630
$a/W=\frac{7}{8}$	h_2	1.03	0.485	0.310	0.196	0.157	0.127	0.109	0.0971	0.0842
	h_3	0.730	0.452	0.313	0.198	0.157	0.127	0.109	0.0973	0.0842

for plane stress. Here Δ is the average load-point displacement defined by

$$\Delta = \frac{1}{2W}\int_{-W}^{W} [u_2(x_1, L) - u_2(x_1, -L)]\, dx_1 \tag{A.2-3}$$

The crack opening displacement at the center of the crack is denoted by δ. With $g_1 = g_4 = a/W$ and $g_2 = g_3 = 1$, the tabulated functions h_1, h_2, and h_3 for plane strain and plane stress appear in Tables A.3 and A.4, respectively.

A.3 Single Edge Notched Specimen

The limit load P_0 per unit thickness for a single edge notched specimen under a uniform remote tension $\sigma = P/W$ (Figure A.3) is

$$P_0 = 1.455\beta b\sigma_y \tag{A.3-1}$$

for plane strain and

$$P_0 = 1.072\beta b\sigma_y \tag{A.3-2}$$

for plane stress where now

$$\beta = [1 + (a/b)^2]^{\frac{1}{2}} - a/b \tag{A.3-3}$$

For this specimen Δ is the load-point displacement at the centerline of the specimen and δ is the crack mouth opening displacement. For the tabulated functions in Tables A.5 and A.6 for plane strain and plane stress, respectively, $g_1 = g_4 = a/W$ and $g_2 = g_3 = 1$.

A.4 Double Edge Notched Specimen

The double edge notched specimen in Figure A.4 is loaded by a uniform remote stress $\sigma = P/2W$. The corresponding limit load per unit thickness is

$$P_0 = (0.72W + 1.82b)\sigma_y \tag{A.4-1}$$

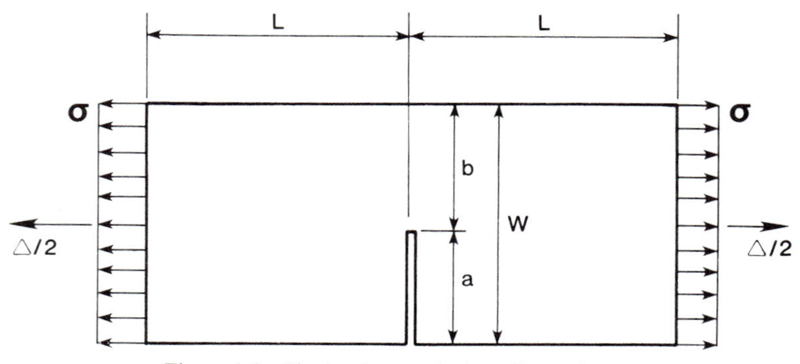

Figure A.3 Single edge notched tensile specimen.

Table A.5 Plane Strain h-Functions for a Single Edge Notched Specimen

		$n = 1$	$n = 2$	$n = 3$	$n = 5$	$n = 7$	$n = 10$	$n = 13$	$n = 16$	$n = 20$
$a/W = \frac{1}{8}$	h_1	4.95	6.93	8.57	11.5	13.5	16.1	18.1	19.9	21.2
	h_2	5.25	6.47	7.56	9.46	11.1	12.9	14.4	15.7	16.8
	h_3	26.6	25.8	25.2	24.2	23.6	23.2	23.2	23.5	23.7
$a/W = \frac{1}{4}$	h_1	4.34	4.77	4.64	3.82	3.06	2.17	1.55	1.11	0.712
	h_2	4.76	4.56	4.28	3.39	2.64	1.81	1.25	0.875	0.552
	h_3	10.3	7.64	5.87	3.70	2.48	1.50	0.970	0.654	0.404
$a/W = \frac{3}{8}$	h_1	3.88	3.25	2.63	1.68	1.06	0.539	0.276	0.142	0.0595
	h_2	4.54	3.49	2.67	1.57	0.946	0.458	0.229	0.116	0.048
	h_3	5.14	2.99	1.90	0.923	0.515	0.240	0.119	0.060	0.0246
$a/W = \frac{1}{2}$	h_1	3.40	2.30	1.69	0.928	0.514	0.213	0.0902	0.0385	0.0119
	h_2	4.45	2.77	1.89	0.954	0.507	0.204	0.0854	0.0356	0.0110
	h_3	3.15	1.54	0.912	0.417	0.215	0.085	0.0358	0.0147	0.00448
$a/W = \frac{5}{8}$	h_1	2.86	1.80	1.30	0.697	0.378	0.153	0.0625	0.0256	0.0078
	h_2	4.37	2.44	1.62	0.0806	0.423	0.167	0.0671	0.0272	0.00823
	h_3	2.31	1.08	0.681	0.329	0.171	0.067	0.0268	0.0108	0.00326
$a/W = \frac{3}{4}$	h_1	2.34	1.61	1.25	0.769	0.477	0.233	0.116	0.059	0.0215
	h_2	4.32	2.52	1.79	1.03	0.619	0.296	0.146	0.0735	0.0267
	h_3	2.02	1.10	0.765	0.435	0.262	0.125	0.0617	0.0312	0.0113
$a/W = \frac{7}{8}$	h_1	1.91	1.57	1.37	1.10	0.925	0.702			
	h_2	4.29	2.75	2.14	1.55	1.23	0.921			
	h_3	2.01	1.27	0.988	0.713	0.564	0.424			

Table A.6 Plane Stress h-Functions for a Single Edge Notched Specimen

		$n = 1$	$n = 2$	$n = 3$	$n = 5$	$n = 7$	$n = 10$	$n = 13$	$n = 16$	$n = 20$
$a/W = \frac{1}{8}$	h_1	3.58	4.55	5.06	5.30	4.96	4.14	3.29	2.60	1.92
	h_2	5.15	5.43	6.05	6.01	5.47	4.46	3.48	2.74	2.02
	h_3	26.1	21.6	18.0	12.7	9.24	5.98	3.94	2.72	2.0
$a/W = \frac{1}{4}$	h_1	3.14	3.26	2.92	2.12	1.53	0.960	0.615	0.400	0.230
	h_2	4.67	4.30	3.70	2.53	1.76	1.05	0.656	0.419	0.237
	h_3	10.1	6.49	4.36	2.19	1.24	0.630	0.362	0.224	0.123
$a/W = \frac{3}{8}$	h_1	2.81	2.37	1.94	1.37	1.01	0.677	0.474	0.342	0.226
	h_2	4.47	3.43	2.63	1.69	1.18	0.762	0.524	0.372	0.244
	h_3	5.05	2.65	1.60	0.812	0.525	0.328	0.223	0.157	0.102
$a/W = \frac{1}{2}$	h_1	2.46	1.67	1.25	0.776	0.510	0.286	0.164	0.0956	0.0469
	h_2	4.37	2.73	1.91	1.09	0.694	0.380	0.216	0.124	0.0607
	h_3	3.10	1.43	0.871	0.461	0.286	0.155	0.088	0.0506	0.0247
$a/W = \frac{5}{8}$	h_1	2.07	1.41	1.105	0.755	0.551	0.363	0.248	0.172	0.107
	h_2	4.30	2.55	1.84	1.16	0.816	0.523	0.353	0.242	0.150
	h_3	2.27	1.13	0.771	0.478	0.336	0.215	0.146	0.100	0.0616
$a/W = \frac{3}{4}$	h_1	1.70	1.14	0.910	0.624	0.447	0.280	0.181	0.118	0.0670
	h_2	4.24	2.47	1.81	1.15	0.798	0.490	0.314	0.203	0.115
	h_3	1.98	1.09	0.784	0.494	0.344	0.211	0.136	0.0581	0.0496
$a/W = \frac{7}{8}$	h_1	1.38	1.11	0.962	0.792	0.677	0.574			
	h_2	4.22	2.68	2.08	1.54	1.27	1.04			
	h_3	1.97	1.25	0.969	0.716	0.591	0.483			

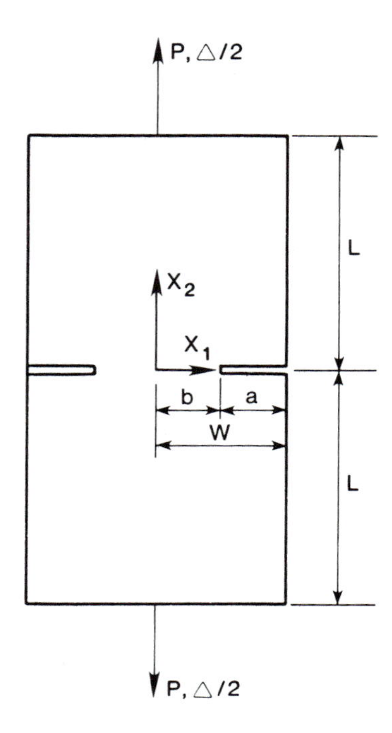

Figure A.4 Double edge notched tensile specimen.

for plane strain and

$$P_0 = 4b\sigma_y/\sqrt{3} \qquad \text{(A.4-2)}$$

for plane stress. Here Δ is the load-point displacement at the centerline of the specimen and δ is the crack mouth opening displacement. Tabulated values of h_1, h_2, and h_3 are presented in Tables A.7 and A.8 for plane strain and plane stress, respectively. These values are associated with $g_1 = g_4 = 1$ and $g_2 = g_3 = W/a - 1$.

A.5 Axially Cracked Pressurized Cylinder

Figure A.5 depicts a long circular cylindrical pressure vessel containing a long axial crack. The crack front is parallel to the axis of the cylinder. The generalized load is the uniform internal pressure p. A lower bound for the limit pressure p_0 is

$$p_0 = 2b\sigma_y/(\sqrt{3}R_c) \qquad \text{(A.5-1)}$$

where $R_c = R_i + a$ is the radial distance from the axis of the cylinder to the. crack front. For $g_1 = a/W$ and $g_2 = 1$ the functions h_1 and h_2 are tabulated in Tables A.9, A.10, and A.11 for $W/R_i = \frac{1}{5}$, $\frac{1}{10}$, and $\frac{1}{20}$, respectively. For this configuration δ is the crack mouth opening displacement.

Table A.7 Plane Strain h-Functions for a Double Edge Notched Specimen

		$n = 1$	$n = 2$	$n = 3$	$n = 5$	$n = 7$	$n = 10$	$n = 13$	$n = 16$	$n = 20$
	h_1	0.572	0.772	0.922	1.13	1.35	1.61	1.86	2.08	2.44
$a/W = \frac{1}{8}$	h_2	0.732	0.852	0.961	1.14	1.29	1.50	1.70	1.94	2.17
	h_3	0.063	0.126	0.200	0.372	0.571	0.911	1.30	1.74	2.29
	h_1	1.10	1.32	1.38	1.65	1.75	1.82	1.86	1.89	1.92
$a/W = \frac{1}{4}$	h_2	1.56	1.63	1.70	1.78	1.80	1.81	1.79	1.78	1.76
	h_3	0.267	0.479	0.698	1.11	1.47	1.92	2.25	2.49	2.73
	h_1	1.61	1.83	1.92	1.92	1.84	1.68	1.49	1.32	1.12
$a/W = \frac{3}{8}$	h_2	2.51	2.41	2.35	2.15	1.94	1.68	1.44	1.25	1.05
	h_3	0.637	1.05	1.40	1.87	2.11	2.20	2.09	1.92	1.67
	h_1	2.22	2.43	2.48	2.43	2.32	2.12	1.91	1.60	1.51
$a/W = \frac{1}{2}$	h_2	3.73	3.40	3.15	2.71	2.37	2.01	1.72	1.40	1.38
	h_3	1.26	1.92	2.37	2.79	2.85	2.68	2.40	1.99	1.94
	h_1	3.16	3.38	3.45	3.42	3.28	3.00	2.54	2.36	2.27
$a/W = \frac{5}{8}$	h_2	5.57	4.76	4.23	3.46	2.97	2.48	2.02	1.82	1.66
	h_3	2.36	3.29	3.74	3.90	3.68	3.23	2.66	2.40	2.19
	h_1	5.24	6.29	7.17	8.44	9.46	10.9	11.9	11.3	17.4
$a/W = \frac{3}{4}$	h_2	9.10	7.76	7.14	6.64	6.83	7.48	7.79	7.14	11.1
	h_3	4.73	6.26	7.03	7.63	8.14	9.04	9.40	8.58	13.5
	h_1	14.2	24.8	39.0	78.4	140.0	341.0	777.0	1570.0	3820.0
$a/W = \frac{7}{8}$	h_2	20.1	19.4	22.7	36.1	58.9	133.0	294.0	585.0	1400.0
	h_3	12.7	18.2	24.1	40.4	65.8	149.0	327.0	650.0	1560.0

Table A.8 Plane Stress h-Functions for a Double Edge Notched Specimen

		$n = 1$	$n = 2$	$n = 3$	$n = 5$	$n = 7$	$n = 10$	$n = 13$	$n = 16$	$n = 20$
	h_1	0.583	0.825	1.02	1.37	1.71	2.24	2.84	3.54	4.62
$a/W = \frac{1}{8}$	h_2	0.853	1.05	1.23	1.55	1.87	2.38	2.96	3.65	4.70
	h_3	0.0729	0.159	0.26	0.504	0.821	1.41	2.18	3.16	4.73
	h_1	1.01	1.23	1.36	1.48	1.54	1.58	1.59	1.59	1.59
$a/W = \frac{1}{4}$	h_2	1.73	1.82	1.89	1.92	1.91	1.85	1.80	1.75	1.70
	h_3	0.296	0.537	0.770	1.17	1.49	1.82	2.02	2.12	2.20
	h_1	1.29	1.42	1.43	1.34	1.24	1.09	0.970	0.873	0.674
$a/W = \frac{3}{8}$	h_2	2.59	2.39	2.22	1.86	1.59	1.28	1.07	0.922	0.709
	h_3	0.658	1.04	1.30	1.52	1.55	1.41	1.23	1.07	0.830
	h_1	1.48	1.47	1.38	1.17	1.01	0.845	0.732	0.625	0.208
$a/W = \frac{1}{2}$	h_2	3.51	2.82	2.34	1.67	1.28	0.944	0.762	0.630	0.232
	h_3	1.18	1.58	1.69	1.56	1.32	1.01	0.809	0.662	0.266
	h_1	1.59	1.45	1.29	1.04	0.882	0.737	0.649	0.466	0.0202
$a/W = \frac{5}{8}$	h_2	4.56	3.15	2.32	1.45	1.06	0.790	0.657	0.473	0.0277
	h_3	1.93	2.14	1.95	1.44	1.09	0.809	0.665	0.487	0.0317
	h_1	1.65	1.43	1.22	0.979	0.834	0.701	0.630	0.297	
$a/W = \frac{3}{4}$	h_2	5.90	3.37	2.22	1.30	0.966	0.741	0.636	0.312	
	h_3	3.06	2.67	2.06	1.31	0.978	0.747	0.638	0.318	
	h_1	1.69	1.43	1.22	0.979	0.845	0.738	0.664	0.614	0.562
$a/W = \frac{7}{8}$	h_2	8.02	3.51	2.14	1.27	0.971	0.775	0.663	0.596	0.535
	h_3	5.07	3.18	2.16	1.30	0.980	0.779	0.665	0.597	0.538

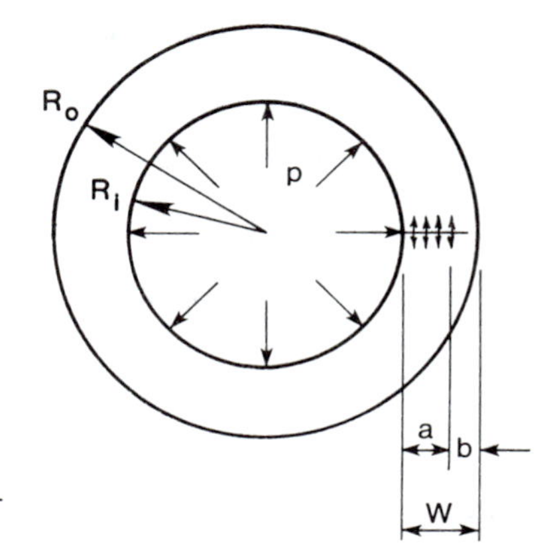

Figure A.5 Axially cracked pressurized cylinder.

Table A.9 h-Functions for an Internally Pressurized, Axially Cracked Cylinder with $W/R_i = \frac{1}{5}$

		$n = 1$	$n = 2$	$n = 3$	$n = 5$	$n = 7$	$n = 10$
$a/W = \frac{1}{8}$	h_1	6.32	7.93	9.32	11.5	13.12	14.94
	h_2	5.83	7.01	7.96	9.49	10.67	11.96
$a/W = \frac{1}{4}$	h_1	7.00	8.34	9.03	9.59	9.71	9.45
	h_2	5.92	8.72	7.07	7.26	7.14	6.71
$a/W = \frac{1}{2}$	h_1	9.79	10.37	9.07	5.61	3.52	2.11
	h_2	7.05	6.97	6.01	3.70	2.28	1.25
$a/W = \frac{3}{4}$	h_1	11.00	5.54	2.84	1.24	0.83	0.493
	h_2	7.35	3.86	1.86	0.556	0.261	0.129

Table A.10 h-Functions for an Internally Pressurized, Axially Cracked Cylinder with $W/R_i = \frac{1}{10}$

		$n = 1$	$n = 2$	$n = 3$	$n = 5$	$n = 7$	$n = 10$
$a/W = \frac{1}{8}$	h_1	5.22	6.64	7.59	8.76	9.34	9.55
	h_2	5.31	6.25	6.88	7.65	8.02	8.09
$a/W = \frac{1}{4}$	h_1	6.16	7.49	7.96	8.08	7.78	6.98
	h_2	5.56	6.31	6.52	6.40	6.01	5.27
$a/W = \frac{1}{2}$	h_1	10.5	11.6	10.7	6.47	3.95	2.27
	h_2	7.48	7.72	7.01	4.29	2.58	1.37
$a/W = \frac{3}{4}$	h_1	16.1	8.19	3.87	1.46	1.05	0.787
	h_2	9.57	5.40	2.57	0.706	0.370	0.232

Table A.11 h-Functions for an Internally Pressurized Axially Cracked Cylinder with $W/R_i = \frac{1}{20}$

		$n = 1$	$n = 2$	$n = 3$	$n = 5$	$n = 7$	$n = 10$
$a/W = \frac{1}{8}$	h_1	4.50	5.79	6.62	7.65	8.07	7.75
	h_2	4.96	5.71	6.20	6.82	7.02	6.66
$a/W = \frac{1}{4}$	h_1	5.57	6.91	7.37	7.47	7.21	6.53
	h_2	5.29	5.98	6.16	6.01	5.63	4.93
$a/W = \frac{1}{2}$	h_1	10.8	12.8	12.8	8.16	4.88	2.62
	h_2	7.66	8.33	8.13	5.33	3.20	1.65
$a/W = \frac{3}{4}$	h_1	23.1	13.1	5.87	1.90	1.23	0.883
	h_2	12.1	7.88	3.84	1.01	0.454	0.240

Table A.12 F and V_1 for Internally Pressurized Axially Cracked Cylinders

		$a/W = \frac{1}{8}$	$a/W = \frac{1}{4}$	$a/W = \frac{1}{2}$	$a/W = \frac{3}{4}$
$W/R_i = \frac{1}{5}$	F	1.19	1.38	2.10	3.30
	V_1	1.51	1.83	3.44	7.50
$W/R_i = \frac{1}{10}$	F	1.20	1.44	2.36	4.23
	V_1	1.54	1.91	3.96	10.4
$W/R_i = \frac{1}{20}$	F	1.20	1.45	2.51	5.25
	V_1	1.54	1.92	4.23	13.5

In the elastic range

$$K_I = \frac{2pR_0^2\sqrt{\pi a}}{R_0^2 - R_i^2} F\left(\frac{a}{W}, \frac{R_i}{R_0}\right) \tag{A.5-2}$$

and

$$\delta_e = \frac{8pR_0^2 a}{(R_0^2 - R_i^2)E'} V_1\left(\frac{a}{W}, \frac{R_i}{R_0}\right) \tag{A.5-3}$$

where the dimensionless functions F and V_1 are presented in Table A.12.

A.6 Circumferentially Cracked Cylinder

An internally, circumferentially cracked circular cylinder subjected to a remote axial tension $\sigma = P/\pi(R_0^2 - R_i^2)$ is depicted in Figure A.6. A lower bound to the limit axial load P_0 is

$$P_0 = 2\pi\sigma_y(R_0^2 - R_c^2)/\sqrt{3} \tag{A.6-1}$$

where $R_c = R_i + a$ is the radial distance from the axis of the cylinder to the crack front. Again δ is the crack mouth opening displacement. For

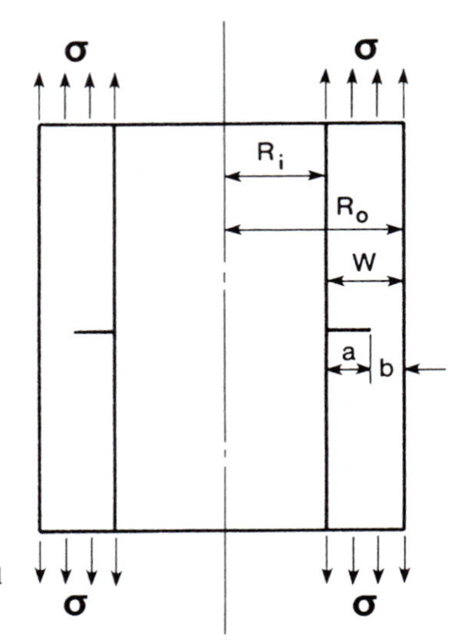

Figure A.6 Circumferentially cracked cylinder.

$g_1 = g_4 = a/W$ and $g_2 = g_3 = 1$, the functions h_1, h_2, and h_3 are presented in Tables A.13, A.14, and A.15 for $W/R_i = \frac{1}{5}, \frac{1}{10}$, and $\frac{1}{20}$, respectively.

In the elastic range

$$K_I = \sigma\sqrt{\pi a}\,F(a/W, R_i/R_0) \quad \text{(A.6-2)}$$

$$\delta_e = 4\sigma a V_1(a/W, R_i/R_0)/E' \quad \text{(A.6-3)}$$

$$\Delta_{ce} = 4\sigma a V_2(a/W, R_i/R_0)/E' \quad \text{(A.6-4)}$$

where the dimensionless functions F, V_1, and V_2 are tabulated in Table A.16.

Table A.13 h-Functions for a Circumferentially Cracked Cylinder in Tension with $W/R_i = \frac{1}{5}$

		$n = 1$	$n = 2$	$n = 3$	$n = 5$	$n = 7$	$n = 10$
	h_1	3.78	5.00	5.94	7.54	8.99	11.1
$a/W = \frac{1}{8}$	h_2	4.56	5.55	6.37	7.79	9.10	11.0
	h_3	0.369	0.700	1.07	1.96	3.04	4.94
	h_1	3.88	4.95	5.64	6.49	6.94	7.22
$a/W = \frac{1}{4}$	h_2	4.40	5.12	5.57	6.07	6.28	6.30
	h_3	0.673	1.25	1.79	2.79	3.61	4.52
	h_1	4.40	4.78	4.59	3.79	3.07	2.34
$a/W = \frac{1}{2}$	h_2	4.36	4.30	3.91	3.00	2.26	1.55
	h_3	1.33	1.93	2.21	2.23	1.94	1.46
	h_1	4.12	3.03	2.23	1.546	1.30	1.11
$a/W = \frac{3}{4}$	h_2	3.46	2.19	1.36	0.638	0.436	0.325
	h_3	1.54	1.39	1.04	0.686	0.508	0.366

Table A.14 h-Functions for a Circumferentially Cracked Cylinder in Tension with $W/R_i = \frac{1}{10}$

		$n = 1$	$n = 2$	$n = 3$	$n = 5$	$n = 7$	$n = 10$
	h_1	4.00	5.13	6.09	7.69	9.09	11.1
$a/W = \frac{1}{8}$	h_2	4.71	5.63	6.45	7.85	9.09	10.9
	h_3	0.548	0.733	1.13	2.07	3.16	5.07
	h_1	4.17	5.35	6.09	6.93	7.30	7.41
$a/W = \frac{1}{4}$	h_2	4.58	5.36	5.84	6.31	6.44	6.31
	h_3	0.757	1.35	1.93	2.96	3.78	4.60
	h_1	5.40	5.90	5.63	4.51	3.49	2.47
$a/W = \frac{1}{2}$	h_2	4.99	5.01	4.59	3.48	2.56	1.67
	h_3	1.555	2.26	2.59	2.57	2.18	1.56
	h_1	5.18	3.78	2.57	1.59	1.31	1.10
$a/W = \frac{3}{4}$	h_2	4.22	2.79	1.67	0.725	0.48	0.300
	h_3	1.86	1.73	1.26	0.775	0.561	0.360

Table A.15 h-Functions for a Circumferentially Cracked Cylinder in Tension with $W/R_i = \frac{1}{20}$

		$n = 1$	$n = 2$	$n = 3$	$n = 5$	$n = 7$	$n = 10$
	h_1	4.04	5.23	6.22	7.82	9.19	11.1
$a/W = \frac{1}{8}$	h_2	4.82	5.69	6.52	7.90	9.11	10.8
	h_3	0.680	0.759	1.17	2.13	3.23	5.12
	h_1	4.38	5.68	6.45	7.29	7.62	7.65
$a/W = \frac{1}{4}$	h_2	4.71	5.56	6.05	6.51	6.59	6.39
	h_3	0.818	1.43	2.03	3.10	3.91	4.69
	h_1	6.55	7.17	6.89	5.46	4.13	2.77
$a/W = \frac{1}{2}$	h_2	5.67	5.77	5.36	4.08	2.97	1.88
	h_3	1.80	2.59	2.99	2.98	2.50	1.74
	h_1	6.64	4.87	3.08	1.68	1.30	1.07
$a/W = \frac{3}{4}$	h_2	5.18	3.57	2.07	0.808	0.472	0.316
	h_3	2.36	2.18	1.53	0.772	0.494	0.330

Table A.16 F, V_1, and V_2 for a Circumferentially Cracked Cylinder in Tension

		$a/W = \frac{1}{8}$	$a/W = \frac{1}{4}$	$a/W = \frac{1}{2}$	$a/W = \frac{3}{4}$
	F	1.16	1.26	1.61	2.15
$W/R_i = \frac{1}{5}$	V_1	1.49	1.67	2.43	3.76
	V_2	0.117	0.255	0.743	1.67
	F	1.19	1.32	1.82	2.49
$W/R_i = \frac{1}{10}$	V_1	1.55	1.76	2.84	4.72
	V_2	0.180	0.290	0.885	2.09
	F	1.22	1.36	2.03	2.89
$W/R_i = \frac{1}{20}$	V_1	1.59	1.81	3.26	5.99
	V_2	0.220	0.315	1.04	2.74

INDEX